COMPUTATIONAL METHODS AND FUNCTION THEORY 1997

APPROXIMATIONS AND DECOMPOSITIONS

Editor-in-Chief: CHARLES K. CHUI

Vol. 1: Wavelets: An Elementary Treatment of Theory and Applications
Tom H. Koornwinder, ed.

Vol. 2: Approximate Kalman Filtering
Guanrong Chen, ed.

Vol. 3: Multivariate Approximation: From CAGD to Wavelets
Kurt Jetter and Florencio I. Utreras, eds.

Vol. 4: Advances in Computational Mathematics: New Delhi, India
H. P. Dikshit and C. A. Micchelli, eds.

Vol. 5: Computational Methods and Function Theory
Proceedings of CMFT '94 Conference, Penang, Malaysia
R. M. Ali, St. Ruscheweyh and E. B. Saff, eds.

Vol. 6: Approximation Theory VIII
Approximation and Interpolation – Vol. 1
Wavelets and Multi-level Approximation – Vol. 2
C. K. Chui and L. L. Schumaker, eds.

Vol. 7: Introduction to the Theory of Weighted Polynomial Approximation
H. N. Mhaskar

Vol. 8: Advanced Topics in Multivariate Approximation
F. Fontanella, K. Jetter and P. J. Laurent, eds.

Vol. 9: Approximate Solutions of Operator Equations
M. J. Chen, Z. Y. Chen and G. R. Chen

Vol. 10: Wavelets & Renormalization
G. Battle

Vol. 11: Computational Methods and Function Theory 1997
N. Papamichael, St. Ruscheweyh and E. B. Saff, eds.

Series in Approximations and Decompositions – Vol. 11

Proceedings of the Third CMFT Conference

COMPUTATIONAL METHODS AND FUNCTION THEORY 1997

13–17 October 1997 Nicosia, Cyprus

edited by

N Papamichael
Department of Mathematics
University of Cyprus, Cyprus

St Ruscheweyh
Mathematisches Institut
Universität Würzburg, Germany

E B Saff
Department of Mathematics
University of South Florida, USA

World Scientific
Singapore • New Jersey • London • Hong Kong

Published by

World Scientific Publishing Co. Pte. Ltd.
P O Box 128, Farrer Road, Singapore 912805
USA office: Suite 1B, 1060 Main Street, River Edge, NJ 07661
UK office: 57 Shelton Street, Covent Garden, London WC2H 9HE

Library of Congress Cataloging-in-Publication Data
International Conference on Computational Methods and Function Theory
(3rd : 1997 : Nicosia, Cyprus)
Computational methods and function theory 1997 : proceedings of the third CMFT conference, 13–17 October 1997, Nicosia, Cyprus / edited by N. Papamichael, St. Ruscheweyh, E.B. Saff.
p. cm. -- (Series in approximations and decompositions ; vol. 11)
The Third International Conference on Computational Methods and Function Theory (CMFT '97) was held in Nicosia, Cyprus, from October 13 to 17, 1997 -- T.p. verso.
ISBN 9810236263 (alk. paper)
1. Numerical analysis -- Congresses. 2. Functions of complex variables -- Congresses. I. Papamichael, N. (Nicolas) II. Ruscheweyh, Stephan. III. Saff, E. B., 1944– . IV. Title. V. Series.
QA297.I473 1997
519.4--dc21 99-14411
CIP

British Library Cataloguing-in-Publication Data
A catalogue record for this book is available from the British Library.

Printed in Singapore by Uto-Print.

Approximations and Decompositions

During the past decade, Approximation Theory has reached out to encompass the approximation-theoretic and computational aspects of several exciting areas in applied mathematics such as wavelets, fractals, neural networks, and computer-aided-geometric design, as well as the modern mathematical development in science and technology. The objective of this book series is to capture this exciting development in the form of monographs, lecture notes, reprint volumes, text books, edited review volumes, and conference proceedings.

Computational Methods and Function Theory (CMFT'97) is the 11th volume of this book series. It is a continuation of Vol. 5, CMFT'94. The 43 carefully refereed and edited contributions in this volume represent both survey articles and state-of-the-art research papers. The central theme of this volume is computational function theory. Topics ranging from hypergeometric functions and orthogonal polynomials, to univalent harmonic exterior maps, to Riccati and Loewner differential equations, and to nonlinear Riemann-Hilbert problems, are covered. The series editor would like to thank the authors for their very fine contributions to this volume, and to congratulate the editors, Professors Papamichael, Ruscheweyh, and Saff, for their outstanding editing work.

World Scientific Series
APPROXIMATIONS AND DECOMPOSITIONS

CHARLES K. CHUI, SERIES EDITOR
Texas A&M University, College Station, Texas

CMFT '97 — University of Cyprus, Nicosia

Preface

The third international conference on Computational Methods and Function Theory (CMFT'97) was held in Nicosia, Cyprus, from October 13 to 17, 1997, under the auspices of the University of Cyprus. Its predecessors CMFT'89 and CMFT'94 were held respectively in Valparaíso, Chile and Penang, Malaysia, but the Cyprus meeting, with about 140 participants from 28 different countries, was by far the largest in the series.

The general theme of the CMFT meetings concerns the various aspects of interaction of complex variables (together with its applications in natural sciences and engineering) and scientific computations. Thus, one of the main goals of the meetings is to bring together researchers from the areas of function theory and computational complex analysis as well as from other associated areas such as approximation theory and numerical analysis.

Another important aim of the CMFT meetings is to assist in the creation and maintenance of contacts between scientists from diverse cultures and, in particular, to encourage and support the participation of eminent researchers from developing institutions with limited resources. This latter aim is also reflected in the choice of venue for the meetings. For instance, the island of Cyprus, traditionally serving as a bridge between East and West, was ideally situated to attract researchers from a variety of different contries.

The University of Cyprus, a newly founded institution celebrating its 5th anniversary in 1997, was an excellent host and provided generous financial assistance, dedicated organizational support and friendly hospitality. On behalf of all participants, the organizers wish to express their gratitude to the University and, in particular, to its Department of Mathematics and Statistics. Of the many local people who contributed to the success of the meeting, special thanks are extended to the members of the Local Organizing Committee; particularly Dr. Nikos Stylianopoulos, who was responsible for the organization and supervision of most technical aspects of the meeting, and to Mrs Naso Mouskoundi who dealt with all the local secretarial work and who was constantly on hand, at the conference reception desk, to offer assistance.

The conference has received financial support, direct or indirect, from several local, regional and international organizations and companies. The organizers wish to express their gratitude to the main local sponsors, which include: the Anastasios G. Levendis Foundation, the Bank of Cyprus and the Cyprus Development Bank Ltd; to the sponsors of special events: ETKO Ltd, KEO Ltd, the Cyprus Institute of Technology and the Cyprus Tourism

Organization; and to IBM Cyprus for providing the conference's computer facilities. Last, but certainly not least, we wish to thank all those agencies which provided travel support for the attendees: The British Council, for its support of British participants, COSTED, Madras, for its support of Indian participants, the Deutsche Forschungsgemeinschaft, for its support of German participants and the National Science Foundation of the USA, for its support of US participants.

Around a hundred research lectures were presented at the conference, including ten invited addresses (see the Appendix for details). This volume of the CMFT'97 Proceedings contains 43 carefully refereed papers (including some detailed survey articles) that cover a broad range of topics from all areas within the general theme of the conference. We are very grateful to all the authors who have contributed to the proceedings, and to all the referees who reviewed the submitted papers.

The production of these proceedings would not have been possible without the dedicated help of Mrs Uta Reyes, Dr. Richard Greiner and Dr. Oliver Roth, all from the Complex Analysis Group at Würzburg University, who have dealt with the typesetting and/or layout processing of all articles. We thank them all! Finally, we would like to thank Professor Charles Chui, the series editor, and World Scientific, the publisher, for accepting these proceedings in their "Approximations and Decompositions" series.

Nicolas Papamichael
Stephan Ruscheweyh
Edward. B. Saff

Obituary

On behalf of all the CMFT'97 participants, the organizers wish to express their deep sorrow for the untimely death of their colleague and co-participant

Professor Vladimir I. Belyĭ,

who passed away only a few weeks after the meeting. We shall always remember with affection this kind man and great mathematician!

Contents

A. Ambroladze, H. Wallin 1
Some open problems in the theory of rational interpolants in the complex plane

G. D. Anderson, S. Qiu, M. K. Vuorinen 9
Bounds for the Hersch-Pfluger and Belinskii distortion functions

V. Andrievskiĭ 23
Locally convex curves and distribution of zeros of polynomials

L. Baratchart 45
Rational and meromorphic approximation in L^p of the circle: system-theoretic motivations, critical points and error rates

V. I. Belyĭ, E. V. Kravchuk 79
Generalized Faber sets and applications

D. Betsakos 87
An extension of the Beurling-Nevanlinna projection theorem

G. Brown, S. Koumandos 91
On a theorem of S. Ruscheweyh

D. Bshouty, W. Hengartner, M.-H. Nicole 99
A constructive method for univalent harmonic exterior maps with Blaschke dilatation

P. Caraman 117
New results about the equality between the p-module and the p-capacity

A. Cuyt, B. Verdonk 135
Extending the qd-algorithm to tackle multivariate problems

A. A. Danielyan 161
M. A. Lavrentyev's problems on pointwise polynomial approximation and related questions

M. J. Dorff 171
Harmonic univalent mappings onto asymmetric vertical strips

P. Duren 177
Robin capacity

H.-J. Fischer 191
On generating orthogonal polynomials with respect to self-similar measures

J.-D. Fournier 203
Complex zeros of random Szegő polynomials

D. Gaier 225
On the relation between $E_n(f)$ and $E_n(f')$

F. W. Gehring, K. Hag 233
A bound for hyperbolic distance in a quasidisk

J. Godula, V. V. Starkov 241
On regularity theorems for linearly invariant families of analytic functions in the unit polydisk

R. Greiner 259
Generalized Jackson kernels in approximation theory

R. Grothmann 267
On the problem of Poreda

M. Ito, M. Shiba 275
Area theorems for conformal mapping and Rankine ovoids

G. Jank, G. Kun 285
Solutions of generalized Riccati differential equations and their approximation

E. A. Karatsuba 303
Fast evaluation of hypergeometric functions by FEE

E. S. Katsoprinakis 315
On the complex Rolle set of a polynomial

C. A. Kokkinos 327
A unified orthonormalization method for the approximate conformal mapping of simply and multiply-connected domains

S. Koumandos 345
Positive trigonometric sums in the theory of univalent functions

P. Kravanja, M. Van Barel, A. Haegemans 359
On computing zeros and poles of meromorphic functions

E. (A. L.) Levin 371
Fast decreasing rational functions and their applications

W. Majchrzak 387
Harmonic univalent mappings of the unit disc onto a vertical strip

D. Minda 397
Euclidean circles of curvature for geodesics of conformal metrics

A. H. M. Murid, M. Z. Nashed, M. R. M. Razali 405
Some integral equations related to the Riemann map

V. Nestoridis 421
An extension of the notion of universal Taylor series

M. Obradović 431
Starlikeness of certain integral transforms

I. E. Pritsker, R. S. Varga 437
Rational approximation with varying weights in the complex plane

D. V. Prokhorov 449
Radii of neighborhoods for coefficient estimates of functions close to the identity

O. Roth 461
A remark on the Loewner differential equation

C. Rudälv 471
Repeated bivariate rational approximants with prescribed poles

G. Schmeisser 485
Real zeros of Bernstein polynomials

G. Schmieder 497
Location of critical points for polynomials

F. Stenger, R. Schmidtlein 505
Conformal maps via Sinc methods

K. Stephenson 551
The approximation of conformal structures via circle packing

E. Wegert 583
Nonlinear Riemann-Hilbert problems — history and perspectives

R. Wegmann 617
Two plane free boundary problems with surface tension

List of Participants 635

Some open problems in the theory of rational interpolants in the complex plane

Amiran Ambroladze
and Hans Wallin

Abstract. We discuss some results about the convergence of rational interpolants with preassigned poles (partly or completely) to analytic functions, and some open problems related to these results. The open problems have relatively elementary formulations and, to our opinion, their solutions would give a better understanding of the subject. We discuss also some computational aspects of this theory.

§1 Problem 1

Let D be a bounded simply connected domain in the open complex plane $\mathbb{C}$, and let f be an analytic function in D. Let also $A_n = \{a_{ni}\}_{i=0}^{n} \subset D$, $n = 0, 1, ...$, be a so called interpolation scheme (a_{ni} are points in D not necessarily different for the same index n). Then for an arbitrary integer $n \geq 0$ there exists a uniquely determined polynomial P_n of degree at most n (*i.e.* $\deg P_n \leq n$) such that

$$(f - P_n)(a_{ni}) = 0, \qquad i = 0, 1, ..., n,$$

counting multiplicities. The polynomials P_n are called interpolating polynomials.

We ask the following question: does there exist (for the domain D) an interpolating scheme guaranteeing that for any analytic function f in D, the following limit relation

$$(f - P_n)(z) \to 0, \qquad \text{as } n \to \infty, \tag{1.1}$$

holds uniformly on compact subsets of D?

(The analogous problem for compact sets (for example for $\overline{D}$) is easier to handle, since any function analytic on a compact set is automatically analytic on a larger open set. For these questions see, for example [10].)

The answer to the question posed above is the following: such an interpolation scheme exists if and only if the boundary ∂D of D is an analytic curve; furthermore, if ∂D is an analytic curve then such an interpolation scheme must satisfy the following (necessary and sufficient) conditions (see [1]):

Computational Methods and Function Theory (CMFT'97)
N. Papamichael, St. Ruscheweyh and E. B. Saff (Eds.), pp. 1–7

ISBN 981-02-3626-3

(i) $\overline{\cup_{n\geq 0} A_n} \subset D$,

(ii) If α is a weak star limit for the sequence of the point counting measures

$$\alpha_n = \frac{1}{n+1} \sum_{i=0}^{n} \delta_{a_{ni}}$$

(or for some of its subsequences), then the sweeping out measure α' of α onto ∂D has to be the equilibrium measure of ∂D.

(Here δ_a denotes the Dirac measure at the point a. For the definitions of equilibrium measure and sweeping out see, for example [9].)

This last condition about α can be rewritten in the following form:

$$\alpha' = \delta'_\infty, \tag{1.2}$$

where the right-hand side denotes the sweeping out measure onto ∂D of the Dirac measure at infinity.

Now some explanations about these two conditions. The behaviour of a function analytic in D can be quite complicated near the boundary of D and consequently the information obtained from the points near to the boundary can be quite erroneous; and so we need to separate interpolation points from the boundary of D. This is done by condition (i).

The second condition (ii) demands a symmetry (about the boundary of D) of the interpolation points a_{ni} and the poles of the interpolating function. In the case of polynomials all the poles lie at infinity.

In terms of the sweeping out technique, the above question (about the existence of an interpolation scheme) is equivalent to the question about existence of a measure α with support (minimal compact set outside of which the measure is zero) in D, and satisfying condition (1.2). And so the answer is that ∂D must be analytic.

To have convergence for general domains we have to restrict the class of analytic functions in D. We now consider the class of bounded analytic functions in D. Then it turns out that for any bounded simply connected domain D there always exists an interpolation scheme guaranteeing the limit relation (1.1) for any bounded analytic function f in D. The boundary of D can be very far from being analytic; as an example we can take the famous Von Koch's snowflake domain. But it is interesting to note that in this case in general (and in particular for the snowflake) the points A_n in the interpolation scheme must go to the boundary of D as n increases, in the sense that the limit measure α, defined above, must be supported on ∂D. This means that even though we have convergence we can not guarantee any geometrical rate for this convergence (since the interpolation points go to the boundary).

But why do the interpolation points have to go to the boundary? The reason is that for the case of bounded analytic functions we still have condition (1.2) as necessary, but for the snowflake, for example, this is impossible unless the measure α is itself the equilibrium measure of ∂D. We will discuss later why this is so, but first we formulate the following problem:

Problem 1. For which bounded simply connected domains $D \subset \mathbb{C}$ does there exist a unit measure α, $\alpha(D) = 1$, such that the sweeping out measure of α onto the boundary of D is the equilibrium measure of the boundary, that is

$$\alpha' = \delta'_\infty?$$

This problem can be put in the following general form:

Problem 1'. For which bounded simply connected domains $D \subset \mathbb{C}$ do there exist unit measures α and β, $\alpha(D) = 1$, $\beta(\overline{\mathbb{C}} \setminus \overline{D}) = 1$ ($\overline{D}$ denotes the closure of D and $\overline{\mathbb{C}} = \mathbb{C} \cup \{\infty\}$), such that

$$\alpha' = \beta',$$

where α' and β' are the sweeping out measures onto ∂D of the measures α and β, respectively?

(The last problem arises naturally if, instead of polynomial interpolation, we consider rational interpolation; more precisely, this problem arises in the so called dual problem (see, for example, [10] or [2]).)

We do not even know if these last two problems are equivalent (like they are when we do not allow the measures α and β to approach the boundary of D, see [1]).

Now let us discuss why for the snowflake domain these problems have negative solution. From the results in [7] it follows that if we take two Dirac measures, one inside the snowflake domain D and the other outside, then their sweeping out measures onto the boundary of D are mutually singular. This assertion assumes a natural generalization for the measures α and β in Problem $1'$. So, for the snowflake domain the question in Problem $1'$ has negative answer. On the other hand, we know that the analyticity of the boundary of D in Problem $1'$ is not a necessary condition for the positive solution of the problem. There is a domain, having the form of diamonds (like in a deck of cards) with four cusps, and for which the problem has positive solution (see [1], Example 1).

We conclude this section by noting that the boundedness of the domain D was assumed just for the sake of discussion convenience. Questions similar to those above can be asked for any simply connected domain in the extended complex plane $\overline{\mathbb{C}}$.

§2 Problem 2

In this section we come back to the problem of interpolation of all analytic functions in a domain D. Let D be simply connected to begin with. We have seen in the previous section that if we want to interpolate by polynomials all analytic functions in D with the same interpolation scheme, then this is possible only in domains with analytic boundaries. It turns out that if we use rational interpolants instead of polynomials, then we can solve this problem for any simply connected domain. To do so we fix a point in D, let us say this is $z = 0$, and then for an analytic function f in D, we consider rational functions P_n/Q_n, $\deg P_n \leq n$, $\deg Q_n = n$, $n \geq 0$, such that

$$(f - \frac{P_n}{Q_n})(z) = O(z^{n+1}), \qquad \text{as } z \to 0. \tag{2.1}$$

Here we fix $Q_n(z)$ by choosing its zeros $\{b_{ni}\}_{i=1}^{n} \subset \partial D$ so that the corresponding zero counting measures

$$\beta_n = \frac{1}{n}\sum_{i=1}^{n} \delta_{b_{ni}}$$

converge (in the weak star sense) to the measure δ_0', which is the harmonic measure of ∂D with respect to the point $z = 0$ (and equal to the sweeping out measure of the Dirac measure δ_0 onto ∂D). The polynomial P_n in (2.1) is uniquely determined as the n-th Taylor polynomial of the function $Q_n f$. Then we have (see, for example, [2]):

$$(f - \frac{P_n}{Q_n})(z) \to 0, \qquad \text{as } n \to \infty, \tag{2.2}$$

uniformly on compact subsets of D.

This assertion is actually true not only for simply connected domains but for any domain D regular with respect to the Dirichlet problem (for the definitions see, for example, [9]). Now the question is what happens if D is irregular. Here we have the following conjecture, which is the second open problem of this paper:

Problem 2 (Conjecture). There exists a sequence of polynomials $\{Q_n\}_{n=1}^{\infty}$, $\deg Q_n = n$, guaranteeing (2.2) for any analytic function$f$ in a domain D, if and only if D is a regular domain.

We have proved this conjecture for the case when ∂D has an isolated point, but even for this case our proof is quite complicated (see [5]).

In this conjecture we may also allow Q_n to have zeros even outside of ∂D, and this makes the problem even more complicated. For the details about this problem see [3].

§3 Problem 3

The last problem of this paper is actually not one problem but a circle of problems which are, to our opinion, worth to be investigated. We start with a concrete example.

Let $f(z) = e^z$. Then the rational interpolants P_n/Q_n, defined in the previous section, with all poles at $z = n$ (that is $Q_n(z) = (z-n)^n$) converge to f uniformly on compact subsets of the comlex plane (see [8]). A similar assertion is true if we consider interpolants $P_n(z)/Q_n(z) = (z+n)^n/Q_n(z)$, with preassigned zeros (instead of poles) at $z = -n$ and free poles, where Q_n is determined from (2.1).

It turns out that both of these approximants accelerate convergence compared with Taylors polynomials of the same order (see [8] or [4]).

A possible explanation of the phenomenon of convergence acceleration is that the function e^z increases to infinity along the positive real axis, and decreases to zero along the negative real axis, and by choosing zeros of the rational interpolants on the negative real axis or poles on the positive real axis, we simulate the behaviour of e^z. The choice $z = -n$ for the zeros and $z = n$ for the poles of the corresponding approximants is probably optimal. As a comparison we recall the fact that the classical diagonal Padé Approximants $[n/n]$ to e^z have all their zeros in the left half-plane and all the poles in the right half-plane, and that they are contained in the annulus $\frac{2}{3}n < |z| < 2n$ (see [6], Theorems 5.7.4 and 5.7.5). This discussion suggests the following problems.

Problem 3A. What can we say about convergence of rational interpolants of the form $R_{2n}(z) = P_n(z)(z+2n)^n/Q_n(z)(z-2n)^n$ to e^z, where P_n and Q_n are free polynomials of degree n determined from the interpolation condition: $e^z - R_{2n}(z) = O(z^{2n+1})$, as $z \to 0$?

Note that in Problem 3A we have fixed zeros where e^z is small and poles where e^z is large. As a motivation for Problem 3A we made numerical experiments with the function e^z. We interpolated at $z = 0$ and evaluated the approximation error at $z = 1$. Here is the result: The Taylor polynomial around zero of degree 40 gives 49 correct decimal digits at $z = 1$. The rational interpolants of the form $P_{40}(z)/(z-40)^{40}$ (giving maximal possible interpolation at zero) gives 55 correct decimal digits at $z = 1$. The classical Padé Approximants $P_{20}(z)/Q_{20}(z)$ gives 60 correct decimal digits at $z = 1$. The rational interpolants of the form $P_{20}(z)(z+40)^{20}/Q_{20}(z)(z-40)^{20}$ (giving maximal possible interpolation at zero) gives 62 correct decimal digits at $z = 1$.

Problem 3B. The function $\sin z$ is bounded on the real axis and converges to infinity along the imaginary axis. What can we say about the convergence of rational interpolants if we preassign zeros on the real axis and/or poles on the imaginary axis?

Questions analogous to those in Problems 3A and 3B can be asked for some other functions as well.

Problem 3C. The functions in Problems 3A and 3B are entire functions and separation of zeros and poles for the approximants were natural. But if we consider for example Markov functions (see, for example [9] for the definitions), then in general they have singularities on the maximal interval I containing the generating measure. This suggests that we should fix some zeros and poles (at the same time) on I. In particular the following question can be asked: What can we say about convergence of rational interpolants $R_{2n}(z) = P_n(z)\tilde{P}_n(z)/Q_n(z)\tilde{Q}_n(z)$ to a Markov function f, where P_n and Q_n have their zeros on I (chosen according to the equilibrium measure of the support of the generating measure of f) and $\tilde{P}_n$ and $\tilde{Q}_n$ are free polynomials of degree $n-1$ and n, respectively, determined by the condition $f(z) - R_{2n}(z) = O(z^{-(2n+1)})$ (as $z \to \infty$)?

To fix both zeros and poles near singularities of f (different from poles) can be a general rule for rational approximants, since such a behaviour is observed by the classical Padé Approximants.

References

[1] Ambroladze, A. and H. Wallin, Balayage and convergence of rational interpolants, *J. Approx. Theory*, 1998 to appear.

[2] Ambroladze, A. and H.Wallin, Rational interpolants with preassigned poles, theory and practice, *Complex Variables* **34** (1997), 399–413.

[3] Ambroladze, A. and H. Wallin, A convergence problem for rational interpolants with preassigned poles: Research Problem 96-1, *Constr. Approx.* **12** (1996), 157–159.

[4] Ambroladze, A. and H. Wallin, Padé type approximants with preassigned zeros, in *Proc. Int. Conf. Approx. and Optim., Aportaciones Matematicas, Mexican Sociaty of Mathematics,* Mexico, 1998 to appear.

[5] Ambroladze, A. and H. Wallin, Convergence of rational interpolants with preassigned poles, *J. Approx. Theory.* **89** No. 2 (1997), 238–256.

[6] Baker, G. A. Jr. and P. R. Graves-Morris, *Padé Approximants, Part 1*, Encycl. Math., vol. 13. Reading, MA: Addison-Wesley, 1981.

[7] Bishop, C. J. et al., Harmonic measures supported on curves, *Pacific J. Math.* **138** No.2 (1989), 233–236.

[8] Le Ferrand, H., Padé type approximants of entire functions and acceleration of convergence: two examples, *J. Comp. Appl. Math.* **57** no.1–2 (1995), 193-202.

[9] Stahl, H. and V. Totik, *General Orthogonal Polynomials*, Cambridge Univ. Press, Cambridge, 1992.

[10] Walsh, J. L., *Interpolation and Approximation by Rational Functions in the Complex Plane*, Fourth edition, Amer. Math. Soc. Colloq. Publ. XX, Amer. Math. Soc., Providence, RI, 1965.

Amiran Ambroladze
Department of Mathematics
Umeå University
S-901 87 Umeå
SWEDEN
amiran.ambroladze@math.lth.se

Hans Wallin
Department of Mathematics
Umeå University
S-901 87 Umeå
SWEDEN
hans.wallin@mathdept.umu.se

Bounds for the Hersch-Pfluger and Belinskii distortion functions

Glen Douglas Anderson

Songliang Qiu

and Matti K. Vuorinen

Abstract. The authors study the Hersch-Pfluger plane quasiconformal distortion function, solving extremal problems, obtaining monotoneity and convexity properties, and finding sharp bounds for it. They also obtain bounds for the Belinskii function, which measures the deviation of a quasiconformal automorphism of the unit disk from the identity map.

§1 Introduction

For $r \in (0,1)$, let $\mu(r)$ denote the modulus of the plane Grötzsch ring $B \backslash [0,r]$, where B is the unit disk. Then [8], p. 60,

$$\mu(r) = \frac{\pi}{2} \frac{\mathcal{K}'(r)}{\mathcal{K}(r)}, \tag{1.1}$$

where

$$\mathcal{K} = \mathcal{K}(r) \equiv \int_0^{\pi/2} (1-r^2 \sin^2 t)^{-1/2} dt, \qquad \mathcal{K}' = \mathcal{K}'(r) \equiv \mathcal{K}(r'), \tag{1.2}$$

$r' = \sqrt{1-r^2}$, $0 < r < 1$, are *complete elliptic integrals of the first kind* [5]. In the sequel, we shall also need the *complete elliptic integrals of the second kind*

$$\mathcal{E} = \mathcal{E}(r) \equiv \int_0^{\pi/2} (1-r^2 \sin^2 t)^{1/2} dt, \qquad \mathcal{E}' = \mathcal{E}'(r) \equiv \mathcal{E}(r'), \tag{1.3}$$

for $r \in [0,1]$, $r' = \sqrt{1-r^2}$.

The function $\mu(r)$ and the distortion function

$$\begin{aligned} \varphi_K(r) &\equiv \mu^{-1}(\mu(r)/K), \\ \varphi_K(0) &= 1 - \varphi_K(1) = 0, \end{aligned} \tag{1.4}$$

Computational Methods and Function Theory (CMFT'97)
N. Papamichael, St. Ruscheweyh and E. B. Saff (Eds.), pp. 9–22

ISBN 981-02-3626-3

for $r \in (0,1)$ and $K \in (0,\infty)$, play an important role in quasiconformal theory [1], [6], [7], [8], [11], [14]. The function $\varphi_K(r)$ has also found applications in other mathematical fields such as number theory [3], [4], [5].

In the past few years, many properties have been derived for the special function $\varphi_K(r)$ [1], [9], [12], [15]. However, the above-mentioned applications motivate us to study this function further. We now state an open problem on its properties.

Problem (P) Given $s = \varphi_K(r)$ and $u = \varphi_K(t)$, for $r, t \in [0,1]$, $K > 1$, find sharp bounds for the functions

$$\frac{1+s}{(1+r)^{1/K}}, \quad \frac{1+s^2}{(1+r^2)^{1/K}}, \quad \text{and} \quad \frac{1+su}{(1+rt)^{1/K}}. \tag{1.5}$$

In the present paper, the authors study the monotoneity and convexity properties of certain combinations of the function $\varphi_K(r)$, from which sharp bounds for this function follow. We shall solve Problem (P) by finding the extremal values

$$\sup_{r,K} \frac{1+s}{(1+r)^{1/K}}, \quad \sup_{r,K} \frac{1+s^2}{(1+r^2)^{1/K}}, \quad \text{and} \quad \sup_{r,t,K} \frac{1+su}{(1+rt)^{1/K}},$$

where $s = \varphi_K(r)$, $u = \varphi_K(t)$, $r, t \in [0,1]$, and $K > 1$.

Throughout this paper, we let $r' = \sqrt{1-r^2}$ for $r \in [0,1]$.

Next, by the well-known quasiconformal Schwarz lemma [8], [7] we have

$$|f(z)| \le \varphi_K(|z|), \quad z \in B, \tag{1.6}$$

where f is a K-quasiconformal mapping of the unit disk B into itself, $K \ge 1$, with $f(0) = 0$. It is well known that $\varphi_1(r) = r$, $0 \le r \le 1$, and that $\varphi_K(0) = 0$, $\varphi_K(1) = 1$, $K \ge 1$. Hence the function $\varphi_K(r) - r$ measures the deviation of a K-quasiconformal automorphism of the unit disk from the identity map, $K > 1, 0 < r < 1$. This function was studied by P. P. Belinskii [2], and explicit estimates were obtained in [13]. In our final result we obtain the following more precise estimates for this function.

Theorem 1. *Let $m(r) = (2/\pi)r'^2\mathcal{K}(r)\mathcal{K}'(r)$ for $r \in (0,1)$. Then for any fixed $r \in (0,1)$, the function $f(K) \equiv \varphi_K(r) - r - (K-1)r^{1/K}m(r)$ is strictly decreasing from $[1,\infty)$ onto $(-\infty, 0]$. Furthermore, for all $K > 1$ and $r \in (0,1)$,*

$$\begin{aligned} \varphi_K(r) - r &< \min\{0.8697 \log K, (K-1)r^{1/K}m(r)\} \\ &< (K-1)\min\{0.8697, r^{1/K}m(r)\}, \end{aligned} \tag{1.7}$$

and

$$\varphi_K(\sqrt{r})^2 - r < a\log K < a(K-1), \tag{1.8}$$

where $a = \mathcal{K}(1/\sqrt{2})^2/\pi = 1.0942\ldots$ *. The coefficient* a *in (1.8) is best possible.*

D. Partyka [10] has proved that the constant $a\log K$ in formula (1.8) can be replaced by $1 - 2\varphi_{1/\sqrt{K}}(1/\sqrt{2})^2$.

We shall frequently employ the following well-known formulas [1]: (5.9), [5], [6], [8] (see also [1], pp. 474-475), for $0 < r < 1,\ \ 0 < K < \infty$:

$$\frac{d\mathcal{K}}{dr} = \frac{\mathcal{E} - r'^2\mathcal{K}}{rr'^2}, \quad \frac{d\mathcal{E}}{dr} = \frac{\mathcal{E} - \mathcal{K}}{r}, \tag{1.9}$$

$$\begin{aligned} &\mathcal{K}\left(\frac{2\sqrt{r}}{1+r}\right) = (1+r)\mathcal{K}(r), \quad \mathcal{K}'\left(\frac{2\sqrt{r}}{1+r}\right) = \frac{1}{2}(1+r)\mathcal{K}'(r), \\ &\mathcal{E}\left(\frac{2\sqrt{r}}{1+r}\right) = \frac{2\mathcal{E}(r) - r'^2\mathcal{K}(r)}{1+r}, \quad \mathcal{E}'\left(\frac{2\sqrt{r}}{1+r}\right) = \frac{\mathcal{E}'(r) + r\mathcal{K}'(r)}{1+r}, \end{aligned} \tag{1.10}$$

$$\mathcal{K}\mathcal{E}' + \mathcal{K}'\mathcal{E} - \mathcal{K}\mathcal{K}' = \frac{\pi}{2}, \tag{1.11}$$

$$\frac{d\mu(r)}{dr} = -\frac{\pi^2}{4rr'^2\mathcal{K}(r)^2}, \tag{1.12}$$

$$\begin{aligned} &\frac{\partial s}{\partial r} = \frac{s}{Kr}\left(\frac{s'\mathcal{K}(s)}{r'\mathcal{K}(r)}\right)^2 = \frac{ss'^2\mathcal{K}(s)\mathcal{K}'(s)}{rr'^2\mathcal{K}(r)\mathcal{K}'(r)}, \\ &\frac{\partial s}{\partial K} = \frac{2}{\pi K}ss'^2\mathcal{K}(s)\mathcal{K}'(s), \end{aligned} \tag{1.13}$$

where $s = \varphi_K(r)$,

$$\varphi_K(r)^2 + \varphi_{1/K}(r')^2 = 1, \quad \varphi_2(r) = \frac{2\sqrt{r}}{1+r}, \tag{1.14}$$

$$\lim_{r\to 0} r^{-1/K}\varphi_K(r) = 4^{1-1/K}. \tag{1.15}$$

We denote

$$m(r) = \frac{2}{\pi}r'^2\mathcal{K}(r)\mathcal{K}'(r), \tag{1.16}$$

for $r \in (0,1)$. Then, by differentiation and (1.11), we have

$$\frac{dm(r)}{dr} = \frac{1}{r} - \frac{4}{\pi r}\mathcal{E}'(r)\mathcal{K}(r). \tag{1.17}$$

§2 Preliminary results

In this section we prove some lemmas that will be needed for the proofs of the main theorems in Sections 3 and 4.

Lemma 1. *There exists a unique* $r_0 \in (0.6084, 0.6085)$ *such that the function* $f(r) \equiv (2/\pi)r(1-r)\mathcal{K}(r)\mathcal{K}'(r) + \log(1+r)$ *is strictly increasing from* $(0, r_0]$ *onto* $(0, f(r_0)]$, *and strictly decreasing from* $[r_0, 1)$ *onto* $(\log 2, f(r_0)]$, *so that*

$$1.003793104 < \sup_{0<r<1} f(r) = f(r_0) < 1.003917448. \tag{2.1}$$

Proof: By differentiation and (1.11), we obtain

$$\frac{\pi(1+r)}{2\mathcal{K}'(r)} f'(r) = g(r) \equiv 2\mathcal{E}(r) - (1+r)\mathcal{K}(r),$$

which is strictly decreasing from $(0,1)$ onto $(-\infty, \pi/2)$, by (1.2) and (1.3), with $g(0.6084) = 0.0002\ldots$ and $g(0.6085) = -0.0002\cdots$. Hence the piecewise monotoneity of f follows.

Next, let $F(r) = r(1-r)\mathcal{K}(r)\mathcal{K}'(r)$. Then, by differentiation and (1.11), we get

$$\frac{\pi(1+r)}{2\mathcal{K}'(r)} F'(r) = G(r) \equiv g(r) - \frac{\pi}{2\mathcal{K}'(r)},$$

which is strictly decreasing from $(0,1)$ onto $(-\infty, \pi/2)$ with $G(0.41) = 0.014\ldots$ and $G(0.42) = -0.020\ldots$. Hence there is a unique $r_1 \in (0.41, 0.42)$ such that F is strictly increasing on $(0, r_1]$, and decreasing on $[r_1, 1)$. Therefore,

$$f(r_0) < \frac{2}{\pi} F(0.6084) + \log 1.6085 = 1.003917447\ldots,$$

$$f(r_0) > \frac{2}{\pi} F(0.6085) + \log 1.6084 = 1.003793104\ldots,$$

and (2.1) holds. ■

Lemma 2.

(1) There exists a unique $r_0 \in (0.736, 0.737)$ *such that the function* $f(r) \equiv 2r^2 m(r)/(1+r^2) + \log(1+r^2)$ *is strictly increasing on* $(0, r_0]$, *and decreasing on* $[r_0, 1)$. *In particular,*

$$1.13842 < \sup_{0<r<1} f(r) = f(r_0) < 1.13891. \tag{2.2}$$

(2) Let

$$g(x,y) = \frac{xy}{1+xy}[m(x) + m(y)] + \log(1+xy)$$

for $(x,y) \in D = (0,1) \times (0,1)$. *Then*

$$1.3842 < \sup_{0<x,y<1} g(x,y) = f(r_0) < 1.13891. \tag{2.3}$$

Proof: (1) By (1.17),

$$\frac{1+r^2}{4r}f'(r) = f_1(r) \equiv \frac{1}{1+r^2}m(r) - \frac{2}{\pi}\mathcal{K}(r)\mathcal{E}'(r) + 1.$$

Since $r'^2\mathcal{K}(r)$ is decreasing with range $(0, \pi/2]$ ([1], Theorem 3.21(7)), clearly f_1 is strictly decreasing from $(0,1)$ onto $(-\infty, \infty)$, so that f_1 has a unique zero $r_0 \in (0,1)$. The piecewise monotoneity of f follows. Because of $f_1(0.736) = 0.0004\ldots$ and $f_1(0.737) = -0.0042\ldots$, it follows that $r_0 \in (0.736, 0.737)$.

Next, let $f_2(r) = f(r) - \frac{1}{4}\log(1+r^2)$. Then f_2 is strictly decreasing on $[0.736, 1)$ since the function

$$f_3(r) \equiv \frac{1+r^2}{4r}f_2'(r) = \frac{1}{1+r^2}m(r) - \frac{2}{\pi}\mathcal{K}(r)\mathcal{E}'(r) + \frac{7}{8}$$

is strictly decreasing on $(0,1)$ and $f_3(0.736) = -0.124\ldots$. Hence, we have the following estimates:

$$f(r_0) < f_2(0.736) + \frac{1}{4}\log(1+0.737^2) = 1.138909\ldots$$

and

$$f(r_0) > f_2(0.737) + \frac{1}{4}\log(1+0.736^2) = 1.138428\ldots,$$

so that (2.2) holds.

(2) Clearly, by (1.16) and the fact that $r'^2\mathcal{K}(r)$ is strictly decreasing with range $(0, \pi/2]$ ([1], Theorem 3.21(7)), we have $g(0^+, y) = g(x, 0^+) = 0$, $g(x, 1^-) = g_1(x) \equiv (2/\pi)x(1-x)\mathcal{K}(x)\mathcal{K}'(x) + \log(1+x)$, and $g(1^-, y) = g_1(y)$. By Lemma 1, $g_1(x)$ and $g_1(y)$ have the upper bound $f(r_0)$ given in (2.1).

Differentiation gives

$$\frac{1+xy}{y}\frac{\partial g}{\partial x} = g_2(x,y) \equiv 2 + \frac{1}{1+xy}[m(x)+m(y)] - \frac{4}{\pi}\mathcal{K}(x)\mathcal{E}'(x)$$

and, by symmetry,

$$\frac{1+xy}{x}\frac{\partial g}{\partial y} = g_2(y,x).$$

It is obvious that g_2 is strictly decreasing in x or in y, separately, from $(0,1)$ onto $(-\infty, \infty)$. Also $g_2(y,x) = g_2(x,y)$ if and only if $\mathcal{K}(x)\mathcal{E}'(x) = \mathcal{K}(y)\mathcal{E}'(y)$, which is true if and only if $x = y$ since $\mathcal{K}\mathcal{E}'$ is increasing. Hence, by letting $\partial g/\partial x = \partial g/\partial y = 0$, we see that g has a unique local maximum at the point (r_0, r_0), where r_0 is as in part (1). Consequently, (2.3) follows from part (1), the limiting values of g, and Lemma 1. ■

§3 Properties of $\varphi_{\mathbf{K}}(\mathbf{r})$

In this section we obtain some properties for $\varphi_K(r)$, the so-called Hersch-Pfluger *φ-distortion function* [7]. In particular, we shall solve Problem (P).

Theorem 2. *Let $D = (0,1) \times (1,\infty)$, and define the function f on D by*

$$f(r,K) = [(1+\varphi_K(r))(1+r)^{-1/K}]^{K/(K-1)}.$$

Then

$$2.728612 < \sup_{(r,K)\in D} f(r,K) < 2.728952. \tag{3.1}$$

In particular, for all $(r,K) \in D$,

$$\frac{1+\varphi_K(r)}{(1+r)^{1/K}} < 2.728952^{1-1/K}, \tag{3.2}$$

and the constant 2.728952 cannot be replaced by 2.728612. For $K \in (0,1)$, (3.2) is reversed.

Proof: Clearly, $f(0,K) - 1 = f(1,K) - 2 = 0$ for all $K > 1$. Let $g(r) = (2/\pi)r(1-r)\mathcal{K}(r)\mathcal{K}'(r) + \log(1+r)$ and $c = 1.003917448$. Then, by l'Hôpital's Rule, we have

$$f(r,1^+) = \exp(g(r)), \quad \lim_{K\to\infty} f(r,K) = 2$$

for all $r \in (0,1)$. Hence, by (2.1),

$$2.728612 < \sup_{(r,K)\in\partial D} f(r,K) < e^c < 2.728952, \tag{3.3}$$

and the lower bound in (3.1) follows.

Next, put $s = \varphi_K(r)$. By logarithmic differentiation and (1.13) we get

$$(1+r)(K-1)\frac{1}{f(r,K)}\frac{\partial f}{\partial r} = K\,\frac{s(1-s)\mathcal{K}(s)\mathcal{K}'(s)}{r(1-r)\mathcal{K}(r)\mathcal{K}'(r)} - 1 \tag{3.4}$$

and

$$\begin{aligned}(K-1)^2\frac{1}{f(r,K)}\frac{\partial f}{\partial K} &= \frac{2}{\pi}(K-1)s(1-s)\mathcal{K}(s)\mathcal{K}'(s)\\ &\quad -\log(1+s)+\log(1+r).\end{aligned} \tag{3.5}$$

Letting $\partial f/\partial r = \partial f/\partial K = 0$, we obtain

$$1+s = (1+r)\exp\left(\frac{2}{\pi}\left(1-\frac{1}{K}\right)r(1-r)\mathcal{K}(r)\mathcal{K}'(r)\right). \tag{3.6}$$

Hence, if f has a local maximum at the point $(r_1, K_1) \in D$, then r_1 and K_1 must satisfy (3.6), so that

$$f(r_1, K_1) = (1 + r_1) \exp\left(\frac{2}{\pi} r_1(1 - r_1)\mathcal{K}(r_1)\mathcal{K}'(r_1)\right) = \exp(g(r_1)) < e^c$$

by (2.1). Hence, by (3.3), we obtain the upper bound in (3.1).

The remaining conclusions are clear. ■

Theorem 3. *Let* $D_1 = (0,1) \times (1,\infty)$ *and* $D_2 = (0,1) \times (0,1) \times (1,\infty)$. *Define functions* f *and* g *on* D_1 *and* D_2 *by*

$$f(r,K) = \left[\frac{1+s^2}{(1+r^2)^{1/K}}\right]^{K/(K-1)} \quad \text{and} \quad g(r,t,K) = \left[\frac{1+su}{(1+rt)^{1/K}}\right]^{K/(K-1)},$$

respectively, where $s = \varphi_K(r)$ *and* $u = \varphi_K(t)$. *Then*

$$3.1218 < \sup_{(r,K)\in D_1} f(r,K) \le \sup_{(r,t,K)\in D_2} g(r,t,K) < 3.1234. \tag{3.7}$$

In particular, for all $r, t \in (0,1)$ *and* $K \in (1,\infty)$,

$$\frac{1+su}{(1+rt)^{1/K}} < 3.1234^{1-1/K}, \tag{3.8}$$

and the constant 3.1234 *cannot be replaced by* 3.1218. *Inequality (3.8) is reversed for* $K \in (0,1)$.

Proof: Since $f(r,K) = g(r,r,K)$, the second inequality in (3.7) holds. Let

$$g_1(r,K) = \left[\frac{1+s}{(1+r)^{1/K}}\right]^{K/(K-1)}, \quad g_2(r,t) = \frac{rt}{1+rt}[m(r)+m(t)] + \log(1+rt).$$

It is easy to find the following limiting values:

$$\left.\begin{aligned} &f(0,K) - 1 = f(1,K) - 2 = \lim_{K\to\infty} f(r,K) - 2 = 0, \\ &f(r,1^+) = (1+r^2)\exp(2r^2 m(r)/(1+r^2)), \end{aligned}\right\} \tag{3.9}$$

$$\left.\begin{aligned} &g(0,t,K) - 1 = g(r,0,K) - 1 = \lim_{K\to\infty} g(r,t,K) - 2 = 0, \\ &g(1,t,K) - g_1(t,K) = g(r,1,K) - g_1(r,K) = 0, \\ &g(r,t,1^+) = \exp(g_2(r,t)). \end{aligned}\right\} \tag{3.10}$$

Here the last limits in (3.9) and (3.10), respectively, follow from l'Hôpital's Rule and (1.13), while the others are clear.

From (3.9), (3.10), Theorem 2, and Lemma 2, it follows that

$$\begin{aligned} 3.1218 < e^{1.13842} < &\sup_{(r,K)\in\partial D_1} f(r,K) \\ \leq &\sup_{(r,t,K)\in\partial D_2} g(r,t,K) < e^{1.13891} < 3.1234. \end{aligned} \tag{3.11}$$

Hence the lower bound in (3.7) follows.

Next, by logarithmic differentiation and (1.13), using symmetry, we have

$$\frac{r(K-1)}{g(r,t,K)}\frac{\partial g}{\partial r} = \frac{su}{1+su}\left(\frac{s'\mathcal{K}(s)}{r'\mathcal{K}(r)}\right)^2 - \frac{rt}{1+rt},$$

$$\frac{t(K-1)}{g(r,t,K)}\frac{\partial g}{\partial t} = \frac{su}{1+su}\left(\frac{u'\mathcal{K}(u)}{t'\mathcal{K}(t)}\right)^2 - \frac{rt}{1+rt},$$

$$\frac{(K-1)^2}{g(r,t,K)}\frac{\partial g}{\partial K} = \frac{(K-1)su}{1+su}[m(s)+m(u)] + \log\frac{1+rt}{1+su}.$$

Letting $\partial g/\partial r = \partial g/\partial t = \partial g/\partial K = 0$ and using the fact that the function $s'\mathcal{K}(s)/(r'\mathcal{K}(r))$ is strictly decreasing ([1], Lemma 10.7 (1),(3)) and definition (1.4), we obtain

$$\left.\begin{aligned} &r = t, \frac{s^2 m(s)}{1+s^2} = \frac{r^2 m(r)}{K(1+r^2)}, \\ &1+s^2 = (1+r^2)\exp\left(\frac{2(K-1)s^2 m(s)}{1+s^2}\right). \end{aligned}\right\} \tag{3.12}$$

Thus, if g has a local maximum at a point $(r,t,K) \in D_2$, then r,t,K must satisfy (3.12), and at this point g takes the value

$$\begin{aligned} g(r,r,K) &= \left[\frac{1+s^2}{(1+r^2)^{1/K}}\right]^{K/(K-1)} \\ &= (1+r^2)\exp\left(\frac{2Ks^2 m(s)}{1+s^2}\right) = \exp(g_3(r)), \end{aligned} \tag{3.13}$$

where $g_3(r) = 2r^2 m(r)/(1+r^2) + \log(1+r^2)$. Hence the upper bound in (3.7) follows from (3.11), (3.13), and Lemma 2 (1).

The remaining conclusions are clear. ■

Theorem 4.

(1) For $K \in [1, \infty)$, define the function f on $[0,1]$ by

$$f(r) = \varphi_K(r) \left[\frac{(1+r)^2}{r} \right]^{1/K}$$

for $r \in (0,1]$, and $f(0) = 4^{1-1/K}$. Then f is strictly increasing from $[0,1]$ onto $[4^{1-1/K}, 4^{1/K}]$ if $K \in [1,2)$, is strictly decreasing from $[0,1]$ onto $[4^{1/K}, 4^{1-1/K}]$ if $K \in (2,\infty)$, and is the constant function $f(r) \equiv 2$ if $K = 2$.

(2) For fixed $r \in (0,1)$ and $K \in (1,\infty)$, define the function g on $[0,1]$ by

$$g(x) = (\varphi_K(r) + x^{1/K})(r+x)^{-1/K}.$$

Then the function g is strictly increasing on $[0, (r/s)^{K/(K-1)}]$, and is decreasing on $[(r/s)^{K/(K-1)}, 1]$, where $s = \varphi_K(r)$. In particular, for $r, x \in (0,1)$ and $K \in (1,\infty)$,

$$\begin{aligned} \min\left\{ \frac{s}{r^{1/K}}, \frac{1+s}{(1+r)^{1/K}} \right\} &< g(x) \leq \left[1 + \left(\frac{s}{r^{1/K}} \right)^{K/(K-1)} \right]^{1-1/K} \\ &< 5^{1-1/K} \end{aligned} \tag{3.14}$$

(3) For $K \geq 2$, the function $h(r) \equiv (r')^{1-1/K}[(1+2r)/r]^{1/K} \varphi_K(r)$ is strictly decreasing from $(0,1)$ onto $(0, 4^{1-1/K})$.

Proof: (1) For $K \in [1,\infty)$, define the function F on $(0,1)$ by

$$F(r) = \frac{s' \mathcal{K}(s)}{(1-r)\mathcal{K}(r)},$$

where $s = \varphi_K(r)$. Then, by logarithmic differentiation, (1.13), and (1.11), we obtain

$$r(r')^2 \mathcal{K}(r)\mathcal{K}'(r)F'(r) = F(r)G(r,K), \tag{3.15}$$

where $G(r,K) = r\mathcal{K}(r)\mathcal{K}'(r) + \mathcal{K}(r)\mathcal{E}'(r) - \mathcal{K}(s)\mathcal{E}'(s)$. Clearly G is strictly decreasing in K on $[1,\infty)$, while $G(r,2) \equiv 0$ by (1.14) and (1.10). Hence, it follows from (3.15) that

$$\left. \begin{aligned} F'(r) \geq 0 \quad &\text{if} \quad K \in [1,2], \\ F'(r) < 0 \quad &\text{if} \quad K \in (2,\infty), \end{aligned} \right\} \tag{3.16}$$

for $r \in (0,1)$, with equality if and only if $K=2$. Differentiation gives

$$\frac{Kr^{1+1/K}}{s(1-r)(1+r)^{2/K-1}} f'(r) = F(r)^2 - 1. \tag{3.17}$$

Since $F(0)=1$, the result for f follows from (3.16), (3.17), and (1.14). We note that $f(0^+) = f(0) = 4^{1-1/K}$ by (1.15).

(2) By differentiation we obtain

$$Kx^{1-1/K}(x+r)^{1+1/K} g'(x) = s\left(\frac{r}{s} - x^{1-1/K}\right),$$

from which the piecewise monotoneity of g follows. The first inequality in (3.14) is clear, the third is implied by the estimate $s < 4^{1-1/K} r^{1/K}$ due to He [6] (*cf.* [8], (3.6), p. 65), and the second follows from

$$\begin{aligned} g(x) &\le g\left(\left(\frac{r}{s}\right)^{K/(K-1)}\right) = \frac{s + (r/s)^{1/(K-1)}}{\left[r + (r/s)^{K/(K-1)}\right]^{1/K}} \\ &= \left[1 + \left(\frac{s}{r^{1/K}}\right)^{K/(K-1)}\right]^{1-1/K}, \end{aligned}$$

where in the last step we have simplified by multiplying numerator and denominator by $(s/r)^{1/(K-1)}$.

(3) Again put $s = \varphi_K(r)$. Then, by logarithmic differentiation and (1.13),

$$Kr\frac{h'(r)}{h(r)} = H(r,K) \equiv \left[\frac{s'\mathcal{K}(s)}{r'\mathcal{K}(r)}\right]^2 - (K-1)\left(\frac{r}{r'}\right)^2 - \frac{1}{1+2r}.$$

Since s is increasing in K by (1.13) and since $s'\mathcal{K}(s)$ is decreasing in s by [1], Theorem 3.21 (7), it is clear that H is strictly decreasing in K on $(1,\infty)$. Since $H(r,2) = -3r^2/[r'^2(1+2r)] < 0$ by (1.14) and (1.10), we see that $H(r,K) < 0$ for all $K \ge 2$ and $0 < r < 1$, so that the monotoneity of h follows.

Clearly, $h(1) = 0$. By (1.15), $h(0^+) = 4^{1-1/K}$. ■

§4 Bounds for the Belinskii function

As we mentioned in the Introduction, the function $\varphi_K(r)$ we have studied in Section 3 has many applications. Theorem 1, which we prove in this section, is an example of these. The proof of this theorem requires the following lemma.

Lemma 3.

(1) For fixed $r \in (0,1)$, the functions $g(K) \equiv \varphi_K(r)(K-1)^{-\frac{1}{2}\log 2} r^{-1/K}$ and $h(K) \equiv \varphi_K(r) K^{-(m(r)+\log r)} r^{-1/K}$ are both strictly decreasing on $(1,\infty)$ with range $(0,\infty)$ and $(0,1)$, respectively, where $m(r) = (2/\pi) r'^2 \mathcal{K}(r)\mathcal{K}'(r)$.

(2) There exists a unique $r_0 \in (0.5251, 0.5252)$ such that the function $G(r) \equiv (2/\pi) r r'^2 \mathcal{K}(r)\mathcal{K}'(r)$ is strictly increasing on $(0, r_0]$ and decreasing on $[r_0, 1)$.

(3) The function $H(r) \equiv r^2 r'^2 \mathcal{K}(r)\mathcal{K}'(r)$ is strictly increasing on $(0, 1/\sqrt{2}]$, and decreasing on $[1/\sqrt{2}, 1)$.

Proof: (1) Let $s = \varphi_K(r)$. Since $K^2/(K-1) \geq 4$ for $K > 1$, since $s > r$, and since the function $r'\mathcal{K}(r)$ is strictly decreasing on $(0,1)$ ([1], Theorem 3.21 (7)), logarithmic differentiation and the fact that $0 < m(r) + \log r < 4$ ([1], Theorem 3.30 (1)) give

$$\begin{aligned}\frac{K^2}{g(K)} g'(K) &= \frac{2}{\pi} s'^2 \mathcal{K}(s)^2 \frac{\mathcal{K}'(r)}{\mathcal{K}(r)} - \frac{K^2}{2(K-1)} \log 2 + \log r \\ &< m(r) + \log(r/4) < 0,\end{aligned}$$

and

$$\begin{aligned}\frac{K^2}{h(K)} h'(K) &= \frac{2}{\pi} s'^2 \mathcal{K}(s)^2 \frac{\mathcal{K}'(r)}{\mathcal{K}(r)} - (m(r) + \log r) K + \log r \\ &< (1-K)(m(r) + \log r) < 0.\end{aligned}$$

Hence the monotoneity properties of g and h follow. The limiting values of g and h are clear.

(2) By differentiation and Legendre's relation (1.11), we get

$$\frac{\pi}{2\mathcal{K}'(r)} G'(r) = G_1(r) \equiv 2\mathcal{E}(r) - (1+r^2)\mathcal{K}(r) - \frac{\pi}{2\mathcal{K}'(r)},$$

which is strictly decreasing from $(0,1)$ onto $(-\infty, \pi/2)$, and hence there exists a unique $r_0 \in (0,1)$ such that G is strictly increasing on $(0, r_0]$ and decreasing on $[r_0, 1)$. Since $G_1(0.5251) = 0.0001\ldots$ and $G_1(0.5252) = 0.0002\ldots$, we see that $r_0 \subset (0.5251, 0.5252)$.

(3) Differentiation and (1.11) give

$$\frac{1}{2r\mathcal{K}'(r)} H'(r) = H_1(r) \equiv \mathcal{E}(r) - r^2\mathcal{K}(r) - \frac{\pi}{4\mathcal{K}'(r)},$$

which is strictly decreasing from $(0,1)$ onto $(-\infty, \pi/2)$ with $H_1(1/\sqrt{2}) = 0$ by Legendre's relation (1.11), and hence the piecewise monotoneity of H follows. ■

Proof of Theorem 1: Set $s = \varphi_K(r)$. Then, by differentiation,

$$\frac{K^2}{s} f'(K) = f_1(K) \equiv \frac{2}{\pi} s'^2 \mathcal{K}(s)^2 \frac{\mathcal{K}'(r)}{\mathcal{K}(r)} - \frac{m(r)}{h(K)} - \frac{m(r)}{g(K)} \log \frac{1}{r},$$

where

$$g(K) = \frac{s}{(K-1)r^{1/K}} = \frac{s}{(K-1)^{\frac{1}{2}\log 2} r^{1/K}} \cdot \frac{1}{(K-1)^{1-\frac{1}{2}\log 2}}$$

and

$$h(K) = \frac{s}{K^2 r^{1/K}} = \frac{s}{K^{m(r)+\log r} r^{1/K}} \cdot \frac{1}{K^{2-(m(r)+\log r)}}.$$

By Lemma 3 and the fact that $m(r) + \log r < \log 4$ ([1], Theorem 3.30 (1)) it follows that g and h are both strictly decreasing on $(1, \infty)$. Hence, f_1 is strictly decreasing from $(1, \infty)$ onto $(-\infty, 0)$, and hence the monotoneity of f follows. The limiting values of f are clear.

The second bound in the first inequality in (1.7) and the second inequality in (1.7) are clear. By (1.13),

$$\frac{ds}{dK} = \frac{2}{\pi K} s s'^2 \mathcal{K}(s)\mathcal{K}'(s) = \frac{1}{K} G(s), \tag{4.1}$$

where G is as in Lemma 3. Since rr' is strictly increasing on $(0, 1/\sqrt{2})$ and since $r'\mathcal{K}(r)$ is strictly decreasing on $(0,1)$ ([1], Theorem 3.21 (7)), it follows from part (2) of Lemma 3 that for all $r \in (0,1)$,

$$\begin{aligned} G(r) &\leq G(r_0) < \frac{2}{\pi} \cdot 0.5252 \cdot C \cdot \mathcal{K}(0.5251) \cdot \mathcal{K}'(0.5251) \\ &= 0.8696\ldots < 0.8697; \qquad C = \sqrt{1-0.5252^2} \cdot \sqrt{1-0.5251^2}. \end{aligned}$$

Hence it follows from (4.1) that $\frac{ds}{dK} < 0.8697/K$. Integrating with respect to K from 1 to K, we get the first bound in the first inequality in (1.7).

Finally, it follows from part (3) of Lemma 3 that

$$\frac{\partial}{\partial K}(\varphi_K(\sqrt{r})^2 - r) = \frac{4}{\pi K} H(\varphi_K(\sqrt{r})) < \frac{a}{K},$$

and hence we obtain (1.8) by integration with respect to K from 1 to K. Since for $r = 1/\sqrt{2}$,

$$\lim_{K \to 1} \frac{\varphi_K(\sqrt{r})^2 - r}{K-1} = \frac{4}{\pi} H(1/\sqrt{2}) = a$$

by l'Hôpital's Rule, the last assertion of Theorem 1 follows. ■

Acknowledgement. This work was completed during the second author's visits to Michigan State University at East Lansing, MI, U.S.A., and to the University of Helsinki, Finland. His visits were supported by grants from the Department of Mathematics at Michigan State University, the Finnish Academy of Science and Letters, the Academy of Finland, and the Commission on Development and Exchanges in Paris. He wishes to express his thanks to these institutions.

References

[1] Anderson, G. G., M. K. Vamanamurthy and M. Vuorinen, *Conformal Invariants, Inequalities, and Quasiconformal Maps*, John Wiley and Sons, New York 1997.

[2] Belinskii, P. P., *General Propcrtics of Quasiconformal Mappings*, Izdat. Nauka. Sibirsk. Otdelenie, Novosibirsk, 1974 (in Russian).

[3] Berndt, B. C., *Ramanujan Notebooks, Part III*, Springer-Verlag, New York, 1991.

[4] Berndt, B. C., S. Bhargava, and F. G. Garvan, Ramanujan's theories of elliptic functions to alternative bases, *Trans. Amer. Math. Soc.* **347** (1995), 4163–4244.

[5] Borwein J. M. and P. B. Borwein, *Pi and the AGM*, John Wiley and Sons, New York (1987).

[6] He, C.-Q., Distortion estimates of quasiconformal mappings, *Sci. Sinica. Ser. A* **27** (1984), 225–232.

[7] Hersch J. and A. Pfluger, Généralization du lemme de Schwarz et du principe de la mesure harmonique pour les fonctions pseudo-analytique, *C. R. Acad. Sci. Paris* **234** (1952), 43–45.

[8] Lehto O. and K. I. Virtanen, *Quasiconformal Mappings in the Plane*, 2nd ed., Grundlehren Math. Wiss., Band 126, Springer-Verlag, Berlin–Heidelberg–New York, 1973.

[9] Partyka, D., Approximation of the Hersch-Pfluger distortion function, *Ann. Acad. Sci. Fenn.* **18** (1993), 343–354.

[10] Partyka, D., The maximal value of the function $[0,1] \ni r \mapsto \Phi_K^2(\sqrt{r}) - r$, *Bull. Soc. Sci. Lett. Łódź Rech. Déform.* **45** (1995), 49–55.

[11] Qiu, S.-L., Distortion properties of K-q.c. maps and a better estimate of Mori's constant, *Acta Math. Sinica* **35**(4) (1992), 492–504.

[12] Qiu, S.-L., Agard's η-distortion function and Schottky's theorem, *Science in China Ser. A*, **26** (8) (1996), 683–690.

[13] Vamanamurthy M. K. and M. Vuorinen, Functional inequalities, Jacobi products and quasiconformal maps, *Illinois J. Math.* **38** (1994), 394–419.

[14] Vuorinen, M., *Conformal Geometry and Quasiregular Mappings*, Lecture Notes in Math., Vol. 1319, Springer-Verlag, Berlin–Heidelberg–New York, 1988.

[15] Zając, J., The distortion function Φ_K and quasihomographies, in *Current Topics in Analytic Function Theory*, H. M. Srivastava and S. Owa (eds.), World Scientific Publ., 1992, 403–428.

Glen Douglas Anderson
Department of Mathematics
Michigan State University
East Lansing, MI 48824
U.S.A.
anderson@math.msu.edu

Songliang Qiu
President's Office
Hangzhou Institute of Electronics Engineering
Hangzhou 310037
P.R. CHINA

Matti K. Vuorinen
Department of Mathematics
P.O. Box 4 (Yliopistonkatu 5)
FIN-00014 University of Helsinki
Helsinki
FINLAND
vuorinen@csc.fi

Locally convex curves and distribution of zeros of polynomials

Vladimir Andrievskiĭ

Abstract. We introduce the class of locally Dini-convex curves which includes convex curves as well as piecewise Dini-smooth curves without cusps. We obtain estimates for the discrepancy between the equilibrium measure μ_L of a locally Dini-convex curve L and the zero-counting measure ν_{p_n} of a monic polynomial p_n of degree $n = 1, 2, \ldots$ in terms of one-sided bounds for the difference $U(\mu_L - \nu_{p_n}, z)$ of their logarithmic potentials. Both situations — when either an upper or a lower bound of $U(\mu_L - \nu_{p_n}, z)$ is known — are treated simultaneously.

§1 Introduction

We are interested in the distribution of the zeros of monic polynomials. To explain the subject of this paper we need to recall a classical theorem of Erdős and Turán.

Let $p(z) = p_n(z) = a_n z^n + \ldots + a_0$ be a polynomial of degree $n = 1, 2, \ldots$. We associate with p the zero-counting measure

$$\nu(A) = \nu_p(A) := \frac{\text{number of zeros of } p \text{ in } A}{n} \qquad (A \subset \mathbb{C}), \tag{1.1}$$

where the zeros are counted with their multiplicities.

Erdős and Turán [8] have proved that if $a_0\, a_n \neq 0$ then the number of zeros of p in the sector

$$S_{\alpha,\beta} := \{re^{i\theta} : r \geq 0, \alpha \leq \theta \leq \beta\} \qquad (0 \leq \alpha < \beta \leq 2\pi)$$

can be estimated as follows:

$$\left| \nu(S_{\alpha,\beta}) - \frac{\beta - \alpha}{2\pi} \right| \leq 16 \sqrt{\frac{\log P}{n}}, \tag{1.2}$$

where

$$P := \frac{1}{\sqrt{|a_n\, a_0|}} \sup_{|z| \leq 1} |p(z)|.$$

Computational Methods and Function Theory (CMFT'97)
N. Papamichael, St. Ruscheweyh and E. B. Saff (Eds.), pp. 23–44

ISBN 981-02-3626-3

By virtue of a simple observation due to Schur, it is well-known that the nontrivial part of the proof of (1.2) concerns only the case when all zeros of p belong to the unit circle.

Many mathematicians have generalized this result under various conditions (for a bibliography, see [5, 6]).

A second result, which is close to the considerations below, concerns the case when all zeros of p are simple and belong to some smooth Jordan curve L. In this case the analogue of (1.2) can be written in terms of lower bounds for $|p'(z)|$ at the zeros of p. The first result of such kind is due to [7]. Using essentially the technique of [10], its authors have proved some discrepancy estimates for the distribution of extreme points in best complex polynomial approximation. A refinement of this result was made in [9].

Our main purpose is to extend the assertions above to the case of Jordan domains which can possess corners (not cusps) on the boundary. It is surprising that the passage from the unit circle or sufficiently smooth curve to a curve with much more complicated geometrical structure does not influence the shape of the results.

We apply essentially some constructions from [4].

§2 Definitions and formulation of the main results

Let $L \subset \mathbb{C}$ be a bounded Jordan curve in the complex plane $\mathbb{C}$. Denote by $G := \text{int } L$ the collection of points interior to L. In addition, set $\Omega :=$ ext $L = \overline{\mathbb{C}} \backslash \overline{G}$, where $\overline{\mathbb{C}} := \mathbb{C} \cup \{\infty\}$ is the extended complex plane. Let Φ denote the Riemann function that conformally and univalently maps Ω onto the exterior $\Delta := \overline{\mathbb{C}} \backslash \overline{\mathbb{D}}$ of the unit disk $\mathbb{D} := \{w : |w| < 1\}$ and which is normalized by the conditions

$$\Phi(\infty) = \infty, \quad \Phi'(\infty) := \lim_{z \to \infty} \frac{\Phi(z)}{z} = (\text{cap } L)^{-1} > 0,$$

where cap L denotes the logarithmic capacity of L [15]. Set $\Psi := \Phi^{-1}$.

By $\mu = \mu_L$ we denote the equilibrium measure for L [15].

In what follows L will be a quasiconformal curve [1]. We recall that a Jordan curve L is called K-quasiconformal, $K \geq 1$, or, briefly, quasiconformal, if there exists a K-quasiconformal mapping of the plane onto itself which carries L into a circle. Ahlfors (see [1]) has established a geometric criterion for the quasiconformality of a curve which can be formulated as follows: A Jordan curve L is quasiconformal if and only if for any pair of points z_1 and $z_2 \in L$ the inequality

$$\min\{\text{diam } L', \text{diam } L''\} \leq c\,|z_1 - z_2|$$

holds with some constant $c = c(L) \geq 1$, where L' and L'' are the two arcs which $L\backslash\{z_1, z_2\}$ consists of.

For more on quasiconformal curves, see [1, 11].

Let $\mathbb{M}_n$, $n = 1, 2, \ldots$, denote the set of monic polynomials $p(z) = p_n(z) = z^n + \ldots$ of degree n. We associate with $p \in \mathbb{M}_n$ the normalized counting measure for its zeros given by (1.1). Consider the logarithmic potentials of the measures μ and ν:

$$\begin{aligned} U(\mu, z) &:= -\int \log|z-\zeta|\,d\mu(\zeta) \\ &= \begin{cases} -\log|\Phi(z)| - \log(\text{cap } L), & \text{if } z \in \Omega\backslash\{\infty\}, \\ -\log(\text{cap } L), & \text{if } z \in \overline{G}, \end{cases} \\ U(\nu, z) &:= -\int \log|z-\zeta|\,d\nu(\zeta) = -\frac{1}{n}\log|p(z)| \end{aligned}$$

and their difference

$$U(\mu - \nu, z) := U(\mu, z) - U(\nu, z).$$

Denote by $Z(p)$ the set of all zeros of p. If $Z(p) \subset L$, then a standard way to measure the discrepancy between the measures μ and ν is by means of in terms of the quantity

$$D[\mu - \nu] := \sup|(\mu - \nu)(J)|,$$

where the supremum is taken over all subarcs $J \subset L$.

Let $z_0 \in G$ be a fixed point. Denote by $w = \varphi(z)$ the conformal mapping of G onto $\mathbb{D}$ with the normalization $\varphi(z_0) = 0$, $\varphi'(z_0) > 0$. Set $\psi := \varphi^{-1}$. The functions $\Phi, \Psi, \varphi, \psi$ can be naturally extended to homeomorphisms between the appropriate closed domains, and we keep the previous notations for these extensions.

Further, set

$$\begin{aligned} L_r &:= \{\zeta : |\Phi(\zeta)| = 1 + r\} \qquad (r \geq 0), \\ L_{-r} &:= \{\zeta : |\varphi(\zeta)| = 1 - r\} \qquad (0 \leq r < 1) \end{aligned}$$

and denote by the symbol $\|\cdot\|_A$ the supremum norm over a set $A \subset \mathbb{C}$. Our main results will concern the case when $Z(p) \subset L$ and will be based on the behavior of the following quantities:

$$\begin{aligned} a_{\sup}(\delta) &:= \sup_{z \in L_\delta \cup L_{-\delta}} U(\mu - \nu, z) \geq 0, \\ a_{\inf}(\delta) &:= -\inf_{z \in L_\delta \cup L_{-\delta}} U(\mu - \nu, z) \geq 0. \end{aligned}$$

More precisely, a monic polynomial $p \in \mathbb{M}_n$ will satisfy one of the following two conditions.

(i) For some $2 < A_n \le e^{n/e}$

$$||p_n||_L \le A_n (\text{cap } L)^n \tag{2.1}$$

and therefore

$$a_{\text{sup}}(\delta) \le \frac{1}{n} \log A_n \quad (\delta > 0). \tag{2.2}$$

(ii) For some $n \le A_n \le e^{n/e}$

$$|p_n'(z)| \ge \frac{1}{A_n} (\text{cap } L)^n \qquad (z \in Z(p)), \tag{2.3}$$

and therefore for $0 < \delta < 1/e$

$$a_{\text{inf}}(\delta) \le c_1 \left(\frac{\log A_n}{n} + \frac{\log 1/\delta}{n} + \delta \right) \tag{2.4}$$

with some constant $c_1 = c_1(L, z_0) > 0$ (for details, see [6]).

Next we are going to explain the notion of a locally Dini-convex curve which is a natural generalization of the notion of a convex curve and a curve consisting of a finite number of Dini-smooth arcs (see [12, 13]) without cusps.

Without additional stipulation we assume below that L is a quasiconformal curve. The necessity of this restriction for the validity of the Erdős-Turán type theorems is explained in [4].

Let h be a positive increasing function satisfying the Dini-condition

$$\int_0^1 \frac{h(x)}{x} < \infty \tag{2.5}$$

and let

$$W(h, \varepsilon) := \{ z = re^{i\theta} : 0 < r < \varepsilon, \, h(r) < \theta < \pi - h(r) \} \quad (0 < \varepsilon \le 1).$$

We say that a subarc $J \subseteq L$ is Dini-convex with respect to E ($= G$ or Ω) if there exist $0 < \varepsilon \le 1$ and a function h satisfying (2.5) such that for any $z \in J$

$$\{\zeta = z + e^{i\theta}\xi : \xi \in W(h, \varepsilon)\} \subset E$$

with some $0 \le \theta = \theta(z) < 2\pi$.

We say that L is locally Dini-convex if there is an $\varepsilon > 0$ such that each subarc $J \subset L$ with diam $J \le \varepsilon$ is either Dini-convex with respect to G or Dini-convex with respect to Ω.

It is easy to see that by our definition each convex curve is Dini-convex with respect to Ω (with $h \equiv 0$). More generally, if there is an $r > 0$ depending only on L such that for each $z \in L$ there is an open disk D_z with radius r such that $D_z \subset E\,(= \Omega$ or $G)$ and $\overline{D}_z \cap L = \{z\}$, then L is Dini-convex with respect to E (with $h(x) = c_2\,x$, $c_2 = c_2(r) = \text{const} > 0$).

At the same time the following curves belong to the class under consideration. Following [13] we call a Jordan curve L Dini-smooth if it is smooth and if the angle $\beta(s)$ of the tangent, considered as a function of the arc length s, satisfies

$$|\beta(s_2) - \beta(s_1)| \le h(s_2 - s_1) \quad (0 < s_2 - s_1 < 1),$$

where h is a function satisfying (2.5).

We call a Jordan arc Dini-smooth if it is a subarc of some Dini-smooth curve.

It is not difficult to see that if L consists of finitely many Dini-smooth arcs and does not have cusps, then L is locally Dini-convex.

Theorem 1. *Let L be an arbitrary locally Dini-convex curve, $z_0 \in \operatorname{int} L$ be fixed. Let the monic polynomial $p = p_n \in \mathbb{M}_n, n \ge 2$, satisfy $Z(p) \subset L$. Then there exist constants $c = c(L, z_0) > 0$ and $0 < \delta_0 = \delta_0(L, z_0)$ such that for $0 < \delta < \delta_0$*

$$D[\mu - \nu] \le c\sqrt{\delta}\left(\frac{2\,a_{\sup}(\delta/2) - U(\mu - \nu, z_0)}{\delta} + 1\right), \tag{2.6}$$

$$D[\mu - \nu] \le c\sqrt{\delta}\left(\frac{2\,a_{\inf}(\delta/2) + U(\mu - \nu, z_0)}{\delta} + 1\right). \tag{2.7}$$

Therefore, the inequality

$$D[\mu - \nu] \le c_3\sqrt{\delta_n} \tag{2.8}$$

holds with

$$\delta_n := \frac{1}{n}\log\frac{A_n^2(\operatorname{cap} L)^n}{|p_n(z_0)|} \tag{2.9}$$

if p satisfies (2.1), and with

$$\delta_n := \frac{1}{n}\log\frac{A_n^{c_4}|p_n(z_0)|}{(\operatorname{cap} L)^n} \tag{2.10}$$

if p satisfies (2.3), where $c_j = c_j(L, z_0) = \text{const} > 1$, $j = 3, 4$.

This theorem generalizes corresponding results in [5, 4].

It will be seen from the proof of Theorem 1 (*cf.* (4.19)) that if L is Dini-convex with respect to Ω then we can omit the term $U(\mu - \nu, z_0)$ in the estimates (2.6) and (2.7). Therefore, in this case relation (2.8) will be valid with

$$\delta_n = \frac{1}{n} \log A_n$$

for any monic polynomial $p \in \mathbb{M}_n$ satisfying either (2.1) or (2.3).

We complete this section with an assertion which shows the sharpness of the last result.

Theorem 2. *Let L be an arbitrary Jordan curve with* cap $L = 1$. *For any $n < A_n \leq e^n$, $n \geq 2$, there exists a monic polynomial $p \in \mathbb{M}_n$ with zeros on L, satisfying condition (2.3) and such that*

$$D[\mu_L - \nu_p] \geq c\sqrt{\frac{\log A_n}{n}}, \quad c = \frac{1}{\sqrt{32\,\pi^3}}. \tag{2.11}$$

Below, we use the following notations:

$$\text{dist}\,(A,B) = \inf_{z \in A, \zeta \in B} |z - \zeta| \quad (A, B \subset \mathbb{C}).$$

By $|\gamma|$ we denote the length of a (rectifiable) Jordan arc γ.

Further, for $\zeta \in \mathbb{C}\backslash\{z_0\}$, $0 < \delta < 1$ and $J \subset L$ we set

$$\begin{aligned} \zeta_L &:= \begin{cases} \Psi\left(\frac{\Phi(\zeta)}{|\Phi(\zeta)|}\right), & \text{if } \zeta \in \Omega\backslash\{\infty\}, \\ \psi\left(\frac{\varphi(\zeta)}{|\varphi(\zeta)|}\right), & \text{if } \zeta \in G\backslash\{z_0\}, \\ \zeta, & \text{if } \zeta \in L, \end{cases} \\ E_\delta &:= (\text{ext}\, L_{-\delta}) \cap (\text{int}\, L_\delta), \\ A_\delta(J) &:= \{\zeta \in E_\delta : \zeta_L \in J\}. \end{aligned}$$

By $L(z,\zeta)$, $z, \zeta \in L$, we denote the smaller (with respect to diameter) connected component of $L \setminus \{z, \zeta\}$.

Further, we denote by $c, c_1, \ldots$ positive constants and by $\varepsilon, \varepsilon_1, \ldots$ sufficiently small positive constants (different each time, in general) that either are absolute or depend on parameters not essential for the arguments; sometimes such a dependence will be indicated. For $a > 0$ and $b > 0$ we use the expression $a \preceq b$ (order inequality) if $a \leq c\,b$ for some $c > 0$. The expression $a \asymp b$ means that $a \preceq b$ and $b \preceq a$, simultaneously.

§3 Some auxiliary facts from geometric function theory

If L is a quasiconformal curve, it is known (see [1], Chapter IV) that the conformal mappings Φ and φ can be extended to quasiconformal mappings of the whole plane onto itself with ∞ as a fixed point. We keep the previous notations for these extensions. Note that the inverse functions $\Psi := \Phi^{-1}$ and $\psi := \varphi^{-1}$ will be quasiconformal, too.

The following result (see [3], p. 97) is useful in the study of metric properties of the mappings $\Phi, \Psi, \varphi, \psi$.

Lemma 1. *Let $w = F(\zeta)$ be a K-quasiconformal mapping of the plane onto itself with $F(\infty) = \infty$, $\zeta_j \in \mathbb{C}$, $w_j := F(\zeta_j)$, $j = 1, 2, 3$, and $|w_1 - w_2| \leq c_1|w_1 - w_3|$. Then $|\zeta_1 - \zeta_2| \leq c_2 |\zeta_1 - \zeta_3|$ and, in addition,*

$$\left|\frac{\zeta_1 - \zeta_3}{\zeta_1 - \zeta_2}\right| \leq c_3 \left|\frac{w_1 - w_3}{w_1 - w_2}\right|^K,$$

where $c_j = c_j(c_1, K)$, $j = 2, 3$.

If a curve L is locally Dini-convex it implies some special distortion properties of the functions Φ and φ.

Lemma 2. *Let L be locally Dini-convex, $z \in L$ belong to some subarc $J \subset L$ which is Dini-convex with respect to Ω, $\tau := \Phi(z)$. Then for any points $\tau_1, \tau_2 \in \overline{\Delta}$ with $|\tau_1 - \tau| \leq |\tau_2 - \tau| \leq 1$ and their preimages $z_j := \Psi(\tau_j)$, $j = 1, 2$, the inequalities*

$$\left|\frac{\tau_2 - \tau}{\tau_1 - \tau}\right| \leq c_1 \left|\frac{z_2 - z}{z_1 - z}\right|, \tag{3.1}$$

$$|z_1 - z| \leq c_2 |\tau_1 - \tau| \tag{3.2}$$

hold with $c_j = c_j(L) > 0$, $j = 1, 2$.

Proof: Note that (3.2) is an obvious consequence of (3.1). Without loss of generality we assume that

$$|z_1 - z| < |z_2 - z| < \varepsilon$$

(otherwise (3.1) follows easily from Lemma 1). Next we introduce the following notations. Denote by $\gamma_z(x) \subset \Omega$, $0 < x < \varepsilon$, that subarc of the circle $\{\xi : |\xi - z| = x\}$ which separates in Ω the point z from ∞. Let $Q_z(\delta, t)$, $0 < \delta < t < \varepsilon$, be the quadrilateral bounded by the arcs $\gamma_z(\delta)$, $\gamma_z(t)$ and the two subarcs of L joining their endpoints. Denote by $\Gamma_z(\delta, t)$ the family of all locally rectifiable arcs in $Q_z(\delta,t)$ which separate the sides $\gamma_z(\delta)$ and $\gamma_z(t)$ and by $m_z(\delta,t)$ the module of $\Gamma_z(\delta,z)$ (see [1, 11]).

For any triplet of points $\xi_1,\xi_2,\xi_3\in\overline{\Omega}$ with $|\xi_1-\xi_2|=|\xi_1-\xi_3|$ we have by Lemma 1

$$|\Phi(\xi_1)-\Phi(\xi_2)|\asymp|\Phi(\xi_1)-\Phi(\xi_3)|.$$

Hence, according to [3, p. 36]

$$\left|\frac{\tau_2-\tau_1}{\tau_1-\tau}\right|\asymp\exp(\pi m_z(|z_1-z|,|z_2-z|)).$$

Therefore, inequality (3.1) is equivalent to

$$m_z(\delta,t)\le\frac{1}{\pi}\log\frac{t}{\delta}+c_3\qquad(0<\delta<t<\varepsilon).\tag{3.3}$$

By our assumption

$$\begin{aligned}Q_z(\delta,t)\quad&\supset\quad Q_z^*(\delta,t)\\ &:=\quad\{\zeta=z+re^{i(\theta+\theta_0)}:\delta<r<t,\,h(r)<\theta<\pi-h(r)\},\end{aligned}$$

where $\theta_0=\theta_0(z)\in[0,2\pi)$ and h satisfies the Dini-condition (2.5).

Let $m(Q_z^*(\delta,t))$ be the module of $Q_z^*(\delta,t)$, *i.e.* the module of the family Γ^* of all locally rectifiable arcs in $Q_z^*(\delta,t)$ which separate the circular subarcs of $\partial Q_z^*(\delta,t)$.

To prove (3.3) it is enough, by virtue of the comparison principle for moduli of families of curves, to show that

$$m(Q_z^*(\delta,t))=m(\Gamma^*)\le\frac{1}{\pi}\log\frac{t}{\delta}+c_4.\tag{3.4}$$

In order to establish (3.4) we assume that $h(\varepsilon)<\pi/4$ and consider the functions

$$\begin{aligned}v(x)\quad&:=\quad\min\{h(e^{\pi}x),h(t)\}\qquad(\delta<x<t),\\ \rho(z)\quad&=\quad\frac{1}{|z|\,(\pi-2\,v(|z|))}\qquad(z\in Q_z^*(\delta,t)).\end{aligned}$$

Note that for any $\gamma\in\Gamma^*$

$$\int_\gamma\rho(z)\,|dz|\ge1.$$

In fact, set

$$r_{\max}:=\sup_{z\in\gamma}|z|,\quad r_{\min}:=\inf_{z\in\gamma}|z|.$$

If $r_{\max} \geq e^{\pi} r_{\min}$, then there exists an arc $\gamma_1 \subset \gamma$ contained in the annulus $\{z : r_{\min} < |z| < r_{\max}\}$ whose endpoints connect the two boundary components of the annulus. Therefore,

$$\int_{\gamma} \rho(z)\,|dz| \geq \frac{1}{\pi}\left|\int_{\gamma_1} \frac{dz}{z}\right| \geq \frac{1}{\pi}\log\frac{r_{\max}}{r_{\min}} \geq 1.$$

If $r_{\max} < e^{\pi} r_{\min}$, then

$$\int_{\gamma} \rho(z)\,|dz| \geq \frac{1}{\pi - 2\,v(r_{\min})}\left|\int_{\gamma} \frac{dz}{z}\right| \geq 1.$$

The definition of the module yields

$$\begin{aligned} m(Q_z^*(\delta,t)) &\leq \int_{Q_z^*(\delta,t)} \rho^2(z)\,dm(z) = \int_{\delta}^{t} \frac{(\pi - 2\,h(r))\,dr}{r(\pi - 2\,v(r))^2} \\ &\leq \frac{1}{\pi}\int_{\delta}^{t}\frac{dr}{r} + 4\int_{\delta}^{t}\frac{v(r)\,dr}{r\,(\pi - 2\,v(r))^2} \\ &\leq \frac{1}{\pi}\log\frac{t}{\delta} + \frac{16}{\pi^2}\int_{e^{2\pi}\delta}^{e^{2\pi}t}\frac{h(r)}{r}\,dr \leq \frac{1}{\pi}\log\frac{t}{\delta} + c_4. \quad \blacksquare \end{aligned}$$

Remark: Changing in the proof of Lemma 2 the places of Ω and G we obtain the following assertion:

Let L be locally Dini-convex, $z \in L$ belong to some subarc $J \subset L$ which is Dini-convex with respect to G, $w := \varphi(z)$. Then for any points $w_1, w_2 \in \overline{\mathbb{D}}$ with $|w_1 - w| \leq |w_2 - w| \leq 1$ and their preimages $z_j := \psi(w_j)$, $j = 1, 2$, the inequalities

$$\begin{aligned} \left|\frac{w_2 - w}{w_1 - w}\right| &\leq c_1\left|\frac{z_2 - z}{z_1 - z}\right|, \\ |z_1 - z| &\leq c_2\,|w_1 - w| \end{aligned}$$

hold with $c_j = c_j(L, z_0) > 0$, $j = 1, 2$.

Let $z \in \mathbb{D}$ and let $l \subset \partial\mathbb{D}$ be an arc. We denote by $\omega(z, \mathbb{D}, l)$ the harmonic measure of l at the point z with respect to $\mathbb{D}$. The next lemma is an immediate consequence of the Poisson formula.

Lemma 3. *For any $z \in \mathbb{D}$ the inequality*

$$\omega(z, \mathbb{D}, l) \leq c_1 \frac{1 - |z|}{d(z, l)}$$

holds with some universal constant $c_1 \geq 1$.

§4 Proof of Theorem 1

We restrict ourselves to the proof of (2.6). The inequality (2.7) can be established in the same way. Estimates (2.8)–(2.10) follow directly from (2.6) and (2.7) via (2.2) and (2.4), respectively.

So let us assume that

$$U^{\mu-\nu}(z) \le a := a_{\sup}(\delta/2) \qquad (z \in \text{ext } L_{\delta/2} \cup \text{int } L_{-\delta/2}). \tag{4.1}$$

Consider an arbitrary subarc $J \subset L$. For our purposes it is enough to establish the inequality

$$(\nu - \mu)(J) \le c\sqrt{\delta}\left(\frac{2a - U(\mu - \nu, z_0)}{\delta} + 1\right) \tag{4.2}$$

for sufficiently small δ and sufficiently small J, such that some fixed neighbourhood of J is Dini-convex either with respect to G or with respect to Ω.

Also, set

$$\begin{aligned} \gamma &:= \Phi(J) = \{e^{i\theta} : \theta_1 \le \theta \le \theta_2\}, \\ \gamma(r) &:= \{e^{i\theta} : \theta_1 - r \le \theta \le \theta_2 + r\} \quad (r > 0), \\ J(r) &:= \Psi(\gamma(r)) \quad (r > 0). \end{aligned}$$

We assume that $\theta_2 - \theta_1 \le \pi/2$ and that for some $0 < \varepsilon < \pi/6$ the arc $J(3\varepsilon)$ is Dini-convex with respect to Ω (the case of Dini-convexity of J with respect to G can be handled in quite the same way).

There are two possibilities:

(i) L is Dini-convex with respect to Ω;

(ii) There is a subarc of L which is Dini-convex with respect to G.

Since the second case is obviously more complicated, we investigate it in detail and explain how we have to change our constructions in case (i).

So suppose that (ii) holds. Let $\zeta_1 := \Psi(e^{i\theta_1})$ and $\zeta_2 := \Psi(e^{i\theta_2})$ be the endpoints of J.

Further, we are going to define points ξ_1 and $\xi_2 \in L\backslash J(2\varepsilon)$ that are "nearest" to ζ_1 and ζ_2, respectively, from such ones at which L becomes convex with respect to G.

This can be done, for example, in the following way. Obviously, there exists a finite number of subarcs $J_k \subset L$, $k = 1, \ldots, m$, such that J_k is Dini-convex either with respect to G or with respect to Ω, and

$$L \setminus J(3\varepsilon) \subset \bigcup_{k=1}^{m} J_k \subset L \setminus J(2\varepsilon). \tag{4.3}$$

Denote by $z_1, \dots, z_l$, $l \le 2m$, the endpoints of those J_k's which are Dini-convex with respect to G (if this set is empty we are in the frame of case (i)).

Let ξ_2 be the first point from $\{z_1, \dots, z_l\}$ when moving from ζ_2 along L in the positive direction and denote by ξ_1 the last point of such kind.

By our construction the subarc of L between the points ξ_1 and ξ_2 containing J is Dini-convex with respect to Ω. At the same time the points ξ_j, $j = 1, 2$, are also endpoints of subarcs Dini-convex with respect to G.

Therefore, by Lemma 1, Lemma 2 and the Remark to Lemma 2 for $z \in E_\delta$ with $z_L = \xi_j$, $j = 1, 2$, we have

$$\text{dist}\,(z, L) \preceq \delta. \tag{4.4}$$

In what follows $\delta > 0$ is "much" smaller than the fixed number ε.

The points $(1-\delta)\varphi(\xi_j), j = 1, 2$, divide the circle $\{\tau : |\tau| = 1 - \delta\}$ into two parts. Denote by S_i' that one which does not include the points $(1-\delta)\ \varphi(\zeta_j)$. Let $S_i := \psi(S_i')$.

Set $t := \sqrt{\delta}$ and consider the function

$$h(z) := \begin{cases} 1, & \text{if } z \in A_{2t}(J(t)), \\ 0, & \text{otherwise in } \mathbb{C}. \end{cases}$$

Let $\rho(z)$, $z \in \mathbb{C}$, be a regularized distance to $A_t(J)$ (see [14]), *i.e.* a function with the following properties

$$\frac{1}{2}\,\text{dist}(z, A_t(J)) \le \rho(z) \le 2\,\text{dist}(z, A_t(J)) \quad (z \in \mathbb{C}), \tag{4.5}$$

$$\rho(z) \in C^\infty(\mathbb{C}), \tag{4.6}$$

$$\left|\frac{\partial^{j+k}}{\partial x^j \partial y^k}\,\rho(x+iy)\right| \le c(j,k)\,\rho(x+iy)^{1-j-k} \quad (j, k = 0, 1, \dots). \tag{4.7}$$

Next, we average the function h in the following way

$$g(z) := \begin{cases} \frac{16}{\rho(z)^2}\int_{\mathbb{C}} h(\zeta) K\left(\frac{4|\zeta - z|}{\rho(z)}\right) dm(\zeta), & \text{if } z \in \mathbb{C}\setminus A_t(J), \\ 1, & \text{if } z \in A_t(J), \end{cases} \tag{4.8}$$

where $K(\zeta)$ is an arbitrary symmetric averaging kernel, *i.e.* $K(z) \in C^\infty(\mathbb{C})$,

$$\begin{aligned} K(z) &= K(|z|) \ge 0 \qquad (z \in \mathbb{C}), \\ K(z) &= 0 \qquad (|z| \ge 1), \\ \int K(z)\,dm(z) &= 1. \end{aligned}$$

Note that, according to (4.6), $g \in C^\infty(\mathbb{C})$. By Lemma 1 there exists $\varepsilon_1 > 0$ such that for $z \in L_\delta$ with $z_L \in J(2\varepsilon_1 t)$,

$$\operatorname{dist}(z, A_t(J)) < \operatorname{dist}(z, \partial A_{2t}(J(t))).$$

Therefore, using (4.5) we obtain

$$g(z) = 1 \qquad (z \in L_\delta,\ z_L \in J(2\varepsilon_1 t)). \tag{4.9}$$

Further, by Lemma 1 there exists a sufficiently large constant $c_1 > 0$ such that for $z \in L_\delta$ with $z_L \in L \setminus J(c_1 t)$,

$$\operatorname{dist}(z, A_{2t}(J(t))) > \frac{1}{2} \operatorname{dist}(z, A_t(J)).$$

Therefore, by (4.5) for such z

$$\rho(z) \le 2 \operatorname{dist}(z, A_t(J)) < 4 \operatorname{dist}(z, A_{2t}(J(t)))$$

and we obtain

$$g(z) = 0 \qquad (z \in L_\delta,\ z_L \in L \setminus J(c_1 t)). \tag{4.10}$$

Note that for the restriction g_S of g to $S := S_i \cup L_\delta$ we have in addition to (4.9) and (4.10) that

$$g_S(z) = 0 \qquad (z \in S_i).$$

Further, if $z = x + iy$, $\xi = x_0 + iy_0 \in L_\delta$ with $z_L, \xi_L \in L(\zeta_3, \zeta_1)$, where $\zeta_3 := \Psi(e^{i(\theta_1 - 3c_1 t)})$, then by the Taylor formula

$$g_S(z) = g_S(\xi) + A(\xi)(x - x_0) + B(\xi)(y - y_0) + r(z, \xi), \tag{4.11}$$

where according to (4.7) we have

$$|A(\xi)| + |B(\xi)| \preceq |\zeta_1 - \zeta_3|^{-1}, \tag{4.12}$$

$$|r(z, \xi)| \preceq \frac{|z - \xi|^2}{|\zeta_1 - \zeta_3|^2}. \tag{4.13}$$

The same relations are valid for $z, \xi \in L_\delta$ with $z_L, \xi_L \in L(\zeta_2, \zeta_4)$, where $\zeta_4 := \Psi(e^{i(\theta_2 + 3c_1 t)})$.

Let f be the function harmonic in $\overline{\mathbb{C}} \setminus S$ with continuous boundary values

$$f(z) := g_S(z) \qquad (z \in S).$$

Set

$$\begin{aligned} \tilde{f}(w) \quad &:= \quad \begin{cases} f(\Psi(w)), & \text{if } w \in \Delta, \\ f(\psi(w)), & \text{if } w \in D, \end{cases} \\ S' \quad &:= \quad S'_i \cup \{w : |w| = 1 + \delta\}. \end{aligned}$$

Then the following estimate holds.

Lemma 4. *Let* $w \in S'_\delta := \{w \in \mathbb{C} : \text{dist}\,(w, S') < \delta\}$. *Then*

$$|\tilde{f}(w) - \tilde{f}(w^*)| \le c_2\, t, \tag{4.14}$$

where w^* *is the point of* S' *nearest to* w.

The proof of Lemma 4 will be given in the next section.

Now, we average the function $\tilde{f}$ separately in $\mathbb{D}$ and Δ in the following way. Let $K(z)$, $z \in \mathbb{C}$, be an arbitrary symmetric averaging kernel. Consider the function

$$\tilde{F}(w) := \begin{cases} \frac{16}{\delta^2} \int \tilde{f}(t) K\left(\frac{4(t-w)}{\delta}\right) dm(t), & \text{if } 1 + \frac{3}{4}\delta \le |w| \le 1 + \frac{5}{4}\delta \\ & \text{or } 1 - \frac{5}{4}\delta \le |w| \le 1 - \frac{3}{4}\delta, \\ \tilde{f}(w), & \text{elsewhere in } \mathbb{C} \setminus \partial\mathbb{D}. \end{cases}$$

Note that $\tilde{F} \in C^\infty(\mathbb{C} \setminus \partial\mathbb{D})$ and that by virtue of (4.14)

$$0 \le \tilde{F}(w) \le 1 \qquad (w \in \mathbb{C} \setminus \partial\mathbb{D}), \tag{4.15}$$

$$|\Delta \tilde{F}(w)| \preceq \frac{t}{\delta^2} \qquad (|w| \ne 1). \tag{4.16}$$

Therefore, the function

$$F(z) := \begin{cases} \tilde{F}(\Phi(z)), & \text{if } z \in \Omega, \\ \tilde{F}(\varphi(z)), & \text{if } z \in G, \\ f(z), & \text{if } z \in L, \end{cases}$$

is of class $C^\infty(\mathbb{C})$. Moreover, by the Green formula

$$\int \Delta F(z)\, dm(z) = 0. \tag{4.17}$$

Next, applying the technique of [5], we can establish the inequality

$$\left| \int F\,(d\mu - d\nu) \right| \preceq \frac{t}{\delta}\, (2\,a - U(\mu - \nu, z_0))\,. \tag{4.18}$$

Indeed, setting

$$\tilde{U}(\mu - \nu, w) := \begin{cases} U(\mu - \nu, \Psi(w)), & \text{if } |w| > 1, \\ U(\mu - \nu, \psi(w)), & \text{if } |w| < 1, \end{cases}$$

and using the representation of the function F by means of the Green formula

$$F(z) = F(\infty) + \frac{1}{2\pi} \int \Delta F(\zeta) \log |z - \zeta| \, dm(\zeta) \quad (z \in \mathbb{C}),$$

we obtain according to (4.1), (4.16) and (4.17) the relations

$$\begin{aligned} \left| \int F \, (d\nu - d\mu) \right| &= \frac{1}{2\pi} \left| \int (a - U(\mu - \nu, \zeta)) \, \Delta F(\zeta) \, dm(\zeta) \right| \\ &\leq \frac{1}{2\pi} \int \left(a - \tilde{U}(\mu - \nu, w) \right) |\Delta \tilde{F}(w)| \, dm(w) \\ &\preceq t \frac{a}{\delta} + t \frac{a - \tilde{U}(\mu - \nu, 0)}{\delta} \\ &= t \frac{2\,a - U(\mu - \nu, z_0)}{\delta} \end{aligned}$$

(for details, see [5]).

The conditions (4.14), (4.15) and (4.18) imply

$$\begin{aligned} (\nu - \mu)(J) &\leq \int F \, (d\nu - d\mu) + \mu(J(c_1 t) \setminus J) \\ &\quad + \int_{L \setminus J(c_1 t)} F \, d\mu + \int_J (1 - F) \, d\nu \\ &\preceq t \left(\frac{2\,a - U(\mu - \nu, z_0)}{\delta} + 1 \right), \end{aligned}$$

which is the assertion of (4.2).

Now, suppose (i) holds, *i.e.* L is Dini-convex with respect to Ω. In this case our constructions are even simpler. The scheme of reasoning in this situation is as follows:

We consider the function g given by (4.8) and its restriction g_R to $R := L_\delta$. Let f be the function, harmonic in $\overline{\mathbb{C}} \setminus R$ with continuous boundary values

$$f(z) := g_R(z) \qquad (z \in R).$$

It is easy to see that the functions

$$\begin{aligned} \tilde{f}(w) &:= f(\Psi(w)) \quad (w \in \Delta), \\ \tilde{F}(w) &:= \begin{cases} \frac{16}{\delta^2} \int \tilde{f}(\tau) \, K\left(\frac{4(\tau - w)}{\delta}\right) dm(\tau), & \text{if } 1 + \frac{3}{4}\delta \leq |w| \leq 1 + \frac{5}{4}\delta, \\ \tilde{f}(w), & \text{if } 1 \leq |w| < \frac{3}{4}\delta \\ & \text{or } \ |w| > 1 + \frac{5}{4}\delta, \end{cases} \end{aligned}$$

$$F(z) = \begin{cases} f(z), & \text{if} \quad z \in \overline{G}, \\ \tilde{F}(\Phi(z)), & \text{if} \quad z \in \Omega, \end{cases}$$

satisfy the same conditions as the corresponding functions in case (ii). Hence, by repeating our reasoning, we come to the even simpler formula

$$(\nu - \mu)(J) \leq c\sqrt{\delta}\left(\frac{a}{\delta} + 1\right). \quad \blacksquare \tag{4.19}$$

§5 Proof of Lemma 4

According to Beurling's theorem (see [2, p. 43]) for all $\zeta \in V := \overline{\mathbb{C}} \backslash S$ with the properties

$$\text{dist}(\zeta, \{z \in L_\delta : z_L \in J(c_1 t)\}) > \varepsilon_2, \quad \text{dist}(\zeta, S) < \varepsilon_2,$$

where ε_2 is an arbitrary constant and c_1 is the constant from (4.10), we have

$$|f(\zeta)| \preccurlyeq \sqrt{\text{dist } (\zeta, S)}. \tag{5.1}$$

Therefore, in the case that $w \in \mathbb{C}$, $|w| \neq 1$, and

$$\zeta := \begin{cases} \Psi(w), & \text{if} \ w \in \Delta, \\ \psi(w), & \text{if} \ w \in \mathbb{D}, \end{cases}$$

satisfy

$$\text{dist}(\zeta, \{z \in L_\delta : z_L \in J(c_1 t)\}) > \varepsilon_2 \,, \quad 1 - \delta \leq |w| \leq 1 + \delta \,,$$

inequality (4.14) is a simple consequence of (4.4), (5.1) and the maximum principle for harmonic functions.

Furthermore, applying Lemma 3 we obtain easily (4.14) for $w = re^{i\theta}$ with

$$1 + \delta < r < 1 + 2\delta \,,$$

$$\theta_1 - \varepsilon_1 t \leq \theta \leq \theta_2 + \varepsilon_1 t \quad \text{or} \quad \theta_2 + 2c_1 t \leq \theta \leq 2\pi + \theta_1 - 2c_1 t \,.$$

If

$$1 + \delta < r < 1 + 2\delta, \quad \theta_1 - 2c_1 t \leq \theta \leq \theta_1 - \varepsilon_1 t \,,$$

we set $R := L_\delta$, $w_R := (1 + \delta)e^{i\theta}$, $\xi := \zeta_R := \Psi(w_R)$ and write the function g_S in the form (4.11).

Lemma 1 and Lemma 2 imply that

$$\left|\frac{\zeta - \zeta_R}{\zeta_1 - \zeta_3}\right| \preceq \left|\frac{\zeta_L - \zeta_R}{\zeta_L - \zeta_1}\right| \preceq \frac{\delta}{t} = t. \tag{5.2}$$

In proceeding, denote by the same symbol $r(z,\xi)$ the harmonic extension to ext R of the function appearing in (4.11) and set

$$\tilde{r}(\tau) := r(\Psi(\tau),\xi) \quad (|\tau| \geq 1+\delta).$$

Since by Lemma 2 for $z \in R$ with $z_L \in L(\zeta_3,\zeta_1)$,

$$\left|\frac{z-\xi}{\zeta_1-\zeta_3}\right| \asymp \left|\frac{z-\xi}{z-\zeta_1}\right| \preceq \frac{|\Phi(z)-w_R|}{t},$$

we have according to (4.13)

$$|\tilde{r}(\tau)| \preceq \frac{|\tau-w_R|^2}{t^2} \quad (|\tau| = 1+\delta,\ |\tau-w_R| \leq c_2\, t). \tag{5.3}$$

At the same time by the definition of the function $\tilde{r}(\tau)$ and (4.11)–(4.13), relation (5.3) remains true for τ such that $|\tau| > 1+\delta$, $|\tau-w_R| = c_2\, t$.

Further, direct computation shows that

$$|\tilde{r}(w)| \preceq t. \tag{5.4}$$

Indeed, let us introduce the auxiliary function $\tilde{R}(\tau)$ which we define to be the harmonic extension to $|\tau| \geq 1+\delta$ of the function

$$\tilde{R}(\tau) := \begin{cases} |\tilde{r}(\tau)|, & \text{if } |\tau| = 1+\delta,\ |\tau-w_R| \leq c_2\, t, \\ c_3, & \text{otherwise for } |\tau| = 1+\delta. \end{cases}$$

It is clear that for sufficiently large c_3 we have

$$|\tilde{r}(\tau)| \leq \tilde{R}(\tau)$$

on the boundary of the domain

$$\{\tau : |\tau| > 1+\delta,\ |\tau-w_R| < c_2\, t\}.$$

Therefore, by the maximum principle for harmonic functions, the Poisson formula and (5.3) we obtain

$$\begin{aligned} |\tilde{r}(w)| &\leq \tilde{R}(w) = \tilde{R}(|w|\, e^{i\theta}) \\ &= \frac{1}{2\pi}\int_0^{2\pi} \tilde{R}((1+\delta)e^{i\eta})\, \frac{|w|^2-(1+\delta)^2}{|w|^2 - 2\,|w|\,(1+\delta)\cos\,(\theta-\eta)+(1+\delta)^2}\, d\eta \\ &\preceq \delta\left(\frac{1}{t^2}\int_{\theta-c_2 t}^{\theta+c_2 t} d\eta + \int_{\theta+c_2 t}^{\theta+\pi} \frac{d\eta}{(\eta-\theta)^2} + \int_{\theta-\pi}^{\theta-c_2 t} \frac{d\eta}{(\eta-\theta)^2}\right) \preceq \frac{\delta}{t} = t. \end{aligned}$$

Comparing (4.11), (4.12), (5.2) and (5.4) we get by virtue of (3.1) the desired inequality (4.14).

The same reasoning gives (4.14) for the case $w = re^{i\theta}$ with

$$1+\delta < r < 1+2\delta\,, \quad \theta_2 + \varepsilon_1 t \le \theta \le \theta_2 + 2c_1 t\,.$$

If $\zeta \in \Omega \cap (\mathrm{int} R)$, $\zeta_L \in L \setminus J(\varepsilon)$, $\arg\Phi(\xi_1) < \arg\Phi(\zeta) < \arg\Phi(\xi_2)$, where ε is the constant from (4.3), then $\mathrm{dist}(\zeta, S) \preceq \delta$ by Lemma 2, and inequality (4.14) follows from the estimate (5.1).

The same elementary arguments based on Beurling's theorem and the maximum principle for harmonic functions are applied to the rest of the possible dispositions of the point w in $\mathbb{D}$.

Thus, in order to complete the proof we have to consider only the case when

$$\zeta \in \Omega \cap \ (\mathrm{int}\ R), \quad \zeta_L \in J(\varepsilon). \tag{5.5}$$

First, we discuss the following auxiliary result. Let $\kappa(z), z \in \overline{\mathbb{C}}$, be a continuous function which is harmonic in V. Set

$$\tilde{\kappa}(\tau) := \kappa(\Psi(\tau)) \quad (\tau \in \Delta).$$

For ζ as in (5.5) we estimate the quantity

$$|\tilde{\kappa}(w) - \tilde{\kappa}(w_R)| = |\kappa(\zeta) - \kappa(\zeta_R)|$$

by some expressions using the notion of harmonic measure and some special cross-cuts of V.

It follows from Lemma 2 that

$$\mathrm{diam}\ (\Omega \cap \{z \in (\mathrm{int}\ R) : z_L = \zeta_L\}) =: d \asymp |\zeta_L - \zeta_R|.$$

Without loss of generality we assume that δ and, consequently, d are sufficiently small. For $d < r < \varepsilon_3 := \mathrm{dist}(J(\varepsilon), L \setminus J(2\varepsilon))$ we denote by $\gamma(r) = \gamma_\zeta(r)$ the arc of the intersection $\{z : |z - \zeta_L| = r\} \cap V$ that separates in V the point ζ from $L \setminus J(2\varepsilon)$. Define $l(r) = l_\zeta(r)$ to be the smaller subarc of R which has the same endpoints as $\gamma(r)$. It follows from the quasiconformality of R that

$$\mathrm{diam}\ l(r) \prec r.$$

Consider the Jordan domain $W = W_s \subset V$, $d < s \le \varepsilon_3$, that includes the point ζ and is bounded by $\partial W = l(s) \cup \gamma(s)$. For $r > 0$ define

$$\begin{aligned} u(r) \quad &:= \sup_{z \in \partial W,\, |z - \zeta_R| \le r} |\kappa(z) - \kappa(\zeta_R)|, \\ m \quad &:= \quad u(\mathrm{diam}\ W). \end{aligned}$$

It is practically shown in [4, p. 131] that for $s > 2d$

$$|\kappa(\zeta) - \kappa(\zeta_R)| \preceq u(2\,c\,d) + \int_{2d}^{s} \frac{u(c\,r)}{r} \exp\left(-\pi \int_d^{r/2} \frac{dx}{|\gamma(x)|}\right) dr$$
$$+m \exp\left(-\pi \int_d^{s/2} \frac{dx}{|\gamma(x)|}\right).$$

By Lemma 1 we have for $d < x < \varepsilon_3$ and some $0 < \beta < 1$

$$\varepsilon_4\, x \leq |\gamma(x)| \leq \pi\,(x + c_4\, x\, h(x) + c_5\, d^\beta x^{1-\beta}),$$

where h is the function from the definition of a locally Dini-convex curve. Hence, by (2.5),

$$\begin{aligned}\pi \int_d^{r/2} \frac{dx}{|\gamma(x)|} &= \int_d^{r/2} \frac{dx}{x} + \int_d^{r/2} \frac{\pi\, x - |\gamma(x)|}{x\,|\gamma(x)|}\, dx \\ &\geq \log\frac{r}{2d} - \frac{c_4}{\varepsilon_4} \int_0^{r/2} \frac{h(x)}{x}\, dx - \frac{c_5}{\varepsilon_4}\, d^\beta \int_d^{r/2} \frac{x^{1-\beta}}{x^2}\, dx \\ &\geq \log\frac{r}{d} - c_6\,,\end{aligned}$$

and, consequently,

$$|\kappa(\zeta) - \kappa(\zeta_R)| \preceq u(2cd) + d \int_{2d}^{s} \frac{u(cr)}{r^2}\, dr + m\,\frac{d}{s}\,. \tag{5.6}$$

It is simple to prove (4.14), if, in addition to (5.5), $\zeta_L \notin J(2c_1 t) \setminus J(\varepsilon_1 t)$. Indeed, let now $\kappa := f$, $s := \varepsilon_5\,|\zeta_L - \zeta_L^*|$, where $\zeta_L^* := \Psi(e^{it}\Phi(\zeta_L))$ and the sufficiently small constant ε_5 is chosen such that $u(c\,s) = 0$. Therefore, by virtue of (5.6) and the obvious inequality

$$\left|\frac{\zeta_L - \zeta_R}{\zeta_L - \zeta_L^*}\right| \preceq \frac{\delta}{t} = t$$

(which is a direct consequence of Lemma 2) we get (4.14).

The situation appears to be more complicated if, in addition to (5.5), $\zeta_L \in J(2c_1 t) \setminus J(\varepsilon_1 t)$. For definiteness, let $\zeta_L \in L(\zeta_3, \zeta_1)$ in addition to the assumptions above. In this case we represent the function g_S in the form (4.11) with $\xi := \zeta_R$ and set $\kappa(z) := r(z, \xi)$ (*i.e.* $\kappa(z)$ is the harmonic extension of $r(z, \xi)$ from S to V), $s := \varepsilon_6\,|\zeta_1 - \zeta_3|$, where ε_6 is chosen so small that for $z \in l(s)$ the function $\kappa(z)$ satisfies (4.13). Since by (4.11) and (4.12),

$$|\kappa(z)| \preceq 1 \quad (z \in \gamma(s)),$$

we have according to (4.13) and (5.6)

$$\begin{aligned} |r(\zeta,\xi)| &\preceq \frac{d^2}{s^2} + \frac{d}{s^2}\int_{2d}^{s} dr + \frac{d}{s} \\ &\asymp \frac{d}{s} \asymp \left|\frac{\zeta_L-\zeta_R}{\zeta_1-\zeta_3}\right| . \end{aligned} \tag{5.7}$$

Comparing (5.7), (4.11) as well as (4.12) and applying Lemma 2 we get

$$|f(\zeta)-f(\zeta_R)| \preceq \left|\frac{\zeta_L-\zeta_R}{\zeta_1-\zeta_3}\right| \asymp \left|\frac{\zeta_L-\zeta_R}{\zeta_L-\zeta_1}\right| \preceq \frac{\delta}{t} = t. \quad \blacksquare$$

§6 Proof of Theorem 2

First we consider the subarc of L given by the formula

$$S = S_a := \{z \in L : a \le \arg \Phi(z) \le 2\pi\},$$

where $a \in (0, \pi/8)$ is a parameter, the choice of which will be made more precise later.

Next, we introduce the conformal mapping Ψ_S of Δ onto $\overline{\mathbb{C}} \setminus S$ normalized by $\Psi_S(\infty) = \infty$, $\Psi_S'(\infty) > 0$.
Let $\varepsilon > 0$ be arbitrary, $R = R(\varepsilon) > 1$ be sufficiently large and such that for ζ with $|\zeta| = R$,

$$|(\Phi \circ \Psi_S)(\zeta)| \le (1+\varepsilon)\, R\, \Psi_S'(\infty) \; =: A\,.$$

We define two families of curves. Denote by Γ_1 the family of all locally rectifiable curves in the annulus

$$\{z : 1 < |z| < R\},$$

that separate its boundary components.

Let Γ_2 be the family of all locally rectifiable curves and arcs (with endpoints on $\{e^{it} : 0 < t < a\}$) in the annulus

$$\{w : 1 \le |w| \le A\},$$

that separate the circle $\{w : |w| = A\}$ from the arc $\{e^{i\theta} : a \le \theta \le 2\pi\}$.

By virtue of the comparison principle for moduli of families of curves [1] we have

$$\frac{1}{2\pi} \log R = m(\Gamma_1) \le m(\Gamma_2). \tag{6.1}$$

Our next aim is to estimate the module $m(\Gamma_2)$ from above. To do this, we consider the functions

$$\rho_1(w) \quad := \quad \begin{cases} \frac{1}{2\pi|w|}, & \text{if } \; 1 \le |w| \le A, \\ 0, & \text{otherwise }, \end{cases}$$

$$\begin{aligned}\rho_2(w) &:= \begin{cases} \frac{1}{\pi}, & \text{if } |w| \ge 1,\ |w-1| \le 2a, \\ 0, & \text{otherwise,} \end{cases} \\ \rho(w) &:= \max\{\rho_1(w), \rho_2(w)\}.\end{aligned}$$

It is clear that for each $\gamma \in \Gamma_2$,

$$\int_\gamma \rho(w)\,|dw| \ge 1.$$

Therefore, by definition of the module [1],

$$\begin{aligned}m(\Gamma_2) &\le \int_{\mathbb{C}} \rho_1^2(w)\,dm(w) + \int_{\mathbb{C}} \rho_2^2(w)\,dm(w) \\ &\le \frac{1}{2\pi}\log A + 4\,a^2.\end{aligned}$$

Comparing the last estimate with (6.1) we get

$$0 \le \frac{1}{2\pi}\log(1+\varepsilon)\,\Psi_S'(\infty) + 4\,a^2.$$

Since ε was arbitrary, we have

$$\exp\{c_1 a^2\}\ \text{cap}\ S \ge 1,$$

where $c_1 := 8\pi$.

Now, consider the n-th Fekete polynomial F_n for the arc S. According to results by Pommerenke [12] at its zeros $z_j \in S$, $j = 1, \ldots, n$,

$$|F_n'(z_j)| \ge (\text{cap}\ S)^{n-1} \ge \exp\{-c_1 a^2 n\}.$$

If now

$$a := \sqrt{\frac{\log A_n}{c_1 n}},$$

then the polynomial F_n (considered on L) satisfies the condition (2.3). At the same time,

$$D[\mu - \nu] \ge \frac{a}{2\pi} = c\sqrt{\frac{\log A_n}{n}},$$

where c is given by (2.11). ■

Acknowledgement. This paper was completed while the author was visiting the Katholische Universität Eichstätt. The author wishes to thank the members of this university for the pleasant mathematical atmosphere they offered him.

References

[1] Ahlfors, L. V., *Lectures on Quasiconformal Mappings*, Van Nostrand, Princeton, 1966.

[2] Ahlfors, L. V., *Conformal Invariants*, Mc Graw-Hill Book Comp., New York, 1973.

[3] Andrievskiĭ, V. V., V. I. Belyĭ and V. K. Dzjadyk, *Conformal Invariants in Constructive Theory of Functions of Complex Variable*, World Federation Publisher, Atlanta, Georgia, 1995.

[4] Andrievskiĭ, V. V. and H.-P. Blatt, Erdős-Turán-type theorems on piecewise smooth curves and arcs, *J. Approx. Th.* **88** (1997), 109–134.

[5] Blatt, H.-P. and R. Grothmann, Erdős-Turán theorems on a system of Jordan curves and arcs, *Constr. Approx.* **7** (1991), 19–47.

[6] Blatt, H.-P. and H. N. Mhaskar, A general discrepancy theorem, *Ark. Mat.* **31** (1993), 219–246.

[7] Blatt, H.-P., E. B. Saff and V. Totik, The distribution of extreme points in best complex polynomial approximation, *Constr. Approx.* **5** (1989), 357–370.

[8] Erdős, P. and P. Turán, On the distribution of roots of polynomials, *Annals of Math.* **51** (1950), 105–119.

[9] Huesing, J., Diskrepanzsätze bei einseitigen Schranken für logarithmische Potentiale, Diplomarbeit, Katholische Universität Eichstätt, 1995.

[10] Kleiner, W., Sur l'approximation de la représentation conforme par la méthode des points extrémaux de M. Leja, *Annales Pol. Math.* **XIV** (1964), 131–140.

[11] Lehto, O. and K. I. Virtanen, *Quasiconformal Mappings in the Plane, 2nd ed.*, Springer-Verlag, New York, 1973.

[12] Pommerenke, Ch., *Univalent Functions*, Vandenhoeck & Ruprecht, Göttingen, 1975.

[13] Pommerenke, Ch., *Boundary Behaviour of Conformal Maps*, Springer-Verlag, New York, 1992.

[14] Stein, E. M., *Singular Integrals and Differentiability Properties of Functions*, Princeton University Press, Princeton, 1970.

[15] Tsuji, M., *Potential Theory in Modern Function Theory*, Chelsea Publ. Co., New York, 1950.

Vladimir Andrievskiĭ
Katholische Universität Eichstätt
Mathematisch-Geographische Fakultät
D-85071 Eichstätt,
GERMANY

and

Institute for Applied Mathematics and Mechanics
Ukrainian Academy of Sciences
ul. Rozy Luxemburg 74
340114 Donetsk,
UKRAINE
mgk002@eo-dec-mathsrv.ku-eichstaett.de

Rational and meromorphic approximation in L^p of the circle: system-theoretic motivations, critical points and error rates

Laurent Baratchart

Abstract. This paper is concerned with best rational or meromorphic approximation of fixed type in L^p of the circle, when poles are constrained to lie in the unit disk. We first discuss some motivations from system theory and n-th root asymptotics for the error when the function to be approximated is analytic in a neighborhood of the circle. We then consider the problem from a differential-topological viewpoint, and derive an index theorem linking together the Morse indices of critical points. We subsequently relate Morse indices and error rates, and apply this to the issue of uniqueness of a critical point in H^2 rational approximation. The approach is illustrated on Markov functions and on the exponential function.

§1 Introduction

This introduction is intended as a guide to the results presented in the following sections. Proofs in this paper are merely outlined.

We will discuss best approximation to an L^p function on the unit circle, by a rational function of fixed type having all its poles in $\{|z| < 1\}$, or by a function that is meromorphic in the unit disk with a prescribed number of poles there, and whose L^p means remain bounded on circles close to the boundary of the disk. The meromorphic problem is conveniently thought of as the limit of the rational one when the degree of the numerator goes to infinity while the degree of the denominator is kept fixed. We may assume at first that $1 \leq p \leq \infty$ and that the function f to be approximated is an arbitrary member of L^p, although the results below really concern the case where $2 \leq p \leq \infty$, f is analytic in a neighborhood of the circle, and the degree of the numerator of the approximant –which is set to $+\infty$ in the meromorphic case– is not less than the number of poles minus 1. Illustrative results from the point of view of function theory specialize even further to $p = 2$, but we shall present them here from a more general perspective.

The constraint that the poles of rational approximants have to lie in the open disk is not completely standard. We shall provide some motivation

Computational Methods and Function Theory (CMFT'97)
N. Papamichael, St. Ruscheweyh and E. B. Saff (Eds.), pp. 45–78

ISBN 981-02-3626-3

for it in the context of control theory and signal processing, where it appears as a stability requirement. In this connection, it is often necessary to restrict further the approximation problem by requiring both the function to be approximated and the approximant to have real Fourier coefficients. In almost every respect, the cases of real and complex coefficients run parallel to each other, but the complex situation does not subsume the real one because best approximants to conjugate-symmetric functions need not be conjugate-symmetric. We shall consider real and complex coefficients together, but the applications to uniqueness given at the end of the paper deal with the real case.

Our point of departure will be a brief discussion of the rate of approximation. Over the past ten years, potential and operator-theoretic methods have led to an accurate appraisal of the degree of rational approximation to analytic functions in the n-th root sense. As a byproduct of work from many contributors, notably [44], [22], [23], [24], [34], [35], [36], [40], [41], one can produce upper bounds both on the *limsup* and the *liminf* of the geometric decay of the error as the degree increases, at least when f is analytic in $|z| \geq 1$. These bounds are nicely expressed in terms of the capacity of the condenser obtained from the unit circle on the one hand and the complement of the domain of analyticity of f on the other hand; for certain domains of analyticity, they are demonstrably sharp when $p \geq 2$ and the degree of the numerator is not less than the degree of the denominator (the so-called super-diagonal case). The more general situation where f is merely analytic in a neighborhood of the circle is less natural, because the requirement that poles lie in the disk may prevent the error from tending to zero unless the degree of the numerator increases faster than the degree of the denominator. When that is the case, bounds may also be obtained on the upper limit of the geometric decay of the error for ray sequences.

Now, it seems only fair to say that estimates on the rate are *a priori* of no help in actually computing best approximants; except when $p = 2$ and no pole is allowed (a trivial situation by the Parseval relation), or else when $m = \infty$ and $p = \infty$ (a case which is settled by the Adamjan-Arov-Krein theory for continuous f), no algorithm is known to compute them. Accordingly, numerical approaches will typically rely on general principles of nonlinear optimization like gradient and Newton methods. These usually converge to *local* best approximants only, and raise the eternal issues whether a local best approximant is globally optimal and what is a good guess to initialize the search. The example shown in Figure 2 of the Appendix indicates how serious the problem may be.

Motivated by this, we aim at initiating the study of *critical points* in this context; here, a critical point is a rational or meromorphic approximant for which the derivatives of the L^p norm of the error, with respect to all the parameters at hand, vanish. To avoid complications, we make this definition for the range $1 < p < \infty$ only, so that the norm itself is differentiable. Local best approximants are critical points of a particular kind, and the most interesting ones from a computational viewpoint, but it is a familiar principle from differential topology that they cannot be globally studied independently from the others. This will be illustrated in the present case by the *index theorem*, our central result, which has been established for the range $2 \leq p < \infty$ when the degree of the numerator is not less than the degree of the denominator minus 1, under the additional assumption that f is analytic in a neighborhood of the circle. To summarize the content of the theorem, let the *Morse index* of a critical point be classically defined as the maximum dimension of a real vector space on which the second derivative of the objective function –in our case the L^p norm of the error– restricts to a negative quadratic form. This number is finite, even in the meromorphic case where the ambient manifold is infinite-dimensional, and we define the *contribution* of a critical point to be 1 or -1 according to whether the Morse index is even or odd. With the afore-mentioned hypotheses, the index theorem asserts that the sum of the contributions of all critical points, provided they are nondegenerate and therefore finite in number, is equal to 1. In particular this sum is independent of the function f to be approximated.

The meaning of the result is perhaps best captured as follows. Let n denote the number of poles that the approximant is allowed to have. The optimal L^p error with fixed poles naturally defines a real-valued function on the set of monic polynomials of degree n, all roots of which lie in the open unit disk. This function extends continuously to the set of monic polynomials having their roots in the *closed* unit disk by cancelling poles of unit modulus with a corresponding zero of the numerator. The boundary of this set consists of those monic polynomials having one root of modulus 1 at least, and the extended function on this boundary takes values corresponding to best approximation with fewer poles, namely those remaining ones we have discarded the poles of modulus 1. In the case of rational approximants, the number of zeros has also decreased since we used some of the degree to perform cancellations. If one notices that no local minimum of the extended function can occur on the boundary by the normality of the approximation problem with free poles, one may intuively view the index theorem as an application of the Poincaré-Hopf relation, once it is understood that the set of monic polynomials of fixed degree whose zeros do not exceed 1 in modulus is topologically a closed ball when endowed with the Euclidean topology on the coefficients. As a matter of fact, a rigorous argument along these lines runs into difficulties due to a lack of smoothness, although the initial proof when $p = 2$ retains much of this line of approach [9]. The full demonstration of the statement uses stratified

Morse theory, and the restrictions on the degree of the numerator and on the range of p, together with the analyticity of f, are used to meet technical requirements of Whitney stratifications on the graph of the L^p error. It is unclear whether the condition $2 \leq p < \infty$ is best possible.

Returning to our approximation problem, we now raise the issue of estimating the number of local minima. From a computational viewpoint, such an estimate is a decisive piece of information to globalize a local search and, in case such a local minimum is unique –and therefore is the global minimum as well– descent methods from numerical optimization will even converge. Of course, such a nice feature as uniqueness cannot happen in general, but we want to enquire for which functions it holds and, if it does, for which values of n. Actually, we shall ask a slightly stronger question, namely *when is a critical point unique?* If it is, then the critical point is necessarily the best approximant we are looking for. This is where the index theorem becomes instrumental, because it allows us to put the question in *local* form. Indeed, since a local minimum has Morse index 0, the theorem entails when all critical points are nondegenerate that there is only one critical point if, and only if, each of them is a local minimum. In other words, it is enough to check that the second derivative at any critical point is positive definite to conclude that uniqueness prevails.

Investigating this second derivative is not so easy in general, and applications of the technique so far concern the case where $p = 2$ and f is analytic in $\{|z| > 1-\varepsilon\}$ including at infinity. The reason for this is that the critical point equation amounts then to an interpolation condition of the multipoint Padé type, thereby allowing us to translate a positivity statement on the Hessian matrix into the language of classical function theory. Another particular feature of this case is that the superdiagonal case reduces to the diagonal one, and even to the type $(n-1, n)$ upon substracting the value at infinity. The first result that was obtained along these lines is the uniqueness of a critical point in each degree for L^2 rational approximation of type $(n - 1, n)$ to Markov functions whose defining measure is supported on a compact interval of $(-1, 1)$ whose endpoints lie within some absolute bounds [14]. The proof depends on specific computations that evaluate in disguise the interpolation constant of the sequence of zeros and poles of the critical points, and contains in essence a more systematic *criterium* for the positiveness of the second derivative at a critical point for functions that need not be of Markov type. This *criterium* expresses in a quantitative manner that the numerator and the denominator of the critical point should be "sufficiently coprime", and was stated as a bound on their Bezout coefficients in H^∞ [10]. The bound in question depends on the error function itself, making for a rather implicit analysis, but one can ensure that it is met if, when n poles are allowed, one finds a multipoint

Padé interpolant of type $(n-2, n-1)$ whose error function is in modulus 3 times bigger, pointwise on the circle, than the error function generated by the considered critical point. Here, one should remember that a critical point is a particular multipoint Padé interpolant of type $(n-1, n)$ (the precise condition is that it should interpolate f with order two at each of the reflection of its poles with respect to the unit circle). That such a condition accounts for coprimeness may be intuitively understood as follows: if zeros and poles are "distinct enough", each of them is effective to approximate f and another interpolant of degree one less should be significantly worse.

The main virtue of the preceding *criterium* is to reduce the positivity analysis of the second derivative at a critical point –and therefore the uniqueness issue by the index theorem– to a classical question in approximation theory: pointwise estimating the error in interpolation. When sharp asymptotics of this error are available, one may check in this way for *asymptotic uniqueness*, that is to say, uniqueness when the number of poles is large enough. This was illustrated in [10], where the case $f(z) = z^{-1}e^{1/z}$ is treated; upon changing z into z^{-1} the result translates into asymptotic uniqueness in best H^2 rational approximation to e^z for the superdiagonal case.

This positivity *criterium* can be refined using Adamjan-Arov-Krein theory, to the effect that the ratio between the error function generated by a multipoint Padé interpolant of type $(n-2, n-1)$ and the error function of the critical point under consideration must remain at distance greater than 2 from the constant 1 on $\mathbb{T}$. This seemingly minor improvement is decisive to prove asymptotic uniqueness of a critical point in L^2 rational approximation of type $(n-1, n)$ to Markov functions whose defining measure is absolutely continuous, is supported on a compact interval of $(-1, 1)$, and satisfies the Szegő condition [11]; in fact, the improved *criterium* is necessary and sufficient for the positivity of the second derivative when the function is of Markov type. The proof makes use of a sharp asymptotic formula for the interpolation error, that was given in [43]. We note the difference with the previously mentioned result on Markov functions: this time the support may be an arbitrary compact interval of $(-1, 1)$, but the measure has to be quite regular and uniqueness is only proved for n large enough depending on f. The result is best possible in some sense, because there exists a Markov function, whose defining measure is singular, having at least two best approximants in each odd degree; and to each odd $n > 1$ there exists a Markov function whose defining measure is absolutey continuous and satisfies the Szegő condition, yet the function has two best approximants in degree n [12]. This, however, leaves room for a conjecture that uniqueness prevails for a suitable subsequence of degrees.

Gathering up the above results, we see that the uniqueness issue in the L^2 superdiagonal case is reasonably understood for Markov functions, and also that some light was shed on the case of the exponential function which is the prototype of an entire function with smoothly decaying Fourier coefficients. In view of the motivations that we mentioned, it is interesting to note that Markov functions do have a system-theoretic *status* as transfer functions of so-called *relaxation systems* [45], [16], [15], [33]. However, much remains to be done for more general families of functions, and we shall conclude the paper trying to open perspectives on classes defined by Cauchy integrals in connection with "orthogonal" polynomials with complex weight.

The general idea we want to convey is the following. After all, it was perhaps not fair to say that error rates are of no help to compute best rational or meromorphic approximants. Somewhat unexpectedly rates appear on the contrary to be deeply connected with the number of local best approximants that may be obtained through numerical searches. However, n-th root asymptotics will not be sufficient, and sharp pointwise estimates on the circle of the *ratio* between two consecutive error functions seem to be needed for this purpose.

§2 Notations

- $\mathbb{T}$: the unit circle.
- $\mathbb{D}$: the open unit disk.
- L^p: the Lebesgue space on $\mathbb{T}$ with norm $\|f\|_p$, where $1 \leq p \leq \infty$.
- H^p: Hardy space of the disk consisting of L^p functions whose Fourier coefficients of negative index all vanish.
- $\bar{H}^p$: Hardy space of the complement of the disk consisting of L^p functions whose Fourier coefficients of positive index do vanish.
- $\bar{H}_0^p$: functions in $\bar{H}^p$ with vanishing mean.
- B_n: Blaschke products of degree at most n:

$$b \in B_n \text{ iff } b(z) = c \prod_{j=1}^{k} \frac{(z-\alpha_j)}{(1-\bar{\alpha_j} z)}, \quad \text{where } k \leq n,\ \alpha_j \in \mathbb{D}, \text{ and } c \in \mathbb{T}.$$

- $\mathcal{P}_m$: algebraic polynomials of degree $\leq m$. We set $\mathcal{P}_{-1} = \{0\}$ by convention.
- $\mathcal{R}_{m,n}$: rational functions of type (m,n), that are lying in $\mathcal{P}_m \mathcal{P}_n^{-1}$.

- $\mathcal{R}^p_{m,n}$: members of $\mathcal{R}_{m,n}$ having no pole in $\mathbb{C} \setminus \mathbb{D}$. The superscript p emphasizes that they are endowed with the L^p norm.
- $\mathcal{R}^p_{\infty,n}$: functions in $B_n^{-1}H^p$. They are those meromorphic functions in $\mathbb{D}$ having at most n poles and bounded L^p means on circles close to $\mathbb{T}$.

H^p functions are often called "analytic" in L^p. Indeed, they are in one-to-one correspondence with a.e. nontangential limits of those f analvtic in $\mathbb{D}$ satisfying:

$$\sup_{0<r<1} \|f(re^{i\theta})\|_p < \infty.$$

This allows us to regard $f \in H^p$ as a function of $z \in \mathbb{D}$. A symmetric remark applies to $\bar{H}^p$, which can be viewed as the space of functions analytic in $\{|z| > 1\}$ whose L^p-means remain bounded on circles centered at 0 with radius greater than 1. With this interpretation, $\bar{H}^p_0$ consists of those H^p functions vanishing at infinity. Members of B_n are those rational functions of degree at most n that are analytic in $\bar{\mathbb{D}}$ and have modulus 1 everywhere on $\mathbb{T}$. The notation $\mathcal{R}^p_{\infty,n}$ is intended to suggest that we may formally regard an analytic function as a polynomial of infinite degree.

The subscript or superscript $\mathbb{R}$ will indicate that we restrict to functions with real Fourier coefficients. We write for instance $H^p_{\mathbb{R}}$, $\mathcal{P}^{\mathbb{R}}_m$, $\mathcal{R}^{p,\mathbb{R}}_{m,n}$, and so on. For rational functions, it means, of course, that they can be represented as a quotient of two polynomials with real coefficients. This restriction to real coefficients is often necessary in applications of approximation theory to signal and systems (but not always, see *e.g.* Figure 3).

§3 A family of approximation problems

To each pair of integers m, n, and each real number p such that

$$0 \le m \le \infty, \quad 0 \le n \le \infty, \quad 1 \le p \le \infty,$$

we associate the following problem:

$\boldsymbol{Pb_{m,n,p}}$
Given $f \in L^p$, find $g_0 \in \mathcal{R}^p_{m,n}$ such that

$$\|f - g_0\|_p = \inf_{g \in \mathcal{R}^p_{m,n}} \|f - g\|_p := \rho(f, m, n, p).$$

There is also a real version:

$\boldsymbol{Pb}^{\mathbb{R}}_{\boldsymbol{m,n,p}}$
Given $f \in L^p_{\mathbb{R}}$, *find* $g_0 \in \mathcal{R}^{p,\mathbb{R}}_{m,n}$ *such that*

$$\|f - g_0\|_p = \inf_{g \in \mathcal{R}^{p,\mathbb{R}}_{m,n}} \|f - g\|_p := \rho_{\mathbb{R}}(f, m, n, p).$$

This formulation allows us to state in a common framework some classical rational and meromorphic approximation problems. Note for instance that:

- the case where $m \leq n$ and $f \in \bar{H}^p$ is that of best rational approximation of type (m, n) in $\bar{H}^p$. When $m = n$, this is equivalent to best rational approximation of type (n, n) to $f(1/z)$ in H^p;
- the case where $m < n$ and $f \in \bar{H}^p_0$ is that of best rational approximation of type (m, n) in $\bar{H}^p_0$. If $m = n - 1$, this is equivalent to best rational approximation of type $(n - 1, n)$ to $z^{-1}f(1/z)$ in H^p;
- the case where $m = \infty$ and $n = 0$ is the classical extremal problem of approximating an L^p function by an H^p function;
- the case $m = \infty$, $n \geq 0$, $p = \infty$ is AAK approximation, namely uniform meromorphic approximation with a prescribed maximum number of poles;
- the case $m = \infty$, $n \geq 0$, $1 \leq p < \infty$ is the natural L^p analog of AAK approximation;

The orthogonality of H^2 and $\bar{H}^2_0$ together with the Parseval relation make the value $p = 2$ a little special because then:

- the cases $n-1 \leq m \leq \infty$ reduce to the case $m = n-1$ with f replaced by $\mathbf{P}_{\bar{H}^2_0}(f)$. In particular, the L^2 analog of AAK approximation is equivalent to the problem of best rational approximation of type $(n-1, n)$ in $\bar{H}^2_0$, and therefore, by a previous remark, to best rational approximation problem of type $(n - 1, n)$ in H^2 upon changing z into $1/z$ and dividing again by z.
- For $f \in \bar{H}^2_0$ and $[n/2] \leq m$, a best approximant from $\mathcal{R}^2_{m,n}$ is also a best approximant from $\mathcal{R}_{m,n}$, namely the restriction that the finite poles should lie in $\mathbb{D}$ is superfluous; here, the brackets designate the integer part. This fact is a little less obvious, but can be derived from the partial fraction expansion of the approximant by differentiating the L^2 norm. Whether the property extends to $0 \leq m < [n/2]$ is apparently unknown.

§4 Motivations from system theory

We will provide some motivation from control and signal theory for some of the preceding problems. The connection between linear control system and function theory has two cornerstones:

- the fact that these systems can be described in the so-called frequency domain as a multiplication operator by the transfer function which belongs to certain Hardy classes if the system has certain stability properties;
- the fact that rational functions are precisely transfer functions of systems having finite-dimensional state-space, namely those that can be designed and handled in practice.

Let us review these notions briefly, beginning with discrete-time systems.

A *discrete control system* is a map $u \to y$ where the input

$$u = (\ldots, u_{k-1}, u_k, u_{k+1}, \ldots)$$

is a real-valued function of the discrete time k, generating an output

$$y = (\ldots, y_{k-1}, y_k, y_{k+1}, \ldots)$$

of the same sort, where y_k depends on u_j for $j \leq k$ only. The system is said to be time-invariant if a shift in time of the input produces a corresponding shift of the output.

Particularly important in applications are the *linear systems*:

$$y_k = \sum_{j=0}^{\infty} f_j u_{k-j},$$

where the output at time k is a linear combination of the past inputs with fixed coefficients $f_j \in \mathbb{R}$. For such systems, function theory enters the picture when signals are encoded by their generating functions:

$$u(z) = \sum_{k\in\mathbb{Z}} u_k z^{-k}, \qquad y(z) = \sum_{k\in\mathbb{Z}} y_k z^{-k}.$$

Indeed, if we define the *transfer function* of the linear control system to be:

$$f(z) = \sum_{k=0}^{\infty} f_k z^{-k},$$

the input-output behaviour can be described as

$$y(z) = f(z)\, u(z).$$

In particular

(i) the system is a bounded operator $l^2 \to l^2$ iff $f \in \bar{H}^\infty_{\mathbb{R}}$, and the operator norm is $\|f\|_\infty$: the system is called (l^2, l^2)-stable;

(ii) the system is a bounded operator $l^2 \to l^\infty$ iff $f \in \bar{H}^2_{\mathbb{R}}$, and the operator norm is $\|f\|_2$: the system is called (l^2, l^∞)-stable.

A linear control system is said to have *finite dimension* n if it can be described as a linear *automata* in terms of a state variable $x_k \in \mathbb{R}^n$ which is updated at each time $k \in \mathbb{Z}$:

$$x_{k+1} = Ax_k + Bu_k, \qquad y_k = Cx_k + Du_k,$$

where A is a real $n \times n$ matrix, B (resp. C) a column (resp. row) vector with n real entries, and D some real number, n being the smallest possible integer for which such an equation holds. The classical result here (see *e.g.* [29], [37]) is that a linear time-invariant system has dimension n iff its transfer-function is rational of degree n and analytic at infinity. The transfer-function is then

$$f(z) = D + C\left(zI_n - A\right)^{-1} B.$$

For finite-dimensional linear systems the requirement that the poles of f should lie in $\{|z| < 1\}$ amounts to $f \in \bar{H}^p$ for some and, in fact, all p, and is equivalent to any reasonable definition of stability. We now mention some specific applications of approximation theory to identification and control of linear dynamical systems.

4.1 Robust $\mathbf{H}^\infty$ control

This example was historically the first to stress a connection between extremal problems in Hardy spaces and control theory [47]. Given a system with transfer function say H, we seek another system, called the *controller*, with transfer function K, such that the *closed-loop system* in Figure 1 has some desired behaviour. This use of a compensator K that automatically generates, from the observed output, an appropriate correction to the input is a fundamental process called *feedback*, see Figure 1.

Some technical assumptions on H are needed.

- First, we assume that the formal power series

$$H = \sum_{k=0}^{\infty} h_k z^{-k} \tag{4.1.1}$$

is *strictly proper*, which means by definition that $h_0 = 0$: there is no direct feedthrough in the system.

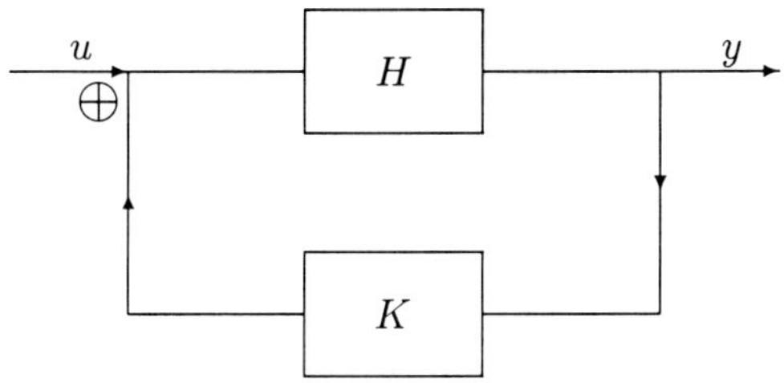

Figure 1. Feedback

- Second, we assume that H may be written as a ratio of the form:

$$H = \frac{n}{d}, \qquad n, d \in \bar{H}^{\infty}_{\mathbb{R}}. \tag{4.1.2}$$

  The quotient in (4.1.2) really means that H is the long division of n by d, so that (4.1.1) needs *not* to be the Fourier expansion of H when the latter is not stable; the hypothesis says, however, that H may be (l^2, l^2)-stabilized if the input signal is pre-filtered by some (l^2, l^2)-stable system (*e.g.* with transfer function d).

- Third, we assume that n and d, generate the unit ideal:

$$an + bd = 1, \qquad a, b \in \bar{H}^{\infty}_{\mathbb{R}},$$

  and we further require that the outer factor of n be invertible in $\bar{H}^{\infty}_{\mathbb{R}}$. These technical requirements assert in some sense that a signal destabilizing $1/d$ destabilizes H, and that the asymptotic response to oscillating signals is never too small.

These hypotheses more or less circumscribe the range of application of H^∞ control, and cover the rational case with no transmission zeros on the circle as well as several infinite-dimensional situations. If H is given as an L^∞ function which is not of the desired form n/d, one way to approximate it by a function of the correct form is to find a meromorphic approximant to $f(1/z)$, namely to solve for the case $p = \infty$ and $m = \infty$ of our family of problems.

As an extra piece of terminology, a controller K is said to *internally stabilize* the system if all transfer functions between two signals in the loop are stable. This notion is justified by the fact that input-output stability is not enough: every signal inside the loop must remain of finite energy to avoid internal damage to the physical system.

Consider now the specific problem where, in the feedback diagram, the controller K must be designed so as to minimize the norm $l^2 \to l^2$ of the closed loop system mapping u to y (rejection of perturbations), subject to the constraint that it internally stabilizes the system. The closed-loop transfer-function is easily computed to be:

$$f = \frac{H}{1 - HK},$$

and it is known after [46], [17] that internally stabilizing controllers can be parametrized as:

$$K(Q) = \frac{dQ + a}{nQ - b},$$

where Q varies freely in $\bar{H}^\infty_{\mathbb{R}}$. Substituting in the expression for the closed loop transfer-function, we have to find:

$$\min_{Q \in \bar{H}^\infty_{\mathbb{R}}} \|n^2 Q - nb\|_\infty.$$

If $n = jw$ is the inner-outer decomposition and since w is invertible in $\bar{H}^\infty$ by assumption, we may take $T = Qw^2$ as a new variable and we end up with a classical extremal problem:

$$\min_{T \in \bar{H}^\infty_{\mathbb{R}}} \|T - nb\bar{j}^2\|_\infty.$$

Upon changing z into $1/z$, this is the case $p = \infty$, $m = \infty$, and $n = 0$ of our family of problems, known as the *Nehari extension*.

This is but an illustration of the connections between H^∞ optimization and control. In practice, one encounters more complicated loop diagrams, other objective *criteria*, and one has to consider matrix-valued transfer-functions. For more details on these non-trivial extensions, we refer the reader to [20] and [18].

4.2 Synthesis

As a sequel to the above example, one usually wants a controller with rational transfer function, for this is what finitely many electronic devices can produce. In the previous problem, K will not be rational in general unless H is, and even then it will be rational of the same degree which may be too high. One is thus often led to approximate the optimal K in $\bar{H}^\infty_{\mathbb{R}}$ by a rational function of type (n, n) with n reasonably small. If K is stable — this happens if the original system is stable and in a number of other cases as well [18] — one does usually want to preserve this precious feature (in case the loops break down) and this is where the requirement that the poles of the approximant should lie in the disk becomes important.

This situation is known as *synthesis*: one wants to design a transfer function which is stable, rational of lowest possible degree, approximating within some prescribed tolerance a given frequency behaviour. We face then $p = \infty$ and $m = n$ in our family of problems.

4.3 Stochastic identification

Consider a discrete time real-valued stationary stochastic process:

$$y = (\ldots, y(k-1), y(k), y(k+1), \ldots).$$

If it is regular (*i.e.* purely non-deterministic in a certain sense), we have the Wold decomposition:

$$y(k) = \sum_{j=0}^{\infty} f_j u(k-j),$$

where u is a white noise called the innovation, and where

$$f_j = \mathbf{E}\left\{ y(k+j)u(k) \right\}$$

is independent of k by stationnarity. We also have, by the Parseval identity, that

$$\sum_{j=0}^{\infty} f_j^2 = \mathbf{E}\left\{ y(k)^2 \right\}.$$

If we set

$$f(z) = \sum_{k=0}^{\infty} f_k z^{-k} \in \bar{H}^2,$$

we see that a regular process is obtained by feeding white noise to an $l^2 \to l^\infty$ stable linear system.

In the special case where f is rational of degree n, then y is called an Auto-Regressive Moving Average process, and such models of a process are seeked because they are efficient for computations. When trying to fit such a model to y, a typical interest is in minimizing the variance of the error between the true output and the prediction of the model. In this way one is led to solve for

$$\min_{g \in \mathcal{R}_{n,n}^{2,\mathbb{R}}} \|f - g\|_2.$$

This is the case $p = 2$ and $m = n$ of our family of problems. This principle can be used to identify a linear system from observed stochastic inputs, although computing f_j is difficult because it requires spectral factorization of the function

$$\sum_{j\in\mathbb{Z}} \mathbf{E}\{y(j+k)y(k)\}$$

whose Fourier coefficients can only be estimated by ergodicity through time averages of the observed sample path of y.

In practice, one would rather use time averages already in the optimization *criterium*, but this can be proved asymptotic to the previous problem [26], [30]. In fact, critical points themselves asymptotically match [6].

To lend perspective to the discussion, let us briefly digress on the more general case where the input is an arbitrary stationary process. Applying the spectral theorem to the shift operator on the Hilbert space of the process allows to compute the squared variance of the output error as a weighted L_2–integral:

$$\frac{1}{2\pi}\int_0^{2\pi} |f-g|^2\,d\mu, \tag{4.3.1}$$

where the positive measure μ is the so-called spectral measure of the input process (that reduces to Lebesgue measure when the latter is white noise), and f now has to belong to the weighted Hardy space $\bar{H}^2(d\mu)$. Though we shall not prove anything concerning this weighted type of approximation, we want to emphasize that the spectral theorem, as applied to shift operators, stresses deep links between time and frequency representations of a stochastic process, and the isometric character of this theorem (that may be viewed as a far-reaching generalization of Parseval's relation) is a fundamental reason why L^2 approximation problems arise in system theory.

There is much more to be said about the connections between classical function theory and stochastic identification. The problem presented above is a special instance of *maximum likelihood identification* where the noise model is fixed. If one looks at a general version of maximum likelihood estimation [26], and if one takes into account the fact that matrix-valued functions have to be considered to handle multiple inputs and outputs, one will find that all this really aims at a matrix-valued and rational extension of the Szegő theory [42]. In this context, the uniqueness problem that we raised is linked to the so-called *consistency issue*.

4.4 Harmonic identification

This example deals with continuous-time rather than discrete-time linear control systems, namely with convolution operators $u(t)\to y(t)$ of the form:

$$y(t)=\int_0^t h(t-\tau)u(\tau)\,d\tau.$$

The function $h : [0, \infty) \to \mathbb{R}$ is called the *impulse response* of the system, as it formally corresponds to the output generated by a delta function. If h and u have exponential growth, so does y and the one-sided Laplace transforms $Y(s)$, $U(s)$ and $H(s)$ are defined on some common half-plane $\{\mathrm{Re}\, z > \sigma\}$. The system operates in this frequency domain as multiplication by the transfer-function H:

$$Y(s) = H(s)U(s).$$

This time, rational transfer-functions of degree n correspond to linear differential operators of order n forced by the input u.

The Hardy spaces involved are now those of the right half-plane, denoted by $\mathcal{H}^p$, and their relation to stability are for instance that

> the system is bounded $L^2[0, \infty) \to L^2[0, \infty)$ iff $H \in \mathcal{H}^\infty_{\mathbb{R}}$;
>
> the system is bounded $L^2[0, \infty) \to L^\infty[0, \infty)$ iff $H \in \mathcal{H}^2_{\mathbb{R}}$;
>
> the system is bounded $L^\infty[0, \infty) \to L^\infty[0, \infty)$ iff $H \in \mathcal{W}_{\mathbb{R}}$, the Wiener algebra of the right half-plane consisting of Laplace transforms of summable functions $[0, \infty) \to \mathbb{R}$.

One of the most popular methods to identify a system which sends bounded inputs to bounded outputs is to plug in a periodic input $u = e^{i\omega t}$ and to observe the steady-state output which is:

$$y(t) = \lambda e^{i\phi} e^{i\omega t},$$

where λ and ϕ are respectively the modulus and the argument of $H(i\omega)$.

In this way, one can estimate the transfer function on the imaginary axis and, for physical as well as computational purposes, one is led to rationally approximate the experimental data one has got. In fact, the practical situation is slightly more complicated, because experiments cannot be performed on the whole axis and usually the system will no longer behave linearly at high frequencies. One way out is to complete a partial model by solving bounded extremal problems [8], [7]. Once a complete model is obtained, one faces an approximation problem on the imaginary axis by a rational function which is analytic in $\{\mathrm{Re}\, s > 0\}$. The $\mathcal{H}^2$ norm is usually better suited, due to the measurement errors. Since

$$H(s) \to f(z) = 2\sqrt{\pi}\frac{(H \circ \varphi)(z)}{z - 1},$$

where

$$\varphi(z) = \frac{z + 1}{z - 1}$$

is an isometry from $\mathcal{H}^2$ onto $\bar{H}_0^2$ preserving rationality and the degree, this approximation issue on the line translates to the case $p = 2$ and $m = n - 1$ of our family of problems. Figure 3 in the appendix shows a numerical example of this identification procedure on a hyperfrequency filter.

All the previous examples deal with $p = 2$ or $p = \infty$, and indeed other values of p are seldom considered. It seems, however, that their study is worth the while as they may offer an interesting trade-off in identification and synthesis between L^2 norms, that tend to oversmooth the data, and L^∞–norms that are put off by irregular samples [39].

Potential applications of approximation theory to control and signal processing go far beyond the few examples we gave. Other kinds of inverse problems than deconvolution may also be approached using meromorphic approximation, for instance Dirichlet-Neuman ones. But we shall stop here with the motivations and return to the study of the problem itself.

§5 Error rates

Let us observe first that a compactness argument easily implies the existence of a solution to $\boldsymbol{Pb_{m,n,p}}$. If we restrict ourselves to the superdiagonal case $m \geq n$ and assume that f is analytic on a neighborhood V of $\bar{\mathbb{C}} \setminus \mathbb{D}$, a case which is already interesting enough, we can get precise asymptotic estimates from above of $\rho(f, m, n, p)$, in the n-th root sense as $n \to \infty$.

To state the result formally, let us define

$$\lambda = \exp\left\{-\frac{1}{C}\right\} < 1,$$

where $C = \mathrm{cap}(\mathbb{T}, \mathbb{C} \setminus V)$ is the condenser capacity of the pair $(\mathbb{T}, \mathbb{C} \setminus V)$ (see *e.g.* [38] for the definition of capacity).

Theorem 1. *For f analytic on $V \supset \bar{\mathbb{C}} \setminus \mathbb{D}$ and $1 \leq p \leq \infty$, we have that*

$$\limsup_{\substack{n\to\infty \\ m\geq n-1}} \rho^{1/n}(f, m, n, p) \leq \lambda, \tag{5.1}$$

$$\liminf_{\substack{n\to\infty \\ m\geq n}} \rho^{1/n}(f, m, n, p) \leq \lambda^2. \tag{5.2}$$

The same estimates apply to $\rho_{\mathbb{R}}(f, m, n, p)$. The limsup estimate is sharp for every $p \geq 2$ in the class of functions analytic on V, and the liminf estimate is sharp at least when V is the complement of a real segment.

Outline of proof: To get (5.1) and (5.2), it is enough to consider $m = n$ and $p = \infty$. The *limsup* estimate follows then from a well-known inequality of

Walsh, which is sharp for all V when the poles are free in the class of analytic functions on V [44], [3]. The *liminf* estimate follows from the positive answer to a conjecture of Gonchar that was given in [34] for the simply connected case which is all we need here (the general case was recently settled in [35]). As to sharpness, it is enough to consider $p = 2$ and subsequently $m = n$. Sharpness of the *limsup* estimate follows from the known sharpness when $p = \infty$ and the Cauchy representation formula of the error function in $|z| \geq 1 + \varepsilon$. Sharpness of the *liminf* when V is the complement of a real segment can be seen in the case of real coefficients by choosing $f \in \bar{H}^2_{0,\mathbb{R}}$ to be a Markov function whose defining measure has positive density a.e. and is supported on some compact subinterval of $(-1, 1)$. Here one uses the fact that $\bar{H}^2_{\mathbb{R}}$ best rational approximants of type (n, n) are particular multipoint Padé approximants whose pole distribution converges weak-star on $[a, b]$ to the equilibrium distribution of the condenser $(\mathbb{T}, [a, b])$ [13]. We then apply classical convergence results on interpolation to Markov functions with a regular defining measure [23], [41]. Sharpness in the case of complex coefficients follows by showing directly from the orthogonality equation (9.1) that a critical point in complex $\bar{H}^2$ rational approximation to a Markov function has in fact real coefficients. ■

We will add two remarks on this result. First, it is possible to give upper bounds of the same type on the *limsup* and the *liminf* of $\rho^{1/(m+n)}(f, m, n, p)$ in the case of subdiagonal *ray sequences*, namely sequences of degrees (m_k, n_k) for which $\lim_{k\to\infty} m_k/n_k = \theta \leq 1$; the capacity must then be defined *via* a modified energy problem where the value of the Green potential at infinity is taken into account with weight $1 - \theta$ [36]. Second, the theorem easily extends to $m = n - 1$ if $f(\infty) = 0$.

If f is merely analytic on a neighborhood V of $\mathbb{T}$, the above estimates do not hold and the requirement that the poles of the approximant lie in $\mathbb{D}$ may even prevent diagonal sequences of best approximants from converging to f. If $m - n$ goes to infinity though, we can at least split f as $f_1 + f_2$ where f_1 is analytic in $V_1 = V \cup \{z;\ |z| \geq 1\}$ and f_2 is analytic in $V_2 = V \cup \mathbb{D}$, and then approximate separately f_1 by a member of $\mathcal{R}_{n,n}$ and f_2 by a member of $\mathcal{P}_{m-n}$. If we let for instance $\lambda > 1$, and if we set

$$C_1 = \mathrm{cap}(\mathbb{T}, \bar{\mathbb{C}} \setminus V_1),$$
$$C_2 = \sup\{|z|^{-1};\ z \in \bar{\mathbb{C}} \setminus V_2\},$$

we get from the previous estimate and the Bernstein-Walsh lemma:

Corollary 1. *For f analytic on $V \supset \mathbb{T}$ and $1 \leq p \leq \infty$, we have that:*

$$\limsup_{\substack{n\to\infty \\ m\geq\lambda n}} \rho^{1/n}(f, m, n, p) \leq \max\left(e^{-1/C_1}, C_2^{\lambda-1}\right).$$

§6 Critical points

Only in very few cases a computational solution to $\mathbf{Pb}_{m,n,p}$ is available. These are:

- $n = 0$ and $p = 2$ which is trivial from Parseval's identity,
- $n \geq 0$ and $p = \infty$ for continuous f, which is solved by AAK theory, [1], [2],
- $m = \infty$ and $n = 0$ for f analytic on $\mathbb{T}$ in the special case where the winding number of the error is -1, see [19], [39].

When $n = 0$ and $1 < p < \infty$, the situation may still be handled to some extent from the numerical viewpoint, because the problem is convex. But when $n \geq 1$, we enter the realm of non-convex optimization, where lots of local minima may impede optimization methods (see Figure 3 of the Appendix).

To understand this increased complexity, the following comparison may be of help. On the one hand, the best $\mathcal{R}^{\infty}_{\infty,n}$ approximant to a continuous f, which is just the AAK approximant with n poles, is given by the formula:

$$g_n = f - \frac{A_f(v_n)}{v_n}, \tag{6.1}$$

where v_n is a maximizing vector of the Hankel operator:

$$A_f : H^2 \to \bar{H}^2_0, \qquad A_f(v) = \mathbf{P}_{\bar{H}^2_0}(fv),$$

when restricted to the linear subspace:

$$\{v_0, v_1, \ldots, v_{n-1}\}^{\perp} \subset H^2.$$

On the other hand, the best $\mathcal{R}^2_{\infty,n}$ approximant to $f \in L^2$, which is equal to $\mathbf{P}_{H^2}(f) + r_{n-1,n}$ where $r_{n-1,n}$ is the best $\bar{H}^2_0$ rational approximant of type $(n-1, n)$ to $\mathbf{P}_{\bar{H}^2_0}(f)$, is again given by (6.1) and v_n still maximizes the Hankel operator, but this time over a highly nonlinear subset which is B_n. And for $2 < p \leq \infty$ the best $\mathcal{R}^p_{\infty,n}$ approximant to f is still given by (6.1) where v_n maximizes the minimal norm of the vectors attained by the Hankel operator over compact symmetric subsets of genus $2m + 2$ on the unit sphere of H^s with $1/s + 1/p = 1/2$ (resp. over compact symmetric subsets of genus $m + 1$ on the unit sphere of $H^s_{\mathbb{R}}$ if one restricts to real Fourier coefficients) [39].

If one wants to handle the problem numerically when $1 < p < \infty$, no doubt, one will take advantage of the differentiability of the norm and use a descent algorithm combined with a Newton-like method in order to solve for a local minimum; and even if $p = \infty$, he will try to solve Kolmogorov type equations that encapsulate in this situation first order variational conditions [28]. Here, a local minimum is a rational or meromorphic function in $\mathcal{R}^p_{m,n}$ that yields the infimum of the function

$$\Phi_{m,n} : \mathcal{R}^p_{m,n} \to \mathbb{R},$$

$$\Phi_{m,n}(g) = \|f - g\|_p^p,$$

on some neighborhood of itself in $\mathcal{R}^p_{m,n}$.

A local minimum is a particular instance of a *critical point*. A critical point of a map is, by definition, a point where the derivative vanishes. In the present case, if we let Δ_n be the set of monic polynomials of degree n whose roots lie in $\mathbb{U}$, a critical point is a pair $(h_c, q_c) \in \mathcal{P}_m \times \Delta_n$ (resp. $H^p \times \Delta_n$ if $m = \infty$) such that the derivative of the map

$$\Phi_{m,n}(h, q) = \|f - \frac{h}{q}\|_p^p$$

vanishes identically at (h_c, q_c). The derivatives are here understood with respect to $h \in \mathcal{P}_m$ (resp. $h \in H^p$ if $m = \infty$), and with respect to the coefficients of $q \in \Delta_n$ (except for the leading one which is fixed and equal to 1). To avoid additional complications, *we assume from now on that* $1 < p < \infty$ *so there is no differentiability question.* Critical points fall into two classes: they are termed *irreducible* if h_c and q_c are coprime, and *reducible* otherwise. When $m \geq n - 1$, it is easily seen that local minima are irreducible unless $f \in \mathcal{R}^p_{m,n}$. *Henceforth we rule this trivial case out,* so that a local minimum corresponds to a well-defined critical point with the above definition.

The *Morse index* of a critical point is the maximal dimension of a real vector space on which the second derivative restricts to a negative quadratic form; it is defined whenever the second derivative in question is nondegenerate.

The partial mapping $\mathcal{P}_m \to \mathbb{R}$ (resp. $H^p \to \mathbb{R}$ if $m = \infty$) given by $h \to \|f - h/q\|_p$ is strictly convex, thus has a unique critical point (a unique point where the derivative vanishes) which is a minimum. The latter, denoted by $h(q)$, is characterized by the "orthogonality" equations:

$$\frac{1}{2i\pi} \int_{\mathbb{T}} \left| f(\xi) - \frac{h(q)}{q}(\xi) \right|^{p-2} \overline{\xi \left(f(\xi) - \frac{h(q)}{q}(\xi) \right)} \frac{\xi^k}{q(\xi)} \frac{d\xi}{\xi} = 0, \quad 0 \leq k \leq m. \tag{6.2}$$

It is an easy consequence of (6.2) and the implicit function theorem that $q \to h(q)$ is smooth $\Delta_n \to \mathcal{P}_m$ (resp. $\Delta_n \to H^p$ if $m = \infty$). Thus every critical point is of the form $(h(q), q)$ where q is critical for the map

$$\Psi_n : \Delta_n \to \mathbb{R}$$

$$q \to \|f - \frac{h(q)}{q}\|_p^p,$$

and it is immediately checked that the Morse indices agree in both cases. In particular, the Morse index of a nondegenerate critical point is finite and at most equal to n, even if $m = \infty$.

The main virtue of the map Ψ_n, as compared to $\Phi_{m,n}$, is that it behaves continuously when roots of q tend to the unit circle. More precisely, if $q_* = q_1 q_2$ where $q_2 \in \Delta_{n-k}$ and q_1 is a monic polynomial of degree k whose roots have modulus 1, it follows from the definition of $h(q)$, the continuity of $\Psi_{n-k} : \Delta_{n-k} \to \mathbb{R}$, and the weak-star compactness of balls in L^p that $\Psi_n(q)$ tends to $\Psi_{n-k}(q_2)$ as q tends to q_*; the convergence of q is here understood coefficientwise. Thus if $\overline{\Delta_n}$ denotes the set of monic polynomials of degree n whose roots lie in $\bar{\mathbb{U}}$, then Ψ_n extends continuously to a map $\Psi : \overline{\Delta_n} \to \mathbb{R}$ upon setting $\Psi(q_*) = \Psi_{n-k}(q_2)$. Note that $\Psi(q)$ depends only on q_2, that is on the roots of q that lie in $\mathbb{U}$. Hence the natural domain of definition for Ψ is the *quotient* of $\overline{\Delta_n}$ obtained by identifying two polynomials if they share the same roots of modulus < 1. This quotient space $\mathcal{Q}$ turns out to be homeomorphic to the projective space $\mathbb{P}^n(\mathbb{C})$ or $\mathbb{P}^n(\mathbb{R})$, according to whether the Fourier coefficients are complex or real. One way to see this is to view $\mathcal{Q}$ as a quotient of B_n where proportional Blaschke products have been identified, and to recognize in the Schur parametrization the cell-decomposition of the projective space [5]. For each $k \in \{0, 1, \ldots, n\}$, there is exactly one (complex or real) k-cell in $\mathcal{Q}$ corresponding to normalized Blaschke products of exact degree k, and the cells partition $\mathcal{Q}$. Henceforth, we always identify $\mathcal{Q}$ with the corresponding projective space. Upon representing a normalized Blaschke product by its monic numerator, the above k-cell identifies naturally with Δ_k, and the restriction of Ψ to it is simply Ψ_k which is smooth. However, Ψ *cannot* be represented as a smooth map on a projective space, because directional derivatives may not exist when we approach Δ_k from Δ_l with $l > k$.

Let $\mathcal{G} \subset \mathcal{Q} \times \mathbb{R}$ be the graph of Ψ, and $\mathbf{p}_2$ be the projection onto the second factor. On $\mathcal{G}$ we get $\mathbf{p}_2(q, \Psi(q)) = \Psi(q)$, so the projection is just a trick to represent Ψ as the restriction of a smooth map to a singular subset of a smooth ambient space that may be taken to be $\mathbb{R}^N \times \mathbb{R}$ ($N = 4n$ will do if the Fourier coefficients are complex and $N = 2n$ is enough if we restrict to real ones). The space $\mathcal{G}$ is naturally stratified with *strata*:

$$\mathcal{S}_k = \{(q, \Psi(q)); \quad q \in \Delta_k\}, \quad k \in \{0, 1, \ldots, n\},$$

and $\mathcal{S}_k$ is diffeomorphic to Δ_k; hence it is a smooth manifold. Let us denote the tangent space to $\mathcal{S}_k$ at $(q, \Psi(q))$ by $\mathcal{T}_q$. A critical point of $\mathbf{p}_2$ on $\mathcal{S}_k$ is, by definition, a pair $(q, \Psi(q)) \in \mathcal{S}_k$ such that $\mathbf{Dp}_2(q, \Psi(q)).v = 0$ for every $v \in \mathcal{T}_q$. Since $\mathbf{Dp}_2(q, \Psi(q)) = \mathbf{p}_2$, this means that q is a critical point of Ψ_k on Δ_k. However, due to the non-smoothness of $\mathcal{G}$, each point $(q, \Psi(q)) \in \mathcal{S}_k$ with $k < n$ has to be endowed with a generalized tangent space $\mathcal{T}_q^*$ comprising not only the members of $\mathcal{T}_q$ but also those vectors that are limits, for some $l > k$, of tangent vectors to $\mathcal{S}_l$ at a sequence of points converging to $(q, \Psi(q))$ in $\mathcal{G}$. A critical point of $\mathbf{p}_2$ on $\mathcal{S}_k$ will be called nondegenerate if it is nondegenerate in the usual sense on the smooth manifold $\mathcal{S}_k$ —this is equivalent to saying that q is a nondegenerate critical point of Ψ_k—and if, in addition, the kernel of $\mathbf{p}_2$ on $\mathcal{T}_q^*$ reduces to $\mathcal{T}_q$. *Henceforth, the word nondegenerate for a critical point of degree $k < n$ will be understood in this extended sense.* We also note that the Morse index of $(q, \Psi(q))$ as a critical point of $\mathbf{p}_2$ on $\mathcal{S}_k$ agrees with the Morse index of q as a critical point of Ψ_k on Δ_k.

For the stratification $\cup_k \mathcal{S}_k$ to be useful from the differential viewpoint, it has to satisfy the Withney conditions [25]:

A if $l > k$ and $\{(q_\ell, \Psi(q_\ell))\}$ is a sequence in $\mathcal{S}_l$ converging in $\mathcal{G}$ to $(q, \Psi(q)) \in \mathcal{S}_k$, and if $\mathcal{T}_{q_\ell}$ converges to $\mathcal{T}$ in the Grassman topology, then $\mathcal{T}_q \subset \mathcal{T}$;

B while the assumptions being as in condition **A**, suppose in addition that $(\chi_\ell, \Psi(\chi_\ell))$ is a sequence in $\mathcal{S}_k$ converging to $(q, \Psi(q))$. If the line through $(q_\ell, \Psi(q_\ell))$ and $(\chi_\ell, \Psi(\chi_\ell))$ converges to $\mathcal{L}$ in the Grassman topology, then $\mathcal{L} \subset \mathcal{T}$.

One can prove that these technical conditions are satisfied when $m \geq n-1$, $p \geq 2$, and f is analytic in a neighborhood of $\mathbb{T}$, although we will not detail the argument here. It is based on the limiting behaviour of the derivative $\mathbf{D}\Psi_l(q)$ when some roots of q tend to the circle, more specifically on the fact that the contour integral on $\mathbb{T}$ evaluating $\mathbf{D}\Psi_l(q)$ can be shrunk to a smaller circle for some appropriate extension of $|f - h(q)/q|^{p-2}(e^{i\theta})$, obtained by solving a complex Dirichlet problem on a neighborhood of $\mathbb{T}$, and using relation (6.2) to get rid of the residues. The important feature of the range $p \geq 2$ for this proof is that the derivative of the L^p norm is then locally Lipschitz.

§7 The index theorem

We shall phrase the result in terms of the map $\Phi_{m,n}$ to stick as much as possible to the initial formulation of the approximation problem.

Theorem 2. *Let $2 \leq p < \infty$, and $m \geq n-1$. Assume f is analytic in a neighborhood of $\mathbb{T}$, $f \notin \mathcal{R}^p_{m,n}$. Let Δ_k denote the set of monic polynomials of degree k whose roots lie in $\bar{\mathbb{D}}$, and let $\mathcal{C}_k$ be, for each $0 \leq k \leq n$, the set of critical points of the map*

$$\Phi_{m-n+k,k} : \mathcal{P}_{m-n+k} \times \Delta_k \to \mathbb{R}$$

(resp. $H^p \times \Delta_k \to \mathbb{R}$ if $m = \infty$), given by

$$\Phi_{m-n+k,k}(h,q) = \|f - \frac{h}{q}\|_p^p.$$

If critical points in $\mathcal{C}_k$ are nondegenerate for $0 \leq k \leq n$ (this is understood in the extended sense if $k < n$), then $\mathcal{C}_n$ is a finite set and

$$\sum_{(h_c,q_c)\in\mathcal{C}_n} (-1)^{M(h_c,q_c)} = 1, \tag{7.1}$$

where $M(h_c,q_c)$ designates the Morse index of the critical point (h_c,q_c). The same result holds if we restrict to real-Fourier coefficients.

Outline of proof: we consider complex Fourier coefficients. Under the stated nondegeneracy conditions, we show by induction on n that the critical points of $\mathbf{p}_2$ on Δ_n satisfy (7.1). We use stratified Morse theory on $\mathcal{G}$. The Withney conditions are satisfied as explained in the previous section, so the Morse data at a critical point of $\mathbf{p}_2$ on $\mathcal{S}_k$ are the product of the tangential Morse data on the stratum and of the normal Morse data. The normality of the approximation problem — namely the fact that the approximation error always decreases if a new pole is added provided that $f \notin \mathcal{R}^p_{m,n}$ — implies that the normal Morse data are homeomorphic to a complex $(n-k)$-cell cross its boundary. Thus, in the process of attaching handles that lead to the Morse inequalities [32], each critical point of index λ on $\mathcal{S}_k$ accounts for attaching one real $2(n-k)+\lambda$-cell along its boundary. Let $C_{\lambda,k}$ denote the number of critical points of $\mathbf{p}_2$ on $\mathcal{S}_k$ with index λ, and set

$$C_k = \sum_{\lambda=0}^{k} (-1)^\lambda C_{\lambda,k}.$$

The Euler-Poincaré formula tells us that

$$\sum_{k=0}^{n}\sum_{\lambda=0}^{k}(-1)^{2(n-k)+\lambda} C_{\lambda,k} = \sum_{k=0}^{n} C_k = \chi,$$

where χ is the Euler characteristic of $\mathcal{G}$. Since the latter is homeomorphic to $\mathbb{P}^n(\mathbb{C})$, we get $\chi = n+1$ and we obtain $C_0 = C_1 = \cdots = C^n = 1$ by induction on n. The argument in the case of real Fourier coefficients is similar. ■

The theorem is stated here for the first time when $p > 2$. The case where $p = 2$ was treated in [4], [9] when $m = n - 1$ or n with a somewhat different proof, and then extended to $m \geq n - 1$ in [10]. It is natural to ask whether the nondegeneracy assumptions are generically satisfied. The answer is yes: functions to which the theorem applies form an open and dense subset of the space of functions analytic in the corona $\{1 - \varepsilon < |z| < 1 + \varepsilon\}$ and continuous on its closure; a proof when $p = 2$ can be found in [4].

§8 Uniqueness in $\mathbf{H}^2$ rational approximation

The results of the previous section apply naturally to the problem of uniqueness of a critical point, hence of a best approximant, as described in the introduction. Uniqueness results are rare in complex approximation, and one should not expect very general statements here: examples where the solution to $\boldsymbol{Pb}_{m,n,p}$ is not unique can be given in the superdiagonal case for each $p < \infty$. However, these examples are highly non-generic, exploiting some symmetry of the function and special values for m, n. Of course, the set $\mathcal{O}_{m,n,p}$ of functions having several nondegenerate critical points in degree (m, n) is open in L^p, and non-void in many cases by what we just said, but it is unclear how large it is nor if

$$\bigcup_{n_0 \in \mathbb{N}} \left\{ \bigcap_{m \geq n-1 \geq n_0} \mathcal{O}_{m,n,p} \right\} \neq \emptyset.$$

Therefore the question as to when and why uniqueness occurs remains very valid.

To prove uniqueness of a critical point in a specific situation, one needs in view of the index theorem, and since local minima have Morse index 0, to establish that:

(i) the critical points are nondegenerate,

(ii) each of them is a local minimum.

We shall illustrate this approach with a few examples that were recently worked out when $p = 2$ and the Fourier coefficients are real. The functions involved lie in $\bar{H}^2_{0,\mathbb{R}}$, so we limit the discussion to $m = n - 1$.

The critical point property for an irreducible rational function $(h(q)/q)$ of type $(n - 1, n)$ in this case is that it should interpolate f with order 2 at the reflected of the roots of q across $\mathbb{T}$. Moreover, the (extended) nondegeneracy condition is fulfilled if the second derivative $\mathbf{D}^2\Psi_n$ is nonsingular at q and if no critical point of Ψ_k in degree $k < n$ has ν extra-interpolation points to f *on* $\mathbb{T}$ with total multiplicity greater than $(n-k)/2-1$, see [14], [10]. A simple computation further shows [14] that the second derivative at some irreducible critical point $h(q)/q \in \mathcal{R}^2_{n-1,n}$ may be expressed in suitable coordinates as a quadratic form on $\mathcal{P}_{n-1}$ by the formula:

$$H(p,p) = \left\|\frac{p}{q}\right\|^2 - 2\langle f - \frac{h(q)}{q}, \frac{\widetilde{q}^2 bp^2}{q^3}\rangle, \qquad p \in \mathcal{P}_{n-1}, \tag{8.1}$$

where $\langle\ ,\ \rangle$ denotes the L^2 scalar product on $\mathbb{T}$, where $\widetilde{q}(z) = z^n q(1/z)$ is the reciprocal of q, and where b is defined by a Bezout equation of the form:

$$bh(q) + cq = 1, \quad b, c \in H^\infty_{\mathbb{R}}. \tag{8.2}$$

Note that (8.1) indeed depends on b *modulo* q only, by the critical point property and the residue formula.

The first case that we shall consider is that of a Markov function:

$$f(z) = \int_{t_1}^{t_2} \frac{d\mu(t)}{t-z}, \tag{8.3}$$

where μ is a positive measure supported on a real segment $[t_1, t_2] \subset (-1, 1)$. It is classical [23] that multipoint Padé approximation to such functions at a conjugate-symmetric set of interpolation conditions is normal, and that the denominator is an orthogonal polynomial in $L^2(d\mu/\widetilde{q}^2)$. In particular, we get in the case of real Fourier coefficients that a critical point has no extra interpolation point to f in $\bar{\mathbb{C}} \setminus [t_1, t_2]$ and that all the poles lie on $[t_1, t_2]$. Thus every critical point is irreducible and we do not have to worry about the extended form of the nondegeneracy condition.

Theorem 3. *If t_1 and t_2 satisfy*

$$(t_2 - t_1)^2 < 2\left(1 - \max\left(t_1^2, t_2^2\right)\right)^2,$$

and if μ has more than n points of increase, on $[t_1, t_2]$, then there is a unique critical point in rational $\bar{H}^2_0$-approximation of type $(n-1, n)$ to f defined in (8.3).

If for instance $t_1 = -t_2$, numerical estimation yields that t_2 should be less than 0.5176... for the theorem to ensure uniqueness. If $t_1 = 0$, the condition improves to $t_2 < \sqrt{1/2}$.

Outline of proof: this is an easy adaptation of [14], Theorem 3, where only $t_1 = -t_2$ is considered; the positivity of $H(p,p)$ is established using Pick's interpolation theorem. ■

A closer look at the proof just quoted shows that the estimate of (8.1) which is given there really amounts to that of the interpolation constant of the zeros of q and $h(q)$, and that the crucial property granted by the restricted support of μ is a "good" coprimeness of q and $h(q)$. In retrospect, the importance of this coprimeness was already apparent from (8.1), whether f is Markov or not, as it is easily checked that $H(p,p)$ is positive as soon as

$$\|b(fq - h(q))\|_\infty < 1/2, \tag{8.4}$$

for some b as in (8.2). The following criterion gives conditions on the interpolation error to f under which (8.4) holds.

Criterion 1. *Let f be analytic in $\{1-\varepsilon < |z| \leq \infty\}$ and $(h(q), q)$ be a coprime critical pair for $\boldsymbol{Pb}^{\mathbb{R}}_{\boldsymbol{n-1,n,2}}$. If there exists a multipoint Padé interpolant to f of type $(n-2, n-1)$, say A/B, such that:*

$$3|f - h(q)/q| < |f - A/B| \qquad \text{everywhere on} \quad \mathbb{T},$$

then $h(q)/q$ is a nondegenerate local minimum of $(h,\chi) \to \|f - h/\chi\|_2^2$.

Criterion 1 is proved by checking that $b = A/(Ah(q) - Bq)$ satisfies (8.4), and we refer the reader to [10] where it is used to prove asymptotic uniqueness of a critical point in $\bar{H}_0^2$ rational approximation of type $(n-1,n)$ to $f = z^{-1}e^{1/z}$. Changing z into $1/z$ and dividing again by z, the result is formally stated in terms of the familiar exponential function as:

Theorem 4. *There exists $n_0 \geq 1$ such that a critical point in rational H^2-approximation of type $(n-1,n)$ to e^z is unique as soon as $n \geq n_0$.*

Outline of proof: Criterion 1 is applied to $f = z^{-1}e^{1/z}$ by choosing A/B to be the familiar Padé interpolant at ∞. The difficult part is to get an upper estimate pointwise on $\mathbb{T}$ of the function associated to a critical point, and this is obtained by carrying over to the circle, using a projection technique and some known estimates in real interpolation to e^x. There is numerical evidence that $n_0 = 1$, but no estimate is available at present. Other applications of Criterion 1 to uniqueness in H^2 approximation on small circles and to Montessus de Ballore-type situations may also be found in [10].

Let us finally return to Markov functions and consider again Theorem 3. On the one hand, it is easy to construct examples of such functions for which uniqueness does not hold, already when $n = 1$, showing that some sort of restriction on the measure μ is indeed necessary for uniqueness to hold irrespectively of n. On the other hand, we may ask whether, given a Markov function, uniqueness holds asymptotically. This turns out to be true if the measure μ is regular enough:

Theorem 5. *Let the positive measure μ be supported on $[t_1, t_2] \subset (-1, 1)$, be absolutely continuous with respect to Lebesgue measure, and assume that its density μ' satisfies the Szegő condition:*

$$\int_{t_1}^{t_2} \frac{\log \mu'(t)}{\sqrt{(t-t_1)(t_2-t)}} dt > -\infty. \tag{8.5}$$

Then, there exists $n_0 = n_0(\mu) \geq 1$ such that a critical point in rational $\bar{H}_0^2$-approximation of type $(n-1, n)$ to f defined by (8.3) is unique as soon as $n \geq n_0$.

Outline of proof: one needs the following refinement of Criterion 1:

Criterion 2. *Let f be analytic $\{1-\varepsilon < |z| \leq \infty\}$ and $(h(q), q)$ be a coprime critical pair for $\boldsymbol{Pb}^{\mathbb{R}}_{\boldsymbol{n-1,n,2}}$. If $A/B \in \mathcal{R}^{\infty}_{n-2,n-1}$ interpolates f at $2n-2$ points in $\mathbb{C} \setminus \mathbb{D}$ and satisfies:*

$$\left| 1 - \frac{(f - A/B)(z)}{(f - h(q)/q)(z)} \right| > 2, \qquad \forall z \in T,$$

then $h(q)/q$ is a nondegenerate local minimum of $(h, \chi) \to \|f - h/\chi\|_2^2$.

The proof of Theorem 5 proceeds by exhibiting some A/B meeting Criterion 2, whose poles are closely connected to those of $h(q)/q$, the pointwise error being controlled through a precise (up to $o(1)$) asymptotic formula in multipoint Padé interpolation to Markov functions given in [43]. The details are carried out in [11]. It is interesting to note that Criterion 1 is not sharp enough to apply here.

As to the proof of Criterion 2 itself, it rests on the fact that $H(p,p)$ can be recomputed as:

$$H(p,p) = 2\langle (I - 2\Gamma_\varphi^\sharp)\, p/q \ ,\ p/q \rangle, \tag{8.6}$$

where

$$\varphi(e^{i\theta}) = e^{i(n-1)\theta}\, \frac{\widetilde{q}}{q}(e^{i\theta})\, \overline{(fq(e^{i\theta}) - h(q)(e^{i\theta}))},$$

and $\Gamma_\varphi^\sharp$ is the pseudo-inverse of the Hankel operator:

$$\Gamma_\varphi : \bar{H}^0_{2,\mathbf{R}}(V) \to \bar{H}^0_{2,\mathbf{R}}(V),$$

$$\Gamma_\varphi(g) = \mathbf{P}_{\bar{H}^0_{2,\mathbf{R}}(V)}(\varphi\, e^{-i\theta}\, \bar{g}),$$

when viewed as an isomorphism from the orthogonal complement of its kernel $\mathcal{P}_{n-1}/q$ onto itself. From (8.6) follows that $H(p,p)$ is positive if Γ_φ has all its nonzero singular values larger than 2 (this is also necessary for Markov functions because Γ_φ is nonnegative in this case), and Criterion 2 follows from the de la Vallée-Poussin lower bound principle for AAK theory [27].

Theorem 5 is in some sense best possible: given $N \geq 1$, there exists a Markov function whose defining measure satisfies the Szegő condition and for which uniqueness fails for at least one $n \geq N$; there is also a Markov function whose defining measure does not satisfy the Szegő condition and for which uniqueness fails in each odd degree [12]. No estimate of $n_0(\mu)$ is available at present.

§9 Conclusion

We have shown in this work how differential topology can be used to approach uniqueness issues in rational and meromorphic approximation in Hardy spaces, and we indicated why these issues are of computational importance in control and identification of linear dynamical systems. We also illustrated how the coprimeness of the numerator and the denominator plays a crucial role in determining the signature of the Hessian matrix at a critical point, thereby allowing us to prove uniqueness in some cases; this was done only for $p = 2$, but Criterion 1 can be generalized to handle the range $2 \leq p < \infty$ and $m \geq n-1$. We finally stressed that pointwise estimates on the multipoint Padé interpolation error on the circle are a mean to control the norm of a Bezout relation between the numerator and the denominator and therefore to adress the coprimarity issue when $n = 2$. Of course, asymptotic uniqueness as in Theorems 4 and 5 is not sufficient from a constructive viewpoint, for n_0 remains to be estimated. It is, however, a step forward into a subject that requires further analysis.

It may seem that the arguments we gave in the case of Markov functions are rather special, but they might be prototypical of more general Cauchy integrals of the form

$$f(z) = \frac{1}{2i\pi}\int_C \frac{w(t)dt}{t-z},$$

where C is a piecewise analytic arc inside the unit disk. For such functions, the critical point equation is

$$\int_C \frac{q}{\widetilde{q}^2}\, w(t)\, t^k\, dt = 0, \qquad 0 \leq k \leq n-1, \tag{9.1}$$

so that q appears to be a "complex" orthogonal polynomial whose behaviour is one of the important issues facing this theory today. If we knew that the zeros of q converge to some compact subset $\mathcal{K}$ of $\mathbb{D}$ as $n \to \infty$ and that there are no extra interpolation condition at critical points, the analysis of [11] would reduce the uniqueness question to a certain Green energy problem on $\bar{\mathbb{C}} \setminus \mathcal{K}$; and even if there were extra interpolation points, the index theorem could still help us to tell how many critical points of given genus there are. If, for instance, C is a real segment and w is continuous and non-vanishing, what we need is a multipoint version, with varying scheme of interpolation, of the uniform convergence over compact subsets of $\bar{\mathbb{C}} \setminus C$ of Padé approximants established in [31]. More generally, for piecewise analytic arcs and analytic w, it follows from [40] that the poles of Padé approximants will converge in some sense to a minimal capacity contour C_m linking the endpoints of the analytic arcs, namely the measure with mass $1/n$ at each pole (counting multiplicities) will converge weak-star to some equilibrium distribution on C_m. However, just like n-th root asymptotics are unable to distinguish between constants like 3 or 2 appearing in Criteria 1 and 2, the weak-star convergence of the counting measure is not sufficient for our purposes. One may hope that this convergence can be sharpened at the possible cost of stronger assumptions on w, in the spirit of [31] where the nonvanishing character of the weight yields the uniform convergence of *all the poles* to the cut when the latter is a real segment.

§10 Appendix

The polynomial in $1/z$ of degree 199 whose coefficients are given below for increasing powers of z^{-1}, ordered from the left to the right and the top to the bottom, generates a phase portrait for the gradient flow in best rational $\bar{H}_0^2$ rational approximation of type $(1, 2)$ which is shown in Figure 2.

9.97196827205383e-14	2.04669203745972e-10	1.14150869986854e-12	-3.28310294979261e-08
5.02747140006639e-13	1.5626962719558e-06	-5.93581611615724e-13	-3.48257915890696e-05
4.70752702081621e-13	0.000441126691824404	-3.57442432371252e-13	-0.00352901353465124
2.73933097187261e-13	0.0190023805714752	-2.16342711788239e-13	-0.0716661210126338
1.75936314088528e-13	0.193920092151961	-1.4611287530278e-13	-0.381035970544271
1.22932164371215e-13	0.544337100777561	-1.04162950166247e-13	-0.559397811075766
8.85211661237948e-14	0.402766423974554	-7.52317390755403e-14	-0.192777091816881
6.38037452222159e-14	0.0550791690905388	-5.39081590540254e-14	-0.00710698956007355
4.5314576668414e-14	7.11637632493339e-15	-3.78518333161695e-14	-9.26255458989828e-15
3.13855653310506e-14	1.06034775033886e-14	-2.58045090071697e-14	-1.13037571037997e-14
2.10118960292336e-14	1.15072220841081e-14	-1.69211511876641e-14	-1.1335991145898e-14
1.34533771966683e-14	1.08922601023105e-14	-1.05358323472808e-14	-1.02596256580072e-14
8.10140099537926e-15	9.5054020241051e-15	-6.08838175688951e-15	-8.68280515511563e-15
4.44034189199018e-15	7.83307079688527e-15	-3.10603125941489e-15	-6.98730777854383e-15
2.03927641405526e-15	6.16840425565104e-15	-1.19880851556569e-15	-5.39235149015959e-15
5.48068070818875e-16	4.6697264839736e-15	-5.49192621897418e-17	-4.00680395411755e-15
-3.08637312696479e-16	3.40648959761016e-15	5.66794262175047e-16	-2.86915407284107e-15
-7.40277611992192e-16	2.39328165175266e-15	8.46696283215339e-16	-1.97602620742456e-15
-9.00858740400251e-16	1.61364144256156e-15	9.15115850601437e-16	-1.30180521558456e-15
-8.99680576233208e-16	1.03592401002207e-15	8.62872874362831e-16	-8.11324882604785e-16
-8.11410928302378e-16	6.23403064731527e-16	7.50649428002061e-16	-4.67735807546012e-16
-6.84771518144866e-16	3.40165118184551e-16	6.1700788373042e-16	-2.36841444514827e-16
-5.49774423191143e-16	1.54250812625137e-16	4.84833749547731e-16	-8.92247551741132e-17
-4.234149924958e-16	3.89407085247104e-17	3.66322348362854e-16	-9.06524518641762e-19
-3.14021787695859e-16	-2.70538925362206e-17	2.66721805838983e-16	4.68242153381848e-17
-2.24432993031929e-16	-6.00195878200703e-17	1.87021647498438e-16	6.80125050474842e-17
-1.54250812625137e-16	-7.19580312271158e-17	1.25816680910526e-16	7.28198119830092e-17
-1.01372422004306e-16	-7.13914534029647e-17	8.05493631867902e-17	6.83216076688376e-17
-6.2976515123356e-17	-6.41337635669843e-17	4.82873231422573e-17	5.92427275488014e-17

-3.61321107662754e-17	-5.39738185401887e-17	2.61819186746124e-17	4.85763704971717e-17
-1.81328731767686e-17	-4.3237433528518e-17	1.17079733154925e-17	3.80924965134374e-17
-6.65781894576063e-18	-3.32356140182992e-17	2.7601871251892e-18	2.8727812299396e-17
1.80579472308645e-19	-2.46039729587523e-17	-2.3353530405731e-18	2.0878789953957e-17
3.85185550945193e-18	-1.75519323400193e-17	-4.85665745254458e-18	1.46116846642669e-17
5.45719023274585e-18	-1.20386535502424e-17	-5.7438216892279e-18	9.80815050163631e-18
5.79177561838649e-18	-7.89237613922632e-18	-5.66307773038656e-18	6.26192269113505e-18
5.40828318201393e-18	-4.88719705629029e-18	-5.06808496933031e-18	3.73918859680356e-18
4.67478332343215e-18	-2.79020383584954e-18	-4.25361610931347e-18	2.01423711605568e-18
3.82389727663401e-18	-1.3873147822926e-18	-3.40006264588995e-18	8.87601084071108e-19
2.9925502220893e-18	-4.95455759970601e-19	-2.608548130431e-18	1.93419476059471e-19
2.25268960078545e-18	3.38885022960359e-20	-1.92756924188005e-18	-1.99878860388783e-19
1.63422622098574e-18	3.1613520240674e-19	-1.37252816010623e-18	-3.92574228808509e-19
1.14147560536947e-18	4.37605913698837e-19	-9.39466783020267e-19	-4.58294608850111e-19
7.6451602311417e-19	4.60519221077882e-19	-6.14379518613943e-19	-4.49119489926767e-19
4.8672344544487e-19	4.28036308346527e-19	-3.79190151492902e-19	-4.00436819898569e-19
2.89478907182398e-19	3.68833558952608e-19	-2.15396871004697e-19	-3.35183072293914e-19
1.54890198347252e-19	3.00982555060606e-19	-1.06066545752997e-19	-2.6734464431156e-19
6.7205661160054e-20	2.35072874484299e-19	-3.67620976506527e-20	-2.04717276153437e-19
1.33617716232308e-20	1.76628651009161e-19	4.20306095252765e-21	-1.50999609036083e-19
-1.6984825003883e-20	1.27902296242185e-19	2.58918122957276e-20	-1.07318841597967e-19

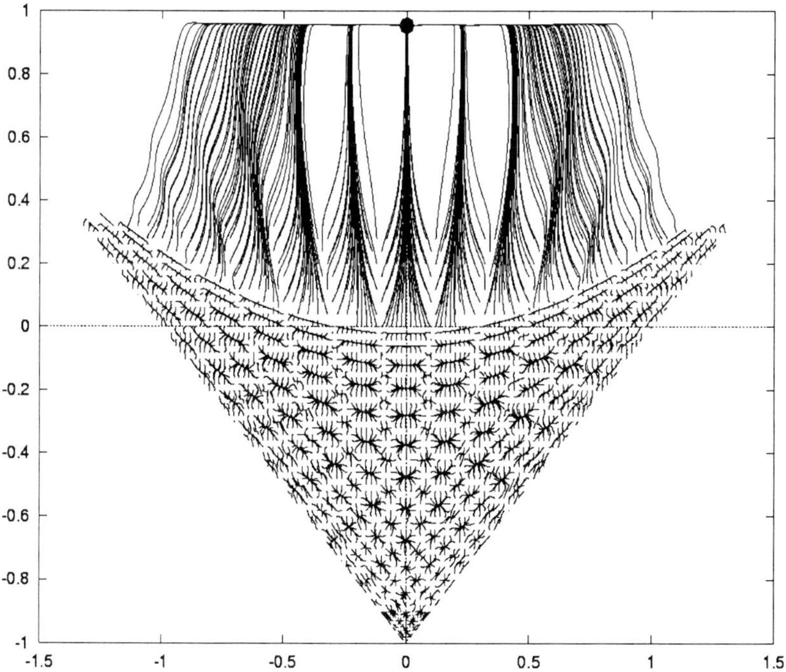

Figure 2. In this diagram, a point with coordinates (x, y) in the plane corresponds to a denominator $q(z) = z^2 + xz + y$. The picture indicates the gradient flow of $\Psi_2^R(q)$ on Δ_2 which is a triangle in these coordinates. The global minimum is the large dot at the top of the figure. Each of the small many legged patterns is a local minimum.

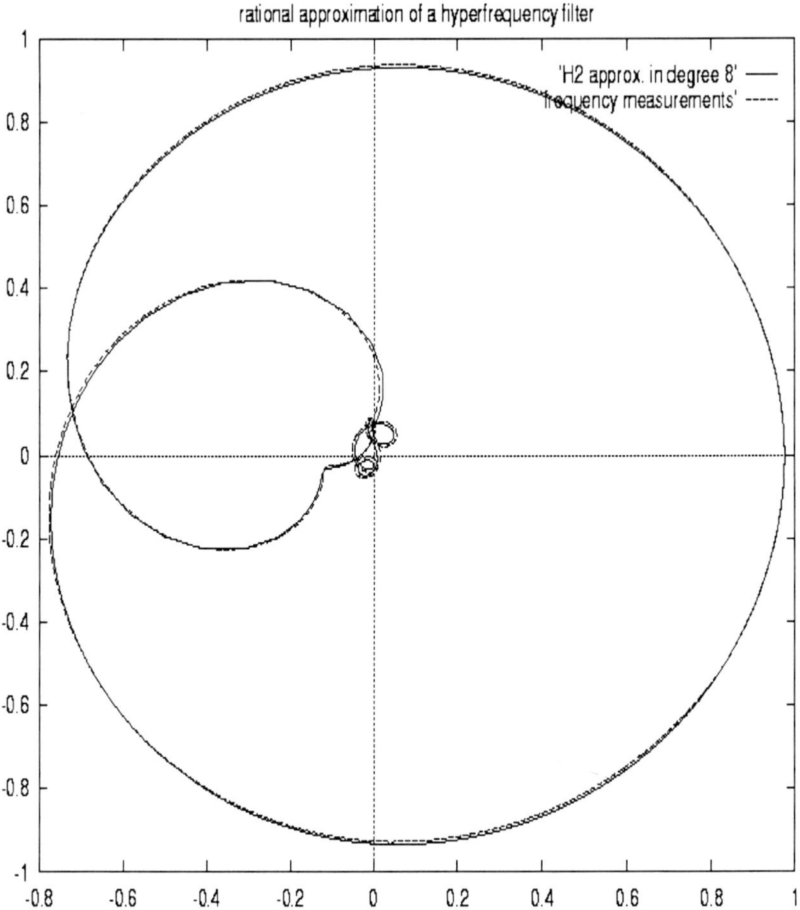

Figure 3. The dotted line in this diagram is the Nyquist plot (*i.e.* the image of the bandwidth on the imaginary axis) of the transfer function of the reflexion of a hyperfrequency filter measured by the French CNES (Toulouse). The data were first completed by solving an $\bar{H}^2$ bounded extremal problem and then approximated by a rational function of degree 8 whose Nyquist plot has been superimposed on the figure. The locus is not conjugate-symmetric because a low-pass transformation sending the central frequency to the origin was performed on the data. This illustrates that approximation with complex Fourier coefficients can be useful in system identification, even though the physical system is real.

Acknowledgement. The autor gratefully ackowledges the help of J. Grimm for deriving the numerical examples using the software *Hyperion.*

References

[1] Adamjan, V. M., D. Z. Arov and M. G. Krein, Infinite Hankel matrices and generalized Caratheodory-Fejer and Riesz problems, *Functional Anal. Appl.* **2** (1968), 1–18.

[2] Adamjan, V. M., D. Z. Arov and M. G. Krein, Analytic properties of Schmidt pairs for a Hankel operator and the generalized Schur-Takagi problem, *Math. USSR-Sb.* **15** (1971), 31–73.

[3] Bagby, T., On interpolation by rational functions, *Duke Math. J.* **36** (1969), 95–104.

[4] Baratchart, L., *Sur l'approximation rationnelle L^2 pour les systèmes dynamiques linéaires*, dissertation, Université de Nice, 1987.

[5] Baratchart, L., On the topological structure of inner functions and its use in identification, in *Progress in Systems and Control Theory* **8**, B. Bonnard, B. Bride, J. P. Gauthier, I. Kupka (eds.), Birkhäuser, 1990, 51–59.

[6] Baratchart, L. and M. Deistler, Differential consistency of maximum likelihood estimates, preprint.

[7] Baratchart, L. and J. Leblond, Hardy approximation to L^p functions on subsets of the circle with $1 \leq p < \infty$, *Constr. Approx.* **14** (1998), 41–56.

[8] Baratchart, L., J. Leblond and J. R. Partington, Hardy approximation to L^∞ functions on subsets of the circle, *Constr. Approx.* **12** (1996), 423–436.

[9] Baratchart, L. and M. Olivi, Index of critical points in l^2 approximation, *Systems & Control Letters* **10** (1988), 167–174.

[10] Baratchart, L., E. B. Saff and F. Wielonsky, A criterion for uniqueness of a critical point in H_2 rational approximation, *J. d'Analyse Math.* **70** (1996), 225–266.

[11] Baratchart, L., H. Stahl and F. Wielonsky, Uniqueness of best H^2 rational approximants to Markov functions, submitted to Constructive Approximation, 1997.

[12] Baratchart, L., H. Stahl and F. Wielonsky, Non-uniqueness of rational best approximants, 1997 to appear in Proc. CONFUN, Trondheim.

[13] Baratchart, L., H. Stahl and F. Wielonsky, Asymptotic error estimates for H^2 rational approximants to Markov functions, preprint.

[14] Baratchart, L. and F. Wielonsky, Rational approximation in the real Hardy space H_2 and Stieltjes integrals: a uniqueness theorem, *Constr. Approx.* **9** (1993), 1–21.

[15] Basu, S. and N. K. Bose, Matrix Stieltjes series and network models, *SIAM J. Math. Anal.* **14** (1983), 209–222.

[16] Brockett, R. W. and P. A. Fuhrmann, Normal symmetric dynamical systems, *SIAM J. Control and Optimization* **14** no.1 (1976), 107–119.

[17] Desoer, C. A., R. W. Liu, J. Murray and R. Saeks, Feedback system design: the fractional representation approach to analysis and synthesis, *IEEE Trans. Aut. Control* AC **25** (1980), 399–412.

[18] Doyle, J. C., B. A. Francis and A. R. Tannenbaum, *Feedback Control Theory*, Macmillan Publishing Company, 1992.

[19] Duren, P. L., *Theory of H^p spaces*, Pure and Applied Math. **38**, Academic Press, New-York, 1970.

[20] Francis, B., *A course in H^∞ control theory*, Lectures notes in control and information sciences **88**, Springer–Verlag, New-York, 1987.

[21] Geronimus, Y. L., *Polynomials orthogonal on a circle and on an interval*, Pergamon Press, Oxford, 1960.

[22] Gonchar, A. A., Rational approximation of analytic functions, in *Linear and Complex Analysis Problem Book*, V.P. Havin et al. (eds.), Lecture Notes in Math. vol. 1043, Springer-Verlag, Berlin, 1984, 471–474.

[23] Gonchar, A. A. and G. López, On Markov's theorem for multipoint Padé approximants, *Mat. Sb.* **105** (**147**) (1978); English transl. in *Math. USSR Sb.* **34** (1978), 449–459.

[24] Gonchar, A. A. and E. A. Rakhmanov, Equilibrium distributions and the degree of rational approximation of analytic functions, *Mat. Sb.* **134** (**176**) (1987), 306–352; English transl. in *Math. USSR Sb.* **62** (1989).

[25] Goresky, M. and R. MacPherson, *Stratified Morse Theory*, Ergebnisse der Mathematik und ihrer Grenzgebiete, 3. Folge, Band **14**, Springer Verlag, Berlin-Heidelberg, 1988.

[26] Hannan, E. J. and M. Deistler, *The statistical theory of linear systems*, Wiley, New York, 1988

[27] Hayashi, E., L. N. Trefethen and M. H. Gutknecht, The CF table, *Constr. Approx.* **6** (1990), 195–223.

[28] Istace, M. P., *On rational Chebyshev approximation in the complex plane*, dissertation, Facultés Universitaires Notre-Dame de la paix, Namur, 1994.

[29] Kalman, R. E. and P. L. Falb and M. A. Arbib, *Topics in mathematical system theory*, Mc Graw Hill, 1969.

[30] Ljung, L., *System identification: Theory for the user*, Prentice–Hall, 1987.

[31] Magnus, A. P., Toeplitz matrix techniques and convergence of complex weight Padé approximants, *J. of Computational and Applied Math.* **19** (1987), 23–38.

[32] Milnor, J., *Morse theory*, Annals of Math. Studies **51**, Princeton University Press, 1963.

[33] Ober, R. J., System theoretic aspects of completely symmetric systems, *Operator Theory, Adv. and Appl.* **87** (1996), 233–262.

[34] Parfenov, O. G., Estimates of the singular numbers of a Carleson operator, *Mat. Sb.* **131** (**173**) (1986), 501–518; English transl. in *Math USSR Sb.* **59** (1988).

[35] Prokhorov, V. A., *Rational approximation of analytic functions*, Russian Acad. Sci. Sb. Math., **78** (**1**), 1994.

[36] Prokhorov, V. A. and E. B. Saff, Rates of best uniform rational of analytic functions by ray sequences of rational functions, to appear in *Constructive Approx.*

[37] Rosenbrock, H. H., *State Space and Multivariable Theory*, Wiley, New York, 1970.

[38] Saff, E. B. and V. Totik, Logarithmic potentials with external fields, Grundlehren der mathematischen Wissenschaften **316**, Springer Verlag, Heidelberg, 1997.

[39] Seyfert, F., Contribution l'approximation méromorphe dans les espaces de Hardy et l'identification fréquentielle, dissertation, Ecole des Mines de Paris, 1998.

[40] Stahl, H., The convergence of Padé approximants to functions with branch points, *J. of Approx. Theory* **91** (1997), 139–204.

[41] Stahl, H. and V. Totik, *General orthogonal polynomials*, Encyclopedia of Mathematics, **43**, Cambridge university press, New York, 1992.

[42] Szegő, G., *Orthogonal polynomials*, Coll. Publ., Vol.23, Amer. Math. Soc., Providence, R. I., 1966.

[43] Totik, V., *Weighted approximation with varying weight*, Lecture notes in Math., 1569, Springer Verlag, Heidelberg, 1994.

[44] Walsh, J. L., *Interpolation and approximation by rational functions in the complex domain*, AMS colloq. Pub. XX, 1969.

[45] Willems, J. C., Dissipative dynamical systems, Part I: general theory, Part II: linear systems with quadratic supply rates, *Arch. Rat. Mech. and Anal.* **45** (1972), 321–351 & 352–392.

[46] Youla, D. C., H. A. Jabr and J. J. Bongiorno, Modern Wiener-Hopf design of optimal controllers: the multivariable case, *IEEE Trans. Aut. Control*, AC **21** (1976), 319–338.

[47] Zames, G., Feedback and optimal sensitivity: model reference transformations, multiplicative semi-norms, and appropriate inverses, *IEEE Trans. Aut. Control*, AC **28** (1981), 585–601.

Laurent Baratchart
INRIA
2004, route des Lucioles
B. P. 93
06902 Sophia Antipolis Cedex
FRANCE
e-mail: baratcha@sophia.inria.fr

Generalized Faber sets and applications

Vladimir I. Belyĭ

and Elena V. Kravchuk

Abstract. Let B be a compact set in the extended complex plane $\mathbb{C}$ with a simply connected complement containing ∞. We consider some generalizations of the Faber transform, namely the operators $\tilde{T} : w^n \mapsto \tilde{F}_n(z)$, $n \in \mathbb{N}$, where $\tilde{F}_n(z)$ are generalized Faber polynomials for B which are generated by the Hadamard composition of Faber polynomials with adjoint kernels. Here the adjoint kernels $\mathcal{F}(z)$ and $\mathcal{F}^*(z)$ are two analytic functions in the unit disk $\mathbb{D}$ with power extensions $\mathcal{F}(z) = \sum_{k=0}^{\infty} b_k z^k$, $b_k \neq 0$, $k \in \mathbb{N}$, $\mathcal{F}^*(z) = \sum_{k=0}^{\infty} b_k{}^{-1} z^k$. It is supposed that $\mathcal{F}(z)$ and $\mathcal{F}^*(z)$ have analytic extensions to $\mathbb{C}$ along an arbirtrary continuous curve Γ which does not intersect the points 1, ∞, 0. We study sufficient geometric conditions for B which imply Γ to be bounded (that is B is a so called "generalized Faber set") and consider some applications to the order estimates of various nth widths for some classes of analytic functions.

§1 Adjoint kernels

Consider two analytic functions $\mathcal{F}$ and $\mathcal{F}^*$ in the unit disk $\mathbb{D}$ with Taylor expansions as follows

$$\mathcal{F}(z) = \sum_{k=0}^{\infty} b_k z^k, \qquad b_k \neq 0, \quad k \in \mathbb{N}, \tag{1.1}$$

$$\mathcal{F}^*(z) = \sum_{k=0}^{\infty} b_k^{-1} z^k. \tag{1.2}$$

It is assumed that series (1.1) and (1.2) both have radius of convergence equal to 1, and that the functions $\mathcal{F}$ and $\mathcal{F}^*$ have analytic extensions to $\mathbb{C}$ along an arbitrary continuous curve Γ which does not intersect the points 1, ∞, 0. It is clear that the point $z = 1$ is a singular point for the functions $\mathcal{F}$ and $\mathcal{F}^*$. It is usual to call the functions $\mathcal{F}$ and $\mathcal{F}^*$ *adjoint* or *adjoint kernels* (see [7]). In order to get some preference ordering, we need to make use of the notion of "admissibility" of adjoint kernels. The adjoint

Computational Methods and Function Theory (CMFT'97)
N. Papamichael, St. Ruscheweyh and E. B. Saff (Eds.), pp. 79–85

ISBN 981-02-3626-3

kernel $\mathcal{F}$ will be called *admissible* in $\mathbb{D}$ for $n \in \mathbb{N}$, if for a given n, all r, $0 < r < 1$, and all $|x| \leq \pi$ the following inequality is true

$$\frac{1}{2} + \operatorname{Re}\left(b_n \sum_{k=1}^{\infty} b_{n+k}^{-1} r^k e^{ikx}\right) \geq 0. \tag{1.3}$$

The admissibility of $\mathcal{F}$ in $\mathbb{D}$ for *all* $n \in \mathbb{N}$ implies the sequence $\{|b_n^{-1}|\}$ is non-increasing. The condition (1.3) was investigated in [3], while a little earlier such a condition was used by Scheik [8] for kernels $\mathcal{F}$ with real nonnegative coefficients [8]. We note in passing that (1.3) implies that the function

$$\phi_n(z) =: 1 + 2b_n \sum_{k=1}^{\infty} b_{n+k}^{-1} z^k$$

belongs to the Herglotz-Riesz class.

§2 Hadamard compositions with adjoint kernels

Let $G \subset \mathbb{C}$ be a bounded and simply connected Jordan region, with $0 \in G$, and consider a function f analytic in G which at the point $z = 0$ has the germ $f(z) = \sum_{k=0}^{\infty} a_k z^k$. Consider also two adjoint kernels $\mathcal{F}$ and $\mathcal{F}^*$. Then by the Hadamard composition of f and $\mathcal{F}$ we mean a new holomorphic function in G defined by extending analytically to G the germ

$$f_*(z) =: (f * \mathcal{F})(z) = \sum_{k=0}^{\infty} a_k b_k z^k.$$

In fact, the constructive realization of such an analytic extension to an arbitrary point $z \in G$ can be made by means of the formula ([7])

$$f_*(z) = \frac{1}{2\pi i} \int_L f(u) \mathcal{F}\left(\frac{z}{u}\right) \frac{du}{u}, \tag{2.1}$$

where L is an arbitrary rectifiable curve in G containing the points z and 0 in its interior. Incidentally, for all $z \in G$,

$$f(z) = \frac{1}{2\pi i} \int_L f_*(u) \mathcal{F}^*\left(\frac{z}{u}\right) \frac{du}{u}. \tag{2.2}$$

Let us remark that formulas (2.1) and (2.2) are not the only means for realizing the desired extension. If the boundary $L_0 = \partial G$ is a rectifiable Jordan curve and if the functions f and f_* belong to the Smirnov class E_1, then we can put $L = L_0$ in (2.1), (2.2) and consider nontangential boundary values of f and f_* on it.

We would like to mention a remarkable similarity: the germ of an analytic function behaves like a gene of a living organism. Thus, Hadamard compositions may be regarded as a form of genetic engineering which can, essentially, change the properties of functions.

§3 Some important special cases

We can get a nice family of adjoint kernels by setting

$$\mathcal{F}(z) =: F(\alpha, 1, \gamma, z), \qquad \mathcal{F}^*(z) =: F(\gamma, 1, \alpha, z), \tag{3.1}$$

where α, γ do not admit the values $\{0, -1, -2, \ldots\} =: \mathbb{Z}_-$, and $F(\alpha, \beta, \gamma, z)$ is a Gauss hypergeometric series, see [7], p. 160,

$$F(\alpha, \beta, \gamma, z) = 1 + \sum_{k=1}^{\infty} \frac{\alpha(\alpha+1)\cdots(\alpha+k-1)\beta(\beta+1)\cdots(\beta+k-1)}{k!\gamma(\gamma+1)\cdots(\gamma+k-1)} z^k,$$

$\beta \notin \mathbb{Z}_-$. If $\alpha = 2$ and $\gamma = 1$ we get the following pair of adjoint kernels

$$\mathcal{F}(z) = \frac{1}{(1-z)^2}, \qquad \mathcal{F}^*(z) = \ln \frac{1}{1-z}.$$

Let us now choose adjoint kernels as follows

$$\mathcal{F}(z) = \sum_{k=0}^{\infty} \frac{\Gamma(k+1)}{\Gamma(k-\alpha+1)} z^k, \qquad \mathcal{F}^*(z) = \sum_{k=0}^{\infty} \frac{\Gamma(k-\alpha+1)}{\Gamma(k+1)} z^k, \tag{3.2}$$

where $\alpha > 0$, $\alpha \notin \mathbb{N}$, is an arbitrary real number. Then

$$f_*(z) = (f * \mathcal{F}_n)(z) = z^\alpha f^{(\alpha)}(z),$$

where $f^{(\alpha)}$ is a fractional αth derivative with the origin $z = 0$ in the sense of Sewell [9].

Finally, let ω be a continuous, non-negative and non-decreasing function on $[0, 1]$, such that the following conditions are satisfied

$$\omega(0) = 1, \qquad \int_0^1 \omega(x)\, dx < +\infty, \qquad \int_r^1 \omega(x)\, dx > 0, \quad r \in [0, 1).$$

By the choice of

$$\Delta_0 = 1, \qquad \Delta_k = k \int_0^1 \omega(x) x^{k-1}\, dx \quad k = 1, 2, \ldots,$$

and

$$\mathcal{F}(z) = \sum_{k=0}^{\infty} \Delta_k z^k, \qquad \mathcal{F}^*(z) = \sum_{k=0}^{\infty} \Delta_k^{-1} z^k \tag{3.3}$$

we have

$$f_*(z) = (f * \mathcal{F})(z) = L^{(\omega)}(f(z)),$$

which is Djarbashian's operator for generalized integration and differentiation ([4]).

Notice that the kernels $\mathcal{F}$ in cases (3.1), (3.2) and (3.3) are admissible for all $n \in \mathbb{N}$ (*cf.* [3]).

§4 Faber operators, generalized Faber operators

Let $G \subset \mathbb{C}$ be a bounded simply connected Jordan region, with $0 \in G$, and let $\overline{G}$ denote the closure of G. Also, let $w = \Phi(z)$ be a Riemann mapping function $\Phi : \mathbb{C}\backslash\overline{G} \longrightarrow \mathbb{C}\backslash\overline{\mathbb{D}}$, and let $z = \Psi(w)$ be its inverse. The functions Φ and $\Psi = \Phi^{-1}$ are normalized by the condition $\lim_{w\to\infty} \Psi(w)/w = \alpha > 0$, $\alpha = \operatorname{cap}(\overline{G})$. Suppose also we have a prescribed function R which is regular for $|w| > 1$ and continuous in $|w| \geq 1$:

$$R(w) = \sum_{k=0}^{\infty} \frac{c_k}{w^k}, \qquad c_0 \neq 0.$$

Consider an operator of a form

$$(\tilde{T}p)(z) = \frac{1}{2\pi i} \int_{|w|=1} p(w)\mathcal{F}^*\left(\frac{z}{\Psi(w)}\right) \frac{\Psi'(w)}{\Psi(w)} R(w)\, dw \tag{4.1}$$

where $\mathcal{F}^*$ is one of two adjoint kernels and, for $|w| = 1$, the derivative $\Psi'(w)$ exists almost everywhere as the nontangential boundary value and belongs to $\mathcal{L}^1$ because of the rectifiability of $L = \partial G$. In fact, it is known [7] that the operator (4.1) being bounded in some spaces changes the degree basis $\{1, w, w^2, \ldots\}$ in $\mathbb{D}$ into the basis $\{\tilde{F}_0(z), \tilde{F}_1(z), \ldots\}$ of generalized Faber polynomials in $\overline{G}$. So

$$\tilde{T} : w^n \to \tilde{F}_n(z), \qquad n \in \mathbb{N}.$$

We have the "canonical" case of the Faber operator if $\mathcal{F}(z) = \mathcal{F}^*(z) = \frac{1}{1-z}$, $R(w) \equiv 1$. This case was investigated by a group of mathematicians: T. Kövari [6], J.-E. Anderson [1], E. M. Dyn'kin, in a series of papers starting with [5], J. M. Anderson and J. Clunie [2] (which contains the latest and strongest result). The set for which the Faber operator is bounded is called the *Faber set*, and if the inverse operator is also bounded, the *inverse*

Faber set. A full description of Faber sets is not yet available. The best result for Faber sets belongs to T. Kövari ([6], see also [2]): closed Jordan regions whose boundary has a bounded boundary rotation; for inverse Faber sets: closed Jordan regions with a bounded boundary rotation without cusps.

§5 Applications

5.1 First application

Generalized Faber sets with admissible adjoint kernels are, generally speaking, a wider class of sets. For example, if $\mathcal{F}$ and $\mathcal{F}^*$ are two adjoint kernels with

$$\mathcal{F}^*(z) = \frac{1}{(1-z)^{1-\alpha}}, \qquad \alpha \in (0;1),$$

then (if, without loss of generality, $R(w) \equiv 1$)

$$(\tilde{T}p)(z) = \frac{1}{2\pi i}\int_{|w|=1} \frac{p(w)\Psi'(w)}{\left(1-\frac{z}{\Psi(w)}\right)^{1-\alpha}} \cdot \frac{dw}{\Psi(w)} = \frac{1}{2\pi i}\int_L \frac{p(\Phi(\zeta))\zeta^{-\alpha}}{(\zeta-z)^{1-\alpha}}\,d\zeta.$$

So, in fact, for the mapping

$$\tilde{T} : A(\overline{\mathbb{D}}) \to A(\overline{G}),$$

an arbitrary Jordan region $\overline{G}$ with a rectifiable boundary is a generalized Faber set.

The following theorem is almost trivial.

Theorem 1 (Analogous to that of J. M. Anderson and J. Clunie). *Let K be a generalized Faber set and suppose that $f \in \tilde{T}\left(A(\overline{\mathbb{D}})\right)$ then*

$$E_n(f;K) \le \text{const} \cdot E_n\left(\tilde{T}^{-1}f;\overline{\mathbb{D}}\right).$$

5.2 Second application

For an inverse Faber set we have

$$E_{n-1}\left(\mathcal{F}\widetilde{A^{K_\rho}},\Lambda^*\right) \asymp \left|b_n^{-1}\right|\rho^n, \qquad 0<\rho<1$$

iff $\mathcal{F}$ is admissible for a given n in $K_\rho \subset \mathbb{D}$.

Theorem 2 (Analogous to that of V. I. Belyĭ, M. Z. Dveirin).

$$E_{n-1}\left(\mathcal{F}E_p(G)\right) \asymp \left|b_n^{-1}\right|, \qquad p \geq 1.$$

5.3 Third application

For an inverse Faber set we have the ordinal estimates for the Kolmogorov, the linear and the S. N. Bernsteinian width, *e.g.* for Kolmogorov width in $E_q(G)$:

$$d_n\left(\mathcal{F}E_p(G), E_q(G)\right) \asymp \left|b_n^{-1}\right|.$$

For other widths we have similar estimates.

References

[1] Anderson, J.-E., On the degree of polynomial and rational approximation of holomorphic functions, dissertation, University of Göteborg, 1973.

[2] Anderson, J. M. and J. Clunie, Isomorphisms of the disk algebra and inverse faber sets, *Math. Z.* **188** (1985), 545–558.

[3] Belyĭ, V. I. and M. Z. Dveirin, *On the Best Linear Methods of Approximation on Classes of Functions Defined by Adjoint Kernels* (in Russian), Metricheskie voprosy teorii functsii i otobrazhenii, II, Naukova Dumka, Kiev, 1971, 37–54.

[4] Djarbashian, M. M., Generalized Riemann-Liouville operator and some of its applications (in Russian), *Izv. AN USSR, ser.matemat.* **32** (1968), 1075–1111.

[5] Dyn'kin, E. M., On the uniform polynomial approximation in a complex domain (in Russian), *Zapiski nauchnyh seminarov LOMI* **47** V (1974), 164–165.

[6] Kövari, T., On the order of polynomial approximation for closed Jordan domains, *Approx. Theory* **5** (1972), 362–373.

[7] Lebedev, N. A. and V. I. Smirnov, *Function of a Complex Variable: Constructive Theory*, The M.I.T. Press, Cambridge, Mass., 1968.

[8] Scheick, I. T., Polynomials of functions analytic in a disk, *Proc. Amer. Math. Soc.* **7** no. 6 (1966), 1238–1243.

[9] Sewell, W. E., Generalized derivatives and approximation by polynomials, *Trans. Amer. Math. Soc.* **41** no. 1 (1937), 84–121.

Vladimir I. Belyĭ
Institute for Applied Mathematics and Mechanics
NAS of Ukraine
Donetsk, 340092
UKRAINE
faber@bvi.donetsk.ua

Elena V. Kravchuk
Department of Economics
Donetsk State University
Donetsk, 340012
UKRAINE
Dmitrii@kravchuk.dgtu.donetsk.ua

An extension of the Beurling-Nevanlinna projection theorem

Dimitrios Betsakos

Abstract. We prove a polarization and a projection result for the harmonic measure on the unit disk. The projection result is a generalization of the Beurling-Nevanlinna projection theorem.

§1 Introduction

Let D be a domain in the complex plane $\mathbb{C}$ and let E be a Borel set on the boundary ∂D of D. The harmonic measure of E at $z \in D$ relative to D is the Perron solution $u(z)$ of the Dirichlet problem in D with boundary values 1 on E and 0 on $\partial D \setminus E$. More precisely, let $\chi_E = 1$ on E and $\chi_E = 0$ on $\partial D \setminus E$. Then

$$u(z) = \sup\{v(z) : \begin{array}{l} v \text{ subharmonic in } D \text{ and} \\ \limsup_{w\to\zeta} v(w) \le \chi_E(\zeta) \text{ for } \zeta \in \partial D\}. \end{array}$$

We shall write

$$u(z) = \begin{cases} \omega(z, E, D) & \text{if } z \in D \cup E \\ 0 & \text{otherwise} \end{cases}.$$

Furthermore, for $A \subset \mathbb{C}$ we set $\overline{A} := \{\bar{z} : z \in A\}$, and if $F \subset \mathbb{C}$ is a closed set then

$$\begin{array}{ll} F_+ := F \cap \{z : \operatorname{Im} z \ge 0\}, & F_- := F \cap \{z : \operatorname{Im} z \le 0\}, \\ F_l := F \cap \{z : \operatorname{Re} z \le 0\}, & F_r := F \cap \{z : \operatorname{Re} z \ge 0\}. \end{array}$$

If O is an open set in $\mathbb{C}$ then

$$\begin{array}{ll} O_+ := O \cap \{z : \operatorname{Im} z > 0\}, & O_- := O \cap \{z : \operatorname{Im} z < 0\}, \\ O_l := O \cap \{z : \operatorname{Re} z < 0\}, & O_r := O \cap \{z : \operatorname{Re} z > 0\}. \end{array}$$

In 1933 Beurling [2] and independently R. Nevanlinna [3] proved a projection theorem which is now a classical result for harmonic measure.

Computational Methods and Function Theory (CMFT'97)
N. Papamichael, St. Ruscheweyh and E. B. Saff (Eds.), pp. 87–90

ISBN 981-02-3626-3

Theorem 1 (The Beurling-Nevanlinna projection theorem). *Let K be a closed set in the unit disk $\mathbb{D}$. Let $K^* = \{-|z| : z \in K\}$. Then*

$$\omega(z, \partial\mathbb{D}, \mathbb{D} \setminus K) \leq \omega(|z|, \partial\mathbb{D}, \mathbb{D} \setminus K^*) \quad \text{for all } z \in \mathbb{D}. \tag{1.1}$$

We will prove the following generalization of Theorem 1.

Theorem 2. *Let $y \in (0,1)$ and let K, K^* be as in Theorem 1. Then*

$$\begin{aligned} &\omega(iy, \partial\mathbb{D}, \mathbb{D} \setminus K) + \omega(-iy, \partial\mathbb{D}, \mathbb{D} \setminus K) \\ &\quad \leq \omega(iy, \partial\mathbb{D}, \mathbb{D} \setminus K^*) + \omega(-iy, \partial\mathbb{D}, \mathbb{D} \setminus K^*). \end{aligned} \tag{1.2}$$

The proof of Theorem 2 is based on *polarization*.

Definition 1. *Let A be a closed or open set. The polarization $\mathbb{P}A$ of A (with respect to $\mathbb{R}$) is $\mathbb{P}A = (A \cup \overline{A})_+ \cup (A \cap \overline{A})_-$.*

We define also the polarization $\mathbb{P}_l A$ of A with respect to any oriented line l.

Definition 2. *Let l be an oriented line and let $T_l : \widehat{\mathbb{C}} \to \widehat{\mathbb{C}}$ be the Möbius transformation that maps ∞ to ∞ and l to $\mathbb{R}$, preserving orientation. The polarization $\mathbb{P}_l A$ of A with respect to l is $\mathbb{P}_l A := T_l^{-1} \mathbb{P} T_l A$.*

The behaviour of harmonic measure under polarization has been studied by Solynin [5] and Betsakos [1]. One of their results is the following:

Theorem 3 (Solynin, Betsakos). *Let K be a compact set in $\mathbb{D}$, and let $z \in \mathbb{D}$. Then*

$$\begin{aligned} &\omega(z, \partial\mathbb{D}, \mathbb{D} \setminus K) + \omega(\bar{z}, \partial\mathbb{D}, \mathbb{D} \setminus K) \\ &\quad \leq \omega(z, \partial\mathbb{D}, \mathbb{D} \setminus \mathbb{P}K) + \omega(\bar{z}, \partial\mathbb{D}, \mathbb{D} \setminus \mathbb{P}K). \end{aligned} \tag{1.3}$$

We will prove a partial extension of Theorem 3.

Theorem 4. *Let $z = re^{i\theta}$, $z' = re^{i(\theta-\pi)}$, $r \in (0,1)$, $\theta \in [\pi/2, \pi]$. Let K be a compact set in $\mathbb{D}$ such that $K = K_1 \cup K_2 \cup K_3 \cup \overline{K_3}$, where $K_1 \subset \mathbb{D}_+$, $K_3 \subset \mathbb{D}_+$, $K_2 \subset \mathbb{D}_-$ and $\overline{K}_1 \cap K_2 = \emptyset$. Assume $K_1 \subset \mathbb{D}_l$ and $K_3 \subset \mathbb{D}_r$. Then*

$$\begin{aligned} &\omega(z, \partial\mathbb{D}, \mathbb{D} \setminus K) + \omega(z', \partial\mathbb{D}, \mathbb{D} \setminus K) \\ &\quad \leq \omega(z, \partial\mathbb{D}, \mathbb{D} \setminus \mathbb{P}K) + \omega(z', \partial\mathbb{D}, \mathbb{D} \setminus \mathbb{P}K). \end{aligned} \tag{1.4}$$

In §2 we prove Theorem 4 and see how it implies Theorem 2.

§2 Proof of the theorems

Proof of Theorem 4: Let $D = \mathbb{D} \setminus K$ and $\widehat{D} = \mathbb{P}D = \mathbb{D} \setminus \overline{\mathbb{P}K}$. By the strong Markov property of harmonic measure (*cf.* [1]) we have

$$\omega(z, \partial\mathbb{D}, D) = \omega(z, (\partial\mathbb{D})_+, D_+) + \int_{-1}^{1} \omega(z, dx, D_+)\omega(x, \partial\mathbb{D}, D), \quad (2.1)$$

$$\omega(z', \partial\mathbb{D}, D) = \omega(z', (\partial\mathbb{D})_-, D_-) + \int_{-1}^{1} \omega(z', dx, D_-)\omega(x, \partial\mathbb{D}, D), \quad (2.2)$$

$$\omega(z, \partial\mathbb{D}, \widehat{D}) = \omega(z, (\partial\mathbb{D})_+, \widehat{D}_+) + \int_{-1}^{1} \omega(z, dx, \widehat{D}_+)\omega(x, \partial\mathbb{D}, \widehat{D}), \quad (2.3)$$

$$\omega(z', \partial\mathbb{D}, \widehat{D}) = \omega(z', (\partial\mathbb{D})_-, \widehat{D}_-) + \int_{-1}^{1} \omega(z', dx, \widehat{D}_-)\omega(x, \partial\mathbb{D}, \widehat{D}). \quad (2.4)$$

So, because of Theorem 3, to prove (1.4) it suffices to prove

$$\begin{aligned} &\omega(z, (\partial\mathbb{D})_+, D_+) + \omega(z', (\partial\mathbb{D})_-, D_-) \\ &\quad \leq \omega(z, (\partial\mathbb{D})_+, \widehat{D}_+) + \omega(z', (\partial\mathbb{D})_-, \widehat{D}_-) \end{aligned} \quad (2.5)$$

and

$$\omega(z, I, D_+) + \omega(z', I, D_-) \leq \omega(z, I, \widehat{D}_+) + \omega(z', I, \widehat{D}_-) \quad (2.6)$$

for every interval I on $(-1, 1)$.

First we prove (2.5). Again by the strong Markov property:

$$\begin{aligned} &\omega(z, (\partial\mathbb{D})_+, D_+) = \\ &\quad \omega(z, (\partial\mathbb{D})_+, \widehat{D}_+) - \textstyle\int_{K_1} \omega(z, dk, D_+)\omega(k, (\partial\mathbb{D})_+, \widehat{D}_+) \end{aligned} \quad (2.7)$$

and

$$\begin{aligned} &\omega(z, (\partial\mathbb{D})_-, D_-) = \\ &\quad \omega(z', (\partial\mathbb{D})_-, D_-) - \textstyle\int_{\overline{K}_1} \omega(z', dk, \widehat{D}_-)\omega(k, (\partial\mathbb{D})_-, D_-). \end{aligned} \quad (2.8)$$

So, to prove (2.5) it suffices to prove

$$\begin{aligned} &\textstyle\int_{\overline{K}_1} \omega(z', dk, \widehat{D}_-)\omega(k, (\partial\mathbb{D})_-, D_-) \\ &\quad \leq \textstyle\int_{K_1} \omega(z, dk, D_+)\omega(k, (\partial\mathbb{D})_+, \widehat{D}_+). \end{aligned} \quad (2.9)$$

But we have

$$\omega(z', \overline{J}, \widehat{D}_-) \leq \omega(z, J, D_+) \quad (2.10)$$

for every Borel set J on the boundary of K. (2.10) holds because of (a) the domain monotonicity of harmonic measure, and (b) the polarization theorem for harmonic measure (Theorem 2.2 in [1]). We also have

$$\omega(k, (\partial\mathbb{D})_-, D_-) \leq \omega(k, (\partial\mathbb{D})_+, \widehat{D}_+) \quad (2.11)$$

by domain monotonicity. (2.10) and (2.11) show that (2.9) holds. Hence (2.5) also holds. The proof of (2.6) is very similar to the proof of (2.5) and we omit it. As we noted above, (2.5) and (2.6) imply (1.4) and Theorem 4 is proved. ■

Proof of Theorem 2: To prove Theorem 2 we use a technique due to Øksendal, [4]. First we note that Theorem 4 and the domain monotonicity of harmonic measure imply:

$$\begin{aligned} &\omega(z, \partial\mathbb{D}, D) + \omega(z', \partial\mathbb{D}, D) \\ &\quad \leq \omega(z, \partial\mathbb{D}, \mathbb{D} \setminus \overline{\mathbb{P}K_-}) + \omega(z', \partial\mathbb{D}, \mathbb{D} \setminus \overline{\mathbb{P}K_-}). \end{aligned} \tag{2.12}$$

We consider successive polarizations of the domain D with respect to the line $\{re^{i\theta} : r \in \mathbb{R},\ \theta = -\pi/2^n\}$ (with the orientation from $e^{-i\theta}$ to $e^{i(\pi-\theta)}$) and then with respect to the line $\{re^{i\theta} : r \in \mathbb{R}, \theta = \pi/2^n\}$ (with the orientation from $e^{i\theta}$ to $e^{i(\pi+\theta)}$), $n = 0, 1, 2, \ldots$. In each step we apply (2.12). In the limit we obtain Theorem 2. ■

References

[1] Betsakos, D., Polarization, conformal invariants, and Brownian motion, *Ann. Acad. Fenn. Ser. A I Math* **23** (1998), 59–82.

[2] Beurling, A., The collected works of Arne Beurling, vol. 1 Complex Analysis, L. Carleson et al. (ed.), Birkhäuser, Boston, 1989.

[3] Nevanlinna, R., Über eine Minimumaufgabe in der Theorie der konformen Abbildung, *Göttinger Nachr.* **I.37** (1933), 103–115.

[4] Øksendal, B., Projection estimates for harmonic measure, *Ark. Mat.* **21** (1983), 191–203.

[5] Solynin, A. Yu., Polarization and functional inequalities, (in Russian) *Algebra i Analiz* **8** no. 6, (1996), 1015–1038.

Dimitrios Betsakos
Anatolikis Romilias 7
50100 Kozani
GREECE
betsakos@compulink.gr

Current address of the author:
Mathematics Department
P.O.B. 4 (Yliopistonkatu 5)
FIN-00014 University of Helsinki
FINLAND

On a theorem of S. Ruscheweyh

Gavin Brown
and Stamatis Koumandos

Abstract. A Theorem of S. Ruscheweyh states that for $\lambda \geq 1/2$, $-1 < x < 1$, and arbitrary nonincreasing sequence a_k, $k = 0, 1, \ldots, n$, we have

$$\sum_{k=0}^{n} a_k \frac{C_k^\lambda(x)}{C_k^\lambda(1)} z^k \neq 0, \qquad |z| \leq 1,$$

where C_k^λ are the Gegenbauer polynomials. We give a new proof of this result by Fourier kernel techniques. We further show that these methods enable us to obtain corresponding results for even polynomial sums of this type and specific sequences a_k. Other related results are also discussed.

§1 Introduction

Consider the polynomials

$$P_n^\lambda(z;\theta) = \sum_{k=0}^{n} \frac{C_k^\lambda(\cos\theta)}{C_k^\lambda(1)} z^k, \qquad 0 < \theta < \pi, \quad n = 1, 2, \ldots, \tag{1.1}$$

where C_k^λ is the Gegenbauer polynomial of degree k and order λ.

Ruscheweyh [10] extended the classical Kakeya-Eneström theorem in a far-reaching way by proving a result on starlike functions and deducing that, for $\lambda \geq 1/2$, the polynomials $P_n^\lambda(z,\theta)$ have no zeros in the interior of the unit disc. The case $\lambda = 1/2$ had been shown by Szegő [11] and, by specializing to the real axis, Ruscheweyh was able to resolve conjectures of Askey [1] and Bustoz [3].

In fact, if the polynomial in (1.1) is never zero in the unit disc it follows that

$$\sum_{k=0}^{n} \frac{C_k^\lambda(\cos\theta)}{C_k^\lambda(1)} > 0, \qquad 0 < \theta < \pi, \quad n = 1, 2, \ldots, \tag{1.2}$$

The best λ in (1.2) was a long standing question now resolved by Brown, Koumandos and Wang [7]. It turns out that (1.2) holds precisely when $\lambda \geq \lambda' = \alpha' + \frac{1}{2}$, where α' is the unique solution of

$$\int_0^{j_{\alpha,2}} t^{-\alpha} J_\alpha(t)\, dt = 0,$$

Computational Methods and Function Theory (CMFT'97)
N. Papamichael, St. Ruscheweyh and E. B. Saff (Eds.), pp. 91–97

ISBN 981-02-3626-3

and J_α is the Bessel function of the first kind of order α with $j_{\alpha,2}$ being its second positive root. Then $\alpha' = -0.269387\ldots$. We conjecture that $P_n^\lambda(z;\theta)$ has no zeros for $|z|<1$ when $\lambda \geq \alpha' + \frac{1}{2}$.

In the spirit of the conference we investigate the interplay of complex and real methods by showing how to derive Ruscheweyh's basic application in a very simple way with the focus on positive sums. The methods trace back to the work of Szegő [11] and the famous posthumous paper of Feldheim [9].

We note also that the results in the cases already discussed are far from best possible. Although the methods described here could in principle be extended to tackle new cases of the zero problem, it would be more interesting to discover whether, for special polynomials such as those in (1.1), there exist ways of transferring real axis estimates to the full disc.

§2 Ruscheweyh's basic result

Theorem 1. *Let $-1<x<1$ and let $a_k>0$ be a nonincreasing sequence, then*

$$Q_n^\lambda = \sum_{k=0}^{n} a_k \frac{C_k^\lambda(x)}{C_k^\lambda(1)} z^k \neq 0, \qquad |z|\leq 1, \quad \lambda \geq \frac{1}{2}. \tag{2.1}$$

Proof: Because the coefficients of the polynomials under discussion are real we can and do restrict attention to $z = re^{i\phi}$, $0\leq\phi\leq\pi$. Let us multiply Q_n^λ by $e^{\frac{i\phi}{2}}$, write $x=\cos\theta$ and take real and imaginary parts to see that it will suffice to prove

$$\sum_{k=0}^{n} a_k \frac{C_k^\lambda(\cos\theta)}{C_k^\lambda(1)} r^k \cos(k+\frac{1}{2})\phi > 0, \qquad 0\leq\phi\leq\theta, \tag{2.2}$$

and

$$\sum_{k=0}^{n} a_k \frac{C_k^\lambda(\cos\theta)}{C_k^\lambda(1)} r^k \sin(k+\frac{1}{2})\phi > 0, \qquad \theta\leq\phi\leq\pi. \tag{2.3}$$

Because $C_k^\lambda(-x) = (-1)^k C_k^\lambda(x)$, we see that (2.2) and (2.3) are, in fact, equivalent.

Let us now argue that it suffices to prove, that for all n,

$$\sum_{k=0}^{n} \frac{C_k^\lambda(\cos\theta)}{C_k^\lambda(1)} \cos(k+\frac{1}{2})\phi > 0, \qquad 0\leq\phi\leq\theta. \tag{2.4}$$

The point is that (2.4) being true for all n allows us to use partial summation to introduce the decreasing sequence (r^k). With the corresponding result established for all n, we may next introduce the decreasing sequence (a_k). Thus (2.2) follows from (2.4).

To prove (2.4) we quote the Dirichlet-Mehler formula (*cf.* [12], p. 86)

$$\frac{C_k^{\frac{1}{2}}(\cos\theta)}{C_k^{\frac{1}{2}}(1)} = \frac{\sqrt{2}}{\pi}\int_\theta^\pi \frac{\sin(k+\frac{1}{2})t}{(\cos\theta-\cos t)^{1/2}}\,dt.$$

It is immediately obvious that it will suffice to prove positivity of

$$S_n = \sum_{k=0}^n \sin(k+\frac{1}{2})t\cos(k+\frac{1}{2})\phi, \qquad \text{for } 0<\phi<t. \tag{2.5}$$

(Note here that $\theta < \pi$).

The fact that $S_n > 0$ in (2.5) is (or, more precisely, in the good old days was) a high school exercise. We note, for example, that

$$\begin{aligned}\frac{S_n}{\sin\frac{1}{2}(t+\phi)} &= \frac{1}{2}\sum_{k=0}^n \frac{\sin(k+\frac{1}{2})(t+\phi)}{\sin\frac{1}{2}(t+\phi)} + \frac{1}{2}\sum_{k=0}^n \frac{\sin(k+\frac{1}{2})(t-\phi)}{\sin\frac{1}{2}(t+\phi)}\\ &= \frac{1-\cos(n+1)(t+\phi)}{4\sin^2\frac{1}{2}(t+\phi)} + \frac{1-\cos(n+1)(t-\phi)}{2(\cos\phi-\cos t)} \geq 0.\end{aligned}$$

To complete the proof of Theorem 1 it remains only to pass from $\lambda = 1/2$ to all greater λ. This follows from a result of Askey and Fitch [2]:

If $\lambda > \nu > 0$, $-1 \leq x \leq 1$. Then there exists a probability measure μ such that

$$\frac{C_k^\lambda(x)}{C_k^\lambda(1)} = \int_{-1}^1 \frac{C_k^\nu(y)}{C_k^\nu(1)}\,d\mu(y),$$

where μ depends on x, but is independent of k. ■

§3 Extensions

Ruscheweyh noted that his methods allowed one to replace k by mk in (1.1) and obtain the conclusion for $\lambda \geq \frac{1}{2}m$. Thus when $m = 2$ his result was $\lambda \geq 1$. We now show how to modify P_n^λ so that we obtain a corresponding conclusion for $\lambda \geq 1/2$.

Theorem 2. *Let* $a_n = (\frac{1}{2})_n/n!$, $b_n = n!/(\frac{3}{2})_n$ *(where, as usual, we use the Pochhammer symbol* $(s)_n = s(s+1)\cdots(s+n-1)$*). Then, for* $0<\theta<\pi$, $n = 1, 2, \ldots$, $\lambda \geq \frac{1}{2}$,

$$Q_n^\lambda(z;\theta) = \sum_{k=0}^n \frac{C_{2k}^\lambda(\cos\theta)}{C_{2k}^\lambda(1)} a_k z^k,$$

$$R_n^\lambda(z;\theta) = \sum_{k=0}^{n} \frac{C_{2k}^\lambda(\cos\theta)}{C_{2k}^\lambda(1)} b_k z^k,$$

have no zeros in the unit disc.

Proof: By following the first part of the proof of Theorem 1 with $Q_n^\lambda(z;\theta)$ in place of $P_n^\lambda(z;\theta)$ we reduce matters to proving positivity of the kernel

$$K_n = \sum_{k=0}^{n} a_k \sin(2k+\frac{1}{2})t\cos(2k+\frac{1}{2})\phi, \qquad \text{for } 0<\phi<t<\pi. \tag{3.1}$$

But

$$2K_n = \sum_{k=0}^{n} a_k \sin(2k+\frac{1}{2})(t+\phi) + \sum_{k=0}^{n} a_k \sin(2k+\frac{1}{2})(t-\phi).$$

It is a Theorem of Vietoris [13] that $\sum_{k=0}^{n} a_k \sin(2k+\frac{1}{2})u > 0$ for $0<u<2\pi$, and this suffices for our purpose.

The corresponding result with coefficients b_k was proved by Brown and Hewitt [4]. As there is no difficulty passing from z^2 to z the proof is complete. ■

Remark 1: Because (a_n), (b_n) are decreasing this is weaker than the corresponding theorem with constant sequences. We would expect that result to be true as well.

Remark 2: The specializations to $z=1$ provide positivity results for sums of Legendre polynomials. In that case the stronger version with constant coefficients is known to be true, *cf.* [6].

Remark 3: The Vietoris result can itself be expressed as a Gegenbauer sum. In fact it states positivity of

$$\sum_{k=0}^{n}(4k+1)a_k \frac{C_{4k}^1(\cos\theta)}{C_{4k}^1(1)}, \qquad \text{for } 0<\theta<\pi.$$

Because $(4k+1)a_k$ is an increasing sequence this is sharper than the corresponding result with constant coefficients.

Although the result just quoted improves on the Ruscheweyh estimate for $m=4$ (*viz.* $\lambda=2$) we cannot resist offering a remarkably simple proof of the latter result.

In fact we quote the Dirichlet-Mehler formula in the form

$$\frac{C_j^\lambda(\cos\theta)}{C_j^\lambda(1)} = \frac{2^\lambda(\sin\theta)^{1-2\lambda}}{B(\lambda,\frac{1}{2})} \int_0^\theta \frac{\cos(j+\lambda)t}{(\cos t-\cos\theta)^{1-\lambda}}\,dt$$

(*cf.* [8], p. 177).

The choice $j = mk$, $\lambda = \frac{1}{2}m$ is then very natural and Ruscheweyh's work implies positivity of

$$\int_0^\theta (\cos t - \cos\theta)^{\frac{m}{2}-1} \frac{\sin(n+1)mt}{\sin\frac{1}{2}mt}\, dt.$$

To prove this directly in the case $m = 4$, we write the integrand as

$$f(t) = \frac{\sin 4(n+1)t}{\sin t}\,(1 - \frac{\cos\theta}{\cos t}).$$

Positivity of the integral is now obvious for $0 < \theta \le \pi/2$ as we are multiplying a sine wave by a positive decreasing function. For $\pi/2 < \theta < \pi$ it is probably simplest to rewrite the integral as

$$\int_0^{\pi-\theta} f(t)\, dt + \int_{\pi-\theta}^{\frac{\pi}{2}} \frac{\sin 4(n+1)t}{2\sin t} \left(\frac{-2\cos\theta}{\cos t}\right) dt$$

and the proof is complete. ■

In fact the best known positivity result for $m = 4$ was established by Brown, Koumandos and Wang in [5], where it is shown that one may take $\lambda = 1/2$.

As noted in the introduction a major breakthrough would be to find a way of converting estimates for $z = 1$ to theorems concerning zeros in the disc. However the interplay between Remarks 1, 2 shows that this is not simple. It may also be useful to consider estimates along rays by considering the sums

$$\sum_{k=0}^{n} \frac{C_k^\lambda(\cos\theta)}{C_k^\lambda(1)} \cos\left[(k+\frac{1}{2})u\theta\right] > 0, \qquad 0 \le u \le 1,$$

bearing in mind that the term $\frac{1}{2}u\theta$ is an artefact of the method used to prove Theorem 1. In that regard recall that $C_k^0(\cos\theta)/C_k^0(1) = \cos k\theta$ and that for $u = 1$ the natural sum is, of course, $\sum_{k=0}^n \cos^2 k\theta$ although, in fact, $\sum_{k=0}^n \cos k\theta \cos(k+\frac{1}{2})\theta$ is also strictly positive.

Using the limit relation

$$\lim_{n\to\infty} \frac{C_n^\lambda(\cos\frac{\theta}{n})}{C_n^\lambda(1)} = \Gamma(\alpha+1)2^\alpha\,\theta^{-\alpha} J_\alpha(\theta), \qquad \alpha = \lambda - \frac{1}{2},$$

we find that

$$\lim_{n\to\infty} \frac{\theta}{n} \sum_{k=0}^{n} \frac{C_k^\lambda(\cos\frac{\theta}{n})}{C_k^\lambda(1)} \cos\left[(k+\frac{1}{2})\frac{\theta}{n}u\right] =$$

$$2^\alpha\Gamma(\alpha+1)\int_0^\theta t^{-\alpha} J_\alpha(t)\cos ut\, dt.$$

The result

$$\int_0^\theta t^{-\alpha} J_\alpha(t) \cos t \, dt > 0, \qquad \text{for } \theta > 0, \quad \alpha \geq -\frac{1}{2},$$

compared with

$$\int_0^\theta t^{-\alpha} J_\alpha(t) \, dt > 0 \qquad \text{for } \theta > 0, \quad \alpha \geq \alpha' = -0.269387\ldots.$$

gives further encouragement to the notion that the case $u = 0$ may be critical for this problem.

References

[1] Askey, R., Positive Jacobi polynomials, *Tôhoku Math. J.* **24** (1972), 109–119.

[2] Askey, R. and J. Fitch, Integral representations for Jacobi polynomials and some applications, *J. Math. Anal. Appl.* **26** (1969), 411–437.

[3] Bustoz, J., Jacobi polynomial sums and univalent Cesáro means, *Proc. Amer. Math. Soc.* **50** (1975), 259–264.

[4] Brown, G. and E. Hewitt, A class of positive trigonometric sums, *Math. Ann.* **268** (1984), 91–122.

[5] Brown, G., S. Koumandos and K. Wang, Positivity of more Jacobi polynomial sums, *Math. Proc. Camb. Phil. Soc.* **119** (1996), 681–694.

[6] Brown, G., S. Koumandos and K. Wang, On the Positivity of some basic Legendre Polynomial sums, *J. London Math. Soc.*, 1998 to appear.

[7] Brown, G., S. Koumandos and K. Wang, Positivity of basic sums of ultraspherical polynomials, submitted.

[8] Erdélyi, A., W. Magnus, F. Oberhettinger and F. G. Tricomi, *Higher transcendental functions* Vol. 2, McGraw-Hill, New York, 1953.

[9] Feldheim, E., On the positivity of certain sums of ultraspherical polynomials. *J. Analyse Math.* **11** (1963), 275–284, (edited with additional notes by G. Szegő), also in G. Szegő, collected papers Vol. 3, Birkhäuser, Boston, 1982, 821-830.

[10] Ruscheweyh, S., On the Kakeya-Eneström theorem and Gegenbauer polynomial sums, *SIAM J. Math. Anal.* **9** no 4 (1978), 682-686.

[11] Szegő, G., Zur Theorie der Legendreschen Polynome, *Jahresbericht Deutsch. Math. Verein.* **40** (1931), 163-166. Also in G. Szegő, collected papers Vol. 2, Birkhäuser, Boston, 1982, 261-264.

[12] Szegő, G., *Orthogonal polynomials*, Amer. Math. Soc. Colloq. Publ. 23 4th ed. Amer. Math. Soc. Providence, R.I., 1975.

[13] Vietoris, L., Über das Vorzeichen gewisser trigonometrischer Summen, *S. B. Öster Akad. Wiss.* **167** (1958), 125–135; Teil II: *Anzeiger Öster. Akad. Wiss.* (1959), 192–193.

Gavin Brown
Vice-Chancellor
The University of Sydney
Sydney, NSW 2006
AUSTRALIA
Vice-Cancellor@vcc.usyd.edu.au

Stamatis Koumandos
Department of Mathematics and Statistics
The University of Cyprus
P.O. Box 537
1678 Nicosia
CYPRUS
skoumand@pythagoras.mas.ucy.ac.cy

A constructive method for univalent harmonic exterior maps with Blaschke dilatation

Daoud Bshouty

Walter Hengartner

and Marc-Hubert Nicole

Abstract. Necessary and sufficient conditions are established which allow the construction of univalent harmonic mappings whose second dilatation function is a given Blaschke product of degree N and which maps the exterior of the unit disk onto a given domain Ω containing infinity. Explicit examples are given for the cases $N = 3$ and $N = 4$.

§1 Introduction

Let f be a univalent harmonic mapping defined on the exterior of the unit disk, $\Delta = \{z : |z| > 1\}$ which keeps infinity fixed.

Without loss of generality, we may assume that f preserves the orientation. Such mappings are solutions of the elliptic partial differential equation

$$\overline{f_{\overline{z}}(z)} = a(z) f_z(z), \qquad z \in \Delta, \tag{1.1}$$

where the second dilatation function $a(z) = \sum_{k=0}^{\infty} \alpha_k z^{-k}$ belongs to $H(\Delta)$, the class of holomorphic functions on Δ, and $|a(z)| < 1$ on Δ. It follows [5] that f admits the representation

$$f(z) = A\left(z - \frac{1}{\overline{z}}\right) + B\left(\overline{z} - \frac{1}{z}\right) + C \log|z|^2 + F(z), \qquad A \neq 0, \tag{1.2}$$

where F is a bounded harmonic function in Δ. Furthermore, substituting (1.2) in (1.1) and comparing the constant and the z^{-1} coefficients in (1.2), we get the following relations between the constants A, B and C,

$$B = \overline{\alpha_0 A} \quad \text{and} \quad C = \frac{\alpha_1 \overline{\alpha_0} A + \overline{\alpha_1 A}}{1 - |\alpha_0|^2}. \tag{1.3}$$

In other words, B and C depend only on the coefficients A, α_0 and α_1 and are completely independent of the boundary values $f^*(e^{it}) = F^*(e^{it})$.

The following result characterizes univalent harmonic mappings from Δ onto a one point punctured plane.

Computational Methods and Function Theory (CMFT'97)
N. Papamichael, St. Ruscheweyh and E. B. Saff (Eds.), pp. 99–115

ISBN 981-02-3626-3

Theorem 1 ([5]). *Let*

$$f(z) = A\left(z - \frac{1}{\overline{z}}\right) + B\left(\overline{z} - \frac{1}{z}\right) + C\log|z|^2 + F(z)$$

be a univalent harmonic and orientation-preserving mapping defined on Δ. Then $f(\Delta) = \overline{\mathbb{C}} \setminus \{q\}$ for some point $q \in \mathbb{C}$ if, and only if f is of the form

$$f(z) = A\left(z - \frac{1}{\overline{z}}\right) + B\left(\overline{z} - \frac{1}{z}\right) + C\log|z|^2 + D.$$

These mappings are solutions of the elliptic differential equation (1.1) where $a(z)$ is of the form

$$a(z) = \frac{\overline{A}}{A}\frac{(1-p_1z)(1-p_2z)}{(z-\overline{p_1})(z-\overline{p_2})}, \qquad |p_1| < 1 \quad \text{and} \quad |p_2| \leq 1. \tag{1.4}$$

Conversely, if $a(z)$ is as in (1.4), then any univalent harmonic mapping of the form (1.2) which is a solution of (1.1) has the property that $f(\Delta) = \overline{\mathbb{C}} \setminus \{q\}$ for some point $q \in \mathbb{C}$.

While Theorem 1 places restrictions on the existence of a univalent harmonic map, the next result, Theorem 2, is a general existence and uniqueness theorem.

Theorem 2 ([5]). *Let Ω be a simply connected domain of $\overline{\mathbb{C}}$ containing infinity whose boundary $\partial\Omega$ is locally connected. Let $a(z) \in H(\Delta)$, $|a(z)| < 1$ on Δ. For a fixed $\arg(A)$, unless $a(z)$ is of the form (1.4), there is a univalent solution of (1.1) with the following properties.*

1. *f is of the form (1.2)*
2. *$f(\Delta) \subset \Omega$.*
3. *The unrestricted limits $f^*(e^{it}) = \lim_{z\to e^{it}} f(e^{it})$ exist on $\partial\Delta \setminus E$ and belong to $\partial\Omega$ where E is a countable set.*
4. *If $e^{it} \in E$, then the cluster set of f at e^{it} is a straight-line segment joining two points of $\partial\Omega$.*
5. *If Ω is the complement of a compact convex set, then $f(\Delta) = \Omega$.*
6. *If $\mathbb{C} \setminus \Omega$ is strictly starlike with respect to a finite point p, and if $|a(z)| \leq k < 1$ for all $z \in \Delta$, then f is uniquely determined.*

Suppose that $\sup_{z\in\Delta}|a(z)| < 1$. Then we have $f(\Delta) = \Omega$ and f extends continuously on $\overline{\Delta}$. Therefore, no collapsing takes place into the interior of Ω and Theorem 2 can be interpreted as the Riemann mapping theorem for harmonic maps. For statement 6 of Theorem 2, a constructive method is developed in [4] which leads to an approximation of the univalent normalized solution of (1.1) mapping Δ onto Ω. The method consists of minimizing the maximum of the Minkowski functional of $f^*(e^{it})$ with respect to $\partial\Omega$. Unlike the case of bounded Jordan domains Ω [3], the corresponding mathematical program is highly nonlinear and the functional to be optimized has many local extrema. Moreover, the requirement that $\sup_{z\in\Delta}|a(z)| < 1$ is an essential condition in order to show that the method converges to the desired mapping.

In this paper, we consider the case where the second dilatation function

$$a(z) = e^{i\gamma}\frac{1}{z^{n_0}}\prod_{k=1}^{N-n_0}\frac{1-p_k z}{z-\overline{p_k}} = \sum_{k=0}^{\infty}\alpha_k z^{-k}, \tag{1.5}$$

$0 \le n_0 \le N$ and $0 < |p_k| < 1$ for $1 \le k \le N - n_0$, is a finite Blaschke product of degree N. We first state in §2 some auxiliary results which are needed in the sequel. In §3, Theorem 6 gives a necessary and sufficient condition for a boundary correspondence in order that the solution of the Dirichlet problem be a univalent solution of (1.1) which maps Δ onto Ω. If in addition, $a(z)$ is a square of a Blaschke product of degree N, then the univalent solutions which map Δ onto Ω generate nonparametric minimal surfaces over Ω which have the property that the image of their Gauss map is the upper half-sphere covered exactly N times. In §4, the necessary conditions of Theorem 6 are expressed in terms of a system of equations and in §5 and §6, we solve explicitly this system for the cases $N = 3$ and $N = 4$ and construct the corresponding mapping functions.

§2 Some auxiliary results

Let Ω be a simply connected domain of $\overline{\mathbb{C}}$ which contains infinity and whose boundary is piecewise continuously differentiable and let $a(z)$ be a finite Blaschke product. We have seen in Theorem 2 that the unrestricted limits $f^*(e^{it}) = \lim_{z\to e^{it}} f(z)$ exist and are on $\partial\Omega$ except for a countable set of points of $\partial\Delta$ where f^* has jumps. Our next result shows that the values of $f^*(e^{it})$ depend strongly on the values of $a(e^{it})$.

Theorem 3 ([1]). *Let $a(z)$ be a finite Blaschke product of degree N and let Ω be a simply connected domain of $\overline{\mathbb{C}}$ which contains infinity and whose boundary is piecewise continuously differentiable. Let f be a univalent* *solution of (1.1) which maps* Δ *onto* Ω. *Then, unless* $f^*(e^{it})$ *is constant on an interval* $\{s : t < s < t+\delta\}$, *we have*

$$\lim_{h\downarrow 0} \arg\left[f^*(e^{i(t+h)}) - f^*(e^{i(t-0)})\right] = -\frac{1}{2}\arg a(e^{it}) \quad \mod \pi. \tag{2.1}$$

The relation (2.1) can be expressed in the differential form

$$df^*(e^{it}) - \overline{a(e^{it})df^*(e^{it})} = 0, \qquad f^*\text{-a.e.} \tag{2.2}$$

or equivalently, by

$$\operatorname{Im}\left(\sqrt{a(e^{it})}df^*(e^{it})\right) = 0. \tag{2.3}$$

Hence, unless $df^*(e^{it}) = 0$, we have

$$\arg df^*(e^{it}) = -\frac{1}{2}\arg a(e^{it}) \quad \mod \pi \qquad f^*\text{-a.e.} \tag{2.4}$$

We shall need the following definitions for the next result.

Definition 1. *Let* $\partial\Omega$ *be a Jordan curve. A point* $q \in \partial\Omega$ *is said to be a point of convexity (with respect to* Ω*) if there is a line segment* L *containing* q *as an interior point such that* $L \setminus \{q\}$ *lies in the exterior of* $f(\Delta)$.

Definition 2. *A simply connected Jordan domain* Ω *of* $\overline{\mathbb{C}}$ *containing infinity is said to be a piecewise concave domain if* $\partial\Omega$ *contains at most finitely many points of convexity.*

In the next result, we shall see that Ω is always a piecewise concave domain. Suppose that $q \in \partial\Omega$ is not an interior point of a jump of f^*. Then $(f^*)^{-1}(q) = J(q) = \{e^{it} : \gamma(q) \leq t \leq \delta\}$. Set $A(t) = \arg a(e^{it})$, $0 \leq t \leq 2\pi$, as a continuous function and define $\triangle A(q) = [A(\delta(q)) - A(\gamma(q))]/2$. If q is an interior point of a jump of f^* at e^{it}, put $\gamma = \delta = t$ and $\triangle A(q) = 0$.

Theorem 4 ([2, Theorem 2.6 and Theorem 3.2]). *Let* Ω *be a simply connected Jordan domain of* $\overline{\mathbb{C}}$ *containing infinity whose boundary is piecewise continuously differentiable and let* $a(z)$ *be a finite Blaschke product of degree* N. *Let* f *be a univalent solution of (1.1) which is of the form (1.2) and which maps* Δ *onto* Ω. *Finally, let* q *be in* $\partial\Omega$ *and denote the opening angle of* $\partial\Omega$ *at* q *as seen from the inside of* Ω *by* $\alpha(q)$. *Then, we have*

1. Ω *is a piecewise concave domain with at most* $N-2$ *points of convexity.*

2. *If* $0 \le \alpha(q) < \pi$, *then* $\alpha(q) = -\triangle A(q)$.

3. *If* $\pi \le \alpha(q) \le 2\pi$, *then either* $\alpha(q) = -\triangle A(q)$ *or* $\alpha(q) = -\triangle A(q) + \pi$.

Theorem 4 states that the total change of $-\arg a(e^{it})/2$ over the interval $J(q) = f^{-1}(q)$ is either equal to the opening angle $\alpha(q)$ as seen from the inside of the domain or, if $\pi \le \alpha \le 2\pi$, it can also be equal to $\alpha(q) - \pi$. In the first case we shall use the following notation.

Definition 3. *We say that* $q \in \partial\Omega$ *is a complete resting point of* f^* *if* $-\triangle A(q) = \alpha(q)$.

Remark 1: If q is an interior point of a linear segment of $f^*(I)$, then either q is an interior point of a jump of f and $\triangle A(q) = 0$ or the inverse image $f^{-1}(q)$ is not a singleton and we have $\triangle A(q) = -\pi$.

Remark 2: If the opening angle at $q \in \partial\Omega$ is strictly less than π then q is a complete resting point of f^*.

Our next result sharpens the first statement of Theorem 4.

Theorem 5 ([2, Theorem 3.2]). *Let* $N \ge 3$. *Under the conditions of Theorem 4,* $\partial\Omega$ *has* $N-2$ *complete resting points of* f^*.

Remark 3: If $N = 1$, then by Theorem 1, there is no univalent solution of (1.1) which maps Δ onto the unbounded component of the complement of a compact continuum.

Remark 4: If $N = 2$, then $f(\Delta)$ is always either a one point punctured plane or the complement of a compact convex set.

§3 The basic result

Recall that a univalent and orientation-preserving harmonic mapping defined on the exterior of the unit disk, Δ, which keeps infinity fixed, is necessarily of the form (1.1), *i.e.*

$$f(z) = A\left(z - \frac{1}{\overline{z}}\right) + B\left(\overline{z} - \frac{1}{z}\right) + C\log|z|^2 + F(z)$$

where F is a bounded harmonic function on Δ. In particular, if f is a solution of (1.1), where $a(z)$ is a Blaschke product of degree $N \ge 2$, then the boundary function f^* satisfies the relations (2.2) or equivalently (2.3) or (2.4). Furthermore, there are $N-2$ points which are complete resting points of f^* and hence, the image domain Ω has at most $N-2$ points of convexity. It is a natural question to ask wether these boundary conditions are sufficient for a solution of the Dirichlet problem to be a univalent harmonic solution of (1.1) which maps Δ onto Ω. Unfortunately the answer is negative. For instance, there is no univalent harmonic mapping $f(z) = Az + O(1), A > 0$, which is a univalent solution of $\overline{f_{\overline{z}}(z)} = z^{-2} f_z(z)$ and maps Δ onto Δ. Moreover, we shall see in §5, that there is no univalent harmonic mapping $f(z) = Az + O(1)$, which is a univalent solution of $\overline{f_{\overline{z}}(z)} = z^{-4} f_z(z)$ and maps Δ onto $\Omega = \{w : |w-1| > 2\} \cap \{w : |w+1| > 1\}$ even if (2.2) is satisfied everywhere on the boundary. Necessary and sufficient conditions are given in [2, Theorem 4.2]. We give here another necessary and sufficient condition which seems to be more practical for applications.

In what follows we need the following property of boundary correspondence.

Definition 4. *Let Ω be a Jordan domain of $\overline{\mathbb{C}}$ which contains infinity and whose boundary is piecewise continuously differentiable. Let ϕ be a conformal univalent mapping from Δ onto Ω keeping infinity fixed. We say that a (orientation-preserving) mapping $f^*(e^{it})$ from $\partial\Delta$ into $\partial\Omega$ is a quasihomeomorphism, if its image contains at least three different non collinear points of $\partial\Omega$ and if $\phi^{-1} \circ f^*(e^{it})$ is the pointwise limit of a sequence of (orientation-preserving) homeomorphisms from $\partial\Delta$ onto $\partial\Delta$. If, in addition, the linear segments from $f^*(e^{i(t-0)})$ to $f^*(e^{i(t+0)})$ are parts of $\partial\Omega$, then we call $f^*(e^{it})$ a quasihomeomorphism from $\partial\Delta$ <u>onto</u> $\partial\Omega$.*

Our main result is

Theorem 6. *Let*

$$a(z) = e^{i\gamma} \frac{1}{z^{n_0}} \prod_{k=1}^{N-n_0} \frac{1 - p_k z}{z - \overline{p_k}} = \sum_{k=0}^{\infty} \alpha_k z^{-k},$$

$0 \le n_0 \le N$ and $0 < |p_k| < 1$ for $1 \le k \le N - n_0$, be a finite Blaschke product of degree N and let Ω be a piecewise concave Jordan domain of $\overline{\mathbb{C}}$ which contains infinity and whose boundary has at most $N-2$ points of convexity. Let $f^*(e^{it})$ be a positively oriented quasihomeomorphism from the unit circle $\partial\Delta$ onto $\partial\Omega$. Then the mapping

$$\begin{aligned} f(z) &= A(z - \frac{1}{\overline{z}}) + B(\overline{z} - \frac{1}{z}) + C \log |z|^2 \\ &\quad - \frac{1}{2\pi} \int_0^{2\pi} \operatorname{Re}\left(\frac{e^{it} + z}{e^{it} - z}\right) f^*(e^{it})\, dt \end{aligned} \tag{3.1}$$

is a univalent solution of

$$\overline{f_{\bar{z}}(z)} = a(z) f_z(z)$$

which maps Δ *onto* Ω *if and only if the following conditions hold.*

1. *The relation (2.3),* $\operatorname{Im}(\sqrt{a(e^{it})}df^*(e^{it})) = 0$, *is satisfied* f^*-*a.e. on* $\partial\Delta$.

2. *The coefficients* A, B *and* C *satisfy the relations (1.3).*

3. $$\frac{\overline{f_{\bar{z}}(z)}^2}{a(z)} \in H(\Delta \setminus \infty). \tag{3.2}$$

4. *If* $n_0 = 2$, *then*

$$\frac{1}{2\pi i}\int_0^{2\pi} e^{-it}\, df^*(e^{it}) = A - \overline{\alpha_2 A}.$$

If $n_0 \geq 3$, *then*

$$\begin{aligned}
\frac{1}{2\pi i}\int_0^{2\pi} e^{-it}\, df^*(e^{it}) &= A, \\
\frac{1}{2\pi i}\int_0^{2\pi} e^{-ikt}\, df^*(e^{it}) &= 0, \qquad 2 \leq k \leq n_0 - 2, \\
\frac{1}{2\pi i}\int_0^{2\pi} e^{-i(n_0-1)t}\, df^*(e^{it}) &= -\overline{\alpha_{n_0} A}.
\end{aligned}$$

Proof: Let f be of the form (1.1). Then $F(\zeta^{-1})$ is a bounded harmonic function of ζ in the unit disk U. Using the Poisson formula with respect to the boundary function F^* of F, $F^* \equiv f^*$, we get for z in Δ,

$$\begin{aligned}
F(z) &= \frac{1}{2\pi}\int_0^{2\pi} \operatorname{Re}\left(\frac{e^{-it} + \frac{1}{z}}{e^{-it} - \frac{1}{z}}\right) f^*(e^{it})\, dt \\
&= -\frac{1}{2\pi}\int_0^{2\pi} \operatorname{Re}\left(\frac{e^{it} + z}{e^{it} - z}\right) f^*(e^{it})\, dt.
\end{aligned}$$

Therefore, f is of the form (3.1) and we have

$$F_z(z) = -\frac{1}{2\pi}\int_0^{2\pi} \frac{e^{it}}{(e^{it} - z)^2} f^*(e^{it})\, dt, \tag{3.3}$$

$$\overline{F_{\bar{z}}(z)} = -\frac{1}{2\pi}\int_0^{2\pi} \frac{e^{it}}{(e^{it} - z)^2} \overline{f^*(e^{it})}\, dt. \tag{3.4}$$

Next, integrating (3.3) by parts, we obtain

$$\begin{aligned} \frac{1}{2\pi i}\int_0^{2\pi}\frac{df^*(e^{it})}{e^{it}-z} &= -\frac{1}{2\pi i}\int_0^{2\pi}\frac{d}{dt}\left(\frac{1}{e^{it}-z}\right)f^*(e^{it})\,dt \\ &= -\frac{1}{2\pi i}\int_0^{2\pi}\frac{-ie^{it}}{(e^{it}-z)^2}f^*(e^{it})\,dt \\ &= \frac{1}{2\pi}\int_0^{2\pi}\frac{e^{it}}{(e^{it}-z)^2}f^*(e^{it})\,dt \qquad (3.5)\\ &= -F_z(z) \\ &= \frac{B}{z^2}+\frac{C}{z}+A-f_z(z). \end{aligned}$$

Similarly, (3.4) reduces to

$$\begin{aligned} \frac{1}{2\pi i}\int_0^{2\pi}\frac{\overline{df^*(e^{it})}}{e^{it}-z} &= \frac{1}{2\pi}\int_0^{2\pi}\frac{e^{it}}{(e^{it}-z)^2}\overline{f^*(e^{it})}\,dt \\ &= -\overline{F_{\overline{z}}(z)} \qquad (3.6)\\ &= \frac{\overline{A}}{z^2}+\frac{\overline{C}}{z}+\overline{B}-\overline{f_{\overline{z}}(z)}. \end{aligned}$$

We now are ready to show the necessity of the conditions. Suppose that (3.1) is a univalent solution of (1.1) which maps Δ onto Ω. Then, the condition (1) of Theorem 6 is a consequence of Theorem 3 and the condition (2) follows by comparing the constant and the z^{-1} coefficients in (1.2). Finally, condition (3) holds since $f_z(z) = \overline{f_{\overline{z}}(z)}/a(z)$ and $\overline{f_{\overline{z}}(z)}$ are analytic on Δ. Next, we write condition (1) (which we have already shown to be necessary,) in the form $\overline{df^*(e^{it})} \equiv a(e^{it})df^*(e^{it})$. Then (3.5) and (3.6) imply that

$$\begin{aligned} \overline{f_{\overline{z}}(z)}-a(z)f_z(z) &= -\frac{1}{2\pi i}\int_0^{2\pi}\frac{\overline{df^*(e^{it})}-a(z)df^*(e^{it})}{e^{it}-z} \\ &\quad +\left(\overline{B}+\frac{\overline{C}}{z}+\frac{\overline{A}}{z^2}\right)-a(z)\left(A+\frac{C}{z}+\frac{B}{z^2}\right) \qquad (3.7)\\ &= -\frac{1}{2\pi i}\int_0^{2\pi}\frac{a(e^{it})-a(z)}{e^{it}-z}df^*(e^{it}) \\ &\quad +\left(\overline{B}+\frac{\overline{C}}{z}+\frac{\overline{A}}{z^2}\right)-a(z)\left(A+\frac{C}{z}+\frac{B}{z^2}\right). \end{aligned}$$

Finally, we deduce the condition (4) by equating (3.7) to zero and by comparing the coefficients of the z^{-k} terms, $2 \le k \le m$. This proves the necessity of the conditions.

To prove that the conditions (1) to (4) are also sufficient, we intend to show that $\overline{f_{\overline{z}}(z)} - a(z)f_z(z) \equiv 0$. It follows then that the harmonic map defined in (3.1) is orientation-preserving and since f^* is an orientation preserving quasihomeomorphism from $\partial\Delta$ onto $\partial\Omega$ we conclude by the argument principle, that f is univalent and that $f(\Delta) = \Omega$.

From (3.7) and the third condition, we conclude that $\overline{f_{\overline{z}}(z)} - a(z)f_z(z)$ is an analytic function on $\overline{\Delta}$. Furthermore, since condition (2) holds, it folllows that near infinity, $\overline{f_{\overline{z}}(z)} - a(z)f_z(z) = O(z^{-2})$. Next we use the facts

$$1 + \frac{2z}{e^{it} - z} = \frac{e^{it} + z}{e^{it} - z}a$$

and

$$\int_0^{2\pi} df^*(e^{it}) = \int_0^{2\pi} a(e^{it})\, df^*(e^{it}) = 0.$$

By (3.5) and (3.6), we have

$$\begin{aligned} \frac{1}{4\pi i}\int_0^{2\pi} \frac{e^{it}+z}{e^{it}-z}\, df^*(e^{it}) &= Az + C + \frac{B}{z} - zf_z(z), \\ \frac{1}{4\pi i}\int_0^{2\pi} \frac{e^{it}+z}{e^{it}-z}\, \overline{df^*(e^{it})} &= \frac{\overline{A}}{z} + \overline{C} + \overline{B}z - z\overline{f_{\overline{z}}(z)}. \end{aligned}$$

Hence, using the first condition, we get

$$\begin{aligned} &\frac{z^2(f_z(z)a(z) - \overline{f_{\overline{z}}(z)})^2}{a(z)} = \\ &\left\{\frac{1}{4\pi i}\int_0^{2\pi} \frac{e^{it}+z}{e^{it}-z}\left(\frac{\overline{df^*(e^{it})}}{\sqrt{a(z)}} - df^*(e^{it})\sqrt{a(z)}\right)\right. \\ &\left. + \left(Az + C + \frac{B}{z}\right)\sqrt{a(z)} - \left(\frac{\overline{A}}{z} + \overline{C} + \overline{B}z\right)\frac{1}{\sqrt{a(z)}}\right\}^2 \end{aligned} \tag{3.8}$$

Suppose first, that $n_0 = 0$ or 1, *i.e.* , that $|\alpha_0| + |\alpha_1| \neq 0$. Then by condition (3), the left hand side of (3.8) is analytic in $\overline{\Delta}$ and is $O(\frac{1}{z^2})$ near infinity. The right hand side shows that its values on the unit circle are real. Hence, $f_z(z)a(z) - \overline{f_{\overline{z}}(z)} = 0$.

Now consider the case $n_0 \geq 2$. By condition (4), we have $\overline{f_{\overline{z}}(z)} - a(z)f_z(z) = O(z^{-(n_0+1)})$ near infinity and the same reasoning as above shows that $f_z(z)a(z) - \overline{f_{\overline{z}}(z)} \equiv 0$. This completes the proof of Theorem 6. ■

§4 The constructive procedure

Let

$$a(z) = e^{i\gamma}\frac{1}{z^{n_0}}\prod_{k=1}^{m}\left[\frac{1-p_k z}{z-\overline{p_k}}\right]^{n_k} = \sum_{k=0}^{\infty}\alpha_k z^{-k},$$

$n_0 \geq 0$, $n_k > 0$, $0 < |p_k| < 1$ for $1 \leq k \leq m$, $p_k \neq p_j$ if $i \neq j$ and $\sum_{k=0}^{m} n_k = N$, be a given Blaschke product of degree N. Let Ω be a domain containing the point at infinity whose boundary is a convex Jordan curve. Theorem 6 provides a constructive way to find a univalent solution of (1.1) with respect to the given dilatation function $a(z)$ which maps Δ onto Ω. The problem is to find the $N-2$ points of $\partial\Omega$ which are complete resting points of f^*. Once these are known, we may find an acceptable boundary correspondence applying continuously the relation (2.3). As we have seen, the constants B and C are completely determined by the constant A and the first two coefficients α_0 and α_1 of the second dilatation function $a(z)$. The equations to solve are divided into two sets, the hypothesis about the behavior of $f_z(z)a(z) - \overline{f_{\overline{z}}(z)}$ at infinity given in condition (3) and condition (4) of Theorem 6.

The first set of equations *EQ 1* is

EQ 1:

If $n_0 = 0$ or 1, then *EQ 1* is void.

If $n_0 = 2$, then

$$\int_0^{2\pi} e^{-it}\, df^*(e^{it}) = 2\pi i(A - \overline{\alpha_2 A}). \tag{4.1}$$

If $n_0 \geq 3$, then

$$\begin{aligned}
\int_0^{2\pi} e^{-it}\, df^*(e^{it}) &= 2\pi i A, \\
\int_0^{2\pi} e^{-ikt}\, df^*(e^{it}) &= 0, \qquad 2 \leq k \leq n_0 - 2, \\
\int_0^{2\pi} e^{-i(n_0-1)t}\, df^*(e^{it}) &= -2\pi i\overline{\alpha_{n_0} A}.
\end{aligned} \tag{4.2}$$

The second set of equations is

EQ 2:

$$\int_0^{2\pi} \frac{e^{ijt}\, df^*(e^{it})}{(e^{it}-\overline{p_k})^j} = 2\pi i\,(j\,\overline{p_k}A + C) + 2\pi i\frac{B}{\overline{p_k}}\delta_{j,1}, \tag{4.3}$$

$1 \leq j \leq \text{int}[(n_k+1)/2]$, $1 \leq k \leq m$, where $\text{int}[x]$ denotes the integer part of the positive real number x and where $\delta_{j,1}$ is the Kronecker symbol. In particular, we have $C = \overline{A}\prod_{k=1}^{m}(-\overline{p_k})^{n_k}$ if $n_0 = 1$ and $C = 0$ if $n_0 > 1$.

Observe that the combined set *EQ* of the above equations is in general not independent. For instance, for $a(z) = z^{-4}$, *EQ 2* is empty and *EQ 1* consists of the three equations

$$\begin{aligned} \int_0^{2\pi} e^{-it}\, df^*(e^{it}) &= 2\pi i A, \\ \int_0^{2\pi} e^{-2it}\, df^*(e^{it}) &= 0, \\ \int_0^{2\pi} e^{-3it}\, df^*(e^{it}) &= -2\pi i \overline{A}. \end{aligned}$$

The third equation is just the conjugate of the first equation and determines the constant A. Indeed, we have

$$\begin{aligned} 2\pi i A &= \overline{\int_0^{2\pi} e^{-3it}\, df^*(e^{it})} = \int_0^{2\pi} e^{3it}\, \overline{df^*(e^{it})} \\ &= \int_0^{2\pi} e^{3it} a(e^{it})\, df^*(e^{it}) = \int_0^{2\pi} e^{-it}\, df^*(e^{it}). \end{aligned}$$

The second equation tells us that the two complete resting points divide the convex curve $\partial\Omega$ into two arcs of equal euclidean length. For more details see [2].

The same set of equations were established by the authors in [2] using a different approach. The following result was also shown there.

Theorem 7 ([2, Theorem 4.4]). *One can replace the conditions (3) and (4) in Theorem 6 by any choice of* $\operatorname{int}[N/2]$ *equations of the set EQ which together with condition (2) are independent.*

§5 The case N = 3

The case where the Blaschke product has only one or two factors has been extensively studied in [2]. One also finds there a discussion about the case

$$a(z) = \frac{1}{z^2}\left(\frac{1}{z} - \frac{pz}{\overline{p}}\right)^2.$$

In this section, we suppose that the second dilatation function $a(z)$ is a Blaschke product with three factors and that Ω is the complement of a compact convex continuum. Applying an appropriate Moebius premapping, we may assume, without loss of generality, that one of the zeros of a is at infinity, *i.e.* , that $n_0 > 0$. Furthermore, we may also assume that the factor $e^{i\gamma}$ in (1.5) is one.

Observe, that there is exactly one complete resting point $\omega_0 \in \partial\Omega$. Furthermore the boundary correspondence has to satisfy (2.3),

$$\operatorname{Im}\left(\sqrt{a(e^{it})}df^*(e^{it})\right) \equiv 0, \qquad f^*\text{-a.e.} \quad \text{on } \partial\Delta$$

and one of the equations given in *EQ 1* and *EQ2*. We distinguish among the following three different cases.

5.1 The case $\mathbf{a(z)} = \frac{1}{\mathbf{z}^3}$

In this case, *EQ 2* is empty and *EQ 1* consists of the two equations

$$\int_0^{2\pi} e^{-it}\, df^*(e^{it}) = 2\pi i A \quad \text{and} \quad \int_0^{2\pi} e^{-2it}\, df^*(e^{it}) = -2\pi i\overline{A}.$$

The same reasoning, as we gave at the end of §4, shows that the second equation is just the conjugate of the first equation.

The construction of f goes as follows. We prescribe the boundary correspondence by applying the relation (2.3) on $t_0 < t < t_0 + 4\pi/3$ and we put $f^*(e^{it}) = \omega_0$ on the interval $t_0 + 4\pi/3 < t < t_0 + 2\pi$. Then

$$\int_{t_0}^{t_0+\frac{4\pi}{3}} e^{-it}\, df^*(e^{it}) = 2\pi i A$$

which determines uniquely the constant A. Observe that $B = C = 0$. Finally, the desired harmonic mapping is obtained by (3.1).

5.2 The case $\mathbf{a(z)} = \frac{1}{\mathbf{z}^2}\left(\frac{1-\mathbf{p}\mathbf{z}}{\mathbf{z}-\overline{\mathbf{p}}}\right)$, $|\mathbf{p}| < 1$

In this case, we have $B = C = 0$ and we may choose one of the following two equations:

$$\int_0^{2\pi} e^{-it}\, df^*(e^{it}) = 2\pi i(A + \overline{pA}) \quad \text{or} \quad \int_0^{2\pi} \frac{e^{it}\, df^*(e^{it})}{e^{it} - \overline{p}} = 2\pi i\overline{p}A$$

The construction of f goes as follows. Fix t_0 and determine τ such that $3\tau - 4\arg(1 + pe^{i(t_0-\tau)}) = 2\pi - 4\arg(1 + pe^{it_0})$. Then prescribe the boundary correspondence by applying the relation (2.3) on the interval $t_0 < t < t_0 + 2\pi - \tau$. On the interval $t_0 + 2\pi - \tau < t < t_0 + 2\pi$, put $f^*(e^{it}) = \omega_0$. Finally, set a to be

$$\frac{1}{2\pi i}\int_{t_0}^{t_0+2\pi-\tau} e^{-it}\, df^*(e^{it})$$

and (3.1) gives the desired mapping.

5.3 The case $\mathbf{a(z) = \frac{1}{z}\frac{1-p_1 z}{z-\overline{p_1}}\frac{1-p_2 z}{z-\overline{p_2}}}$, $\mathbf{|p_1| < 1,\ |p_2| < 1}$

In this case, we have $B = 0.$ and $C = \overline{p_1 p_2 A}$. The set *EQ 1* is empty and *EQ 2* contains the two equations

$$\int_0^{2\pi} \frac{e^{it}\, df^*(e^{it})}{e^{it} - \overline{p_1}} = 2\pi i(\overline{p_1}A + \overline{p_1 p_2 A})$$

and

$$\int_0^{2\pi} \frac{e^{it}\, df^*(e^{it})}{e^{it} - \overline{p_2}} = 2\pi i\ (\overline{p_2}A + \overline{p_1 p_2 A}).$$

We may choose any one of these equations and the construction of f goes the same way as explained in the previous case.

Summarizing we have shown the following existence and uniqueness result.

Theorem 8. *Let $a(z)$ be a Blaschke product consisting of three factors and let Ω be the complement of a compact convex continuum with piecewise continuously differentiable boundary. Then, for each fixed resting point of $\partial\Omega$, there exist three univalent solutions of (1.1) which map Δ onto Ω and which keep infinity fixed.*

Remark 5: The above theorem also holds for the case where $\partial\Omega$ contains a single convexity point ω_0. In this case ω_0 is a complete resting point and the above procedure applies. The length τ of the interval $(f^*)^{-1}(\omega_0)$ is given by the equation

$$\frac{1}{2}\left(\arg a(e^{i(t_0-\tau)}) - \arg a(e^{it_0})\right) = \alpha(\omega_0).$$

Remark 6: For the complete resting point $\omega_0 \in \partial\Omega$, the above construction automatically yields to (see §2, Definition 2)

$$\frac{1}{2}\Big(\arg(a(e^{i\gamma})) - \arg(a(e^{i\delta}))\Big) = \alpha(\omega_0)$$

5.4 Examples

We conclude this section with some examples.

Example 1: In our first example, we consider the case where $a(z) = z^{-3}$, and $\Omega = \Delta$. Then there is a complete resting point q of f^* such that $q = f^*(e^{it})$ for all t in the interval $t_0 - 2\pi/3 < t < t_0$. As pointed out in Section 5.1, the boundary correspondence has to satisfy

$$\operatorname{Im}\left(\sqrt{a(e^{it})}df^*(e^{it})\right) \equiv 0, \qquad f^*\text{-a.e.} \quad \text{on } \partial\Delta$$

and the equation

$$\int_0^{2\pi} e^{-it}\, df^*(e^{it}) = 2\pi i A.$$

Since $\Omega = \Delta$, we have $f^*(e^{it}) = \pm i e^{3it_0/2}$ for all t, $t_0 - 2\pi/3 \le t \le t_0$ and $f^*(e^{it}) = \pm i e^{3it/2}$ for all t, $t_0 \le t \le t_0 + 4\pi/3$. If we take the positive sign, then

$$A = \frac{-3}{4\pi i}\int_{t_0}^{t_0+4\pi/3} e^{it/2}\, dt = -\frac{3\sqrt{3}}{4\pi}\left(1 + i\sqrt{3}\right) e^{it_0/2}.$$

Hence, by (3.1), the desired harmonic mapping is

$$f(z) = -\frac{3\sqrt{3}}{4\pi}\left(1 + i\sqrt{3}\right) e^{it_0/2}\left(z - \frac{1}{\bar{z}}\right) - \frac{1}{2\pi}\int_0^{2\pi} \mathrm{Re}\left(\frac{e^{it}+z}{e^{it}-z}\right) f^*(e^{it})\, dt.$$

Example 2: In our next example, we suppose that Ω is the complement of the horizontal real line segment $[-1, 1]$ and we choose $a(z) = z^{-3}$. Note that Ω is not a Jordan domain. However, the boundary can be considered as a Jordan curve by distinguishing between the two sides of the line segment $(-1, 1)$. Then $\partial\Omega$ has no points of convexity and the above theory applies for this domain as well. Hence, there is a t_0 such that $q = f^*(e^{it})$, $t_0 - 2\pi/3 < t < t_0$, is a complete resting point of f^*. As in the previous example the boundary correspondence has to satisfy

$$\mathrm{Im}\left(\sqrt{a(e^{it})}df^*(e^{it})\right) \equiv 0, \qquad f^*\text{-a.e.} \quad \text{on } \partial\Delta$$

and

$$\int_0^{2\pi} e^{-it}\, df^*(e^{it}) = 2\pi i A.$$

Since $\partial\Omega$ is the real interval $[-1, 1]$, df^* is real and hence, $f^*(e^{it})$ has to jump at the three points $1, e^{2i\pi/3}$ and $e^{4i\pi/3}$. Furthermore, it has to be constant on each one of the three intervals of $\partial\Delta$ bounded by these points. Note that t_0 is one of these points and the other two are 1 and -1. Choose $\omega_0 \in [-1, 1]$. Then the possible boundary correspondences are given by the six functions

$$f_1^*(e^{it}) = \begin{cases} \omega_0 & \text{if} \quad 0 < t < \frac{2\pi}{3} \\ -1 & \text{if} \quad \frac{2\pi}{3} < t < \frac{4\pi}{3} \\ 1 & \text{if} \quad \frac{4\pi}{3} < t < 2\pi \end{cases},$$

$$f_2^*(e^{it}) \quad = \quad f_1^*\left(e^{i(t+2\pi/3)}\right),$$

$$\begin{aligned} f_3^*(e^{it}) &= f_1^*\left(e^{i(t+4\pi/3)}\right), \\ f_4^*(e^{it}) &= f_1^*\left(e^{-it}\right), \\ f_5^*(e^{it}) &= f_1^*\left(e^{-i(t+2\pi/3)}\right), \\ f_6^*(e^{it}) &= f_1^*\left(e^{-i(t+4\pi/3)}\right). \end{aligned}$$

Each f_j^* leads to a univalent solution of (1.1) which maps Δ onto Ω. By symmetry, we need only to determine the constant A for the functions $f_1(z)$ and we get

$$A = \frac{\sqrt{3}}{4\pi}\left[(3+\omega_0+i(1-\omega_0)\sqrt{3}\right].$$

§6 The case N = 4

Of particular interest is the case where $a(z)$ is a complete square due to its relation to minimal surfaces. In fact

$$a(z) = \frac{1}{z^2}\left(\frac{1-pz}{z-\overline{p}}\right)^2$$

yields to nonparametric minimal surfaces over Ω whose Gauss map image covers the upper half-sphere twice. This follows from the fact that $\sqrt{a(\Delta)}$ is the unit disk covered twice.

Let Ω be the complement of a compact convex continuum. Then there exist two points of $\partial\Omega$, ω_0 and ω_1, which are complete resting points. For the case where $\partial\Omega$ is a line segment, see Example 2 for the interpretation of ω_0 and ω_1. Fixing t_0, $0 \le t_0 < 2\pi$, we are looking for three t's, $t_0 < t_1 < t_2 < t_3 < t_0 + 2\pi$, such that

$$f_1^*(e^{it}) = \begin{cases} \omega_0 & \text{if} \quad t_3 < t < t_0 + 2\pi \\ \omega_1 & \text{if} \quad t_1 < t < t_2 \end{cases},$$

and otherwise, f^* is defined as a sense-preserving continuous homeomorphism satisfying (2.3). Note that t_3 is completely determined by t_0 and $a(z)$ given by Remark 1 in §5. Similarly, given t_2, then t_1 is uniquely determined by $a(z)$. The system of equations *EQ 1* and *EQ 2* reduces to

$$\frac{1}{2\pi i}\int_0^{2\pi} \frac{e^{it}}{e^{it}-\overline{p}}\, df^*(e^{it}) = \overline{p}A \tag{6.1}$$

and

$$\frac{1}{2\pi i}\int_0^{2\pi} e^{-2it}\, df^*(e^{it}) = 2(1-|p|^2)\overline{pA} \tag{6.2}$$

which determines uniquely t_2 and A. Given ω_0, there are four possible choices for t_0 and therefore we have

Theorem 9. *Let $a(z)$ be the square of a Blaschke product of degree two and let Ω be the complement of a compact convex continuum. Then for each fixed $\omega_0 \in \partial\Omega$, there are four univalent solutions $f(z)$ of (1.1) which map Δ onto Ω keeping infinity fixed and having ω_0 as a complete resting point for f^*.*

Example 3: Let $0 < p < 1$ and $\Omega = \Delta$. Fix $t_0 = 0$ and $\omega_0 = 1$. It then follows that $t_3 = \pi + \arccos(p)$ and for given $t_2 < t_3$, t_1 is uniquely determined by the equation

$$-t_1 + \arg(1 - pe^{it_1}) = \frac{\pi}{2} - t_2 + \arg(1 - pe^{it_2})$$

where $t_0 < t_1 < t_2$. Eliminating A from (6.1) and (6.2) yields

$$\int_0^{2\pi} \frac{df^*(e^{it})}{(e^{it} - p)^2} = 0. \tag{6.3}$$

Since $f^*(e^{it}) = i\sqrt{a} = ie^{it}\frac{1-pe^{-it}}{p-e^{-it}}$, for $t_0 < t < t_1$ and $t_2 < t < t_3$, we conclude that

$$df^*(e^{it}) = \frac{-2(1 - p\cos(t))}{(e^{-it} - p)^2}\, dt \equiv K(t)\, dt$$

on (t_0, t_1) and $df^*(e^{it}) = -K(t)\, dt$ on (t_2, t_3). Summarizing we have

$$H(t_2) = \int_0^{t_1(t_2)} \frac{1 - p\cos(t)}{|e^{it} - p|^4}\, dt - \int_{t_2}^{\pi+\arccos(p)} \frac{1 - p\cos(t)}{|e^{it} - p|^4}\, dt = 0.$$

Note that t_2 is uniquely determined. Indeed, $H(t)$ is a monotonic decreasing function of t with the property that $H(t) > 0$ for $t = t_3$ and $H(t) < 0$ for $t_1(t) = t_0 = 1$. Finally, A is determined from (6.1).

Numerical calulations for $p = 0, 0.25, 0.5, 0.75$ yield the following table:

p	t_1	t_2	t_3	A
0	1.5708	3.1416	4.7124	$-0.637 + i0.637$
0.25	0.5977	2.1641	4.4597	$-0.184 + i0.712$
0.5	01427	1.3526	4.1888	$0.106 + i0.547$
0.75	0.0184	0.7991	3.8643	$0.244 + i0.390$

Table 1.

Acknowledgements. This work was initiated while the second author was visiting the Technion supported by the Wenger Grant of The Canadian Technion Society. It was also supported in parts by the Promotion of Research Grant (Technion) and grants from the NSERC (Canada) and FCAR (Quebec).

References

[1] Bshouty, D. and W. Hengartner, Boundary values versus dilatations of harmonic mappings, *J. Analyse Math.* **72** (1998), 141–164.

[2] Bshouty, D. and W. Hengartner, Exterior univalent harmonic mappings with finite Blaschke dilatations, preprint.

[3] Bshouty, D., N. Hengartner and W. Hengartner, A constructive method for starlike harmonic mappings, *Numer. Math.* **54** (1988), 167–178.

[4] Hengartner, W. and L. Nadeau, Univalent harmonic exterior mappings as solutions of an optimization problem. *Mat. Vesnik.* **40** (1988), 233–240.

[5] Hengartner, W. and G. Schober, Univalent harmonic exterior and ring mappings, *J. Math. Anal. Appl.* **156** (1991), 154–171.

Daoud Bshouty
Department of Mathematics
Technion
Haifa
ISRAEL
daoud@techunix.technion.ac.il

Walter Hengartner
Département des Mathématiques
Université Laval
Québec City
CANADA
walheng@mat.ulaval.ca

Marc-Hubert Nicole
Département des Mathématiques
Université Laval
Québec City
CANADA

New results about the equality between the p-module and the p-capacity

Petru Caraman

Abstract. In this paper, continuing our earlier work ([2]–[5], [7], [10]–[19],[22]–[36]), we establish some new cases of equality between the p-module and the p-capacity of a condenser, *i.e.* of a tripple of sets (E_0, E_1, D), where $D \subset \overline{\mathbb{R}^n}$ is a domain, $E_0, E_1 \subset \hat{D}$, $\overline{E}$ is the closure of E with respect to $\mathbb{R}^n$ and $\hat{E}$ is its closure with respect to $\overline{R^n}$. Finally, we improve a result of this kind obtained previously in [18].

§1 Introduction

Let $\overline{\mathbb{R}^n}$ be the one-point compactification of the Euclidean $n-$space $\mathbb{R}^n$; E, E_0, E_1 arbitrary sets; $\overline{E}$ the closure of E with respect to $\mathbb{R}^n$; $\hat{E}$ its closure with respect to $\overline{\mathbb{R}^n}$; F, F_0, F_1 closed sets with respect to $\mathbb{R}^n$; and $\hat{F}, \hat{F}_0, \hat{F}_1$ closed sets with respect to $\overline{\mathbb{R}^n}$. Finally, let ∂D be the boundary of D with respect to $\mathbb{R}^n$ and $\hat{\partial D}$ its boundary with respect to $\overline{\mathbb{R}^n}$. A *condenser* is a triple of sets (E_0, E_1, D), where D is a domain of $\overline{\mathbb{R}^n}$ and $E_0, E_1 \subset \hat{D}$.

The *p-capacity of* $E_0, E_1 \subset \hat{D} \subset \overline{\mathbb{R}^n}$ *relatively to* D is

$$\operatorname{cap}_p(E_0, E_1, D) = \inf_{u \in U} \int_D |\nabla u|^p dm,$$

where

$$U = \{u : D \cup E_0 \cup E_1 \to [0,1] \text{ continuous, } u_{|D-\{\infty\}} \text{ locally Lipschitz,}$$
$$u_{|E_0} = 0, u_{|E_1} = 1\},$$

$p > 1$, and ∇u is the gradient of u. In the particular case $p = n$,

$$\operatorname{cap}(E_0, E_1, D) = \operatorname{cap}_n(E_0, E_1, D)$$

is called *conformal capacity of the condenser* (E_0, E_1, D). By means of the conformal capacity, F. W. Gehring [39], [40] defined the class of quasiconformal mappings (a generalization of the class of conformal mappings)

Computational Methods and Function Theory (CMFT'97)
N. Papamichael, St. Ruscheweyh and E. B. Saff (Eds.), pp. 117–133

ISBN 981-02-3626-3

An *arc* is the homeomorphic image of the open set $(0,1)$. The *p-module of an arc family* Γ is

$$M_p\Gamma = \inf_{\rho\in F(\Gamma)} \int \rho^p\, dm,$$

$$F(\Gamma) = \{\rho : \mathbb{R}^n \to [0,\infty] \text{ Borel measurable and } \int_\gamma \rho dH^1 \geq 1\ \forall\gamma\in\Gamma\},$$

where $p > 1$, m is the n-dimensional Lebesgue measure and H^1 the linear Hausdorff measure. $M\Gamma = M_n\Gamma$ is called the *module* of Γ. The concepts of module and p-module were applied to the study of quasiconformal mappings *e.g.* by F. W. Gehring [38], J. Hesse [41], H. M. Reimann [42], B. V. Šabat [43], [44], G. D. Suvorov [45], J. Väisälä ([46]–[48]) and W. Ziemer ([50]–[52]), while, for the quasiregular mappings, by M. Vuorinen [49] (a quasi-conformal mapping is a homeomorphic quasiregular one).

In this paper we continue our earlier work ([2]–[5], [7], [10]–[19], [22]–[36]) by improving some previous results of [18] and establishing some cases of equality between the p-module and the p-capacity. It should be noted that the module of an arc family is an important tool in studying the boundary behaviour of quasiconformal mappings (*cf. e.g.* [6], [8], [9], [20], [21]), of quasiregular mappings of bounded Dirichlet integrals ([49] and [45]). Note that the p-module (and, in particular, the module) is a concept of geometric function theory, while the p-capacity (and, in particular, the conformal capacity) is a notion of potential theory. Thus, cases of equality between the two are very interesting, because they provide more flexibility and technical advantages in the arrangement of proofs. Some other cases of equality were obtained for example by J. Hesse [41] and W. Ziemer ([50]-[52]).

In what follows we shall use "Proposition" and "Theorem" to refer to well-known and new results respectively.

§2 Preliminaries

Proposition 1 (B. Fuglede [37], Theorem 1a,b). *If* $\Gamma \subset \bigcup_{k=1}^{\infty} \Gamma_k$ *then*

$$M_p\Gamma \leq \sum_{k=1}^{\infty} M_p\Gamma_k, \qquad (p>1).$$

Proposition 2 (J. Hesse [41], Lemma 5.2). *If* $\hat{F}_0, \hat{F}_1 \subset \hat{D} \subset \overline{\mathbb{R}^n}$ *and* $\hat{F}_0 \cap \hat{F}_1 = \emptyset$ *then*

$$M_p\Gamma(\hat{F}_0, \hat{F}_1, D\cap\mathbb{R}^n) \leq \operatorname{cap}_p(\hat{F}_0, \hat{F}_1, D\cap\mathbb{R}^n), \qquad (p>1).$$

Proposition 3 (P. Caraman [32], Theorem 1). *If $\hat{F}_0, \hat{F}_1 \subset \hat{D} \subset \overline{\mathbb{R}^n}$ then*

$$M_p\Gamma(\hat{F}_0, \hat{F}_1, D) \leq \text{cap}_p(\hat{F}_0, \hat{F}_1, D), \qquad (p > 1). \tag{2.1}$$

Theorem 1. *If $E_0, E_1 \subset \overline{\mathbb{R}^n}$ then*

$$M_p\Gamma(E_0, E_1, \overline{\mathbb{R}^n}) \leq \text{cap}_p(E_0, E_1, \overline{\mathbb{R}^n}), \qquad (p > 1).$$

Proof: Since the function u involved in the definition of the p-capacity is continuous in its domain of definition ($D = \overline{\mathbb{R}^n}$), from Propositions 1 and 3, it follows that

$$\begin{aligned} M_p\Gamma(E_0, E_1, \overline{\mathbb{R}^n}) &\leq M_p\Gamma(\hat{E}_0, \hat{E}_1, \overline{\mathbb{R}^n}) \leq \text{cap}_p(\hat{E}_0, \hat{E}_1, \overline{\mathbb{R}^n}) \\ &= \text{cap}_p(E_0, E_1, \overline{\mathbb{R}^n}), \qquad (p > 1), \end{aligned}$$

as desired. ■

Proposition 4 (P. Caraman [32], Theorem 2). *If $F_0, F_1 \subset \overline{D} \subset \mathbb{R}^n$ then*

$$M_p\Gamma(F_0, F_1, D) \leq \text{cap}_p(F_0, F_1, D), \qquad (p > 1). \tag{2.2}$$

Theorem 2. *If $E_0, E_1 \subset \mathbb{R}^n$ then*

$$M_p\Gamma(E_0, E_1, \mathbb{R}^n) \leq \text{cap}_p(E_0, E_1, \mathbb{R}^n), \qquad (p > 1).$$

Proof: Propositions 1, 4 (where $D = \mathbb{R}^n$) and the definition of the p-capacity imply that

$$\begin{aligned} M_p\Gamma(E_0, E_1, \mathbb{R}^n) &\leq M_p\Gamma(\overline{E_0}, \overline{E_1}, \mathbb{R}^n) \leq \text{cap}_p(\overline{E_0}, \overline{E_1}, \mathbb{R}^n) \\ &= \text{cap}_p(E_0, E_1, \mathbb{R}^n), \qquad (p > 1), \end{aligned}$$

as desired. ■

An arc family Γ is said to be *p-exceptional* ($p > 0$) if $M_p\Gamma = 0$.

Proposition 5 (B. Fuglede [37], Theorem 5). *If $\Gamma_0 = \{\gamma \subset \mathbb{R}^n; x_0 \in \gamma\}$ then*

$$M_p\Gamma_0 = 0, \qquad (1 < p \leq n).$$

Lemma 1. *If $D \subset \mathbb{R}^n$ is a domain and $E_0, E_1 \subset \overline{D}$, then,*

$$M_p\Gamma(E_0, E_1, D) = M_p^{\circ\Delta}\Gamma(E_0, E_1, D) \stackrel{\text{def}}{=} \inf_{\rho \in F^{\circ\Delta}(\Gamma)} \int \rho^p dm, \qquad (2.3)$$

where $\Delta = D - (\overline{E_0} \cup \overline{E_1})$, $F^{\circ\Delta}(\Gamma)$ is the class of all admissible functions $\rho \in F(\Gamma)$, continuous in Δ and 0 in $\complement\Delta$ ($\complement\Delta$ is the complement of Δ).

The proof of Lemma 1 is similar to the proof of the Theorem of [19], p. 904.

Remark: In the theorem mentioned above, as well as in the lemma on page 831 of [13], the functions $\rho \in \tilde{F}(\Gamma)$ were erroneously supposed to be also bounded in $\mathbb{R}^n$. We now show that $M_p\Gamma(E_0, E_1, D) \neq m_p$, where $m_p \equiv \inf_{\rho \in F_b(\Gamma)} \int \rho^p dm$ and $F_b(\Gamma)$ is the class of bounded functions in $F(\Gamma)$. In order to see this, let $E_0 = \{0, 1 - \frac{1}{2}, \ldots, 1 - \frac{1}{k}, \ldots\} \subset \mathbb{R}, F_1 = S$ (the unit sphere of $\mathbb{R}^n$) and D the domain with boundary $\partial D = E_0 \cup F_1$ (which clearly is a closed set). It is easy to see, on account of Propositions 1 and 5, that

$$M_p\Gamma(E_0, F_1, D) \leq \sum_{k=1}^{\infty} M_p\Gamma(\{1 - \frac{1}{k}\}, S, D) = 0,$$

but at the same time the arcs joining the points $\{1 - \frac{1}{k}, 0, ..., 0\}$ to S may have a length as small as one wishes. In such a case, the admissible functions must be unbounded. Another counter-example may be obtained choosing for example in $\mathbb{R}^2, F_0 = \{(1,1), (2, \frac{1}{2}), \ldots, (k, \frac{1}{k}), \ldots\}$, while $F_1 = \{x \geq 0, y = 0\}$ and $D = \mathbb{R}^2 - (F_0 \cup F_1)$ with the boundary $\partial D = F_0 \cup F_1$. This time F_0 and F_1 are both closed (with respect to $\mathbb{R}^2$) and disjoint, $M_p\Gamma(F_0, F_1, D) = 0$, while the arcs joining F_0 and F_1 have a length which can be as small as one wishes. Thus, again, there is no admissible bounded ρ (*i.e.* satisfying the condition $\int_\gamma \rho dH^1 \geq 1$ for all $\gamma \in \Gamma(F_0, F_1, D)$), so that $M_p\Gamma(F_0, F_1, D) = \infty$. Finally, the preceding two examples of this remark, allow us to conclude that the infimum in the definition of $M_p\Gamma(F_0, F_1, D)$ may not be obtained only with respect to bounded admissible ρ. Hence, it follows that the Corollary of [19] asserting that "$d_D(E_0, E_1) = 0 \Longrightarrow M_p\Gamma(E_0, E_1, D) = \infty$", where the distance $d_D(E_0, E_1)$ between $E_0, E_1 \subset \overline{D}$ relatively to D means the infimum of the length of the arcs $\gamma \in \Gamma(E_0, E_1, D)$, is also wrong, because, in the preceding two examples, the distance $d_D(E_0, E_1) = 0$, while $M_p\Gamma(E_0, E_1, D) = 0$.

By the same arguments as in Lemma 1, and with the same notations, we obtain

Corollary 1. *With the hypotheses of Lemma 1,*

$$M_p\Gamma(E_0, E_1, D) = M_p^{\circ D}\Gamma(E_0, E_1, D) \stackrel{\text{def}}{=} \inf_{\rho \in F^{\circ D}(\Gamma)} \int \rho^p dm,$$

where $F^{\circ D}(\Gamma)$ is the class of all admissible funtions $\rho \in F(\Gamma)$, continuous in D and 0 in $\complement D$.

Corollary 2. *Lemma 1 and Corollary 1 still hold if D is supposed to be only open.*

Corollary 3. *With the hypotheses of Lemma 1, if $\Gamma_1 \subset \Gamma$ such that $M_p\Gamma_1 = M_p\Gamma$, then $M_p^{\circ\Delta}\Gamma_1 = M_p\Gamma_1 = M_p\Gamma = M_p^{\circ\Delta}\Gamma$.*

Corollary 4. *With the hypotheses of Corollary 1, if $\Gamma_1 \subset \Gamma$ such that $M_p\Gamma_1 = M_p\Gamma$, then $M_p^{\circ D}\Gamma_1 = M_p\Gamma_1 = M_p\Gamma = M_p^{\circ D}\Gamma$.*

Lemma 2. *If $D \subset \mathbb{R}^n$ is a domain and $\Gamma = \Gamma(E_1, E_1, D)$, where $E_0, E_1 \subset \overline{D}$, then*

$$M_p^*\Gamma \stackrel{\text{def}}{=} \inf_{\rho \in F^*(\Gamma)} \int \rho^p dm = M_p^{*\circ\Delta}\Gamma \stackrel{\text{def}}{=} \inf_{\rho \in F^{*\circ\Delta}(\Gamma)} \int \rho^p dm, \tag{2.4}$$

where $F^*(\Gamma)$ is the family of admissible functions $\rho \in F(\Gamma)$ such that $\int_\gamma \rho dH^1 < \infty$ for all $\gamma \in \Gamma$ rectifiable, while $F^{*\circ\Delta}(\Gamma)$ is the family of the functions $\rho \in F^*(\Gamma)$, continuous in Δ and 0 in $\complement\Delta$.

Proof: Clearly, $M_p^*\Gamma \leq M_p^{*\circ\Delta}\Gamma$, so that, in order to prove the opposite inequality, we may suppose $M_p^*\Gamma < \infty$, since otherwise, there is nothing to prove. Given $\varepsilon > 0$ and $\rho \in F^*(\Gamma)$ such that $\int \rho^p dm \leq (1+\varepsilon)M_p^*\Gamma$, let us consider the function

$$\tilde{\rho}(x) = \begin{cases} \frac{1+\varepsilon}{\Omega_n} \displaystyle\int_B \rho[x + \varepsilon r(x)y]dm(y) & \text{if } x \in \Lambda, \\ 0 & \text{if } x \in \complement\Delta, \end{cases}$$

where $y \in B$ (unit ball in $\mathbb{R}^n$), $\Omega_n = mB$ and $r(x) = \inf(|x - \xi|; \xi \in \partial D \cup E_0 \cup E_1)$. By the same argument as in the theorem of V. V. Aseev [1], it follows that $\tilde{\rho} \in F^*(\Gamma)$ is continuous in Δ and 0 in $\complement\Delta$. Let us show that $\int_\gamma \tilde{\rho} dH^1 < \infty$ for all rectifiable $\gamma \in \Gamma$. If $z(x) = x + \varepsilon r(x)y$, then,

$|dz| = |dx + \varepsilon r'(x) y dx| > |dx|(1-\varepsilon)$ and since $\rho \in F^*(\Gamma)$, it follows that $\int_\gamma \rho dH^1 = K(\gamma) < \infty$. This allows us to conclude that

$$\begin{aligned}
\int_\gamma \tilde{\rho}(x) d\gamma(x) &= \int_\gamma \left\{ \frac{1+\varepsilon}{\Omega_n} \int_B \rho[x + \varepsilon r(x) y] dm(y) \right\} d\gamma(x) \\
&= \frac{1+\varepsilon}{\Omega_n} \int_B \left\{ \int_\gamma \rho[z(x)] d\gamma(x) \right\} dm(y) \\
&\leq \frac{1+\varepsilon}{(1-\varepsilon)\Omega_n} \int_B \left\{ \int_\gamma \rho(z) d\gamma(z) \right\} dm(y) \\
&= \frac{1+\varepsilon}{(1-\varepsilon)\Omega_n} \int_B K(\gamma) dm(y) = \frac{1+\varepsilon}{1-\varepsilon} K(\gamma) < \infty.
\end{aligned}$$

Taking into account also that $1 - \varepsilon \leq |J(x)| \leq 1 + \varepsilon$, we deduce (arguing as in Aseev's theorem mentioned above) that

$$\int \tilde{\rho}(x)^p dm(x) \leq \frac{(1+\varepsilon)^{p+1}}{1-\varepsilon} M_p^* \Gamma,$$

and, by taking the infimum over all $\tilde{\rho} \in F^{*\circ\Delta}(\Gamma)$, we obtain

$$M^{*\circ\Delta}\Gamma \leq \frac{(1+\varepsilon)^{p+1}}{1-\varepsilon} M_p^* \Gamma.$$

Hence, since $\varepsilon > 0$ may be taken arbitrarily small, we get

$$M_p^{*\circ\Delta}\Gamma \leq M_p^*\Gamma,$$

which, on account of the opposite inequality from above, yields (2.4), as desired. ■

With the same notations and by the same argument as above, but with $r(x) = d(x, \partial D)$, we obtain:

Corollary 5. *If $E_0, E_1 \subset \overline{D} \subset \mathbb{R}^n$, then,*

$$M_p^*\Gamma = M_p^{*\circ D}\Gamma = \inf_{\rho \in F^{*\circ D}(\Gamma)} \int \rho^p dm,$$

where $F^{*\circ D}(\Gamma)$ is the family of functions $\rho \in F^*(\Gamma)$, continuous in D and 0 in $\complement D$.

Proposition 6 (B. Fuglede [37], Theorem 2). *An arc family Γ is p-exceptional iff there exists a function $\rho \in L^p, \rho \geq 0$, such that $\int_\gamma \rho dH^1 = +\infty$ for all $\gamma \in \Gamma$.*

Corollary 6. *For every $p > 0$ $p-$almost every bounded arc is rectifiable.*

Remark 1: In the particular case $p = n$, Corollary 6 was obtained by J. Väisälä ([46] Theorem 3.2, [48] Theorem 6.9).

Remark 2: In the proof of Lemma 5 of [18], there was a mistake. We give a correct proof in:

Lemma 3. *Let $D \subset \mathbb{R}^n$ be a domain, $E_0, E_1 \subset \overline{D}$, $\Gamma = \Gamma(E_0, E_1, D)$, $\Gamma' = \Gamma(\overline{E_0}, \overline{E_1}, D)$ and $\{\rho_k\}$ a sequence of functions $\rho_k \in F(\Gamma)$ such that*

$$M_p\Gamma = \lim_{k\to\infty} \int \rho_k^p dm < \infty. \tag{2.5}$$

Let further $\Gamma_k = \{\gamma \in \Gamma; \int_\gamma \rho_k ds = \infty\}$, $\Gamma_0 = \cup_{k=1}^{\infty}\Gamma_k$, $\Gamma'' = \Gamma - \Gamma_0$, $\Gamma^* = \{\gamma' \in \Gamma'; \gamma' \subset \gamma'' \in \Gamma''\}$, and $\Delta = D - (\overline{E_0} \cup \overline{E_1}) \neq \emptyset$. Then

$$M_p\Gamma = M_p\Gamma'' = M_p\Gamma^* = M_p^*\Gamma^* = M_p^{\circ\Delta}\Gamma = M_p^{\circ\Delta}\Gamma'' = M_p^{\circ\Delta}\Gamma^*. \tag{2.6}$$

Proof: By means of B. Fuglede's Proposition 6 and by the same argument as in Lemma 5 of [18], it follows that

$$M_p\Gamma = M_p\Gamma' = M_p\Gamma'' = M_p^*\Gamma'' = M_p^*\Gamma^*.$$

From Corollary 3, since $\Gamma'' \subset \Gamma^* \subset \Gamma'$, and $M_p\Gamma'' \leq M_p\Gamma^* \leq M_p\Gamma' = M_p\Gamma''$, we deduce that

$$M^*\Gamma^* = M_p^*\Gamma'' = M_p\Gamma'' = M_p^{\circ\Delta}\Gamma'' = M_p^{\circ\Delta}\Gamma^*,$$

yielding (2.6), as desired. ■

Lemma 4. *With the same notations as in Lemma 3, where D is an open set, we obtain*

$$\begin{aligned} M_p\Gamma &= M_p\Gamma'' = M_p^*\Gamma'' = M_p^{*\circ\Delta}\Gamma'' = M_p^{*\circ D}\Gamma'' = M_p^*\Gamma^* \\ &= M_p^{*\circ\Delta}\Gamma^* = M_p^{*\circ D}\Gamma^*. \end{aligned} \tag{2.7}$$

Proof: By the same argument as in Lemma 5 of [18], we obtain

$$M_p\Gamma - M_p\Gamma'' = M_p^*\Gamma'' = M_p^*\Gamma^*,$$

and, by the same argument as in Lemma 2 and Corollary 5, it follows that $M_p^*\Gamma^* = M_p^{*\circ\Delta}\Gamma^* = M_p^{*\circ D}\Gamma^*$ and that $M_p^*\Gamma'' = M_p^{*\circ\Delta}\Gamma'' = M_p^{*\circ D}\Gamma''$, yielding (2.7), as desired. ■

Corollary 7. *Let D be a domain, $F_0, F_1 \subset \overline{D}, \Gamma = \Gamma(F_0, F_1, D)$ and $\{\rho_k\}$ a sequence of functions $\rho_k \in F(\Gamma)$ such that (2.5) holds. Let further $\Gamma_k = \{\gamma \in \Gamma; \int_\gamma \rho_k dH^1 = \infty\}$, $\Gamma_0 = \cup_{k=1}^\infty \Gamma_k$, $\Gamma'' = \Gamma - \Gamma_0$, $\Delta = D - (F_0 \cup F_1) \neq \emptyset$, then*

$$M_p\Gamma = M_p\Gamma'' = M_p^*\Gamma'' = M_p^{*\circ D}\Gamma'' = M_p^{\circ\Delta}\Gamma'' = M_p^{\circ\Delta}\Gamma. \tag{2.8}$$

§3 Main result

Theorem 3. *Suppose that $D \subset \mathbb{R}^n$ is a domain, $F_0, F_1 \subset \overline{D}$ such that $d(F_0, F_1) > 0$ (where d is the distance between two sets) and that for each $\xi \in \partial D \cap F_i$, with $i = 0$ or $i = 1$, we have*

$$\lim_{x \to \xi, x \in D} \inf_\gamma H^1[\gamma(F_i, x)] = 0, \tag{3.1}$$

where the infimum is taken over all $\gamma = \gamma(F_i, x)$ joining x to F_i in D. Then

$$M_p\Gamma(F_0, F_1, D) = \operatorname{cap}_p(F_0, F_1, D), \qquad (p > 1). \tag{3.2}$$

Proof: From Proposition 4, it follows that (2.2) holds. In order to prove the opposite inequality, since, on account of (2.8), $M_p\Gamma = M_p^{*\circ D}\Gamma''$, it is enough to establish that

$$\operatorname{cap}_p(F_0, F_1, D) \leq \int \rho^p dm \tag{3.3}$$

holds for every $\rho \in F^{*\circ D}\Gamma''$. Now, assume that (3.1) holds for each $\xi \in \partial D \cap F_0$ and let

$$u(x) = \inf_\gamma \int_\gamma \rho dH^1,$$

where the infimum is taken over all $\gamma \in \Gamma(F_0, x, D)$ (*i.e.* all $\gamma \subset D$ joining x and F_0). Since $\rho \in F^{*\circ D}(\Gamma'')$, it is easy to see that u has boundary value 0 on $F_0 \cap \overline{D}$. Indeed, if $\xi \in F_0 \cap D$, then, there exists a compact rectifiable arc $\gamma_0(x_0) \subset D$ joining x_0 to ξ. From the continuity of ρ in D, it follows that ρ is bounded on such an arc, *i.e.* $K_0 = \sup_{x \in \gamma_0(x_0)} \rho(x) < \infty$, so that

$$\begin{aligned} \lim_{x \to \xi, x \in D} u(x) &= \lim_{x \to \xi, x \in D} \inf_\gamma \int_\gamma \rho dH^1 \leq \lim_{x \to \xi, x \in \gamma_0(x_0)} \int_{\gamma_0(x)} \rho dH^1 \\ &\leq K_0 \lim_{x \to \xi, x \in \gamma_0(x_0)} \int_{\gamma_0(x)} dH^1 = 0, \end{aligned}$$

where the infimum is taken over all $\gamma \in \Gamma(F_0, x, D)$. Next, suppose that $\xi \in F_0 \cap \partial D$ and $\gamma_1(x) \in \Gamma(F_0, x, D)$, where $\gamma_1(x)$ is a subarc joining F_0

and x of an arc $\gamma_1 \in \Gamma''$. Then, we have

$$\lim_{x\to\xi, x\in D} u(x) = \lim_{x\to\xi, x\in D} \inf_\gamma \int_\gamma \rho dH^1 \leq \lim_{x\to\xi, x\in\gamma_1} \int_{\gamma_1(x)} \rho dH^1 = 0,$$

in view of (3.1), since $\rho \in F^{*\circ D}(\Gamma'')$ and $\int_{\gamma_1} \rho dH^1 < \infty$ for all $\gamma_1 \in \Gamma''$.

It is easy to see that $u(x) < \infty$ in D. Indeed, from $u(x) \to 0$ as $x \to \xi$ and $x \in D$, it follows that, for $x_0 \in D$ sufficiently close to $\xi, u(x_0) < \infty$. Hence, there exists an open arc $\gamma_0 \subset D$, joining x_0 to F_0, such that $\int_{\gamma_0} \rho dH^1 < \infty$. If $\tilde{x}$ is an arbitrary point of D, then, there is a rectifiable arc $\gamma_1 = \gamma(x_0, \tilde{x}) \subset D$ joining x_0 to $\tilde{x}$. Indeed, $\overline{\gamma}_1 = \gamma_1 \cup \{x_0, \tilde{x}\} \subset D$ is compact and then, $d(\overline{\gamma}_1, \partial D) = r_0 > 0$. Next, let us consider a covering of $\overline{\gamma}_1$ by balls $B(x, r)$ with $x \in \overline{\gamma}_1, r < r_0$. Since $\overline{\gamma}_1$ is compact, we can extract a finite covering $\{B(x_k, r_k)\}$ of $\overline{\gamma}_1$ by balls with $x_k \in \overline{\gamma}_1, r_k < r_0$ $(k = 1, ..., q)$. Clearly, these q balls can be ordered along $\overline{\gamma}_1$ from x_0 towards $\tilde{x}$ and the intersection Δ_k of two successive balls can be an open non-empty domain. In each $\Delta_k = B(x_k, r_k) \cap B(x_{k+1}, r_{k+1})$, we choose a point y_k; the polygonal line λ corresponding to the points $x_0, y_1, ..., y_{q-1}, \tilde{x}$, is contained in D and $H^1(\lambda) < 2(q+1)r_0 < \infty$. Now, let $\gamma \subset \gamma_0 \cup \lambda$ bc an arc joining F_0 to $\tilde{x}_1$. Then, since $\rho \in F^{*\circ D}(\Gamma'')$, it follows that $\sup_\lambda \rho(x) = M_\lambda < \infty$, and we obtain

$$u(\tilde{x}) \leq \int_\gamma \rho dH^1 \leq \int_{\gamma_0} \rho dH^1 + \int_\lambda \rho dH^1 \leq \int_{\gamma_0} \rho dH^1 + 2(q+1)r_0 M_\lambda < \infty,$$

as desired.

Let us show that u is locally Lipschitz in D. Indeed, if $x_0 \in D$ and $x_1, x_2 \in B(x_0, r) \subset D, \gamma_1 \in \Gamma(F_o, x_1, D)$, and λ_0 is the line segment joining x_1 to x_2, then,

$$u(x_2) \leq \int_{\gamma_1} \rho dH^1 + \int_{\lambda_0} \rho dH^1.$$

Thus, taking the infimum over all the arcs joining x_1 and F_0 in D, we obtain

$$\begin{aligned} u(x_2) &\leq \inf_{\gamma_1} \int_{\gamma_1} \rho dH^1 + \int_{\lambda_0} \rho dH^1 \leq u(x_1) + M_{\lambda_0} \int_{\lambda_0} dH^1 \\ &= u(x_1) + M_{\lambda_0} |x_1 - x_2|. \end{aligned}$$

Then, on account of the symmetrical role of x_1, x_2, $|u(x_1) - u(x_2)| \leq M_{\lambda_0}|x_1 - x_2|$, *i.e.* u is locally Lipschitz in D, as desired.

Next, let us prove that

$$|\nabla u(x)| \leq \rho(x) \tag{3.4}$$

a.e. in D. Firstly, we observe that u is differentiable a.e. in D and, at a point of differentiability,

$$|\nabla u(x)| = \sup_s \left|\frac{\partial u(x)}{\partial s}\right|, \tag{3.5}$$

where

$$\frac{\partial u(x)}{\partial s} = \lim_{t\to 0} \frac{u(x+te_s)-u(x)}{t}$$

is the directional derivative of u. Clearly, $u(x+te_s) \le \int_\gamma \rho dH^1 + \int_{\lambda_t} \rho dH^1$, where $\gamma \in \Gamma(F_0, x, D)$, and λ_t is the line segment joining x and $x+te_s$. This inequality yields

$$u(x+te_s) \le \inf_{\gamma(F_0,x)} \int_{\gamma(F_0,x)} \rho dH^1 + \int_{\lambda_t} \rho dH^1 = u(x) + \int_{\lambda_t} \rho dH^1.$$

Hence, in the case $u(x) \le u(x+te_s)$, we have

$$\left|\frac{u(x+te_s)-u(x)}{t}\right| = \frac{u(x+te_s)-u(x)}{t} \le \frac{1}{t}\int_{\lambda_t} \rho dH^1.$$

Arguing as above, we get also

$$u(x) \le \inf_{\gamma(F_0,x+te_s)} \int_{\gamma(F_0,x+te_s)} \rho dH^1 + \int_{\lambda_t} \rho dH^1 = u(x+te_s) + \int_{\lambda_t} \rho dH^1,$$

so that, for $u(x+te_s) \le u(x)$, we deduce

$$\left|\frac{u(x+te_s)-u(x)]}{t}\right| = \frac{u(x)-u(x+te_s)}{t} \le \frac{1}{t}\int_{\lambda_t} \rho dH^1.$$

Thus, in both cases,

$$\left|\frac{u(x+te_s)-u(x)]}{t}\right| \le \frac{1}{t}\int_{\lambda_t} \rho dH^1,$$

and hence

$$\left|\frac{\partial u(x)}{\partial s}\right| = \lim_{t\to 0}\left|\frac{u(x+te_s)-u(x)}{t}\right| \le \lim_{t\to 0}\frac{1}{t}\int_{\lambda_t} \rho dH^1 = \rho(x)$$

a.e. in D, since ρ is supposed to be continuous in D and the points of continuity are Lebesgue points. This implies the last part of the preceding relation, which, by (3.5), yields (3.4), as desired.

Next, we remark that $h = \inf_{x \in F_1} u(x) \geq 1$. We consider the truncation

$$u^*(x) = \begin{cases} h & \text{for } u(x) \geq h, \\ u(x) & \text{for } 0 \leq u(x) \leq h, \end{cases}$$

and observe that u^*/h is admissible for $\text{cap}_p(F_0, F_1, D)$. Indeed, $0 \leq u^*(x)/h \leq 1$, $u^*/h_{|F_0} = 0$ since $u_{|F_0} = 0$ and $u^*/h_{|F_1} = 1$ because, by the definition of h, $u(x) \geq h$ for all $x \in F_1$ and u^* is the truncation of u at the level h. Further, since u is locally Lipschitzian in D with Lipschitz constant M_λ (as it was proved above), it is easy to see that u^*/h is also locally Lipschitzian in D with Lipschitz constant M_λ/h. Indeed, if $x, y \in D_h = \{x \in D; u(x) < h\}$, then, clearly,

$$\left| \frac{u^*(x)}{h} - \frac{u^*(y)}{h} \right| = \frac{1}{h}|u(x) - u(y)| \leq \frac{M_\lambda}{h}|x - y|,$$

if $x, y \in D \cap \complement D_h$, then, $u^*(x) = u^*(y) = h$, so that

$$\left| \frac{u^*(x)}{h} - \frac{u^*(y)}{h} \right| = 0 \leq \frac{M_\lambda}{h}|x - y|,$$

and, if $x \in D_h, y \in D \cap \complement D_h$, then,

$$\left| \frac{u^*(x)}{h} - \frac{u^*(y)}{h} \right| = \left| \frac{u(x)}{h} - 1 \right| \leq \left| \frac{u(x)}{h} - \frac{u(y)}{h} \right| \leq \frac{M_\lambda}{h}|x - y|.$$

Now, let us also verify that

$$|\nabla u^*(x)| = \sup_s \left| \frac{\partial u^*}{\partial s} \right| \leq \rho(x) \tag{3.6}$$

at each point of differentiability of u^*, *i.e.* a.e. in D. Indeed, suppose that x is such a point of differentiability. If $x \in D_h$, then it is evident that

$$\frac{\partial u^*(x)}{\partial s} = \lim_{t \to 0} \frac{u^*(x + te_s) - u^*(x)}{t} = \lim_{t \to 0} \frac{u(x + te_s) - u(x)}{t} = \frac{\partial u(x)}{\partial s}$$

for every direction s. Hence, $|\nabla u^*(x)| = |\nabla u(x)|$. If $x \in \complement \overline{D}_h$, then, for t sufficiently small, $u^*(x + te_s) = u^*(x) = h$, yielding $|\frac{\partial u^*(x)}{\partial s}| = 0$ for all s and then, $|\nabla u^*(x)| = 0$. Now, if $x \in \partial D_h \cap D$, then

$$\begin{aligned} \left| \frac{\partial u^*(x)}{\partial s} \right| &= \lim_{t \to 0} \left| \frac{u^*(x + te_s) - u^*(x)}{t} \right| = \lim_{t \to 0} \left| \frac{u^*(x + te_s) - u(x)}{t} \right| \\ &\leq \lim_{t \to 0} \left| \frac{u(x + te_s) - u(x)}{t} \right| = \left| \frac{\partial u(x)}{\partial s} \right|. \end{aligned}$$

Hence, $|\nabla u^*(x)| \leq |\nabla u(x)|$ a.e. in D and, on account of (3.4),

$$|\nabla u^*(x)| \leq \rho(x) \tag{3.7}$$

a.e. in D.

Since u^*/h, with $h \geq 1$, is admissible for $\text{cap}_p(F_0, F_1, D)$ and taking into account (3.6), we obtain

$$\text{cap}_p(F_0, F_1, D) \leq \frac{1}{h^p} \int_D |\nabla u^*|^p dm \leq \frac{1}{h^p} \int \rho^p dm \leq \int \rho^p dm.$$

Here ρ is an arbitrary function of $F^{*\circ D}(\Gamma'')$, so that taking the infimum over all such ρ, we deduce that

$$\text{cap}_p(F_0, F_1, D) \leq M_p^{*\circ D}\Gamma'' = M_p\Gamma,$$

which, together with (2.1), yields (3.2) if (3.1) is supposed to hold for F_0. If (3.1) is verified for $i = 1$, we repeat the argument from above with F_1 instead of F_0 and, finally, we consider the function $v = 1 - \frac{u^*}{h,}$ which is admissible for $\text{cap}_p(F_0, F_1, D)$. ■

§4 Concluding remarks

According to Lemma 1, in the definition of the p-module, the corresponding infimum may be taken with respect to the subclass of the admissible functions ρ, which are supposed to be not only Borel measurable, but also continuous in $\Delta = D - (F_0 \cap F_1)$ and 0 in $\complement D$. However, as was pointed out in the remark following Lemma 1, we do not obtain, in general, the p-module if the infimum is taken with respect to the subclass of bounded admissible functions (this infimum may be strictly larger). Our proof of Theorem 1 of [18], is not correct because we used there (as in the proof of Theorem 3 of this paper) the relation $M_p\Gamma = M_p^{*\circ D}\Gamma''$, where the infimum in the definition of $M_p^{*\circ D}\Gamma''$ was taken there with respect to the functions $\rho \in F(\Gamma'')$, supposed to be not only continuous in D, 0 in $\complement D$, but also bounded (which is not correct, according to the remark following Lemma 1) and also because we used there the inequality $M_p\Gamma(E_0, E_1, D) \leq \text{cap}_p(E_0, E_1, D)$ for arbitrary sets E_0, E_1 with $d(E_0, E_1) > 0$, (which is still an open question). This is why it was necessary to give here a correct, although a little more restrictive version of that theorem.

References

[1] Aseev, V. V. , Ob odnom cvoĭstve modulja, *Dokl. Akad. Nauk SSSR* **200** (1971), 513–514 (Russian).

[2] Caraman, P., Quasiconformality and extremal length, in *Proc. of the Romanian-Finnish Seminar on Teichmüller spaces and quasiconformal mappings*, Edit. Acad. Române, Braşov România 1969, Bucureşti 1971, 111–145.

[3] Caraman, P., About the characterization of the quasiconformality (QCf) by means of the moduli of q-dimensional surface families, *Rev. Roumaine Math. Pures Appl.* **16** (1971), 1329–1348.

[4] Caraman, P., Characterization of quasiconformality by arc families of extremal length zero, *Ann. Acad. Sci. Fenn. Ser. A I* **528** (1973), 1–10.

[5] Caraman, P., *n-dimensional Quasiconformal Mappings*, Edit. Acad. Bucureşti România and Abacus Press, Tunbridge Wells, Kent, England 1974, 553p.

[6] Caraman, P., Exceptional sets for the boundary correspondence of quasiconformal mappings, in *Proc. of the scientific jubilee session for* 25th *anniversary of the Institute of Mathematics*, Edit. Acad. Bucureşti România 1975, 117–128.

[7] Caraman, P., p-capacity and p-modulus, *Symp. Math.* **18** (1976), 455–484.

[8] Caraman, P., Quasiconformality and boundary correspondence, *Mathematica - Rev. Anal. Numér. Théorie Approximation (Cluj)* **5** (1976), 117–126.

[9] Caraman, P., About a conjecture of F. W. Gehring on the boundary correspondence by quasiconformal mappings, *Ann. Polonici Math.* **33** (1976), 21–33.

[10] Caraman, P., Relations between capacities, Hausdorff h-measures and p-modules, *Mathematica - Rev. Anal. Numér. Théorie Approximation (Cluj)* **7** (1978), 13–49.

[11] Caraman, P., Estimate of exceptional sets for quasiconformal mappings in space, in *Proc. of the* III^d *Romanian-Finnish Seminar, Bucharest 27. VI - 2. VII. 1979*, Lecture Notes in Math. 743, Springer-Verlag Berlin-Heidelberg-New York 1979, 66–107.

[12] Caraman, P., About capacities and moduli in infinite-dimensional spaces, in *Proc of the IVth Romanian-Finnish Seminar on Complex Analysis and of the Colloquium on Complex Analysis, Joensuu, Finland 24-27. VIII. 1978*, Lecture Notes in Math. 747, Springer-Verlag, Berlin-Heidelberg-New York 1979, 73–91.

[13] Caraman, P., Le p-module et la p-capacité du cylindre, *C.R. Acad.Sci. Paris* **290** (1980), 829–832.

[14] Caraman, P., p-capacity and conformal capacity in infinite-dimensional spaces, in *Proc. of the Conference on Analytic Functions, Kozubnik 1979*, Lecture Notes in Math. 798, Springer-Verlag, Berlin-Heidelberg-New York 1980, 68–108.

[15] Caraman, P., Estimates for an exceptional set for quasiconformal mappings in $\mathbb{R}^n$, in *Kongress und Tagungsberichte der Martin Luther Univ. Halle - Wittenberg 'Komplexe Analysis und ihre Anwendungen auf Partielle Differentialgleichungen'*, VEB Kongress- und Werbedruck 9273 Oberlungwitz 1980, Teil 3, 210–221.

[16] Caraman, P., About the equality between the p-module and the p-capacity in $\mathbb{R}^n$, *Institutul Naţional pentru Creaţie Ştiinţifică şi Tehnică - Institutul de Matematică. Preprints Series in Math.* **106** (1981), 1–58.

[17] Caraman, P., Module and p-module in an abstract Wiener space, *Rev. Roumaine Math. Pure Appl.* **37** (1982), 551–599.

[18] Caraman, P., About the equality between the p-module and the p-capacity in $\mathbb{R}^n$, in *Analytic functions. Blażejewko 1982. Proc. of a Conference held in Blażejewko, Poland, August 19–27, 1982*, Lecture Notes in Math. 1039, Springer-Verlag Berlin-Heidelberg-New York-Tokyo 1983, 32–83.

[19] Caraman, P., Une propriété du module des familles d'arcs, *C.R. Acad. Sci. Paris.* **299** (1984), 903–905.

[20] Caraman, P., Boundary behaviour of quasiconformal mappings in normed spaces, *Ann. Polonici Math.* **46** (1985), 35–54.

[21] Caraman, P., Boundary behaviour of quasiconformal mappings in normed spaces, *Acad. R.S.R. Memoriile Secţiilor Ştiinţifice (IV) 9 (1986)*, Bucureşti 1989, 87–112.

[22] Caraman, P., New cases of equality between p-module and p-capacity, *Ann. Polonici Math.* **55** (1991), 37–56.

[23] Caraman, P., p-module and p-capacity of a topological cylinder, *Rev. Roumaine Math. Pures Appl.* **36** (1991), 333–343.

[24] Caraman, P., p-module and p-capacity in $\mathbb{R}^n$, *Rev. Roumaine Math. Pure Appl.* **36** (1991), 345–353.

[25] Caraman, P., p-capacità e p-modulo, in *Simpozionul al VI-lea Tiraspolian de Topologie Generală şi Aplicaţii. Vadul lui Vodă, 9-17.IX, 1991, Chişinău*, 1991, 48–49.

[26] Caraman, P., Rezultate noi despre p-modul şi p-capacitate, in *Lucrările Conferinţei Pregătitoare pentru Congresul Matematicienilor din România 19-21.XI.1992*, vol. II, Bucureşti 1993, 119–122 (Romanian).

[27] Caraman, P., Relations between p-capacity and p-module (I),(II), *Rev. Roumaine Math. Pures Appl.* **39** (1994), 509–553 and 555–577.

[28] Caraman, P., p-module for Grötzsch and Teichmüller rings, *Rev. Roumaine Math. Pures Appl.* **39** (1994), 579–585.

[29] Caraman, P., Some properties of the p-module and the p-capacity, in *Technical Univ. of Iaşi. Preprints of the Department of Math. Proc. of the Memorial Conf. of the Depart. of Math. Iaşi, December* 21st *1991. Session B 1, Paper No. 9/92*, 1995 B.1–B.6.

[30] Caraman, P., The problem of equality between the p-capacity and the p-module, *Bull. de la Société des Sciences et des Lettres de Łodz* **45** (1995); Série: "Recherches sur les déformations." 20, 79–89.

[31] Caraman, P., Some new cases of equality between p-module and p-capacity, in *Symposium Septimum Tiraspolense Generalis Topologiae et suae Applicationum, Chişinău*, 1996, 63–66.

[32] Caraman, P., Some remarks on the equality between the p-module and the p-capacity, *Academia Română. Memoriile Secţiilor Ştiinţifice. (IV)* **17** 1996 (1997), 21–27.

[33] Caraman, P., About the strict inequality between p-module and p-capacity, *Scripta Scientiarum Matematicarum* **1** (1997), 37–50.

[34] Caraman, P., Nuovi risultati sulla p-capacità ed il p-modulo, *Scripta Scientiarum Matematicarum* **1** (1997), 51–66.

[35] Caraman, P., Some comments on the equality between the p-module and the p-capacity of a condenser, *Scripta Scientiarum Matematicarum*, 1998 to appear.

[36] Caraman, P., About cases of equality between the p-module and the p-capacity, in *Analysis and topology*, C. Andreian-Cazacu, O. Lehto and Th. M. Rassias (eds.), World Scientific, 1998 to appear.

[37] Fuglede, B., Extremal length and functional completion, *Acta Math.* **98** (1957), 171–219.

[38] Gehring, F. W., Extremal length definition of the conformal capacity of rings in space, *Michigan Math. J.* **9** (1962), 137–150.

[39] Gehring, F. W., Rings and quasiconformal mappings in space, *Trans. Amer. Math. Soc.* **103** (1962), 353–93.

[40] Gehring, F. W., Quasiconformal mappings in space, *Bull. Amer. Math. Soc.* **69** (1963), 146–164.

[41] Hesse, J., A p-extremal length and p-capacity, *Ark. Mat.* **13** (1975), 131–144.

[42] Reimann, H. M., Über harmonische Kapazität und quasikonforme Abbildungen im Raum, *Math. Helv.* **44** (1969), 284–304.

[43] Šabat, B. V., Metod moduleĭ v prostranstve, *Dokl. Akad. Nauk SSSR* **130** (1960), 1210–1213 (Russian).

[44] Šabat, B. V., Ekstremal'nye zadači teorii kvazikonformnyh otobraženiĭ v prostranstve, in *Sovieto-American joint Symposium on partial differential equations*, Novosibirsk, August 1963 (Russian).

[45] Suvorov, G., *Generalized length-area principle in the theory of mappings*, Kiev, Naukova Dumka, 1985 (Russian).

[46] Väisälä, J., Quasiconformal mappings in space, *Ann. Acad. Sci. Fenn. Ser. A I Math* **298** (1961), 1–36.

[47] Väisälä, J., Two new characterizations for quasiconformality, *Ann. Acad. Sci. Fenn. Ser. A I Math.* **362** (1965), 1–12.

[48] Väisälä, J., *Lectures on n-dimensional quasiconformal mappings*, Lecture Notes in Math. 229, Springer-Verlag, Berlin-Heidelberg-New York, 1971, 144p.

[49] Vuorinen, M., *Conformal geometry and quasiregular mappings*, Lecture Notes in Math. 1319, Springer-Verlag, Berlin-Heidelberg-New York-London-Paris-Tokyo, 1988, 209p.

[50] Ziemer, W., Extremal length and conformal capacity, *Trans. Amer. Math. Soc.* **126** (1967), 460–473.

[51] Ziemer, W., Extremal length and p-capacity, *Michigan Math. J.* **16** (1969), 43–51.

[52] Ziemer, W., Extremal length as a capacity, *Michigan Math. J.* **17** (1970), 117–123.

Petru Caraman
Mathematical Institute "Octav Mayer"
Romanian Academy - Branch of Iaşi
Str. Palade 12
6600 Iaşi
ROMÂNIA
caraman@tuiasi.ro

Extending the qd-algorithm to tackle multivariate problems

Annie Cuyt

and Brigitte Verdonk

Abstract. A lot has been said and done about the *qd*-algorithm [13, 9, 7, 5, 14]. Our main interest is to analyze how the original algorithm and its various improvements can be generalized for use in several multivariate applications. The present paper recalls the known univariate results in the Sections 2.1 and 3, and discusses their multivariate generalization in the Sections 2.2 and 4, but without going into all the multivariate details. We just make everything multivariate-ready for implementation in floating-point polynomial arithmetic (covering additional difficulties not encountered in exact polynomial arithmetic). The reader who is only familiar with the properties of the univariate *qd*-algorithm and does not have an extensive knowledge of the multivariate theory, can easily follow the analysis.

§1 Introduction: A state of the art in *qd*-algorithms

When studying the literature on the *qd*-algorithm, the results have to be classified mainly into two groups. A number of papers discuss the application of the *qd*-algorithm to the moments of a meromorphic function, for the calculation of its poles and multiplicities. This technique will be summarized in Section 2. Another collection of publications discuss the use of the *qd*-algorithm applied to an underlying rational function, for the computation of the eigenvalues or singular values of certain tri- or bidiagonal matrices. These computational schemes are recalled in Section 3. We stress that the main difference lies in the fact that the latter applications rely on the fact that the underlying function is rational and hence has a terminating continued fraction expansion.

In [11] the authors describe an ingenious way to transform a problem of the first kind into one of the second kind, at the expense of great computational complexity though. Our interest is in problems of the first kind, mainly with the ambition to generalize the application of the *qd*-algorithm to multivariate functions in order to obtain information on the location of their pole curves. Such methods, that decipher information on the poles of a

Computational Methods and Function Theory (CMFT'97)
N. Papamichael, St. Ruscheweyh and E. B. Saff (Eds.), pp. 135–159

ISBN 981-02-3626-3

function from its Taylor series development, are also closely related to the computation of Padé approximants or the formally orthogonal Hadamard polynomials. The Padé approximant $r_{n,m}(z)$ for a function $f(z)$ is known to be the irreducible form of a rational function $p(z)/q(z)$ of degree n in the numerator and m in the denominator satisfying $(fq-p)(z) = O(z^{n+m+1})$. Here we normalize the Padé approximant so that the denominator is monic.

By performing an equivalence transformation on the underlying continued fraction, the computation of the qd-table can be combined with that of the Hadamard polynomials into one scheme [7]. The main interest of this rewrite is in the fact that this last form of the qd-algorithm can also be applied to multivariate functions for the computation of their poles [6]. In Section 4 we describe a new implementation of this symbolic qd-algorithm in floating-point polynomial arithmetic, instead of exact polynomial arithmetic, thus creating the possibility to work with real-life numeric data. This gives rise to many new applications that we intend to discuss in future papers, such as numeric–symbolic factorization techniques for multivariate polynomials, numeric pole estimation of multivariate meromorphic functions etc.

§2 Meromorphic function case

2.1 The progressive qd-algorithm.

We immediately present the more stable progressive form of the qd-algorithm [10, p. 614]. Let the function $f(z)$ be known by its formal series expansion

$$f(z) = \sum_{i=0}^{\infty} c_i z^i \tag{2.1.1}$$

For arbitrary integers n and for integers $m \geq 0$ we define the Hankel determinants

$$H_m^{(n)} = \begin{vmatrix} c_n & c_{n+1} & \cdots & c_{n+m-1} \\ c_{n+1} & c_{n+2} & \cdots & c_{n+m} \\ \vdots & & & \vdots \\ c_{n+m-1} & c_{n+m} & \cdots & c_{n+2m-2} \end{vmatrix}$$

with $c_i = 0$ for $i < 0$ and $H_0^{(n)} = 1$. The series (2.1.1) is termed *k-normal* if $H_m^{(n)} \neq 0$ for $m = 0, 1, \ldots, k$ and $n \geq 0$. It is called *ultimately k-normal* if for every $0 \leq m \leq k$ there exists an $n(m)$ such that $H_m^{(n)} \neq 0$ for $n > n(m)$. With (2.1.1) as input we can define the qd-scheme as follows.

(a) The start values are given by

$$
\begin{aligned}
q_k^{(-k+1)} &= 0, & k &= 2, 3, \ldots, \\
e_k^{(-k+1)} &= \frac{w_{k+1}}{w_k}, & k &= 1, 2, \ldots, \\
e_0^{(k)} &= 0, & k &= 1, 2, \ldots, \\
q_1^{(0)} &= \frac{c_1}{c_0},
\end{aligned}
$$

where $w_0, w_1, \ldots$ are the Taylor coefficients of the function $1/f$. In other words,

$$
\begin{aligned}
w_0 &= \frac{1}{c_0}, \\
w_k &= -\frac{(c_1 w_{k-1} + \ldots + c_k w_0)}{c_0}, \qquad k = 1, 2, \ldots.
\end{aligned}
$$

(b) The rhombus rules for continuation of the scheme are given by

$$
\begin{aligned}
q_m^{(-m+n)} &= e_m^{(-m+n-1)} - e_{m-1}^{(-m+n)} + q_m^{(-m+n-1)}, \\
& \qquad\qquad\qquad\qquad\qquad m = 1, 2, \ldots. \\
& \qquad\qquad\qquad\qquad\qquad n = 2, 3, \ldots. \\
e_m^{(-m+n)} &= \frac{q_{m+1}^{(-m+n-1)}}{q_m^{(-m+n)}} e_m^{(-m+n-1)},
\end{aligned}
$$

Usually the values $q_m^{(n)}$ and $e_m^{(n)}$ are arranged in a table where subscripts indicate columns and superscripts downward sloping diagonals and each continuation rule links four elements in a rhombus:

$$
\begin{array}{ccccccccc}
 & q_1^{(0)} & & q_2^{(-1)} & & q_3^{(-2)} & & \cdots & \\
e_0^{(1)} & & e_1^{(0)} & & e_2^{(-1)} & & e_3^{(-2)} & & \cdots \\
 & q_1^{(1)} & & q_2^{(0)} & & q_3^{(-1)} & & \cdots & \\
e_0^{(2)} & & e_1^{(1)} & & e_2^{(0)} & & e_3^{(-1)} & & \cdots \\
 & q_1^{(2)} & & q_2^{(1)} & & q_3^{(0)} & & \cdots & \\
e_0^{(3)} & & e_1^{(2)} & & e_2^{(1)} & & e_3^{(0)} & & \cdots \\
 & q_1^{(3)} & & q_2^{(2)} & & \ddots & & & \\
e_0^{(4)} & & e_1^{(3)} & & e_2^{(2)} & & & & \\
 & q_1^{(4)} & & q_2^{(3)} & & \ddots & & & \\
e_0^{(5)} & \vdots & e_1^{(4)} & \vdots & e_2^{(3)} & & & & \\
\vdots & & \vdots & & \vdots & & & &
\end{array}
$$

The following property can be found in [10].

Theorem 1 ([10, pp. 612–613]). *Let (2.1.1) be the Taylor series at $z = 0$ of a function f meromorphic in the disk $B(0, R) = \{z : |z| < R\}$ and let the poles z_i of f in $B(0, R)$ be numbered such that*

$$z_0 = 0 < |z_1| \leq |z_2| \leq \ldots < R,$$

each pole occuring as many times in the sequence $\{z_i\}_{i \in \mathbb{N}}$ as indicated by its order. If f is ultimately k-normal for some integer $k > 0$, then the qd-scheme associated with f has the following properties (put $z_{k+1} = \infty$ if f has only k poles):

(a) For each m with $0 < m \leq k$ and $|z_{m-1}| < |z_m| < |z_{m+1}|$,

$$\lim_{n \to \infty} q_m^{(n)} = z_m^{-1}. \tag{2.1.2}$$

(b) For each m with $0 < m \leq k$ and $|z_m| < |z_{m+1}|$,

$$\lim_{n \to \infty} e_m^{(n)} = 0. \tag{2.1.3}$$

Any index m such that the strict inequality

$$|z_m| < |z_{m+1}|$$

holds, is called a critical index. It is clear that the critical indices of a function do not depend on the order in which the poles of equal modulus are numbered. The theorem above states that if m is a critical index and f is ultimately m-normal, then

$$\lim_{n\to\infty} e_m^{(n)} = 0.$$

Thus the qd-table of a meromorphic function is divided into subtables by those e-columns tending to zero. This property motivated Rutishauser [13] to apply the rhombus rules satisfied by the q- and e-values, namely

$$\begin{aligned} q_m^{(n+1)} e_m^{(n+1)} &= e_m^{(n)} q_{m+1}^{(n)} \\ e_{m-1}^{(n+1)} + q_m^{(n+1)} &= q_m^{(n)} + e_m^{(n)} \end{aligned}$$

in their progressive form as formulated above. When computing the q-values from the top down rather than from left to right, one avoids divisions by possibly small e-values that can inflate rounding errors. Any q-column corresponding to a simple pole of isolated modulus is flanked by such e-columns and converges to the reciprocal of the corresponding pole. If a subtable contains $j > 1$ columns of q-values, the presence of j poles of equal modulus is indicated. In [10, p. 642] it is also explained how to determine these poles if $j > 1$.

Theorem 2 ([10, p. 642]). *Let m and $m+j$ with $j > 1$ be two consecutive critical indices and let f be $(m+j)$-normal. Let the polynomials $v_k^{(n)}$ be defined by*

$$\begin{aligned} v_0^{(n)}(z) &= 1, \\ v_{k+1}^{(n)}(z) &= z v_k^{(n+1)}(z) - q_{m+k+1}^{(n)} v_k^{(n)}(z), \qquad n \geq 0, \quad k = 0, 1, \ldots, j-1. \end{aligned}$$

Then there exists a subsequence $\{n(\ell)\}_{\ell\in\mathbb{N}}$ such that

$$\lim_{\ell\to\infty} v_j^{(n(\ell))}(z) = (z - z_{m+1}^{-1}) \cdots (z - z_{m+j}^{-1}).$$

The polynomials $v_k^{(n)}$ are closely related to the formally orthogonal Hadamard polynomials [10, p. 629] which can be defined as follows. Let $A_m^{(n)}$ denote the matrix

$$A_m^{(n)} = \begin{pmatrix} q_1^{(n)} + e_0^{(n)} & q_1^{(n)} e_1^{(n)} & & & 0 \\ 1 & q_2^{(n)} + e_1^{(n)} & q_2^{(n)} e_2^{(n)} & & \\ & \ddots & \ddots & \ddots & \\ & & 1 & q_{m-1}^{(n)} + e_{m-2}^{(n)} & q_{m-1}^{(n)} e_{m-1}^{(n)} \\ 0 & & & 1 & q_m^{(n)} + e_{m-1}^{(n)} \end{pmatrix}.$$

Then the Hadamard polynomials $p_m^{(n)}$ are given by

$$p_0^{(n)}(z) = 1, \qquad p_m^{(n)}(z) = \det(zI - A_m^{(n)}), \qquad m = 1, 2, \ldots, \tag{2.1.4}$$

and under the conditions of Theorem 1 they satisfy, for each critical index m,

$$\lim_{n\to\infty} p_m^{(n)} = (z - z_1^{-1})(z - z_2^{-1})\ldots(z - z_m^{-1}). \tag{2.1.5}$$

From the above theorems the *qd*-scheme seems to be an ingenious tool for determining, under certain conditions, the poles of a meromorphic function f directly from its Taylor series at the origin. This result is very much related to the convergence theorem of de Montessus de Ballore for Padé approximants of meromorphic functions [1, pp. 246–254]. The reason is that the q- and e-values appear in the partial numerators and denominators of the corresponding continued fraction for f

$$f(z) = c_0 + \sum_{i=1}^{\infty}\left(\frac{-q_i^{(1)}z|}{|1} + \frac{-e_i^{(1)}z|}{|1}\right). \tag{2.1.6}$$

The $(2m)^{th}$ convergent of (2.1.6) is the Padé approximant $r_{m,m}(z)$ for f of degree m in numerator and denominator. The separate treatment of poles of equal modulus is now easy to understand from the theorem of de Montessus de Ballore. Under the conditions of Theorem 1, the poles of the Padé approximants with denominator degree m are approximations for the poles z_i of the function f. Now the *qd*-algorithm computes approximations for the values $1/z_i$ which are, again in the limit, the zeroes of the Hadamard polynomials of degree m. Since convergence of Padé approximants is obtained on compact subsets of a disk of meromorphy, poles equidistant from the origin (or the point around which f is developed into a Taylor series) cannot be treated separately. Increasing the radius of the disk (in *qd*-terminology moving to the next critical index) includes all the next poles of equal modulus simultaneously.

The next lemma gives an explicit formula for the q- and e-values in terms of the determinants $H_m^{(n)}$ which apparently play a crucial role.

Lemma 1 ([10, pp. 610–613]). *Let f be given by its formal Taylor series expansion (2.1.1). If there exists a positive integer k such that f is k-normal, then the values $q_m^{(n)}$ and $e_m^{(n)}$ exist for $m = 1, \ldots, k$ and $n \geq 0$ and they are given by*

$$q_m^{(n)} = \frac{H_m^{(n+1)}H_{m-1}^{(n)}}{H_m^{(n)}H_{m-1}^{(n+1)}},$$

$$e_m^{(n)} = \frac{H_{m+1}^{(n)}H_{m-1}^{(n+1)}}{H_m^{(n+1)}H_m^{(n)}}.$$

2.2 The symbolic *qd*-algorithm.

By imposing another form than (2.1.6) for the corresponding continued fraction for f,

$$f(z) = c_0 + \sum_{i=1}^{\infty} \left(\frac{-Q_i^{(1)}(z)}{1 + Q_i^{(1)}(z)} \Big| + \frac{-E_i^{(1)}(z)}{1 + E_i^{(1)}(z)} \Big| \right) \tag{2.2.1}$$

we obtain slightly different computation rules for the $Q_m^{(n)}(z)$ and $E_m^{(n)}(z)$ which are now rational functions of z, *cf.* [7]:

(a) help entries $g_{0,m}^{(n)}$ are given by

$$g_{0,m}^{(n)} = -c_{n-m+1} z^{n-m+1}, \qquad m \geq 1, \quad n \geq 1, \tag{2.2.2}$$

$$g_{r,m}^{(n)} = \frac{g_{r-1,m}^{(n)} g_{r-1,r}^{(n+1)} - g_{r-1,m}^{(n+1)} g_{r-1,r}^{(n)}}{g_{r-1,r}^{(n+1)} - g_{r-1,r}^{(n)}}, \tag{2.2.3}$$
$$1 \leq r \leq m-1, m \geq 2, n \geq 1,$$

where the values $g_{r,m}^{(n)}$ are stored as in table 1,

(b) and the symbolic *qd*-like algorithm is defined by

$$Q_1^{(n)}(z) = \frac{c_{n+1}}{c_n} z \frac{g_{0,1}^{(n)}}{g_{0,1}^{(n)} - g_{0,1}^{(n+1)}}, \quad n \geq 1, \tag{2.2.4}$$

$$E_m^{(n)}(z) + 1 = \frac{g_{m-1,m}^{(n+m-1)} - g_{m-1,m}^{(n+m)}}{g_{m-1,m}^{(n+m-1)}} \left(Q_m^{(n+1)}(z) + 1 \right), m, n \geq 1, \tag{2.2.5}$$

$$Q_m^{(n)}(z) = \frac{E_{m-1}^{(n+1)}(z) Q_{m-1}^{(n+1)}(z)}{E_{m-1}^{(n)}(z)} \frac{g_{m-2,m-1}^{(n+m-2)} - g_{m-2,m-1}^{(n+m-1)}}{g_{m-2,m-1}^{(n+m-2)}} \times \tag{2.2.6}$$

$$\times \frac{g_{m-1,m}^{(n+m-1)}}{g_{m-1,m}^{(n+m-1)} - g_{m-1,m}^{(n+m)}}, \qquad m \geq 2, \qquad n \geq 1.$$

$$
\begin{array}{ccccccccccccc}
g_{0,1}^{(1)} & | & g_{0,2}^{(1)} & & | & | & g_{0,r}^{(1)} & & & | & | & g_{0,m}^{(1)} & \\
 & | & & g_{1,2}^{(1)} & | & | & & g_{1,r}^{(1)} & & | & | & & \ddots \\
g_{0,1}^{(2)} & | & g_{0,2}^{(2)} & & | & | & g_{0,r}^{(2)} & & \ddots & | & | & g_{0,m}^{(2)} & g_{m-1,m}^{(1)} \\
 & | & & g_{1,2}^{(2)} & | & | & & g_{1,r}^{(2)} & g_{r-1,r}^{(1)} & | & | & \vdots & \vdots \\
g_{0,1}^{(3)} & | & g_{0,2}^{(3)} & \vdots & | & \cdots | & g_{0,r}^{(3)} & \vdots & \vdots & | \cdots & | & & \\
\vdots & | & \vdots & & | & | & \vdots & & g_{r-1,r}^{(n+m-r+1)} & | & | & & g_{m-1,m}^{(n+1)} \\
 & | & & & | & | & & \iddots & & | & | & & \\
 & | & & g_{1,2}^{(n+m-1)} & | & | & & g_{1,r}^{(n+m-1)} & & | & | & \iddots & \\
g_{0,1}^{(n+m)} & | & g_{0,2}^{(n+m)} & & | & | & g_{0,r}^{(n+m)} & & & | & | & g_{0,m}^{(n+m)} &
\end{array}
$$

Table 1

The continued fraction (2.2.1) is especially interesting because it has the same form as the one underlying the general multivariate Padé approximation theory. The recursive computation rules (2.2.2)–(2.2.6) are no more than the univariate special case of the multivariate qd-like algorithm that was presented in [5]. The functions $Q_m^{(n)}(z)$ and $E_m^{(n)}(z)$ can be arranged in a table in the same way as the values $q_m^{(n)}$ and $e_m^{(n)}$ above. Symbolic or polynomial computation is necessary here because the $Q_m^{(n)}$ and $E_m^{(n)}$ in (2.2.1) are not merely coefficients anymore but rational functions. A progressive form of the computation rules (2.2.4)–(2.2.6) can be given but does not seem useful at first sight because (2.2.6) retains the rational function $Q_m^{(n)}(z)$ as the quotient of two polynomials. As can be deduced from the next lemma and as will be detailed in Section 4.2, a straightforward implementation of (2.2.4)–(2.2.6) involves a large number of greatest common divisor (GCD) computations and is not very economical. The implementation of (2.2.4)–(2.2.6) presented in Section 4 takes into account all possible simplifications that remain valid in the multivariate version of this qd-like algorithm. Further simplifications that only hold in the univariate case would of course bring us back to the classical progressive qd-algorithm, and that is not the issue here.

Let us now first generalize Lemma 1 and give explicit determinant representations for the rational expressions $Q_m^{(n)}(z)$ and $E_m^{(n)}(z)$. In addition to $H_m^{(n)}$ we define the determinants

$$
H_{1,0}^{(n)} = 1, \qquad H_{1,m}^{(n)}(z) = \begin{vmatrix} z^m & \cdots & 1 \\ c_n & \cdots & c_{n+m} \\ \vdots & & \vdots \\ c_{n+m-1} & \cdots & c_{n+2m-1} \end{vmatrix}, \quad m \geq 1, \tag{2.2.7}
$$

$$H_{3,0}^{(n)}(z) = z^{n-1}, \; H_{3,m}^{(n)}(z) = \begin{vmatrix} z^m & \dots & 1 \\ z^m \sum_{k=0}^{n-1} c_k z^k & \dots & \sum_{k=0}^{n+m-1} c_k z^k \\ c_n & \dots & c_{n+m} \\ \vdots & & \vdots \\ c_{n+m-2} & \dots & c_{n+2m-2} \end{vmatrix}, \quad m \geq 1, \quad (2.2.8)$$

and, for $s \geq 1$, $m \geq 2$,

$$H_m^{(n,s)} = \begin{vmatrix} c_{n-s} & \dots c_{n+m-1-s} \\ c_n & \dots & c_{n+m-1} \\ \vdots & & \vdots \\ c_{n+m-2} & \dots & c_{n+2m-3} \end{vmatrix} \quad (2.2.9)$$

$$= \begin{vmatrix} c_{n-s} & c_{n+1-s} & \dots & c_{n+m-1-s} \\ c_n & & & \\ \vdots & & H_{m-1}^{(n+1)} & \\ c_{n+m-2} & & & \end{vmatrix}.$$

It is easy to verify the following relations between the $H_{3,m}^{(n)}(z)$, the $H_m^{(n,s)}$ and the Hankel determinants $H_m^{(n)}$:

$$\begin{aligned} H_m^{(n,1)} &= H_m^{(n-1)}, \\ H_{3,m}^{(n)}(z) &= z^{n+2m-1} H_m^{(n)}. \end{aligned} \quad (2.2.10)$$

By means of Sylvester's and Schweins' determinant identities we can prove the following lemma.

Lemma 2 ([5]). *Let f be given by its formal Taylor series expansion (2.1.1). If there exists a positive integer k such that f is k-normal, then for $m = 1, \dots, k$ and $n \geq 0$ the following values exist and are given by:*

$$Q_m^{(n)}(z) = -\frac{H_m^{(n+1)} H_{1,m-1}^{(n)}(z) H_{3,m}^{(n)}(z)}{H_m^{(n)} H_{1,m}^{(n)}(z) H_{3,m-1}^{(n+1)}(z)}, \quad (2.2.11)$$

$$E_m^{(n)}(z) = -\frac{H_{m+1}^{(n)} H_{1,m-1}^{(n+1)}(z) H_{3,m}^{(n+1)}(z)}{H_m^{(n+1)} H_{1,m}^{(n+1)}(z) H_{3,m}^{(n)}(z)}, \quad (2.2.12)$$

$$g_{m-1,m}^{(n)}(z) = -z^n \frac{H_m^{(n-m+1)}}{H_{1,m-1}^{(n-m+2)}(z)}, \quad (2.2.13)$$

$$g_{r,m}^{(n)}(z) = -z^{n+1-m+r} \frac{H_{r+1}^{(n-r+1,m-r)}}{H_{1,r}^{(n-r+1)}(z)}, \quad 1 \leq r \leq m-1. \quad (2.2.14)$$

Let us now consider how the Theorems 1 and 2 generalize into one powerful theorem for the new qd-like algorithm (2.2.4)–(2.2.6). To this end we introduce the monomial

$$\tilde{E}_m^{(n)}(z) = -\frac{H_{m+1}^{(n)} H_{3,m}^{(n+1)}(z) H_{m-1}^{(n+2)}}{H_m^{(n+1)} H_{3,m}^{(n)}(z) H_m^{(n+2)}} \tag{2.2.15}$$

that can be extracted from $E_m^{(n)}(z)$. We do not simplify the expression (2.2.15) for $\tilde{E}_m^{(n)}(z)$ further using (2.2.10), because in the multivariate case this is the way to specify $\tilde{E}_m^{(n)}(z)$. We also draw the reader's attention to the denominator factor $H_{1,m}^{(n)}(z)$ in $Q_m^{(n)}(z)$.

Theorem 3 ([7]). *Let (2.1.1) be the Taylor series at $z = 0$ of a function f meromorphic in the disk $B(0, R) = \{z : |z| < R\}$ and let the poles z_i of f in $B(0, R)$ be numbered such that*

$$z_0 = 0 < |z_1| \leq |z_2| \leq \ldots < R,$$

each pole occuring as many times in the sequence $\{z_i\}_{i\in\mathbb{N}}$ as indicated by its order. If f is ultimately k-normal for some integer $k > 0$, then the symbolic qd-like scheme (2.2.4)–(2.2.6) associated with f has the following properties (put $z_{k+1} = \infty$ if f has only k poles):

(a) *For each m with $0 < m \leq k$ and $|z_m| < |z_{m+1}|$,*

$$\lim_{n\to\infty} \frac{H_{1,m}^{(n)}(z)}{H_m^{(n+1)}} = (z - z_1)\ldots(z - z_m).$$

(b) *For each m with $0 < m \leq k$ and $|z_m| < |z_{m+1}|$,*

$$\lim_{n\to\infty} \tilde{E}_m^{(n)}(z) = 0,$$

both uniformly on compact subsets of $B(0, |z_{m+1}|)$.

From the above it is clear that, when trying to decipher pole information from Taylor coefficients, one is in fact interested in the functions $H_{1,m}^{(n)}(z)/H_m^{(n+1)}$ each time $|z_m| < |z_{m+1}|$, or in other words, for each m for which $\tilde{E}_m^{(n)}(z) \to 0$. We therefore still call every index m for which $\tilde{E}_m^{(n)}(z) \to 0$ a critical index. The information contained in $1/q_m^{(n)}$ in the classical qd-algorithm is now contained in the polynomial $H_{1,m}^{(n)}(z)$, which can be found in the denominator of the symbolic expression $Q_m^{(n)}(z)$. Similarly, the information contained in $e_m^{(n)}$ in the classical *qd*-algorithm is contained in $\tilde{E}_m^{(n)}(z)$ in the symbolic *qd*-algorithm. In fact it is easy to verify that

$$\tilde{E}_m^{(n)}(z) = -\frac{e_m^{(n)}}{q_m^{(n+1)}} z.$$

It is also apparent that equimodular poles do not need a separate treatment such as in Theorem 2 anymore: they are automatically retrieved in $H_{1,m}^{(n)}(z)/H_m^{(n+1)}$ from one critical index to another.

§3 Rational function case

If f is a rational function, even more properties of the *qd*-algortihm can be proven than is the case for a meromorphic f. Assume f has numerator degree n and denominator degree $m \le n$, with finite poles $z_1, \ldots, z_m$ each pole occurring as many times as indicated by its multiplicity, just like above. Assume also that the series (2.1.1) is m-normal.

Theorem 4. *Let (2.1.1) be the Taylor series at $z = 0$ of a rational function of degree n in the numerator and $m \le n$ in the denominator. Then if the series f is m-normal,*

(a) all the Hadamard polynomials $p_m^{(n-m+h)}(z)$ are identical:

$$p_m^{(n-m+h)}(z) = (z - z_1^{-1})(z - z_2^{-1}) \ldots (z - z_m^{-1}), \qquad h > 0,$$

(b) and the mth e-column equals zero:

$$e_m^{(n-m+h)} = 0, \qquad h > 0,$$

thus terminating the computation of further q- and e-values.

The above theorem is again easy to understand when linking it to the theory of Padé approximants. For a rational function f of degree n in the numerator and m in the denominator, all Padé approximants $r_{n+h,m+h}(z)$ with $h \ge 0$ are equal [8, p. 68–72]. This property translates to the normalized Padé denominators of $r_{n+h,m}(z)$ being equal. As has already been hinted at in (2.1.5) and as will become clear from Lemma 3 and (4.1.4), there is a one-to-one link between the normalized Padé denominators and the Hadamard polynomials. Hence also all the Hadamard polynomials $p_m^{(n-m+h)}(z), h \ge 1$, are equal. Moreover, the consistency property of Padé approximants states that if f is a rational function of degree n in the numerator and m in the denominator, all $r_{n+h,m+h}(z) = f(z)$ [2, p. 30]. This then implies that the corresponding continued fraction (2.1.6) terminates, because it is a rational function, and hence that from a certain column on all q- and e-values are zero.

Since the polynomial $p_m^{(n-m+h)}(z)$, defined by (2.1.4), is the characteristic polynomial of the matrix $A_m^{(n-m+h)}$, the theorem above tells us that all the matrices $A_m^{(k)}$ with $k > n - m$, have the same eigenvalues. It is also easy to prove that the matrix $A_m^{(n-m+h)}$ has the same eigenvalues as the matrix $B_m^{(n-m+h)}$ where $B_m^{(n)}$ is defined by

$$B_m^{(n)} = \begin{pmatrix} q_1^{(n)} + e_0^{(n)} & -q_1^{(n)} & & & 0 \\ -e_1^{(n)} & q_2^{(n)} + e_1^{(n)} & -q_2^{(n)} & & \\ & \ddots & \ddots & \ddots & \\ & & -e_{m-2}^{(n)} & q_{m-1}^{(n)} + e_{m-2}^{(n)} & -q_{m-1}^{(n)} \\ 0 & & & -e_{m-1}^{(n)} & q_m^{(n)} + e_{m-1}^{(n)} \end{pmatrix}.$$

Hence the qd-algorithm is an ingenious way to compute the eigenvalues of tridiagonal matrices $A_m^{(n-m+h)}$ or $B_m^{(n-m+h)}$. For distinct eigenvalues for instance [13, pp. 466–469], all indices $j = 1, \dots, m-1$ are critical and hence $\lim_{h\to\infty} e_j^{(n-m+h)} = 0$ for $j = 1, \dots, m-1$. In that case the values $q_j^{(n-m+h)}$ converge to the zeroes $1/z_j$ of the polynomial in Theorem 4(a) and the eigenvalues can be read from the diagonal of $B_m^{(n-m+h)}$, for h sufficiently large.

In addition, if all q- and e-values in $A_m^{(n)}$ are positive, then computing the eigenvalues of $A_m^{(n)}$ is also equivalent to computing the eigenvalues of $T_m^{(n)}$ where

$$T_m^{(n)} = \begin{pmatrix} q_1^{(n)} + e_0^{(n)} & \sqrt{q_1^{(n)} e_1^{(n)}} & & & 0 \\ \sqrt{q_1^{(n)} e_1^{(n)}} & q_2^{(n)} + e_1^{(n)} & \sqrt{q_2^{(n)} e_2^{(n)}} & & \\ & \ddots & \ddots & \ddots & \\ & & & q_{m-1}^{(n)} + e_{m-2}^{(n)} & \sqrt{q_{m-1}^{(n)} e_{m-1}^{(n)}} \\ 0 & & & \sqrt{q_{m-1}^{(n)} e_{m-1}^{(n)}} & q_m^{(n)} + e_{m-1}^{(n)} \end{pmatrix}.$$

This is in turn equivalent to computing the singular values of the bidiagonal matrix

$$R_m^{(n)} = \begin{pmatrix} \sqrt{q_1^{(n)}} & \sqrt{e_1^{(n)}} & & & 0 \\ 0 & \sqrt{q_2^{(n)}} & \sqrt{e_2^{(n)}} & & \\ & \ddots & \ddots & \ddots & \\ & & & \sqrt{q_{m-1}^{(n)}} & \sqrt{e_{m-1}^{(n)}} \\ 0 & & & 0 & \sqrt{q_m^{(n)}} \end{pmatrix},$$

where $T_m^{(n)} = {R_m^{(n)}}^T R_m^{(n)}$ is the Cholesky decomposition of $T_m^{(n)}$.

Several variants of the progressive qd-algorithm have been developed in the framework of computing the eigenvalues of a tridiagonal matrix and the singular values and vectors of a bidiagonal matrix. In particular, we refer to the differential qd-algorithm [13, pp. 505–506], the qd-algorithm with shift [13, pp. 472–474], the stationary qd-algorithm [13, pp. 508–513] and the orthogonal qd-algorithm [9, 14]. The main aim of these variants is either to avoid the pitfalls of a floating-point implementation or to accelerate the convergence of the algorithm or both. These improved schemes, however, rely on the fact that the matrix elements satisfy certain properties (such as positivity). Such properties can not be assumed in general when using the qd-algorithm to locate the poles of a meromorphic function from its Taylor series expansion. The generalization of these variants will be the subject of a future investigation in the framework of applying multivariate qd-schemes to the solution of parameterized eigenvalue problems. For the moment there is no need to incorporate these improvements in the numeric-symbolic qd-algorithm which is implemented in the next section for the detection of pole curves of meromorphic functions.

§4 Towards a numeric-symbolic implementation

4.1 More symbolic computation rules.

Let us first exhibit the relation between the Hadamard polynomials from Section 2.1, that play an important role through Theorem 2, and the polynomials $H_{1,m}^{(n)}(z)/H_m^{(n+1)}$ from Section 2.2 that according to Theorem 3 contain valuable information.

Lemma 3. *The Hadamard polynomial $p_m^{(n)}(z)$ equals*

$$p_m^{(n)}(z) = (-1)^m z^m H_{1,m}^{(n)}(1/z)/H_m^{(n)}, \qquad m \geq 0, \qquad n \geq 0.$$

Proof: The proof is very easy if we make use of the fact that, as indicated in [10, p. 625], the Hadamard polynomials are monic and admit the determinant representation

$$p_0^{(n)} = 1,$$

$$p_m^{(n)}(z) = \frac{1}{H_m^{(n)}} \begin{vmatrix} c_n & \cdots & c_{n+m-1} & 1 \\ \vdots & & \vdots & z \\ c_{n+m} & \cdots & c_{n+2m-1} & z^m \end{vmatrix}, \quad n = 0,1,\ldots, m = 1,2,\ldots.$$

Rewriting this determinant formula as

$$\begin{aligned} p_m^{(n)}(z) &= \frac{(-1)^m z^m}{H_m^{(n)}} \begin{vmatrix} z^{-m} & c_n & \cdots & c_{n+m-1} \\ z^{-m+1} & & & \\ \vdots & \vdots & & \vdots \\ 1 & c_{n+m} & \cdots & c_{n+2m-1} \end{vmatrix} \\ &= \frac{(-1)^m z^m}{H_m^{(n)}} H_{1,m}^{(n)}(1/z) \end{aligned}$$

concludes the proof. ■

From Theorem 3 in Section 2.2 it is apparent that in the symbolic version of the *qd*-algorithm, one is mainly interested in the functions $H_{1,m}^{(n)}(z)$ (for pole information) and $\tilde{E}_m^{(n)}(z)$ (for critical index information). Extracting this information from the $Q_m^{(n)}(z)$ and $E_m^{(n)}(z)$ is only possible in exact polynomial arithmetic [6], [7], whereas here we are interested in obtaining this information also when dealing with inexact data. It can be observed from (2.2.13) that the pole information $H_{1,m}^{(n)}(z)$ can be directly found in the denominator of $g_{m,m+1}^{(n+m-1)}(z)$. In order to obtain the $\tilde{E}_m^{(n)}(z)$ with as little computational effort as possible, we introduce the functions

$$\Delta_m^{(n)}(z) = \frac{H_{m+1}^{(n)} H_{3,m}^{(n+1)}(z)}{H_{1,m}^{(n)}(z) H_{1,m}^{(n+1)}(z)}. \tag{4.1.1}$$

These satisfy a simpler recurrence relation than the $E_m^{(n)}(z)$.

Theorem 5. $\Delta_m^{(n)}(z)$ *satisfies the recurrence relation*

$$\Delta_0^{(n)}(z) = c_n z^n, \quad n \geq 1, \tag{4.1.2}$$

$$\Delta_m^{(n)}(z) = g_{m-1,m}^{(n+m)} \left(\frac{\Delta_{m-1}^{(n+1)}(z)}{g_{m-1,m}^{(n+m)} - g_{m-1,m}^{(n+m-1)}} - \frac{\Delta_{m-1}^{(n+2)}(z)}{g_{m-1,m}^{(n+m+1)} - g_{m-1,m}^{(n+m)}} \right), \quad n \geq 1, \quad m \geq 1. \qquad (4.1.3)$$

Proof: A proof can again be obtained through the link with Padé approximation theory. Under the condition of m-normality, we can give a determinant representation of the Padé approximants for $f(z)$ in terms of the determinants $H_{1,m}^{(n)}(z)$ and the determinants $H_{2,m}^{(n)}(z)$ defined by

$$H_{2,0}^{(n)} = \sum_{k=0}^{n-1} c_k z^k, \quad H_{2,m}^{(n)}(z) = \begin{vmatrix} z^m \sum\limits_{k=0}^{n-1} c_k z^k & \dots & \sum\limits_{k=0}^{n+m-1} c_k z^k \\ c_n & \dots & c_{n+m} \\ \vdots & & \vdots \\ c_{n+m-1} & \dots & c_{n+2m-1} \end{vmatrix}, \quad m \geq 1.$$

The Padé approximant $r_{n,m}(z)$ for f is given by [8]

$$r_{n,m}(z) = \frac{H_{2,m}^{(n-m+1)}(z)/H_m^{(n-m+2)}}{H_{1,m}^{(n-m+1)}(z)/H_m^{(n-m+2)}}. \qquad (4.1.4)$$

An application of Sylvester's determinant identity shows that

$$\Delta_m^{(n)}(z) = r_{n+m,m}(z) - r_{n+m-1,m}(z). \qquad (4.1.5)$$

Using the E-algorithm [4, 3] one can show that

$$r_{n,m}(z) = \frac{r_{n,m-1}(z) g_{m-1,m}^{(n+1)} - r_{n+1,m-1}(z) g_{m-1,m}^{(n)}}{g_{m-1,m}^{(n+1)} - g_{m-1,m}^{(n)}}.$$

Substituting this into (4.1.5) easily yields the desired recurrence relation for the $\Delta_m^{(n)}(z)$. ■

4.2 A numeric-symbolic GCD–free implementation.

The computation rules (2.2.2)–(2.2.3), and (4.1.2)–(4.1.3) will allow us to compute pole information and critical indices for a function f given by its Taylor series expansion. In order to retrieve this information from the rational functions $g_{m-1,m}^{(n)}$ and $\Delta_m^{(n)}$, the computed rational functions must be irreducible and normalized. Let us introduce the following notations to denote the numerator and denominator of the normalized irreducible rational functions $g_{r,m}^{(n)}$ and $\Delta_m^{(n)}$, where the normalization is such that the denominator polynomials of $g_{r,m}^{(n)}$ and $\Delta_m^{(n)}$ are monic:

$$\hat{g}_{r,m}^{(n)}(z) = -z^{n+1-m+r}\frac{H_{r+1}^{(n-r+1,m-r)}}{H_r^{(n-r+2)}}, \tag{4.2.1}$$

$$\overline{g}_{r,m}^{(n)}(z) = \frac{H_{1,r}^{(n-r+1)}(z)}{H_r^{(n-r+2)}}, \tag{4.2.2}$$

$$\hat{\Delta}_m^{(n)}(z) = \frac{H_{m+1}^{(n)}H_{3,m}^{(n+1)}(z)}{H_m^{(n+1)}H_m^{(n+2)}}, \tag{4.2.3}$$

$$\overline{\Delta}_m^{(n)}(z) = \frac{H_{1,m}^{(n)}(z)H_{1,m}^{(n+1)}(z)}{H_m^{(n+1)}H_m^{(n+2)}}. \tag{4.2.4}$$

The computation of pole and critical index information is reduced to computing $\hat{g}_{m-1,m}^{(n)}$, $\overline{g}_{m-1,m}^{(n)}$ and $\hat{\Delta}_m^{(n)}(z)$ as the following lemma indicates.

Lemma 4.

$$\frac{H_{1,m}^{(n)}(z)}{H_m^{(n+1)}} = \overline{g}_{m,m+1}^{(n+m-1)}(z),$$

$$\tilde{E}_m^{(n)}(z) = \frac{\hat{\Delta}_m^{(n)}(z)}{z^m\hat{g}_{m-1,m}^{(n+m-1)}(z)}.$$

Proof: The result is an obvious consequence of (4.2.1)–(4.2.3) and (2.2.15), where we have used (2.2.10). ■

Our aim is now to implement the recursive computation rules (2.2.2)–(2.2.3) for $g_{m-1,m}^{(n)}(z)$ and (4.1.2)–(4.1.3) for $\Delta_m^{(n)}(z)$ in such a way that the computed rational functions are, as required, irreducible and normalized. Moreover, this will be achieved without GCD computations, in contrast to a straightforward symbolic implementation of (2.2.2)–(2.2.6). Using the notations (4.2.1)–(4.2.4), it is easy to rewrite the recursive rule (2.2.3) as

$$\frac{\hat{g}_{r,m}^{(n)}}{\overline{g}_{r,m}^{(n)}} = \frac{\hat{g}_{r-1,m}^{(n)}\hat{g}_{r-1,r}^{(n+1)} - \hat{g}_{r-1,m}^{(n+1)}\hat{g}_{r-1,r}^{(n)}}{\hat{g}_{r-1,r}^{(n+1)}\overline{g}_{r-1,r}^{(n)} - \hat{g}_{r-1,r}^{(n)}\overline{g}_{r-1,r}^{(n+1)}}.$$

Let us point out that the left hand side of this equation is relatively prime while the right hand side is not. By applying Sylvester's identity to the numerator and denominator of the right hand side, one can verify that the GCD of this rational function is $\hat{g}_{r-2,r-1}^{(n+1)}/z$. Hence

$$\frac{\hat{g}_{r,m}^{(n)}}{\overline{g}_{r,m}^{(n)}} = \frac{z\left(\hat{g}_{r-1,m}^{(n)}\hat{g}_{r-1,r}^{(n+1)} - \hat{g}_{r-1,m}^{(n+1)}\hat{g}_{r-1,r}^{(n)}\right) \Big/ \hat{g}_{r-2,r-1}^{(n+1)}}{z\left(\hat{g}_{r-1,r}^{(n+1)}\overline{g}_{r-1,r}^{(n)} - \hat{g}_{r-1,r}^{(n)}\overline{g}_{r-1,r}^{(n+1)}\right) \Big/ \hat{g}_{r-2,r-1}^{(n+1)}}.$$

Let us also point out that in the left hand side the denominator $\overline{g}_{r,m}^{(n)}$ is monic while the denominator in the right hand side is not. We shall denote by $\gamma_r^{(n)}$ the highest degree coefficient of

$$z\left(\hat{g}_{r-1,r}^{(n+1)}\overline{g}_{r-1,r}^{(n)} - \hat{g}_{r-1,r}^{(n)}\overline{g}_{r-1,r}^{(n+1)}\right) \Big/ \hat{g}_{r-2,r-1}^{(n+1)}.$$

Then

$$\begin{aligned}(\hat{g}_{r,m}^{(n)}, \overline{g}_{r,m}^{(n)}) &= \texttt{Normalize}\left(\frac{z}{\hat{g}_{r-2,r-1}^{(n+1)}}\left(\hat{g}_{r-1,m}^{(n)}\hat{g}_{r-1,r}^{(n+1)} - \hat{g}_{r-1,m}^{(n+1)}\hat{g}_{r-1,r}^{(n)}\right),\right. \\ &\qquad \left.\frac{z}{\hat{g}_{r-2,r-1}^{(n+1)}}\left(\hat{g}_{r-1,r}^{(n+1)}\overline{g}_{r-1,r}^{(n)} - \hat{g}_{r-1,r}^{(n)}\overline{g}_{r-1,r}^{(n+1)}\right)\right). \qquad (4.2.5)\end{aligned}$$

The function `Normalize` is applied to make the denominator polynomial monic. In other words, it divides both its arguments by the highest degree coefficient of the second argument. In a similar way, we can deduce from (4.1.3) separate formulas for the numerator and denominator of $\Delta_m^{(n)}$. Starting from (4.1.3), we find

$$\begin{aligned}\frac{\hat{\Delta}_m^{(n)}}{\overline{\Delta}_m^{(n)}} &= \frac{\hat{\Delta}_{m-1}^{(n+1)}\overline{g}_{m,m+1}^{(n+m)}\dfrac{\hat{g}_{m-2,m-1}^{(n+m+1)}}{z}\gamma_m^{(n+m)} - \hat{\Delta}_{m-1}^{(n+2)}\overline{g}_{m,m+1}^{(n+m-1)}\dfrac{\hat{g}_{m-2,m-1}^{(n+m)}}{z}\gamma_m^{(n+m-1)}}{\overline{g}_{m-1,m}^{(n+m)}\overline{g}_{m,m+1}^{(n+m)}\dfrac{\hat{g}_{m-2,m-1}^{(n+m+1)}}{z}\overline{g}_{m,m+1}^{(n+m-1)}\dfrac{\hat{g}_{m-2,m-1}^{(n+m)}}{z}\gamma_m^{(n+m)}\gamma_m^{(n+m-1)}} \times \\ &\quad \times \hat{g}_{m-1,m}^{(n+m)}.\end{aligned}$$

Using (4.2.4), we can deduce that the GCD of the rational function in the right-hand side of the above expression is

$$\frac{\overline{g}_{m-1,m}^{(n+m)}\hat{g}_{m-2,m-1}^{(n+m+1)}\hat{g}_{m-2,m-1}^{(n+m)}\gamma_m^{(n+m)}\gamma_m^{(n+m-1)}}{z^2}$$

and hence

$$
\begin{aligned}
(\hat{\Delta}_m^{(n)}, \overline{\Delta}_m^{(n)}) = \texttt{Normalize}\Bigg(& \frac{z^2 \hat{g}_{m-1,m}^{(n+m)}}{\overline{g}_{m-1,m}^{(n+m)} \hat{g}_{m-2,m-1}^{(n+m+1)} \hat{g}_{m-2,m-1}^{(n+m)} \gamma_m^{(n+m)} \gamma_m^{(n+m-1)}} \times \\
& (\hat{\Delta}_{m-1}^{(n+1)} \overline{g}_{m,m+1}^{(n+m)} \frac{\hat{g}_{m-2,m-1}^{(n+m+1)}}{z} \gamma_m^{(n+m)} - \hat{\Delta}_{m-1}^{(n+2)} \overline{g}_{m,m+1}^{(n+m-1)} \frac{\hat{g}_{m-2,m-1}^{(n+m)}}{z} \gamma_m^{(n+m-1)}), \\
& \overline{g}_{m,m+1}^{(n+m)} \overline{g}_{m,m+1}^{(n+m-1)} \Bigg) \\
= \Bigg(& \frac{z^2 \hat{g}_{m-1,m}^{(n+m)}}{\overline{g}_{m-1,m}^{(n+m)} \hat{g}_{m-2,m-1}^{(n+m+1)} \hat{g}_{m-2,m-1}^{(n+m)} \gamma_m^{(n+m)} \gamma_m^{(n+m-1)}} \times \\
& \Big(\hat{\Delta}_{m-1}^{(n+1)} \overline{g}_{m,m+1}^{(n+m)} \frac{\hat{g}_{m-2,m-1}^{(n+m+1)}}{z} \gamma_m^{(n+m)} - \hat{\Delta}_{m-1}^{(n+2)} \overline{g}_{m,m+1}^{(n+m-1)} \frac{\hat{g}_{m-2,m-1}^{(n+m)}}{z} \gamma_m^{(n+m-1)}\Big), \\
& \overline{g}_{m,m+1}^{(n+m)} \overline{g}_{m,m+1}^{(n+m-1)} \Bigg).
\end{aligned}
\tag{4.2.6}
$$

The main outcome of implementing the computation rules (2.2.3) and (4.1.3) as (4.2.5) and (4.2.6) is that the computation of GCDs at each application of the rules (2.2.3) and (4.1.3), in order to obtain the irreducible forms $\hat{g}_{r,m}^{(n)}/\overline{g}_{r,m}^{(n)}$ and $\hat{\Delta}_m^{(n)}/\overline{\Delta}_m^{(n)}$, is replaced by mere polynomial divisions in (4.2.5) and (4.2.6). By the analysis above we have shown that we know beforehand what the GCDs are, from the use of some determinant identities. We cannot emphasize enough the reduction in computational complexity between a true GCD computation and a mere polynomial division, where it is moreover theoretically known that the polynomial remainder equals zero.

Let us take a closer look at the degrees and orders of the polynomials occurring in the right hand side of the equations (4.2.5) and (4.2.6).

Lemma 5. *Let f be given by its formal Taylor series expansion (2.1.1). If there exists a positive integer k such that f is k-normal, then for $m = 1, \ldots, k-1$ and $n \geq 0$ the following values exist and we have*

$$
\begin{aligned}
&\hat{\Delta}_m^{(n)} \text{ is a monomial of degree } n + 2m, \\
&\overline{\Delta}_m^{(n)} \text{ is a polynomial of degree } 2m, \\
&\hat{g}_{m-1,m}^{(n)} \text{ is a monomial of degree } n, \\
&\overline{g}_{m-1,m}^{(n)} \text{ is a polynomial of degree } m-1.
\end{aligned}
$$

Proof: The proof is an obvious consequence of the determinant formulas (2.2.7)–(2.2.9) and definition (4.2.1)–(4.2.4), where we have also used the fact that

$$H_{3,m}^{(n)}(z) = z^{n+2m-1} H_m^{(n)}. \quad \blacksquare$$

Information on the order and the degree of the polynomials occurring in the right-hand side of the computation rules (4.2.5) and (4.2.6) is important for error correction and validation when computing in floating-point polynomial arithmetic, as we shall indicate now. Since the univariate $\hat{g}^{(n)}_{m-1,m}$ are monomials of degree n, it appears that only one polynomial division must be applied in the above computation rules for $(\hat{\Delta}^{(n)}_m, \overline{\Delta}^{(n)}_m)$ and $(\hat{g}^{(n)}_{r,m}, \overline{g}^{(n)}_{r,m})$, namely the division by $\overline{g}^{(n+m)}_{m-1,m}$ for the computation of $\hat{\Delta}^{(n)}_m$. We shall now see that, moreover, this polynomial division further reduces to the simple division of highest degree coefficients. Using Lemma 5, it can easily be verified that the first argument of `Normalize` in (4.2.6) can be written as

$$\frac{\rho^{(n)}_m(z)}{\overline{g}^{(n+m)}_{m-1,m}(z)}, \tag{4.2.7}$$

where $\rho^{(n)}_m(z)$ is of the form

$$\rho^{(n)}_m(z) = \sigma^{(n)}_m z^{n+2m-1} \sum_{i=0}^{m} \tau_i z^i,$$

while, also according to Lemma 5, the expression (4.2.7) for $\hat{\Delta}^{(n)}_m$ has to reduce to a monomial of degree $n+2m$. Taking into account that $\overline{g}^{(n+m)}_{m-1,m}$ is a polynomial of degree $m-1$, this implies

$$\tau_0 = 0 \qquad \text{and} \qquad \overline{g}^{(n+m)}_{m-1,m}(z) | \sum_{i=1}^{m} \tau_i z^i, \tag{4.2.8}$$

and hence

$$\frac{\rho^{(n)}_m(z)}{\overline{g}^{(n+m)}_{m-1,m}(z)} = \sigma^{(n)}_m z^{n+2m-1} \frac{\sum_{i=1}^{m} \tau_i z^i}{\overline{g}^{(n+m)}_{m-1,m}(z)} = \sigma^{(n)}_m \tau_m z^{n+2m}. \tag{4.2.9}$$

In other words, the polynomial division by $\overline{g}^{(n+m)}_{m-1,m}(z)$ has reduced to determining the highest degree coefficient of a polynomial, where we have taken into account that $\overline{g}^{(n+m)}_{m-1,m}(z)$ is monic. We shall denote by `HDC` the function which, when given a polynomial as input, returns the highest degree coefficient of that polynomial.

Whereas (4.2.8) certainly holds when all computations are performed in exact polynomial arithmetic, this may no longer be the case when dealing with real-life numeric data or when performing the computations in floating-point polynomial arithmetic. Data and rounding errors may have as effect that

$$\tau_0 \neq 0 \qquad \text{or} \qquad \overline{g}^{(n+m)}_{m-1,m}(z) \not| \sum_{i=1}^{m} \tau_i z^i.$$

Applying (4.2.9) to compute $\hat{\Delta}^{(n)}_m$ then in fact amounts to approximating the computed polynomial $\sum_{i=0}^{m} \tau_i z^i$ by its best polynomial approximation of degree m and order one in the ℓ_1–sense, in other words setting $\tau_0 = 0$.

Taking the above into account, we obtain the following algorithm in floating-point polynomial arithmetic for the computation of the poles and critical indices of the function f.

`Given :`

$$f(z) = \sum_{i=0}^{\infty} c_i z^i$$

`Initialization :`

$$(\hat{g}^{(n)}_{0,1}, \overline{g}^{(n)}_{0,1}) = (-c_n z^n, 1)$$

$$\hat{\Delta}^{(n)}_0 = c_n z^n \qquad \texttt{n = 1,...,nmax}$$

`For m = 1,2, ...`

- `Initialization of` $(\hat{g}^{(n)}_{0,m+1}, \overline{g}^{(n)}_{0,m+1})$:

$$(\hat{g}^{(n)}_{0,m+1}, \overline{g}^{(n)}_{0,m+1}) = (-c_{n-m} z^{n-m}, 1) \qquad \texttt{n = 1,...,nmax}$$

- `Computation of` $\hat{g}^{(n)}_{m,m+1}$ `and` $\overline{g}^{(n)}_{m,m+1}$:

$$(\hat{g}^{(n)}_{r,m+1}, \overline{g}^{(n)}_{r,m+1}) =$$

$$\texttt{Normalize}\left(\frac{z}{\hat{g}^{(n+1)}_{r-2,r-1}} \left(\hat{g}^{(n)}_{r-1,m+1} \hat{g}^{(n+1)}_{r-1,r} - \hat{g}^{(n+1)}_{r-1,m+1} \hat{g}^{(n)}_{r-1,r} \right), \frac{z}{\hat{g}^{(n+1)}_{r-2,r-1}} \left(\hat{g}^{(n+1)}_{r-1,r} \overline{g}^{(n)}_{r-1,r} - \hat{g}^{(n)}_{r-1,r} \overline{g}^{(n+1)}_{r-1,r} \right) \right)$$

$$\gamma^{(n)}_r = \texttt{HDC}\left[\frac{z\left(\hat{g}^{(n+1)}_{r-1,r} \overline{g}^{(n)}_{r-1,r} - \hat{g}^{(n)}_{r-1,r} \overline{g}^{(n+1)}_{r-1,r} \right)}{\hat{g}^{(n+1)}_{r-2,r-1}} \right]$$

`r = 1,...,m and n = 1,...,nmax-r`

- `Computation of` $\hat{\Delta}_m^{(n)}$:

$$\hat{\Delta}_m^{(n)} = \frac{z^{2n+3m}\hat{g}_{m-1,m}^{(n+m)}}{\hat{g}_{m-2,m-1}^{(n+m+1)}\hat{g}_{m-2,m-1}^{(n+m)}\gamma_m^{(n+m)}\gamma_m^{(n+m-1)}} \times$$

$$\text{HDC}\left[\hat{\Delta}_{m-1}^{(n+1)}\overline{g}_{m,m+1}^{(n+m)}\frac{\hat{g}_{m-2,m-1}^{(n+m+1)}}{z}\gamma_m^{(n+m)} - \hat{\Delta}_{m-1}^{(n+2)}\overline{g}_{m,m+1}^{(n+m-1)}\frac{\hat{g}_{m-2,m-1}^{(n+m)}}{z}\gamma_m^{(n+m-1)}\right]$$

`n = 1,...,nmax-2m`

- `Critical index information:`

$$\tilde{E}_m^{(n)}(z) = \frac{\hat{\Delta}_m^{(n)}(z)}{z^m\hat{g}_{m-1,m}^{(n+m-1)}} \qquad \text{n=0,1,..., nmax-2m}$$

- `Pole information (if` m `is a critical index):`

$$\overline{g}_{m,m+1}^{(n+m-1)}(z) = \frac{H_{1,m}^{(n)}(z)}{H_m^{(n+1)}} \qquad \text{n=0,1,..., nmax-2m+1}$$

All essential ideas discussed here for the univariate case can be carried over to the multivariate case. The multivariate algorithm analogously consists of an initialization phase with terms from the multivariate Taylor series expansion, followed by the computation of the multivariate $\overline{g}_{m,m+1}^{(n)}$, $\hat{g}_{m,m+1}^{(n)}$ and $\hat{\Delta}_m^{(n)}$. The latter two give information on which indices are critical in the multivariate case, while the first gives information on the pole curves of the multivariate function if the index m is critical. The multivariate exposition of the algorithm is burdened by the fact that what are monomials in the univariate case, become multivariate polynomials of homogeneous degree in the multivariate case. Also, it is necessary in the multivariate case to work with the notion of partial degree, which indicates the degree of the multivariate polynomial per variable. A reader interested in the multivariate case is referred to [5] for the multivariate equivalent of all the determinants given here, while the actual multivariate algorithm will be described in a future paper.

§5 Numerical example

We conclude the paper by applying the classical *qd*-algorithm, the symbolic *qd*-algorithm and the numeric–symbolic *qd*-algorithm to compute the poles (with multiplicities) of the meromorphic function

$$f(z) = \frac{e^z}{(z-1)(z-2)(z+2)}$$

given by its Taylor series expansion.

`INPUT` classical *qd*-algorithm:

$$\text{floating} - \text{point representation of } (c_0, c_1, \ldots, c_{18})$$

`OUTPUT` :

$$\begin{aligned} q_1^{(17)} &= 0.1000004\text{E}{+}01 \\ e_1^{(16)} &= -0.3674957\text{E}{-}05 \\ q_2^{(15)} &= 0.4479084\text{E}{+}00 \\ e_2^{(14)} &= -0.1102231\text{E}{+}00 \\ q_3^{(13)} &= -0.5581391\text{E}{+}00 \\ e_3^{(12)} &= -0.3003665\text{E}{-}07 \end{aligned}$$

`CONCLUSION` (see Theorems 1 and 2):

$$\begin{aligned} z_1 &\approx \frac{1}{q_1^{(17)}} = 0.9999960\text{E}{+}00 \\ z_i^{-2} &- (q_2^{(14)} + q_3^{(13)}) z_i^{-1} + q_2^{(13)} q_3^{(13)} \approx 0, \qquad i = 2, 3 \\ z_2 &\approx 0.2000095\text{E}{+}01 \\ z_3 &\approx -0.2000032\text{E}{+}01 \end{aligned}$$

We now display the output of the symbolic *qd*-like algorithm (2.2.4)–(2.2.6), which relies on exact polynomial arithmetic and involves many GCD computations, for the same meromorphic function. Here we will denote by $\tilde{H}_{1,m}^{(n)}(z)$ the monic polynomial $H_{1,m}^{(n)}(z)/H_m^{(n+1)}$.

`INPUT` symbolic *qd*-like algorithm :

$$\sum_{i=0}^{18} c_i z^i \text{ with exact } c_i$$

`OUTPUT` :

$$\begin{aligned} Q_1^{(17)}(z) &= \frac{-z}{z - 0.9999960} \\ E_1^{(16)}(z) &= \frac{3.674943\text{E}{-}06 * z}{z - 0.9999960} \\ Q_2^{(15)}(z) &= \frac{-z * (z - 0.9999842)}{(z - 2.232566) * (z - 0.9999990)} \end{aligned}$$

$$\begin{aligned}
E_2^{(14)}(z) &= \frac{0.2460840\text{E}{+}00 * z * (z - 0.9999842)}{(z - 2.232566) * (z - 0.9999990)} \\
Q_3^{(13)}(z) &= \frac{-z * (z - 0.9999961) * (z - 2.232565)}{(z - 1.000000) * (z - 2.000000) * (z + 2.000000)} \\
E_3^{(12)}(z) &= \frac{-5.387212\text{E}{-}08 * z * (z - 0.9999961) * (z - 2.232565)}{(z - 1.000000) * (z - 2.000000) * (z + 2.000000)}
\end{aligned}$$

`CONCLUSION` (see Theorem 3):

$$\begin{aligned}
\tilde{E}_1^{(16)}(z) &= 3.674943\text{E}{-}06 * z \\
\tilde{H}_{1,1}^{(17)}(z) &= z - 0.9999960 \\
\tilde{E}_2^{(14)}(z) &= 0.2460840\text{E}{+}00 * z \\
\tilde{E}_3^{(12)}(z) &= -5.387212\text{E}{-}08 * z \\
\tilde{H}_{1,3}^{(13)}(z) &= (z - 1.000000) * (z - 2.000000) * (z + 2.000000)
\end{aligned}$$

It is important to emphasize that in the above algorithm all computations are performed in exact (rational) polynomial arithmetic. Only the final display of the results is done in floating-point format. This is in contrast to the new numeric–symbolic *qd*-algorithm presented in this paper, of which we display the output next.

`INPUT` numeric–symbolic *qd*-algorithm:

$$\text{floating} - \text{point representation of } (c_0, c_1, \ldots, c_{18})$$

`OUTPUT:`

$$\begin{aligned}
\frac{\hat{\Delta}_1^{(16)}}{z\hat{g}_{0,1}^{(16)}} &= \frac{3.32979\text{E}{-}07 * z^{18}}{0.90608 * z^{17}} \\
\bar{g}_{1,2}^{(17)} &= z - 0.9999960 \\
\frac{\hat{\Delta}_2^{(14)}}{z^2\hat{g}_{1,2}^{(15)}} &= \frac{3.27768\text{E}{-}06 * z^{18}}{1.33193\text{E-}05 * z^{17}} \\
\frac{\hat{\Delta}_3^{(12)}}{z^3\hat{g}_{2,3}^{(14)}} &= \frac{-7.06301\text{E}{-}13 * z^{18}}{1.31107\text{E}{-}05 * z^{17}} \\
\bar{g}_{3,4}^{(15)} &= 4.0000 - 4.0000z - z^2 + z^3
\end{aligned}$$

`CONCLUSION` (see Lemmas 4 and 5):

$$\tilde{E}_1^{(16)}(z) = 3.674940\text{E}{-}06 * z$$

$$\begin{aligned}
\tilde{H}_{1,1}^{(17)}(z) &= z - 0.9999960 \\
\tilde{E}_2^{(14)}(z) &= 0.2460840\text{E}{+}00 * z \\
\tilde{E}_3^{(12)}(z) &= -5.387212\text{E}{-}08 * z \\
\tilde{H}_{1,3}^{(13)}(z) &= 4.0000 - 4.0000 * z - z^2 + z^3
\end{aligned}$$

References

[1] Baker, G. A. and P. Graves-Morris, *Padé Approximants Part I: Basic Theory*, Encycl. Math. 13, Addison Wesley, London, 1981.

[2] Baker, G. A. Jr. and P. Graves-Morris, *Padé Approximants (2nd Ed.)*, Cambridge University Press, 1996.

[3] Brezinski, C., A general extrapolation algorithm, *Numer. Math.* **35** (1980), 175–187.

[4] Cuyt, A., A recursive computation scheme for multivariate rational interpolants, *SIAM J. Numer. Anal.* **24** (1987), 228–238.

[5] Cuyt, A., A multivariate qd-like algorithm, *BIT* **28(86-15)** (1988), 98–112.

[6] Cuyt, A., Floating-point versus symbolic computations in a multivariate *qd*-like algorithm, *Communications Anal. Th. Cont. Fr.* **IV** (1995), 84–90.

[7] Cuyt, A., Floating-point versus symbolic computations in the qd-algorithm, *J. Symbolic Computation* **24** (1997), 695–703.

[8] Cuyt, A. and L. Wuytack, *Nonlinear Methods in Numerical Analysis*, North-Holland, Amsterdam, 1987.

[9] Vince Fernando, K. and B. N. Parlett, Accurate singular values and differential *qd*-algorithms, *Numer. Math.*, **67** (1994), 191–229.

[10] Henrici, P., *Applied and Computational Complex Analysis I*, John Wiley, New York, 1974.

[11] Kravanja, P. and A. Haegemans, Computing zeros of analytic functions via the qd-algorithm, Technical Report TW 233, KU Leuven, 1995.

[12] Perron, O., *Die Lehre von den Kettenbrüchen II*, Teubner, Stuttgart, 1977.

[13] Rutishauser, H., *Lectures on Numerical Mathematics*, Birkhäuser, Boston, 1990.

[14] Von Matt, U., The orthogonal QD-algorithm, *SIAM J. Sci. Statist. Comput.* **18** (1997), 1163–1186.

Annie Cuyt
Research Director FWO–Vlaanderen
Department of Mathematics and Computer Science
Universiteit Antwerpen (UIA)
Universiteitsplein 1
B–2610 Wilrijk
BELGIUM
email: cuyt@uia.ua.ac.be

Brigitte Verdonk
Postdoctoral Fellow FWO–Vlaanderen
Department of Mathematics and Computer Science
Universiteit Antwerpen (UIA)
Universiteitsplein 1
B–2610 Wilrijk
BELGIUM
email: verdonk@uia.ua.ac.be

M. A. Lavrentyev's problems on pointwise polynomial approximation and related questions

Arthur A. Danielyan

Dedicated to the memory of Professor Lee A. Rubel

Abstract. A function $f(z)$ defined in a domain D of the complex plane is called a limit function, if there exists a polynomial sequence converging to $f(z)$ everywhere in D. The set E of singular points of the limit function is nowhere dense in D. In 1936 M. A. Lavrentyev posed a problem on the mutual connection between the values of a limit function in different domains complementary to E. The main purpose of the article is to give a complete solution to this problem.

§1 Introduction

The problems of pointwise polynomial approximation in the complex plane were first studied by Montel, Osgood, Farrell, Hartogs, Rosenthal, Lavrentyev, Keldysh, and Mergelyan (*cf.* [6]–[7]).

Following Lavrentiev [6], we call a function $f(z)$ defined in a domain D a limit function if there exists a polynomial sequence converging to $f(z)$ everywhere in D. It is well-known that the set E of singular points of a limit function is nowhere dense in D (Osgood). In the general case, the set E can have a complicated character; for example, each portion of E can separate the plane. For this reason Lavrentyev [6] posed in 1936 a problem on the mutual connection between the values of limit functions in different components of the complement of E. A similar question (also mentioned by Lavrentyev [6]) for limit functions of harmonic polynomials remained open as well. We give complete solutions to these problems and also consider a few related questions.

Let us recall the scheme of construction of a set in the complex plane $\mathbb{C}$ given by Lavrentyev [6].

Let C be a circle of radius r and z_0 be a fixed point on C. Let z_1 be the midpoint of the arc $C \setminus \{z_0\}$, and C_1 be the circle of radius $r/2$, outside C which is tangent to C at the point z_1. Let z_2, z_3 be the midpoints of the two arcs $C \setminus \{z_0, z_1\}$ and C_2, C_3 be circles of radii $r/2^2$, which lie outside C and which are tangent to C at the points z_2 and z_3, respectively. Continuing this process indefinitely (taking circles of radii $r/2^k$ on the step k), we shall get a certain sequence of circles $\{C_n\}$ tangent to C. Evidently $\{C_n\}$ is completely defined by the start circle C and its point z_0.

Computational Methods and Function Theory (CMFT'97)
N. Papamichael, St. Ruscheweyh and E. B. Saff (Eds.), pp. 161–170

ISBN 981-02-3626-3

Definition 1. *$\{C_n\}$ is called a standard sequence of circles, corresponding to a given circle C and its point z_0.*

Let $C_0 = \Gamma_0$ be the unit circle $|z| = 1$. Consider the standard sequence of circles corresponding to C_0 and its point $z_0 = 1$. Let Γ_1 be the union of all these new circles of the standard sequence. For each circle C from Γ_1 consider the standard sequence of circles corresponding to C and its point at which C is tangent to C_0. Let Γ_2 be the union of all new circles arising this way. Let us consider for each circle C from Γ_2 the standard sequence of circles corresponding to C and its point at which C is tangent to a circle from Γ_1. Let Γ_3 be the union of all the circles obtained this way. By induction one can get the set Γ_k for each natural number k. Set $\Gamma = \bigcup_{k=0}^{\infty} \Gamma_k$. The closure $\bar{\Gamma}$ of this set is the desired set of Lavrentyev. $\bar{\Gamma}$ coincides with the boundary of a domain containing ∞, and $\bar{\Gamma}$ is a set of singular points for limit functions [6]. Evidently the set of bounded components for the complement of $\bar{\Gamma}$ coincides with the collection of all open disks bounded by circles from Γ. Every neighbourhood of each point of $\bar{\Gamma}$ contains circles from Γ (that is, each portion of $\bar{\Gamma}$ separates the plane).

The following problem was posed by Lavrentyev [6], §19:

Let $\{D_{n_k}\}$ and $\{D_{m_k}\}$ be two infinite subsequences of bounded components of the complement of $E = \bar{\Gamma}$ and such that every neighbourhood of each point of E contains domains from both subsequences. Let two limit functions f_1 and f_2 be defined in a disk U containing E, which are regular on the bounded components of the complement of E and coincide in each domain from $\{D_{n_k}\}$. Do there exist, under these conditions, domains from $\{D_{m_k}\}$, where f_1 and f_2 coincide identically?

Below, we give a negative answer to this question (Corollary 1). We also give a similar solution for the corresponding problem for *real limit functions of harmonic polynomials* mentioned as well by Lavrentyev [6], after Theorem 28.

For this purpose we first prove a topological lemma, which gives a Baire first class function $f(z)$ on the set of Lavrentyev. Then we use the characterization (obtained in [1]) of functions representable on an arbitrary compact set with connected complement as a limit of converging and uniformly bounded sequences of polynomials. This shows that $f(z)$ is a limit function (on the set of Lavrentyev). To complete the proof we need to use results from the theory of uniform approximation by polynomials and here we refer to the general theorem of Mergelyan.

Note that Corollary 2 of the main result follows directly from Lemma 1 and the theorem of Walsh-Lebesgue on uniform approximation by harmonic polynomials. Thus the proof of Lemma 1 is, in fact, enough to resolve the question of Lavrentyev in the case of harmonic functions.

§2 Statement of results

In the following main theorem we use the notations given in the problem of Lavrentyev formulated above.

Theorem 1. *There exists a function $f(z)$ defined in the disk U such that the following properties hold:*

1. *Everywhere in U the function $f(z)$ is the limit of a converging sequence of polynomials;*
2. *$f(z)$ is regular in the complement of $\bar{\Gamma}$ ($\bar{\Gamma}$ is a set of singular points of $f(z)$);*
3. *$f(z)$ equals zero in each domain from $\{D_{n_k}\}$ and is different from zero in each domain from $\{D_{m_k}\}$.*

Taking $f_1(z) = f(z)$ and $f_2(z) = 2f(z)$ we get the solution of the problem of Lavrentyev, that is:

Corollary 1. *There exist two limit functions $f_1(z)$ and $f_2(z)$ defined in U, which are regular in the complement of $\bar{\Gamma}$ ($\bar{\Gamma}$ is the set of singular points for $f_1(z)$ and $f_2(z)$), coincide in each domain from $\{D_{n_k}\}$ and have different values everywhere in each domain from $\{D_{m_k}\}$.*

Note that $f_1(z)$ and $f_2(z)$ have different values everywhere in each domain from $\{D_{m_k}\}$, which is even more than the requirement of the problem (not to be equal identically in each domain from $\{D_{m_k}\}$).

Next, because the values of $f(z)$ are real (see the proof of Theorem 1 below), from Corollary 1 we have:

Corollary 2. *There exist two real limit functions (for harmonic polynomials) $f_1(x,y)$ and $f_2(x,y)$ defined in U, which are harmonic in the complement of $\bar{\Gamma}$ ($\bar{\Gamma}$ is the set of singular points for $f_1(x,y)$ and $f_2(x,y)$), coincide in each domain from $\{D_{n_k}\}$ and have different values everywhere in each domain from $\{D_{m_k}\}$.*

§3 Proofs

We need some notation concerning Lavrentyev's set $\bar{\Gamma}$. Let C be a circle from Γ_k ($k \geq 1$). We shall denote by $\Gamma_{k+1}(C)$ all the circles from Γ_{k+1} that are tangent to C, and by $\Gamma_{k+2}(C)$ all the circles from Γ_{k+2} that are tangent to a circle from $\Gamma_{k+1}(C)$; similarly, one can define $\Gamma_m(C)$ for every $m > k$. The set $\bigcup_{m=k+1}^{\infty} \Gamma_m(C)$ is the collection of all circles "arising" from C.

We start our proofs by establishing the following lemma.

Lemma 1. *There exists a Baire first class function $f(z)$ defined on the set $\bar{\Gamma}$, such that:*

1. $0 \leq f(z) \leq 1$ *on* $\bar{\Gamma}$*;*

2. *on* ∂D_{n_k} $(k = 1, 2, 3, \ldots)$ $f(z) = 0$ *except, perhaps, for a countable set;*

3. *on* ∂D_{m_k} $(k = 1, 2, 3, \ldots)$ $f(z) = r_k > 0$ *except, perhaps, for one point* (r_k *is the radius of* D_{m_k}*).*

Proof: If there exist bounded components (disks) of the complement of $\bar{\Gamma}$ which belong neither to $\{D_{n_k}\}$ nor to $\{D_{m_k}\}$, we include without loss of generality such disks in $\{D_{m_k}\}$. Let us define a function $f(z)$ on $\bar{\Gamma}$ as follows.

If the domain bounded by the circle $\Gamma_0 = C_0$ is in the sequence $\{D_{m_k}\}$, we put $f(z)$ equal to the radius of C_0, *i.e.* 1. Otherwise we put $f(z) = 0$ at all points of C_0 that are not points of tangency with circles from Γ_1, (that is $f(z) = 0$ on $C_0 \setminus \Gamma_1$). At those points of $C_0 \cap \Gamma_1$ that are points of tangency with circles from $\{D_{n_k}\}$, we put again $f(z) = 0$. At each of the remaining points of the set $C_0 \cap \Gamma_1$ we define the value of $f(z)$ to be equal to the radius of the circle from Γ_1 that is tangent to C_0 at that point. Thus, on $\Gamma_0 = C_0$ the function $f(z)$ is defined.

Let $f(z)$ be defined on $\Gamma_0, \Gamma_1, \ldots, \Gamma_k$ (*i.e.* on $\bigcup_{j=0}^{k} \Gamma_j$), we now define $f(z)$ on Γ_{k+1}. It is enough to define $f(z)$ on an arbitrary circle C from Γ_{k+1}. At a point t of C that is a point tangency of C with a corresponding circle from Γ_k, the value of the function is already defined. If the domain bounded by C is in the sequence $\{D_{m_k}\}$, on $C \setminus \{t\}$ we put $f(z)$ equal to the radius of C. Otherwise we define $f(z) = 0$ at all points from $C \setminus \{t\}$ that are not points of tangency with circles from Γ_{k+2}. At those points of $C \cap \Gamma_{k+2}$ which are points of tangency with circles from $\{D_{n_k}\}$ we put again $f(z) = 0$. And at each of the remaining points of the set $C \cap \Gamma_{k+2}$ we define the value of $f(z)$ to be equal to the radius of the circle from Γ_{k+2} tangent to C at that point. Thus $f(z)$ is defined on C. So $f(z)$ is defined on $\Gamma = \bigcup_{k=0}^{\infty} \Gamma_k$. We complete the definition of $f(z)$ on $\bar{\Gamma}$ by taking $f(z) = 0$ on $\bar{\Gamma} \setminus \Gamma$.

Evidently, conditions 1–3 of Lemma 1 are fulfilled. Also, $0 \leq f(z) \leq 2^{1-k}$, $z \in \Gamma_k$ $(k \geq 1)$. It follows from this inequality that $f(z)$ is continuous at each point of $\bar{\Gamma} \setminus \Gamma$. By construction of $f(z)$, it is not difficult to see that $f(z)$ is also continuous at each point t of Γ at which the value of the function is 0.

Let us prove that $f(z)$ belongs to Baire first class (on $\bar{\Gamma}$). By the theorem of Baire it is enough to show that for an arbitrary non-empty closed subset F of $\bar{\Gamma}$ the restriction $f|_F$ of f to F has a point of continuity.

Let the closed set F be such that $\Gamma \cap F \neq \emptyset$ (the case $\Gamma \cap F = \emptyset$ is trivial, because then we will have $F \subset \bar{\Gamma} \setminus \Gamma$). There exists a smallest number $k_1 \geq 0$, such that $\Gamma_{k_1} \cap F \neq \emptyset$. Let C_1 be a circle from Γ_{k_1}, such that $C_1 \cap F \neq \emptyset$ (in particular, C_1 may coincide with the single circle C_0 from Γ_0).

For the circle C_1, consider the set $\Gamma_k(C_1)$, $k > k_1$ (see the notation above). If there exists $k > k_1$ such that $\Gamma_k(C_1) \cap (F \setminus C_1) \neq \emptyset$, we denote by k_2 the smallest of these numbers; $k_2 > k_1$. Let C_2 be a circle from $\Gamma_{k_2}(C_1)$ such that $C_2 \cap (F \setminus C_1) \neq \emptyset$.

We do the same also for C_2. Generally, suppose that the circle C_n from $\Gamma_{k_n}(C_{n-1})$ ($k_n > k_{n-1}$) with $C_n \cap (F \setminus C_{n-1}) \neq \emptyset$ has been found and consider the set $\Gamma_k(C_n)$, $k > k_n$. If there exists $k > k_n$ such that $\Gamma_k(C_n) \cap (F \setminus C_n) \neq \emptyset$, let k_{n+1} be the smallest of these numbers; $k_{n+1} > k_n$. Let C_{n+1} be a circle from $\Gamma_{k_{n+1}}(C_n)$ such that $C_{n+1} \cap (F \setminus C_n) \neq \emptyset$.

If one can continue this process indefinitely, one will have a sequence of circles $\{C_n\}$. It is clear by construction of $\{C_n\}$ that one can "connect" in a unique manner C_n and C_{n+1} with the help of a finite number of intermediate circles (if C_n and C_{n+1} are tangent, then they are already "connected"). As a result we get an infinite chain of circles from Γ which has C_1 as a first circle and in which there exists exactly one circle from each Γ_k ($k \geq k_1$). Evidently, the sequence of centers of these circles converges to some point t_0. Note that $t_0 \notin \Gamma$, because otherwise we will have contradicted the fact that $\bar{\Gamma}$ is the boundary of a domain containing ∞. Thus $t_0 \in \bar{\Gamma} \setminus \Gamma$, and hence t_0 is a point of continuity of $f(z)$ (on $\bar{\Gamma}$). On the other hand each circumference from $\{C_n\}$ has a non-empty intersection with F, that is $t_0 \in F$. Hence t_0 is a point of continuity for $f|_F$.

If the above mentioned process of construction of circles terminates after a finite number of steps, that is if we have circles $C_1, \ldots, C_n$, but C_{n+1} does not exist, then (in case $n > 1$) this means the following: for C_n from $\Gamma_{k_n}(C_{n-1})$, $k_n > k_{n-1}$, we have

$$C_n \cap (F \setminus C_{n-1}) \neq \emptyset \tag{3.1}$$

$$\Gamma_k(C_n) \cap (F \setminus C_n) = \emptyset \quad \text{for all } k > k_n. \tag{3.2}$$

In the case $n = 1$ the circle C_2 already does not exist, which means that for C_1 from Γ_{k_1} we have $C_1 \cap F \neq \emptyset$, as well as the relation (3.2) with $n = 1$.

If there exists $z \in F \cap (\bar{\Gamma} \setminus \Gamma)$, then z is a point of continuity of $f|_F$. Hence one needs only consider the case when $F \subset \Gamma$. From the last relation and (3.2), follows that

$$F \setminus C_n \subset \Gamma \setminus \bigcup_{k=k_n+1}^{\infty} \Gamma_k(C_n). \tag{3.3}$$

We now consider the possible cases when C_{n-1} and C_n are tangent, or not tangent, as well as the case $n = 1$.

1) Let C_{n-1} and C_n be tangent to each other $(n > 1)$. Denote by τ the point of tangency of C_{n-1} and C_n. By (3.1) there exists a point t such that $t \in C_n$, $t \in F$ and $t \notin C_{n-1}$. Thus, there exists a point $t \in F$ such that $t \in C_n \setminus \{\tau\}$.

Because the set $\bar{\Gamma}$ is the boundary of a domain containing ∞, it is not difficult to see that each point of the set $C_n \setminus \{\tau\}$ can be a limit point only for those points of the set Γ which belong to $\bigcup_{k=k_n+1}^{\infty} \Gamma_k(C_n)$. In other words, points of $C_n \setminus \{\tau\}$ are not limit points for $\Gamma \setminus \bigcup_{k=k_n+1}^{\infty} \Gamma_k(C_n)$. This assertion, together with (3.3), means that points of the set $(C_n \setminus \{\tau\}) \cap F$ (and in particular t) are not limit points for $F \setminus C_n$.

If the disk bounded by C_n is in $\{D_{m_k}\}$, then $f(z)$ is constant on $C_n \setminus \{\tau\}$ and hence t is a point of continuity for $f|_{C_n \cap F}$. Because t is not a limit point for $F \setminus C_n$, the point t is a point of continuity also for $f|_F$.

If the disk bounded by C_n is in $\{D_{n_k}\}$, then there exists a closed arc $\gamma \subset C_n$ with endpoints τ_1 and τ_2, such that $t \in \gamma$, $\tau \notin \gamma$ and $f(\tau_1) = f(\tau_2) = 0$. If the set $\gamma \cap F \neq \emptyset$ is perfect, then it must contain some point z_1 such that $f(z_1) = 0$ (on C_n the function $f(z)$ equals to 0 except perhaps for a countable set and $\gamma \cap F$ has cardinality of the continuum). The point z_1 is a point of continuity for $f(z)$ (on $\bar{\Gamma}$) and hence also for $f|_F$. In case $\gamma \cap F$ is not a perfect set, it has an isolated point z_2 and evidently z_2 is a point of continuity for $f|_{\gamma \cap F}$. If in addition $f(z_2) = 0$ (in particular, if z_2 coincides with τ_1 or τ_2) we will have again that z_2 is a point of continuity for $f|_F$. Otherwise $f(z_2) \neq 0$ and z_2 is an inner point of the arc γ. Because z_2 is not a limit point for $F \setminus C_n$ in a certain (circular) neighbourhood of z_2 the function $f|_F$ coincides identically with $f|_{\gamma \cap F}$. This means that z_2 is a point of continuity also for $f|_F$.

2) Now suppose C_{n-1} and C_n are not tangent to each other $(n > 1)$.

Evidently $C_n \cap F = C_n \cap (F \setminus C_{n-1})$, and by (3.1) there exists $t \in C_n \cap F$. Let C' be a circle from $\Gamma_{k_n - 1}(C_{n-1})$ tangent to C_n, and τ the point of tangency. By definition of the number k_n the relation $\tau \in F$ is impossible. Hence $\tau \notin F$, $C_n \cap F \subset C_n \setminus \{\tau\}$ and, in particular, $t \in C_n \setminus \{\tau\}$.

Thus it remains to repeat the argument of case 1). Owing to the condition $C_n \cap F \subset C_n \setminus \{\tau\}$ the final part of the proof (when we assumed that the disk bounded by C_n is in $\{D_{n_k}\}$) becomes even simpler here.

3) Finally, let us consider the case $n = 1$, which only formally differs from previous cases.

We have C_1 from Γ_{k_1} such that $C_1 \cap F \neq \emptyset$, and we have the relation (3.2) for $n = 1$. Let C' be the circle from $\Gamma_{k_1 - 1}$ tangent to C_1, and τ be the point of tangency. As above, $\tau \notin F$, $C_1 \cap F \subset C_1 \setminus \{\tau\}$ and points of the set $(C_1 \setminus \{\tau\}) \cap F = C_1 \cap F$ are not limit points for $F \setminus C_1$. Hence one can prove the existence of a point of continuity for the function $f|_F$ as in case 2). In the trivial case when C_1 coincides with $\Gamma_0 = C_0$ (then a circle C' does not exist) we will have $F \neq \emptyset$, $F \subset C_0$ and, evidently, there exists a point of continuity for $f|_F$ (as above, we need only to distinguish the cases when F is perfect or has an isolated point).

Lemma 1 is proved. ■

Proof of Theorem 1: If there exist bounded components (disks) of the complement of $\bar{\Gamma}$ which are neither in $\{D_{n_k}\}$ nor in $\{D_{m_k}\}$, as above, we include such disks in $\{D_{m_k}\}$. Set $K = \bar{\Gamma} \cup \bigcup_{n=1}^{\infty} D_n$, where $\{D_n\}$ is the sequence of all disks in the complement of $\bar{\Gamma}$. Then K is a compact set whose complement is connected. Evidently, there exist Jordan domains G_n with rectifable boundaries, and such that $K \subset G_{n+1} \subset \bar{G}_{n+1} \subset G_n \subset \bar{G}_n \subset U$, $\bigcap_{n=1}^{\infty} G_n = K$.

Let $H_n = \bar{U} \setminus G_n$, $B_n = \{z = x + iy \in H_n : x > 0, 0 < y < 1/n\}$ and $L_n = H_n \setminus B_n$. Clearly $K \cap L_n = \emptyset$ and $K \cup \bigcup_{n=1}^{\infty} L_n = \bar{U}$. Also L_n and $K \cup L_n$ are compact sets, each with connected complement.

Define the function $f(z)$ on $\bar{U}$ as follows. On $\bar{\Gamma}$ it is just the function given in Lemma 1. In each domain D_{m_k} of $\{D_{m_k}\}$ we put $f(z)$ equal to the radius of the disk D_{m_k}. In each domain of $\{D_{n_k}\}$ we put $f(z) = 0$. Thus, $f(z)$ is defined on K. Put $f(z) = -1$ on $\bar{U} \setminus K$. Clearly, $\bar{\Gamma}$ is the set of singular points of $f(z)$. Note that the values of $f(z)$ are real.

The function $f(z)$ is bounded on K, analytic in $\bigcup_{n=1}^{\infty} D_n$, belongs to Baire first class on $\bar{\Gamma}$, and in each domain D_n the boundary values of $f(z)$ coincide with its values on ∂D_n almost everywhere. By Theorem 2 of [1] these conditions are necessary and sufficient for the representation of $f(z)$ as a limit (everywhere on K) of a uniformly bounded on K sequence of polynomials $\{Q_n(z)\}$.

By the theorem of Mergelyan on uniform approximation by polynomials, for each n there exists a polynomial $P_n(z)$ such that

$$|f(z) - P_n(z)| < \frac{1}{n} \quad \text{on } L_n \qquad \text{and} \qquad |Q_n(z) - P_n(z)| < \frac{1}{n} \quad \text{on } K.$$

Thus, $\{P_n(z)\}$ converges to $f(z)$ everywhere in U, and Theorem 1 is proved. ■

Note that $\{P_n(z)\}$ is uniformly bounded on $\bar{\Gamma}$.

Now we give a new direct proof of Corollary 2.

Proof of Corollary 2: Let $f(x, y) = f(z)$, $(z = x + iy)$ be the function defined in the beginning of the proof of Theorem 1, and L_n be the set defined there. By Lemma 1, $f(x, y)$ is a bounded Baire first class function on $\bar{\Gamma}$, and hence there exists a uniformly bounded on Γ sequence of continuous real functions $\{f_n(x,y)\}$ converging to $f(x,y)$ everywhere on $\bar{\Gamma}$. By the theorem of Walsh-Lebesgue (see [4]) there exists a harmonic polynomial $u_n(x,y)$ such that $|f(x,y)-u_n(x,y)| < \frac{1}{n}$ on ∂L_n and $|f_n(x,y)-u_n(x,y)| < \frac{1}{n}$ on $\bar{\Gamma}$. Clearly $\{u_n(x,y)\}$ converges to $f(x,y)$ everywhere in U, and the functions $f_1(x,y) = f(x,y)$ and $f_2(x,y) = 2f(x,y)$ in U are limit functions for harmonic polynomials. They have all the properties noted in Corollary 2. ■

Finally we give a discussion of closely related problems and results.

Discussion: A problem posed by Montel ("Montel's problem", see [6]) seeks to find necessary and sufficient conditions for a function $f(z)$ defined in a domain D to be a limit function in D. Significant progress towards the solution of this problem was provided by a theorem of Lavrentyev [6], Theorem 24, later extended by Mergelyan [7], Theorem 6, to functions defined on F_σ sets. A solution to the problem (in terms of inner properties of $f(z)$) was found in a joint paper [8] of Mergelyan and the author.

Concerning pointwise approximation problems, Lavrentyev [6] introduced the classes M and M^* of closed sets, where $M^* \subset M$. For each $E \in M^*$ Theorem 27 of [6] shows that an arbitrary function $f(z)$, which is regular outside of E and belongs to Baire first class on E, is a limit function in the whole plane $\mathbb{C}$ (see also [7], Theorem 10 and page 168). Hence, the values of $f(z)$ in domains D_n complementary to E and on E are totally independent: an arbitrary function $f(z)$ will be a limit function if only $f(z)$ is regular in each domain D_n and belongs to Baire first class on E.

A similar result for $E \in M \setminus M^*$ is not true (see [6], Theorems 25 and 26). However, each such set E is a set of singular points for a limit function $f(z)$ (defined, for example, in an arbitrary disk $U \supset E$), and there arise questions:

(i) on the mutual connection between the values of $f(z)$ in different domains D_n complementary to E and

(ii) on the connection between the values of $f(z)$ on E and in the domains D_n.

Both questions are mentioned in [6], §19. For question (i), in case of an arbitrary $E \in M \setminus M^*$, Lavrentyev did not mention any further formulation and only posed a problem for the case when $E = \bar{\Gamma}$, which is solved in this article (note that, of course, $\bar{\Gamma} \in M \setminus M^*$). Because already in case of the set $\bar{\Gamma}$ the question has a negative answer, one can say that generally for $E \in M \setminus M^*$ there is no mutual connection (in a sense noted by Lavrentyev) between the values of $f(z)$ in different domains complementary to E.

Concerning question (ii), for a limit function $f(z)$ Lavrentyev [6], Theorem 25, proved the existence of a subsequence $\{D_{n_k}\}$ of domains complementary to E, such that almost everywhere on the boundary ∂D_{n_k} (in the conformal mapping sense [6], §6), the values of f are equal to the boundary values of a function $f(z)$ regular and bounded in D_{n_k} $(k = 1, 2, 3, \dots)$. Theorem 25 gives a necessary condition for $f(z)$ to be a limit function (that is, for the solution of Montel's problem), but the question of the sufficiency of this condition with remaining necessary conditions was open. Note that from the above mentioned result of [8] it follows, in particular, a positive answer to this question. Thus, the noted connection between the values of f on E and in the domains D_{n_k} is one of the necessary and sufficient conditions for $f(z)$ to be a limit function. This answers question (ii) (compare with the corresponding case for harmonic functions [6], after Theorem 28).

Unfortunately, for a technical reason, in the author's article [1] a reference to Lavrentyev [6] was omitted. Actually the necessary conditions of the main theorems proved in [1] were formulated in [6]. We remark that pointwise polynomial approximation results were applied to certain problems of simultaneous approximation in the complex plane (see [1],[2]) investigated by Rubel, Stray, Davie and other mathematicians. One related recent result due to Rubel and the author appears in [3].

Acknowledgements. The author wishes to thank Professor S. N. Mergelyan for helpful discussions. He also wishes to thank the organizers of CMFT'97 for the support of his participation in that conference. This research was partially supported by grants from the Russian Foundation for Fundamental Research (grant no. 96-01-01240 and no. 97-01-10920).

References

[1] Danielyan, A. A., On the representation of functions by sequences of uniformly bounded polynomials on compact sets of the complex plane, *Soviet Journ. of Contemporary Math. Analysis* **21** (1986), 30–43.

[2] Danielyan, A. A., Certain problems arising from Rubel's problem of simultaneous approximation, *Doklady Math.* **51** (1995), 164–165.

[3] Danielyan, A. A. and L. A. Rubel, Uniform approximation by entire functions that are all bounded on a given set, *Constructive Approximation* **14** (1998), 469–473.

[4] Gamelin, T. W., *Uniform Algebras*, Prentice-Hall, 1969.

[5] Hartogs, F., and A. Rosenthal, Über Folgen analytischer Functionen, *Math. Ann.* **100** (1928), 212–263.

[6] Lavrentyev, M., *Sur les functions d'une variable complexe representables par des series de polynomes*, Hermann, Paris, 1936.

[7] Mergelyan, S. N., On certain classes of sets and their applications (in Russian), in *Nekotorye problemy matematiki i mekhaniki*, Novosibirsk, 1961, 133–172.

[8] Mergelyan, S. N. and A. A. Danielyan, On the sequences of polynomials converging on F_σ sets (in Russian), *Doklady Akad. Nauk Armyan. SSR* **86** (1988), 54–57.

Arthur A. Danielyan
Dept. of Math. (Chair 806)
Moscow Aviation Institute
4, Volokolamskoe shosse
125871 Moscow
RUSSIA
daniel@k806.mainet.msk.su

Harmonic univalent mappings onto asymmetric vertical strips

Michael John Dorff

Abstract. Let Ω_α be the asymmetrical vertical strips defined by $\Omega_\alpha = \{w : \frac{\alpha-\pi}{2\sin\alpha} < \text{ Re } w < \frac{\alpha}{2\sin\alpha}\}$, where $\pi/2 \leq \alpha < \pi$, and let $\mathbb{D}$ be the unit disk. We characterize the class $S_H(\mathbb{D}, \Omega_\alpha)$ of univalent harmonic orientation-preserving functions f which map $\mathbb{D}$ onto Ω_α and are normalized by $f(0) = 0$, $f_{\overline{z}}(0) = 0$, and $f_z(0) > 0$. Then we use this characterization to demonstrate a few other results.

§1 Introduction

Let S_H be the class of complex-valued harmonic functions f which are univalent and orientation-preserving mappings of the open unit disk $\mathbb{D} = \{z : |z| < 1\}$ and are normalized by $f(0) = 0$ and $f_z(0) = 1$. Clunie and Sheil-Small [1] showed that functions in such a class have the form

$$f = h + \overline{g},$$

where

$$h(z) = z + \sum_{k=2}^{\infty} a_k z^k \quad \text{and} \quad g(z) = \sum_{k=1}^{\infty} b_k z^k$$

are analytic in $\mathbb{D}$. They also showed that the orientation-preserving condition implies that $|b_1| < 1$ and so $(f - \overline{b_1 f})/(1 - |b_1|^2) \in S_H$. Hence it is customary to just consider the subclass

$$S_H^0 = \{f \in S_H \text{ with } f_{\overline{z}}(0) = 0\}.$$

The uniqueness result of the Riemann Mapping Theorem does not extend to these classes of harmonic functions, and several authors have studied the subclass of functions that map $\mathbb{D}$ onto specific domains. In particular, Hengartner and Schober [3] considered the strip $\Omega = \{w : | \text{ Im } w| < \pi/4\}$. We will apply their results to derive a family of functions that includes all mappings in S_H^0 from $\mathbb{D}$ onto vertical strip domains that are asymmetric with respect to the imaginary axis. Using this, we will characterize all mappings in S_H^0 whose image is either a right-half plane or the entire plane minus a slit lying on the negative real axis.

Computational Methods and Function Theory (CMFT'97)
N. Papamichael, St. Ruscheweyh and E. B. Saff (Eds.), pp. 171–175

ISBN 981-02-3626-3

§2 Asymmetric vertical strip mappings

In [3], Hengartner and Schober investigated the family $S_H(\mathbb{D},\Omega)$ of normalized harmonic univalent mappings from the unit disk $\mathbb{D}$ onto the horizontal strip domain $\Omega = \{w : |\text{ Im } w| < \pi/4\}$. By the use of a rotation and a composition on their family of functions, we derive analogous results about the family of normalized univalent mappings from $\mathbb{D}$ onto the vertical asymmetric strips.

In particular, let $f \in S_H(\mathbb{D},\Omega_\alpha)$, the family of all normalized univalent mappings from $\mathbb{D}$ onto the vertical asymmetric strip domains $\Omega_\alpha = \{w : \frac{\alpha-\pi}{2\sin\alpha} < \text{ Re }(w) < \frac{\alpha}{2\sin\alpha}\}$, where $\frac{\pi}{2} \leq \alpha < \pi$. Recall that $f = h + \overline{g}$, where h, g are in the space of analytic functions, $H(\mathbb{D})$, on $\mathbb{D}$, and that $|a(z)| = |g'(z)/h'(z)| < 1$. Now, $f =$ Re $(h+g) + i$ Im $(h-g)$. So

$$\begin{aligned} h(z) - g(z) &= \int \frac{h'(z)-g'(z)}{h'(z)+g'(z)}[h'(z)+g'(z)]dz \\ &= \int \frac{1-a(z)}{1+a(z)}\varphi'(z)dz, \end{aligned}$$

where $\varphi(z) = h(z) + g(z)$.

Now φ is the conformal map from $\mathbb{D}$ onto Ω_α, normalized by $\varphi(0) = 0$ and $\varphi'(0) > 0$. To see this, note that if we consider the map $F(w) = \zeta = \xi + i\eta = \varphi(f^{-1}(w))$, then f consists of the successive transformations $(u,v) \to (w,\overline{w}) \to (z,\overline{z}) \to (\varphi,\overline{\varphi}) \to (\xi,\eta)$ so that

$$\begin{pmatrix} \frac{\partial\xi}{\partial u} & \frac{\partial\xi}{\partial v} \\ \frac{\partial\eta}{\partial u} & \frac{\partial\eta}{\partial v} \end{pmatrix} = \begin{pmatrix} \frac{1}{2} & \frac{1}{2} \\ \frac{1}{2i} & -\frac{1}{2i} \end{pmatrix} \begin{pmatrix} h'+g' & 0 \\ 0 & \overline{h'+g'} \end{pmatrix} \cdot \begin{pmatrix} \frac{\overline{h'}}{\Delta} & -\frac{\overline{g'}}{\Delta} \\ -\frac{g'}{\Delta} & \frac{h'}{\Delta} \end{pmatrix} \begin{pmatrix} 1 & i \\ 1 & -i \end{pmatrix}$$

and thus $\frac{\partial\xi}{\partial u} = 1$, $\frac{\partial\xi}{\partial v} = 0$ and $\frac{\partial\eta}{\partial v} = \frac{|h'+g'|^2}{\Delta}$ where $\Delta = |h'|^2 - |g'|^2 =$ Re $[\overline{(h'+g')}(h'-g')]$. Therefore, φ is a univalent map from $\mathbb{D}$ onto a vertical strip. Because of the normalization of φ, we see that φ is the map $z \mapsto (1/2)\log[(1+z)/(1-z)]$ rotated by $-iz$, composed with the Möbius transformation $z \mapsto (z+p)/(1+\overline{p}z)$, where $0 < p < 1$, and normalized.

Hence, any map f in $S_H(\mathbb{D},\Omega_\alpha)$ is of the form

$$\begin{aligned} f(z) &= \text{Re } \varphi(z) + i \text{ Im } \int \frac{1-a(z)}{1+a(z)}\varphi'(z)dz. \\ &= \varphi(z) - 2i \text{ Im } \int \frac{a(z)}{1+a(z)}\varphi'(z)dz. \qquad (2.1) \end{aligned}$$

Since a is in $H(\mathbb{D})$, $|a(z)| < 1$ on $\mathbb{D}$, and $a(0) = 0$, we have

$$\frac{1-a(z)}{1+a(z)} = \int_{|\eta|=1} \frac{1+\eta z}{1-\eta z} d\mu(\eta),$$

where $\mathcal{P}$ is the set of probability measures on the Borel sets of $|\eta| = 1$.

Definition 1. *For $z \in \mathbb{D}$ and $|\eta| = 1$, define the kernel*

$$K(z,\eta) = \int_0^z \frac{1+\eta w}{1-\eta w} \frac{1}{(1+we^{i\alpha})(1+we^{-i\alpha})} dw$$

$$= \begin{cases} \dfrac{\cos\alpha}{2\sin^2\alpha} \log\left(\dfrac{1+ze^{i\alpha}}{1+ze^{-i\alpha}}\right) \\ +\dfrac{1}{i\sin\alpha}\left(\dfrac{ze^{i\alpha}}{1+ze^{i\alpha}}\right), & \text{if } \eta = -e^{i\alpha} \\ \\ \dfrac{\cos\alpha}{2\sin^2\alpha} \log\left(\dfrac{1+ze^{-i\alpha}}{1+ze^{i\alpha}}\right) \\ -\dfrac{1}{i\sin\alpha}\left(\dfrac{ze^{-i\alpha}}{1+ze^{-i\alpha}}\right), & \text{if } \eta = -e^{-i\alpha} \\ \\ \dfrac{1}{2i\sin\alpha}\left(\dfrac{1-\eta e^{i\alpha}}{1+\eta e^{i\alpha}}\right) \log\left(\dfrac{1-\eta z}{1+ze^{-i\alpha}}\right) \\ -\dfrac{1}{2i\sin\alpha}\left(\dfrac{1-\eta e^{-i\alpha}}{1+\eta e^{-i\alpha}}\right) \log\left(\dfrac{1-\eta z}{1+ze^{i\alpha}}\right), & \text{if } \eta \neq -e^{\pm i\alpha} \end{cases}$$

Define the family

$$\mathcal{F}_\alpha = \left\{ f : f(z) = \ Re \left[\frac{1}{2i\sin\alpha} \log\left(\frac{1+ze^{i\alpha}}{1+ze^{-i\alpha}}\right)\right] \right.$$
$$\left. +i \ Im \int_{|\eta|=1} K(z,\eta)\, d\mu(\eta), \quad \mu \in \mathcal{P} \right\}.$$

From our discussion above, we obtain an isomorphism between the family $S_H(\mathbb{D},\Omega)$ from Hengartner and Schober [3] and the class $S_H(\mathbb{D},\Omega_\alpha)$. Hence we have the following theorem.

Theorem 1. *The following properties hold:*

1. *If f is a univalent harmonic and orientation preserving map from the unit disk $\mathbb{D}$ onto $\Omega_\alpha = \{w : \frac{\alpha-\pi}{2\sin\alpha} < Re\,(w) < \frac{\alpha}{2\sin\alpha}\}$ such that $f(0) = 0$ and $f_z(0) > 0$, then $f_z(0) = 1$.*

2. *The set* $S_H(\mathbb{D}, \Omega_\alpha) \subset \mathcal{F}_\alpha$ *with* $\overline{S_H(\mathbb{D}, \Omega_\alpha)} = \mathcal{F}_\alpha$.

3. *If* $f \in \mathcal{F}_\alpha$, *then* $f(\mathbb{D})$ *is either the strip* Ω_α, *a halfstrip, a triangle, or a trapezium.*

§3 Consequences

The results from the previous section yield a few nice consequences.

Theorem 2. *Every right-half plane mapping* $f \in S_H^0$ *can be expressed as a limit of functions in* $\mathcal{F}_\alpha$. *In particular,* f *maps* $\partial\mathbb{D}$ *into the line* $Re\ w = -\frac{1}{2}$.

Proof: This follows from the normality of the family S_H^0 and an approximation theorem (theorem 3.7 in [1]). ■

Corollary 1. *Let* $f = h + \overline{g} \in S_H^0$ *be a right-half plane mapping. Then*

$$f(z) = h(z) + g(z) - 2i\ Im\ g(z) = \frac{z}{1-z} - 2i\ Im \int_0^{2\pi} K(z,t)\, d\mu(t),$$

where

$$K(z,\eta) = \begin{cases} \dfrac{-\frac{1}{2}z^2}{(1-z)^2}, & \text{if } \eta = 1 \\[2ex] \dfrac{z}{(1-\overline{\eta})(1-z)} + \dfrac{1}{(1-\overline{\eta})(1-\eta)} \log\left(\dfrac{1-z}{1-\eta z}\right), & \text{if } \eta \neq 1 \end{cases}$$

Proof: Let $f \in S_H(\mathbb{D}, \Omega)$, where f is of the form in (2.1). The result follows from taking the limit of f as $\alpha \to \pi$. ■

Corollary 1 provides a general description for right-half plane mappings in S_H^0, so that in such cases we know that $h(z) + g(z) = z/(1-z)$. In a similar fashion, it has been shown that all slit mappings in S_H^0 whose slit lie on the negative real axis have the property that $h(z) - g(z) = 1/(1-z)^2$ ([2] or see [4]). Corollary 2 provides another proof of this.

Corollary 2. *Let* $f = h + \overline{g} \in S_H^0$ *be a slit mapping whose slit lies on the negative real axis. Then*

$$h(z) - g(z) = \frac{z}{(1-z)^2}.$$

Proof: Sheil-Small (remark 7 in [5]) showed that if $f = h + \overline{g} \in S_H^0$ is starlike, then $\hat{f} = \hat{h} - \overline{\hat{g}}$ is convex in S_H^0, where

$$\hat{h}(z) = \int_0^z \frac{h(w)}{w}\,dw \quad \text{and} \quad \hat{g}(z) = \int_0^z \frac{g(w)}{w}\,dw.$$

Let $f = h + \overline{g} \in S_H^0$ be a slit mapping whose slit lies on the negative real axis. Then $\hat{f}$ is convex. In particular, $\hat{f}$ is a right-half plane mapping since the process $\hat{f}(z) = \int_0^z f(w)/w\,dw$ makes the boundary of $\hat{f}$ normal to the boundary of f. Hence, by Corollary 1

$$\frac{z}{1-z} = \hat{h}(z) - \hat{g}(z) = \int_0^z \frac{h(w) - g(w)}{w}\,dw.$$

Therefore,

$$\frac{z}{(1-z)^2} = h(z) - g(z). \quad \blacksquare$$

Acknowledgement. The author wishes to thank W. Hengartner and T. Suffridge for their helpful suggestions.

References

[1] Clunie, J. and T. Sheil-Small, Harmonic univalent functions, *Ann. Acad. Sci. Fenn. Ser. A.I Math.* **9** (1984), 3–25.

[2] Hengartner, W., personal correspondence.

[3] Hengartner, W. and G. Schober, Univalent harmonic functions, *Trans. Amer. Math. Soc.* **299** (1987), 1–31.

[4] Livingston, A. E., Univalent harmonic mappings, *Ann. Polon. Math.* **57** (1992), 57–70.

[5] Sheil-Small, T., Constants for planar harmonic mappings, *J. London Math. Soc.* **42** (1990), 237–248.

Michael John Dorff
Dept. of Math. and Stat.
Univ. of Missouri–Rolla
Rolla, MO 65409–0020
U.S.A.
mdorff@umr.edu

Robin capacity

Peter Duren

Dedicated to the memory of Menahem Max Schiffer (1911-1997)

Abstract. This is a survey of recent work on Robin capacity. With respect to a given domain in the complex plane, Robin capacity measures the size of a subset of the boundary. Its potential-theoretic definition is given in terms of the Robin function of the domain, so named by Bergman and Schiffer [3] for the French mathematical physicist Gustave Robin (1855-1897). The Robin function is a generalization of Green's function and plays a similar role of resolvent kernel in the solution of mixed boundary-value problems. Like ordinary capacity, the Robin capacity can be defined in a variety of equivalent ways: through potential theory, extremal length, and least-energy considerations. In addition, it has a basic connection with conformal mapping.

§1 Transfinite diameter and capacity

Let us begin by recalling the classical definition of transfinite diameter. Given a compact subset $E \subset \mathbb{C}$ of the complex plane, choose a system of n points $z_1, z_2, ..., z_n$ in E and form the homogenized product

$$\left\{ \prod_{1 \le j < k \le n} |z_j - z_k| \right\}^{\frac{2}{n(n-1)}}$$

of distances between the $n(n-1)/2$ pairs of points z_j. Let $d_n(E)$ be the maximum of these products as the points z_j range over E. Note that $d_2(E)$ is the diameter of E, while $d_3(E)$ measures its "spread". The quantity $d_n(E)$ is called the *n-diameter* of E. It can be shown (see Goluzin [13], Ch. VII) that $d_{n+1}(E) \le d_n(E)$. Thus as n tends to infinity, $d_n(E)$ decreases to a limit $d(E) \ge 0$, known as the *transfinite diameter* of E. The transfinite diameter of a closed disk (or of a circle) is found to be its radius; while the transfinite diameter of a linear segment is one-quarter of its length.

The capacity of E is defined as follows. Let $\Omega = \hat{\mathbb{C}} \backslash E$ be the domain in the extended complex plane complementary to E, and suppose that Ω

Computational Methods and Function Theory (CMFT'97)
N. Papamichael, St. Ruscheweyh and E. B. Saff (Eds.), pp. 177–190

ISBN 981-02-3626-3

is bounded by finitely many Jordan curves. Let $G(z,\zeta)$ denote Green's function of Ω with pole at a point $\zeta \in \Omega$. This is the function harmonic in $\Omega\backslash\{\zeta\}$ with $G(z,\zeta) - \log\frac{1}{|z-\zeta|}$ harmonic near ζ, and $G(z,\zeta) = 0$ on $\partial\Omega$. For $\zeta = \infty$ the definition is modified by requiring $G(z,\infty) - \log|z|$ to be harmonic near infinity. Then the *capacity* of E is $e^{-\gamma(E)}$, where

$$\gamma(E) = \lim_{z\to\infty}\{G(z,\infty) - \log|z|\}$$

is the *Robin constant* of E. By a remarkable theorem of Szegő [21], capacity is equal to transfinite diameter: $d(E) = e^{-\gamma(E)}$. A proof can be found in Goluzin [13].

Note that Green's function $G(z,\infty)$ is invariant under precomposition with an arbitrary conformal mapping of Ω of the form

$$w = f(z) = z + b_0 + b_1 z^{-1} + b_2 z^{-2} + \ldots$$

near infinity. Such mappings will be called *admissible*. It follows that the capacity of a compact set E is invariant under all admissible mappings f of $\Omega = \hat{\mathbb{C}}\backslash E$; that is, $d(f(E)) = d(E)$. If E is the disk $|z| \leq R$, then Green's function $G(z,\infty)$ of Ω is simply $\log|z| - \log R$, and so $d(E) = R$. From this the capacities of many other sets can be found by conformal mapping. Some examples are given in the next section.

§2 Robin function and Robin capacity

Suppose now that $\Omega = \hat{\mathbb{C}}\backslash E$ is a Jordan domain whose boundary $\partial\Omega$ consists of a finite number of analytic Jordan curves. Let the boundary be partitioned into a pair of disjoint sets A and B, each consisting of a finite number of subarcs of boundary curves. With respect to such a partition, the *Robin function* $R(z,\zeta)$ of Ω with pole at ζ is defined by the following properties. It is harmonic in Ω except at ζ, where $R(z,\zeta) - \log\frac{1}{|z-\zeta|}$ is harmonic. It is continuous together with its first partial derivatives up to the boundary of Ω. The boundary function satisfies $R(z,\zeta) = 0$ for $z \in A$; while its inner normal derivative has the property $\frac{\partial R}{\partial n}(z,\zeta) = 0$ for $z \in B$. For $\zeta = \infty$ the definition is again modified to require that $R(z,\infty) - \log|z|$ be harmonic at infinity.

The Robin function can be used to solve the mixed boundary-value problem for harmonic functions. This is the problem of finding the function $u(z)$ harmonic in Ω and of class C^1 in $\overline{\Omega}$, with prescribed values on A and prescribed normal derivative on B. The solution is given by the integral formula

$$u(\zeta) = \frac{1}{2\pi}\int_A \frac{\partial R}{\partial n}(z,\zeta)\, u(z)\, |dz| - \frac{1}{2\pi}\int_B R(z,\zeta)\, \frac{\partial u}{\partial n}(z)\, |dz|.$$

The Robin function $R(z,\infty)$ is again invariant under precomposition with admissible conformal mappings of Ω. The conformal invariance allows a definition of the Robin function even for domains with rough boundary, where no normal direction is defined. (Successive applications of the Riemann mapping theorem show that an arbitrary finitely connected domain with nondegenerate boundary components can be mapped conformally onto a Jordan domain with analytic boundary.) The Robin function may be viewed as a generalization of Green's function, to which it reduces when the set B is empty.

The *Robin capacity* of A with respect to Ω is now defined as $\delta(A) = e^{-\rho(A)}$, where

$$\rho(A) = \lim_{z\to\infty} \{R(z,\infty) - \log|z|\}$$

is the Robin constant of A with respect to Ω. It can be shown [9] with the help of Hopf's lemma that $R(z,\zeta) \geq G(z,\zeta)$ for each domain Ω and each set $A \subset \partial\Omega$. It follows in particular that $\delta(A) \leq d(A)$.

Here are two examples. (See [6] for further information.)

Example 1: Let E be the closed disk $|z| \leq 1$, and let A be an arc on the boundary which subtends an angle α at the center. Then it can be shown by conformal mapping that $d(A) = \sin\alpha/4$, while $\delta(A) = \sin^2\alpha/4$. Thus $\delta(B) = \cos^2\alpha/4$, so that $\delta(A) + \delta(B) = 1$. On other hand $d(A \cup B) = d(E) = 1$.

Example 2: Let E be the upper semidisk $|z| \leq 1$, $\mathrm{Im}\{z\} \geq 0$. Let A be the upper semicircle, so that B is the real segment $[-1,1]$. Then $d(A) = 1/\sqrt{2}$, $\delta(A) = 1/\sqrt{3}$, $d(B) = 1/2$ and $\delta(B) = 1/3\sqrt{3}$; while $d(A \cup B) = d(E) = 4/3\sqrt{3}$. Observe that again $\delta(A) + \delta(B) = d(A \cup B)$.

On the basis of these examples, a general principle emerges. Let C be an arbitrary Jordan curve, and let E be the closure of the domain inside it. Let A and B be complementary arcs of the boundary. Then $\delta(A) + \delta(B) = d(C)$. The proof simply appeals to the conformal invariance of all quantities under admissible mappings, together with the truth of the assertion in the special case of Example 1. The relation breaks down when the subsets A and B are not single arcs of the boundary, but it is still true that $\delta(A) \leq d(A) \leq d(C)$ and $\delta(B) \leq d(C)$, so that $\delta(A) + \delta(B) \leq 2d(C)$. That the constant 2 is best possible is seen by taking C to be the unit circle and A the union of n arcs of length π/n centered at the nth roots of unity. Then $\delta(A) = \delta(B) = 2^{-1/n} \to 1$ as $n \to \infty$. (See [6] for details.)

Example 1 also illustrates the following general relation. If $\Omega = \{z \in \hat{\mathbb{C}} : |z| > 1\}$ is the exterior of the unit disk and A is any nice subset of the boundary, then $\delta(A) = d(A)^2$; that is, the Robin capacity of A with respect to Ω is the square of the ordinary capacity. To see this, observe that if $g(z,\zeta)$

is Green's function $\hat{\mathbb{C}}\backslash A$, then for $\zeta \in \Omega$, $G(z,\zeta) = g(z,\zeta) - g(1/\bar{z},\zeta)$ is Green's function of Ω and $R(z,\zeta) = g(z,\zeta) + g(1/\bar{z},\zeta)$ is Robin's function of Ω. Hence $\log|z| = g(z,\infty) - g(1/\bar{z},\infty)$ and $R(z,\infty) = g(z,\infty) + g(1/\bar{z},\infty)$. Addition of this last pair of equations gives the result. (See [10] for further details. The relation is also implicit in [20].)

§3 Connection with Conformal Mapping

Suppose once again that Ω is a Jordan domain whose boundary consists of a finite number of analytic Jordan curves. Although the capacity of the full boundary is invariant under normalized conformal mappings, the capacity of any fixed subset $A \subset \partial\Omega$ may well be distorted. It was shown in [10] that the minimum of the capacity $d(f(A))$ under all admissible mappings f is precisely the Robin capacity $\delta(A)$. This explains and generalizes a result of Pommerenke [16], who had shown that $d(f(A)) \geq d(A)^2$ for subsets of the unit circle. In fact, as noted above, $\delta(A) = d(A)^2$ in this case.

The proof that $\delta(A)$ is the minimum of $d(f(A))$ goes by a variational method. Here is a sketch of the argument. Suppose, after suitable conformal mapping, that Ω is actually an extremal domain. Choose a point $z_0 \in B$, and construct a boundary variation [18, 5]

$$z* = z + \frac{a\rho^2}{z - z_0} + O(\rho^3), \qquad \rho \to 0.$$

It acts as an interior variation on the domain $\hat{\Omega} \supset \Omega$ with $\partial\hat{\Omega} = A$. The variational formula for the capacity of the boundary is ([19])

$$d(A^*) = d(A)\left[1 - \operatorname{Re}\left\{a\rho^2[\hat{p}'(z_0)]^2\right\} + O(\rho^3)\right],$$

where $\hat{p}(z)$ is the analytic completion of Green's function $\hat{G}(z,\infty)$ for the domain $\hat{\Omega}$. Since the domain Ω was taken to be extremal, it follows that $d(A^*) \geq d(A)$, so that

$$\operatorname{Re}\left\{a\rho^2[\hat{p}'(z_0)]^2\right\} + O(\rho^3) \leq 0.$$

An appeal to Schiffer's theorem (see [5], p. 297) shows that B consists of analytic arcs lying on trajectories of the quadratic differential $[\hat{p}'(z)]^2\,dz^2 > 0$. Thus the tangential derivative of $\hat{p}(z)$ is real, and it follows from the generalized Cauchy-Riemann equations that $\frac{\partial}{\partial n}\hat{G}(z,\infty) = 0$ on B. In other words, Green's function of $\hat{\Omega}$ is Robin's function of Ω, showing that $d(A) = \delta(A)$ in the extremal case. This establishes the sharp inequality $d(f(A)) \geq \delta(A)$ for all admissible mappings f of Ω. In particular, $\delta(A) \leq d(A)$.

This last inequality also follows immediately from alternate definitions of capacity and Robin capacity in terms of extremal length, or reduced extremal distance. Further details appear in the next section. However, it must be acknowledged that the proof of sharpness of this inequality, as given in [10], contains a gap. The difficulty lies in proving the existence of an extremal domain. There is no problem if Ω is simply connected. In that case, an extremal domain can be described explicitly. It is the domain bounded by the circle $|w| = \delta$ to which exterior radial segments (images of the arcs of B under the associated admissible mapping f) are attached. Green's function of the domain $|w| > \delta$ is then $\hat{G}(w, \infty) = \log|w| - \log\delta$, and it is easily verified that this is the Robin function of the image domain $f(\Omega)$. The actual conformal mapping f onto this extremal domain has the form

$$w = f(z) = \delta\ e^{R(z,\infty)+i\tilde{R}(z,\infty)},$$

where $R(z,\infty)$ is the Robin function of Ω and $\tilde{R}(z,\infty)$ is its harmonic conjugate.

Similar considerations show that if Ω is multiply connected and the set A consists of a single boundary component, an extremal domain will result from mapping A onto the circle $|w| = \delta$ and each of the other boundary components onto a radial segment. The existence of an extremal domain can also be proved if Ω is multiply connected and the subsets A and B are unions of entire boundary components. In that case an appeal to Koebe's *Kreisnormierungssatz* reduces the convergence question to that of a sequence of domains bounded by circles, and kernel convergence means that the sequences of centers and radii are convergent. Details will appear in a forthcoming paper [7].

§4 Extremal length

Alternate descriptions of capacity and Robin capacity can be given in terms of extremal length. These results are essentially due to Ahlfors and Beurling, who invented and developed the concept of extremal length around 1950. Good introductory accounts may be found in the books by Ahlfors [1] and Fuchs [11].

Extremal length is defined as follows. Let Γ be a family of locally rectifiable curves in a domain Ω. An *admissible metric* for Γ is a Borel measurable function $\rho(z) \geq 0$ with the property

$$\int_\gamma \rho(z)|dz| \geq 1 \quad \text{for all} \quad \gamma \in \Gamma.$$

The *extremal length* $\lambda(\Gamma)$ is then defined by

$$\frac{1}{\lambda(\Gamma)} = \inf \iint_\Omega \rho(z)^2\, dxdy, \qquad z = x + iy,$$

where the infimum is taken over all admissible metrics ρ. The importance of extremal length in function theory is due mainly to its conformal invariance.

Now let A and B be two disjoint subsets of the boundary of a domain Ω, and let Γ be the family of curves in Ω which connect A to B. Then the extremal length $\lambda(\Gamma)$ is called the *extremal distance* from A to B and is denoted by $\lambda_\Omega(A, B)$. The following theorem is due to Ahlfors and Beurling (see [1], p. 65).

Theorem 1. *Let Ω be a bounded domain whose boundary consists of a finite number of analytic Jordan curves. Let A and B be disjoint subsets of $\partial\Omega$, each consisting of a finite number of closed arcs. Let u be harmonic in Ω and continuous in $\overline{\Omega}$, with $u(z) = 1$ on A, $u(z) = 0$ on B, and $\frac{\partial u}{\partial n}(z) = 0$ on $\partial\Omega\backslash(A \cup B)$. Then*

$$\frac{1}{\lambda_\Omega(A, B)} = \iint\limits_{\Omega} |\nabla u|^2 \, dxdy.$$

With the help of this theorem, it is possible to derive an expression for the capacity of a compact set $E \subset \mathbb{C}$. Let $\Omega = \hat{\mathbb{C}}\backslash E$ and suppose that $A = \partial\Omega$ consists of a finite number of components. Let C_r be a circle of radius r centered at the origin and large enough to enclose the set A. Let Ω_r be the part of Ω which lies inside C_r. Then the capacity of E is given by the formula $d(E) = e^{-\gamma(E)}$, where

$$\gamma(E) = \lim_{r\to\infty} \{2\pi \, \lambda_{\Omega_r}(A, C_r) - \log r\}.$$

Robin capacity can be expressed in a similar way. Suppose once again that $\Omega \subset \hat{\mathbb{C}}$ is a domain containing infinity whose boundary $\partial\Omega$ consists of a finite number of analytic Jordan curves, and let A and B be complementary subsets of $\partial\Omega$, each consisting of finitely many arcs. Again surround $\partial\Omega$ by a sufficiently large circle C_r, and let $\Omega_r = \Omega \cap \{|z| < r\}$. Then the Robin capacity of A with respect to Ω is given by $\delta(A) = e^{-\rho(A)}$, where

$$\rho(A) = \lim_{r\to\infty} \{2\pi \, \lambda_{\Omega_r}(A, C_r) - \log r\}.$$

Note that in the case of Robin capacity, the family of curves joining A to C_r is restricted to lie in the domain Ω_r, and so is in general smaller than the family of curves in the larger domain

$$\hat{\Omega}_r = \{|z| < r\}\backslash A,$$

which join A to C_r. Since a smaller family of curves gives rise to a larger collection of admissible metrics, the definition of extremal length shows that

$\lambda_{\Omega_r}(A, C_r) \geq \lambda_{\hat{\Omega}_r}(A, C_r)$, so that $\rho(A) \geq \gamma(A)$ and hence $\delta(A) \leq d(A)$. Thus if we are given the descriptions in terms of extremal length, it is a matter of simple logic to show that Robin capacity is majorized by ordinary capacity.

On the basis of the extremal-length definition of Robin capacity, it is easy to establish some comparison theorems which are not so accessible from the potential-theoretic description. For instance, if $A \subset \tilde{A} \subset \partial\Omega$, then $\delta(A) \leq \delta(\tilde{A})$. The proof is straightforward and is very similar to the proof just given that $\delta(A) \leq d(A)$.

It is also of interest to ask how the Robin capacity of a boundary set is affected by a change of domain. Here it is convenient to adopt the more specific notation $\delta_\Omega(A)$ for the Robin capacity of a boundary set A with respect to the domain Ω. The following results were noted by Duren and Pfaltzgraff [6]; again the proofs are simple logical arguments based on the extremal-length description of Robin capacity.

Theorem 2. *Let Ω and $\tilde{\Omega}$ be smoothly bounded Jordan domains with $\Omega \subset \tilde{\Omega}$. Then*

(i) If $A \subset (\partial\Omega) \cap (\partial\tilde{\Omega})$, then $\delta_\Omega(A) \leq \delta_{\tilde{\Omega}}(A)$.

(ii) If $B \subset (\partial\Omega) \cap (\partial\tilde{\Omega})$, $A = (\partial\Omega)\backslash B$, $\tilde{A} = (\partial\tilde{\Omega})\backslash B$, then $\delta_{\tilde{\Omega}}(\tilde{A}) \leq \delta_\Omega(A)$.

§5 Quadrilaterals

A *quadrilateral* Q is a Jordan domain $\Omega \subset \hat{\mathbb{C}}$ together with four distinct boundary points z_1, z_2, z_3, z_4 (called the *vertices* of Q) arranged in the positive sense with respect to Q. Any quadrilateral can be mapped conformally onto the interior of a rectangle with vertices going to vertices, and the ratio of side-lengths is a conformal invariant known as the *modulus* of Q.

Suppose now that there is a conformal mapping of Q onto the *exterior* of a rectangle with side-lengths a and b, carrying vertices to vertices and making the boundary arcs (z_1, z_2) and (z_3, z_4) correspond to the sides of length a. We then define the quantity $m(Q) = a/b$, which we call the *exterior modulus* of Q. It can be shown that the exterior modulus is again a conformal invariant.

If a quadrilateral Q contains the point at infinity, we take A to be the union of the two arcs (z_1, z_2) and (z_3, z_4) of $\partial\Omega$, and let $B = \partial\Omega\backslash A$ be the union of the two complementary arcs. Then if $\delta(A)$ and $\delta(B)$ denote the respective Robin capacities of A and B with respect to Ω, we define the *Robin modulus* of Ω by $\mu(Q) = \delta(A)/\delta(B)$. It is clear that the Robin modulus is invariant under conformal mappings that preserve infinity.

The relation between the exterior modulus and the Robin modulus of a quadrilateral is given by the following theorem, found by Duren and Pfaltzgraff [6], except for the numerical values of the constants, which were established in [2].

Theorem 3. *Let Q be a quadrilateral over a Jordan domain containing the point at infinity. Suppose that Q is conformally equivalent to a quadrilateral over the outside of a rectangle, under a mapping that preserves infinity. Then the exterior modulus $m(Q)$ and the Robin modulus $\mu(Q)$ are related by the sharp inequalities*

$$\frac{\pi}{4}\, m(Q) \le \mu(Q)^2 \le \frac{4}{\pi}\, m(Q).$$

Proof: Since all quantities are conformally invariant, we may suppose without loss of generality that Q is the outside of a rectangle. In order to compute the Robin capacity of a pair of opposite sides, we fix an angle α, $0 < \alpha < \pi/2$, and construct a Schwarz-Christoffel mapping from the domain Δ outside the unit circle to the outside of a rectangle, carrying the four points $e^{i\alpha}, e^{-i\alpha}, -e^{i\alpha}, -e^{-i\alpha}$ onto the vertices and preserving the point at infinity. This was done in [6]. In terms of the parameter $r = \tan^2(\alpha/2)$, it was found that the resulting rectangle R has sidelengths

$$2J = \frac{1}{r}[E - (1-r)K] \quad\text{and}\quad L = \frac{1}{2r}[E' - rK'],$$

where $K = K(r)$ and $E = E(r)$ are the standard complete elliptic integrals of first and second kinds, respectively; while $K'(r) = K(r')$, $E'(r) = E(r')$, and $r' = (1-r^2)^{1/2}$. The capacity of R was found to be $d(R) = (1+r)/8r$. More symmetric formulas can be obtained, however, with the change of parameter $k = \frac{1-r}{1+r}$, or $r = \frac{1-k}{1+k}$. Then $k = \cos\alpha$ and standard transformations give

$$\begin{aligned} K(r) &= \frac{1+k}{2}\, K(k'), & K(r') &= (1+k)K(k), \\ E(r) &= \frac{1}{1+k}\, [E(k') + kK(k')], & E(r') &= \frac{1}{1+k}[2E(k) - k'^2 K(k)]. \end{aligned}$$

Thus the rectangle R is found to have side-lengths

$$2J = \frac{1}{1-k}\, [E' - k^2 K'] \quad\text{and}\quad L = \frac{1}{1-k}\, [E - k'^2 K],$$

where now $K = K(k)$, $K' = K(k')$, $E = E(k)$, $E' = E'(k)$. Note that R is a square when $\alpha = \pi/4$, so that $k = k' = 1/\sqrt{2}$. The capacity of the rectangle is $d(R) = \frac{1}{4(1-k)}$.

Now let A be the union of the two opposite sides of R of length $2J$, and let B be the union of the sides of length L. Referring the calculations back to the disk and appealing to conformal invariance, we find (see [6]) that the Robin capacities are

$$\begin{aligned} \delta(A) &= \frac{1}{4(1-k)} \sin\alpha = \frac{k'}{4(1-k)}, \\ \delta(B) &= \frac{1}{4(1-k)} \cos\alpha = \frac{k}{4(1-k)}. \end{aligned}$$

Thus the Robin modulus is $\mu(R) = k'/k$, while the exterior modulus is

$$m(R) = \frac{2J}{L} = \frac{E' - k^2 K'}{E - k'^2 K}.$$

The ratio

$$\rho(k) = \frac{\mu^2}{m} = \frac{k'^2}{k^2} \frac{E - k'^2 K}{E' - k^2 K'}$$

now has the property $\rho(k') = 1/\rho(k)$, reflecting the geometrically obvious symmetries of the moduli m and μ under interchange of A and B. A simple argument (see [2]) now shows that $\rho(k)$ increases from $\pi/4$ to $4/\pi$ as k increases from 0 to 1. This concludes the proof. ■

It is interesting to note that the apparatus developed here can be applied to calculate the extremal length of the family of curves *outside* a given rectangle and joining a pair of opposite sides. Let the rectangle have side-lengths a and b, and let Γ be the family of curves running outside the rectangle between the two sides of length b. Now choose the parameter k $(0 < k < 1)$ so that

$$\frac{a}{b} = \frac{2J}{L} = \frac{E' - k^2 K'}{E - k'^2 K}.$$

Then the extremal length of Γ is found to be $\lambda(\Gamma) = K/K'$.

§6 Minimum energy

A physical interpretation of the capacity of a set $E \subset \mathbb{C}$ proceeds as follows (*cf.* Tsuji [25]). Imagine that E is a conducting plate and that a unit charge distribution $\mu \geq 0$ is placed on it. Consider the *energy integral*

$$I(\mu) = \int_E\int_e \log \frac{1}{|z-\zeta|}\, d\mu(z)\, d\mu(\zeta),$$

which measures the potential energy of the charge distribution in the presence of a logarithmic potential. The charge will distribute itself so as to minimize the energy integral. (If E has several components, we can imagine that they are connected by thin wires to allow a free flow of charge.) Let $V = \inf\ I(\mu)$ be the minimum energy. It can be shown that there exists a unique minimizing measure μ, called the *equilibrium distribution.* Then

$$U(z) = \int_E \log \frac{1}{|z-\zeta|}\, d\mu(\zeta) = -\log|z| + O(1/z)$$

is called the *conductor potential.* It turns out that $U(z) = V$ *quasi-everywhere* on E; that is, except for a set of capacity zero. Thus $G(z,\infty) = V - U(z)$ is Green's function of $\Omega = \hat{\mathbb{C}}\backslash E$, and so $d(E) = e^{-V}$ is the capacity of E. Furthermore, if E is bounded by a finite number of smooth Jordan curves, it can be shown (see [1], pp. 25-27) that the equilibrium measure is supported on the boundary and has the form $d\mu(z) = \frac{1}{2\pi}\frac{\partial G}{\partial n}(z,\infty)|dz|$, where $|dz|$ denotes arclength measure on the boundary.

We now turn to a corresponding physical interpretation of Robin capacity. Let $\Omega \subset \hat{\mathbb{C}}$ be a finitely connected domain containing the point at infinity, bounded by analytic Jordan curves $C_1, C_2, ..., C_n$. Let $A = C_1 \cup \ldots \cup C_k$ and $B = C_{k+1} \cup \ldots \cup C_n$, where $1 \le k < n$. Let $N(z,\zeta)$ be the *Neumann function* of the larger domain $\hat{\Omega}$ bounded only by B. This is the function harmonic $\hat{\Omega}$ except at ζ and infinity, with the properties

(i) $N(z,\zeta) - \log\frac{1}{|z-\zeta|}$ is harmonic near ζ,

(ii) $N(z,\zeta) + \log|z|$ is harmonic near infinity,

(iii) $\lim_{z\to\infty}\{N(z,\zeta) + \log|z|\} = 0$, and

(iv) $\frac{\partial N}{\partial n}(z,\zeta) = 0$ for $z \in B$.

(See for instance Henrici [14], p. 270 or Garabedian [12], p. 245. The terminology is not standard.) For example, it can be verified that Neumann's function for the domain outside the unit disk has the form

$$N(z,\zeta) = \log\ \frac{|z\zeta|}{|z-\zeta|\,|1-\bar{\zeta}z|}.$$

The Neumann function can be viewed as the potential at z of a point-charge at ζ in the presence of perfectly insulating "islands" with boundary B. As in the case of Green's function, an elementary calculation shows that the Neumann function is symmetric: $N(z,\zeta) = N(\zeta,z)$. If $\delta(A)$ is the Robin capacity of A with respect to Ω, then $-\log\delta(A)$ can be interpreted as the minimum energy of a unit charge distribution on the set E bounded by A, in the presence of insulation on B. This is the main content of the following theorem, found by Duren, Pfaltzgraff, and Thurman [8].

Theorem 4. *Let Ω be a finitely connected Jordan domain with analytic boundary, containing the point at infinity, and let its boundary be partitioned as above into two sets $A = C_1 \cup \ldots \cup C_k$ and $B = C_{k+1} \cup \ldots \cup C_n$. Let $N(z,\zeta)$ be the Neumann function of the domain $\hat{\Omega}$ bounded only by B. Suppose that $\mu \geq 0$ is a unit charge distribution on the compact set E with boundary A, and let*

$$J(\mu) = \int_E \int_E N(z,\zeta)\, d\mu(z)\, d\mu(\zeta)$$

denote its energy integral with respect to the Neumann potential. Moreover, let $W = \inf J(\mu)$ be the minimum energy. Then there is a unique unit measure μ such that $J(\mu) = W$, and its conductor potential

$$U(z) = \int_E N(z,\zeta)\, d\mu(\zeta)$$

has the property that $U(z) = W$ quasi-everywhere on E. Furthermore, the Robin capacity of A with respect to Ω is $\delta(A) = e^{-W}$.

Proof: As for the logarithmic potential, a variational argument shows that the conductor potential $U(z) \equiv W$ on E. By property (iv) of the Neumann function,

$$\frac{\partial U}{\partial n}(z) = \int_E \frac{\partial N}{\partial n}(z,\zeta)\, d\mu(\zeta) = 0, \qquad z \in B.$$

Observe also that

$$U(z) = -\log|z| + o(1) \qquad \text{as} \quad z \to \infty,$$

by property (iii) of the Neumann function. The function $F(z) = W - U(z)$ therefore has the properties $F(z) = 0$ on A and $\frac{\partial F}{\partial n}(z) = 0$ on B; while $F(z) - \log|z|$ is harmonic near infinity. Thus $F(z) = R(z,\infty)$, the Robin function of Ω with pole at infinity. Consequently,

$$W = \lim_{z\to\infty} \{F(z) - \log|z|\}$$

is the Robin constant of A with respect to Ω, and $\delta(A) = e^{-W}$ is its Robin capacity. In fact, the equilibrium distribution μ is supported on the boundary $A = \partial E$ and has the form $d\mu(z) = \frac{1}{2\pi}\frac{\partial R}{\partial n}(z,\infty)|dz|$, a generalization of the structure already noted for the logarithmic potential. ∎

§7 Further developments

Robin capacity arises as the minimum of the capacity of the image of a specified subset A of the boundary under normalized conformal mappings of the domain. The question arises to describe the *maximum* distortion of capacity under the same class of mappings. Thurman [22, 23, 24] has investigated this question in some detail, giving a formula for the maximum capacity in terms of standard conformal invariants such as harmonic measures and the Riemann matrix of periods of their harmonic conjugates. A new phenomenon arises for the maximum problem: if the subset A contains even a single point of a given boundary component, it may as well contain the entire boundary component. Thurman [23] has given an extremal-length description of maximum capacity in terms of a notion of "bridged extremal distance".

Betsakos [4] has studied the effect on Robin capacity of a certain type of symmetrization known as polarization. Here the difficulty is to devise methods of symmetrization which will respect the given partition of the boundary. It seems likely that further results in this direction await discovery.

References

[1] Ahlfors, L. V., *Conformal Invariants*, McGraw-Hill, New York, 1973.

[2] Anderson, G. D., P. Duren, and M. K. Vamanamurthy, An inequality for complete elliptic integrals, *J. Math. Anal. Appl.* **182** (1994), 257–259.

[3] Bergman S. and M. Schiffer, *Kernel Functions and Elliptic Differential Equations in Mathematical Physics*, Academic Press, New York, 1953.

[4] Betsakos, D., Polarization, conformal invariants, and Brownian motion, *Ann. Acad. Sci. Fenn.*, to appear.

[5] Duren, P., *Univalent Functions*, Springer-Verlag, New York, 1983.

[6] Duren P. and J. Pfaltzgraff, Robin capacity and extremal length, *J. Math. Anal. Appl.* **179** (1993), 110–119.

[7] Duren P. and J. Pfaltzgraff, Hyperbolic capacity and its distortion under conformal mapping, preprint

[8] Duren, P., J. Pfaltzgraff, and R. E. Thurman, Physical interpretation and further properties of Robin capacity, *Algebra i Analiz* **9** No. 3 (1997), 211–219 = *St. Petersburg Math. J.* **9** (1997), 607–614.

[9] Duren, P. and M. M. Schiffer, Robin functions and energy functionals of multiply connected domains, *Pacific J. Math.* **148** (1991), 251–273.

[10] Duren, P and M. M. Schiffer, Robin functions and distortion of capacity under conformal mapping, *Complex Variables Theory Appl.* **21** (1993), 189–196.

[11] Fuchs, W. H. J., *Topics in the Theory of Functions of One Complex Variable*, Van Nostrand, Princeton, N. J., 1967.

[12] Garabedian, P. R., *Partial Differential Equations*, Second Edition, Chelsea Publ. Co., New York, 1986.

[13] Goluzin, G. M., *Geometric Theory of Functions of a Complex Variable*, Second edition, Izdat. "Nauka", Moscow, 1966; English translation, American Mathematical Society, Providence, R.I., 1969.

[14] Henrici, P., *Applied and Computational Complex Analysis*, Volume 3, Wiley-Interscience, New York, 1986.

[15] Nehari, Z., *Conformal Mapping* , McGraw-Hill, New York, 1952; Dover reprint, 1975.

[16] Pommerenke, Ch., On the logarithmic capacity and conformal mapping, *Duke Math. J.* **35** (1968), 321–25.

[17] Pommerenke, Ch., *Univalent Functions*, Vandenhoeck & Ruprecht, Göttingen, 1975.

[18] Schiffer, M. M., A method of variation within the family of simple functions, *Proc. London Math. Soc.* **44** (1938), 432–449.

[19] Schiffer, M. M., Hadamard's formula and variation of domain-functions, *Amer. J. Math.* **68** (1946), 417–448.

[20] Schiffer, M. M., Some distortion theorems in the theory of conformal mapping, *Atti Accad. Naz. Lincei Mem.* **10** (1970), 1–19.

[21] Szegő, G., Bemerkungen zu einer Arbeit von Herrn M. Fekete: Über die Verteilung bei gewissen algebraischen Gleichungen mit ganzzahligen Koeffizienten", *Math. Z.* **21** (1924), 203–208.

[22] Thurman, R. E., Upper bound for distortion of capacity under conformal mapping, *Trans. Amer. Math. Soc.* **346** (1994), 605–616.

[23] Thurman, R. E., Bridged extremal distance and maximal capacity, *Pacific J. Math.* **176** (1996), 507–528.

[24] Thurman, R. E., Maximal capacity, Robin capacity, and minimum energy, *Indiana Univ. Math. J.*, to appear.

[25] Tsuji, M., *Potential Theory in Modern Function Theory*, Maruzen, Tokyo, 1959; Chelsea reprint.

Peter Duren
Department of Mathematics
University of Michigan
Ann Arbor
MI 48109-1109
U.S.A.
duren@umich.edu

On generating orthogonal polynomials with respect to self-similar measures

Hans-Jürgen Fischer

Abstract. In the present paper, we derive an algorithm for computing the recurrence coefficients of orthogonal polynomials with respect to self-similar measures. This means that the cumulative distribution function of the measure is a fractal interpolation function in the sense of Barnsley. As examples show, this notion is very flexible. But here we consider only real fractals, as in the case of Cantor measure or of some special Riesz-Nagy measures.

§1 Introduction

Let σ be some measure with compact support $S(\sigma) \subset [a, b]$. We are interested in the uniquely defined family of monic polynomials $\tilde{\pi}_j$ with

$$\int \tilde{\pi}_l(x)\tilde{\pi}_j(x)\, d\sigma(x) = 0 \quad \text{for} \quad l < j$$

and

$$\int \tilde{\pi}_j^2(x)\, d\sigma(x) = \gamma_j^{-2} > 0, \quad j = 0, 1, \ldots$$

They satisfy a three-term recurrence

$$\tilde{\pi}_{j+1}(x) = (x - \alpha_j)\tilde{\pi}_j(x) - \beta_j\tilde{\pi}_{j-1}(x) \quad \text{for} \quad j = 0, 1, \ldots, \tag{1.1}$$

if we set $\tilde{\pi}_{-1}(x) \equiv 0$. We can assume the orthogonal polynomials to be known if we know the coefficients α_j and β_j. Thus the problem of determining these recurrence coefficients is crucial. For classical measures this problem is solved analytically. In the case of practically important measures, several numerical procedures were developed. They depend on how the measure is given. Since we consider orthogonality of polynomials, we must have an algorithm to compute the integral $\int q(x)\, d\sigma(x)$ for any polynomial q.

Computational Methods and Function Theory (CMFT'97)
N. Papamichael, St. Ruscheweyh and E. B. Saff (Eds.), pp. 191–201

ISBN 981-02-3626-3

Definition 1. *The set of all polynomials of degree* $\leq D$ *will be denoted by* $\mathcal{P}_D$.

A representation of the form

$$q(x) = \sum_{j=0}^{D} c_j(q) p_j(x)$$

for polynomials $q \in \mathcal{P}_D$ with some appropriate base $(p_j)_{j=0,\ldots,D} \subset \mathcal{P}_D$ leads to a procedure of modified moments.

However, the measure can be defined by some invariance property of the measure as well. This means that an equation like

$$\int h(x)\, d\sigma(x) = \int Ah(x)\, d\sigma(x) \quad \text{for} \quad h \in \mathbf{H} \tag{1.2}$$

with some linear function space $\mathbf{H}$ and some linear operator $A : \mathbf{H} \to \mathbf{H}$ is satisfied. This kind of equation can be obtained from integration by parts, if the weight function of the measure satisfies some linear differential equation with polynomial coefficients. But it can be derived from a self-similarity of the measure, too.

§2 Formulas for recurrence coefficients

Principially, simple explicit formulas for the coefficients α_j and β_j do exist (due to Darboux and Stieltjes), see *e.g.* [3], eq. (2.2): We have

$$\alpha_j = \frac{\int x\tilde{\pi}_j^2(x)\, d\sigma(x)}{\int \tilde{\pi}_j^2(x)\, d\sigma(x)} \quad \text{for} \quad j = 0, 1, \ldots, \tag{2.1}$$

and

$$\beta_j = \frac{\int \tilde{\pi}_j^2(x)\, d\sigma(x)}{\int \tilde{\pi}_{j-1}^2(x)\, d\sigma(x)} \quad \text{for} \quad j = 1, 2, \ldots. \tag{2.2}$$

These equations can be applied to the computation of α_j and β_j, since in virtue of the recurrence relation (1.1) the polynomials $\tilde{\pi}_j$ and $\tilde{\pi}_{j-1}$ are well-defined by the coefficients up to α_{j-1} and β_{j-1}. The problem, of course, is in the computation of the integrals in these formulas. The approximation by quadrature sums leads to an algorithm, which Gautschi calls *discretized Stieltjes procedure*. Interestingly, there is another way to exploit the formulas (2.1) and (2.2) for the computation of the recurrence coefficients: In some cases from the invariance (1.2) we can obtain recurrence relations between the integrals in (2.1) and (2.2). This will be made more precise in the next section. First we need a simple auxiliary result.

Lemma 1. *The integral* $\int x\tilde{\pi}_j(x)\tilde{\pi}_l(x)\,d\sigma(x)$ *can be evaluated as*

$$\int x\tilde{\pi}_j(x)\tilde{\pi}_l(x)\,d\sigma(x) = \begin{cases} 0 & \textit{for} \quad l < j-1 \\ \gamma_j^{-2} & \textit{for} \quad l = j-1 \\ \alpha_j\gamma_j^{-2} & \textit{for} \quad l = j \end{cases}. \tag{2.3}$$

Proof: In the first case $l < j-1$, the polynomial $x\tilde{\pi}_l(x)$ has degree $< j$ and thus is orthogonal to $\tilde{\pi}_j(x)$. If $l = j-1$, then we have from equation (1.1)

$$x\tilde{\pi}_{j-1}(x) = \tilde{\pi}_j(x) + \alpha_{j-1}\tilde{\pi}_{j-1}(x) + \beta_{j-1}\tilde{\pi}_{j-2}(x),$$

the second and third terms on the right-hand side of this equation being orthogonal to $\tilde{\pi}_j(x)$, and consequently

$$\int x\tilde{\pi}_j(x)\tilde{\pi}_{j-1}(x)\,d\sigma(x) = \int \tilde{\pi}_j^2(x)\,d\sigma(x) = \gamma_j^{-2}.$$

The last case $l = j$ follows analogously from the equation

$$x\tilde{\pi}_j(x) = \tilde{\pi}_{j+1}(x) + \alpha_j\tilde{\pi}_j(x) + \beta_j\tilde{\pi}_{j-1}(x).$$

Here the first and third terms of the right-hand side are orthogonal to $\tilde{\pi}_j(x)$, and we obtain

$$\int x\tilde{\pi}_j^2(x)\,d\sigma(x) = \int \alpha_j\tilde{\pi}_j^2(x)\,d\sigma(x) = \alpha_j\gamma_j^{-2}. \quad \blacksquare$$

We only remark that actually equation (2.1) is just another notation of the last case in (2.3). Equation (2.2) could be derived simply by multiplying the recurrence relation (1.1) by $\tilde{\pi}_{j-1}(x)$ and integrating.

§3 Self-similar measures

Now we will specify some types of operators A, for which the invariance (1.2) together with results like (2.3) lead to recurrence relations for the integrals $\int \tilde{\pi}_j^2(x)\,d\sigma(x)$ and $\int x\tilde{\pi}_j^2(x)\,d\sigma(x)$. For classical measures, the weights of which satisfy simple differential equations, the operator A will be some differential operator. Indeed, this leads to recursion formulas for α_j and β_j, as was shown in the literature long ago.

We will consider here two other types of operators closely related to *fractal interpolation functions* (see Barnsley and Harrington [1]).

3.1 Fractal interpolation functions

In their paper [1], Barnsley and Harrington give the following definition of fractal interpolation functions:

Definition 2. *Let $x_0 < x_1 < \ldots < x_N$. Let L_i be the affine map satisfying $L_i(x_0) = x_{i-1}$, $L_i(x_N) = x_i$, $i = 1, 2, \ldots, N$. Let $y_0, y_1, \ldots, y_N \in \mathbf{R}$. Let $-1 < \lambda_i < 1$, $i = 1, 2, \ldots, N$. Let $X = [x_0, x_N]$. Let $F_i : X \times \mathbf{R} \to \mathbf{R}$ satisfy*

$$|F_i(x, y_1) - F_i(x, y_2)| \le |\lambda_i||y_1 - y_2|, \quad x \in X, y_1, y_2 \in \mathbf{R}. \tag{3.1.1}$$

$$F_i(x_0, y_0) = y_{i-1}, \quad F_i(x_N, y_N) = y_i, \quad i = 1, 2, \ldots, N. \tag{3.1.2}$$

The fractal interpolation function, abbreviated FIF, associated with $\{L_i(x), F_i(x, y)\}_{i=1}^N$ is the unique function $f : X \to \mathbf{R}$ satisfying

$$f(L_i(x)) = F_i(x, f(x)), \quad i = 1, 2, \ldots, N, x \in X \tag{3.1.3}$$

As in [1], we deal with FIFs where

$$F_i(x, y) = \lambda_i y + q_i(x). \tag{3.1.4}$$

Thus, we assume that the cumulative distribution function of our measure σ (which for simplicity we will denote by $\sigma(x)$) is a FIF.

Lemma 2. *Let the cumulative distribution function of our measure σ be a FIF defined in Definition 2 and let F_i be defined by (3.1.4). Then σ satisfies the invariance property (1.2), where the operator A is defined via*

$$Ah(x) = \sum_{i=1}^N \left(\lambda_i h(L_i(x)) + \gamma_0^2 \int h(L_i(y))\, dq_i(y) \right) \tag{3.1.5}$$

for $h \in \mathbf{H}$ and we can assume

$$\mathbf{H} = \bigcup_{\mathbf{D}} \mathcal{P}_{\mathbf{D}}.$$

Proof: From the definitions (with σ instead of f) we obtain

$$\int h(x)\, d\sigma(x) = \sum_{i=1}^N \int_{[x_{i-1}, x_i]} h(x)\, d\sigma(x) = \sum_{i=1}^N \int h(L_i(x))\, d\sigma(L_i(x))$$

$$= \sum_{i=1}^N \left(\lambda_i \int h(L_i(x))\, d\sigma(x) + \int h(L_i(x))\, dq_i(x) \right),$$

and this can be written as

$$\int h(x)\, d\sigma(x) = \int \sum_{i=1}^N \left(\lambda_i h(L_i(x)) + \gamma_0^2 \int h(L_i(y))\, dq_i(y) \right) d\sigma(x)$$

for $h \in \mathbf{H}$. This, however, is an equation of type (1.2) with an operator A defined by (3.1.5). ■

Remark: From Definition 2 follows that the linear functionals L_i are of the form

$$L_i(x) = a_i x + b_i \tag{3.1.6}$$

with $a_i > 0$. This guarantees the continuity of the FIF, *i.e.* the absence of atomic parts in our measure. If we drop the assumption $a_i > 0$, our equations still make sense. Simply our distribution function will no longer be continuous, and our measure will be atomic. In the degenerate case, where $a_i = 0$ and $q_i \equiv 0$ for $i = 1, \ldots, N$, we obviously get a discrete measure with weight $\gamma_0^{-2}\lambda_i$ on $\{b_i\}$.

Obviously, the operator A converts polynomials to polynomials of the same degree, *i.e.* if $h \in \mathcal{P}_D$ then $Ah \in \mathcal{P}_D$, too.

Lemma 3. *There exists a representation*

$$\tilde{\pi}_j(L_i(x)) = \sum_{k=0}^{j} c_{jk}^{(i)} \tilde{\pi}_k(x) \tag{3.1.7}$$

with appropriate real coefficients $c_{jk}^{(i)}$. These can be determined recursively from their initial values $c_{00}^{(i)} = 1$ via the relation

$$c_{j+1,k}^{(i)} = a_i c_{j,k-1}^{(i)} + (a_i\alpha_k + b_i - \alpha_j)c_{jk}^{(i)} + a_i\beta_{k+1}c_{j,k+1}^{(i)} - \beta_j c_{j-1,k}^{(i)} \tag{3.1.8}$$

valid for $0 \leq k \leq j+1$ (we assume $c_{jk}^{(i)} = 0$ for $k < 0$ or $k > j$).

Proof: This can be obtained immediately from the relation (1.1): If we substitute the affine functions

$$L_i(x) = a_i x + b_i$$

instead of x and compare the coefficients of $\tilde{\pi}_k(x)$, then after an easy calculation we obtain the equation (3.1.8). ■

One easily observes that we can compute all coefficients $c_{jk}^{(i)}$, knowing only α_k, β_k up to $\alpha_{j-1}, \beta_{j-1}$. Now we can use the invariance property (3.1.5) with $h(x) = \tilde{\pi}_j^2(x)$ and $h(x) = x\tilde{\pi}_j^2(x)$ to derive recurrences for α_j and β_j.

Theorem 1. *Let the measure σ satisfy equation (1.2) with an operator A defined by (3.1.5), where the functionals L_i have the explicit form (3.1.6), and let the coefficients $c_{jk}^{(i)}$ be defined via Equation (3.1.7). Then γ_j^{-2} and α_j can be calculated from the equations*

$$\gamma_j^{-2}\left(1-\sum_{i=1}^{N}\lambda_i c_{jj}^{(i)2}\right)=\sum_{i=1}^{N}\sum_{k=0}^{j-1}\lambda_i c_{jk}^{(i)2}\gamma_k^{-2}+r_j \tag{3.1.9}$$

and

$$\begin{aligned}\alpha_j\gamma_j^{-2}&\left(1-\textstyle\sum_{i=1}^{N}\lambda_i a_i c_{jj}^{(i)2}\right)\\ &=\sum_{i=1}^{N}\lambda_i\left\{\sum_{k=0}^{j-1}[b_i c_{jk}^{(i)}+a_i(\alpha_k c_{jk}^{(i)}+2c_{j,k-1}^{(i)})]c_{jk}^{(i)}\gamma_k^{-2}\right.\\ &\qquad\left.+[b_i c_{jj}^{(i)}+2a_i c_{j,j-1}^{(i)}]c_{jj}^{(i)}\gamma_j^{-2}\right\}+s_j,\end{aligned} \tag{3.1.10}$$

where

$$r_j=\sum_{i=1}^{N}\int\tilde{\pi}_j^2(L_i(x))\,dq_i(x)$$

and

$$s_j=\sum_{i=1}^{N}\int L_i(x)\tilde{\pi}_j^2(L_i(x))\,dq_i(x).$$

Proof: In the first case $h(x)=\tilde{\pi}_j^2(x)$ we have

$$\int h(L_i(x))\,d\sigma(x)=\int\left(\sum_{k=0}^{j}c_{jk}^{(i)}\tilde{\pi}_k(x)\right)^2 d\sigma(x)=\sum_{k=0}^{j}c_{jk}^{(i)2}\gamma_k^{-2},$$

using the relations (2.3). Thus, (3.1.5) may be written as

$$\gamma_j^{-2}=\sum_{i=1}^{N}\sum_{k=0}^{j}\lambda_i c_{jk}^{(i)2}\gamma_k^{-2}+r_j.$$

If we solve the last equation for γ_j^{-2}, we have Equation (3.1.9).

Analogously, in the second case $h(x)=x\tilde{\pi}_j^2(x)$ using the formula

$$\int x\left(\sum_{k=0}^{j}c_{jk}^{(i)}\tilde{\pi}_k(x)\right)^2 d\sigma(x)=\sum_{k=0}^{j}c_{jk}^{(i)}(\alpha_k c_{jk}^{(i)}+2c_{j,k-1}^{(i)})\gamma_k^{-2}$$

we obtain

$$\alpha_j\gamma_j^{-2} = \sum_{i=1}^{N}\lambda_i\sum_{k=0}^{j}[b_i c_{jk}^{(i)} + a_i(\alpha_k c_{jk}^{(i)} + 2c_{j,k-1}^{(i)})]c_{jk}^{(i)}\gamma_k^{-2} + s_j.$$

This gives finally Equation (3.1.10). ■

Now it is clear how to design an algorithm for the recursive computation of the coefficients α_j and β_j:

- The functions q_i (if non-constant) are by assumption polynomials, so we can generate with standard methods appropriate discretizations of the integrals in the quantities r_j and s_j in the preceding theorem.
- If we have already computed all α_l and β_l with $l < j$ (*i.e.* none, if $j = 0$), we can calculate the new coefficients $c_{jk}^{(i)}$ from Equation (3.1.8) or take their initial values for $j = 0$.
- The polynomials $\tilde{\pi}_j$ are determined by the coefficients already computed, so we are able to calculate r_j and s_j.
- We compute γ_j^{-2} from (3.1.9) and $\alpha_j\gamma_j^{-2}$ from (3.1.10),
- and finally we obtain $\alpha_j = \alpha_j\gamma_j^{-2}/\gamma_j^{-2}$ and (if $j > 0$) $\beta_j = \gamma_j^{-2}/\gamma_{j-1}^{-2}$.

§4 Examples

In the following we give two examples illustrating the actual behaviour of the algorithm developed in this paper. The error magnification tabulated there is defined as the maximum of the absolute errors in the α_j and the relative errors in the β_j, divided by the machine eps. Of course, the errors can not always be calculated exactly — we have explicit formulas for the coefficients only in special cases. Thus, we estimate the errors by comparison with the results of the same algorithm computed with higher precision.

One remark is in place here: If we want to calculate *many* of the coefficients α_j and β_j with this algorithm (*e.g.* we could compute some thousands of them to get some feeling for the asymptotic behaviour of these numbers), then we must ensure that our computer can represent floating point numbers in a large range of magnitude. It's well-known that γ_j^{-2} in sufficiently regular cases behaves like a geometric sequence with ratio cap$S(\sigma)$, and thus for really large j numerical overflow or underflow occurs. We can avoid this to some extent by an appropriate scaling of the support of our measure to make its capacity near to 1. Of course, this is impossible if the capacity is zero, and it's difficult if we don't know the capacity of the support.

4.1 Special Riesz-Nagy measures

We consider here only the special case of Lemma 4.1 in the paper of Goh and Wimp [6]: The distribution function σ of the measure (with support $[0,1]$ and total weight ${\gamma_0}^{-2} = 1$) satisfies the equations

$$\frac{1-\tau}{1+\tau}\sigma(x) + \sigma\left(\frac{1}{2}\right) = \sigma\left(x + \frac{1}{2}\right) \quad \text{for} \quad x \in \left[0, \frac{1}{2}\right] \tag{4.1.1}$$

and

$$\sigma\left(\frac{x}{2}\right) = \frac{1+\tau}{2}\sigma(x) \quad \text{for} \quad x \in [0,1] \tag{4.1.2}$$

(we wrote down Equations 4.4 and 4.5 of the cited paper in our notations). If we substitute $\frac{x}{2}$ instead of x in (4.1.1) and use (4.1.2), we obtain the equation

$$\sigma\left(\frac{x+1}{2}\right) = \frac{1-\tau}{2}\sigma(x) + \sigma\left(\frac{1}{2}\right) \quad \text{for} \quad x \in [0,1]. \tag{4.1.3}$$

If we now compare the last two equations with Definition 2, we see that σ is indeed a FIF, where $N = 2$, $X = [0,1]$, $x_0 = 0$, $x_1 = \frac{1}{2}$, $x_2 = 1$, $L_1(x) = \frac{x}{2}$ and $L_2(x) = \frac{x+1}{2}$. Moreover, we have $\lambda_1 = \frac{1+\tau}{2}$ and $\lambda_2 = \frac{1-\tau}{2}$, whereas q_1 and q_2 are constant and thus can be ignored in our algorithm (the quantities r_j and s_j in Equations (3.1.9) and (3.1.10) are all equal to 0).

We remark here that the real parameter τ in [6] is assumed to be $\in (0,1)$. The above equations, however, suggest that the natural interval for τ should be $(-1,1)$. Of course, for $\tau = 0$ the distribution function is not a fractal, and the measure σ is not singular — in fact, this measure is simply Lebesgue measure restricted to $[0,1]$. For $\tau \neq 0$, of course, no explicit results are known, so we have to estimate errors. The numerical results show that our algorithm seems to be stable.

n	error magnification
10	3.00000000000000E+0000
20	1.25000000000000E+0001
50	4.00000000000000E+0001
100	4.30000000000000E+0001
200	6.34658203125000E+0001

Table 1. Error magnification for Riesz-Nagy measure with $\tau = 0.5$

Interestingly, there is some numerical evidence that the measure belongs to Nevai's class $M(\frac{1}{2}, \frac{1}{8})$ (see the famous Memoir [8] for a definition) not only for $\tau = 0$, but for all $\tau \in (-1, 1)$. However, the author is not able to give a proof for this conjecture.

4.2 Cantor measure

Cantor measure is defined by its cumulative distribution function — the well-known Cantor function. Standard properties of this function are mentioned in the paper [6]: The function is continuous, non-decreasing and satisfies the equations

$$\sigma\left(\frac{x}{3}\right) = \frac{1}{2}\sigma(x) \quad \text{for} \quad x \in [0, 1]$$

and

$$\sigma\left(\frac{x+2}{3}\right) = \frac{1}{2} + \frac{1}{2}\sigma(x) \quad \text{for} \quad x \in [0, 1].$$

Again, we see easily that this is a FIF: Here $N = 3$, $X = [0, 1]$, $L_1(x) = \frac{x}{3}$, $L_3(x) = \frac{x+2}{3}$ (the missing $L_2(x)$ would be $\frac{x+1}{3}$, but can be omitted, since $\lambda_2 = 0$ and q_2 is constant), $\lambda_1 = \lambda_3 = \frac{1}{2}$ and q_1 and q_3 constant. Again, only estimates of the errors in the algorithm are available. There are no explicit formulas for the recurrence coefficients, and as far as we are aware this is the first attempt to calculate numerically *many* of these coefficients. Recursive equations for the moments of Cantor measure based on self-similarity, of course, are well- known (see [6]). But numerical methods using moments are severely ill-conditioned. The same is true in this case for computations using modified moments — the orthogonal polynomials with respect to Cantor measure are very large in the gaps of the support, and the theory predicts bad condition of our problem (see the thorough discussion in the articles of Gautschi [4] or [5] and the alternative approach in our paper [2]). In their paper [6], Goh and Wimp tried to compute recurrence coefficients from ordinary moments using exact (rational) arithmetic. This avoids rounding errors, but one can expect rapid growth of the length of numerators and denominators of the rationals in the algorithm, making this method impractical.

The algorithm proposed here is reasonably fast, and the (estimated) error magnification reported below indicates stability. Of course, the recurrence coefficients β_j show a very irregular behaviour due to the gaps in the support of the measure (but we have $\alpha_j \equiv \frac{1}{2}$ by symmetry of the measure).

Note added in proof: As we learned only after finishing the work on this article, Giorgio Mantica in his 1996 paper [7] has designed a very

n	error magnification
10	1.97265625000000E+0000
20	4.86962890625000E+0000
50	2.25175781250000E+0001
100	3.15063476562500E+0001
200	7.98388671875000E+0001

Table 2. Error magnification for Cantor measure

similar method for computing the recurrence coefficients in the case considered here. However, he works with normalized polynomials and uses the invariance (3.1.5) with other polynomials to obtain equations for the recurrence coefficients. This makes his algorithm slightly more complicated, but he is able to prove numerical stability of the method. Moreover the use of normalized polynomials avoids the problem of overflow or underflow in the γ_j.

References

[1] Barnsley, M. F. and A. N. Harrington, The calculus of fractal interpolation functions, *Journal of Approximation Theory* **57** (1989), 14–34.

[2] Fischer, H.-J., On the condition of orthogonal polynomials via modified moments, *Zeitschrift für Analysis und ihre Anwendungen* **15** (1996), 223–244.

[3] Gautschi, W., On generating orthogonal polynomials, *SIAM J. Sci. Stat. Comp.* **3** (1982), 289–317.

[4] Gautschi, W., Computational aspects of orthogonal polynomials, in *Orthogonal Polynomials, Theory and Practice*, Paul Nevai (ed.), NATO ASI Series C, Vol. 294, NATO, 1990, 181–216.

[5] Gautschi, W., Orthogonal polynomials: applications and computations, in *Acta Numerica 1996*, Cambridge University Press, 1996, 45–119.

[6] Goh, W. and J. Wimp, Asymptotics for the moments of singular distributions, *Journal of Approximation Theory* **74** (1993), 301–334.

[7] Mantica, G., A stable Stieltjes technique for computing orthogonal polynomials and Jacobi matrices associated with a class of singular measures, *Constr. Approx.* **12** (1996), 509–530.

[8] Nevai, P., *Orthogonal polynomials*, Mem. Amer. Math. Soc. **18** (1978), No. 213.

Hans-Jürgen Fischer
Technische Universität Chemnitz
Fakultät für Mathematik
D-09107 Chemnitz
GERMANY
hfischer@mathematik.tu-chemnitz.de

Complex zeros of random Szegő polynomials

Jean-Daniel Fournier

Abstract. Orthogonal Szegő polynomials are built using a measure supported by the unit circle. Here its density is taken as the energy spectrum of a *random* real signal $X(m)$, with disrete m, $0 \leq m < N$. Given some probabilistic assumptions on the signal, one is after the statistics of the zeros of the associated polynomials. For *e.g.* white gaussian signal, the probability distribution of their moduli is extremely peaked at $\Lambda = N^{-1/(2p)}$, where p is the degree of the polynomial and N and p are large. Angularly, there are also $2p$ preferred sites. Other new akin crystallization phenomena take plake under less restricted probability hypotheses on the signal. A variety of results is described here, rigorously or on the basis of numerical experiments or asymptotic estimates.

§1 Introduction

We are concerned with the statistical properties of the zeros of polynomials of degree p in the complex variable z, the coefficients of the polynomial being real random numbers. Specifically one may consider the positions in $\mathbb{C}$ of the local maxima of the expectation value of the density of the zeros and study the "motion" of those maxima as the degree p varies and/or when some parameter of the probability distribution function (pdf) of the coefficients, *e.g.* its variance, changes (the mean being assumed zero). This in turn allows a classification of the pdf's of the coefficients of the polynomials into classes.

The introduction of probability in this classical problem produces elegant phenomena. The singular role of the real axis, which, under certain hypotheses on the real coefficients, attracts a finite fraction of the zeros, although it has a zero Lebesgue measure in the complex plane, is one famous example discussed by Mark Kac. These phenomena are often more simple than their counterparts in the deterministic case (if any) and simpler to tackle mathematically — at least at the level of asymptotic estimates considered here (we do not present proofs of convergence in the senses used in the stochastic processes studies). This is largely due to the powerful role of the central limit theorem and in turn to the particular properties of the gaussian law, for which we obtained some proofs (§4). Otherwise

Computational Methods and Function Theory (CMFT'97)
N. Papamichael, St. Ruscheweyh and E. B. Saff (Eds.), pp. 203–223

ISBN 981-02-3626-3

we present numerical evidences and/or scaling derivations and asymptotic estimates.

The motivation for introducing probability in this classical complex analysis problem is not only the mathematical interest — stressed above — of the problem. Random polynomials appear as models in quantum mechanics and statistical mechanics. Random orthogonal polynomials play a role in signal processing and data analysis methods. This is the way we entered the field. Most of this expository paper is devoted to the zeros of random orthogonal Szegő polynomials.

§2 Szegő polynomials

Classical orthogonal polynomial families are built using an orthogonality relation defined with the aid of a measure whose support is a segment (a, b) of the real line. Suppose this measure has a density, often denoted $w(x)$. Then the p zeros of the polynomial with degree p, belonging to the family, are simple and lie all inside the segment (a, b). When $w(x)$ has "peaks" or local maxima, the zeros tend to accumulate towards these maxima when p increases. Usually, the polynomials are built using a recurrence relation which involves the two preceding terms of the family. Sometimes closed formulae in terms of special functions of the variable x are known.

As far as I know, G. Szegő was the first to study orthogonal polynomials built using a measure whose support lies on a line in the complex. This class of polynomials are called after his name. We consider here a real positive measure whose support is the unit circle in the complex and assume that this measure has a density, which we denote $\rho(\theta)$, $\theta \in [-\pi, \pi]$, the points on the unit circle being denoted $e^{i\theta}$. One introduces the scalar product between two analytic functions

$$\langle f, g \rangle = \int_{-\pi}^{\pi} f(e^{i\theta})\overline{g(e^{i\theta})}\rho(\theta)\, d\theta, \tag{2.1}$$

where the bar denotes complex conjugation.

The family of orthogonal Szegő polynomials $S_p(z)$ is then entirely prescribed by the conditions

$$S_p(z) = \sum_{q=0}^{p} g_{(p-q)}^{(p)} z^q, \quad \langle S_p, S_r \rangle \propto \delta_{p,r}, \quad g_{(0)}^{(p)} = 1. \tag{2.2}$$

The last condition (coefficient of the highest degree term equals unity) replaces a normalisation condition on the polynomial. Such polynomials are called "monic" in english (and "unitaire" in french). This condition arises naturally from the signal processing origin of the problem. It is an important hypothesis in the sense that later on, all the g's shall be random except $g_{(0)}^{(p)} = 1$.

The usual two-term recurrence relation is replaced here by a matrix recurrence relation

$$\begin{bmatrix} S_p(z) \\ S_p^*(z) \end{bmatrix} = \begin{bmatrix} z & \delta_p \\ z\delta_p & 1 \end{bmatrix} \begin{bmatrix} S_{p-1}(z) \\ S_{p-1}^*(z) \end{bmatrix}, \tag{2.3}$$

where $S_p^*(z)$ is defined by

$$S_p^*(z) = z^p S_p(z^{-1}) \tag{2.4}$$

and the seed of the recurrence is

$$S_0(z) = S_0^*(z) = 1. \tag{2.5}$$

The real number δ_p is called Levinson reflection factor, for reasons due to the history of algorithmics and wave propagation theory. It is given by

$$\delta_p = -\frac{\langle zS_{p-1}, 1\rangle}{\langle S_{p-1}^*, 1\rangle} \tag{2.6}$$

and is such that $|\delta_p| < 1$, which is the technical condition which ensures that all zeros of $S_p(z)$ lie strictly inside the unit disk.

Finally a closed determinantal formula is known,

$$S_p(z) = \begin{vmatrix} C(0) & C(1) & \dots & C(p-1) & C(p) \\ C(1) & C(0) & \dots & C(p-2) & C(p-1) \\ \vdots & \vdots & \ddots & \vdots & \vdots \\ C(p-1) & C(p-2) & \dots & C(0) & C(1) \\ 1 & z & \dots & z^{p-1} & z^p \end{vmatrix}, \tag{2.7}$$

which is a sort of complemented Toeplitz matrix. The C's are the trigonometric moments of the measure

$$C(k) = \int_{\pi}^{\pi} e^{-ik\theta} \rho(\theta)\, d\theta. \tag{2.8}$$

Both (2.3) and (2.7) are useful in building proofs, both in the deterministic and random case. Note however that both can be extremely heavy to manipulate in the deterministic case, while the introduction of probability allows to obtain approximate but very simple forms for the Szegő polynomials in some asymptotic cases (see §4). (Of course there remains then the question of the localisation of their zeros.)

Let us mention that using (2.1), (2.2), (2.6), (2.8), it is an easy matter to give the following expression for the Levinson factor

$$\delta_p = -\frac{\sum_{q=0}^{p-1} g_{(p-q-1)}^{(p-1)} C(q+1)}{\sum_{q=0}^{p-1} g_{(p-q-1)}^{(p-1)} C(p-q-1)}. \tag{2.9}$$

It is interesting to note that any reference to the measure has disappeared. Only its trigonometric moments matter to build the S_p's with (2.3), (2.4), (2.5), (2.9). Thus the whole setting of the polynomials being orthogonal is apparently gone. Indeed some signal processing textbooks (where the C's are viewed as correlation factors, see §3) just ignore this aspect, and many proofs can be obtained in one language or the other, although some times more easily in one of the two. Also, the orthogonal polynomial point of view sheds more light on the behavior of the zeros, which is our primary interest here. Finally the theory of orthogonal polynomials has roots in common with other mathematical questions: special functions, Padé approximants, etc., some of them of major importance to data analysis and signal processing.

§3 Connection with signal processing; random Szegő polynomials

In the sequel we will be interested in the zeros of random Szegő polynomials. By this we mean that the density $\rho(\theta)$ will be taken randomly from a given probability space. By (2.1), (2.3), (2.4), (2.5), (2.6), the g's are then themselves random, and so also the (in general complex) zeros of the S_p's. One then can ask questions about their mean location, their correlation, etc. It turns out that, although in the purely deterministic case only partial answers can be obtained, in the purely random case the phenomena are more clear-cut; specifically the location of the maxima of the expectation value of the density of presence of the zeros exhibit remarkable patterns and scenarios.

We note in passing that such a problem where the state of a system is defined using a measure which is itself taken randomly is called a "disorder problem" by physicists and a "double stochasticity" problem by people from signal processing.

Finally, for technical reasons and because of the way we entered the problem, the probabilistic hypothesis will not be formulated in terms of the density $\rho(\theta)$ itself, but in terms of a signal which has the C's (see above (2.7), (2.8)) as correlation factors. We need thus a slight detour in the concepts of signal processing.

3.1 Connection with signal processing and building the measure

We consider a real signal X from which one takes N successive values, at equally spaced instants denoted $m = 0, 1, \ldots, N-1$. To this (discretized real time) signal, one associates its autocorrelation factors

$$C(k) = \sum_{m=0}^{N-1} X(m)X(m+k), \qquad k = 0, 1, \ldots, N-1. \tag{3.1.1}$$

(Outside the sampling time window, X is considered to vanish). Parity allows construction of $C(k)$ for negative k

$$C(k) = C(|k|), \qquad k = -N-1, \ldots, -1. \tag{3.1.2}$$

Of course, these C's of §3 are the same as the C's of §2. One then introduces the discrete Fourier transform of the C's

$$\mathcal{E}_N(\theta) = \frac{1}{2\pi} \sum_{k=-(N-1)}^{k=N-1} C(k)e^{ik\theta}. \tag{3.1.3}$$

Trivial manipulations on (3.1.3) using (3.1.1), (3.1.2) show that

$$\mathcal{E}_N(\theta) = \frac{1}{2\pi} \left| \sum_{m=0}^{N-1} X(m)e^{-im\theta} \right|^2, \tag{3.1.4}$$

hence the notation. $\mathcal{E}_N(\theta)$ is the squared modulus of the Fourier transform of the signal itself and is thus often called the energy spectrum of the signal. From (3.1.4) it is clear that $\mathcal{E}_N(\theta)$ is a real positive quantity, periodic of period 2π. It is thus a good candidate to be the density of a measure "sitting" on the unit circle. Indeed, it is nothing else than our previous $\rho(\theta)$. Of course, for (3.1.3) one has a reciprocal formula, which was given in (2.8). Formula (3.1.3) can also be read as the squared modulus of the z-transform of the sequence of numbers $\{X(m)\}$, namely $T_z\{X\} = \sum_{m=0}^{N-1} X(m)z^{-m}$, taken for z on the unit circle. The showing-up of z-transforms can be viewed as the origin of the role of complex analysis in signal processing. We have no space here to dwell on this matter, developed in some textbooks and previous introductory papers by members of this CMFT community.

Not only the measure has a signal processing interpretation, but the polynomial S_p's themselves: they can be viewed as the denominator of the transfer function of a linear Wiener filter (*i.e.* the z-transform of its discrete output to an impulse input) and as such serve to perform the harmonic analysis of stochastic signals. They can also be viewed as solution of the extremalisation problem in the l_2 norm set down by N. Wiener (and solved by him, Levinson and Szegő) in his formalization of the prediction problem (weather forecasting, stock exchange prediction). Again we cannot dwell more on this here.

The connection between signal processing and orthogonal polynomials was already presented in this series of Conferences. (See also the Conference Paper in Columbus by Ph. Delsarte and Y. Génin.)

Now, of course, the signal X can be purely deterministic, purely random, or a mixture of the two. The first case served as an introduction to the domain. In the nice series of papers by L. Pakula and W. B. Jones, O. Njåstad and E. B. Saff, one considers the case where $X(t)$ is a finite sum of cosine functions, with amplitudes, frequencies and phases being fixed real quantities and the frequencies ω_l, $l = 1, \ldots, b$, being incommensurate. Numerical indications and theorems give a picture of the phenomenon of "attraction" of $2b$ zeros of the polynomial $S_p(z)$ by the "phase-points" $e^{\pm i\omega_l}$ sitting *on* the unit circle. These results aroused our interest for the "motion" of zeros in the second case, the purely random case. I give below some new results. Finally, I am not aware of any publication in the academic literature on the third case, a subject which we plan to tackle in the future and which is not only mathematically appetizing but obviously the most important in practice.

3.2 Random Szegő polynomials

In the sequel, the $\{X(m)\}$, $m = 0, \ldots, N-1$, will be real random numbers. Later on we will specify the hypotheses on their probability law. It is clear that the Szegő polynomials built from the corresponding measure will have real random coefficients. They form a particular family among all possible polynomials with real random coefficients. Our goal is the investigation of the statistical behavior of their zeros, mainly the complex ones.

Given a probability law for the X's, one has two parameters at hand: the degree p of the polynomial and the length N of the sample of the signal. In this theoretical study, we are interested in the limit where both N and p are large. However one cannot for example fix p and let N go to infinity: usually all zeros will collapse at the origin of the complex plane. One thus has to do asymptotic expansions, with some ordering. We will usually assume

$$1 \ll p \ll N, \tag{3.2.1}$$

with some additional prescription for the position of $\ln N$ with respect to p.

Also for simplicity we assume that the variables X have zero mean, and specifically that their probability laws are even: $\mathcal{P}(X) = \mathcal{P}(-X)$.

§4 The white noise Gaussian case

I discuss this case for reference for the sequel and to show the key points of the proofs. Part of the material of this chapter was published in Europhysics Letters and Journal of Statistical Physics. In the latter paper, a formula equivalent to (4.3.14) was given for $\mathcal{N}_{int}(R)$, the mean number of roots inside a disk of radius R, along with its proof. However no visual illustration was given *e.g.* of the extremely sharp behavior of this function in the asymptotic regime, nor of the very good confirmation by numerical experiments. This is why we illustrate the present chapter with Figures 1, 2, 3 and Table 1.

4.1 Most of the zeros sit on a circular quasi-crystal

In this section, we mention the hypotheses and the most interesting results. The theory and the numerical findings are to be found in the subsequent sections and the figures.

We assume here that the X's are taken randomly independently. In other words the process $X(t)$ is a white noise, that is the expectation value of the autocorrelation is a Dirac at the origin and the expectation value of the energy spectrum is flat. (Of course there are fluctuations around those means in each particular "realization"). Moreover the variables $X(m)$ are taken randomly according to a given probability law, the same for all X's. One speaks of i.i.d. (independent identically distributed) random variables. In the present chapter, this identical probability law is the Laplace-Gauss distribution, with zero mean and variance unity.

In spite of the absence of structure in the expectation value of the density $\mathcal{E}_N(\theta)$, the expectation value of the density of zeros exhibits a considerable structure, which we have termed a "circular quasi-crystal". First, the expectation value of the modulus of the zeros is extremely peaked (in the regime (3.2.1)) around $N^{-\tau/p}$, where τ tends to one-half as one goes deeper and deeper in the regime (3.2.1). With the additional assumption

$$\ln N \ll p,$$

the circle on which most of the zeros lie approches the unit circle. Second, the expectation value of the argument of the zeros exhibits $2p$ peaks, located at the $2p$th's roots of unity. Most of the probability of presence is concentrated near these points. Loosely speaking, a small region (in the complex) near any of these points has a probability $1/2p$ for containing a zero of the random $S_p(z)$. In a mathematically suggestive language, we say that each location contains in the mean "one-half zero". We note that p zeros have to be set down in $2p$ prefered sites, so that, in a typical realization, half of the sites are occupied and half are empty — under the constraints of complex conjugacy and of the slight attracting power of the real line. Thus, in a physically suggestive language, we say that each realization is a crystal with defects, or a quasi-crystal.

4.2 The low disorder expansion

For small or moderate N, it is difficult to say anything about the coupling of the correlation factors, the C's, among themselves (respectively: the δ's, the g's) and their order of magnitude. On the contrary, for large N it is known that the first $C's$ decouple and that their order of magnitude and fluctuations are described by the central limit theorem formulae, which with our hypotheses read

$$C(k) \simeq \begin{cases} N-(2N)^{1/2}\hat{\delta}_0, & k=0 \\ -N^{1/2}\hat{\delta}_k, & k\neq 0 \end{cases}, \quad \text{and} \quad N \gg 1. \tag{4.2.1}$$

Here the $\hat{\delta}$'s are independent random gaussian variables with zero mean and variance unity. (The minus sign in (4.2.1) is a convenient notation for the sequel. It has absolutely no meaning since the gaussian law is even.) Within the same order of approximation, one can then show by recurrence that

$$\delta_k \simeq -\frac{C(k)}{C(0)} \simeq N^{-1/2}\hat{\delta}_k \simeq g^{(p)}_{(k)}, \qquad 0 < k \leq p,$$

while of course $g^{(p)}_{(0)}$ remains equal to unity. Thus in this regime each of the coefficients g's is expressed in terms of only one Levinson factor, and an explicit asymptotic form of the random Szegő polynomial is obtained,

$$\begin{aligned} S_p(z) \simeq \tilde{S}_p(z) \quad \equiv \quad & z^p + N^{-1/2}\hat{\delta}_1 z^{p-1} + \ldots \\ & + N^{-1/2}\hat{\delta}_{p-1} z + N^{-1/2}\hat{\delta}_p, \qquad N \gg 1. \end{aligned} \tag{4.2.2}$$

As to the terminology, we point out that once one has understood that the δ's have a small dispersion, one can perform, to the lowest order of approximation, the iteration of the recurrence (2.3), in neglecting any non-linear combination of δ's. Physicists would call this a low disorder expansion and it leads to the same result (4.2.2).

We have thus arrived to a well defined auxiliary probabilistic problem which deserves a treatment of its own. The problem becomes trivial if N goes to infinity while p is fixed. We concentrate here on the asymptotic regime where N is large but fixed and p moderate, with of course $1 < p < N$.

4.3 An auxiliary probabilistic problem

(i) The scaling laws and the quasi-crystal, heuristically

Given the definition of the $\hat{\delta}$'s, it is not difficult to guess that the typical length scale in the distribution of the zeros of $\tilde{S}_p(z)$, defined in (4.2.2), is

$$\Lambda = N^{-1/2p}, \tag{4.3.1}$$

at least if Λ is small, that is if

$$p \ll \ln N. \tag{4.3.2}$$

Indeed, if one performs a change of variable $z = \Lambda y$, the problem of finding the random zeros of $\tilde{S}_p(z)$ is mapped into the problem of finding the y_α's, which are the random roots of

$$\begin{aligned} Q_p(y) &= \Lambda^{-p}\tilde{S}_p[z(y)] \\ &= y^p + \Lambda^{p-1}\hat{\delta}_1 y^{p-1} + \ldots \\ &\quad +\Lambda^{p-q-1}\hat{\delta}_q y^{p-q-1} + \ldots + \Lambda\hat{\delta}_{p-1}y + \hat{\delta}_p. \end{aligned} \tag{4.3.3}$$

Here, only the first and the last term have coefficients whose magnitude is of order 1. All others have extra Λ factors and the sum of their typical orders of magnitude is easily evaluated, to first non trivial order, to $\Lambda(1+\Lambda)$ (if $p > 2$). We retain only the two dominant terms. We recall that $\hat{\delta}_p$ is a real random variable with a typical order of magnitude 1, but with both signs possible. (The large deviations — unlikely within the gaussian hypothesis — would anyway be erased in taking the power $1/p$ to calculate the zeros. In other words they are flattened out at the level of other terms already neglected in this approximation.) The picture of a crystal with $2p$ sites on a circle of radius Λ then emerges for the most likely locations of the random zeros of $S_p(z)$. It comes out here as a result of the gaussianity of the signal X and of the assumed asymptotics (*cf.* (3.2.1) and (4.3.2)) and seems to depend crucially on the interplay between the first (deterministic) term and the last (random) term of $S_p(z)$. It is however not dramatically modified when one takes into account properly all the intermediate terms in (4.2.2) or (4.3.3) and even if one relaxes (4.3.2), as shall be seen in the next subsection. The influence of the gaussianity will again be discussed at the end of this chapter and in §5.

(ii) Some exact results and an outline of their proof

We first extend the problem (4.2.2) in considering generalized random polynomials

$$F(z) = \Phi(z) + \sum_{q=1}^{p} \hat{\delta}_q f_q(z),$$

where $\Phi(z)$ and the f's are real-analytic functions ($f(\bar{z}) = \overline{f(z)}$) and the other notations are as before. (In spite of the higher rigor of this subsection, I do not elaborate here on the probabilistic language, to keep in with the spirit of the rest of the paper. Of course, here every quantity should be thought of as depending on ω, where $\omega \in \Omega$, and Ω is some probability space, etc.) The extreme case $\Phi \equiv 0$ will be called homogeneous. The Szegő case is recovered for

$$\Phi(z) = N^{1/2} z^p, \qquad f_q(z) = z^{p-q}. \tag{4.3.4}$$

I will concentrate here on the distribution of complex (non real) random zeros of $F(z)$. The real zeros need a special treatment because the distribution is singular on the real axis. One introduces the counting measure

$$\sigma(z) = \sum_\alpha \delta^{(2)}(z - z_\alpha), \tag{4.3.5}$$

where the z_α's are the zeros of a given realization of $F(z)$. Here $\delta^{(2)}$ is the Dirac distribution on $\mathbb{R}^2$. Introducing the proper Jacobian and calculating it, using the Cauchy-Riemann relations, (4.3.5) may be rewritten

$$\sigma(z) = \left|\dot{F}(z)\right|^2 \delta^{(2)}[F(z)]. \tag{4.3.6}$$

(Here and in the sequel of this subsection, the dot stands for derivation with respect to the complex variable z). The goal is now to prove properties of the random $\sigma(z)$ described by (4.3.6), for example to obtain its moments and, to start with, its expectation value

$$D(z) = \mathbb{E}\{\sigma(z)\}.$$

The gaussianity hypothesis on the $\hat{\delta}$'s enters in here in a crucial way. Because gaussianity is preserved by linear transformation, $F(z)$ and $\dot{F}(z)$ are, for any given fixed z, correlated random *gaussian* variables. One can thus at once write their joint probability distribution

$$\mathcal{P}(F, \dot{F}; z) = \left(\frac{1}{2\pi}\right)^2 \frac{1}{\sqrt{\det \Delta}} \exp\left\{-\frac{1}{2}\left(F, \dot{F}\right) \Delta^{-1} \begin{pmatrix} F \\ \dot{F} \end{pmatrix}\right\}, \tag{4.3.7}$$

where Δ is the 4×4 correlation matrix. Combining (4.3.6) and (4.3.7), the mean density of the zeros obtains

$$D(z) = \int \left|\dot{F}\right|^2 \mathcal{P}\left(0, \dot{F}; z\right) d^{(2)}\dot{F}. \tag{4.3.8}$$

As any gaussian integral, (4.3.8) can be expressed in terms of the invariants of the matrix Δ or its inverse. The elements of Δ are the cross-correlations of the elements of the 4-vector $(\operatorname{Re} F, \operatorname{Im} F, \operatorname{Re} \dot{F}, \operatorname{Im} \dot{F})$. For example

$$\Delta_{11} = \mathbb{E}\{\operatorname{Re} F \cdot \operatorname{Re} F\}, \ldots . \tag{4.3.9}$$

To lighten the formalism, I shall from now on consider the homogeneous case. The monic case, which clearly is our primary interest, only adds technicalities which can be found elsewhere. With this assumption, we carry on with the example above,

$$\begin{aligned} \Delta_{11} &= \mathbb{E}\left\{\left[\sum_q \hat{\delta}_q \operatorname{Re} f_q\right]\left[\sum_r \hat{\delta}_r \operatorname{Re} f_r\right]\right\} \\ &= \sum_{q,r} \delta_{q,r} \operatorname{Re} f_q \operatorname{Re} f_r = \sum_q \left[\operatorname{Re} f_q\right]^2, \ldots . \end{aligned}$$

One sees that one can write Δ as a block matrix, made of four 2×2 matrices. Symbolically, one writes

$$\Delta = \begin{bmatrix} (ff) & (f\dot{f}) \\ (\dot{f}f) & (\dot{f}\dot{f}) \end{bmatrix}. \tag{4.3.10}$$

Similarly Δ^{-1} can be seen as made of four blocks. To perform the gaussian integral (4.3.8), one needs only to know the lower right block of Δ^{-1}, the 2×2 matrix which stands at the $(2,2)$ corner of the block matrix Δ^{-1}. The properties of Δ ensure that this block is itself invertible and we denote it by its inverse M. That is, symbolically,

$$[\Delta^{-1}]_{2,2} = M^{-1}.$$

The integral (4.3.8) can then be rewritten

$$\begin{aligned} D(z) &= \frac{1}{2\pi}\frac{\sqrt{\det M}}{\sqrt{\det \Delta}} \\ &\quad \cdot \int \frac{1}{2\pi}\frac{u_1^2+u_2^2}{\sqrt{\det M}} \exp\left\{-\frac{1}{2}(u_1,u_2)M^{-1}\begin{pmatrix}u_1\\u_2\end{pmatrix}\right\} du_1 du_2, \end{aligned}$$

where the two-variables gaussian integral is equal to $\operatorname{Tr} M$. Thus

$$D(z) = \frac{1}{2\pi}\frac{\sqrt{\det M}}{\sqrt{\det \Delta}} \operatorname{Tr} M. \tag{4.3.11}$$

At this point, there are two steps left. First, some linear algebra lemmas allow to link the invariants of Δ and of M, which enter in (4.3.11), to the invariants of the building blocks of Δ. Specifically we use a sort of block-Cholesky triangular decomposition of Δ and Δ^{-1} to prove that

$$M = (\dot{f}\dot{f}) - (\dot{f}f)(ff)^{-1}(f\dot{f}),$$

and

$$\det \Delta = \det(f, f) \det M,$$

where the notations of (4.3.10) are used. With the same notations,

$$D(z) = \frac{1}{2\pi} \frac{\operatorname{Tr} M}{\sqrt{\det(f, f)}}. \tag{4.3.12}$$

Second, the invariants entering in (4.3.12) have to be evaluated for the given choice of functions f_q's. Even in the case of simple monomials mentioned in (4.3.4), it involves heavy trigonometric manipulations which can not be reproduced here. Once it is done, it is still difficult to figure out the content of the formula, except in resorting to numerical plots (the computer playing there just the role of a pocket calculator) or in studying some asymptotic regime — to which the above calculations are not restricted in principle.

We close this subsection with some results for the monic-Szegő case (4.3.4). From (4.3.9) and on, a heavier calculation can be done along the same lines. (The functions Φ and $\dot{\Phi}$ play the role of a mean field in the gaussian integrals.) A numerical plot of $D(z)$ as a function of r and θ, where $z = re^{i\theta}$, provides a theoretical confirmation of the quasi-crystallisation phenomenon in the regime (3.2.1). The additional assumption (4.3.2) used in the heuristic derivation, is not necessary for it, and it survives even in the opposite regime

$$\ln N \ll p, \tag{4.3.13}$$

that is when the concentration circle approaches the unit circle. Nevertheless, when p becomes very large (within (3.2.1)), the oscillations in $D(r, \theta)$, as a function of θ, are progressively washed out — except for the real line: nothing surprising. Finally, if one integrates the expectation value $D(r, \theta)$ of the density of zeros over a disk of radius R, one finds that the expectation value $\mathcal{N}_{int}(R)$ of the number of zeros contained in this disk is represented by

$$\mathcal{N}_{int}(R) \simeq p - \left[\frac{1}{2\ln R} - \frac{p}{R^{2p} - 1}\right] \exp\left(\frac{-2N \ln R}{1 - R^{-2p}}\right) \tag{4.3.14}$$

in the regime (3.2.1) and (4.3.13). This function has an extremely sharp behavior, as can be seen on Figure 1 and on Table 1, based on the values $N = 1024$ and $p = 50$ taken as example.

The radial concentration effect is very strong and very well described by the simple scaling law (4.3.1) obtained heuristically and in numerical

R	$R_- = 0.901$	$\Lambda = 0.933$	$R_* = 0.94727$	$R_+ = 0.965$
$\mathcal{N}_{int}(R)$	5.1	12.7	25	45.6

Table 1. Numerical illustration of the behavior of the mean number of zeros inside a disk of radius R, $\mathcal{N}_{int}(R)$, according to (4.3.14). The parameters are as for Figure 1: $N = 1024$, $p = 50$. Λ is the heuristic prediction, $\Lambda = N^{-1/2p}$, for the concentration radius. Note that $R_* \simeq 1.015\Lambda$, $R_- = \Lambda - 0.032$, $R_+ = \Lambda + 0.032$.

experimemts. The radius R_* at which half of the zeros has been absorbed differs from Λ only by a factor 1.015 (see Table 1. Note the very elongated scale on the x-axis of Figure 1). The theoretical prediction even underestimates the concentration: on Figure 1, the theoretical curve is above the numerical experiment curve for $R < \Lambda$ and the reverse for $R > \Lambda$.

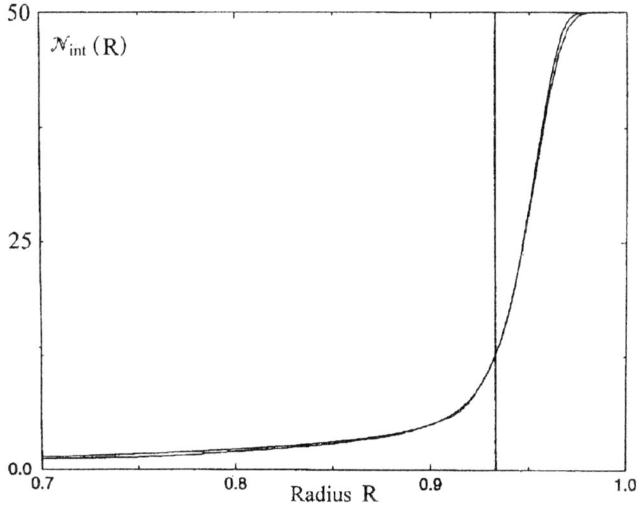

Figure 1. Expectation value $\mathcal{N}_{int}(R)$ of the number of complex zeros inside a disk of radius R, centered at the origin. The random Szegő polynomials of degree $p = 50$ are built out of random gaussian sequences of length $N = 1024$. The asymptotic prediction of the probabilistic theory (formula (4.3.14)) coincides with the result of numerical experiments in the concentration region, where most of the zeros lie. The vertical bar marks the approximate evaluation of the position of this region, $\Lambda = N^{-1/2p}$ (formula (4.3.1)), which turns out to be accurate within $1.5 \cdot 10^{-2}$ (see Table 1 and Section 4.3). The theoretical curve is above the numerical experiment curve for $R < \Lambda$ and the reverse for $R > \Lambda$.

4.4 Significance of the findings and numerical experiments

We have performed numerical experiments on this problem: one generates the random X's, one builds iteratively the polynomial $S_p(z)$, and one performs a careful search of its zeros. Finally after repeating $\mathcal{M}$ times ($\mathcal{M} = 10000$ *e.g.*) the process, some statistical treatments are carried out: radial and angular histograms, integrated mean density, check of the scaling law (4.3.1). We discussed just above the resulting picture given on Figure 1 for the experimental $\mathcal{N}_{int}(R)$, using independent gaussian X's. Radial and angular histograms are given in Figure 2, taken from an other experiment where X's with a certain level of correlation were used, hence the name "pink noise". They again demonstrate the features of the circular quasi-crystal. We have also performed extensive experiments using independent X's but taken according to a large class of probability laws $\mathcal{P}$'s other than gaussian (but the same for all X's). If $\mathcal{P}$ does not allow large deviations, that is if its second moment is finite, we have found evidence that the findings concerning the gaussian case always extend to those new cases. We believe it is indeed rigorously true because of the role of the central limit theorem. In other words a more elaborate derivation in Section 4.2 ought to map also the problem with the non-gaussian (but not too much so) laws into the same auxiliary problem (4.2.2). Then the considerations of Section 4.3 (ii) apply.

§5 Relaxing the Gaussian hypothesis: Levy classes

This is ongoing work with G. Mantica. One figure in a particular case has been included in the Proceedings of the Chamonix School. I give here other preliminary numerical and theoretical results.

We still consider white noise, with the same probability law $\mathcal{P}(X)$ for each variable $X(m)$, in other words the X's are still i.i.d. variables. The law $\mathcal{P}(X)$ is still such that it is normalizable and $\mathcal{P}(X) = \mathcal{P}(-X)$, so the mean of X is zero. However, now we choose our $\mathcal{P}$'s in the Levy classes, thus leaving the realm of the quasi-gaussian laws discussed at the end of the preceding chapter. Precisely, we assume an algebraic decreasing of $\mathcal{P}$ for large modulus of X

$$\mathcal{P}(X) \sim \frac{1}{X^{1+\mu}} \quad \text{for} \quad X \gg 1 \quad \text{with} \quad 0 < \mu \leq 2.$$

This is in contrast with the very fast decrease of the gaussian law and allows large excursions for X in any given realization. This slow decrease forbids the existence of the second moment of $\mathcal{P}$ and modifies the large N behavior of the correlation factor C's. However, we found numerically that

(a)

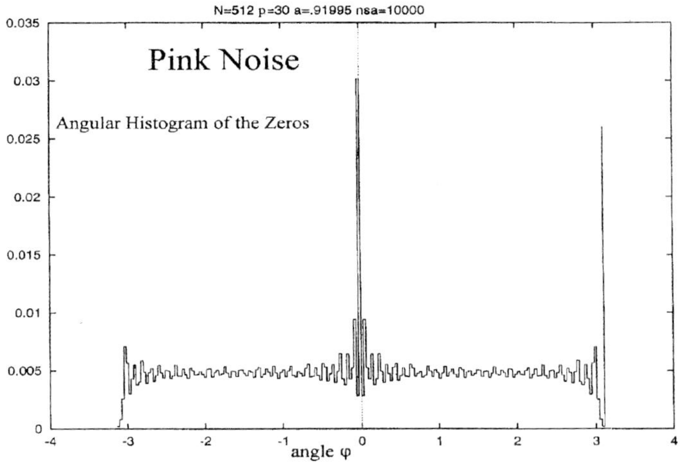

(b)

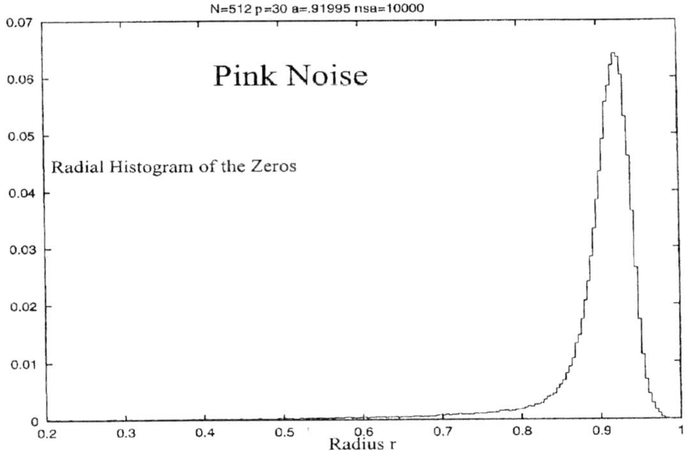

Figure 2. Statistical treatment of the results of a numerical experiment. See Sections 4.3, 4.4 and §6. The number of independent lots (denoted $\mathcal{M}$ in the body of the paper and *nsa* on the figures) is 10^4. Each sample is a realization $\{X(m)\}$ of an $AR(1)$ process of length $N = 512$ and one studies the random zeros of the polynomial of degree $p = 30$. There is some correlation among the X's, measured by the parameter a. In the present experiment, the statistical behavior is akin to the white case. On the angular histogram (Figure 2 (a)), there are 58 peaks in the complex, around the sites of the crystal, and two peaks exactly in the real, with slightly more than their share, as expected. All the radii (Figure 2 (b)) are concentrated near Λ.

the circular quasi-crystallization phenomenon persists. Only the value τ entering the radius $N^{-\tau/p}$ of this circle is changed. Numerically, it definitely does not settle to one-half. Clearly the fluctuations, due to a finite length N of the signal and a finite number $\mathcal{M}$ of lots taken randomly in the computer, are now larger and make a precise conclusion more difficult. However, we have, within reasonable error bars, a numerical agreement with a very heuristic (too heuristic to be presented here!) suggestion according to which

$$\tau = \frac{1}{\mu}.$$

Actually, after a discussion with D. Talay (Stochastic Processes group, Inria, Sophia-Antipolis, France) and with V. Petersen at the CMFT meeting at Cyprus, I got convinced that to stabilize the problem numerically and to have a better setting mathematically one should first introduce a weight on certain of the X's, in other words process a pre-filtered signal. We have some hope in this direction.

§6 Relaxing the whiteness hypothesis: correlated noises

This is ongoing work with G. Mantica and Th. Ayguesparsses. An internal report with preliminary results has been written.

We noticed in §4 that although the expectation value of the measure in the uncorrelated case has no structure, there appears a lot of structure in the expectation value of the distribution of zeros, namely the circular quasi-crystallization.

A puzzling question was: what would survive of this in the case of correlated noise? In the modeling of noises appearing in technology or natural phenomena, it is customary to consider random variables $X(m)$ such that a linear combination of $X(m)$ and of a few X's taken at previous times (an auto regressive combination) is equal to a linear combination of successive independant gaussian variables (a moving average source). We took such a model, where

$$X(m) - aX(m-1) = W(m), \qquad m \geq 1,$$

with $X(0)$ and $W(m)$, $m = 1, \ldots, N-1$, being independent centered gaussian variables. The parameter a is a real number with $|a| < 1$. In the above terminology, this process is called an AR(1) model (one retarded term in the left hand side and no retarded term in the right hand side). Clearly the X's are no more independent. Some physicists and meteorologists call this model the red noise, because there is correlation in the large time scales. Precisely, let us normalize the energy spectrum, $\mathcal{E}_N(\theta)$, by the expectation value of the first auto-correlation factor, $\mathbb{E}\{C(0)\}$, thus introducing

$$\hat{\mathcal{E}}_N(\theta) = \frac{\mathcal{E}_N(\theta)}{\mathbb{E}\{C(0)\}}.$$

The expectation value of this normalized spectrum takes asymptotically (large N) the form

$$\mathbb{E}\{\hat{\mathcal{E}}_N(\theta)\} \simeq \frac{1}{2\pi}\frac{1-a^2}{1-2a\cos\theta + a^2}, \qquad N \gg 1,$$

which indeed exhibits a peak at low frequencies, that is for small θ.

There is plenty of room for various phenomena in this problem with three parameters (N, p, a). Let us just describe the scenario we found for N and p fixed and taken in the asymptotic regime (3.2.1), and increasing a, with $0 < a < 1$. The interesting result is that there is no "melting" of the circular quasi-crystal, but more a rearrangement of it. The radius of the circle is the same as in §4, but there is a critical range of values of a for which the two sites of the crystal closest to the positive real axis $(e^{\pm i\pi/p})$ empties onto the positive real axis. As a increases, the quasi-crystal rearranges itself. It has $2p-2$ sites for $p-1$ zeros. In the mean time, the additional real zeros tend towards the unit circle, creating a "precursor peak" in the expectation value of the distribution of the modulus of the zeros. The transition between the two regimes is illustrated by the two groups of Figures 2 and 3. In the first two, a (or N) is not large enough for the correlation between the X's to be "felt" by the zeros. They behave statistically like in the white noise case described in §4, although the mean energy spectrum of the experiment is definitely not flat. Hence the name "pink noise". In the second group of figures, the transition has taken place. A new quasi-crystal shows up in the complex. On the real line, there is, in the mean, "one half" root at $-\Lambda$, "one half" root at Λ, and "one" precursor root between Λ and 1. The color or correlation of the "red noise" is now acting.

Numerically, the location of the precursor peak tends to the point 1 as a keeps increasing. We still have to find the correct asymptotic ordering on N, p, a to prove a mathematical equivalent of this phenomenological statement

§7 Comments

The only problem treated rigorously to this point is the auxiliary probabilistic problem (4.2.2) — at least at the level of the calculation of expectation values. It would be of interest to calculate the correlations etc between the zeros and also to obtain rigorous convergence results in one of the senses used in probability (convergence in law *e.g.*). This concerns the problem

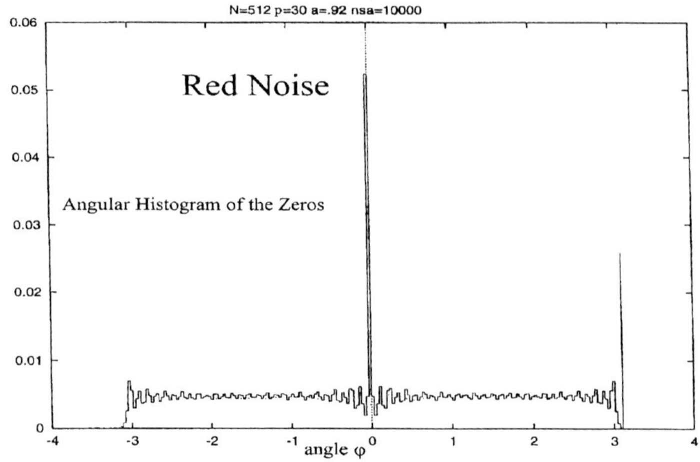

(a)

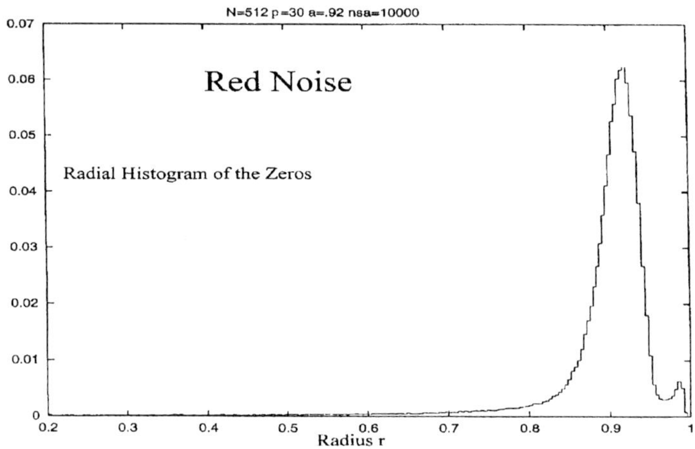

(b)

Figure 3. Same as Figure 2 but with a higher value of the correlation parameter (see §6). The attraction of the peak of the energy spectrum, located at 1, has distorted the crystal, which eventually has lost two sites. The crystal then undergoes a re-arrangement (Figure 3 (a)): there are 56 peaks in the complex and one on the real negative line. The real positive line supports in the mean one additionnal root. The modulus of the latter is larger than Λ, giving rise to the precursor peak which can be seen on Figure 3 (b).

of §4. Similar questions may be raised concerning §5 and §6 in which many other open problems have been also mentioned.

On the other hand, we are now (joint work with A. Noullez and F. Benady) starting a study of a family of stochastic processes, with a parameter which allows to go continuously from a periodic deterministic signal, of the class studied by other authors (see Section 3.1), to a stochastic stationary ergodic Markov process. Here also many phenomena remain to be disclosed and proven, and among them those concerning the connection between the deterministic and the random case.

Acknowledgements. I mention with gratitude that most of this work was done in collaboration with colleagues from Nice, Saclay/Atlanta, Como and Bucuresti. In fact, the present report includes unpublished work with Th. Ayguesparsses, F. Benady, G. Mantica and A. Noullez with some reference to published work with F. Aaron, D. Bessis, G. Mantica and G. Mezincescu.

Finally, I want to thank all the organizers of the Cyprus meeting. I thank also many colleagues, some of whom I met there for the first time, for their interest in this and related work on data analysis (random Padé), for their comments, for their suggestions for a proper mathematical setting, and for sending me related papers. Special mention should be made of E. B. Saff and of the Control Theory group (Inria, Sophia-Antipolis, France) for scientific remarks and for introducing me to the CMFT community. And of the people who made this travel possible for me, particularly Mrs Limouzis, administrative assistant of the Inria group. The anonymous referee had a useful impact on this paper. Finally, I take the opportunity to express my gratitude to J.-P. Kahane for his continuing interest in this work.

§8 Bibliography

8.1 Basic books

- Szegő, G., *Orthogonal Polynomials*, Am. Math. Soc., Providence, 1939.

- Wiener, N. (with an Appendix by N. Levinson), *Extrapolation, Interpolation and Smoothing of Stationary Time Series*, Technology Press of MIT, Cambridge, and Wiley, New-York, 1949.

- Kac, M., *Probability and Related Topics in Physical Sciences*, Wiley-

Interscience, New-York, 1959.

8.2 Connection between Szegő polynomials and frequency analysis: deterministic quasi-periodic case

- Delsarte, Ph. and Y. Génin, On the rôle of orthogonal polynomials on the unit circle in digital signal processing applications, in *Orthogonal Polynomials: Theory and Practice*, P. Nevai (ed.), NATO ASI Ser., Ser. C 294, 1990, 115–133.

- Jones, W. and O. Njåstad, Applications of Szegő polynomials to digital signal processing, *Rocky Mt. J. Math.* **21** No. 1 (1991), 387–436.

- Jones, W., O. Njåstad and E. Saff, Szegő polynomials associated with Wiener-Levinson filters, *J. Comput. Appl. Math.* **32** No. 3 (1990), 387–406.

- Jones, W., O. Njåstad, W. J. Thron and H. Waadeland, Szegő polynomials applied to frequency analysis, *J. Comp. Appl. Math.* **46** (1993), 217–228.

- Njåstad, O. and H. Waadeland, Generalized Szegő theory in frequency analysis, *J. of Math. Analysis and Applications* **206** No. 1 (1997), 280–307.

- Pakula, L., Asymptotic zero distribution of orthogonal polynomials in sinusoidal frequency estimation, *IEEE Trans. Inf. Theory* **IT-33** (1987), 569–576.

- Pan, K. and E. B. Saff, Asymptotics for zeros of Szegő polynomials associated with trigonometric polynomial signals, *J. Approximation Theory* **71** No. 3 (1992), 239–251.

- Petersen, V., contribution in *Comm. in the Analytic Theory of Continued Fractions V*, **27**, (1996).

- Petersen, V., On measures in frequency analysis, Contribution to the Proceedings of a Conference held in Trondheim, Norway, June 1997.

8.3 Orthogonal polynomials with varying measures

- Li, X., Asymptotics of columns in the table of orthogonal polynomials with varying measures, *Methods and Applications of Analysis* **2**, 2 1995, 222–236.

8.4 Publications and reports (J.-D. F. and colleagues) related to the work presented in this contribution

- Meszincescu, G. A., D. Bessis, J.-D. Fournier and G. Mantica and F. D. Aaron, Distribution of roots of random real generalized polynomials, *J. of Stat. Phys.* **86**, 3/4 (1997), 675–705.

- Ayguesparsses, Th., Analyse nouvelle de signaux stochastiques corrélés, Internal Report, Dept. G. D. Cassini (Mémoire de DEA), Observatoire de la Côte d'Azur, 1996.

- Benady, F., Polynômes de Szegő aléatoires associés à un processus de Markov, Internal Report, Dept. G. D. Cassini (Mémoire de DEA), Observatoire de la Côte d'Azur, 1997.

- Fournier, J.-D., A., G. Mantica, A. Mezincescu and D. Bessis, Universal statistical behavior of the complex zeros of Wiener transfer functions, *Europhysics Letters* **22**, 5 (1993), 325ff.

- Fournier, J.-D., G. Mantica, A. Mezincescu and D. Bessis, Statistical properties of the zeros of the transfer functions in signal processing, in *Chaos and Diffusion in Hamiltonian Systems*, D. Benest and Cl. Froesché (eds.), Proceedings of a School held in Chamonix, Publ. Editions Frontières, Orsay, 1996, 257–271.

Jean-Daniel Fournier
(member of C.N.R.S.)
Dêpartement G. D. Cassini: Observatoire de Nice
B.P. 4229
06304 NICE Cedex 4
FRANCE
fournier@obs-nice.fr

On the Relation between $E_n(f)$ and $E_n(f')$

Dieter Gaier

Abstract. Let G be a Jordan domain, let f be holomorphic in G and continuous on $\overline{G}$, and assume that also f' has a continuous extension to $\overline{G}$. We investigate the relation between the errors of approximation by polynomials, $E_n(f)$ and $E_n(f')$, under certain assumptions on G. Similarly, we study this relation when f and f' are continuous on a Jordan arc.

§1 Introduction

We study the degree of polynomial approximation of a function f given on a Jordan domain G or on a Jordan arc Γ. The assumptions are on $f^{(p)}$, for some $p = 0, 1, 2, \ldots$, and we want to estimate the minimal error $E_n(f, \overline{G})$ or $E_n(f, \Gamma)$, the error being measured in the uniform norm.

To obtain such estimates, two types of theorems are needed. If $p = 0$, we need a theorem of

$$\text{Type 1 (Jackson type):} \qquad E_n(f) \leq C\omega_f\left(\frac{1}{n}\right) \qquad (n = 1, 2, \ldots).$$

If $p = 1, 2, \ldots$, we need a theorem of

$$\text{Type 2 (Recursion type):} \qquad E_n(f) \leq \alpha_n E_{n-1}(f') \qquad (n = 1, 2, \ldots).$$

For example, if $\Gamma = [-1, 1]$, one has the estimates $E_n(f, \Gamma) \leq 12\omega_f(\frac{1}{n})$ and $E_n(f, \Gamma) \leq \frac{12}{n} E_{n-1}(f', \Gamma)$; see Natanson [10], p. 112. Combining estimates of Type 1 and Type 2, we get

$$\begin{aligned} E_n(f) \leq \alpha_n E_{n-1}(f') &\leq \alpha_n \alpha_{n-1} E_{n-2}(f'') \\ &\leq \alpha_n \alpha_{n-1} \cdots \alpha_{n-p+1} E_{n-p}(f^{(p)}) \\ &\leq \alpha_n \alpha_{n-1} \cdots \alpha_{n-p+1} C\omega_{f^{(p)}}\left(\frac{1}{n-p}\right) \end{aligned}$$

which gives our desired estimate.

In this paper we are mainly concerned with theorems of Type 2. We shall say that $f \in A(\overline{G})$ if f is holomorphic in G and continuous on $\overline{G}$, and we shall say that $f \in C(\Gamma)$ if f is continuous on Γ.

Computational Methods and Function Theory (CMFT'97)
N. Papamichael, St. Ruscheweyh and E. B. Saff (Eds.), pp. 225–231

ISBN 981-02-3626-3

§2 Approximation on a domain G

We assume that G is bounded by a Jordan curve Γ, and that ψ is the conformal mapping of $\{w : |w| > 1\}$ onto the exterior of Γ, normalized by

$$\psi(w) = cw + c_0 + \frac{c_1}{w} + \ldots \qquad \text{with } c > 0, \text{ for } |w| > 1.$$

As is well known, ψ has a continuous extension to $\{w : |w| \geq 1\}$.

2.1 An approximation theorem of Type 1

We make the following assumptions.

Assumptions on G: Let G be a Faber domain, and let $\psi \in \operatorname{Lip}\alpha$ on $|w| = 1$, where $0 < \alpha \leq 1$.

Assumptions on f: Let $f \in A(\overline{G})$ and $f \in \operatorname{Lip}_1 1$ on Γ, *i.e.* the inequality $|f(z_1) - f(z_2)| \leq |z_1 - z_2|$ holds for z_1, z_2 on Γ.

Theorem 1. *Under these assumptions there exists a constant C depending on Γ but not on f such that*

$$E_n(f, \overline{G}) \leq \frac{C}{n^\alpha} \qquad \textit{for } n = 1, 2, \ldots. \tag{2.1.1}$$

Remark 1: Geometric conditions for G to be a Faber domain are well known; see for example Gaier [7], p. 51. Sufficient is that ∂G is piecewise convex; cusps are permitted.

Remark 2: Sufficient for $\psi \in \operatorname{Lip}\alpha$ is that Γ satisfies an exterior α-wedge condition. That means that for each $\zeta \in \Gamma$ there is a fixed wedge of opening $\alpha\pi$ and with tip at ζ, lying in the complement of G. For $0 < \alpha < 1$ see Lesley [9]. For $\alpha = 1$ Γ is convex and hence $|\psi'(w)| < 2$ for $|w| > 1$ (Grötzsch; see Pommerenke [11], p. 48).

Since (2.1.1) is essentially known, we only sketch the proof of Theorem 1. We consider $h := f \circ \psi$ on $\partial\mathbb{D}$. Then $h \in \operatorname{Lip}\alpha$ and also its Fourier conjugate $\tilde{h} \in \operatorname{Lip}\alpha$ on $\partial\mathbb{D}$, provided that $0 < \alpha < 1$. More precisely, $\tilde{h}$ has a modulus of continuity $\omega_{\tilde{h}}$ on $|w| = 1$ which satisfies $\omega_{\tilde{h}}(t) \leq const \cdot t^\alpha$ with a constant that depends only on Γ. Hence the analytic part h^+ of h is in $\operatorname{Lip}\alpha$ on $\partial\mathbb{D}$, so that by classical results (see Timan [12], p. 260)

$$E_n(h^+, \partial\mathbb{D}) \leq c_1\, \omega_{h^+}\left(\frac{1}{n}\right) \leq c_1 c_2 \cdot \frac{1}{n^\alpha} \qquad (n = 1, 2, \ldots); \tag{2.1.2}$$

the constants depend only on Γ. But since G is a Faber domain, its Faber operator T is bounded and hence

$$E_n(f, \overline{G}) \leq \|T\| \cdot E_n(h^+, \partial\mathbb{D}) \qquad \text{with } \|T\| < \infty$$

(see Korevaar [8], p. 287), and this gives (2.1.1). In the case $\alpha = 1$, the functions $\tilde{h}$ and h^+ are in the Zygmund class Z, and (2.1.2) holds with $\alpha = 1$.

Remark 3: Polynomials P_n of degree n approximating f to the degree given in (2.1.1) can be constructed by using certain means of the Faber series of f. Our method of proving (2.1.1) is therefore more constructive than that used by the Ukrainian school of mathematicians who apply approximations of the Cauchy kernel $\frac{1}{z-\zeta}$ to prove Jackson type theorems. See for example Dzjadyk's book [5], chapter IX, §6–10.

2.2 An approximation theorem of Type 2

We make the following assumptions.

Assumptions on G: Let G be a Faber domain with rectifiable boundary, and let $\psi \in \mathrm{Lip}\,\alpha$ on $|w| = 1$, where $0 < \alpha \leq 1$.
(H): If $z_1, z_2 \in \partial G$, there is a Jordan arc $\Gamma(z_1, z_2) \subset \overline{G}$ connecting the point z_1 to z_2 with length $L(z_1, z_2) \leq const \cdot |z_1 - z_2|$.

Assumption on f: Let $f' \in A(\overline{G})$.

Theorem 2. *Under these assumptions, $f \in A(\overline{G})$, and there exists a constant C depending on Γ but not on f such that*

$$E_n(f, \overline{G}) \leq \frac{C}{n^\alpha} \cdot E_{n-1}(f', \overline{G}) \qquad \text{for } n = 1, 2, \ldots. \tag{2.2.1}$$

Remark 4: The class of domains which satisfy condition (H) was introduced by Andrievskii ([2] and [3]). Note that (H) permits angles and outward cusps but no inward cusps.

Remark 5: It is not difficult to see that $f' \in A(\overline{G})$ implies $f \in A(\overline{G})$ whenever ∂G is rectifiable. For, if $z_n \in G$ and $z_n \to \zeta \in \partial G$, the sequence $\{f(z_n)\}$ will be a Cauchy sequence. To see this, we write

$$|f(z_n) - f(z_m)| = \left| \int_{\gamma_{n,m}} f'(t)\,dt \right| \leq ML(\gamma_{n,m})$$

where $\gamma_{n,m}$ is any rectifiable arc connecting z_n to z_m. It is easy to see that we can always find such an arc with $L(\gamma_{n,m}) \to 0$ $(n, m \to \infty)$, if ∂G is rectifiable.

In fact, knowing $f \in A(\overline{G})$, we even find that $f \in \mathrm{Lip}\,1$ on ∂G provided that condition (H) holds:

$$|f(z_1) - f(z_2)| = \left| \int_{\Gamma(z_1,z_2)} f'(t)\,dt \right| \leq M \cdot L(z_1, z_2) \overset{(H)}{\leq} const \cdot |z_1 - z_2|$$

for $z_1, z_2 \in \partial G$.

Proof of Theorem 2: Choose the polynomials P_n of degree $\leq n$ such that

$$|f'(z) - P_{n-1}(z)| = E_{n-1}(f', \overline{G}) \qquad \text{for } z \in \overline{G} \text{ and } n = 1, 2, \ldots.$$

Choose Q_n so that $Q_n' = P_{n-1}$ to get

$$|f'(z) - Q_n'(z)| = E_{n-1}(f', \overline{G}) \qquad \text{for } z \in \overline{G}, \tag{2.2.2}$$

and consider $F := f - Q_n \in A(\overline{G})$. We claim that $F \in \operatorname{Lip} 1$ on ∂G:

$$\begin{aligned} |F(z_1) - F(z_2)| &= \left| \int_{\Gamma(z_1, z_2)} F'(t)\, dt \right| \\ &\overset{(2.2.2)}{\leq} E_{n-1}(f', \overline{G}) \cdot L(z_1, z_2) \\ &\overset{(H)}{\leq} E_{n-1}(f', \overline{G}) \cdot c_1 \cdot |z_1 - z_2|. \end{aligned}$$

Using (2.1.1) from Section 2.1, we get

$$E_n(F, \overline{G}) \leq c_1 c_2 E_{n-1}(f', \overline{G}) \cdot \frac{1}{n^\alpha},$$

so that there is a polynomial R_n of degree $\leq n$ with

$$\|f - Q_n - R_n\| \leq c_1 c_2 E_{n-1}(f', \overline{G}) \cdot \frac{1}{n^\alpha},$$

and this proves (2.2.1). ■

§3 Approximation on an arc Γ

3.1 An auxiliary problem

Let Γ be a Jordan arc, and assume that f is given on Γ with the property that $f'(z)$ exists for all $z \in \Gamma$ and that

$$|f'(z)| \leq M \qquad \text{for } z \in \Gamma. \tag{3.1.1}$$

Question: When do we have

$$|f(z_1) - f(z_2)| \leq CM|z_1 - z_2| \qquad \text{for } z_1, z_2 \in \Gamma \tag{3.1.2}$$

with a constant C depending only on Γ?

Not always does (3.1.1) imply (3.1.2). Take for example $\Gamma = \Gamma_1 \cup \Gamma_2$ where Γ_1 is the arc $\{x + ix^2 : 0 \le x \le 1\}$ and Γ_2 is the interval $[0,1]$, and let

$$f(z) := \begin{cases} z^{5/4} & \text{for} \quad z \in \Gamma_1 \\ z^{3/2} & \text{for} \quad z \in \Gamma_2 \end{cases} .$$

Then f' exists everywhere on Γ and is continuous on Γ, but if $z_1 = x$ and $z_2 = x + ix^2$ we see that

$$\frac{|f(z_1) - f(z_2)|}{|z_1 - z_2|} \to \infty \qquad \text{as } x \to 0+ .$$

Notice that Γ is piecewise smooth but has a cusp at $z = 0$.

Lemma 1. *If Γ is piecewise smooth without cusps, then inequality (3.1.1) implies (3.1.2).*

Proof: If Γ is smooth, we can represent Γ by $z = z(t)$ where z' is continuous on $[0,1]$ and $z'(t) \ne 0$ there. It follows that

$$0 < c_1 \le \left| \frac{z(t_1) - z(t_2)}{t_1 - t_2} \right| \le c_2 < \infty \qquad \text{for } t_1, t_2 \in [0,1].$$

Hence, with $h(t) := f(z(t))$, we have $|h'(t)| \le c_2 M$ and consequently

$$\begin{aligned} |f(z_1) - f(z_2)| &= |h(t_1) - h(t_2)| = \left| \int_{t_1}^{t_2} h'(t)\, dt \right| \\ &\le c_2 M |t_1 - t_2| \le \frac{c_2}{c_1} M |z(t_1) - z(t_2)| = \frac{c_2}{c_1} M |z_1 - z_2|, \end{aligned}$$

so (3.1.2) is true.

If Γ_1 and Γ_2 are two smooth arcs meeting at z' and $z_1 \in \Gamma_1$, $z_2 \in \Gamma_2$, then

$$|f(z_1) - f(z')| \le C_1 M |z_1 - z'| \qquad \text{and} \qquad |f(z_2) - f(z')| \le C_2 M |z_2 - z'|$$

so that

$$|f(z_1) - f(z_2)| \le (C_1 + C_2) M \left(|z_1 - z'| + |z_2 - z'| \right) .$$

If z' is not a cusp of Γ, the last bracket is $\le$ *const* $\cdot |z_1 - z_2|$, so (3.1.2) is true again. We proceed similarly when $\Gamma = \bigcup \Gamma_j$ with a finite number of smooth arcs Γ_j. ■

We do not know whether (3.1.1) implies (3.1.2) whenever Γ is a quasiconformal arc. Remembering the Hardy-Littlewood property for functions holomorphic in domains with quasiconformal boundary, it might even be true that quasiconformal arcs Γ can be *characterized* by the property that (3.1.1) implies (3.1.2) for all functions f which have a derivative f' everywhere on Γ.

3.2 An approximation theorem of Type 1

For a Jordan arc Γ, we denote by ψ again the conformal mapping of $\{w : |w| > 1\}$ onto the complement of Γ, normalized as in the beginning of §2. Several Jackson type theorems have been proved by Andrievskii ([1] and [4]); we only need a special case.

Assumptions on Γ:	Let Γ be a quasiconformal arc, and let $\psi \in \operatorname{Lip}\alpha$ on $\lvert w\rvert = 1$, where $0 < \alpha \leq 1$.
Assumption on f:	Let $f \in \operatorname{Lip}_1 1$ on Γ, *i.e.* $\lvert f(z_1) - f(z_2)\rvert \leq \lvert z_1 - z_2\rvert$ for z_1, z_2 on Γ.

Theorem 3 (Andrievskii). *Under these assumptions there exists a constant C depending on Γ but not on f such that*

$$E_n(f,\Gamma) \leq \frac{C}{n^\alpha} \qquad \textit{for } n = 1, 2, \ldots. \tag{3.2.1}$$

Remark 6: Recently Frerick and Müller [6] have extended the notion of Faber sets to general compact sets, in particular to Jordan arcs. Zero angles are permitted there, so that their results offer a new approach to Jackson type theorems for arcs.

3.3 An approximation theorem of Type 2

To be able to apply Lemma 1, we make the following assumptions.

Assumptions on Γ:	Let Γ be piecewise smooth and without cusps, and let $\psi \in \operatorname{Lip}\alpha$ on $\lvert w\rvert = 1$, where $0 < \alpha \leq 1$.
Assumption on f:	Let f' exist everywhere on Γ and be continuous on Γ.

Theorem 4. *Under these assumptions, f is continuous on Γ, and there exists a constant C depending on Γ but not on f such that*

$$E_n(f,\Gamma) \leq \frac{C}{n^\alpha} E_{n-1}(f',\Gamma) \qquad \textit{for } n = 1, 2, \ldots.$$

The proof uses Lemma 1 but otherwise is similar to that of Theorem 2.

Added in proof: The referee pointed out to the author that the statement made in Remark 5 is already contained in S. N. Mergelyan, Certain questions of the constructive theory of functions, (Russian), *Trudy mat. Inst. Steklov.* **37** (1951).

References

[1] Andrievskii, V. V., Direct theorems of the theory of approximation on quasiconformal arcs, (Russian), *Izv. Akad. Nauk SSSR Ser. Mat.* **44** (1980), no. 2, 243–261.

[2] Andrievskii, V. V., Geometric properties of V.K. Dzjadyk domains, (Russian), *Ukrain. Mat. Zh.* **33** (1981), no. 6, 723–727.

[3] Andrievskii, V. V., Geometric structure of domains and direct theorems of the constructive theory of functions, (Russian), *Mat. Sbornik (N.S.)* **126 (168)** (1985), no. 1, 41–58.

[4] Andrievskii, V. V., Approximation of functions on arcs with zero angles, (Russian), *Ukrain. Mat. Zh.* **37** (1985), no. 5, 547–551.

[5] Dzjadyk, V. K., *Introduction to the Theory of Uniform Approximation of Functions by Polynomials*, (Russian), Moscow, 1977.

[6] Frerick, L. and J. Müller, Polynomial approximation and maximal convergence on Faber sets, *Constr. Approx.* **10** (1994), 427–438.

[7] Gaier, D., *Vorlesungen über Approximation im Komplexen*, Birkhäuser Verlag, Basel, 1980.

[8] Korevaar, J., Polynomial and rational approximation in the complex plane, in: *Aspects of Contemporary Complex Analysis*, D. A. Brannan and J. G. Clunie (eds.), Academic Press, New York, 1980, 251-292.

[9] Lesley, F. D., Conformal mappings of domains satisfying a wedge condition, *Proc. Amer. Math. Soc.* **93** (1985), 483–488.

[10] Natanson, I. P., *Konstruktive Funktionentheorie*. Akademie Verlag, Berlin, 1955.

[11] Pommerenke, Ch., *Univalent functions*, Vandenhoeck und Ruprecht, Göttingen, 1975.

[12] Timan, A. F., *Theory of Approximation of Functions of a Real Variable*, Oxford, 1963.

Dieter Gaier
Mathematisches Institut
Justus-Liebig-Universität Giessen
Arndtstraße 2
D-35392 Giessen
GERMANY
dieter.gaier@math.uni-giessen.de

A bound for hyperbolic distance in a quasidisk

Frederick W. Gehring

and Kari Hag

Abstract. A simply connected proper subdomain D of the complex plane $\mathbb{C}$ is a K-quasidisk if and only if the hyperbolic distance $h_D(z_1, z_2)$ between z_1 and z_2 in D is bounded above by $a\, j_D(z_1, z_2) + b$ where a and b are constants and j_D a certain function of the ratios of the euclidean distance between z_1 and z_2 and the euclidean distances from z_1 and z_2 to ∂D. We derive here a distortion theorem for quasiconformal mappings which yields simple upper bounds for a and b as functions of K.

§1 A distortion theorem

For $0 < r < 1$ let $\mu(r)$ denote the modulus of the ring domain bounded by the segment $[0, r]$ and the unit circle $|z| = 1$ and set

$$\phi_K(r) = \mu^{-1}\left(\frac{\mu(r)}{K}\right) \tag{1.1}$$

for $K \geq 1$. Then

$$\phi_K(r) < 4^{1-1/K}\, r^{1/K} \tag{1.2}$$

for $0 < r < 1$ and

$$\lim_{r \to 0} r^{-1/K} \phi_K(r) = 4^{1-1/K}. \tag{1.3}$$

See, for example, pp. 63–65 in [8].

If we combine (1.2) and (1.3) with results due to Agard [1] and Teichmüller [11], we obtain the following surprisingly simple distortion theorem for quasiconformal mappings.

Theorem 1. *If f is a K-quasiconformal self mapping of $\overline{\mathbb{C}}$ which fixes 0, 1 and ∞, then*

$$|f(z)| + 1 \leq 16^{K-1}\,(|z| + 1)^K \tag{1.4}$$

for $z \in \overline{\mathbb{C}}$. The coefficient 16^{K-1} in (1.4) cannot be replaced by a smaller constant.

Computational Methods and Function Theory (CMFT'97)
N. Papamichael, St. Ruscheweyh and E. B. Saff (Eds.), pp. 233–240

ISBN 981-02-3626-3

Proof: Fix $z \in \overline{\mathbb{C}}$, $z \neq 0, 1, \infty$, and let h_G denote the hyperbolic metric with curvature -1 in $G = \mathbb{C} \setminus \{0, 1, \infty\}$. Then

$$h_G(f(z), z) \leq \log K \tag{1.5}$$

by a theorem of Teichmüller; see pp. 29–31 in [11]. Next

$$h_G(-|f(z)|, -|z|) \leq h_G(f(z), z) \tag{1.6}$$

by a result due to Lehto, Virtanen and Väisälä [9]. See also Agard [1] and Hempel [7]. Finally

$$\left| \log \mu \left(\frac{1}{\sqrt{|f(z)| + 1}} \right) - \log \mu \left(\frac{1}{\sqrt{|z| + 1}} \right) \right| = h_G(-|f(z)|, -|z|) \tag{1.7}$$

by a theorem due to Agard; see (2.8) in [1]. Hence

$$\mu \left(\frac{1}{\sqrt{|f(z)| + 1}} \right) \leq K \, \mu \left(\frac{1}{\sqrt{|z| + 1}} \right)$$

by (1.5), (1.6) and (1.7). Then since $\mu(r)$ is decreasing, we conclude from (1.1) and (1.2) that

$$\frac{1}{\sqrt{|z| + 1}} \leq \phi_K \left(\frac{1}{\sqrt{|f(z)| + 1}} \right) \leq 4^{1-1/K} \left(\frac{1}{\sqrt{|f(z)| + 1}} \right)^{1/K}$$

from which (1.4) follows.

To show that the coefficient 16^{K-1} in (1.4) is best possible, fix $K > 1$ and choose c so that $0 < c < 16^{K-1}$. Then $c^{1/2K} < 4^{1-1/K}$ and by (1.3) we can choose $t < 0$ so that

$$\phi_K \left(\frac{1}{\sqrt{|t| + 1}} \right) > c^{1/2K} \left(\frac{1}{\sqrt{|t| + 1}} \right)^{1/K} . \tag{1.8}$$

Next since $h_G(t, z) \to \infty$ as $z \to 0$, we can choose $s \in (t, 0)$ so that

$$h_G(-|t|, -|s|) = h_G(t, s) = \log K.$$

Then by the theorem of Teichmüller cited above, there exists a K-quasiconformal self mapping f of $\overline{\mathbb{C}}$ which fixes $0, 1, \infty$ such that $f(s) = t$. Hence

$$\mu \left(\frac{1}{\sqrt{|f(s)| + 1}} \right) = K \, \mu \left(\frac{1}{\sqrt{|s| + 1}} \right)$$

by (1.7) with $z = s$,

$$\frac{1}{\sqrt{|s|+1}} = \phi_K\left(\frac{1}{\sqrt{|f(s)|+1}}\right) > c^{1/2K}\left(\frac{1}{\sqrt{|f(s)|+1}}\right)^{1/K}$$

by (1.8) since $f(s) = t$ and we obtain $|f(s)| + 1 > c\,(|s|+1)^K$. ■

Theorem 1 yields the following simple bound for the change of the cross-ratio under a quasiconformal mapping.

Corollary 1. *If f is a K-quasiconformal self mapping of $\overline{\mathbb{C}}$, then*

$$|(f(z_1), f(z_2), f(z_3), f(z_4))| + 1 \le 16^{K-1}(|(z_1, z_2, z_3, z_4)| + 1)^K \qquad (1.9)$$

for each quadruple of points $z_1, z_2, z_3, z_4 \in \overline{\mathbb{C}}$.

Proof: Suppose that z_1, z_2, z_3, z_4 are distinct points and define the function $\phi(z) = h \circ f \circ g^{-1}(z)$ where

$$g(z) = (z, z_2, z_3, z_4), \qquad h(w) = (w, f(z_2), f(z_3), f(z_4)).$$

Then ϕ is a K-quasiconformal self mapping of $\overline{\mathbb{C}}$ which fixes $0, 1, \infty$ and

$$\begin{aligned} |(f(z_1), f(z_2), f(z_3), f(z_4))| + 1 &= |\phi(g(z_1))| + 1 \\ &\le 16^{K-1}(|g(z_1)| + 1)^K \\ &= 16^{K-1}(|(z_1, z_2, z_3, z_4)| + 1)^K \end{aligned}$$

by (1.4). ■

§2 Hyperbolic distance in a quasidisk

Suppose that h_D is the hyperbolic metric with curvature -1 in a simply connected proper subdomain D of $\mathbb{C}$. Then

$$h_D(z_1, z_2) \ge \frac{1}{4}\, j_D(z_1, z_2) \qquad (2.1)$$

for $z_1, z_2 \in D$ where

$$j_D(z_1, z_2) = \log\left(\frac{|z_1 - z_2|}{\operatorname{dist}(z_1, \partial D)} + 1\right)\left(\frac{|z_1 - z_2|}{\operatorname{dist}(z_2, \partial D)} + 1\right), \qquad (2.2)$$

a function of the ratios of the euclidean distance between z_1 and z_2 and the euclidean distances from these points to ∂D. See [4] and [5]. The function j_D also yields an upper bound for h_D, namely

$$h_D(z_1, z_2) \leq a\, j_D(z_1, z_2) + b \tag{2.3}$$

for $z_1, z_2 \in D$ if and only if D is a K-quasidisk, the image of a disk or half plane under a K-quasiconformal self mapping of $\overline{\mathbb{C}}$. In this case $a = a(K)$ and $b = b(K)$. Inequality (2.3) implies that

$$h_D(z_1, z_2) \leq c\, j_D(z_1, z_2), \qquad c = a + \max(b, \sqrt{b}) \tag{2.4}$$

by [4]. Moreover,

$$h_D(z_1, z_2) \leq j_D(z_1, z_2) \tag{2.5}$$

if D is a disk or half plane [4].

Corollary 1 allows us to obtain some simple estimates for $a(K)$ and $b(K)$.

Theorem 2. *If f is a K-quasiconformal self mapping of $\mathbb{C}$, then for each proper subdomain D of $\mathbb{C}$,*

$$j_{f(D)}(f(z_1), f(z_2)) \leq K j_D(z_1, z_2) + 2(K-1)\log 16 \tag{2.6}$$

for $z_1, z_2 \in D$.

Proof: Fix $z_1, z_2 \in D$ and choose $w_1, w_2 \in \partial D$ so that

$$|f(z_k) - f(w_k)| = \operatorname{dist}(f(z_k), \partial f(D))$$

for $k = 1, 2$. Since f has a K-quasiconformal extension to $\overline{\mathbb{C}}$ which fixes ∞,

$$\begin{aligned}
\frac{|f(z_1) - f(z_2)|}{\operatorname{dist}(f(z_1), \partial f(D))} + 1 &= |(f(z_1), \infty, f(z_2), f(w_1))| + 1 \\
&\leq 16^{K-1}(|(z_1, \infty, z_2, w_1)| + 1)^K \\
&\leq 16^{K-1}\left(\frac{|z_1 - z_2|}{\operatorname{dist}(z_1, \partial D)} + 1\right)^K
\end{aligned}$$

by Corollary 1. Similarly

$$\frac{|f(z_1) - f(z_2)|}{\operatorname{dist}(f(z_2), \partial f(D))} + 1 \leq 16^{K-1}\left(\frac{|z_1 - z_2|}{\operatorname{dist}(z_2, \partial D)} + 1\right)^K$$

and we conclude that

$$\begin{aligned} j_{f(D)}(f(z_1), f(z_2)) &= \\ &= \log\left(\frac{|f(z_1)-f(z_2)|}{\mathrm{dist}(f(z_1), \partial f(D))}+1\right)\left(\frac{|f(z_1)-f(z_2)|}{\mathrm{dist}(f(z_2), \partial f(D))}+1\right) \\ &\leq \log\left(16^{K-1}\right)^2\left(\frac{|z_1-z_2|}{\mathrm{dist}(z_1, \partial D)}+1\right)^K\left(\frac{|z_1-z_2|}{\mathrm{dist}(z_2, \partial D)}+1\right)^K \\ &= K j_D(z_1, z_2) + 2(K-1)\log 16. \quad \blacksquare \end{aligned}$$

Corollary 2. *If D is a domain in $\mathbb{C}$ and if there exists a K-quasiconformal self mapping f of $\mathbb{C}$ which maps D conformally onto a disk or half plane, then*

$$h_D(z_1, z_2) \leq K j_D(z_1, z_2) + 2(K-1)\log 16. \tag{2.7}$$

for $z_1, z_2 \in D$.

Proof: If $z_1, z_2 \in D$, then

$$\begin{aligned} h_D(z_1, z_2) &= h_{f(D)}(f(z_1), f(z_2)) \\ &\leq j_{f(D)}(f(z_1), f(z_2)) \\ &\leq K j_D(z_1, z_2) + 2(K-1)\log 16 \end{aligned}$$

by the conformal invariance of hyperbolic distance, (2.5) and (2.6). ■

Corollary 2 yields the following bound for hyperbolic distance in a quasidisk.

Theorem 3. *If $D \subset \mathbb{C}$ is a K-quasidisk, then*

$$h_D(z_1, z_2) \leq K^2 j_D(z_1, z_2) + 2(K^2-1)\log 16. \tag{2.8}$$

for $z_1, z_2 \in D$. In particular,

$$h_D(z_1, z_2) \leq c\, j_D(z_1, z_2) \tag{2.9}$$

for $z_1, z_2 \in D$ where $c = c(K) \to 1$ as $K \to 1$.

Proof: By hypothesis, there exists a K-quasiconformal self mapping f of $\overline{\mathbb{C}}$ which maps D onto a disk or half plane.

If D is bounded, then by composing f with a Möbius transformation we may assume that f fixes ∞ and $f(D)$ is the unit disk B. The existence theorem for the Beltrami equation implies there exists a K-quasiconformal self mapping $g : B \to B$ which fixes 0 such that $g \circ f$ is conformal in D. Reflection in ∂B extends g to a K-quasiconformal self mapping of $\mathbb{C}$. Then $h = g \circ f$ is K^2-quasiconformal and we can apply Corollary 2 to obtain (2.8).

If D is unbounded, then D is the union of an increasing sequence of bounded K-quasidisks D_n which contain z_1, z_2 and

$$h_D(z_1, z_2) = \lim_{n\to\infty} h_{D_n}(z_1, z_2), \qquad j_D(z_1, z_2) = \lim_{n\to\infty} j_{D_n}(z_1, z_2).$$

Hence (2.8) follows from what was proved above.

Finally inequality (2.9) follows from (2.4) with

$$a = K^2, \qquad b = 2(K^2 - 1)\log 16. \quad \blacksquare$$

§3 An example

We conclude with an example which yields lower bounds for the coefficients of $j_D(z_1, z_2)$ in (2.7) and (2.8).

Example 1: If D is the angular sector

$$D = \{z = re^{i\theta} : 0 < r < \infty,\ |\theta| < \pi\alpha/2\}, \qquad 0 < \alpha \le 1, \tag{3.1}$$

then

$$\lim_{x\to\infty} \frac{h_D(x,1)}{j_D(x,1)} = \frac{1}{\alpha}. \tag{3.2}$$

Proof: Since $f(z) = z^\alpha$ maps the right half plane $H = \{z : \mathrm{Re}(z) > 0\}$ conformally onto D,

$$h_D(x,1) = h_H(x^{1/\alpha}, 1) = \frac{1}{\alpha}\log x$$

while

$$j_D(x,1) = \log\left(\frac{x-1}{\beta x} + 1\right)\left(\frac{x-1}{\beta} + 1\right), \qquad \beta = \sin(\pi\alpha/2),$$

for $x > 1$. Hence

$$\lim_{x\to\infty}\left(h_D(x,1) - \frac{1}{\alpha}\, j_D(x,1)\right) = \frac{1}{\alpha}\log\left(\frac{\beta^2}{\beta+1}\right) = \gamma, \qquad |\gamma| < \infty,$$

whence

$$\lim_{x\to\infty}\left(\frac{h_D(x,1)}{j_D(x,1)}-\frac{1}{\alpha}\right)=\frac{\gamma}{\lim\limits_{x\to\infty} j_D(x,1)}=0. \quad \blacksquare$$

Remark 1: The coefficient K of $j_D(z_1,z_2)$ in (2.7) cannot be replaced by a constant less than $(K+1)/2$. The coefficient K^2 of $j_D(z_1,z_2)$ in (2.8) cannot be replaced by a constant less than $(K^2+1)/2$.

Proof: Choose $0<\alpha\le\beta\le 1$, set $f(re^{i\theta})=r^\beta\, e^{i\phi(\theta)}$, where $\phi(\theta)$ is odd with

$$\phi(\theta)=\begin{cases}\alpha\theta & \text{if } 0\le\theta\le\pi/2,\\[2ex] (2-\alpha)\,\theta+(\alpha-1)\,\pi & \text{if } \pi/2\le\theta\le\pi,\end{cases}$$

and let $f(\infty)=\infty$. Then f is a K-quasiconformal self mapping of $\overline{\mathbb{C}}$ where

$$K=\max\{\beta/\alpha,(2-\alpha)/\beta\} \tag{3.3}$$

and f maps the right half plane H onto the angular sector D in (3.1). Suppose that $h_D(z_1,z_2)\le a\; j_D(z_1,z_2)+b$ in D. If $\beta=\alpha$, then $K+1=2/\alpha$, f is conformal in H and

$$a\ge\lim_{x\to\infty}\frac{h_D(x,1)}{j_D(x,1)}=(K+1)/2$$

by (3.2). Next if $\beta^2=2\alpha-\alpha^2$, then $K^2+1=2/\alpha$ and

$$a\ge\lim_{x\to\infty}\frac{h_D(x,1)}{j_D(x,1)}=(K^2+1)/2. \quad \blacksquare$$

Acknowledgement. Research supported in part by grants from the U. S. National Science Foundation, the Norwegian University of Science and Technology and the Technion, the Israel Institute of Technology.

References

[1] Agard, S. B., Distortion theorems for quasiconformal mappings, *Ann. Acad. Sci. Fenn. AI* **413** (1968), 1–11.

[2] Gehring, F. W., Characteristic properties of quasidisks, *Les Presses de l'Université de Montréal* (1982), 1–97.

[3] Gehring, F. W., Characterizations of quasidisks, *Proceedings of the FPU Summer School - Lublin*, 1997 to appear.

[4] Gehring, F. W. and K. Hag, Hyperbolic geometry and disks, *Proceedings of the Conference on Continued Fractions and Geometric Function Theory - Trondheim*, 1998 to appear.

[5] Gehring, F. W. and B. P. Palka, Quasiconformally homogeneous domains, *J. d'Analyse Math.* **30** (1976), 172–199.

[6] Hag, K., What is a disk?, *Proceedings of the FPU Summer School - Lublin*, 1997 to appear.

[7] Hempel, J. A., The Poincaré metric on the twice punctured plane and the theorems of Landau and Schottky, *J. London Math. Soc.* **20** (1979), 435–445.

[8] Lehto O. and K. I. Virtanen, *Quasiconformal mappings in the plane*, Springer-Verlag, 1973.

[9] Lehto, O., K. I. Virtanen and J. Väisälä, Contributions to the distortion theory of quasiconformal mappings, *Ann. Acad. Sci. Fenn. AI* **273** (1959), 1–14.

[10] Martin, G. J., The distortion theorem for quasiconformal mappings, Schottky's theorem and holomorphic motions, *Proc. Amer. Math. Soc.* **125** (1997), 1095–1103.

[11] Teichmüller, O., Extremale quasikonforme Abbildungen und quadratische Differentiale, *Abh. Preuss. Akad. Wiss.* **22** (1940), 1–197.

Frederick W. Gehring
University of Michigan
Ann Arbor, MI 48109
U.S.A.
FGehring@math.lsa.umich.edu

Kari Hag
Norwegian University of Science and Technology
N-7034 Trondheim
NORWAY
kari@math.ntnu.no

On regularity theorems for linearly invariant families of analytic functions in the unit polydisk

Janusz Godula

and Victor V. Starkov

§1 Introduction

In [4] Pommerenke has introduced and studied the notion of a linearly invariant family of functions holomorphic in the unit disk $\mathbb{D} = \{z \in \mathbb{C} : |z| < 1\}$. Linearly invariant families play an important role in the theory of conformal mappings. Furthermore,these families are of interest because of their connections with the Bloch class, *cf.* [2]. In [1] and [5] the regularity theorems were obtained for such families. In [2] we defined linearly invariant families of functions analytic in the unit polydisk $\mathbb{D}^m \subset \mathbb{C}^m$, $m \geq 1$. In this paper we continue the study of the regularity theorem for linearly invariant families of functions defined on the unit polydisk $\mathbb{D}^m$. As we will see, the effect of higher dimensions leads to problems that differ from those for $m = 1$, and consequently for $m > 1$ we obtain more complete results. Let $\mathbb{T} = \{z \in \mathbb{C} : |z| = 1\}$ and $\mathbb{T}^m$ be the unit torus. We will consider the class $\mathcal{H}(\mathbb{D}^m)$ of all functions $f : \mathbb{D}^m \to \mathbb{C}$ analytic in $\mathbb{D}^m$. For $z = (z_1, \dots, z_m) \in \mathbb{C}^m$ we define the norm $\|z\| := \max_{1 \leq j \leq m} |z_j|$. Let $\mathbf{0} = (0, \dots, 0) \in \mathbb{C}^m$. For $a = (a_1, \dots, a_m) \in \mathbb{D}^m$ by the formula $\phi_a(z) = (\phi_1(z_1), \dots, \phi_m(z_m))$, where $\phi_j(z_j) = \frac{z_j + a_j}{1 + \overline{a_j} z_j}, j = 1, \dots, m$, we define the Möbius map ϕ_a of $\mathbb{D}^m$ onto $\mathbb{D}^m$.

Definition 1. *Let $l \in \{1, \dots, m\}$ be fixed. A l-linearly invariant family $\mathfrak{M}_l$ is a class of all functions $f \in \mathcal{H}(\mathbb{D}^m)$, such that*

(1) $f(\mathbf{0}) = 0$, $\frac{\partial f}{\partial z_l}(\mathbf{0}) = 1$, $\frac{\partial f}{\partial z_l}(z) \neq 0$, *for* $z \in \mathbb{D}^m$,

(2) for all $f \in \mathfrak{M}_l$ *and* $\theta = (\theta_1, \dots, \theta_m) \in \mathbb{R}^m$, $f(ze^{i\theta})e^{-i\theta_l} \in \mathfrak{M}_l$, *where* $ze^{i\theta} = (z_1 e^{i\theta_1}, \dots, z_m e^{i\theta_m})$,

(3) for all $f \in \mathfrak{M}_l$ *and* $a = (a_1, \dots, a_m) \in \mathbb{D}^m$, $f(a, z) := \frac{f(\phi_a(z)) - f(\phi_a(\mathbf{0}))}{\frac{\partial f}{\partial z_l}(a)(1 - |a_l|^2)}$ *belongs to* $\mathfrak{M}_l$.

The following definition extends Pommerenke's notion of the order of a function, *cf.* [4]. Let $\frac{\partial f}{\partial z_l}(z) = 1 + c_1(f) z_1 + \ldots + c_m(f) z_m + o(\|z\|)$.

Computational Methods and Function Theory (CMFT'97)
N. Papamichael, St. Ruscheweyh and E. B. Saff (Eds.), pp. 241–257

ISBN 981-02-3626-3

Definition 2. *Suppose that f satisfies (1) of Definition 1. The order of the function f is defined as follows:*

$$\operatorname{ord} f := \sup_{a\in\mathbb{D}^m} \frac{1}{2}\left\|\nabla\frac{\partial f(a,\mathbf{0})}{\partial z_l}\right\| = \frac{1}{2}\sup_{a\in\mathbb{D}^m}\|(c_1(f(a,\cdot)),\ldots,c_m(f(a,\cdot)))\|.$$

Then we define the order of a linearly invariant family $\mathfrak{M}_l$ as

$$\operatorname{ord}\mathfrak{M}_l := \sup_{f\in\mathfrak{M}_l}\operatorname{ord} f.$$

Definition 3. *The universal l-linearly invariant family $\mathcal{U}_\alpha^l$ of order α is defined as*

$$\mathcal{U}_\alpha^l := \bigcup\{\mathfrak{M}_l : \operatorname{ord}\mathfrak{M}_l \le \alpha\}.$$

It is a very important and interesting problem to study the behaviour of functions "near" the torus $\mathbb{T}^m$. Let us denote $\mathbf{r} = (r_1,\ldots,r_m)$, $\theta = (\theta_1,\ldots,\theta_m)\in\mathbb{R}^m$, $\mathbf{1}^- = (1^-,\ldots,1^-)$. Moreover let

$$M(r,p) = \max_{\|z\|\le r}|p(z)|$$

and let

$$\Phi_\theta(\mathbf{r}) = \left|\frac{\partial f}{\partial z_l}(\mathbf{r}e^{i\theta})\right|\prod_{k=1}^m(\frac{1-r_k}{1+r_k})^\alpha(1-r_l^2).$$

Theorem 1 (Regularity Theorem, [3]).

(1) *For any function $f\in\mathcal{U}_\alpha^l$ and fixed θ, the functions $\max_{\theta\in\mathbb{R}^m}\Phi_\theta(\mathbf{r})$ and $\Phi_\theta(\mathbf{r})$ are nonincreasing with respect to every $r_k\in(0,1)$ and $k=1,\ldots,m$. Moreover the function $M\left(r,\frac{\partial f}{\partial z_l}\right)\frac{(1-r)^{\alpha m+1}}{(1+r)^{\alpha m-1}}$ is nonincreasing with respect to $r\in(0,1)$.*

(2) *There exist $d_0\in[0,1]$ and $\theta^0\in[0,2\pi)^m$ such that for every $k = 1,\ldots,m$*

$$\begin{aligned} d_0 &= \lim_{r\to 1^-}\left[M\left(r,\frac{\partial f}{\partial z_l}\right)\frac{(1-r)^{\alpha m+1}}{(1+r)^{\alpha m-1}}\right] \\ &= \lim_{\mathbf{r}\to\mathbf{1}^-}\max_\theta\Phi_\theta(\mathbf{r}) \\ &= \lim_{\mathbf{r}\to\mathbf{1}^-}\Phi_{\theta^0}(\mathbf{r}) \\ &= \lim_{\mathbf{r}\to\mathbf{1}^-}\left[\max_\theta\left|\frac{\partial^2 f}{\partial z_l\partial z_k}(z)\right|\frac{(1-r_k^2)(1-r_l^2)}{2(\alpha+\delta_k^l r_k)}\prod_{k=1}^m\left(\frac{1-r_k}{1+r_k}\right)^\alpha\right], \end{aligned}$$

where δ_k^l is the Kronecker symbol.

(3) $d_0 = 1$ *iff*

$$f_\theta(z) = \frac{e^{i\theta_l}}{2\alpha}\left[\prod_{k=1}^{m}(\frac{1+z_k e^{-i\theta_k}}{1-z_k e^{-i\theta_k}})^\alpha - 1\right] + Q(\ldots, z_{l-1}, z_{l+1}, \ldots),$$

where Q is a function analytic in $\mathbb{D}^{m-1}$ such that $Q(\mathbf{0}) = 0$.

Throughout the paper we will use the notions from Theorem 1. We will call a vector θ^0 from Theorem 1 the *direction of the maximal growth* (d.m.g.) of the function f, and δ_0 the *Hayman number* of the function f. A vector $\theta \in [0, 2\pi)^m$ is called the *direction of the intensive growth* (d.i.g.) of a function $f \in \mathcal{U}^l_\alpha$ if $\lim_{\mathbf{r}\to\mathbb{I}^-} \Phi_\theta(\mathbf{r}) = \delta_\theta > 0$; moreover $\delta_\theta \in (0, 1]$ is called the *Hayman number* for the d.i.g. θ.

Let $\mathcal{U}^l_\alpha(\delta)$ denote the family of functions $f \in \mathcal{U}^l_\alpha$ with corresponding fixed Hayman number δ. If $f \in \mathcal{U}^l_\alpha$ then there exists an infinite set of functions

$$\begin{aligned}\Psi(z) &= \int_0^{z_l} \frac{\partial f}{\partial z_l}(\ldots, z_{l-1}, s, z_{l+1}, \ldots)\, ds \\ &\quad + Q(\ldots, z_{l-1}, z_{l+1}, \ldots) \\ &\in \mathcal{U}^l_\alpha,\end{aligned} \tag{1.1}$$

where $Q(0, \ldots, 0) = 0$, Q-analytic in $\mathbb{D}^{m-1}$ and such that $\frac{\partial \Psi}{\partial z_l} \equiv \frac{\partial f}{\partial z_l}$. Let us consider one of the functions of the above form, namely

$$F(z) = \int_0^{z_l} \frac{\partial f}{\partial z_l}(z_1, \ldots, z_{l-1}, s, z_{l+1}, \ldots, z_m)\, ds. \tag{1.2}$$

For this function the following, higher dimensional counterpart of the regularity theorem from [5], holds.

Theorem 2. *Let $f \in \mathcal{U}^l_\alpha(\delta_0)$ and θ be d.m.g. of the function f. Then*

$$\begin{aligned}&\lim_{\mathbf{r}\to\mathbf{1}^-}\left[\int_0^{r_l}\left|\frac{\partial f}{\partial z_l}(\ldots, r_{l-1}e^{i\theta_{l-1}}, \rho e^{i\theta_l}, r_{l+1}e^{i\theta_{l+1}}, \ldots)\right| d\rho \right.\\ &\qquad \left. \cdot\, 2\alpha \prod_{k=1}^{m}\left(\frac{1-r_k}{1+r_k}\right)^\alpha\right] \\ &= \lim_{\mathbf{r}\to\mathbf{1}}\left[\int_0^{r_l} \max_{\psi\in\mathbb{R}^m}\left|\frac{\partial f}{\partial z_l}(\ldots, r_{l-1}e^{i\phi_{l-1}}, \rho e^{i\phi_l}, r_{l+1}e^{i\phi_{l+1}}, \ldots)\right| d\rho \right.\\ &\qquad \left. \cdot\, 2\alpha \prod_{k=1}^{m}\left(\frac{1-r_k}{1+r_k}\right)^\alpha\right] \\ &= \lim_{\mathbf{r}\to\mathbf{1}^-}\left[\left|F(\mathbf{r}e^{i\theta})\right| 2\alpha \prod_{k=1}^{m}\left(\frac{1-r_k}{1+r_k}\right)^\alpha\right] = \delta_0.\end{aligned}$$

Remark: Since the function Q in (1.1) can grow arbitrary quickly as z is reaching $\mathbb{T}^m$, Theorem 2 does not hold for every function Ψ of the form (1.1).

Before proceeding to the proof of Theorem 2 we provide two lemmas. Let $\dot{\mathcal{U}}^l_\alpha(\delta) := \{\frac{\partial f}{\partial z_l} : f \in \mathcal{U}^l_\alpha(\delta)\}$.

Lemma 1. *Let $h_\delta \in \dot{\mathcal{U}}^l_\alpha(\delta)$ and let $h_\delta(z) \to h_1(z)$ locally uniformly in $\mathbb{D}^m$ as $\delta \to 1^-$. Then there exists $\theta \in \mathbb{R}^m$ such that $h_1(z) = \frac{\partial f_\theta}{\partial z_l}(z)$, where f_θ is a function from Theorem 1 (3).*

Proof: By [2], Theorem 4.2, the family

$$\dot{\mathcal{U}}^l_\alpha = \bigcup_{\delta\in[0,1]} \dot{\mathcal{U}}^l_\alpha(\delta)$$

is sequentially compact in the topology of locally uniform convergence in $\mathbb{D}^m$. Thus $h_1 \in \dot{\mathcal{U}}^l_\alpha(\delta_0)$ for some $\delta_0 \in [0,1]$.

Let us suppose that the thesis of Lemma 1 is not true and $h_1 \neq \frac{\partial f_\theta}{\partial z_l}$ for every $\theta \in \mathbb{R}^m$. Then, by Theorem 1, there is the Hayman number δ_0 of h_1 which is less than one. Let us fix $\varepsilon \in (0, (1-\delta_0)/4)$. By Theorem 1 there exists $r(\varepsilon) > 0$ such that for all $r \in (r(\varepsilon), 1)$

$$M(r, h_1)\frac{(1-r)^{\alpha m+1}}{(1+r)^{\alpha m-1}} < \delta_0 + \varepsilon.$$

Let us fix $r_0 \in (r(\varepsilon), 1)$. Since $h_\delta \to h_1$ locally uniformly in $\mathbb{D}^m$ there exists $\mu \in (0,1)$ such that

$$|M(r_0, h_\delta) - M(r_0, h_1)|\frac{(1-r_0)^{\alpha m+1}}{(1+r_0)^{\alpha m-1}} < \varepsilon$$

for all $\delta \in (\mu, 1)$. Consequently

$$M(r_0, h_\delta)\frac{(1-r_0)^{\alpha m-1}}{(1+r_0)^{\alpha m+1}} < \delta_0 + 2\varepsilon < \frac{1+\delta_0}{2}$$

for all $\delta \in (\mu, 1)$. Choose $\delta > (1+\delta_0)/2$. By Theorem 1 (1)

$$\delta = \lim_{r\to 1-}\left[M(r, h_\delta)\frac{(1-r)^{\alpha m+1}}{(1+r)^{\alpha m+1}}\right] \leq M(r_0, h_\delta)\frac{(1-r_0)^{\alpha m+1}}{(1+r_0)^{\alpha m-1}} < \frac{1+\delta_0}{2},$$

which is impossible. ■

Lemma 2. *Let $f \in \mathcal{U}_\alpha^l(\delta_0)$, $\delta_0 > 0$, and let $\gamma = (\gamma_1, \dots, \gamma_m)$ be any d.i.g. of the function f. Then for every fixed $r_0 \in (0,1)$*

$$\frac{\frac{\partial f}{\partial z_l}\left(\frac{z_1+\rho_1 e^{i\gamma_1}}{1+\rho_1 e^{-i\gamma_1} z_1}, \dots, \frac{z_m+\rho_m e^{i\gamma_m}}{1+\rho_m e^{-i\gamma_m} z_m}\right)}{\frac{\partial f}{\partial z_l}(\rho e^{i\gamma})} - \frac{\frac{\partial f_\gamma}{\partial z_l}\left(\frac{z_1+\rho_1 e^{i\gamma_1}}{1+\rho_1 e^{-i\gamma_1} z_1}, \dots, \frac{z_m+\rho_m e^{i\gamma_m}}{1+\rho_m e^{-i\gamma_m} z_m}\right)}{\frac{\partial f_\gamma}{\partial z_l}(\rho e^{i\gamma})}$$

tends to 0 *as* $\rho = (\rho_1, \dots, \rho_m) \to \mathbf{1}^-$ *uniformly in the polydisk* $\{z \in \mathbb{C}^m : \|z\| \le r_0\}$.

Proof: If $f \in \mathcal{U}_\alpha^l(1)$ then by Theorem 1 (3) $f(z) = f_\gamma(z)$ and our lemma is proved in this case. Now, let $\delta_0 \in (0,1)$. Let δ_γ be the Hayman number of a function f which corresponds to the d.i.g. γ; $0 < \delta_\gamma \le \delta_0 < 1$. In [3, Lemma 2] it was shown that γ is d.i.g. of a function f if and only if $\gamma(a)$ is d.i.g. of $f(a,\cdot)$ from Definition 1, where γ and $\gamma(a)$ are connected by the formulas

$$e^{i\gamma_k(a)} = \frac{e^{i\gamma_k} - a_k}{1 - \overline{a_k} e^{i\gamma_k}}$$

for $k = 1, \dots, m$. Moreover, the numbers δ_γ and $\delta_{\gamma(a)}$ are connected by the formula (8) from [3]

$$\delta_{\gamma(a)} = \frac{\delta_\gamma}{|\frac{\partial f}{\partial z_l}(a)|} \prod_{k=1}^m \frac{(1-|a_k|^2)^\alpha}{|1-\overline{a_k}e^{i\gamma_k}|^{2\alpha}} \frac{1}{1-|a_l|^2}. \tag{1.3}$$

Let $\rho = (\rho_1, \dots, \rho_m) \in (0,1)^m$ and $a = \rho e^{i\gamma}$. Then (1.3) has the following form

$$\delta_{\gamma(a)} = \frac{\delta_\gamma}{|\frac{\partial f}{\partial z_l}(a)|} \prod_{k=1}^m \left(\frac{1+\rho_k}{1-\rho_k}\right)^\alpha \frac{1}{1-\rho_l^2}.$$

Consequently by Theorem 1 (2) it follows that $\delta_{\gamma(a)} \to 1$ as $\rho \to \mathbf{1}^-$. Since $\dot{\mathcal{U}}_\alpha^l$ is sequentially compact it follows, by Lemma 1, that there exists $\rho^{(n)} = (\rho_1^{(n)}, \dots, \rho_m^{(n)}) \in (0,1)^m$ such that $\rho^{(n)} \to \mathbf{1}^-$ and $\frac{\partial f}{\partial z_l}(\rho^{(n)} e^{i\gamma}, z) \to \frac{\partial f_\theta}{\partial z_l}(z)$ locally uniformly in $\mathbb{D}^m$, as $n \to \infty$, with some vector $\theta \in [0, 2\pi)^m$. Next we will show that $\theta = \gamma$. Let $R_k^{(n)} = \frac{r_k + \rho_k^{(n)}}{1 + r_k \rho_k^{(n)}}$, $k = 1, \dots, m$, $R^{(n)} = (R_1^{(n)}, \dots, R_m^{(n)})$. Since

$$\lim_{n\to\infty} \frac{1-\rho_k^{(n)}}{1-R_k^{(n)}} = \frac{1+r_k}{1-r_k},$$

using Theorem 1, with fixed $\mathbf{r} \in [0,1)^m$ one can get

$$\left|\frac{\partial f_\theta}{\partial z_l}(\mathbf{r}e^{i\gamma})\right|$$
$$= \lim_{n\to\infty}\left\{\frac{\left|\frac{\partial f}{\partial z_l}(R^{(n)}e^{i\gamma})\right|\prod_{k=1}^m\left(\frac{1-R_k^{(n)}}{1+R_k^{(n)}}\right)^\alpha\left(1-\left(R_l^{(n)}\right)^2\right)}{\left|\frac{\partial f}{\partial z_l}(\rho^{(n)}e^{i\gamma})\right|\prod_{k=1}^m\left(\frac{1-\rho_k^{(n)}}{1+\rho_k^{(n)}}\right)^\alpha\left(1-(\rho_l^{(n)})^2\right)\left(1+\rho_l^{(n)}r_l\right)^2}\right.$$
$$\left.\cdot\frac{1-\rho_l^{(n)}}{1-R_l^{(n)}}\prod_{k=1}^m\left(\frac{1-\rho_k^{(n)}}{1-R_k^{(n)}}\right)^\alpha\right\}$$
$$= \frac{1}{1-r_l^2}\prod_{k=1}^m\left(\frac{1+r_k}{1-r_k}\right)^2.$$

Therefore $\frac{\partial f_\theta}{\partial z_l}(\mathbf{r}e^{i\gamma}) \to \infty$ as $r_k \to 1^-$ for every $k = 1,\ldots,m$. And this is possible in the case $\theta = \gamma$ only. Because the above arguments were used for any sequence $\rho^{(n)} \to \mathbf{1}^-$ such that $\frac{\partial f}{\partial z_l}(\rho^{(n)}e^{i\gamma}, z)$ is locally uniformly convergent in $\mathbb{D}^m$, we get:

$$\frac{\partial f}{\partial z_l}(\rho e^{i\gamma}, z) \to \frac{\partial f_\gamma}{\partial z_l}(z), \tag{1.4}$$

in $\mathbb{D}^m$, as $\rho \to \mathbf{1}^-$. In [2], Theorem 2.1, we showed that functions from the family $\dot{\mathcal{U}}_\alpha^l$ are locally uniformly bounded in $\mathbb{D}^m$. Therefore one can use Vitali's Theorem for the functions $\frac{\partial f}{\partial z_l}(\rho e^{i\gamma}, z)$. Thus, in (1.4) the convergence is locally uniform in $\mathbb{D}^m$. In particular, for every fixed $r_0 \in (0,1)$

$$\frac{\frac{\partial f}{\partial z_l}\left(\frac{z_1+\rho_1 e^{i\gamma_1}}{1+\rho_1 e^{-i\gamma_1}z_1},\ldots,\frac{z_m+\rho_m e^{i\gamma_m}}{1+\rho_m e^{-i\gamma_m}z_m}\right)}{\frac{\partial f}{\partial z_l}(\rho e^{i\gamma})\left(1+\rho_l e^{-i\gamma_l}z_l\right)^2} \to \frac{1}{1-z_l^2e^{-2i\gamma_l}}\prod_{k=1}^m\left(\frac{1+z_ke^{-i\gamma_k}}{1-z_ke^{-i\gamma_k}}\right)^\alpha$$

uniformly in the polydisk $\{z \in \mathbb{C}^m : \|z\| \le r_0\}$, as $\rho \to \mathbf{1}^-$. Thus the functions

$$\frac{\frac{\partial f}{\partial z_l}\left(\frac{z_1+\rho_1 e^{i\gamma_1}}{1+\rho_1 e^{-i\gamma_1}z_1},\ldots,\frac{z_m+\rho_m e^{i\gamma_m}}{1+\rho_m e^{-i\gamma_m}z_m}\right)}{\frac{\partial f}{\partial z_l}(\rho e^{i\gamma})} \tag{1.5}$$

and

$$\frac{\left(1+\rho_l e^{-i\gamma_l}z_l\right)^2}{1-z_l^2e^{-2i\gamma_l}}\prod_{k=1}^m\left(\frac{1+z_ke^{-i\gamma_k}}{1-z_ke^{-i\gamma_k}}\right)^\alpha = \frac{\frac{\partial f_\gamma}{\partial z_l}\left(\frac{z_1+\rho_1 e^{i\gamma_1}}{1+\rho_1 e^{-i\gamma_1}z_1},\ldots\right)}{\frac{\partial f_\gamma}{\partial z_l}(\rho e^{i\gamma})}$$

converge uniformly in $\{z \in \mathbb{C}^m : \|z\| \le r_0\}$ to the same analytic function in $\mathbb{D}^m$, as $\rho \to \mathbf{1}^-$. ■

Proof of Theorem 2: Let us fix $\phi = (\phi_1, \dots, \phi_m) \in \mathbb{R}^m$ and $\varepsilon > 0$. From Theorem 1 it follows that there exists $\mathbf{r}^0 = (r_1^0, \dots, r_m^0) \in [0,1)^m$ such that for every $r_k \in (r_k^0, 1)$, $k = 1, \dots, m$,

$$\delta_0 - \varepsilon < \left|\frac{\partial f}{\partial z_l}(\mathbf{r}e^{i\theta})\right| \prod_{k=1}^{m} \left(\frac{1-r_k}{1+r_k}\right)^{\alpha} (1-r_l^2) < \delta_0 + \varepsilon, \tag{1.6}$$

$$\max_{\phi \in \mathbb{R}^m} \left|\frac{\partial f}{\partial z_l}(\mathbf{r}e^{i\phi})\right| \prod_{k=1}^{m} \left(\frac{1-r_k}{1+r_k}\right)^{\alpha} (1-r_l^2) < \delta_0 + \varepsilon. \tag{1.7}$$

Now let

$$W = \prod_{k \neq l} \left(\frac{1+r_k}{1-r_k}\right)^{\alpha} \int_{r_l^0}^{r_l} \frac{(1+\rho)^{\alpha-1}}{(1-\rho)^{\alpha+1}}\, d\rho.$$

Then from (1.6) one can get the following trivial inequality:

$$(\delta_0 - \varepsilon) W \leq (\delta_0 + \varepsilon) W. \tag{1.8}$$

But

$$\int_{r_l^0}^{r_l} \frac{(1+\rho)^{\alpha-1}}{(1-\rho)^{\alpha+1}}\, d\rho = \frac{1}{2\alpha}\left[\left(\frac{1+r_l}{1-r_l}\right)^{\alpha} - \left(\frac{1+r_l^0}{1-r_l^0}\right)^{\alpha}\right]$$

and

$$\int_{r_l^0}^{r_l} \left|\frac{\partial f}{\partial z_l}\right| d\rho = \int_0^{r_l} \left|\frac{\partial f}{\partial z_l}\right| d\rho - \int_0^{r_l^0} \left|\frac{\partial f}{\partial z_l}\right| d\rho, \tag{1.9}$$

where

$$\int_0^{r_l^0} \left|\frac{\partial f}{\partial z_l}\right| d\rho \leq r_l^0 \prod_{k \neq l} \left(\frac{1+r_k}{1-r_k}\right)^{\alpha} \frac{(1+r_l^0)^{\alpha-1}}{(1-r_l^0)^{\alpha+1}}, \tag{1.10}$$

by ([2], Theorem 2.1)

$$\left|\frac{\partial f}{\partial z_l}(\dots, r_{l-1}e^{i\theta_{l-1}}, \rho e^{i\theta_l}, r_{l+1}e^{i\theta_{l+1}}, \dots)\right| \leq \prod_{k \neq l} \left(\frac{1+r_k}{1-r_k}\right)^{\alpha} \frac{(1+\rho)^{\alpha-1}}{(1-\rho)^{\alpha+1}}.$$

Multiplying (1.8) by $2\alpha \prod_{k=1}^{m} (\frac{1-r_k}{1+r_k})^{\alpha}$ and using (1.9), (1.10) we get for $\mathbf{r} \to \mathbf{1}^-$

$$\begin{aligned}
\delta_0 - \varepsilon &\leq \liminf_{\mathbf{r}\to\mathbf{1}^-} \int_0^{r_l} \left|\frac{\partial f}{\partial z_l}(\dots, \rho e^{i\theta_l}, \dots)\right| d\rho \cdot 2\alpha \prod_{k=1}^{m} \left(\frac{1-r_k}{1+r_k}\right)^{\alpha} \\
&\leq \limsup_{\mathbf{r}\to\mathbf{1}^-} \int_0^{r_l} \left|\frac{\partial f}{\partial z_l}(\dots, \rho e^{i\theta_l}, \dots)\right| d\rho \cdot 2\alpha \prod_{k=1}^{m} \left(\frac{1-r_k}{1+r_k}\right)^{\alpha} \\
&\leq \delta_0 + \varepsilon.
\end{aligned}$$

Since ε is arbitrary we obtain the first equality in our Theorem. Analogously by (1.7) we get

$$\begin{aligned}\int_{r_l^0}^{r_l} \max_{\phi\in\mathbb{R}^m}\left|\frac{\partial f}{\partial z_l}\left(\ldots,r_{l-1}e^{i\theta_{l-1}},\rho e^{i\theta_l},r_{l+1}e^{i\theta_{l+1}},\ldots\right)\right| d\rho \\ \leq (\delta_0+\varepsilon)\prod_{k\neq l}\left(\frac{1+r_k}{1-r_k}\right)^{\alpha}\int_{r_l^0}^{r_l}\frac{(1+\rho)^{\alpha-1}}{(1-\rho)^{\alpha+1}}\,d\rho.\end{aligned} \tag{1.11}$$

Similarly as above, multiplying (1.11) by $2\alpha\prod_{k=1}^m(\frac{1-r_k}{1+r_k})^\alpha$, we obtain

$$\limsup_{\mathbf{r}\to\mathbf{1}^-}\int_0^{r_l}\max_{\theta\in\mathbb{R}^m}\left|\frac{\partial f}{\partial z_l}\left(\ldots,\rho e^{i\theta_l},\ldots\right)\right| d\rho\cdot 2\alpha\prod_{k=1}^m\left(\frac{1-r_k}{1+r_k}\right)^\alpha \leq \delta_0.$$

Observe that

$$\begin{aligned}\delta_0 &= \lim_{\mathbf{r}\to\mathbf{1}^-}\left[\int_0^{r_l}\left|\frac{\partial f}{\partial z_l}\left(\ldots,\rho e^{i\theta_l},\ldots\right)\right| d\rho\cdot 2\alpha\prod_{k=1}^m\left(\frac{1-r_k}{1+r_k}\right)^\alpha\right]\\ &\leq \limsup_{\mathbf{r}\to\mathbf{1}^-}\left[\int_0^{r_l}\max_{\phi\in\mathbb{R}^m}\left|\frac{\partial f}{\partial z_l}\left(\ldots,\rho e^{i\phi_l},\ldots\right)\right| d\rho\cdot 2\alpha\prod_{k=1}^m\left(\frac{1-r_k}{1+r_k}\right)^\alpha\right]\\ &\leq \delta_0.\end{aligned}$$

Thus we get the second equality in Theorem 2.

Now let us fix $\phi\in\mathbb{R}^m$ and $\varepsilon>0$ again. From (1.10), (1.11) it follows that for all $r_k\in(r_k^0,1)$, $k=1,\ldots,m$, the following inequalities hold:

$$\begin{aligned}&\max_{\phi\in\mathbb{R}^m}\left|F(\mathbf{r}e^{i\phi})\right|\\ &\quad\leq\prod_{k\neq l}\left(\frac{1+r_k}{1-r_k}\right)^\alpha\left[r_l^0\frac{(1+r_l^0)^{\alpha-1}}{(1-r_l^0)^{\alpha+1}}\right.\\ &\qquad\left.+(\delta_0+\varepsilon)\frac{1}{2\alpha}\left(\left(\frac{1+r_l}{1-r_l}\right)^\alpha-\left(\frac{1+r_l^0}{1-r_l^0}\right)^\alpha\right)\right].\end{aligned}$$

As above we multiply the last inequality by $2\alpha\prod_{k=1}^m\left(\frac{1-r_k}{1+r_k}\right)^\alpha$. Then, for $\mathbf{r}\to\mathbf{1}^-$, we obtain

$$\limsup_{\mathbf{r}\to\mathbf{1}^-}\left[\left|F\left(\mathbf{r}e^{i\theta}\right)\right|2\alpha\prod_{k=1}^m\left(\frac{1-r_k}{1+r_k}\right)^\alpha\right]\leq\delta_0+\varepsilon.$$

Since ε is an arbitrary positive number we proved our theorem for $\delta_0=0$. Next we will show that there exists the limit

$$\lim_{\mathbf{r}\to\mathbf{1}^-}\left[\left|F\left(\mathbf{r}e^{i\theta}\right)\right|2\alpha\prod_{k=1}^m\left(\frac{1-r_k}{1+r_k}\right)^\alpha\right]=\delta_0$$

for $\delta_0 > 0$. Let us suppose that this is not true. Then thereexists a sequence $\rho^{(n)} = (\rho_1^{(n)}, \ldots, \rho_m^{(n)}) \to \mathbf{1}^-$ such that

$$\lim_{n\to\infty} \left[\left| F(\rho^{(n)} e^{i\theta}) \right| 2\alpha \prod_{k=1}^{m} \left(\frac{1-\rho_k^{(n)}}{1+\rho_k^{(n)}} \right)^{\alpha} \right] = \delta_1 < \delta_0.$$

Let $r_0 \in (0,1)$ be fixed. Denote $R_l^{(n)} = \frac{\rho_l^{(n)} - r_0}{1 - r_0 \rho_l^{(n)}}$, $a^{(n)} = (a_1^{(n)}, \ldots, a_m^{(n)}) = \rho^{(n)} e^{i\theta}$. Note that

$$\lim_{n\to\infty} \frac{1-\rho_l^{(n)}}{1-R_l^{(n)}} = \frac{1-r_0}{1+r_0}.$$

One can assume that there exists

$$\begin{aligned}\delta_2 &= \lim_{n\to\infty} \Bigg[\left| F\left(\ldots, a_{l-1}^{(n)}, R_l^{(n)} e^{i\theta_l}, \ldots \right) \right| \\ &\qquad \cdot 2\alpha \left(\frac{1-R_l^{(n)}}{1+R_l^{(n)}} \right)^{\alpha} \prod_{k\neq l} \left(\frac{1-\rho_k^{(n)}}{1+\rho_k^{(n)}} \right)^{\alpha} \Bigg].\end{aligned}$$

If the above limit does not exist we can consider an appropriate subsequence of $\rho^{(n)}$. Because of this we suppose that there exist $\psi_0,\ \psi_1,\ \psi_2 \in \mathbb{R}$ and the limit

$$\begin{aligned}&\lim_{n\to\infty} \frac{F\left(\ldots, a_{l-1}^{(n)}, \frac{-r_0 e^{i\theta_l} + a_l^{(n)}}{1-\bar{a}_l^{(n)} r_0 e^{i\theta_l}}, a_{l+1}^{(n)}, \ldots \right) - F\left(a^{(n)}\right)}{\frac{\partial F}{\partial z_l}(a^{(n)}) \left(1 - (\rho_l^{(n)})^2 \right)} \\ &= \frac{\delta_2 e^{i\psi_2} \left(\frac{1-r_0}{1+r_0} \right)^{\alpha} - \delta_1 e^{i\psi_1}}{2\alpha\delta_0 e^{i\psi_0}}.\end{aligned} \tag{1.12}$$

If in (1.12), instead of $F \in \mathcal{U}_\alpha^l(\delta_0)$, we put

$$f_\theta(z) = \frac{e^{i\theta_l}}{2\alpha} \left[\prod_{k=1}^{m} \left(\frac{1+z_k e^{-i\theta_k}}{1-z_k e^{-i\theta_k}} \right)^{\alpha} - 1 \right] \in \mathcal{U}_\alpha^l(1), \tag{1.13}$$

then we get

$$\begin{aligned}&\lim_{n\to\infty} \frac{\frac{e^{i\theta_l}}{2\alpha} \prod_{k\neq l} \left(\frac{1+\rho_k^{(n)}}{1-\rho_k^{(n)}} \right)^{\alpha} \left[\left(\frac{1+R_l^{(n)}}{1-R_l^{(n)}} \right)^{\alpha} - \left(\frac{1+\rho_l^{(n)}}{1-\rho_l^{(n)}} \right)^{\alpha} \right]}{\prod_{k=1}^{m} \left(\frac{1+\rho_k^{(n)}}{1-\rho_k^{(n)}} \right)^{\alpha}} \\ &= \frac{e^{i\theta_l}}{2\alpha} \left[\left(\frac{1-r_0}{1+r_0} \right)^{\alpha} - 1 \right].\end{aligned} \tag{1.14}$$

Since θ is the d.i.g. of F (by Lemma 2) then for every $\varepsilon > 0$ there is $N \in (0,1)$ such that for all $\rho \in (0,1)^m$, $\|\rho\| \in (N,1)$ and for all $z_l \in \mathbb{D}$, $|z_l| \le r_0$ the following inequality holds:

$$\left| \frac{\frac{\partial F}{\partial z_l}\left(\ldots, \frac{z_l+\rho_l e^{i\theta_l}}{1+\rho_l e^{-i\theta_l} z_l}, \ldots\right)}{\frac{\partial F}{\partial z_l}(\rho e^{i\theta})(1-\rho_l^2)} - \frac{\frac{\partial f_\theta}{\partial z_l}\left(\ldots, \frac{z_l+\rho_l e^{i\theta_l}}{1+\rho_l e^{-i\theta_l} z_l}, \ldots\right)}{\frac{\partial f_\theta}{\partial z_l}(\rho e^{i\theta})(1-\rho_l^2)} \right| < \varepsilon. \tag{1.15}$$

Integrating of (1.15) with respect to z_l over the interval $[0, z_l]$ gives

$$\left| \frac{F\left(\ldots, \frac{z_l+\rho_l e^{i\theta_l}}{1+\rho_l e^{-i\theta_l} z_l}, \ldots\right) - F(\rho e^{i\theta})}{\frac{\partial F}{\partial z_l}(\rho e^{i\theta})(1-\rho_l^2)} - \frac{f_\theta\left(\ldots, \frac{z_l+\rho_l e^{i\theta_l}}{1+\rho_l e^{-i\theta_l} z_l}, \ldots\right) - f_\theta(\rho e^{i\theta})}{\frac{\partial f_\theta}{\partial z_l}(\rho e^{i\theta})(1-\rho_l^2)} \right| < \varepsilon. \tag{1.16}$$

Thus the left side of (1.16) tends to 0 as $\rho \to \mathbf{1}^-$. Therefore, if we put $\rho^{(n)}$ and $-r_0 e^{i\theta_l}$ instead of ρ and z_l, respectively, from (1.12) and (1.14) we get

$$\frac{\delta_2 e^{i\psi_2}\left(\frac{1-r_0}{1+r_0}\right)^\alpha - \delta_1 e^{i\psi_1}}{2\alpha\delta_0 e^{i\psi_0}} = \frac{e^{i\theta_l}}{2\alpha}\left[\left(\frac{1-r_0}{1+r_0}\right)^\alpha - 1\right]$$

as $n \to \infty$. Consequently $\delta_0 = \delta_1 = \delta_2$, $e^{i(\psi_2-\psi_0)} = e^{i(\psi_1-\psi_0)} = e^{i\theta_l}$. ■

Theorem 3. *Let $f \in \mathcal{U}_\alpha^l(\delta_0)$, $\delta_0 > 0$ and let $\gamma = (\gamma_1, \ldots, \gamma_m)$ be any d.i.g. of f to which corresponds a Hayman number $\delta \in (0, \delta_0]$. Denote*

$$\Phi(\zeta) := \arg \frac{\partial f}{\partial z_l}(\rho e^{i\gamma}),$$

where $\rho = (\rho_1, \ldots, \rho_m)$,

$$\begin{aligned} \rho_k &= \rho_k(\zeta_k) = \sqrt{\frac{(1-r_0^2)^2}{4r_0^4 c_k^2} - \frac{1}{r_0^2}} - \frac{1}{2c_k}\left(1 - \frac{1}{r_0^2}\right), \\ c_k &= \operatorname{Re}(\zeta_k e^{-i\gamma_k}) - \tan\eta |\operatorname{Im}(\zeta_k e^{-i\gamma_k})|, \end{aligned}$$

$k = 1, \ldots, m$, $r_0 = \sin\eta$. Then for every $n = 0, 1, 2, \ldots$, and every positive integer $k_1, \ldots, k_n \in [1, m]$ and for every fixed $\eta \in (0, \frac{\pi}{2})$:

$$e^{-i\Phi(\zeta)} \frac{\frac{\partial^{n+1} f}{\partial z_l \partial z_{k_1} \ldots \partial z_{k_n}}(\zeta)}{\frac{\partial^{n+1} f_\gamma}{\partial z_l \partial z_{k_1} \ldots \partial z_{k_n}}(\zeta)} \to \delta$$

locally uniformly in the angular domain

$$\mathbb{D}^m(R,\eta) \quad := \quad \{\zeta = (\zeta_1,\ldots,\zeta_m) \in \mathbb{D}^m : |\arg(1-\zeta_k e^{-i\gamma_k})| < \eta, \text{ for all } k, R < \|\zeta\| < 1\}$$

as $R \to 1^-$.

Proof: Let us fix $r_0 \in (0,1)$, $\rho \in (0,1)^m$. The function

$$\zeta_k = \frac{z_k + \rho_k e^{i\gamma_k}}{1+\rho_k e^{-i\gamma_k} z_k}$$

univalently maps the disk $\{z_k \in \mathbb{D} : |z_k| < r_0\}$ onto the disk with center $c_k(\rho) = e^{i\gamma_k}\rho_k \frac{1-r_0^2}{1-\rho_k^2 r_0^2}$ and radius $r_k(\rho) = \frac{r_0(1-\rho_k^2)}{1-\rho_k^2 r_0^2}$. By Lemma 2 we obtain

$$\frac{\frac{\partial f}{\partial z_l}(\zeta)}{\frac{\partial f}{\partial z_l}(\rho e^{i\gamma})} - \frac{\frac{\partial f_\gamma}{\partial z_l}(\zeta)}{\frac{\partial f_\gamma}{\partial z_l}(\rho e^{i\gamma})} \to 0 \tag{1.17}$$

uniformly in the polydisk

$$K_\rho(r_0) = \{\zeta \in \mathbb{D}^m : |\zeta_k - c_k(\rho)| \le r_k(\rho),\ k = 1,\ldots,m\}$$

as $\rho \to \mathbf{1}^-$. Because of

$$\lim_{\rho\to\mathbf{1}^-} \frac{\frac{\partial f_\gamma}{\partial z_l}(\zeta)}{\frac{\partial f_\gamma}{\partial z_l}(\rho e^{i\gamma})} \ne 0$$

in the polydisk $\{z \in \mathbb{C}^m : \|z\| \le r_0\}$ (see the proof of Lemma 2), the convergence in (1.17) is equivalent to

$$\frac{\frac{\partial f}{\partial z_l}(\zeta)\ \frac{\partial f_\gamma}{\partial z_l}(\rho e^{i\gamma})}{\frac{\partial f_\gamma}{\partial z_l}(\zeta)\ \frac{\partial f}{\partial z_l}(\rho e^{i\gamma})} \to 1$$

uniformly in $K_\rho(r_0)$ as $\rho \to \mathbf{1}^-$. Thus, using Theorem 1 we get

$$\frac{\frac{\partial f}{\partial z_l}(\zeta)}{\frac{\partial f_\gamma}{\partial z_l}(\zeta)} e^{-i\Phi(\zeta)} \to \delta$$

uniformly in $K_\rho(r_0)$, as $\rho \to \mathbf{1}^-$. Hence, for every $\varepsilon > 0$ there is a $\rho_* = \rho_*(\varepsilon) \in (0,1)$ such that

$$\left| \frac{\frac{\partial f}{\partial z_l}(z)}{\frac{\partial f_\gamma}{\partial z_l}(\zeta)} e^{-i\Phi(\zeta)} - \delta \right| < \varepsilon \tag{1.18}$$

for all $\rho \in (0,1)^m$, $\|\rho\| \in (\rho_*, 1)$ and for every $\zeta \in K_\rho(r_0)$.

Let $2\beta_k$ be the measure of an angle with the vertex in $e^{i\gamma_k}$, $k = 1, \ldots, m$, and with the arms tangential to the circle $\{\zeta_k \in \mathbb{C} : |\zeta_k - c_k(\rho)| \le r_k(\rho)\}$. Then

$$\sin \beta_k = \frac{r_k(r_0)}{1 - |c_k(r_0)|} = \frac{r_0(1+\rho_k)}{1+\rho_k r_0^2} = q(\rho_k).$$

The function $q(x)$ is increasing. Consequently the set of polydisks $K_\rho(r_0)$ fills up any subset of $\mathbb{D}^m(R,\eta)$ for some R and η. One can take $\arcsin r_0$ (a value corresponding to $\rho = \mathbf{0}$) instead of η. Therefore η can be arbitrary from the interval $(0, \frac{\pi}{2})$, if we choose a suitable r_0. Thus (1.18) holds in $\mathbb{D}^m(R,\eta)$, where $\eta \in (0, \frac{\pi}{2})$ is fixed and R depends of ε. Then for every $\zeta \in \mathbb{D}^m(R,\eta)$ by (1.18) there is $\Phi(\zeta) = \arg \frac{\partial f}{\partial z_l}(\rho e^{i\gamma})$ (not necessary one), where ρ is such that $\zeta \in K_\rho(r_0)$. Consequently, for $\zeta \in \mathbb{D}^m(R,\eta)$ we can choose $\rho = (\rho_1, \ldots, \rho_m)$ such that for every $k = 1, \ldots, m$ the point ζ_k lies on a radius of the disk $\{\zeta_k \in \mathbb{C} : |\zeta_k - c_k(\rho)| \le r_k(\rho)\}$ which is orthogonal to one of the sides of the angular domain $\{\zeta_k \in \mathbb{C} : |\arg(1 - \zeta_k e^{-i\gamma_k})| < \eta\}$. After calculations we get

$$|c_k| = |c_k(\rho)| = \operatorname{Re}(\zeta_k e^{-i\gamma_k}) - \tan\eta |\operatorname{Im}(\zeta_k e^{-i\gamma_k})|.$$

Since $|c_k(\rho)| = \rho_k \frac{1-r_0^2}{1-\rho_k^2 r_0^2}$ we get

$$\rho_k = \rho_k(\zeta_k) = \sqrt{\frac{(1-r_0^2)^2}{4r_0^4|c_k|^2} - \frac{1}{r_0^2}} - \frac{1}{2|c_k|}\left(1 - \frac{1}{r_0^2}\right),$$

where $r_0 = \sin\eta$. This proves Theorem 3 in the case $n = 0$.

Now, let $n \ge 1$. For functions analytic in $\mathbb{D}^m$ holds Weierstraß's Theorem on sequences of functions locally uniformly convergent in $\mathbb{D}^m$. Therefore we can differentiate the relationship from Lemma 2 with respect to $z_{k_1}, \ldots, z_{z_n}$, $1 \le k_1, \ldots, k_n \le m$, preserving uniform convergence in $\{z \in \mathbb{C}^m : \|z\| \le r_0\}$. After every differentiation with respect to z_u we multiply the result by $(1 + z_u \rho_u e^{-i\gamma_u})^2$. In this way

$$\frac{\frac{\partial^{n+1} f}{\partial z_l \partial z_{k_1} \ldots \partial z_{k_n}}\left(\frac{z_1+\rho_1 e^{i\gamma_1}}{1+\rho_1 e^{-i\gamma_1} z_1}, \ldots, \frac{z_m+\rho_m e^{i\gamma_m}}{1+\rho_m e^{-i\gamma_m} z_m}\right)\prod_{j=1}^n (1-\rho_{k_j}^2)}{\frac{\partial f}{\partial z_l}(\rho e^{i\gamma})}$$
$$- \frac{\frac{\partial^{n+1} f_\gamma}{\partial z_l \partial z_{k_1} \ldots \partial z_{k_n}}\left(\frac{z_1+\rho_1 e^{i\gamma_1}}{1+\rho_1 e^{-i\gamma_1} z_1}, \ldots, \frac{z_m+\rho_m e^{i\gamma_m}}{1+\rho_m e^{-i\gamma_m} z_m}\right)\prod_{j=1}^n (1-\rho_{k_j}^2)}{\frac{\partial f_\gamma}{\partial z_l}(\rho e^{i\gamma})}$$
$$\to 0, \tag{1.19}$$

uniformly in $\{z \in \mathbb{C}^m : \|z\| \le r_0\}$, as $\rho \to \mathbf{1}^-$. In this polydisk the right side of (1.5) is separated from 0, for $\rho \to \mathbf{1}^-$. We differentiate (1.5), step by step, with respect to z_u, $u = k_1, \ldots, k_n$, and multiply by $(1 + z_u e^{-i\gamma_u}\rho_u)^2$. Each time we obtain an expression bounded away from 0 for $\rho \to \mathbf{1}^-$. Thus

$$\frac{\frac{\partial^{n+1} f_\gamma}{\partial z_l \partial z_{k_1} \ldots \partial z_{k_n}}\left(\frac{z_1+\rho_1 e^{i\gamma_1}}{1+\rho_1 e^{-i\gamma_1} z_1}, \ldots, \frac{z_m+\rho_m e^{i\gamma_m}}{1+\rho_m e^{-i\gamma_m} z_m}\right) \prod_{j=1}^n (1-\rho_{k_j}^2)}{\frac{\partial f_\gamma}{\partial z_l}(\rho e^{i\gamma})}$$

is bounded away from 0, as $\rho \to \mathbf{1}^-$. Therefore (1.19) is equivalent to

$$\frac{\frac{\partial^{n+1} f}{\partial z_l \partial z_{k_1} \ldots \partial z_{k_n}}(\zeta) \frac{\partial f_\gamma}{\partial z_l}(\rho e^{i\gamma})}{\frac{\partial^{n+1} f_\gamma}{\partial z_l \partial z_{k_1} \ldots \partial z_{k_n}}(\zeta) \frac{\partial f}{\partial z_l}(\rho e^{i\gamma})} \to 1$$

uniformly in $K_\rho(r_0)$, as $\rho \to \mathbf{1}^-$. Next, similarly as in the case $n = 0$,

$$e^{-i\Phi(\zeta)} \frac{\frac{\partial^{n+1} f}{\partial z_l \partial z_{k_1} \ldots \partial z_{k_n}}(\zeta)}{\frac{\partial^{n+1} f_\gamma}{\partial z_l \partial z_{k_1} \ldots \partial z_{k_n}}(\zeta)} \to \delta,$$

uniformly in $\mathbb{D}^m(R, \eta)$, as $R \to 1^-$. ■

Theorem 4. *Let $f \in \mathcal{U}_\alpha^l$, and let γ be a d.i.g. of f. Let δ be the corresponding Hayman number. Then for every nonnegative integers $q_1, \ldots, q_m$ such that $n = q_1 + \ldots + q_m$ there exists the limit*

$$\lim_{\mathbf{r} \to \mathbf{1}^-} \left[\left| \frac{\partial^{n+1} f}{\partial z_l \partial z_1^{q_1} \ldots \partial z_m^{q_m}}(\mathbf{r} e^{i\gamma}) \right| (1-r_l) \prod_{k=1}^m (1-r_k)^{\alpha+q_k} \right]$$
$$= \delta \lim_{\mathbf{r} \to \mathbf{1}^-} \left[\frac{\partial^{n+1} g}{\partial r_l \partial r_1^{q_1} \ldots \partial r_m^{q_m}}(\mathbf{r})(1-r_l) \prod_{k=1}^m (1-r_k)^{\alpha+q_k} \right]$$
$$= \delta 2^{\alpha m - 1} \left(1 + \frac{q_l}{\alpha}\right) \prod_{k=1}^m [\alpha(\alpha+1) \cdots (\alpha + q_k - 1)],$$

where $g(\mathbf{r}) = \frac{1}{2\alpha} \prod_{k=1}^m (\frac{1+r_k}{1-r_k})^\alpha$. If $q_j = 0$ for some j then the corresponding expression in the brackets is equal to 1.

Proof: The function $\log \frac{1+\xi}{1-\xi}$, $\zeta \in \mathbb{D}$, has nonnegative Taylor coefficients. Therefore, for $\alpha > 0$, the function $(\frac{1+\xi}{1-\xi})^\alpha$ has nonnegative coefficients also. Consequently, the derivatives of arbitrary order of the function $(\frac{1+r}{1-r})^\alpha$ increase with respect to $r \in [0, 1)$. Thus the function $g(\mathbf{r})$ as well as its derivatives of arbitrary order are increasing with respect to an arbitrary variable $r_k \in [0,1)$, $k = 1, \ldots, m$. By Theorem 3 there exists the limit

$$\lim_{\mathbf{r}\to 1^-} \left| \frac{\frac{\partial^{n+1} f}{\partial z_l \partial z_1^{q_1} \ldots \partial z_m^{q_m}}(\mathbf{r}e^{i\gamma})}{\frac{\partial^{n+1} f_\gamma}{\partial z_l \partial z_1^{q_1} \ldots \partial z_m^{q_m}}(\mathbf{r}e^{i\gamma})} \right| = \delta,$$

which is equivalent to

$$\lim_{\mathbf{r}\to 1^-} \left| \frac{\frac{\partial^{n+1} f}{\partial z_l \partial z_1^{q_1} \ldots \partial z_m^{q_m}}(\mathbf{r}e^{i\gamma})}{\frac{\partial^{n+1} g}{\partial r_l \partial r_1^{q_1} \ldots \partial r_m^{q_m}}(\mathbf{r})} \right| = \delta.$$

Since

$$\lim_{\mathbf{r}\to 1^-} \left[g(\mathbf{r}) \prod_{k=1}^{m} (1 - r_k)^\alpha \right] = \frac{2^{\alpha m - 1}}{\alpha},$$

by [3], Lemma 1, we obtain

$$\lim_{\mathbf{r}\to 1^-} \left[\left| \frac{\partial^{n+1} f}{\partial z_l \partial z_1^{q_1} \ldots \partial z_m^{q_m}}(\mathbf{r}e^{i\gamma}) \right| (1 - r_l) \prod_{k=1}^{m} (1 - r_k)^{\alpha + q_k} \right]$$
$$= \frac{2^{\alpha m - 1}}{\alpha} \prod_{k=1}^{m} [\alpha \cdots (\alpha + q_k - 1)](\alpha + q_l). \quad \blacksquare$$

One can ask about "largeness" of the set of the directions of the intensive growth for a given function $f \in \mathcal{U}_\alpha^l$. From Privalov's Uniqueness Theorem it follows that the measure of this set on $\mathbb{T}^m$ is equal 0. The set can be empty (for example for bounded functions from $\mathcal{U}_\alpha^l$). For the function f_θ given by (1.13), there is only one d.i.g.. In the case $m = 1$, for every positive integer $n \geq 2$, examples of functions $g_{n,\alpha} \in \mathcal{U}_\alpha$, $\alpha \geq \alpha_n$ and $g_{\infty,\alpha} \in \mathcal{U}_\alpha$, $\alpha \geq 1 + 2/(e-1)$, are given in [6], where

$$\alpha_2 = 1.241\ldots, \qquad \alpha_n = \frac{\sqrt{n^2 e^2 - (e+1)^2} + n}{\sqrt{n^2 e^2 - (e+1)^2} - n},$$

for $n \geq 3$. For $g_{n,\alpha}$ there exist exactly n d.i.g.'s and for $g_{\infty,\alpha}$ — a countable set of d.i.g.'s. That means that in the case $m = 1$, $\alpha \geq \alpha_0$, there exist examples of functions from $\mathcal{U}_\alpha$ with arbitrary, at most countable, sets of d.i.g.'s. Using the functions $g_{n,\alpha}$ and $g_{\infty,\alpha}$ we will construct, in the case $m \geq 2$, examples of functions from $\mathcal{U}_\alpha^l$ which have any countable set of d.i.g.'s.

Let us denote

$$G_*(z) := \prod_{k\neq l}\left(\frac{1+z_k}{1-z_k}\right)^{\alpha}(g_{*,\alpha}(z_l)+1)-1,$$

$z\in\mathbb{D}^m$, where $*$ denotes ∞ or a positive integer greater than 1. We will show, that $G_*\in\mathcal{U}^l_\alpha$ for $\alpha\geq\alpha_*$. Since $G_*(\mathbf{0})=0$, $\frac{\partial G_*}{\partial z_l}(z)\neq 0$ in $\mathbb{D}^m$, $\frac{\partial G_*}{\partial z_l}(\mathbf{0})=1$, it suffices to prove that ord $G_*=\alpha$. As it was shown in [3], Theorem 1.1,

$$\text{ord } G_* = \max_{1\leq k\leq m}\sup_{z\in\mathbb{D}^m}\left|\frac{\frac{\partial^2 G_*}{\partial z_l\partial z_k}(z)}{\frac{\partial G_*}{\partial z_l}(z)}\frac{1-|z_k|^2}{2}-\overline{z_k}\delta^l_k\right|.$$

For $k\neq l$ we have

$$\sup_{z\in\mathbb{D}^m}\left|\frac{\frac{\partial^2 G_*}{\partial z_l\partial z_k}(z)}{\frac{\partial G_*}{\partial z_l}(z)}\frac{1-|z_k|^2}{2}\right| = \sup_{z_k\in\mathbb{D}}\left|\alpha\frac{1-|z_k|^2}{1-z_k^2}\right|=\alpha,$$

and

$$\sup_{z\in\mathbb{D}^m}\left|\frac{\frac{\partial^2 G_*}{\partial z_l^2}(z)}{\frac{\partial G_*}{\partial z_l}(z)}\frac{1-|z_l|^2}{2}-\overline{z_l}\right| = \sup_{z_l\in\mathbb{D}}\left|\frac{g''_{*,\alpha}(z_l)}{g'_{*,\alpha}(z_l)}\frac{1-|z_l|^2}{2}-\overline{z_l}\right|=\alpha,$$

where $\alpha\geq\alpha_*$, see [6]. Thus ord $G_*=\alpha$. Now, let us note that if $\tilde{\gamma}$ is a d.i.g. of $g_{*,\alpha}$ then the vector $(0,\ldots,\tilde{\gamma},\ldots,0)$ (where $\tilde{\gamma}$ is at position l) is a d.i.g. of G_*. That is G_* has as many d.i.g.'s as $g_{*,\alpha}$. The next theorem gives a relationship between $\mathcal{U}^l_\alpha(\delta)$ for various δ.

Theorem 5. *For every $f\in\mathcal{U}^l_\alpha(\delta_0)$, $\delta_0\in[0,1]$ and every function $\delta^\star(\lambda)$, $\lambda\in(0,1)$, with values in $[0,\delta_0]$ there exists a family of functions $\psi_\lambda\in\mathcal{U}^l_\alpha(\delta^\star(\lambda))$, $\lambda\in(0,1)$, such that $\frac{\partial\psi_\lambda}{\partial z_l}(z)\to\frac{\partial f}{\partial z_l}(z)$ locally uniformly in $\mathbb{D}^m$, as $\lambda\to 0$.*

Proof: Let us denote $\delta(\lambda)=\delta_0(\frac{\delta^\star(\lambda)}{\delta_0})^{1/\lambda}\in[0,\delta_0]$ for $\lambda\in(0,1)$. From [3], Theorem 2, and the above example of the function $G_*\in\mathcal{U}^l_\alpha$ it follows that $\mathcal{U}^l_\alpha(\delta(\lambda))\neq\emptyset$. For every $\lambda\in(0,1)$ let us choose $f_\lambda\in\mathcal{U}^l_\alpha(\delta(\lambda))$. One can assume that the function f_λ has the same d.m.g. as the function f. This is possible because one can use a rotation $e^{-i\theta}f(ze^{i\theta})$. Let us consider one of the functions ψ_λ which satisfy the following conditions:

$$\psi_\lambda(\mathbf{0})=0,\quad \frac{\partial\psi_\lambda}{\partial z_l}(z)=\left(\frac{\partial f}{\partial z_l}(z)\right)^{1-\lambda}\left(\frac{\partial f_\lambda}{\partial z_l}(z)\right)^{\lambda}.$$

From [2], Theorem 1.1, it follows that for h_0, $h_1 \in \mathcal{U}_\alpha^l$ and $\lambda \in [0,1]$ the function $h_\lambda \in \mathcal{U}_\alpha^l$ if

$$h_\lambda(\mathbf{0}) = 0, \quad \frac{\partial h_\lambda}{\partial z_l}(z) = \left(\frac{\partial h_0}{\partial z_l}(z)\right)^{1-\lambda} \left(\frac{\partial h_1}{\partial z_l}(z)\right)^{\lambda}.$$

Thus $\psi_\lambda \in \mathcal{U}_\alpha^l$ and its Hayman number is equal to $\delta_0^{1-\lambda}\delta^\lambda(\lambda) = \delta^\star(\lambda)$. Thus $\psi_\lambda \in \mathcal{U}_\alpha^l(\delta^\star(\lambda))$. Since (by [2], Theorem 2.1) for $f \in \mathcal{U}_\alpha^l$ we have

$$\frac{1}{1-|z_l|^2}\prod_{k=1}^m \left(\frac{1-|z_k|}{1+|z_k|}\right)^\alpha \le \left|\frac{\partial f}{\partial z_l}(z)\right| \le \frac{1}{1-|z_l|^2}\prod_{k=1}^m \left(\frac{1+|z_k|}{1-|z_k|}\right)^\alpha,$$

one gets

$$\left(\frac{\frac{\partial f_\lambda}{\partial z_l}(z)}{\frac{\partial f}{\partial z_l}(z)}\right)^\lambda \to 1$$

locally uniformly in $\mathbb{D}^m$ as $\lambda \to 0$. Therefore

$$\frac{\partial \psi_\lambda}{\partial z_l}(z) = \frac{\partial f}{\partial z_l}(z)\left(\frac{\frac{\partial f_\lambda}{\partial z_l}(z)}{\frac{\partial f}{\partial z_l}(z)}\right)^\lambda \to \frac{\partial f}{\partial z_l}(z)$$

locally uniformly in $\mathbb{D}^m$ as $\lambda \to 0$. ■

If $\delta^\star(\lambda) \equiv \delta \in [0, \delta_0]$ then from Theorem 5 we get

Corollary 1. *For every $f \in \mathcal{U}_\alpha^l(\delta_0)$ $\delta_0 \in [0,1]$ and $\delta \in [0,\delta_0]$ there is a family of functions $\psi_\lambda \in \mathcal{U}_\alpha^l(\delta)$, $\lambda \in (0,1)$, such that $\frac{\partial \psi_\lambda}{\partial z_l}(z) \to \frac{\partial f}{\partial z_l}(z)$ locally uniformly in $\mathbb{D}^m$, as $\lambda \to 0$.*

References

[1] Campbell, D. M., Applications and proof of a uniqueness theorem for linear invariant families of finite order, *Rocky Mountain J. Math.* **4**, 4 (1974), 621–634.

[2] Godula, J. and V. V. Starkov, Linearly invariant families of holomorphic functions in the unit polydisk, *Generalizations of Complex Analysis, Banach Center Publ.* **37** (1996), 115–127.

[3] Godula, J. and V. V. Starkov, Regularity theorem for linearly invariant families of functions in a polydisk (in Russian), *Izviestia V. U. Z.* **8** (1995), 21–31.

[4] Pommerenke, Chr., Linear-invariante Familien analytischer Funktionen I, *Math. Ann.* **155** (1964), 108–154.

[5] Starkov, V. V., Theorems of regularity for universal linearly invariant families of functions (in Russian), *Serdika (Bulg. Math.Soc.)* **11** (1985), 299–318.

[6] Starkov, V. V., Directions of intensive growth of locally univalent functions, in *Complex Analysis and Appl.'87*, Sofia, 1989, 517–522.

Janusz Godula
Institute of Mathematics
Maria Curie-Skłodowska University
20-031 Lublin
POLAND
godula@golem.umcs.lublin.pl

Victor V. Starkov
Department of Mathematics
University of Petrozavodsk
185640 Petrozavodsk
RUSSIA
starkov@mainpgu.karelia.ru

Generalized Jackson kernels in approximation theory

Richard Greiner

Abstract. We study approximative identities

$$I_n(f)(x) = \frac{1}{2\pi}\int_{-\pi}^{\pi} \chi_n(u) f(x-u)\,du$$

with positive even trigonometric polynomials χ_n of degree n as kernels. It is well known that in general the approximants $I_n(f)$ are norm convergent to f with order at most n^{-2} and that convergence for the test function $f(x) = (2\sin(x/2))^2$ at $x = 0$ implies norm convergence for a wide class of functions f.

We discuss the order of convergence at $x = 0$ for $(2\sin(x/2))^{2r}$ where $r \geq 1$ is an integer and characterize kernels χ_n which produce best convergence. This leads to estimates of the type

$$\int_{-\pi}^{\pi} \chi_n(u)|u|^{2r}\,du = \mathcal{O}(n^{-2r})$$

which arise *e.g.* in the investigation of the degree of approximation of differentiable functions. Generalized Jackson kernels will be shown to have the same (best possible) rate of convergence. All the kernels discussed are suitable for approximative identities. Moreover, their special properties find applications, also in complex approximation theory [1], [2], [3], [7], [10].

§1 Introduction

Let $r \geq 1$ be a natural number. We consider the singular integral

$$I_n(r,x) = \frac{1}{2\pi}\int_{-\pi}^{\pi} \chi_n(u) s_r(x-u)\,du, \tag{1.1}$$

where $s_r(x) = (2\sin(x/2))^{2r}$ and

$$\chi_n(u) = 1 + 2\sum_{k=1}^{n} \Theta_n(k)\cos ku \geq 0 \tag{1.2}$$

Computational Methods and Function Theory (CMFT'97)
N. Papamichael, St. Ruscheweyh and E. B. Saff (Eds.), pp. 259–266

ISBN 981-02-3626-3

is a positive even kernel with real coefficients $\Theta_n(k)$. We denote the set of these kernels by $\mathcal{T}_n^{e+}$. If $r=1$ then

$$I_n(1,0) \geq \pi^2 n^{-2} + o(n^{-2}) \tag{1.3}$$

and equality is obtained iff χ_n is the Fejér-Korovkin kernel

$$k_{1,n}(x) = \begin{cases} \frac{2\sin^2(\pi/(n+2))}{n+2}\left[\frac{\cos((n+2)x/2)}{\cos(\pi/(n+2))-\cos x}\right]^2, & x \neq \pm\frac{\pi}{n+2}+2j\pi \\ (n+2)/2 & , \quad x = \pm\frac{\pi}{n+2}+2j\pi \end{cases},$$

an element of $\mathcal{T}_n^{e+}$. These results go back to Fejér [8], Korovkin and Nikolskiĭ [15]. For the general framework see *e.g.* [5], Chapter 1.

In this paper we will discuss properties of the best (*i.e.* smallest) asymptotic constants N_r such that

$$I_n(r,0) \geq N_r n^{-2r} + o(n^{-2r}). \tag{1.4}$$

Notice that this is equivalent with

$$\int_{-\pi}^{\pi} \chi_n(u)|u|^{2r}\, du = \mathcal{O}(n^{-2r}). \tag{1.5}$$

The results mentioned above state $N_1 = \pi^2$ and show that the corresponding kernel $k_{1,n}$ is unique. We will show that the order $-2r$ of approximation is maximal and determine those functions in $\mathcal{T}_n^{e+}$ for which equality holds.

Since these kernels may not be computed easily we will discuss generalized Jackson kernels. These kernels as well as (1.5) arise in the investigation of the degree of approximation of differentiable functions, *cf.* [14].

For generalized Jackson kernels we also have equality in (1.4) but with N_r replaced by some larger number which we will use to estimate N_r. On the other hand, the results obtained promote the use of generalized Jackson kernels, *cf.* [3], [7]. Positive kernels with property (1.4) have been used in complex approximation theory, too, *cf.* [1], [2], [10].

In the proofs we will switch between the integrals $I_n(r,0)$ and Toeplitz matrices $T_{r,n} \in \mathbb{R}^{(n+1)\times(n+1)}$ defined by

$$T_{r,n} := \begin{pmatrix} \binom{2r}{r} & -\binom{2r}{r-1} & \cdots & (-1)^r\binom{2r}{0} & & & 0 \\ -\binom{2r}{r-1} & \ddots & & & \ddots & & \\ \vdots & & \ddots & & & \ddots & \\ (-1)^r\binom{2r}{0} & & & \ddots & & & (-1)^r\binom{2r}{0} \\ & \ddots & & & \ddots & & \vdots \\ & & \ddots & & & \ddots & -\binom{2r}{r-1} \\ 0 & & & (-1)^r\binom{2r}{0} & \cdots & -\binom{2r}{r-1} & \binom{2r}{r} \end{pmatrix}.$$

In what follows we will need and will obtain informations about their eigenvalues.

The kernels discussed will be suitable for approximating functions using singular integrals, since $I_n(r,0) = \mathcal{O}(n^{-2r})$ implies $I_n(1,0) = \mathcal{O}(n^{-2})$ by the Hölder inequality. This, in turn implies norm convergence with the same rate of convergence of the singular integral

$$I_n(f)(x) := \frac{1}{2\pi}\int_{-\pi}^{\pi} \chi_n(u) f(x-u)\,du \to f(x) \tag{1.6}$$

for any function f in the generalized Lipschitz class $\mathbf{Lip}^*(\mathbf{X}_{2\pi};2) \subset \mathbf{X}_{2\pi}$, since $||I_n(f) - f||_{\mathbf{X}_{2\pi}}$ may be estimated by $I_n(1,0)$. Here, $\mathbf{X}_{2\pi}$ denotes the spaces of continuous functions or the space of functions which are Lebesgue integrable to the pth power, endowed with their usual norms. For the details see again *e.g.* [5].

Further, a well known theorem of Bohman-Korovkin [4], [11] for kernels $\chi_n \in \mathcal{T}_n^{e+}$ states that norm convergence in (1.6) takes place for any function $f \in \mathbf{X}_{2\pi}$ if it already takes place for the single test function $f = s_1$.

Notice, that for f non-constant we cannot expect anything better than $||I_n(f) - f||_{\mathbf{X}_{2\pi}} = \mathcal{O}(n^{-2})$ if the singular integral (1.6) is generated by a kernel of type (1.2), *cf.* [6], [12]. This, however, does not rule out the better rate of convergence (1.4) at the point $x = 0$.

§2 Statement of results

Let us denote the eigenvalues of $T_{r,n}$ by $\lambda_{r,n,\mu}$, $\mu = 0,1,\ldots,n$. The numbers

$$\lambda_{1,n,\mu} = \left(2\sin\frac{(\mu+1)\pi}{2(n+2)}\right)^2 \tag{2.1}$$

have already been computed by Fejér [8]. Hence $T_{1,n}$ is a positive definite matrix. This remains valid for $T_{r,n}$ since this matrix may be obtained from the rth power of $T_{1,n+2r-2}$ by deleting the leading and tailing $r-1$ rows and columns. An inclusion theorem of Cauchy for eigenvalues of symmetric matrices now gives

$$(\lambda_{1,n+2r-2,\mu})^r \le \lambda_{r,n,\mu} \le (\lambda_{1,n+2r-2,\mu+2r-2})^r \tag{2.2}$$

proving that $0 < \lambda_{r,n,\mu} < 4^r$. Here we arranged the eigenvalues by $\lambda_{r,n,0} \le \lambda_{r,n,1} \le \cdots \le \lambda_{r,n,n}$. The minimal eigenvalues $\lambda_{r,n,0}$ are neatly related with the asymptotic constants N_r and the corresponding normalized eigenvectors $x_{r,n,0}$ with the best kernels.

Theorem 1. *If $\chi_n \in \mathcal{T}_n^{e+}$ then*

$$I_n(r,0) \geq N_r n^{-2r} + o(n^{-2r}), \qquad N_r = \limsup_{n\to\infty}(n^{2r}\lambda_{r,n,0}) < \infty. \tag{2.3}$$

The asymptotic constant N_r is positive. The exponent $-2r$ may not be replaced by any smaller number. Equality holds in (2.3) iff the coefficients $\Theta_n(k)$ of χ_n in (1.2) are obtained by the formula

$$\Theta_n(k) = x_0 x_k + x_1 x_{k+1} + \cdots + x_{n-k} x_n \tag{2.4}$$

from the coordinates $x_0, x_1, \ldots, x_n$ of an eigenvector $x_{r,n,0}$ for the minimal eigenvalue $\lambda_{r,n,0}$ of $T_{r,n}$ normalized by ${x_0}^2 + {x_1}^2 + \cdots + {x_n}^2 = 1$.

Let $k_{r,n}$ denote the kernel in $\mathcal{T}_n^{e+}$ with coefficients $\Theta_n(k)$ defined by (2.4). For $r = 1$ we obtain $N_1 = \pi^2$ using (2.1). The corresponding normalized eigenvector $x_{1,n,0}$ is unique up to a sign and may be computed explicitely, *cf.* [8]. Therefore $k_{1,n}$ is unique and, after some computations, appears to be the Fejér-Korovkin kernel. Therefore (1.3) is contained in Theorem 1.

The asymptotic constants N_r may be estimated by evaluating $I_n(r,0)$ for the generalized Jackson kernel of order r

$$j_{r,n}(x) = \begin{cases} a_{r,n}\left(\frac{\sin(nx/2)}{\sin(x/2)}\right)^{2r+2}, & x \neq 0 \\ a_{r,n} n^{2r+2}, & x = 0 \end{cases}$$

where the constant $a_{r,n}$ is determined by $\int_{-\pi}^{\pi} j_{r,n}(u)\,du = 2\pi$. (See Lemma 1 below on how to compute $a_{r,n}$.) Notice, that $j_{r,n} \in \mathcal{T}^{e+}_{(r+1)(n-1)}$. For $r = 1$ we obtain the Jackson kernel, the case $r = 0$ corresponds to the Fejér kernel.

Theorem 2. *The asymptotic constants N_r are increasing with increasing r and satisfy the estimate*

$$\begin{aligned} N_r \;\leq\; N_{1,r} &= \frac{(r+1)^{2r}(2r+1)!\binom{2r}{r}}{\sum_{k=0}^{r}(-1)^k\binom{2r+2}{k}(r+1-k)^{2r+1}} \\ \leq\; N_{2,r} &= \sqrt{\frac{\pi}{3}}\frac{(r+1)^{2r+5/2}\binom{2r}{r}}{r^2 + \frac{77}{40}r + \frac{4131}{4480}}. \end{aligned} \tag{2.5}$$

Due to [9] we have $N_2 = \eta^4$ where $\eta \doteq 4.73004$ is the least positive solution of the equation $\cos x \cosh x = 1$.

r	${N_r}^{1/2r}$	${N_{1,r}}^{1/2r}$	${N_{2,r}}^{1/2r}$	I_r
1	$\pi \doteq 3.14159$	3.46410	3.46955	
2	$\eta \doteq 4.73004$	5.45216	5.45240	
3		7.44939	7.44942	$(6.252, 6.283)$
4		9.44868	9.44869	$(7.774, 7.819)$
5		11.44853	11.44854	$(9.258, 9.348)$

Table 1. Estimates for N_r.

Table 1 compares the upper bounds ${N_{1,r}}^{1/2r}$ and ${N_{2,r}}^{1/2r}$ in (2.5) with intervals I_r for ${N_r}^{1/2r}$ obtained by numerical calculations of the least eigenvalues $\lambda_{r,n,0}$ of $T_{r,n}$. The results are a striking argument in favour of generalized Jackson kernels which now appear to be close to the best possible kernels. As an immediate consequence of Theorem 1, Theorem 2, and (2.2) we obtain

Corollary 1. *The asymptotic behaviour of the least eigenvalue* $\lambda_{r,n,0}$ *of* $T_{r,n}$ *is described by* $\lambda_{r,n,0} = N_r n^{-2r} + o(n^{-2r})$ *with* $\pi^{2r} \leq N_r \leq N_{1,r}$.

§3 Proofs

Proof of Theorem 1: The proof basically uses a nice characterization of positive trigonometric polynomials due to Fejér [8]. It states that the coefficients $\Theta_n(k)$ of a kernel $\chi_n \in \mathcal{T}_n^{e+}$ in (1.2), and only of those, may be written in the form (2.4) with certain real numbers $x_0, x_1, \ldots, x_n$ satisfying ${x_0}^2 + {x_1}^2 + \cdots + {x_n}^2 = 1$. On the other hand we may compute

$$\begin{aligned} I_n(r,0) &= \frac{1}{2\pi}\int_{-\pi}^{\pi} \chi_n(u) s_r(u)\, du \\ &= \frac{1}{2\pi}\int_{-\pi}^{\pi} \left(1 + 2\sum_{k=1}^{n} \Theta_n(k)\cos ku\right) \times \\ &\qquad \left(\binom{2r}{r} + 2\sum_{k=0}^{r-1}(-1)^{r-k}\binom{2r}{k}\cos(r-k)u\right) du \\ &= \binom{2r}{r} + \sum_{k=1}^{\min\{r,n\}} (-1)^k \binom{2r}{r-k}\Theta_n(k) \end{aligned}$$

for those χ_n. Using (2.4) we obtain

$$I_n(r,0) = x^T T_{r,n} x \tag{3.1}$$

where $x \in \mathbb{R}^{n+1}$ is a normalized vector with coordinates $x_0, x_1, \ldots, x_n$. From (3.1) we now readily deduce $I_n(r,0) \geq {x_{r,n,0}}^T T_{r,n} x_{r,n,0} = \lambda_{r,n,0}$. Equality holds iff χ_n corresponds via (2.4) to a normalized eigenvector $x_{r,n,0}$ for the eigenvalue $\lambda_{r,n,0}$ of $T_{r,n}$.

It remains to show $\lambda_{r,n,0} = N_r n^{-2r} + o(n^{-2r})$ with an asymptotic constant $N_r = \limsup_{n\to\infty}(n^{2r}\lambda_{r,n,0}) \in (0,+\infty)$. But this already follows from (2.2) and (2.1). ■

Now we provide a technical lemma.

Lemma 1. *For $p = 0, 1, \ldots, r$*

$$\frac{1}{2\pi}\int_{-\pi}^{\pi}\left(\frac{\sin(nu/2)}{\sin(u/2)}\right)^{2r+2} s_p(u)\,du$$
$$= \frac{(-1)^p}{4}\sum_{k=0}^{R}(-1)^k\binom{2r+2}{k}\binom{n(r+1-k)+r-p}{2(r+1-p)-1}$$

where R denotes the largest integer less or equal $r+1-\frac{r+1-p}{n}$.

Proof: By the residue theorem we calculate

$$\frac{1}{2\pi}\int_{-\pi}^{\pi}\left(\frac{\sin(nu/2)}{\sin(u/2)}\right)^{2r+2}(2\sin(u/2))^{2p}\,du$$
$$= \frac{2^{2p}}{2\pi}\int_{-\pi}^{\pi}\frac{(\sin nx)^{2r+2}}{(\sin x)^{2(r+1-p)}}\,dx$$
$$= \frac{(-1)^p}{2p}\frac{-i}{4}\int_{|z|=1}\frac{(z^n - z^{-n})^{2r+2}}{(z-z^{-1})^{2(r+1-p)}}\frac{dz}{z}$$
$$= \frac{(-1)^p}{4}\operatorname{res}_{z=0}\left\{\left(\sum_{k=0}^{2r+2}\binom{2r+2}{k}z^{nk}(-z^{-n})^{2(r+1)-k}\right)\times \frac{z^{2(r+1-p)}}{(1-z^2)^{2(r+1-p)}}\frac{1}{z}\right\}$$
$$= \frac{(-1)^p}{4}\operatorname{res}_{z=0}\left\{\left(\sum_{k=0}^{2r+2}(-1)^k\binom{2r+2}{k}z^{2n(k-r-1)}\right)\times \left(\sum_{k=0}^{\infty}\binom{2(r+1-p)+k-1}{k}z^{2(r+1+k-p)}\right)\frac{1}{z}\right\}$$
$$= \frac{(-1)^p}{4}\sum_{k=0}^{R}(-1)^k\binom{2r+2}{k}\binom{n(r+1-k)+r-p}{2(r+1-p)-1}. \qquad \blacksquare$$

Proof of Theorem 2: By an application of the Hölder inequality and $\int_{-\pi}^{\pi}\chi_n(u)\,du = 2\pi$ we get $N_r \leq N_s$ if $r < s$.

Using Lemma 1 with $p = r$ and $p = 0$ we compute for $n > r+1$

$$\frac{1}{2\pi}\int_{-\pi}^{\pi} j_{r,n}(u)s_r(u)\,du = \frac{\frac{1}{2\pi}\int_{-\pi}^{\pi}\left(\frac{\sin(nu/2)}{\sin(u/2)}\right)^{2r+2} s_r(u)\,du}{\frac{1}{2\pi}\int_{-\pi}^{\pi}\left(\frac{\sin(nu/2)}{\sin(u/2)}\right)^{2r+2} du}$$
$$= \frac{(-1)^r\sum_{k=0}^{r}(-1)^k\binom{2r+2}{k}(r+1-k)n}{\sum_{k=0}^{r}(-1)^k\binom{2r+2}{k}\binom{n(r+1-k)+r}{2r+1}}.$$

Both numerator and denominator of this fraction are polynomials in n of degree 1, $2r+1$, resp.; the leading coefficients may be computed as $a = \binom{2r}{r}$ and $b = \frac{1}{(2r+1)!}\sum_{k=0}^{r}(-1)^k\binom{2r+2}{k}(r+1-k)^{2r+1}$. The later sum is known to be the Eulerian number $A(2r+1, r+1)$. By a result due to Lesieur and Nicolas [13] it may be estimated from below by

$$\begin{aligned} b &= \frac{1}{(2r+1)!}\sum_{k=0}^{r}(-1)^k\binom{2r+2}{k}(r+1-k)^{2r+1} \\ &= \frac{1}{(2r+1)!}A(2r+1,r+1) \qquad (3.2) \\ &\geq \sqrt{\frac{3}{\pi(r+1)}}\left(1-\frac{3}{40(r+1)}-\frac{13}{4480(r+1)^2}\right). \end{aligned}$$

Recall that $j_{r,n} \in \mathcal{T}^{e+}_{(r+1)(n-1)}$. Hence, $N_{1,r} = (r+1)^{2r}a/b$ gives an upper bound for N_r. By (3.2) we obtain the estimate for $N_{2,r}$. ■

Acknowledgement. The author would like to thank Reinhold Küstner who drew my attention to the connection between the Eulerian numbers and $\sum_{k=0}^{r}(-1)^k\binom{2r+2}{k}(r+1-k)^{2r+1}$.

References

[1] Andrievskiĭ, V. V. and H.-P. Blatt, A discrepancy theorem on quasi-conformal curves, *Constr. Approx.* **13** (1997), 363–379.

[2] Andrievskiĭ, V. V. and St. Ruscheweyh, Maximal polynomial subordination to univalent functions in the unit disk *Constr. Approx.* **10** (1994), 131–144.

[3] Belyĭ, V. I., A. P. Golub, I. A. Shevchuk, Dzyadyk's research on the theory of approximation of functions of a complex variable, *Ukr. Math. J.* **41** (1989), 384–395.

[4] Bohman, H., On approximation of continuous and of analytic functions, *Ark. Mat.* **2** (1952), 43–56.

[5] Butzer, P. L. and R. J. Nessel, *Fourier Analysis and Approximation, Vol.I*, Birkhäuser Verlag, Basel and Stuttgart, (1971).

[6] Curtis, P. C., Jr., The degree of approximation of a function by certain Fourier-integrals, *Michigan Math. J.* **12** (1965), 155–160.

[7] Dzjadyk, V. K., On the approximation of continuous functions in closed domains with corners, and on a problem of S.M. Nikolskiĭ, *Izv. Akad. Nauk SSSR Ser. Mat.* **26** (1962), 797–824; Engl. translation in *Amer. Math. Soc. Transl. (2)* **53** (1966).

[8] Fejér, L., Über trigonometrische Polynome, *J. Reine Angew. Math.* **146** (1916), 53–87.

[9] Greiner, R., *Zur Güte der Approximation schlichter Abbildungen durch maximal subordinierende Polynomfolgen*, Diploma thesis, Würzburg, 1993.

[10] Greiner, R., On a theorem of Andrievskii and Ruscheweyh, Proceedings of the Ashkelon Workshop on Complex Function Theory 1996, *Israel Math. Conf. Proc.* **11**, Bar-Ilan Univ., Ramat Gan, (1997), 83–90.

[11] Korovkin, P. P., On convergence of linear positive operators in the space of continuous functions, (Russ.), *Dokl. Akad. Nauk SSSR* **90** (1953), 961–964.

[12] Korovkin, P. P., On the order of the approximation of functions by linear positive operators (Russ.), *Dokl. Akad. Nauk SSSR* **114** (1957), 1158–1161.

[13] Lesieur, L. and J.-L. Nicolas, On the Eulerian numbers $M_n = \max_{1\le k\le n} A(n,k)$, *European J. Combin.* **13** no. 5 (1992), 379–399.

[14] Lorentz, G., *Approximation of functions*, Holt, Rinehart and Winston, New York, (1966).

[15] Nikolskiĭ, S. M., Sur l'allure asymptotique du reste dans l'approximation au moyen des sommes de Fejér des fonctions vérifiant la condition de Lipschitz (Russ.), *Izv. Akad. Nauk SSSR Ser. Mat.* **4** (1940), 501–508.

Richard Greiner
Universität Würzburg
Mathematisches Institut
Am Hubland
D-97074 Würzburg
GERMANY
greiner@mathematik.uni-wuerzburg.de

On the problem of Poreda

Rene Grothmann

Abstract. We investigate a problem posed by Poreda on the behaviour of the strong uniqueness constant with increased polynomial degree. It has been conjectured by Bartelt and McLaughlin that this constant tends to zero for all non-polynomial functions. In this paper, we give evidence for this and prove a special result, which we conjecture to be a worst case result.

§1 Introduction to the Problem

Let us first introduce the necessary notations. We denote by

$$e_n(f) = d(f, \mathcal{P}_n) := \min_{p_n \in \mathcal{P}_n} \|f - p_n\|$$

the error of best approximation with respect to the space $\mathcal{P}_n$ of polynomials of degree n, the norm being the sup-norm

$$\|g\| = \|g\|_{[a,b]} := \sup_{x \in [a,b]} |g(x)|,$$

and denote the best approximation by $P_n^* f$:

$$\|f - P_n^* f\| = e_n(f).$$

Then Freud proved that for some smallest constant $L_n(f)$

$$\frac{\|P_n^* f - P_n^* g\|}{\|f - g\|} \leq L_n(f), \qquad \text{for all } g \in C[a,b],$$

and, fore some largest constant $\gamma_n(f)$, w have the following due to Newman and Shapiro [9]

$$\|f - q_n\| \geq \|f - P_n^*(f)\| + \gamma_n(f) \|q_n - P_n^* f\|, \qquad \text{for all } q_n \in \mathcal{P}_n.$$

Furthermore

$$L_n(f) \leq \frac{2}{\gamma_n(f)}.$$

The proof of this and other details may be found in the book of Cheney [3].

Computational Methods and Function Theory (CMFT'97)
N. Papamichael, St. Ruscheweyh and E. B. Saff (Eds.), pp. 267–273

ISBN 981-02-3626-3

Poreda [8] has posed the following question: "How does the strong uniqueness constant behave as $n \to \infty$?" Bartet and McLaughlin [2] conjectured that

$$\liminf_{n\to\infty} \gamma_n(f) = 0$$

for all non-polynomial f.

We do not have enough space here to cover all the results which throw light on this conjecture. The simplest and yet most useful estimate is due to Blatt [1]. He proved that

$$\gamma_n(f) \le \frac{1}{n+1},$$

if $f - p_n^*(f)$ has exactly $n+2$ alternation points. This can be extended somethat to extremal error sets with slightly more points (see [7]). Moreover, using a real Carathéodory-Fejér method of Gutknecht and Trefethen, one can use this result to prove that the conjecture is true for functions analytic in a certain neighborhood of $[-1, 1]$ (see [5] and [6]).

We will need the characterization of Barlelt and McLaughlin

$$\begin{aligned}\gamma_n(f) \quad = \quad & \inf\{h \in \mathbb{R} : \text{there is a } p \in \mathcal{P}_n, \\ & \|p\| = 1 \text{ with } (s_n f(x))p(x) \le h \text{ for } x \in E_n(f)\},\end{aligned}$$

where $s_n f(x) = \text{sign } (f - p_n^*)(x)$ and $E_n(f)$ is the extremal set of the error function $f - p_n^*$. For a subsequence, $E_n(f)$ will consist of exactly $n+2$ subsets $E_0 < \ldots < E_{n+1}$, such that the error function oscillates in sign on these subsets. We choose points

$$x_0 = -1 \le E_0 < x_1 < E_2 < \ldots < E_{n+1} \le x_{n+2} = 1$$

and fix a function σ_n with

$$\sigma_n(x) = (-1)^i, \qquad x \in (x_i, x_{i+1}).$$

Then 0 is the best approximation of σ_n with respect to $\mathcal{P}_n$ and

$$\gamma_n(f) \le \gamma_n(\sigma_n)$$

by the characterization above. Thus we only investigate such extremal functions. We will call such a function a signum function of order $n+2$.

Conjecture 1.

$$\gamma_n(f) \le \frac{C}{\log n}$$

for some constant $C > 0$ independent of n and f.

Strong uniqueness does hold for approximation in the space $C_{2\pi}$ of 2π-periodic continuous functions with respect to $\mathcal{T}_n$, the set of all trigonometric polynomials of degree n. We denote the corresponding strong uniqueness constant by $\hat{\gamma}(f)$.

Let us explain why it suffices to prove the conjecture for the trigonometric case. For functions $g : [-1, 1] \to \mathbb{R}$ we set

$$\hat{g}(t) = g(\cos t), \qquad t \in [0, 2\pi).$$

We start with a signum function σ_n of order $n + 2$ on $[-1, 1]$. Then the best approximation to $\hat{\sigma}_n$ with respect to $\mathcal{T}_n$ on $[0, 2\pi)$ is 0, because $\hat{\sigma}_n$ has $2n + 2$ alternating extremal points. By the characterization above, we find a function $v \in \mathcal{T}_n$ with $\|v\| = 1$ and

$$v(t)\hat{\sigma}_n(t) \leq \hat{\gamma}(\hat{\sigma}_n).$$

Take $w(t) = (v(t) + v(-t))/2$, and $p \in \mathcal{P}_n$ such that $\hat{p} = w$. Then

$$\|p\| \geq \frac{1 - \hat{\gamma}(\hat{\sigma}_n)}{2},$$

and

$$p(x)\sigma_n(x) \leq \hat{\gamma}(\hat{\sigma}_n).$$

We obtain the following Lemma.

Lemma 1. *For any signum function σ_n on $[-1, 1]$ of order n, we get*

$$\gamma_n(\sigma_n) \leq \frac{2\hat{\gamma}(\hat{\sigma}_n)}{1 - \hat{\gamma}(\hat{\sigma}_n)}.$$

§2 Results

The following result takes care of an extremal case.

Theorem 1. *Let s_n be the trigonometric sign function, defined as*

$$s_n(t) = (-1)^\nu, \qquad t \in \left[\frac{\nu\pi}{n+1}, \frac{(\nu+1)\pi}{n+1}\right),$$

for $t \in [0, 2\pi)$. Then

$$\frac{C_1}{\log n} \leq \hat{\gamma}_n(s_n) \leq \frac{C_2}{\log n}$$

for constants $0 < C_1 < C_2$.

The rest of this section is devoted to the proof of this Theorem.

Proof: Let $v_n \in \mathcal{T}_n$ be the trigonometric polynomial of maximal norm, such that

$$v_n(t)s_n(t) \le 1$$

for all t. We may assume that this polyomial takes its norm at a point $t_0 \in [0, \pi/(n+1)]$, where it is negative. Then the polynomial will maximize $v(t_0)$ under the restriction $v_n(t)s_n(t) \le 1$. By the theory of semi-infinite optimization, there must be points $t_\nu \in [\nu\pi/(n+1), (\nu+1)\pi/(n+1)]$ for $\nu = 1, \dots, 2n+1$ such that $v_n(t_\nu) = (-1)^\nu$. A simple zero counting argument shows that this maximization problem has a unique solution, which is independent of the point t_0. A symmetry argument shows that it is symmetric to $\pi/(2n+2)$, where it takes its norm.

We will now investigate

$$w_n(t) = -v_n\left(t + \frac{\pi}{2n+2}\right), \qquad t \in [-\pi, \pi].$$

This is an even polynomial and takes its norm in 0. It maximizes $w_n(0)$ among all polynomials with

$$(-1)^\nu w_n(t) \ge -1, \qquad t \in I_\nu := \left[\frac{\nu\pi}{2n+2}, \frac{(\nu+2)\pi}{2n+2}\right], \quad \nu \text{ odd}.$$

We will study w_n on $[0, \pi]$. On this interval, w_n belongs to $\mathcal{C}_n$, the space of all cosine polynomials of maximal degree n. We compare w_n to the polynomial $v_n \in \mathcal{C}_n$ defined by the $n+1$ interpolation properties

$$v_n(\tau_\nu) = (-1)^{\frac{\nu-1}{2}}, \qquad \tau_\nu = \frac{\nu\pi}{2n+1}, \quad \nu = 1, 3, 5, \dots, 2n+1. \tag{2.1}$$

Note that $\tau_\nu \in I_\nu$. By a simple zero counting estimate, we have

$$v_n(0) > w_n(0).$$

This will yield an upper estimate for w_n and thus a lower estimate for $\hat{\gamma}(s_n)$.

We mention that we could now use well known results about the best interpolation norm in the trigonometric case. But, in any case, we need more detailed information anyway.

Consider the polynomial $\phi_n \in \mathcal{T}_n$ defined by

$$\phi_n(t) = \operatorname{Re} \frac{z^{2n+1} - 1}{(2n+1)z^n(z-1)} = \operatorname{Re} \frac{z^{-n} + \dots + z^n}{2n+1}$$

with $z = e^{it}$. Setting

$$\phi_{\nu,n}(z) = \phi_n(z - \tau_\nu) + \phi_n(z + \tau_\nu),$$

we see that we have just constructed the Lagrange polynomials for the interpolation problem (2.1). To estimate $|\phi_{\nu,n}(0)|$, we need to compute ϕ_n at odd multiples of $d_n := \pi/(2n+1)$. Setting

$$z_{\nu,n} = e^{i\nu d_n},$$

we get for ν odd

$$|\phi_n(\nu d_n)| = \frac{2}{2n+1}\left|\text{Re}\left(\frac{z_{\nu,n}^n}{z_{\nu,n}-1}\right)\right| \leq \frac{1}{(2n+1)|z_{\nu,n}-1|}.$$

Since

$$|z_{\nu,n}-1| \geq \frac{|\nu d_n|}{2},$$

we obtain

$$|\phi_n(\nu d_n)| \leq \frac{2}{\nu\pi}.$$

Thus

$$|\phi_{\nu,n}(0)| = 2|\phi_n(\tau_\nu)| \leq \frac{4}{\nu\pi}.$$

Summing up the Lagrange polynomials, we get

$$\begin{aligned} w_n(0) &\leq v_n(0) \\ &= |\phi_{1,n}(0)| + |\phi_{3,n}(0)| + \ldots + |\phi_{2n+1,n}(0)| \\ &\leq D_1 \log(n) \end{aligned}$$

with a constant $D_1 > 0$, which does not depend on n.

We can also derive a lower estimate for $v_n(0)$. Since $z_{\nu,n}^n \to i$ for $n \to \infty$, we get for n large enough and $|\nu d_n| \leq \pi/2$

$$\begin{aligned} |\phi_n(\nu d_n)| &\geq \frac{\text{Im}\,(z_{\nu,n}-1)}{(2n+1)|z_{\nu,n}-1|^2} \\ &\geq \frac{1}{2(2n+1)|\sin(\nu d_n)|} \\ &\geq \frac{1}{2(2n+1)|\nu d_n|} \\ &= \frac{1}{2\pi\nu}. \end{aligned}$$

Thus we see that

$$v_n(0) \geq D_2 \log(n),$$

and that our estimate for $v_n(0)$ was sharp.

To get a lower bound for $w_n(0)$, we really need to construct a polynomial $h_n \in C_n$, which satisfies the conditions

$$(-1)^\nu h_n(t) \leq 1, \tag{2.2}$$

for

$$t \in \left[\frac{(2\nu+1)\pi}{2n+2}, \frac{(2\nu+3)\pi}{2n+2}\right] \cap [0,\pi],$$

$\nu = -1, \ldots, n$. $h_n(0)$ will then be a lower estimate for $w_n(0)$.

We could use the polynomial v_n, properly scaled, for this purpose, but the proofs will get very technical. Thus, we use instead the polyomials

$$u_n(t) = \sum_{\nu=1}^{n_\alpha} \phi_{\nu,n}(t)$$

with $n_\alpha = [n^\alpha]$ (using the Gaussian bracket) for some $0 < \alpha < 1$ and n large enough. Then with the same estimates as above, we still get

$$u_n(0) \geq D_3 \log(n)$$

for some constant $D_3 > 0$. Morevover, if we take some $0 < \alpha < \beta < 1$, we have by the representation of ϕ_n

$$|u_n(t)| \leq D_4 n^{\alpha-\beta}, \qquad \frac{\pi}{n^\beta} \leq |t| \leq \pi.$$

Clearly

$$u_n(t) \geq 1, \qquad t \in \left[0, \frac{\pi}{2n+2}\right].$$

By elementary calculations, we see that absolute values of the derivatives of ϕ_n, taken at the zeros of ϕ_n

$$s_\nu = \frac{2\nu\pi}{2n+1}, \qquad \nu \neq 0,$$

of ϕ_n, are monotonically decreasing. Using this, we obtain

$$u_n'(\tau_1) < 0, \quad u_n'(\tau_2) > 0, \quad \ldots$$

With a zero counting argument applied to u_n', we see that we only need to estimate u_ν in the intervals $[\nu\pi/(2n+2), \nu\pi/(2n+1)]$ for odd ν with $\nu \leq n^\beta$. On the right endpoint of this interval, we have values ± 1 or 0 by the interpolation property. Repeating the same estimate as above, we can prove

$$\|u_n\|_{[0,\pi]} \leq D_5 \log n$$

for some constant $D_5 > 0$. By the Markov inequality, we finally see that $u_n/2$, for n big enough, satisfies the inequalities (2.2).

This completes the proof of the Theorem. ■

References

[1] Blatt, H.-P., Exchange algorithms, error estimations, and strong unicity in convex programming and Chebyshev approximation, in *Approximation Theory and and Spline Functions*, NATO ASI series, Series C, Math. and Phys. Sciences, Vol. 136, Dordrecht, 1984.

[2] Bartelt, M. W. and H. W. McLaughlin, Characterizations of strong unicity in approximation theory, *J. Approx. Theory* **28** (1973), 255–260.

[3] Cheney, E. W., *Introduction to Approximation Theory*, McGraw-Hill, New York, 1966.

[4] Freud, G., Eine Ungleichung für Tschebyscheff'sche Approximationsprobleme, *Acta Scientarium Math.* **19** (1958), 162–164.

[5] Grothmann, R., Zur Größenordnung der starken Eindeutigkeitskonstanten von holomorphen Funktionen, Dissertation, Kath. Univ. Eichstätt, 1986.

[6] Grothmann, R., On the real CF-method for polynomial approximation and strong unicity constants, *J. Approx. Theory* **55** (1988), 86–103.

[7] Grothmann, R., A note on strong uniqueness constants, *J. Approx. Theory* **58** (1989), 358–360.

[8] Poreda, S. J., Counterexamples on best approximation, *Proc. Americ. Soc.* **56** (1976), 167–171.

[9] Newman, D. J. and H. S. Shapiro, Some theorems on Chebyshev approximation, *Duke Math J.* **30** (1963), 673-681.

Rene Grothmann
Kath. Univ. Eichstätt
D-85072 Eichstätt
GERMANY
grothm@ku-eichstaett.de

Area theorems for conformal mapping and Rankine ovoids

Masaaki Ito
and Masakazu Shiba

Abstract. We study the Rankine ovoids in hydrodynamics in connection with the theory of conformal mapping of the unit disk. We show that for any given complex number κ with $|\kappa| < 1$ there is an essentially unique Rankine ovoid outside of which is precisely the image of a conformal mapping of the open unit disk with the expansion $1/z + \kappa z + o(|z|)$ near $z = 0$. The area of the Rankine ovoid is computed and compared with the maximum value of the possible area for the parameter κ which is attained by an ellipse.

§1 Introduction

Every conformal mapping of a given plane domain defines a conformal embedding of the domain in the Riemann sphere, essentially the single compact Riemann surface of the genus zero. In other words, through conformal mappings one can observe various realizations of the ideal boundary of the domain. From this point of view the closed set on the sphere which is not covered by the image domain is important. Some typical examples will be given by the classical theorems due to Koebe, de Possel, Grötzsch, Grunsky and Schiffer (see [1], [2], [3], [7]). One of the authors recently investigated the outer area of the closed sets in detail and obtained several interesting properties ([9]). For instance, a close connection of the Poincaré metric and the area has been revealed in [9], which led us to the study of the interrelation between various metrics ([4]). The size of the realized ideal boundary has been also studied ([8]).

In the present note we discuss the relation between conformal mapping and the Rankine ovoids which appear in fluid dynamics. A Rankine ovoid is a figure in the plane which is brought forth by the superposition of a flow with a pair of sink and source (of the same strength) and a uniform flow. We then prove that in the class of (normalized) conformal mappings of the unit disk there is one which maps the unit disk conformally outside of a Rankine ovoid. While by our previous work ([9]) we know that no Rankine ovoids attain the maximum area in the class, we show below that their area is very close to the maximum which is attained by an ellipse. The area of the Rankine ovoid is obtained numerically with the help of computers.

Computational Methods and Function Theory (CMFT'97)
N. Papamichael, St. Ruscheweyh and E. B. Saff (Eds.), pp. 275–283

ISBN 981-02-3626-3

§2 Preliminaries

Let G be a domain in the complex plane $\hat{\mathbb{C}}$. It may be unbounded as well as infinitely connected. Let ζ be a fixed point in G, which we may assume is not equal to ∞. We consider the set $\mathcal{F}(G,\zeta)$ of normalized univalent meromorphic functions f on G which have a single simple pole at ζ with residue 1 and a vanishing constant term. Each function $f \in \mathcal{F}(G,\zeta)$ has then a Laurent expansion

$$f(z) = \frac{1}{z-\zeta} + \kappa_f(z-\zeta) + \cdots \qquad \text{about } \zeta,$$

where $\kappa_f \in \mathbb{C}$.

Let $\mathrm{K}(G,\zeta)$ denote the set of coefficients κ_f:

$$\mathrm{K}(G,\zeta) = \{\kappa \in \mathbb{C} \mid \kappa = \kappa_f \text{ for some } f \in \mathcal{F}(G,\zeta)\}.$$

We recall that the class $\mathcal{F}(G,\zeta)$ corresponds to the class $\Sigma'(G)$ in [3] (see Def. 1.3). (If G is the domain $\hat{\mathbb{C}} \setminus \{|z| \le 1\}$ and the point ζ is chosen to be ∞, our class would coincide with Σ_0 in [6]; *cf.* p. 13).

Let $\delta[f]$ be the complementary area of $f(G)$ — *i.e.* the *outer* area of $\hat{\mathbb{C}} \setminus f(G)$ —, which is always finite. For any fixed $\kappa \in \mathrm{K}(G,\zeta)$, we consider the class $\mathcal{F}_\kappa(G,\zeta)$ of functions $f \in \mathcal{F}(G,\zeta)$ such that $\kappa_f = \kappa$.

We can gather the classical theorems due to Koebe, de Possel, Grötzsch, Grunsky, Schiffer and our recent results in the following:

Theorem 1.

(I) $\mathrm{K}(G,\zeta)$ *is a closed disk:*

$$\mathrm{K}(G,\zeta) = \{\kappa \in \mathbb{C} \mid \, |\kappa - \kappa^*| \le \rho\} \qquad \textit{with } \kappa^* \in \mathbb{C} \textit{ and } \rho \ge 0.$$

(II) Let $\kappa(t) = \kappa^* + \rho e^{\pi i t}$, $-1 < t \le 1$, *be a parametrization of the boundary circle* $\partial\mathrm{K}(G,\zeta)$. *Then, for each* $t \in (-1,1]$ *there exists one and only one function* $f_{\kappa(t)} \in \mathcal{F}(G,\zeta)$ *with* $\kappa_{f_{\kappa(t)}} = \kappa(t)$. *Furthermore,* $\hat{\mathbb{C}} \setminus f_{\kappa(t)}(G)$ *is a compact plane set of null area which consists of parallel segments making an angle* $\pi t/2$ *with the real axis.*

(III) The radius ρ *of* $\mathrm{K}(G,\zeta)$ *vanishes if and only if* G *belongs to* O_{AD}.

(IV) For any $\kappa \in \mathrm{K}(G,\zeta)$ *there exists a unique element* f_κ *in* $\mathcal{F}_\kappa(G,\zeta)$ *which maximizes* $\delta[f]$ *in the class* $\mathcal{F}_\kappa(G,\zeta)$.

(V) *The maximum δ_κ of $\delta[f]$ in the class $\mathcal{F}_\kappa(G,\zeta)$ depends on the distance $r = r(\kappa) := |\kappa - \kappa^*|$ only: $\delta_\kappa = \pi(\rho^2 - r^2)/\rho$.*

We also recall some facts from hydrodynamics. Let a, m be positive real numbers and consider the plane flow of an ideal fluid on the Z-plane caused by a sink and a source, of strength m, at $Z = a$ and $Z = -a$. This flow is given by the analytic function

$$\Phi_1(Z; a, m) := m \log \frac{Z+a}{Z-a}.$$

We superpose this flow and the uniform flow $\Phi_0(Z) = Z$ on the plane to obtain a new plane flow defined by

$$\Phi(Z; a, m) := \Phi_0(Z) + \Phi_1(Z) = Z + m \log \frac{Z+a}{Z-a}.$$

The stagnation points of this flow are $Z = \pm\sqrt{a^2 + 2am}$ and the stream lines through these stagnation points are given by

$$Y = 0 \qquad \text{and} \qquad Y = m \arg \frac{Z-a}{Z+a},$$

where $Z = X + iY$ and where the branch of $\arg\zeta$ is chosen such that $\arg\zeta \in (-\pi, \pi)$ on the plane with the negative real axis removed.

By the mapping property of

$$Z \mapsto \zeta = \frac{Z-a}{Z+a},$$

we see that the function $m \arg \zeta$ is a decreasing (resp. increasing) function on $\{Z \,|\, X = 0,\ Y > 0\}$ (resp. $\{Z \,|\, X = 0,\ Y < 0\}$). Hence, for each $m > 0$ there exists a unique Y which satisfies

$$Y = m \arg \frac{Z-a}{Z+a}, \qquad 0 < Y < \frac{\pi}{2m} \qquad (\text{ resp. } 0 > Y > -\frac{\pi}{2m}).$$

There are exactly two stream arcs γ_1, γ_2, one of which emanates from one stagnation point $-\sqrt{a^2 + 2am}$, passes through the upper or lower half plane, and ends at the other stagnation point $\sqrt{a^2 + 2am}$.

The domain which contains the origin and is bounded by γ_1, γ_2 is called the *Rankine ovoid* or *Rankine solid* (see [5] and [10]) with sink and source of strength m at the points a and $-a$ respectively. We remind the reader of the fact that the Joukowski transformation

$$w = J(z) := z + \frac{1}{z}$$

corresponds to a superposition of a uniform flow and a dipole flow. The unit disk $\mathbb{D}$ can be thereby considered as an extreme case of Rankine ovoids.

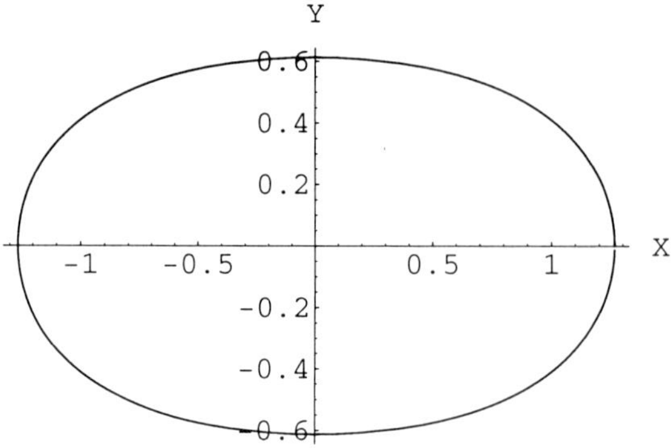

Figure 1. Rankine ovoid for $a = 1, m = 0.2$

§3 Conformal mapping of the unit disk and the Rankine ovoids

Hereafter we restrict ourselves to the case of a simply connected domain G. More specifically, we take $\mathbb{D}$ as the domain G and the center of $\mathbb{D}$ as the reference point ζ. Indeed, this choice does not impose any essential restriction. Now, we easily see that the set $\mathrm{K} := \mathrm{K}(\mathbb{D}, 0)$ of coefficients is the unit disk in the complex κ plane and the extremal function f_κ in the class $\mathcal{F}_\kappa := \mathcal{F}_\kappa(\mathbb{D}, 0)$ is given by

$$w = f_\kappa(z) = \frac{1}{z} + \kappa z.$$

The image domain $f_\kappa(\mathbb{D})$ is thus the exterior of the ellipse

$$(1 - |\kappa|)^2 u^2 + (1 + |\kappa|)^2 v^2 = (1 - |\kappa|^2)^2, \qquad w = u + iv,$$

so that the area of the complement is equal to

$$S(\kappa) := \pi(1 + |\kappa|)(1 - |\kappa|) = \pi(1 - |\kappa|^2).$$

Now we can show the following:

Theorem 2. *For each $\kappa \in \mathrm{K}$ there exist positive real numbers a, m and a function $\check{f}_\kappa \in \mathcal{F}_\kappa$ such that $\hat{\mathbb{C}} \setminus \check{f}_\kappa(\mathbb{D})$ is the Rankine ovoid with sink and source of strength m at the points a and $-a$ respectively.*

For the proof of this theorem, we may assume that κ is real and positive, since the general case can be treated easily by a suitable rotation of the disk.

In the sequel we set

$$k := |\kappa|.$$

We also need:

Lemma 1. *For each k with $0 < k < 1$ the equation*

$$\sqrt{1+2\mu} + \mu \log \frac{\sqrt{1+2\mu}+1}{\sqrt{1+2\mu}-1} = 2\sqrt{\frac{2}{1-k}\,\mu}$$

has a unique solution $\mu = \mu(k)$ with $0 < \mu < (1-k)(2k)$. The solution $\mu = \mu(k)$ is strictly decreasing for $0 < k < 1$. If k tends to 1 then μ tends to 0, and if k tends to 0 then μ tends to ∞.

Proof: We fix a number k $(0 < k < 1)$ once and for all. For simplicity we set

$$C := 2\sqrt{\frac{2}{1-k}}$$

and consider the function

$$\chi(\mu) := \frac{\sqrt{1+2\mu}}{\mu} + \log \frac{\sqrt{1+2\mu}+1}{\sqrt{1+2\mu}-1} - \frac{C}{\sqrt{\mu}}, \qquad \mu > 0.$$

We first note that $\lim_{\mu\to+0} \chi(\mu) = \infty$ and $\lim_{\mu\to+\infty} \chi(\mu) = 0$. By a simple computation we have

$$\frac{d\chi}{d\mu} = \frac{C\sqrt{\mu} - 2\sqrt{1+2\mu}}{2\mu^2}.$$

Hence, in case of nonvanishing k, we have

$$\frac{d\chi}{d\mu} \begin{cases} > 0 & \text{if } \mu > \dfrac{1-k}{2k}, \\ = 0 & \text{if } \mu = \dfrac{1-k}{2k}, \\ < 0 & \text{if } \mu < \dfrac{1-k}{2k}. \end{cases}$$

Thus, for nonvanishing k there is one and only one solution μ of our equation.

The solution $\mu = \mu(k)$ satisfies $0 < \mu < \dfrac{1-k}{2k}$ and

$$\frac{d\mu}{dk} = \frac{\sqrt{2\mu} + \sqrt{(1-k)(2\mu+1)}}{\mu - \dfrac{1-k}{2k}} < 0.$$

In particular, $\mu = \mu(k)$ is strictly decreasing for $0 < k < 1$. The rest of the assertion is easy to prove. ∎

Remark 1: It is clear from the above argument that the value

$$\Omega(k) := \log\frac{1+\sqrt{k}}{1-\sqrt{k}} - \frac{2\sqrt{k}}{1-k}$$

of the function χ at $(1-k)(2k)$ is negative for all k $(0 < k < 1)$. We can show this directly. In fact, the function $\Omega(k)$ satisfies

$$\lim_{k\to+0}\Omega(k) = 0 \qquad \text{and} \qquad \frac{d\Omega}{dk} = -\frac{2\sqrt{k}}{(1-k)^2} < 0.$$

Proof of Theorem 2: To complete the proof of the theorem, we need only set

$$a = a(k) := \sqrt{\frac{1-k}{2\mu}} \qquad \text{and} \qquad m = m(k) := \sqrt{\frac{1-k}{2}\mu}$$

and consider the function

$$w = \Phi(z) = \Phi(Z; a, m) = Z + m\log\frac{Z+a}{Z-a}.$$

The stagnation points of the flow defined by this function are

$$\pm\sqrt{a^2+2am} = \pm\sqrt{\frac{2\mu+1}{2\mu}(1-k)},$$

so that by Lemma 1 we have $\Phi(\pm\sqrt{a^2+2am}; a, m) = \pm 2$. It is now easy to show that $Z = \check{f}_k(z) := \Phi^{-1}\circ J(z)$ is the required function. ■

Remark 2:

(1) Observe that a tends to 0 and m tends to ∞ if k tends to 0. We see that the stagnation points tend to 1 and -1 respectively. Hence, the Rankine ovoid becomes the unit disk if $k \to 0$.

(2) We can also see that if k tends to 1, then a tends to 2 and m tends to 0. The stagnation points tend to 2 and -2 respectively. In other words, the final stage would be the Joukowski mapping as could be easily guessed. The Rankine ovoid reduces to the segment of length 4 on the real axis.

§4 The area of the Rankine ovoid

We now study the area of Rankine ovoids. In fact, the area involves only one parameter k.

Theorem 3. *The area $s(k)$ of the Rankine ovoid $\hat{\mathbb{C}} \setminus \check{f}_k(\mathbb{D})$ is*

$$s(k) = \frac{2(1-k)}{\mu(k)} \int_0^\tau \sqrt{F[t;k]}\, dt,$$

where $F[t;k]$ is the function defined by

$$F[t;k] := 1 - t^2 + \frac{2t}{\tan \dfrac{t}{\mu(k)}},$$

μ is the function obtained in Lemma 1, and τ is the solution of

$$\tan \frac{t}{\mu(k)} = \frac{2t}{t^2 - 1},$$

with $0 < \tau < \dfrac{\mu(k)\pi}{2}$.

Proof: Let $a = a(k)$, $m = m(k)$ be as before and let $\tau(a)$ be the solution of

$$\tan \frac{t}{m} = \frac{2at}{t^2 - a^2}$$

in the interval $0 < \tau < m\pi/2$. The Rankine ovoid obtained in Theorem 2 (with sink and source of strength m at $a, -a$ respectively) is described by the equation

$$\tan \frac{Y}{m} = \frac{2aY}{X^2 + Y^2 - a^2},$$

so that its area is given by

$$4 \int_0^{\tau(a)} \sqrt{F(t;a,m)}\, dt,$$

where

$$F(t;a,m) := a^2 - t^2 + \frac{2at}{\tan \dfrac{t}{m}}.$$

By the change of variables $t \mapsto a^{-1}t$ we easily have the result. ∎

The area $s(k)$ can be numerically computed as a function of k (see Figure 2). As was noted earlier, the maximum area for the functions in the class $\mathcal{F}_k$ is attained by an ellipse and is equal to $S(k) = \pi(1 - k^2)$. The deviation $d(k) := S(k) - s(k)$, $0 < k < 1$, which is shown in Figure 3, is so small that it is not worth to present the graph of $S(k)$. The function $d(k)$ is continuous and vanishes at the endpoints. Hence it takes a maximum value somewhere on the interval $(0, 1)$. We do not know, however, where the function $d(k)$ is maximal.

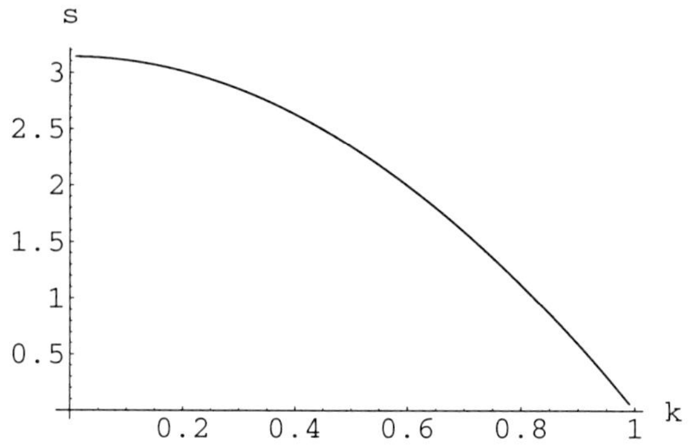

Figure 2. The area function $s(k), k = |\kappa|$

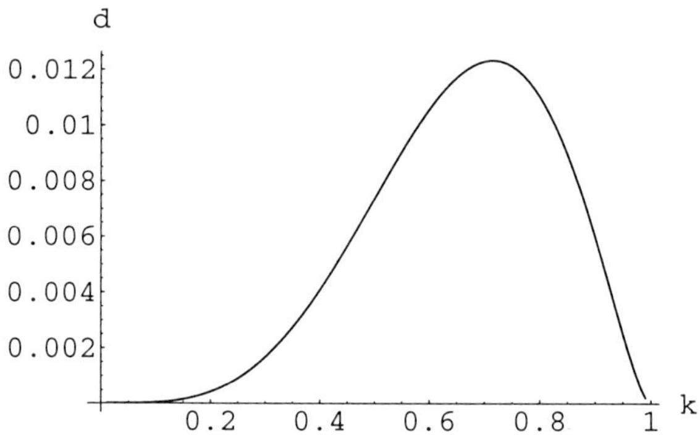

Figure 3. The deviation $d(k) = S(k) - s(k)$

Acknowledgement. The authors thank K. Kamihisa and S. Miyazaki for their interest. They also thank Professors Nick Papamichael and Stephan Ruscheweyh for their kind advice and suggestions.

The authors also thank the Ministry of Education, Government of Japan, for the financial support (Scientific Program, Contract Nr. 09640193).

References

[1] Grötzsch, H., Zum Parallelschlitztheorem der konformen Abbildung schlichter unendlich zusammenhängender Bereiche, *Ber. Verh. Sächs. Akad. Wiss. Leipzig, math.-phys. Kl.* **83** (1931), 185–200.

[2] Grunsky, H., *Lectures on Theory of Functions in Multiply Connected Domains*, Vandenhoeck & Ruprecht, Göttingen, 1978.

[3] Jenkins, J. A., *Univalent Functions and Conformal Mapping*, Springer-Verlag, Berlin-Göttingen-Heidelberg, 1958.

[4] Masumoto, M. and M. Shiba, Intrinsic disks on a Riemann surface, *Bull. London Math. Soc.* **27** (1995), 371–379.

[5] Milne-Thomson, L. M., *Theoretical Hydrodynamics, Fifth edition*, Macmillan Press, London and Basingstoke, 1968 (Reprint: 1979).

[6] Pommerenke, Chr., *Univalent Functions*, Vandenhoeck & Ruprecht, Göttingen, 1975.

[7] Schiffer, M., The span of multiply connected domains, *Duke Math. J.* **10** (1943), 209–216.

[8] Schmieder, G. and M. Shiba, Realizations of the ideal boundary of a planar Riemann surface in the sphere, *Nonlinear Analysis, Theory, Methods and Applications* **30** (1997), 5267–5274.

[9] Shiba, M., The euclidean, hyperbolic, and spherical spans of an open Riemann surface of low genus and the related area theorems, *Kodai Math. J.***16** (1993), 118–137.

[10] Wieghardt, K., *Theoretische Strömungslehre*, B. G. Teubner, Stuttgart, 1974.

Masaaki Ito
Department of Applied Mathematics
Hiroshima University
Higashi-Hiroshima 739-8527
JAPAN
ito@amath.hiroshima-u.ac.jp

Masakazu Shiba
Department of Applied Mathematics
Hiroshima University
Higashi-Hiroshima 739-8527
JAPAN
shiba@amath.hiroshima-u.ac.jp

Solutions of generalized Riccati differential equations and their approximation

Gerhard Jank

and Gábor Kun

Abstract. We study the solution of generalized matrix Riccati differential equations as they appear in differential games theory with various information structures. We propose a numerical approximation for the generalized Riccati differential equation arising from an optimally controlled differential game with memoryless feedback. This method makes use of the determination of the exact solution of associated nonsymmetric Riccati differential equations. Furthermore, we present an example of a generalized algebraic Riccati equation having several stabilizing positive semidefinite solutions.

§1 Introduction

In this paper we study first the matrix Riccati differential equation (RDE) (or $(RDE)_{\mathbb{R}}$)

$$\dot{W} = B_{21} + B_{22}W - WB_{11} - WB_{12}W, \qquad t \in \mathbb{C} \quad (\text{or } t \in \mathbb{R}), \tag{1.1}$$

with the terminal value $W(t_f) = W_f$ and the corresponding algebraic Riccati equation (ARE)

$$0 = B_{21} + B_{22}W - WB_{11} - WB_{12}W, \tag{1.2}$$

where $W(t)$ (or W) is a complex $m \times n$ matrix and where B_{11}, B_{12}, B_{21}, B_{22} are constant complex matrices of dimensions $n \times n$, $n \times m$, $m \times n$ and $m \times m$ respectively. In the next section we present explicit formulas for solutions of (1.1) and discuss some of their function theoretic and asymptotic properties. These solutions are then used to create a numerical algorithm to solve a more involved generalized Riccati differential equation of the following type

$$\begin{aligned} \dot{K} &= \tilde{A}^T K - K\tilde{A} - \tilde{Q} + K\tilde{S}K + JK\tilde{S}JK \\ &\quad + KJ\tilde{S}KJ - JKJS_0JKJ, \end{aligned} \tag{1.3}$$

with terminal values $K(t_f) = K_f, K_f^T = K_f$. This equation describes a coupled system of Riccati differential equations appearing in Nash Games

Computational Methods and Function Theory (CMFT'97)
N. Papamichael, St. Ruscheweyh and E. B. Saff (Eds.), pp. 285–302

ISBN 981-02-3626-3

where the information structure of the players is supposed to be of memoryless feedback type (see [21]). If for simplicity we assume a two player game then the matrices in (1.3) are as follows:

$$\tilde{A} = \begin{pmatrix} A & 0 \\ 0 & A \end{pmatrix}, \qquad \tilde{Q} = \begin{pmatrix} Q_1 & 0 \\ 0 & Q_2 \end{pmatrix}, \qquad \tilde{S} = \begin{pmatrix} S_1 & 0 \\ 0 & S_2 \end{pmatrix},$$

$$S_0 = \begin{pmatrix} S_{12} & 0 \\ 0 & S_{21} \end{pmatrix}, \qquad K = \begin{pmatrix} K_1 & 0 \\ 0 & K_2 \end{pmatrix}, \qquad K_f = \begin{pmatrix} K_{1f} & 0 \\ 0 & K_{2f} \end{pmatrix},$$

$$J = \begin{pmatrix} 0 & I_n \\ I_n & 0 \end{pmatrix},$$

where $A \in \mathbb{R}^{2n\times 2n}$ and $\tilde{Q}, \tilde{S}, S_0 \in \mathbb{R}^{2n\times 2n}$ are symmetric and one is only interested in symmetric solutions $K(t) \in \mathbb{R}^{2n\times 2n}$. Since the differential equation (1.3) is symmetric, this means that one has to stick to terminal value problems with symmetric data.

In several other applications there exist also generalized Riccati differential equations similar to (1.3) which could be treated in an analogous way, but here we will confine ourselves to numerical calculations concerning solutions of (1.3).

The matrix Riccati equation (especially $(RDE)_{\mathbb{R}}$ and (ARE) for $m = n$) plays an important role in many branches of applied mathematics, notably in variational theory and the allied areas of optimal control and filtering, invariant imbedding, spectral factorization and dynamic programming and scattering theory (see [5], [12], [17], [13]); nonsquare matrix Riccati equations appear for example in Nash and Stackelberg control problems, where the properties of Riccati equations determine the existence of the optimal open-loop strategies (see [1], [16], [23] for further references).

On the other hand (RDE) is also of mathematical interest, since it is the description in local coordinates of the differential equation on a Graßmann manifold, whose flow is given by the action of a one-parameter subgroup of $Gl(n+m, \mathbb{C})$ (see [11], [18], [19]).

In §2 we give a fundamental representation formula (2.4) for the solutions of (RDE); this formula contains the complete information on the phase portrait of (RDE) and in particular on the structure of the set Γ of all solutions of (ARE). In §3 we refer to some recently obtained sufficient conditions such that no finite escape time phenomena can appear in a given interval. In §4 we propose a numerical method for the integration of the generalized Riccati differential equation (1.3). This method is based on a piecewise approximation by solutions of an appropriate (RDE), that appear in differential games with discrete time memoryless feedback information structure. The last section deals with numerical results on a given autonomous generalized Riccati equation.

§2 Representation of solutions

To explain the function theoretical and asymptotic properties of solutions and for the application in the numerical algorithm we recall here a representation formula for the solutions of (1.1) as already derived in [7].

We first describe the connection between (RDE) and the corresponding linear differential equation (2.1) (see [17] p. 11), and we then derive a detailed formula for the general solution of (RDE).

Let

$$B = \begin{pmatrix} B_{11} & B_{12} \\ B_{21} & B_{22} \end{pmatrix}, \qquad Y = \begin{pmatrix} Q \\ P \end{pmatrix}$$

with complex $n \times n$ and $m \times n$ matrix functions Q and P, respectively. Then (RDE) and the following linear system of differential equations

$$\dot{Y} = BY, \qquad t \in \mathbb{C}, \tag{2.1}$$

are equivalent in the following sense:

(i) Let $I_n \in \mathbb{C}^{n \times n}$ be the unit matrix, let W be a solution of (RDE). If Q is the unique solution of the initial value problem

$$\dot{Q} = (B_{11} + B_{12}W(t))Q, \qquad Q(t_0) = I_n, \quad t_0 \in G \subset \mathbb{C}$$

and $P(t) = W(t)\,Q(t)$, then $Y := \binom{Q}{P}$ is a solution of (2.1).

(ii) If $Y = \binom{Q}{P}$ is a solution of (2.1) such that $Q(t)$ is regular for $t \in G \subset \mathbb{C}$, then

$$W : G \to \mathbb{C}^{m \times n}, \qquad t \mapsto P(t)Q^{-1}(t) = W(t), \tag{2.2}$$

is a solution of (RDE).

This well known relationship between linear systems and Riccati differential equations gave rise in [7] to a general representation formula for solutions of (RDE) from which many important properties of (RDE)'s can be read almost directly by inspection.

As many of these properties can also be seen in the case that the matrix B is diagonalizable, here for the sake of simplicity we shall only discuss this case. In some respect this can be considered as a generic case since by appropriate small perturbations of B it is always possible to obtain simple eigenvalues.

Let $V = (v_1, \ldots, v_{n+m}) \in \mathbb{C}^{(n+m)\times(n+m)}$ be the matrix defined by a basis of eigenvectors of B such that

$$V^{-1}BV = \operatorname{diag}(\lambda_1, \ldots, \lambda_{n+m}) = \Delta$$

and, without loss of generality, assume that

$$\mathrm{Re}(\lambda_1) \leq \mathrm{Re}(\lambda_2) \leq \ldots \leq \mathrm{Re}(\lambda_{n+m}).$$

If

$$C = \begin{pmatrix} c_1 \\ \vdots \\ c_{n+m} \end{pmatrix} = \begin{pmatrix} c_{11} & \ldots & c_{1n} \\ \vdots & & \vdots \\ c_{n+m,1} & \ldots & c_{n+m,n} \end{pmatrix} \in \mathbb{C}^{(n+m)\times n},$$

then an arbitrary $(n+m) \times n$ matrix-solution Y of (2.1) has the form

$$Y(t;C) = Y(t) = Ve^{\Delta t}C = \begin{pmatrix} Q(t) \\ P(t) \end{pmatrix} = \begin{pmatrix} Q(t;C) \\ P(t;C) \end{pmatrix} \tag{2.3}$$

with $Q(t) \in \mathbb{C}^{n\times n}$, $P(t) \in \mathbb{C}^{m\times n}$ and $Y(0) = VC = \binom{Q_0(C)}{P_0(C)}$.

Let $v_\nu(t) = \begin{pmatrix} v_{\nu,1}(t) \\ \vdots \\ v_{\nu,n+m}(t) \end{pmatrix}$, $1 \leq \nu \leq n+m$, and for $1 \leq j \leq n$ and $1 \leq \ell \leq m$, set

$$\tilde{v}_\nu := \begin{pmatrix} v_{\nu 1} \\ \vdots \\ v_{\nu n} \end{pmatrix}, \qquad v_\nu(\ell, j) := (v_{\nu 1}, \ldots, v_{\nu,j-1}, v_{\nu,n+\ell}, v_{\nu,j+1}, \ldots, v_{\nu n})^T.$$

Then we obtain from [7], Theorem 1, the following fundamental representation theorem.

Theorem 1 (Fundamental representation formula for solutions of (RDE)). *Let $Y(\cdot;C), Q_0(C)$ and $P_0(C)$ be defined by (2.3). Further let $Q_0(C)$ be regular and $W_C := P_0(C)Q_0(C)^{-1}$. Then*

$$\begin{aligned} W(\cdot;W_C) : \mathbb{C} \setminus \{t \in \mathbb{C} | \det Q(t) = 0\} &\rightarrow \mathbb{C}^{m\times n}, \\ t &\mapsto P(t)Q(t)^{-1} = W(t;W_C) \end{aligned}$$

is a solution of (RDE) with $W(0;W_C) = W_C$ and

$$w_{\ell\alpha}(t;W_C) = \frac{\displaystyle\sum_{1\leq k_1<\ldots<k_n\leq n+m} e^{(\lambda_{k_1}+\ldots+\lambda_{k_n})t} |v_{k_1}(\ell,\alpha), \ldots, v_{k_n}(\ell,\alpha)| \begin{vmatrix} c_{k_1} \\ \vdots \\ c_{k_n} \end{vmatrix}}{\displaystyle\sum_{1\leq k_1<\ldots<k_n\leq n+m} e^{(\lambda_{k_1}+\ldots+\lambda_{k_n})t} |\tilde{v}_{k_1}, \ldots, \tilde{v}_{k_n}| \begin{vmatrix} c_{k_1} \\ \vdots \\ c_{k_n} \end{vmatrix}}, \tag{2.4}$$

$1 \leq \ell \leq m, 1 \leq \alpha \leq n.$

Remark 1: Obviously any solution $W(\cdot; W_C)$ defines a meromorphic matrix function with poles at most in the zeros of $\det Q(t)$. Hence it is continuous (in chordal metric) in the complex plane. $W(\cdot; W_C)$ is called the solution corresponding to $V_0 = span(VC)$, the column space of VC.

The determinants

$$\begin{vmatrix} c_{k_1} \\ \vdots \\ c_{k_n} \end{vmatrix}$$

appearing in (2.4) as coefficients are the so-called *Plücker coordinates* of C (or of the n-dimensional subspace $span(C)$); these coefficients and also the coefficients $|\tilde{v}_{k_1}, \ldots, \tilde{v}_{k_n}|$ and $|v_{k_1}(\ell, \alpha), \ldots, v_{k_n}(\ell, \alpha)|$ are coupled (see [7] Lemma 1 and Lemma 3), and this has a strong influence on the behaviour of the solutions of (RDE) and on the structure of the set of all solutions of (ARE).

Remark 2: Since (2.4) represents all solutions of (1.1), it also represents all solutions of (1.2), *i.e.* of (ARE). If for instance for a tuple of indices $(k_1, \ldots, k_n)$, $|\tilde{v}_{k_1}, \tilde{v}_{k_2}, \cdots, \tilde{v}_{k_n}| \neq 0$ then, by choosing C such that all determinants with rows from C are zero except that $\left|(c_{k_1}^T, \cdots, c_{k_n}^T)^T\right| \neq 0$, we obtain a solution of (ARE) by

$$w_{\ell\alpha} = \frac{|v_{k_1}(\ell, \alpha), \ldots, v_{k_n}(\ell, \alpha)|}{|\tilde{v}_{k_1}, \ldots, \tilde{v}_{k_n}|}, \tag{2.5}$$

$1 \leq \ell \leq m, 1 \leq \alpha \leq n$.

Remark 3: As it was mentioned in [7], Section 3.3, all (complex) solutions of (ARE) may be described in different ways. Either by all constant matrix functions W given by (2.4), or by limits (in chordal metric) as $t \to \infty$ in certain sectors of the complex plane of all functions defined by (2.4), or by utilizing the B-invariant subspaces.

In the sequel we shall study a particular situation: a so-called dichotomically separated situation. It turns out that this is the most important case for practical applications and also for numerical calculations, since in this case the system is not sensitive with respect to parameter variations.

Theorem 2 (Dichotomic solution).

(i) Assume that the eigenvalues of B are such that

$$\operatorname{Re} \lambda_1 \leq \operatorname{Re} \lambda_2 \leq \cdots \leq \operatorname{Re} \lambda_n < \operatorname{Re} \lambda_{n+1} \leq \cdots \leq \operatorname{Re} \lambda_{n+m}. \tag{2.6}$$

Then (1.1), or the associated system (2.1), is said to be (negative)-dichotomically separated. If furthermore $|\tilde{v}_1, \ldots, \tilde{v}_n| \neq 0$, *then the (negative) dichotomic solution* $W^- = (w_{\ell\alpha})$, *where*

$$w_{\ell\alpha} = \frac{|v_1(\ell,\alpha), \ldots, v_n(\ell,\alpha)|}{|\tilde{v}_1, \ldots, \tilde{v}_n|},$$

$1 \le \ell \le m$, $1 \le \alpha \le n$, *exists.*

(ii) *If the dichotomic solution* W^- *exists, then for any solution* $W(\cdot, W_C)$ *of (1.1),*

$$W(t, W_C) \longrightarrow W^- \qquad \text{as } t \to -\infty,$$

(with respect to the chordal metric) as long as the first n rows of C are linearly independent, i.e. provided that $\left|(c_1^T, \ldots, c_n^T)^T\right| \neq 0$.

(iii) *The complementary set of the basin of attraction (in chordal metric) of* W^- *(if* W^- *exists) has zero measure in* $\mathbb{C}^{m\times n}$. *The convergence of a solution* $W(\cdot, W_C)$ *towards the dichotomic solution* W^- *takes place at an exponential rate.*

(iv) W^- *is stable with respect to small perturbations of the system, i.e. taking* $\tilde{B} = B + \Delta B$ *instead of* B, *where* $||\Delta B||$ *is sufficiently small.*

Proof: Part (i) follows from (2.5). Part (ii) follows from the representation formula (2.4), by noticing that for a dichotomically separated system there is just one exponential term, namely $e^{(\lambda_1+\ldots+\lambda_n)t}$, growing asymptotically faster than all the others, as $t \to -\infty$. Part (iii) is a direct consequence of (ii), formula (2.4) and the dichotomy condition (2.6). Part (iv) follows from the continuous dependence of the eigenvalues from the entries of a matrix. ■

Remark 4: In a similar way one can define (positive) dichotomically separated systems and obtain similar results regarding $t \to \infty$. If for instance

$$\operatorname{Re}\lambda_1 \le \ldots \le \operatorname{Re}\lambda_{n-1} < \operatorname{Re}\lambda_n = \operatorname{Re}\lambda_{n+1} < \operatorname{Re}\lambda_{n+2} \le \ldots \le \operatorname{Re}\lambda_{n+m},$$

then either $\operatorname{Im}\lambda_n = \operatorname{Im}\lambda_{n+1}$ and (ARE) (1.2) has infinitely many solutions, or $\operatorname{Im}\lambda_n \neq \operatorname{Im}\lambda_{n+1}$ and then (RDE) can (1.1) has a periodic/quasiperiodic limit cycle. For a further discussions on the structure of the set of solutions of (ARE) and the phase portrait of (RDE); see [7], [19].

§3 Lyapunov-type functions

As it was explained in the previous section the solutions of a Riccati differential equation always exist globally as meromorphic functions. Most of the applications, however, require the boundedness of a solution in a specific given interval on the real line. Therefore, determining appropriate invariant domains for (RDE) therefore always leads us to sufficient conditions for boundedness. In classical Control Theory, for instance, the symplectic (RDE)

$$\dot{K} = -A^T K - KA - Q + KSK, \qquad Q \geq 0, \quad S \geq 0,$$

appears. It is well known and remarkably important in Control Theory that in this case the cone of positive semidefinite matrices $K \geq 0$ is negative invariant (see [8]). This behaviour changes drastically if for instance S is indefinite (see [3] Theorem 1). In the general non-symplectic (non-symmetric) case there are no such strong results. Here from [9] we cite two different approaches in order to characterize invariant domains. The first approach works for equations admitting symmetric solutions (not necessarily linearizable) and, in particular, for equations of type (1.3); the second works for all linearizable Riccati differential equations (RDE).

Theorem 3. *Assume that in equation*

$$\begin{aligned} \dot{K} &= -A(t)^T K - KA(t) - Q(t) + KS(t)K \\ &\quad -f_1(t,K) + Kf_2(t,K) + f_2^T(t,K)K, \end{aligned} \qquad K(t_f) = K_f \qquad (3.1)$$

where for all $t \in (-\infty, t_f]$:

(i) *$A(t), Q(t), S(t) \in \mathbb{R}^{n\times n}$ are piecewise continuous,*

(ii) *$Q(t)$, $S(t)$ and K_f are symmetric,*

(iii) *$f_1(t,K)$ and $f_2(t,K)$ map matrices $K \in \mathbb{R}^{n\times n}$ into $\mathbb{R}^{n\times n}$,*

(iv) *$(f_1(t,K))^T = f_1(t,K)$ and,*

(v) *f_1 and f_2 are piecewise continuous in t and Lipschitz-continuous w.r.t. $K \in \mathbb{R}^{n\times n}$.*

If for some positive α and $\forall x \in \mathbb{R}^n$

$$x^T(f_1(K) + Q(t))x \geq \alpha ||x||^2,$$

for $t \in (-\infty, t_f]$ and $K \geq 0$, then any solution of (3.1) with positive initial data $K_0 > 0$ stays positive for $t \leq t_f$.

Theorem 4. *If for some $\varepsilon > 0$, $\alpha, \beta \in \mathbb{R}$, $\alpha < \beta$, for all $t \leq t_f$ and $y \in \mathbb{R}^n$, $||y|| = 1$, the following inequalities hold*

$$\begin{aligned} y^T(-2\alpha A(t) + \alpha^2 S(t) - Q(t))y &\leq -P(|\alpha|) - \varepsilon, \\ y^T(-2\beta A(t) + \beta^2 S(t) - Q(t))y &\geq P(|\beta|) + \varepsilon, \end{aligned}$$

where

$$||f_1(K) - Kf_2(K) - f_2^T(K)K|| \leq P(||K||)$$

for all symmetric matrices $K \in \mathbb{R}^{n\times n}$, then the set

$$I(\alpha, \beta) = \{K = K^T \in \mathbb{R}^{n\times n} \,|\, \alpha I_n \leq K \leq \beta I_n\}$$

is negative invariant.

For the proof of this theorem see [9].

Remark 5: It should be pointed out that, in practice, in the case of (RDE) Theorem 4 can only provide reasonable results if a dichotomic solution exists. Then it is quite clear, from Theorem 2 and from the continuous dependence of the existence interval on the initial data (see [6], Section 1, Theorem 7.1), that there is a small neighbourhood of the dichotomic solution which is negative invariant. Theorem 4 then gives some effective bounds for that neighbourhood.

Remark 6: The perturbation term in equation (3.1) was chosen in this particular form, with the functions f_1, f_2, in order to cover important generalized Riccati differential equations from applications, like for instance (1.3). Clearly the method will also work for other types of generalized equations.

In the next theorem, which (for the sake of completeness) we cite here from [10], we use the following abbreviation

$$L(t) = \begin{pmatrix} CB_{11} + DB_{21} & CB_{12} + B_{11}^T D + DB_{22} \\ 0 & B_{12}^T D \end{pmatrix}.$$

Theorem 5. *Let $B_{11}, B_{12}, B_{21}, B_{22}$ be piecewise continuous on the interval $(-\infty, T]$. If for some matrices $C \in \mathbb{R}^{n\times n}$, with $C^T = C$, $D \in \mathbb{R}^{n\times m}$,*

$$L(t) + L^T(t) \leq 0$$

for all $t \leq t_f \leq T$ and if for some $W_f \in \mathbb{R}^{m\times n}$

$$C + DW_f + W_f^T D^T > 0,$$

then the solution $W(t, W_f)$ of (1.1) with $W(t_f, W_f) = W_f$ exists for all $t \leq t_f$.

§4 The "sampled-data" information structure

There is a strong relationship between the matrix Riccati differential equation (1.1) and the generalized Riccati differential equation arising from a differential game with open loop information structure (see [4]):

$$\dot{K} = -\tilde{A}^T - K\tilde{A} - \tilde{Q} + K\tilde{S}K + KJ\tilde{S}KJ, \tag{4.1}$$

where the matrices $\tilde{A}$, $\tilde{Q}$, $\tilde{S}$, J and K are defined as in equation (1.3). Using the notations

$$B_{11} = A, \qquad B_{12} = \begin{pmatrix} -A^T & 0 \\ 0 & -A^T \end{pmatrix}, \qquad W = \begin{pmatrix} K_1 \\ K_2 \end{pmatrix},$$
$$B_{21} = \begin{pmatrix} -Q_1 \\ -Q_2 \end{pmatrix}, \qquad B_{22} = \begin{pmatrix} S_1 & S_2 \end{pmatrix},$$

the equivalence between equations of type (4.1) and (1.1) is obvious.

As shown in §2 the (RDE) (and therefore even a slightly modified one arising from a differential game with *open-loop* information structure (4.1)) have the nice property of being linearizable. On the other hand, the memoryless feedback (RDE), that arises from a differential game with continuous state-feedback (see [21]) or from an optimal-controlled jump-linear system (see for instance [22] or [2]), do not have this property. This makes equations of type (1.3) particularly difficult to solve. In fact, as far as we are aware, there is no efficient technique for solving them.

Besides a perfectly continuous state-feedback, a discrete-time memoryless feedback information structure could also be considered. This means, that the state of the system can only be measured at predefined times. (Such information structures appear in computer science, where the frequency of sampling is bounded by the speed of the processor and also in economics, where most of the statistical data is calculated monthly or even annually.)

In this context a differential game with a discrete-time memoryless-feedback (or shorter: *sampled-data*) information structure is nothing else but a sequence of *open-loop* differential games, where the duration of each game is given and the terminal value of the previous game represents the initial value of the current one. This means, that we have one *open-loop* Riccati-type (4.1) and one Lyapunov-type (4.2) differential equations to solve for each sampling-interval $[t_{j-1}, t_j]$, $j = 1 \dots h$ (see also [20]):

$$\begin{aligned} \dot{K}_j &= -\tilde{A}^T K - K\tilde{A} - \tilde{Q} \\ &\quad + K\tilde{S}K + KJ\tilde{S}KJ, \end{aligned} \qquad K_j(t_j) = \Lambda_{j+1}(t_j), \tag{4.2}$$

$$\begin{aligned}\dot{\Lambda}_j &= -\tilde{A}^T\Lambda - \Lambda\tilde{A} - \tilde{Q} + \Lambda\tilde{S}K \\ &\quad +K^T\tilde{S}\Lambda + JK^T\tilde{S}J\Lambda + \Lambda J\tilde{S}KJ \qquad \Lambda_j(t_j) = \Lambda_{j+1}(t_j). \\ &\quad -K^T\tilde{S}K - JK^TJS_0JKJ,\end{aligned} \tag{4.3}$$

Hereby take $\Lambda_{h+1}(t_f) = K_f$ and $t_h = t_f$ (see (1.3)).

After carefully studying (4.3) and (1.3), it is clear that for the case $\Lambda_j = K_j$ the two terminal-value problems are equivalent. If we denote the r.h.s. of (4.3) by $g(\Lambda, K)$, the r.h.s. of (1.3) by $\hat{g}(K)$, we can state the following (see also [14]).

Theorem 6. *Under the assumption that the memoryless-feedback (RDE) has a unique bounded solution (i.e. no blow-up) over the interval $[t_0, t_f]$, the solution of the sampled-data information structure converges to this function as the largest sampling-time tends to zero.*

Remark 7: Simaan and Cruz (see [20]) also claim the convergence of the solution of differential games with sampled-data information structure to the solution of those with memoryless-feedback information structure. However, neither a proof of this statement is given nor it is clear what happens if the solution of the differential game using the memoryless-feedback information structure has finite escape time.

Proof of Theorem 6: Denote the length of the largest sampling-interval by δ, the solution of the memoryless feedback Riccati differential equation (1.3) by $\hat{K}(t)$ and the solution of the Lyapunov-type differential equation (4.3), which is continuous in the interval $[t_0, t_f]$, by $\Lambda(t)$. Furthermore denote the piecewise continuous function over $[t_0, t_f]$, being assembled from the segments K_j that solve the open-loop RDE (4.2) for $[t_{j-1}, t_j]$, by $K(t)$.

Since for each sampling interval $[t_{j-1}, t_j]$ the terminal values of both Λ and K are the same and because they are (at least for a small enough δ) analytical, we can state that

$$\|\Lambda - K\| \leq \max_{t_{j-1}\leq t\leq t_j} \left(\|\dot{\Lambda} - \dot{K}\|\right) \cdot \delta \leq M_j \cdot \delta \qquad \text{for } t \in [t_{j-1}, t_j]$$

for some $M_j > 0$. Furthermore,

$$\|g(\Lambda, K) - g(\Lambda, \Lambda)\| \leq \tilde{M}_j \cdot \|\Lambda - K\| \leq N_j \cdot \delta \qquad \text{for } t \in [t_{j-1}, t_j]$$

holds for some $N_j > 0$.

With the notation $N := \max_{j=1\ldots h} N_j$ the following holds

$$\|g(\Lambda, K) - \hat{g}(\Lambda)\| \leq N \cdot \delta \qquad \text{for } t \in [t_0, t_f].$$

Because the linearizable equations (4.2) and (4.3) have over a finite interval bounded solutions, the function $g(\Lambda, K)$ is bounded and furthermore

$$\limsup_{\delta \to 0} \left(\max_{j=1\ldots h} \tilde{M}_j \right) < \infty.$$

This means that

$$\limsup_{\delta \to 0} \|g(\Lambda, K) - \hat{g}(\Lambda)\| = 0$$

and hence that the Lyapunov-type equation (4.3) can be considered as a perturbed memoryless-feedback (RDE), where the perturbation term vanishes as the sampling time tends to zero. This implies (see Chapter 2.4 in [6]) that the function Λ and therefore also the function K converge to $\hat{K}$. ■

With the help of the previous proposition, an approximation method for the solution of the memoryless-feedback (RDE) (1.3) can be constructed as follows:

1. Let h be a sufficiently large fixed positive integer and $T := \frac{t_f - t_0}{h}$. Further let $\tilde{\Lambda} = K_f$ and set $j = h$.
2. Define $t_k := t_0 + kT$ for each $k = 0, \ldots, h$.
3. Let K be the (exact) solution of the *open-loop* RDE (4.2), according to Theorem 2.1, over the interval $[t_{j-1}, t_j]$ for the terminal-value problem $K(t_j) = \tilde{\Lambda}$.
4. Calculate $K_0 = K\left(\frac{t_{j-1}+t_j}{2}\right)$.
5. Determine the (exact) solution of the Lyapunov-type differential equation (4.3) over $[t_{j-1}, t_j]$ for the terminal-value problem $\Lambda(t_j) = \tilde{\Lambda}$, by substituting $K(t) \cong K_0$.
6. Calculate $\tilde{\Lambda} = \frac{1}{2}(\Lambda(t_{j-1}) + \Lambda^T(t_{j-1}))$, as well as $\hat{K}(t) = \frac{1}{2}(K(t) + K^T(t))$, for $t \in [t_{j-1}, t_j]$.
7. For the case $j > 1$, decrease j by one and go to step 3.

Remark 8: The symmetrization in step 6 serves for accelerating the convergence. It is based on the a priori knowledge that the solution of the memoryless-feedback (RDE) is symmetric if the coefficients and initial data are symmetric.

Remark 9: It should be pointed out that the above procedure is not only a solution technique for generalized Riccati differential equations, but also a realisation of an information structure used in differential systems with discrete-time feedback.

§5 Numerical results

In this last section numerical results are presented on the sampled-data approximation and on the boundedness-criteria of the solution of the memoryless feedback RDE. We also give an example of a generalized algebraic Riccati equation having more than one positive definite solutions. In the case of standard Riccati differential equations, as they appear in Control Theory, under stabilizability and controllability conditions there is exactly one positive semidefinite solution (see [12], Theorem 10.6). It was not noticed by several authors (see for instance [15]) that this property does not hold for generalized Riccati matrix differential equations.

Consider the autonomous generalized RDE of the memoryless feedback type (1.3) with coefficients

$$\tilde{A} = \begin{bmatrix} 20 & 50 & 0 & 0 \\ -25 & 15 & 0 & 0 \\ 0 & 0 & 20 & 50 \\ 0 & 0 & -25 & 15 \end{bmatrix}, \qquad \tilde{Q} = \begin{bmatrix} 8 & 2 & 0 & 0 \\ 2 & 4 & 0 & 0 \\ 0 & 0 & 8 & 2 \\ 0 & 0 & 2 & 14 \end{bmatrix},$$

$$\tilde{S} = \begin{bmatrix} 18 & 18 & 0 & 0 \\ 18 & 18 & 0 & 0 \\ 0 & 0 & 4 & 8 \\ 0 & 0 & 8 & 16 \end{bmatrix}, \qquad S_0 = \begin{bmatrix} 0 & 0 & 0 & 0 \\ 0 & 0 & 0 & 0 \\ 0 & 0 & 0 & 0 \\ 0 & 0 & 0 & 0 \end{bmatrix},$$

and terminal value problem

$$K(0) = \begin{bmatrix} 2.5 & 0 & 0 & 0 \\ 0 & 1 & 0 & 0 \\ 0 & 0 & 5 & 0 \\ 0 & 0 & 0 & 3 \end{bmatrix}.$$

As a first question, we can study the existence of negative invariant sets. Using $f_1(K) = JKJS_0JKJ = 0$ and $f_2(K) = J\tilde{S}KJ$, in Theorem 3 we can state that *the set of positive definite matrices is negative invariant.* However, our hope for also obtaining (with the help of Theorem 4) an upper bound for the norm of the solution is not fulfilled. Nevertheless, the statement of Theorem 4 also holds for a slightly weakened assumption (see also [14]):

Theorem 7. *If for some* $\epsilon > 0$, $\kappa > 0$ *and for all* $t \le t_f$

$$x(t)^T(-2\kappa\tilde{A}(t) + \kappa^2\tilde{S}(t) - \tilde{Q}(t))x(t) \ge P(\kappa) + \epsilon,$$

where $x(t)$ *is the unit eigenvector belonging to the biggest eigenvalue of* $K(t)$ *and* $P(\|K\|) \ge 2\|\tilde{S}(t)\|\,\|K\|^2$, *then the set* $\{K \mid \|K\| \le \kappa\}$ *is negative invariant for the RDE*

$$\dot{K} = -\tilde{A}^T K - K\tilde{A} - \tilde{Q} + K\tilde{S}K + JK\tilde{S}JK + KJ\tilde{S}KJ.$$

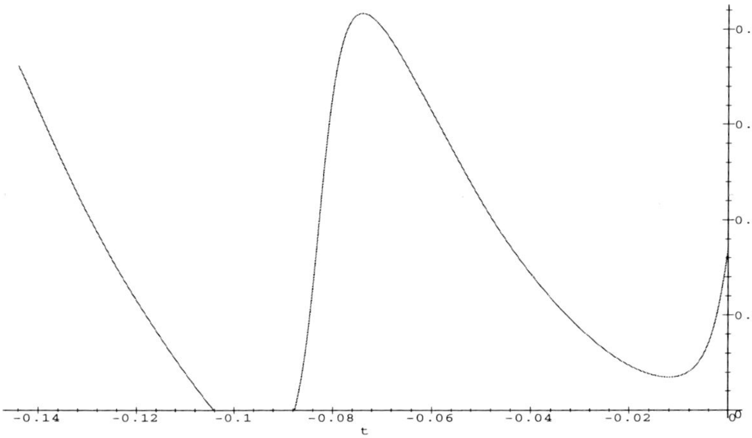

Figure 1. Maximal invariant set using Theorem 7

Although this weaker condition is difficult to check in general, it is more advantageous in numerical approximations. Rearranging the terms in the inequality in Theorem 7 and using $P(\kappa) \ge 2\|\tilde{S}(t)\|\kappa^2$ we get

$$\kappa^2(x^T\tilde{S}(t)x - 2\|\tilde{S}(t)\|) - 2\kappa x^T\tilde{A}(t)x - x^T\tilde{Q}(t)x - \epsilon \ge 0.$$

Since $\|x\| = 1$, it is $x^T\tilde{S}(t)x - 2\|\tilde{S}(t)\| < 0$ as long as $\tilde{S} \ne 0$. This shows that the possible values for κ have an upper bound. Figure 1 represents the maximal value of κ over $t \in [-0.14, 0]$. As it can be seen, for $t \approx -0.1$ not even the weakened condition of Theorem 7 is fulfilled. This means that we cannot obtain an invariant set by this theorem.

Two approximate solutions of the equation, obtained by using a general *Runge-Kutta method* of order 4-5 and the *method of upwind-discretization*, are presented in Figures 2 and 3, respectively. Because of the block-diagonal structure of the coefficient matrices, the solution must also be of the same form. The diagrams represent, respectively, the eigenvalues of the first and second block, *i.e.* K_1, K_2 in (1.3), for the interval $[-1,0]$. Since no error estimate or convergence result is known for these two methods in the case of the (RDE), the qualitative behaviour of the solution cannot be determined easily.

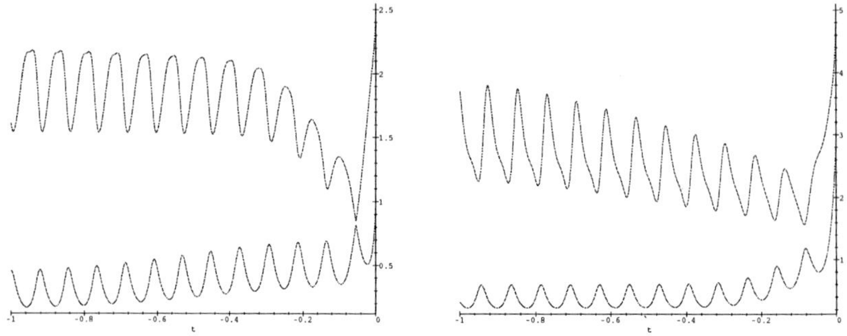

Figure 2. Eigenvalues of the solution with the Runge-Kutta method. $\Delta t = 10^{-3}$

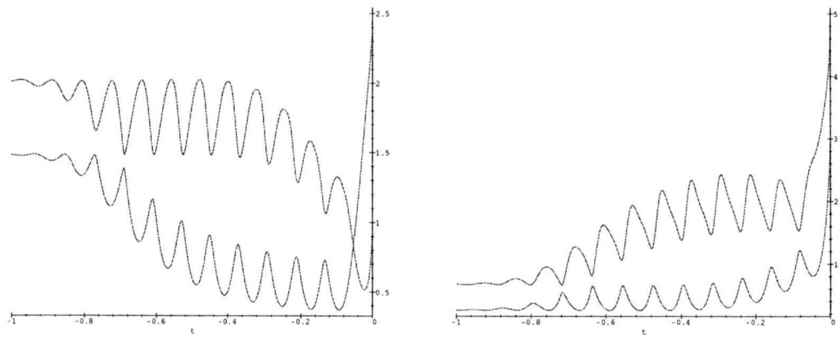

Figure 3. Eigenvalues of the solution with upwind-discretization. $\Delta t = 3.3 \cdot 10^{-4}$

In Figures 4–6 we present the approximate solutions obtained by means of the sampled-data method described in §4. Theorem 6 indicates, that *if the solution of (1.3) exists and is bounded*, then the functions on Figures 4, 5 and 6 converge to it. Thus the qualitative behaviour of the given solution is best reflected in Figures 2 and 6.

Finally, we present some solutions of the generalized (ARE) of the memoryless feedback type

$$0 = -\tilde{A}^T K - K\tilde{A} - Q + K\tilde{S}K + JK\tilde{S}JK + KJ\tilde{S}KJ - JKJS_0JKJ$$

with the same coefficients as before. There are up to four different solutions

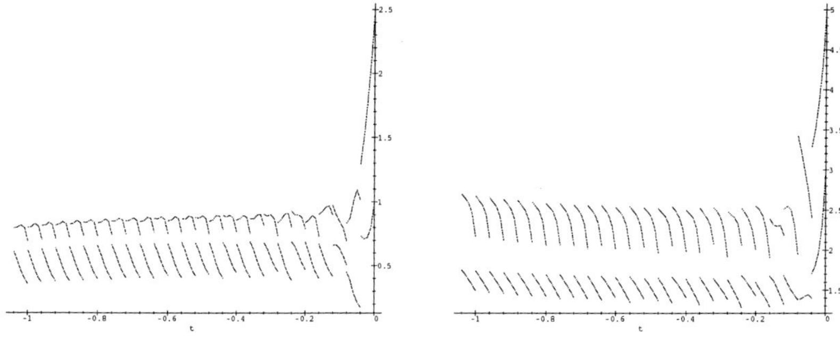

Figure 4. Eigenvalues using the sampled-data approximation. $T = 0.04$

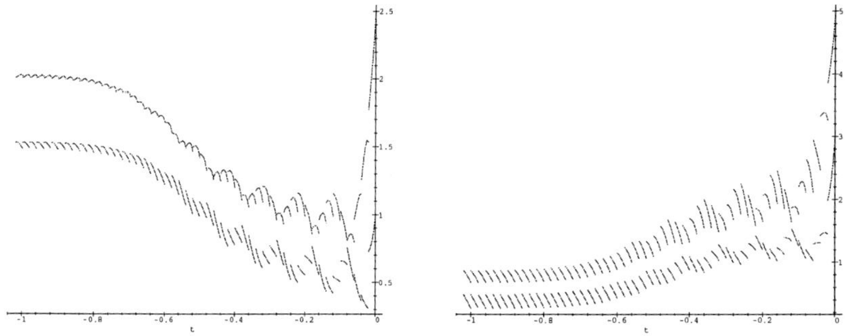

Figure 5. Eigenvalues using the sampled-data approximation. $T = 0.02$

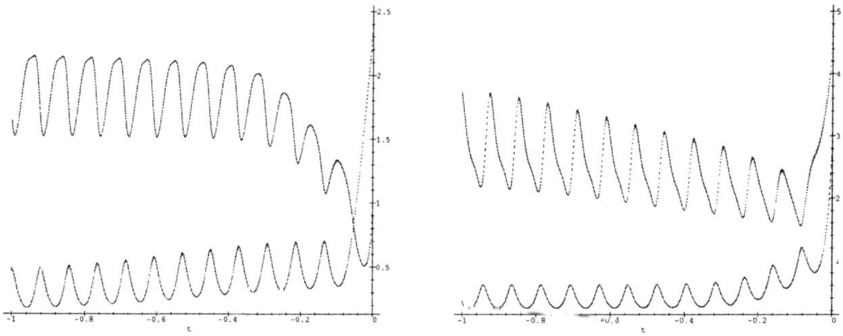

Figure 6. Eigenvalues using the sampled-data approximation. $T = 10^{-3}$

known, two of which are positive definite ($K^{(1)}$ and $K^{(2)}$), one negative definite ($K^{(3)}$) and one indefinite ($K^{(4)}$):

$$K^{(1)} \approx \begin{bmatrix} 2.0139 & .07042 & 0 & 0 \\ .07042 & 1.4959 & 0 & 0 \\ 0 & 0 & .63421 & -.11875 \\ 0 & 0 & -.11875 & .29936 \end{bmatrix}$$

$$K^{(2)} \approx \begin{bmatrix} .29894 & .068680 & 0 & 0 \\ .068680 & .13624 & 0 & 0 \\ 0 & 0 & 3.1659 & .43560 \\ 0 & 0 & .43560 & 2.3835 \end{bmatrix}$$

$$K^{(3)} \approx \begin{bmatrix} -.14501 & .037882 & 0 & 0 \\ .037882 & -.18551 & 0 & 0 \\ 0 & 0 & .-.17930 & .00018170 \\ 0 & 0 & .00018170 & -.34354 \end{bmatrix}$$

$$K^{(4)} \approx \begin{bmatrix} -.11991 & -.0013110 & 0 & 0 \\ -.0013110 & 2.6932 & 0 & 0 \\ 0 & 0 & -.60076 & -1.2321 \\ 0 & 0 & -1.2321 & .084834 \end{bmatrix}$$

If we study the asymptotic behaviour of the games being controlled by the constant controlling laws that arise from the above solutions, then we can state that *both positive definite solutions are asymptotically stabilizing* (*i.e.* the matrix $A - S_1K_1 - S_2K_2$ is stable), the negative definite produces an unstable linear differential system with all eigenvalues having positive real parts and the indefinite solution $K^{(4)}$ leads to a linear differential system with two eigenvalues having positive and two having negative real parts.

Acknowledgement. The authors are indebted to the referees and the editors, in particular to N. Papamichael, for their careful reading and valuable suggestions for improvements of this paper.

References

[1] Abou-Kandil, H., G. Freiling and G. Jank, Necessary conditions for constant solutions of coupled Riccati equations in Nash games, *System & Control Letters*, **21** (1993), 295–306.

[2] Abou-Kandil, H., G. Freiling and G. Jank, Solution and asymptotic behaviour of coupled Riccati equations in jump linear systems, *IEEE Transactions on Automatic Control*, **39** (8) 1994.

[3] Abou-Kandil, H., G. Freiling, G. Jank and S.-R. Lee, On the dependence of the solutions of algebraic and differential game Riccati equations on the parameter μ^*, *European Journal of Control* **2** (1996), 69–78.

[4] Başar,T. and G. J. Olsder, *Dynamic Noncooperative Game Theory*, Academic Press, 1995.

[5] Brockett, R. W., *Finite Dimensional Linear Systems*, John Wiley, New York, 1970.

[6] Coddington, E. A. and N. Levinson, *Theory of Ordinary Differential Equations*, McGraw-Hill, New York, 1955.

[7] Freiling, G. and G. Jank, Nonsymmetric matrix Riccati equations, *Journal for Analysis and its Applications*, **14** (1995), 259–284.

[8] Freiling, G. and G. Jank, Existence and comparison theorems for algebraic Riccati equations and Riccati differential and difference equations, *Journal of Dynamical and Control Systems* **2** (4) (1996), 529–547.

[9] Freiling, G., G. Jank and A. Sarychev, Non-blow-up conditions for Riccati-type matrix differential and difference equations, to appear.

[10] Freiling, G., G. Jank and A. Sarychev, Lyapunov-type functions and invariant sets for Riccati matrix differential equations, *European Control Conference*, 1997, Brussels, Belgium.

[11] Hermann, R., and C. Martin, Lie and Morse theory for periodic orbits of vector fields and matrix Riccati equations, I: general Lie-theoretic methods, *Math. Systems Theory* **15** (1982), 277–284.

[12] Knobloch, H. W., and H. Kwakernaak, *Lineare Kontrolltheorie*. Springer-Verlag, Berlin, Heidelberg, New York, Tokio, 1985.

[13] Kuiper, H. J., Positive invariance and asymptotic stability of solutions to certain Riccati equations, *Dynamics and Stability of Systems* **9** (4) (1994), 331–344.

[14] Kun, G., Verallgemeinerte Riccati-Gleichungen, Diplomarbeit in Mathematik, RWTH Aachen, Lehrstuhl II für Mathematik, July 1997.

[15] Li, T.-Y. and Z. Gajic, Lyapunov iterations for solving coupled agebraic Riccati equations of Nash differential games and algebraic Riccati equations of zero-sum games, in *New Trends in Dynamic Games and Applications*, G. J. Olsder (ed.), Birkhäuser, Boston, 1995, 333–351.

[16] Medanic, J., Geometric properties and invariant manifolds of the Riccati equation, *IEEE Trans. Automat. Control* **AC-27** (1982), 670–677.

[17] Reid, W. T., *Riccati Differential Equations*, Academic Press, New York, 1972.

[18] Schneider, C. R., Global aspects of the matrix Riccati equation, *Math. Systems Theory* **7** (1973), 281–286.

[19] Shayman, M. A., Phase portrait of the matrix Riccati equation, *SIAM J. Control and Opt.* **24** (1986), 1–65.

[20] Simaan, M. and Jr. J. B. Cruz, Sampled-data Nash controls in non-zero-sum differential games, *Int. J. Control* **17** (6) (1973), 1201–1209.

[21] Starr, A. W. and Y. C. Ho, Nonzero-sum differential games, *J. Optimization Theory & Applications* **3** (3) (1968), 184–206.

[22] Sworder, D. D., Feedback control of a class of linear systems with jump parameters, *IEEE Transactions on Automatic Control* **AC-14** (1969), 9–14.

[23] Telford, A. J. and J. B. Moore, On the existence of solutions to non-symmetric algebraic Riccati equations, *Lect. Notes of the Workshop on 'The Riccati Equation in Control, Systems and Signal'*, S. Bittanti (ed.), Pitagora Editrice, Bologna, 1989, 83–86.

Gerhard Jank
Lehrstuhl II für Mathematik,
RWTH Aachen
Templergraben 55
D-52056 Aachen
GERMANY
jank@math2.rwth-aachen.de

Gábor Kun
Lehrstuhl C für Mathematik
RWTH Aachen
Templergraben 55
D-52056 Aachen
GERMANY

Fast evaluation of hypergeometric functions by FEE

Ekatharine A. Karatsuba

Abstract. An algorithm for fast calculation of hypergeometric functions for algebraic values of argument and parameters is proposed. The computational complexity is near to optimal. The merit of the method is the possibility of its implementation in parallel processing.

§1 Introduction and statement of the result

In a series of earlier papers ([6]–[9]) we proposed a new method for the fast computation of transcendental functions. Since all these functions are known in Number Theory as Siegel E-functions, we call this method FEE (Fast E-function Evaluation). It was proved that by using FEE one could calculate quickly any elementary transcendental function, the classical constants e, π, and the Euler constant γ, as well as higher transcendental functions such as the gamma function, Bessel functions and other special functions for algebraic values of argument and parameters. The computational complexity is near to optimal.

We shall assume that numbers are written in binary notation.

By the complexity of multiplication of two n-digit integers we shall mean the total number of elementary (bit) operations $M(n)$ sufficient for the computation of the product of these two n-digit integers.

The total number of bit-operations (further bit-operations, for the sake of brevity, will simply be called operations) that are sufficient for the evaluation of the function $y = f(z)$ with accuracy 2^{-n} at the point $z = z_0$ of the domain of a function is denoted by $s_f(n)$ and is called the complexity of the evaluation of the function $f(z)$ at the point $z = z_0$.

It was proved that the complexity of calculation by the FEE of the aforementioned elementary and higher transcendental functions and constants is estimated by

$$s_f(n) = O(M(n)\log^2 n).$$

Here and in what follows $\log z$ denotes the base 2 logarithm of the number z. The history of fast computations began in the 1950's , when A. N. Kolmogorov stated the problem *of estimating from above the number of operations sufficient for calculating the product of two n-digit integers.* Since the previously known methods of multiplication had the complexity upper bound $M(n) \leq cn^2$, $M(n)$ was believed to have such lower bound too.

Computational Methods and Function Theory (CMFT'97)
N. Papamichael, St. Ruscheweyh and E. B. Saff (Eds.), pp. 303–314

ISBN 981-02-3626-3

In 1960 A. A. Karatsuba [4] proposed the method for fast multiplication of two n-digit integers with the complexity bound

$$M(n) \leq cn^{\log_2 3},$$

where c is a constant.

These investigations were continued by other authors and in 1971 A. Schönhage and V. Strassen [11] constructed an algorithm for fast multiplication, that used the Fast Fourier Transform (FFT), with the near to optimal complexity

$$M(n) \leq cn \log n \log \log n,$$

where c is a constant. Different algorithms for fast multiplications and their efficiencies are described in detail in [10].

In 1975 Yu. V. Benderskij [1] proposed an algorithm for the fast evaluation of elementary algebraic functions with complexity

$$O(M(n)).$$

In 1976 R. P. Brent [3] constructed iterative algorithms, using the AGM-method of Gauss, for fast evaluation of elementary transcendental functions with the near to optimal complexity

$$O(M(n) \log n).$$

The 1986 book "π and AGM" [2] by J. M. Borwein and P. B. Borwein gives more complete AGM-algorithms for fast computations of different functions and classical constants.

Here we consider the use of the FEE-method for the fast calculation of the hypergeometric function $y = F(a, b, c, z)$ and prove the following theorem.

Theorem 1. *Let $y = F(a, b, c, z)$ be a hypergeometric function, a, b, c, z algebraic numbers, $|z| \leq 1$. Then*

$$s_F(n) = O(M(n) \log^2 n).$$

§2 Proof of Theorem 1

By definition

$$F(a, b, c, z) = 1 + \frac{ab}{1!c} z + \frac{a(a+1)b(b+1)}{2!c(c+1)} z^2 + \dots . \tag{2.1}$$

We assume, that a, b, c are non-negative integers or zero. Then for $|z| < 1$ the series (2.1) converges absolutely, and for $|z| > 1$ it diverges. The satisfaction of the inequality $a + b < c$ is necessary and sufficient for absolute convergence of the series (2.1) for $|z| = 1$ ([12], p. 38).

We consider the case of the positive parameters a, b, c and positive argument $z = x$, $0 < x < 1$ and rewrite (2.1) in the form:

$$F(a,b,c,x) = F_r(a,b,c,x) + \delta F_r(a,b,c,x), \tag{2.2}$$

where

$$\begin{aligned} F_r(a,b,c,x) &= 1 + \frac{ab}{1!c}x + \dots \\ &+ \frac{a(a+1)\cdots(a+r-1)b(b+1)\cdots(b+r-1)}{r!c(c+1)\cdots(c+r-1)}x^r, \end{aligned} \tag{2.3}$$

$$\delta F_r(a,b,c,x) = \frac{a(a+1)\cdots(a+r)b(b+1)\cdots(b+r)}{(r+1)!c(c+1)\cdots(c+r)}x^{r+1} + \cdots. \tag{2.4}$$

First we estimate $\delta F_r(a,b,c,x)$ from above. We set

$$\begin{aligned} \alpha(k) &= (a+1)(a+2)\cdots(a+k), \\ \beta(k) &= (b+1)(b+2)\cdots(b+k), \\ \gamma(k) &= (c+1)(c+2)\cdots(c+k), \end{aligned} \tag{2.5}$$

$$\varphi(k) = \frac{ab}{c}\frac{\alpha(k)\beta(k)}{(k+1)!\gamma(k)}. \tag{2.6}$$

Using the Euler-Maclaurin summation formula ([5], p. 10) for $\alpha(k)$ with $k \geq 1$ we obtain:

$$\begin{aligned} \ln\alpha(k) &= \sum_{i=1}^{k}\ln(a+i) \\ &= \int_{\frac12}^{k+\frac12}\ln(a+x)\,dx + \rho\left(k+\frac12\right)\ln\left(a+k+\frac12\right) \\ &\quad \rho\left(\frac12\right)\ln\left(a+\frac12\right) \\ &\quad +\sigma\left(\frac12\right)\frac{1}{a+\frac12} - \sigma\left(k+\frac12\right)\frac{1}{a+k+\frac12} \\ &\quad + \int_{\frac12}^{k+\frac12}\sigma(x)\left(-\frac{1}{(a+x)^2}\right)dx, \end{aligned} \tag{2.7}$$

where

$$\rho(x) = \frac{1}{2} - \{x\}, \qquad \sigma(x) = \int_0^x \rho(u)\, du. \tag{2.8}$$

(Here $\{x\}$ denotes the fractional part of x. ln denotes the natural logarithm or the base e logarithm). From (2.8) we see, that $|\rho(x)| \le \frac{1}{2}$, $x > 0$, and $|\sigma(x)| \le \frac{1}{8}$. Taking into account that $\rho\left(\frac{1}{2}\right) = \rho\left(k + \frac{1}{2}\right) = 0$, $\sigma\left(\frac{1}{2}\right) = \sigma\left(k + \frac{1}{2}\right) = \frac{1}{8}$, we have from (2.7):

$$\ln \alpha(k) = (a + x)\ln(a + x) - (a + x)\Big|_{\frac{1}{2}}^{k+\frac{1}{2}} + \theta_1, \tag{2.9}$$

where $\theta_1 = \theta_1(k)$ satisfies

$$|\theta_1| \le \frac{1}{4}. \tag{2.10}$$

One can also write for $\beta(k)$, $\gamma(k)$, $(k + 1)!$ expressions that are analogous to (2.9)–(2.10). Then we get from (2.6):

$$\begin{aligned} \varphi(k) &= \frac{ab}{c} \exp\left[\left(a + k + \frac{1}{2}\right) \ln\left(a + k + \frac{1}{2}\right) - k\right. \\ &- \left(a + \frac{1}{2}\right) \ln\left(a + \frac{1}{2}\right) + \left(b + k + \frac{1}{2}\right) \ln\left(b + k + \frac{1}{2}\right) - k \\ &- \left(b + \frac{1}{2}\right) \ln\left(b + \frac{1}{2}\right) - \left(c + k + \frac{1}{2}\right) \ln\left(c + k + \frac{1}{2}\right) + k \\ &+ \left.\left(c + \frac{1}{2}\right) \ln\left(c + \frac{1}{2}\right) + (k + 1) - \left(k + \frac{3}{2}\right) \ln(k + 1) + \theta_2\right], \end{aligned} \tag{2.11}$$

where $\theta_2 = \theta_2(k)$ satisfies

$$|\theta_2| \le 1. \tag{2.12}$$

We put

$$\max\left(a + \frac{1}{2}, b + \frac{1}{2}, c + \frac{1}{2}\right) = A, \tag{2.13}$$

and assume that

$$k \ge r \ge 2A. \tag{2.14}$$

Then

$$\ln\left(k + d_j + \frac{1}{2}\right) = \ln k + \frac{d_j + \frac{1}{2}}{k} + \varepsilon_j \left(\frac{d_j + \frac{1}{2}}{k}\right)^2, \qquad j = 1, 2, 3, 4, \tag{2.15}$$

where $d_1 = a$, $d_2 = b$, $d_3 = c$, $d_4 = \frac{1}{2}$,

$$|\varepsilon_j| \le \frac{1}{2}, \qquad j = 1, 2, 3, 4. \tag{2.16}$$

Substituting (2.15) for $j = 1, 2, 3, 4$ in (2.11), we obtain

$$\begin{aligned}\varphi(k) = \; & \frac{ab}{c}\exp\left[(a+b-c-1)\ln k + \left(\left(a+\frac{1}{2}\right)^2 + \left(b+\frac{1}{2}\right)^2\right.\right. \\ & \left. - \left(c+\frac{1}{2}\right)^2 - 1\right)\frac{1}{k} + \left(a+\frac{1}{2}\right)\left(1-\ln\left(a+\frac{1}{2}\right)\right) \\ & + \left(b+\frac{1}{2}\right)\left(1-\ln\left(b+\frac{1}{2}\right)\right) \\ & \left. - \left(c+\frac{1}{2}\right)\left(1-\ln\left(c+\frac{1}{2}\right)\right) + \theta_3\right],\end{aligned} \tag{2.17}$$

where

$$\theta_3 = \theta_3(k) = \theta_2 + \varepsilon_1\left(\frac{a+\frac{1}{2}}{k}\right)^2 + \varepsilon_2\left(\frac{b+\frac{1}{2}}{k}\right)^2 + \varepsilon_3\left(\frac{c+\frac{1}{2}}{k}\right)^2 + \varepsilon_4\frac{1}{k^2}.$$

From (2.12)–(2.16) we have

$$|\theta_3| \leq 2.$$

We next rewrite (2.17) in the form

$$\varphi(k) = \frac{ab}{c}\eta_1(a,b,c,k)\eta_2(a,b,c,k)\eta_3(a,b,c)k^{a+b-c-1}, \tag{2.18}$$

where

$$\eta_1(a,b,c,k) = e^{\theta_3} \leq e^2 = \eta_1,$$

$$\begin{aligned}\eta_2(a,b,c,k) &= \exp\left[\left(\left(a+\frac{1}{2}\right)^2 + \left(b+\frac{1}{2}\right)^2 - \left(c+\frac{1}{2}\right)^2 - 1\right)\frac{1}{k}\right] \\ &\leq \exp\left[\frac{1}{2}(a+b+1)\right] = \eta_2, \\ \eta_3(a,b,c) &= \exp\left[\left(a+\frac{1}{2}\right)\left(1-\ln\left(a+\frac{1}{2}\right)\right)\right. \\ &\qquad + \left(b+\frac{1}{2}\right)\left(1-\ln\left(b+\frac{1}{2}\right)\right) \\ &\qquad \left. - \left(c+\frac{1}{2}\right)\left(1-\ln\left(c+\frac{1}{2}\right)\right)\right] = \eta_3.\end{aligned} \tag{2.19}$$

From (2.4)–(2.6) and (2.18)–(2.19) we have

$$\delta F_r(a,b,c,x) = \sum_{k\geq r}\varphi(k)x^{k+1} \leq \frac{ab}{c}\eta_1\eta_2\eta_3\sum_{k\geq r}k^{a+b-c-1}x^{k+1}. \tag{2.20}$$

We will assume that

$$r \geq 3, \tag{2.21}$$

and let r be an integer such that

$$\frac{r}{\ln r} \geq 2\frac{a+b-c-1}{\ln \frac{1}{x}}. \tag{2.22}$$

Then for $k \geq r$ the function $f(k) = k^{a+b-c-1}x^{k/2}$ decreases monotonically. Therefore

$$k^{a+b-c-1}x^{k/2} \leq r^{a+b-c-1}x^{r/2} < 1,$$

$$\sum_{k\geq r} k^{a+b-c-1}x^{k+1} \leq \sum_{k\geq r} x^{k/2+1} = \frac{x^{r/2+1}}{1-\sqrt{x}}. \tag{2.23}$$

From (2.14), (2.20)–(2.23) we obtain

$$\delta F_r(a,b,c,x) < \frac{ab}{c}\eta_1\eta_2\eta_3\frac{x^{r/2+1}}{1-\sqrt{x}},$$

where η_1, η_2, η_3 are defined by (2.19), r satisfies conditions (2.14), (2.21), (2.22), $a > 0$, $b > 0$, $c > 0$, $0 < x < 1$. Let r be the least integer so that

$$\frac{ab}{c}\eta_1\eta_2\eta_3\frac{x^{r/2+1}}{1-\sqrt{x}} \leq 2^{-n}.$$

It is clear that for fixed a, b, c, x the value of r has order $O(n)$ and then

$$|\delta F_r(a,b,c,x)| < 2^{-n}. \tag{2.24}$$

Now we apply directly the algorithm. We represent (2.3) in the form

$$F_r(a,b,c,x) = 1 + S, \tag{2.25}$$

where

$$S = \sum_{j=1}^{r} \frac{a(a+1)\cdots(a+j-1)b(b+1)\cdots(b+j-1)}{j!c(c+1)\cdots(c+j-1)}x^j. \tag{2.26}$$

First we consider the case for rational numbers a, b, c, x: $x = \frac{p}{q}$, $(p,q) = 1$, $0 < p < q$, $a = \frac{a_1}{a_2}$, $(a_1,a_2) = 1$, $b = \frac{b_1}{b_2}$, $(b_1,b_2) = 1$, $c = \frac{c_1}{c_2}$, $(c_1,c_2) = 1$. We take

$$r = i + 1 = 2^m + 1, \qquad m \geq 1, \tag{2.27}$$

terms of the series (2.25) and let the numbers $S_{i-\nu}(0)$, $\nu = 0, 1, \ldots, i-1$, satisfy the equalities

$$S_{i-\nu}(0) = \frac{a(a+1)\ldots(a+i-1-\nu)b(b+1)\cdots(b+i-1-\nu)}{(i-\nu)!c(c+1)\cdots(c+i-1-\nu)}x^{i-\nu}.$$

By the definition of S from (2.26), we have

$$S = S_1(0) + S_2(0) + \ldots + S_i(0).$$

We perform the computation of S in m steps of the FEE process, namely, combining the summands of S in each step successively in pairs and taking the "obvious" common multiplier out of the parentheses, we compute the integer value of the expression in the parentheses.

At the first step we have:

$$S = S_1(1) + S_2(1) + \ldots + S_{i_1-1}(1) + S_{i_1}(1), \quad i_1 = 2^{-1}i,$$

where for $\nu = 0, 1, 2, \ldots, i_1 - 1$,

$$\begin{aligned} S_{i_1-\nu}(1) = \;& S_{i-2\nu}(0) + S_{i-(2\nu+1)}(0) \\ = \;& \frac{a_1(a_1+a_2)\cdots(a_1+a_2(i-2\nu-1))b_1(b_1+b_2)\cdots(b_1+b_2(i-2\nu-1))}{(i-2\nu)!c_1(c_1+c_2)\cdots(c_1+c_2(i-2\nu-1))}\left(\frac{pc_2}{qa_2b_2}\right)^{i-2\nu} \\ & + \frac{a_1(a_1+a_2)\cdots(a_1+a_2(i-(2\nu+1)-1))b_1(b_1+b_2)\cdots(b_1+b_2(i-(2\nu+1)-1))}{(i-(2\nu+1))!c_1(c_1+c_2)\cdots(c_1+c_2(i-(2\nu+1)-1))} \\ & \cdot\left(\frac{pc_2}{qa_2b_2}\right)^{i-(2\nu+1)} \\ = \;& \frac{a_1(a_1+a_2)\cdots(a_1+a_2(i-2\nu-2))b_1(b_1+b_2)\cdots(b_1+b_2(i-2\nu-2))}{(i-2\nu)!c_1(c_1+c_2)\cdots(c_1+c_2(i-2\nu-1))} \\ & \cdot\frac{(pc_2)^{i-2\nu-1}}{(qa_2b_2)^{i-2\nu}}\alpha_{i_1-\nu}(1). \end{aligned}$$

At the first step the integers $\alpha_{i_1-\nu}(1)$, $\nu = 0, 1, 2, \ldots, i_1 - 1$ are calculated as follows:

$$\begin{aligned} \alpha_{i_1-\nu}(1) = \;& (a_1 + a_2(i-2\nu-1))(b_1 + b_2(i-2\nu-1))pc_2 \\ & +(i-2\nu)(c_1 + c_2(i-2\nu-1))qa_2b_2 \end{aligned} \tag{2.28}$$

After j steps of the process we obtain in the $j+1$-th step, $j+1 \le m$,

$$S = S_1(j+1) + S_2(j+1) + \ldots + S_{i_{j+1}-1}(j+1) + S_{i_{j+1}}(j+1),$$

where

$$i_{j+1} = 2^{-1}i_j = 2^{-(j+1)}i,$$

and where for $\nu = 0, 1, 2, \ldots, i_{j+1} - 1$,

$$\begin{aligned}
S_{i_{j+1}-\nu}(j+1) = &\ S_{i_j-2\nu}(j) + S_{i_j-(2\nu+1)}(j) \\
= &\ \frac{a_1(a_1+a_2)\cdots(a_1+a_2(i-2^j2\nu-2^j))b_1\cdots(b_1+b_2(i-2^j2\nu-2^j))}{(i-2^j2\nu)!c_1(c_1+c_2)\cdots(c_1+c_2(i-2^j2\nu-1))} \\
&\cdot\frac{(pc_2)^{i-2^j2\nu-2^j+1}}{(qa_2b_2)^{i-2^j2\nu}}\alpha_{i_j-2\nu}(j)+ \\
&\cdot\frac{a_1\cdots(a_1+a_2(i-2^j(2\nu+1)-2^j))b_1(b_1+b_2)\cdots(b_1+b_2(i-2^j(2\nu+1)-2^j))}{(i-2^j(2\nu+1))!c_1(c_1+c_2)\cdots(c_1+c_2(i-2^j(2\nu+1)-1))} \\
&\cdot\frac{(pc_2)^{i-2^j(2\nu+1)-2^j+1}}{(qa_2b_2)^{i-2^j(2\nu+1)}}\alpha_{i_j-(2\nu+1)}(j) \\
= &\ \frac{a_1(a_1+a_2)\cdots(a_1+a_2(i-2^{j+1}\nu-2^{j+1}))b_1\cdots(b_1+b_2(i-2^{j+1}\nu-2^{j+1}))}{(i-2^{j+1}\nu)!c_1(c_1+c_2)\cdots(c_1+c_2(i-2^{j+1}\nu-1))} \\
&\cdot\frac{(pc_2)^{i-2^{j+1}\nu-2^{j+1}+1}}{(qa_2b_2)^{i-2^{j+1}\nu}}\alpha_{i_{j+1}-\nu}(j+1).
\end{aligned}$$

At the $(j+1)$-th step the integers $\alpha_{i_{j+1}-\nu}(j+1)$, $\nu = 0, 1, 2, \ldots, i_{j+1} - 1$, $i_{j+1} = 2^{-(j+1)}i$ are calculated as follows:

$$\begin{aligned}
&\alpha_{i_{j+1}-\nu}(j+1) \\
&\quad = (a_1 + a_2(i - 2^{j+1}\nu - 2^j))(a_1 + a_2(i - 2^{j+1}\nu - 2^j - 1)) \\
&\qquad \cdots (a_1 + a_2(i - 2^{j+1}\nu - 2^{j+1} + 1)) \\
&\qquad (b_1 + b_2(i - 2^{j+1}\nu - 2^j))(b_1 + b_2(i - 2^{j+1}\nu - 2^j - 1)) \\
&\qquad \cdots (b_1 + b_2(i - 2^{j+1}\nu - 2^{j+1} + 1))(pc_2)^{2^j}\alpha_{i_j-2\nu}(j) \\
&\qquad +(i - 2^{j+1}\nu)(i - 2^{j+1}\nu - 1)\ldots(i - 2^{j+1}\nu - 2^j + 1) \\
&\qquad (c_1 + c_2(i - 2^{j+1}\nu - 1))(c_1 + c_2(i - 2^{j+1}\nu - 2)) \\
&\qquad \cdots (c_1 + c_2(i - 2^{j+1}\nu - 2^j))(qa_2b_2)^{2^j}\alpha_{i_j-(2\nu+1)}(j).
\end{aligned} \tag{2.29}$$

And so on. After m steps of the process we obtain the sum S,

$$\begin{aligned}
&S = S_1(m) = \frac{pc_2a_1b_1\alpha_{i_m}(m)}{(qa_2b_2)^i i! c_1(c_1+c_2)\ldots(c_1+c_2(i-1))}, \\
&i_m = 2^{-m}i = 1,
\end{aligned} \tag{2.30}$$

by dividing with accuracy 2^{-n} the integer numerator of the fraction (2.30) by the integer denominator of this fraction.

Now we count the number of operations, that is sufficient for the computation of the integers $\alpha_{i_{j+1}-\nu}(j+1)$, $\nu = 0, 1, 2, \ldots, i_{j+1}-1$, $i_{j+1} = 2^{-(j+1)}i$, at the $(j+1)$-th step, assuming that the integers $\alpha_\mu(j)$ have already been calculated. Let $\alpha(j) = \max_\mu \alpha_\mu(j)$. The order of the integers participating in the calculations at the $(j+1)$-th step can be estimated from (2.28), (2.29) and from (2.13), (2.14), (2.21) as follows:

$$\alpha(j+1) \leq (Bi)^{2^{j+2}}, \tag{2.31}$$

where $B = 2pqa_1a_2b_1b_2c_1c_2 = const.$
In order to compute the product

$$\prod_{i,\nu}(j+1) = (i - 2^{j+1}\nu)(i - 2^{j+1}\nu - 1)\cdots(i - 2^{j+1}\nu - 2^j + 1)$$

for $\nu = 0, 1, 2, \ldots, i_{j+1} - 1$, $i_{j+1} = 2^{-(j+1)}i$, (we assume, that the products $\prod_\mu(l)$, $l = 1, 2, \ldots, j$, have already been computed) we represent it in the form

$$\prod_{i,\nu}(j+1) = \prod_{2^1\nu}(j) \prod_{2^2\nu+2-1}(j-1)\cdots \prod_{2^{j-1}\nu+2^{j-2}-1}(2) \prod_{2^j\nu+2^{j-1}-1}(1)(i - 2^{j+1}\nu - 2^j + 1)$$

and multiply the factors in the right-hand side of this identity consecutively, starting with the last position. The total number of operations sufficient for evaluating $\prod_{i,\nu}(j+1)$ is

$$O\left(\sum_{k=1}^{j} M(2^k \log i)\right). \tag{2.32}$$

The complexity of the calculations of the products

$$\prod_{c,\nu} = (c_1 + c_2(i - 2^{j+1}\nu - 1))\cdots(c_1 + c_2(i - 2^{j+1}\nu - 2^j))$$

and

$$\prod_{d,\nu} = (d_1 + d_2(i - 2^{j+1}\nu - 2^j))\cdots(d_1 + d_2(i - 2^{j+1}\nu - 2^{j+1} + 1)),$$

$d = a, b$ is estimated analogously By (2.29), (2.31) and (2.32) for the computation of $\alpha_{i_{j+1}-\nu}(j+1)$,

$$K(j+1) = O\left(\sum_{k=1}^{j} M(2^k \log i) + M(2^{j+1} \log i)\right) \tag{2.33}$$

operations are sufficient. For the calculation of $\alpha_1(m) = \alpha_{r_m}(m)$,

$$K = \sum_{j=0}^{m-1} i_{j+1} K(j+1) \tag{2.34}$$

operations are sufficient. From (2.33), (2.34) and (2.30) we obtain that for the calculation of the sum S

$$O(M(i)\log^2 i) \tag{2.35}$$

operations are sufficient. Then by (2.2), (2.24), (2.25), (2.27), (2.35) the complexity of the evaluation of the hypergeometric function with the accuracy 2^{-n} for rational positive parameters a, b, c and rational argument x, $0 < x < 1$, is

$$O(M(n)\log^2 n). \tag{2.36}$$

If any of the numbers x, a, b, c is negative or $x = 1$ and $a + b < c$, then both the estimation of the expression (2.4) and the calculation of the expression (2.3) are conducted just the same way. If the parameters a, b, c and the argument $x = z$ are algebraic numbers, then the algorithm is modified.

Let λ_k, $k = 1, 2, 3, 4$, be the degree of the polynomial with integer coefficients, whose root is a, b, c, z, $0 < z < 1$,

$$\lambda_k \geq 2, \qquad k = 1, 2, 3, 4. \tag{2.37}$$

At the $(j+1)$-th step, $j + 1 \leq m$, $2^j < \lambda_k$, $k = 1, 2, 3, 4$, in the formula for $\alpha_{i_{j+1}-\nu}(j+1)$ we have

$$\begin{aligned}\alpha_{i_{j+1}-\nu}(j+1) \;&=\; (a+i-2^{j+1}\nu-2^j)\cdots(a+i-2^{j+1}\nu-2^{j+1}+1)\\ &\cdot(b+i-2^{j+1}\nu-2^j)\cdots(b+i-2^{j+1}\nu-2^{j+1}+1)x^{2^j}\alpha_{i_j-2\nu}(j)\\ &+(i-2^{j+1}\nu)(i-2^{j+1}\nu-1)\cdots(i-2^{j+1}\nu-2^j+1)\\ &\cdot(c+i-2^{j+1}\nu-1)\cdots(c+i-2^{j+1}\nu-2^j)\alpha_{i_j-(2\nu+1)}(j),\end{aligned}$$

the integer coefficients of the powers of a, b, c, x in the polynomials

$$\begin{aligned}A(a) &= (a+i-2^{j+1}\nu-2^j)\cdots(a+i-2^{j+1}\nu-2^{j+1}+1),\\ B(b) &= (b+i-2^{j+1}\nu-2^j)\cdots(b+i-2^{j+1}\nu-2^{j+1}+1),\\ C(c) &= (c+i-2^{j+1}\nu-1)\cdots(c+i-2^{j+1}\nu-2^j),\\ X(x) &= x^{2^j},\end{aligned} \tag{2.38}$$

are calculated. Let the number j_k, $k = 1, 2, 3, 4$, be defined by the inequality $2^{j_k - 1} < \lambda_k \leq 2^{j_k}$. By (2.37) $j_k \geq 1$. After the j_k-th step of the FEE-algorithm one of the polynomials (2.38) becomes a polynomial of degree 2^{j_k} with integer coefficients of the number a, b, c, or z which is the root of the polynomial of degree λ_k. Before the $(j_k + 1)$-th step, we reduce this polynomial modulo polynomial of degree λ_k with integer coefficients, whose root is the number associated with index k. This process is different from the process which is described in [6], [8] only in the need to keep track of the increase in the degrees of four polynomials (2.38) simultaneously and to reduce each of them separately.

To obtain then the polynomial with integer coefficients, we multiply – if necessary – the remainder polynomial by the corresponding integer coefficient. Since the numbers λ_k, $k = 1, 2, 3, 4$, and the coefficients in the polynomials of degree λ_k are absolute constants, then with these reductions the values of the coefficients used in the computation may increase only a fixed number of times. Then the complexity (2.36) is also valid in the case of an algebraic argument z and algebraic parameters a, b, c. The theorem is proved.

Acknowledgement. The author wishes to acknowledge Edward Saff and Peter Borwein for their attention and support when preparing the paper.

References

[1] Benderskij, Yu. V., Fast computations, *Doklady Akad. Nauk SSSR* **223** (1975), 1041–1043.

[2] Borwein, J. M. and P. B. Borwein, *π and AGM*, Wiley, New York, 1987.

[3] Brent, R. P., Fast multiple-precision evaluation of elementary functions, *Journal ACM* **23 (2)** (1976), 242–251.

[4] Karatsuba A. A. and Yu. P. Ofman, Multiplication of multiplace numbers by automata, *Dokl. Akad. Nauk SSSR* **145 (2)** (1962), 293–294.

[5] Karatsuba, A. A., *Elements of Analytic Number Theory* (in Russian), Nauka, Moscow, 1983.

[6] Karatsuba, E. A., Fast evaluations of transcendental functions, *Probl. Peredachi Inf.* **27 (4)** (1991), 87–110.

[7] Karatsuba, E. A., Fast calculation of $\zeta(3)$, *Problems of Inform. Transmission* **29 (1)** (1993), 58–62.

[8] Karatsuba, E. A., Fast evaluation of Bessel functions, *Integral Transforms and Special Functions* **1 (4)** (1993), 269–276.

[9] Karatsuba, E. A., Fast calculation of the Riemann zeta function $\zeta(s)$ for integer values of the argument s, *Problems of Information Transmission* **31 (4)** (1995), 353–362.

[10] Schönhage, A., A. F. Grotefeld and E. Vetter, *Fast Algorithms: A Multitape Turing Machine Implementation*, BI-Wiss.-Verl., Mannheim-Zürich, 1994.

[11] Schönhage, A. and V. Strassen, Schnelle Multiplikation großer Zahlen, *Computing* **7** (1971), 281–292.

[12] Wittaker, E. T. and J. N. Watson, *A Course of Modern Analysis* (in Russian), Fizmatgiz, Moscow, 1963.

Ekatharine A. Karatsuba
Computer Centre of RAS
Vavilova 40
117967 Moscow
RUSSIA
ekar@ext1.ccas.ru

On the complex Rolle set of a polynomial

Emmanuel S. Katsoprinakis

Abstract. Let $p(z)$ be a complex polynomial of degree $n \geq 2$ with zeros z_j and let w_k be the zeros of its derivative. If $z_1 + \cdots + z_n = 0$, then I. J. Schoenberg called the set $E = \{z_j, w_k\}$ the "complex Rolle set" of $p(z)$ and conjectured that:

$$|w_1|^2 + \cdots + |w_{n-1}|^2 \leq \frac{n-2}{n}(|z_1|^2 + \cdots + |z_n|^2).$$

Here we verify special cases of this conjecture, using a generalization of Van den Berg's Theorem and the Theory of Majorization. Moreover, we arrive at a new conjecture, which implies the above conjecture and also another Schoenberg-type conjecture proposed by M. G. de Bruin and A. Sharma.

§1 Introduction

In this article we assume that $n \geq 2$ and let

$$p(z) := z^n + c_{n-1}z^{n-1} + \cdots + c_0 = \prod_{j=1}^{n}(z - z_j),$$

be a polynomial with complex coefficients. We also consider its derivative $p'(z)$ and let

$$p'(z) = n\prod_{k=1}^{n-1}(z - w_k).$$

If $z_j = x_j + iy_j$ and $w_k = a_k + ib_k$, then we denote by $p^*(z)$ the real polynomial

$$p^*(z) = \prod_{j=1}^{n}(z - x_j)$$

and by a_k^*, $k = 1, 2, ..., n-1$, the zeros of its derivative. We further assume that

$$a_1 \leq a_2 \leq \cdots \leq a_{n-1} \quad \text{and} \quad a_1^* \leq a_2^* \leq \cdots \leq a_{n-1}^*.$$

If $c_{n-1} = 0$ (*i.e.* $z_1 + \ldots + z_n = w_1 + \ldots + w_{n-1} = 0$), then in [6] I. J. Schoenberg called the set $E := \{z_j, w_k : j = 1, 2, \ldots, n;\ k = 1, 2, \ldots, n-1\}$ the *complex Rolle set of* $p(z)$ and proved, among other things, that:

Computational Methods and Function Theory (CMFT'97)
N. Papamichael, St. Ruscheweyh and E. B. Saff (Eds.), pp. 315–326

ISBN 981-02-3626-3

(i) If we turn the set E about 0 by an angle ϑ, the new set $E(\vartheta) = \{z_j e^{i\vartheta}, w_k e^{i\vartheta}\}$ is also a complex Rolle set, namely of $e^{in\vartheta}p(ze^{-i\vartheta})$.

(ii) $w_1^2 + \ldots + w_{n-1}^2 = \frac{n-2}{n}(z_1^2 + \ldots + z_n^2)$.

(iii) If all w_k are on a line, then,

$$|w_1|^2 + \ldots + |w_{n-1}|^2 \leq \frac{n-2}{n}(|z_1|^2 + \ldots + |z_n|^2),$$

with the equality sign if and only if all z_j are collinear.

Further, he conjectured that the conclusion of (iii) is true for any complex Rolle set, and verified it in the cases of binomial polynomials and of polynomials of degree $n = 3$. Thus, we have the following conjecture:

Conjecture 1. *For any complex Rolle set $E = \{z_j, w_k\}$ we have the inequality*

$$\sum_{k=1}^{n-1} |w_k|^2 \leq \frac{n-2}{n} \sum_{j=1}^{n} |z_j|^2,$$

with the equality sign if and only if all z_j are collinear.

Remark 1: We observe that in order to admit all polynomials $p(z)$ (*i.e.* without the restriction $c_{n-1} = 0$) we can translate Conjecture 1 using $z_j - G$ and $w_k - G$, instead of z_j and w_k respectively, where G is the common centroid of z_j and w_k, *i.e.*

$$G = \frac{1}{n-1} \sum_{k=1}^{n-1} w_k = \frac{1}{n} \sum_{j=1}^{n} z_j,$$

which we may assume real. This, by a simple calculation, gives

$$\sum_{k=1}^{n-1} |w_k|^2 \leq |G|^2 + \frac{n-2}{n} \sum_{j=1}^{n} |z_j|^2,$$

and so, instead of Conjecture 1, we have the following:

Conjecture 2. *For any polynomial $p(z)$ of degree $n \geq 2$ we have the inequality*

$$\sum_{k=1}^{n-1} |w_k|^2 \leq |G|^2 + \frac{n-2}{n} \sum_{j=1}^{n} |z_j|^2,$$

with the equality sign if and only if all z_j lie on a straight line.

During our investigation of Conjecture 1 we arrived at a new conjecture (Conjecture 3 below) which, as we shall see, implies, among other things, the above conjectures.

At first we give the following definition:

Definition 1. *With our notation, we say that the polynomial $p(z)$ fulfills the majorization condition if the vector $(a_1, a_2, \dots, a_{n-1})$ is majorized by $(a_1^*, a_2^*, \dots, a_{n-1}^*)$. In symbols:*

$$(a_1, a_2, \dots, a_{n-1}) \prec (a_1^*, a_2^*, \dots, a_{n-1}^*).$$

This means, according to one of the characterizations of Majorization (cf. [5]), that

$$\sum_{k=1}^{m} a_k^* \leq \sum_{k=1}^{m} a_k, \qquad m = 1, 2, \dots, n-2,$$
$$\sum_{k=1}^{n-1} a_k^* = \sum_{k=1}^{n-1} a_k.$$

Conjecture 3. *Any polynomial $p(z)$ fulfills the majorization condition.*

Roughly speaking, we conjecture that the real parts of the zeros of the derivative of $p(z)$ are *less spread out* or *more nearly equal* than are the zeros of the derivative of the real polynomial $p^*(z)$. Further, one can say, that, in some sense, this generalizes the Root Dragging Theorem in the complex case (see *Amer. Math. Monthly* **100**, 864–866), but we do not discuss this subject here.

On the other hand, J. Szynal informed me (during the CMFT'97 Conference), that there are some articles on the above and on similar, Schoenberg-type, conjectures and he sent me one of these. In that article (see [2]) M. G. de Bruin and A. Sharma considered a polynomial $p(z)$ with $c_{n-1} = 0$ and defined the *quadratic mean radius* and *quartic mean radius* of $p(z)$ by the relations:

$$R(p) := \left(\frac{1}{n} \sum_{j=1}^{n} |z_j|^2 \right)^{1/2}, \qquad S(p) := \left(\frac{1}{n} \sum_{j=1}^{n} |z_j|^4 \right)^{1/4},$$

respectively. Further, they conjectured that:

Conjecture 4. *For any complex Rolle set $E = \{z_j, w_k\}$ we have the inequality*

$$\sum_{k=1}^{n-1} |w_k|^4 \leq \frac{n-4}{n} \sum_{j=1}^{n} |z_j|^4 + \frac{2}{n^2} \left(\sum_{j=1}^{n} |z_j|^2 \right)^2 ,$$

or equivalently

$$\frac{n-4}{n-1} S(p)^4 + \frac{2}{n-1} R(p)^4 \geq S(p')^4,$$

with the equality sign if and only if all z_j are collinear.

In §2 the main results concerning these conjectures will be given, followed in §3, §4 and §5 by their proofs and some further remarks and comments.

§2 Main results

First, for our Conjecture 3 we shall prove the following:

Proposition 1.

(a) *Conjecture 3 is true for polynomials with at most 4 distinct zeros.*

(b) *If $p(z)$ is a polynomial of degree n with at most 5 distinct zeros then there is some angle ϑ such that the polynomial $q(z) = e^{in\vartheta} p(ze^{-i\vartheta})$ fulfills the majorization condition.*

Next, for the other conjectures, the following propositions contain some known and some new results:

Proposition 2. *Conjecture 1 is true in the following cases:*

(a) *For all polynomials with w_k on a line.*

(b) *Degree $p(z) = 3$.*

(c) *For all polynomials of the form $z^n + c_{n-k} z^{n-k}$, $2 \leq k \leq n$.*

(d) *For all polynomials with $|z_1| = |z_2| = \cdots = |z_n|$.*

(e) *For all polynomials with zeros opposite in pairs.*

(f) *For all polynomials of the form $(z^s - a^s)^k (z^s - b^s)^m$, $s \geq 2$, $k, m \geq 1$.*

(g) *For any polynomial $p(z)$ for which there is some angle ϑ such that the polynomial $q(z) = e^{in\vartheta} p(ze^{-i\vartheta})$ fulfills the majorization condition.*

(h) *For all polynomials with at most 5 distinct zeros.*

Proposition 3. *Conjecture 4 is true in the following cases:*

(a) *For all polynomials with w_k on a line.*

(b) *Degree $p(z) = 3$.*

(c) *For all polynomials of the form $z^n(z^3 - a^3)^m$, $m, n \geq 1$.*

(d) *For all polynomials of the form $(z^s - a^s)^k(z^s - b^s)^m$, $s \geq 2$, $k, m \geq 1$.*

(e) *For all polynomials of the form $(z^2 + 1)^{m+1} - z^{2m+2}$, $m \geq 1$.*

(f) *For all polynomials with $|z_1| = |z_2| = \cdots = |z_n|$.*

(g) *For all polynomials $p(z)$ with the property that the majorization condition is fulfilled by all the rotations $q(z) = a^{-n}p(az)$ $(|a| = 1)$.*

(h) *For all polynomials with at most 4 distinct zeros.*

§3 Proof of Proposition 1

We have not verified Conjecture 3 in general, but, as we shall see, in the cases (a) and (b) of Proposition 1, it is an immediate consequence of an old theorem (see Theorem 2 below).

More precisely, in [6] Schoenberg noticed that, although the case (b) of Proposition 2 is a corollary of the case (a) (see also (iii) in §1), there is still, in that case (degree $n = 3$), another direct proof derived easily by (as named by Schoenberg) Van den Berg's Theorem, which states that:

Theorem 1. *Let C be the Steiner ellipse of the triangle $T = (z_1, z_2, z_3)$, where $z_1 \neq z_2 \neq z_3 \neq z_1$ are the zeros of $p(z)$, which is of degree $n = 3$. Then the zeros of $p'(z)$ are identical with the foci w_1 and w_2 of C.*

There are several generalizations of Theorem 1 (one of which is due to Van den Berg). See Theorem (4.2) in the classical book [4]. In [3] B. Z. Linfield proved the following Theorem, which we shall need here:

Theorem 2. *For any polynomial $p(z)$ we consider the x-intercepts of lines through z_j parallel to any direction as the zeros of a new real polynomial $q(z)$. Then, those w_k which are distinct from z_j lie between the two lines parallel to the same direction and passing through the least and the greatest zero of the derivative $q'(z)$.*

This, in our notation, simply says that

$$a_1^* \leq a_1 \qquad \text{and} \qquad a_{n-1} \leq a_{n-1}^*, \tag{3.1}$$

which is exactly what Schoenberg proved in the case $n = 3$.

Now, let $p(z)$ be of degree $n = 4$ and assume that is has 4 distinct zeros. By (3.1) and the fact that the centroids of a_j and a_j^* coincide we have:

$$a_1^* \le a_1, \quad a_1^* + a_2^* \le a_1 + a_2, \quad a_1^* + a_2^* + a_3^* = a_1 + a_2 + a_3.$$

Thus,

$$(a_1, a_2, a_3) \prec (a_1^*, a_2^*, a_3^*).$$

Furthermore, if $p(z)$ has 4 distinct zeros and is of arbitrary degree n, then in each member of the above relations we have to add the same quantity every time we pass through a multiple root of $p(z)$. Thus we conclude again that

$$(a_1, a_2, \ldots, a_{n-1}) \prec (a_1^*, a_2^*, \ldots, a_{n-1}^*).$$

Since the cases of at most 2 or 3 distinct zeros are more trivial, we see that Theorem 2 yields at once the case (a) of Proposition 1.

Similarly, in order to prove case (b) it is evident from the above discussion that it suffices to prove it when the degree $n = 5$ and $p(z)$ has 5 distinct zeros. Then we have to show that there is some angle ϑ such that the polynomial $q(z) = e^{in\vartheta} p(ze^{-i\vartheta})$ fulfills the majorization condition. This means that we must find a direction, for which, if we consider it as the real axis, we have that:

$$\begin{aligned} a_1^* &\le a_1, \\ a_1^* + a_2^* &\le a_1 + a_2, \\ a_1^* + a_2^* + a_3^* &\le a_1 + a_2 + a_3. \end{aligned}$$

and

$$a_1^* + a_2^* + a_3^* + a_4^* = a_1 + a_2 + a_3 + a_4.$$

In order to find this direction we consider two cases: First, if all w_k lie on a straight line L, we choose the desired direction to be perpendicular to L, so that $a_1 = \ldots = a_4 = 0$. In the other case, there is obviously a line L passing through, say, w_2 and w_3, which strictly separates w_1 from w_4. Choosing as above the desired direction to be perpendicular to L, we have $a_1 < a_2 = a_3 < a_4$. Thus, in all cases, because of (3.1) and the equation

$$a_1^* + a_2^* + a_3^* + a_4^* = a_1 + a_2 + a_3 + a_4$$

we have the above listed relations except of the second inequality $a_1^* + a_2^* \le a_1 + a_2$. But, if $a_1^* + a_2^* > a_1 + a_2$, then we obtain $a_2^* > a_2$ and $a_3^* < a_3$. Thus, $a_2 < a_2^* \le a_3^* < a_3$, while in all cases we have $a_2 = a_3$, which is a contradiction. Thus the proof of Proposition 1 is complete. ■

§4 Proof of Proposition 2

As we have mentioned before, the proofs of the cases (a), (b) and (c) are contained in [6].

case (d): We need the following lemma:

Lemma 1. *If the zeros z_j, $j = 1, 2, \ldots, n$, of a polynomial $p(z)$ lie on a circle centered at the centroid G of the z_j, then $p'(G) = 0$.*

Proof: We have $p(G) \neq 0$ and

$$\frac{p'(G)}{p(G)} = \sum_{j=1}^{n} \frac{1}{G - z_j} = \sum_{j=1}^{n} \frac{\overline{G - z_j}}{|G - z_j|^2} = 0. \quad \blacksquare$$

Now, in our case (d), we have that $G = 0$ and so, according to Lemma 1, at least one w_k is zero. Say $w_{n-1} = 0$. Then, using in addition the Gauss-Lucas Theorem, we have

$$\sum_{k=1}^{n-1} |w_k|^2 = \sum_{k=1}^{n-2} |w_k|^2 \leq (n-2)|z_1|^2 = \frac{n-2}{n} \sum_{j=1}^{n} |z_j|^2.$$

Moreover, equality can only arise when there is only one w_k distinct from z_j (in our case $w_{n-1} = 0$) and, obviously, this can happen only when z_j lie on a line. $\blacksquare$

case (e): Here we need the following theorem (see [1]):

Theorem 3. *For any polynomial $p(z)$ of degree $n \geq 2$ we have, if $m \geq 1$,*

$$\sum_{k=1}^{n-1} |w_k|^m \leq \frac{n-1}{n} \sum_{j=1}^{n} |z_j|^m.$$

There is equality in the following two cases only: (a) if $m = 1$ and all z_j lie on the same half line with endpoint 0, and (b) if $z_1 = z_2 = \ldots = z_n$.

Remark 2: We notice that if we consider the second term in the inequality in Conjecture 2, then, by the Cauchy-Schwarz inequality we obtain that

$$|G|^2 + \frac{n-2}{n} \sum_{j=1}^{n} |z_j|^2 \leq \frac{n-1}{n} \sum_{j=1}^{n} |z_j|^2,$$

which means that Conjecture 2 (or equivalently Conjecture 1) implies the case $m = 2$ of Theorem 3.

We return now to our case (e) and consider a polynomial of the form

$$p(z) = \prod_{j=1}^{r} (z^s - v_j^s), \quad rs = n.$$

The derivative of $p(z)$ is of the form

$$p'(z) = nz^{s-1} \prod_{k=1}^{r-1} (z^s - u_k^s)$$

and, using the transformation $z^s = t$ and Theorem 3, we obtain that

$$\sum_{k=1}^{r-1} |u_k|^{sm} \leq \frac{r-1}{r} \sum_{j=1}^{r} |v_j|^{sm}, \quad m \geq 1.$$

Since the zeros of $p(z)$ are of the form $v_j \exp(2i\pi q/s)$ and those of $p'(z)$ are either 0 or of the form $u_k \exp(2i\pi q/s)$, $q = 0, 1, \ldots, s-1$, the last inequality gives that

$$\sum_{k=1}^{n-1} |w_k|^{sm} \leq \frac{n-s}{n} \sum_{j=1}^{n} |z_j|^{sm}, \quad m \geq 1,$$

where z_j and w_k denote the zeros of $p(z)$ and $p'(z)$ respectively.

Observe now that if $s = 2$ and $m = 1$, then the last inequality is exactly what we wanted to show. There is equality if and only if all z_j lie on a line according to (a) and (b) of Theorem 3. Furthermore, we note that for $m \geq 1$, we have $sm \geq 2$ and the following more general inequality holds in this case

$$\sum_{k=1}^{n-1} |w_k|^{\mu} \leq \frac{n-2}{n} \sum_{j=1}^{n} |z_j|^{\mu}, \quad \mu \geq 2.$$

Moreover, if $s \geq 2$ we have some new inequalities and it is interesting to consider under what conditions these are true in general. ∎

case (f): The proof of this case is very similar to the proof of case (c) in Theorem 2 in [2] (see Proposition 3(d)) and so we do not present it here. ∎

case (g): We assume that there is some angle ϑ such that the polynomial $q(z) = e^{in\vartheta} p(ze^{-i\vartheta})$ fulfills the majorization condition. Since the zeros of $p(z)$ are $z_j e^{i\vartheta}$ and those of $p'(z)$ are $w_k e^{i\vartheta}$, it suffices to prove Conjecture 1 for the polynomial $q(z)$. Moreover, we may assume that $\vartheta = 0$. Then by (ii) in §1, we easily have that

$$A := \sum_{k=1}^{n-1} a_k^2 - \frac{n-2}{n}\sum_{j=1}^{n} x_j^2 = \sum_{k=1}^{n-1} b_k^2 - \frac{n-2}{n}\sum_{j=1}^{n} y_j^2,$$

and thus

$$\sum_{k=1}^{n-1} |w_k|^2 - \frac{n-2}{n}\sum_{j=1}^{n} |z_j|^2 = 2A.$$

Moreover, by (iii) in §1, (and, essentially, by the case of equality) we also have

$$\sum_{k=1}^{n-1} (a_k^*)^2 = \frac{n-2}{n}\sum_{j=1}^{n} x_j^2.$$

This gives that

$$A = \sum_{k=1}^{n-1} a_k^2 - \sum_{k=1}^{n-1} (a_k^*)^2.$$

Now, we have to show that $A \leq 0$. Indeed, we know, by the Theory of Majorization [5], Chapter 3, Sections A and C, that for any Schur-convex function (*i.e.* any real function preserving the ordering of majorization) and in particular any real convex function (for example, in our case, the function $|t|^m$, $m \geq 1$), we have

$$\sum_{k=1}^{n-1} |a_k|^m \leq \sum_{k=1}^{n-1} |a_k^*|^m.$$

If $m = 2$, the last inequality gives $A \leq 0$, with equality if and only if $a_k = a_k^*$. This means, according to Theorem 2, that all $y_j = 0$, as we wished. ■

case (h): This case is a corollary of case (g), Proposition 1 (b), and the fact that the inequality under question is rotation invariant. ■

Remark 3: We saw in case (g) of Proposition 2 that in order to prove Conjecture 1 (and thus Conjecture 2) for some polynomial $p(z)$, it suffices to know that Conjecture 3 is true not for $p(z)$ itself, but for the polynomial $q(z) = a^{-n}p(az)$, for some a with $|a| = 1$. One of the reasons behind this is the fact that all the exponents appearing in Conjecture 1 as well as in the equation (ii) of §1, are equal to 2. Let us see under what assumptions we can prove the same inequality as in Conjecture 1, but with an exponent $m > 2$, i.e. the inequality

$$\sum_{k=1}^{n-1} |w_k|^m \leq \frac{n-2}{n} \sum_{j=1}^{n} |z_j|^m, \; m > 2.$$

We need to assume that:

(I) we know that this inequality is true whenever all z_j lie on a straight line; and

(II) the majorization condition is fulfilled for all the rotations $a^{-n}p(az)$, ($|a| = 1$), of $p(z)$.

Now, using our notation, (I) implies that

$$\sum_{k=1}^{n-1} |a_k^*|^m \leq \frac{n-2}{n} \sum_{j=1}^{n} |x_j|^m.$$

Furthermore, by (II) we conclude, as in the proof of case (g) of Proposition 2, that

$$\sum_{k=1}^{n-1} |a_k|^m \leq \sum_{k=1}^{n-1} |a_k^*|^m.$$

Thus, we finally obtain that the real parts of the zeros of $p(z)$ and $p'(z)$ satisfy the inequality

$$\sum_{k=1}^{n-1} |\mathrm{Re}\; w_k|^m \leq \frac{n-2}{n} \sum_{j=1}^{n} |\mathrm{Re}\; z_j|^m.$$

Next, we observe that because of (I) and (II), we can apply this inequality to the polynomial $a^{-n}p(az)$, $|a| = 1$ and thus, we can replace $|\mathrm{Re}\; z|$ in this inequality by $|\mathrm{Re}(az)|$ or by $|\mathrm{Im}(az)|$ (*cf.* [1]). Using this observation and the fact that the distance of the point z to the line through the origin 0 making an angle ϑ with the positive real axis is

$$|\cos\vartheta\, \mathrm{Im}\; z - \sin\vartheta\, \mathrm{Re}\; z| = |e^{-i\vartheta}\mathrm{Im}\; z|,$$

we get

$$\sum_{k=1}^{n-1} |e^{-i\vartheta}\mathrm{Im}\; w_k|^m \leq \frac{n-2}{n} \sum_{j=1}^{n} |e^{-i\vartheta}\mathrm{Im}\; z_j|^m.$$

Integrating this inequality over $[0, \pi]$ we have

$$\sum_{k=1}^{n-1} \int_0^{\pi} |e^{-i\vartheta} \text{Im } w_k|^m d\vartheta \le \frac{n-2}{n} \sum_{j=1}^{n} \int_0^{\pi} |e^{-i\vartheta} \text{Im } z_j|^m d\vartheta,$$

which gives immediately the desired inequality. ■

§5 Proof of Proposition 3

The proofs of the cases (b), (c), (d) and (e) are contained in [2]. In case (a) we observe that, if we start with the identity

$$\sum_{k=1}^{n-1} w_k^4 = \frac{n-4}{n} \sum_{j=1}^{n} z_j^4 + \frac{2}{n^2} \left(\sum_{j=1}^{n} z_j^2 \right)^2,$$

then, the proof is the same as the proof of Theorem 2 in [5]. We notice that this identity is an easy consequence of the relations between the elementary symmetric functions of the zeros of $p(z)$ and $p'(z)$ and their coefficients (see [2], Section 1). ■

case (f): Using Lemma 1 and the same reasoning as in case (d) of Proposition 2 we have

$$\sum_{k=1}^{n-1} |w_k|^4 = \sum_{k=1}^{n-2} |w_k|^4 \le (n-2)|z_1|^4 = \frac{n-4}{n} \sum_{j=1}^{n} |z_j|^4 + \frac{2}{n^2} \left(\sum_{j=1}^{n} |z_j|^2 \right)^2,$$

with equality if and only if all z_j lie on a line. ■

case (g): We assume that the majorization condition is fulfilled for all the rotations $a^{-n}p(az)(|a| = 1)$ of $p(z)$. From this and the above case (a) the two assumptions (I) and (II) are satisfied. Then, the proof of this case is the same as the proof of the last inequality in Remark 3 and so we omit it. ■

case (h): This is a corollary of case (g) and Proposition 1 (a). ■

Acknowledgement. The author wishes to thank Professor Jan Szynal who kindly informed him about the results of M. G. de Bruin and A. Sharma. He also wishes to thank the referee for his very helpful comments and suggestions.

References

[1] de Bruijn, N. G. and T. A. Springer, On the zeros of a polynomial and of its derivative II, *Nederl. Acad. Wetensch* **50** (1947), 458–464 and *Indag. Math.* **9** (1947), 264–270.

[2] de Bruin, M. G. and A. Sharma, On a Schoenberg-type conjecture, preprint.

[3] Linfield, B. Z., On certain polar curves with their application to the location of the roots of the derivatives of a rational function, *T. A. M. S.* **25** (1923), 239–258.

[4] Marden, M., *Geometry of Polynomials*, Math. Surv. Monogr., Number 3, A.M.S., Providence, 1966.

[5] Marshall, A. and J. Olkin, (ed.), *Inequalities: Theory of majorization and its applications*, Academic Press, Boston, 1979.

[6] Schoenberg, I. J., A conjectured analogue of Rolle's Theorem for polynomials with real or complex coefficients, *Amer. Math. Monthly* **93** (1986), 8–13.

Emmanuel S. Katsoprinakis
Department of Mathematics
University of Crete
714 09 Heraklion - Crete
GREECE
katsopr@talos.cc.uch.gr

A unified orthonormalization method for the approximate conformal mapping of simply and multiply-connected domains

Christodoulos A. Kokkinos

Abstract. We present a unified method for computing the following conformal maps: (a) The mapping of a finite N-connected domain bounded by $N \geq 2$ ($N \geq 1$) closed piecewise analytic Jordan curves onto a circular ring (disc) slit along $N-2$ ($N-1$) concentric circular arcs. (b) The mapping of a domain exterior to $N \geq 1$ closed piecewise analytic Jordan curves onto the exterior of a circle (the whole complex plane) slit along $N-1$ (N) concentric circular arcs.

§1 Introduction

Let Ω be a finite or infinite N-connected domain in $\mathbb{C}$ with boundary

$$\partial\Omega = \bigcup_{\nu=1}^{N} \Gamma_\nu,$$

where Γ_ν, $\nu = 1, 2, \ldots, N$, are closed piecewise analytic Jordan curves, and let f denote a conformal map of Ω onto some canonical domain S. This paper is concerned with the study of an orthonormalization method (ONM) for the numerical approximation of f.

The use of the ONM has been studied already in [3, 6, 9] in connection with the approximation of the conformal map f in cases where Ω is finite, has connectivity $N \geq 2$ and the domain S is one of the standard canonical domains that are discussed frequently in the literature [1], Ch. VI, [2], Ch. V, and [5], Ch. VII. More precisely, in [9] the method is studied for the case where Ω is doubly-connected ($N = 2$) and S is a circular ring, [3] is concerned with the cases where Ω is N-connected with $N \geq 2$ and S is either a circular ring with circular slits or a disc with circular slits and [6] contains a general survey for the mapping of N-connected domains with $N \geq 2$. For the case $N = 1$, *i.e.* for the mapping of a simply connected domain onto a disc, the closely related Bergman kernel method [4, 7, 8, 11] is generally preferred. However, as observed in [13], pp. 20–22, there are circumstances under which the use of the ONM might present certain advantages over the Bergman kernel method.

Computational Methods and Function Theory (CMFT'97)
N. Papamichael, St. Ruscheweyh and E. B. Saff (Eds.), pp. 327–344

ISBN 981-02-3626-3

The main purpose of this paper is to present the ONM in a unified framework and, at the same time, to extend somewhat its area of application by showing that it can also be used for computing approximations to f in cases where Ω is infinite. More specifically, our main purpose is to show that essentially the same method can be used for computing approximations to the following conformal maps:

(a) the mapping of a finite N-connected domain bounded by $N \geq 2$ ($N \geq 1$) closed piecewise analytic Jordan curves onto a circular ring (disc) with $N-2$ ($N-1$) concentric circular slits,

(b) the mapping of a domain exterior to $N \geq 1$ closed piecewise analytic Jordan curves onto the exterior of a circle (the whole complex plane) slit along $N-1$ (N) concentric circular arcs.

The paper is organized as follows: In §2 we give a brief description, in unified form, of the numerical method and also of the theory on which the method is based. In §3 we distinguish the two cases where Ω is finite and infinite and, in each case, we give the necessary approximating formulae. Finally in §4 we present several numerical examples illustrating the application of the method.

§2 The orthonormalization method

With the notations introduced in §1, assume that the origin 0 lies either in Ω or in the interior of one of the boundary curves Γ_ν, $\nu = 1, 2, \ldots, N$, and let $L_2(\Omega)$ be the Hilbert space of all square integrable functions which are analytic in Ω and possess a single valued indefinite integral in Ω, with respect to the inner product

$$\langle g, h \rangle := \iint_\Omega g(z)\overline{h(z)}\, dx dy.$$

Also let

$$A(z) := \log f(z) - \log z,$$

so that

$$f(z) = z \exp(A(z)),$$

and note that A is analytic and single-valued in Ω, as z and $f(z)$ travel once around the boundary curves, and its derivative H is given by

$$H(z) := A'(z) = \frac{f'(z)}{f(z)} - \frac{1}{z}. \tag{2.1}$$

Then, the main theoretical results on which the ONM is based, can be summarized as follows (see *e.g.* [1], pp. 59, 96, [2], p. 118, [5], p. 210, and [6], pp. 22, 34).

Complex form of Green's formula: Let $u, v \in L_2(\Omega)$ be continuous on $\partial\Omega$ apart, possibly, from a finite number of branch point singularities of the form

$$(z - z_c)^a, \qquad z_c \in \partial\Omega, \quad -\frac{1}{2} < a < 0. \tag{2.2}$$

Then

$$\langle u, v\rangle = \frac{1}{2i}\int_{\partial\Omega} u(z)\overline{\mu(z)}\,dz, \qquad \text{where } \mu'(z) = v(z). \tag{2.3}$$

Inner product of H: For each function $u \in L_2(\Omega)$ which is also continuous on $\partial\Omega$ apart, possibly, from a finite number of singularities of the form (2.2)

$$\langle u, H\rangle = i\int_{\partial\Omega} u(z)\log|z|\,dz. \tag{2.4}$$

This last property shows that the determination of $\langle u, H\rangle$, $u \in L_2(\Omega)$, does not require the knowledge of H.

Let $\{\eta_j\}_1^\infty$ be a complete set of funtions for the space $L_2(\Omega)$ and assume that the basis functions η_j satisfy the additional continuity requirement needed for (2.3) and (2.4) to hold. Then the ONM involves the following three main computational steps:

(i) Orthonormalize the first n basis functions $\{\eta_j\}_1^n$ by means of the Gram-Schmidt process to obtain the orthonormal set $\{\eta_j^*\}_1^n$.

(ii) Approximate the function H by the Fourier sum

$$H_n(z) = \sum_{j=1}^{n}\langle H, \eta_j^*\rangle\eta_j^*(z),$$

where the Fourier coefficients $\langle H, \eta_j^*\rangle$ are known by means of (2.4). (That is, H_n is the least squares approximation to H in the subspace $\mathrm{span}(\eta_1, \eta_2, \ldots, \eta_n)$).

(iii) Approximate the mapping function f by

$$f_n(z) = z\exp(A_n(z)),$$

where from (2.1)

$$A_n(z) = \int_\zeta^z H_n(t)\,dt$$

and $\zeta \in \overline{\Omega}$ is some normalizing point so that

$$A(\zeta) = 0.$$

It follows from the above that the approximations to f are of the form

$$f_n(z) = z \exp\left(\sum_{j=1}^{n} a_j \int_{\zeta}^{z} \eta_j(t)\, dt \right),$$

where $\{\eta_j\}$ is the set of basis functions. In fact, the choice of the basis set plays a very decisive role in numerical computations. More specifically, the use of the "classical" monomial basis sets, *i.e.* sets that consist of positive and negative monomial powers, is not satisfactory for numerical work. The reason for this is that:

(a) the resulting approximating sequences $\{H_n\}$ converge slowly when the function H has singularities either on the boundary $\partial\Omega$ or in $\mathbb{C} \setminus \overline{\Omega}$ but close to $\partial\Omega$,

(b) the slow convergence of $\{H_n\}$, coupled with the instability of the Gram-Schmidt process, makes it difficult (and often impossible) to compute approximations of acceptable accuracy.

In order to overcome the above difficulties it is essential that the basis set $\{\eta_j\}$ contains singular functions that reflect the main singular behaviour of H on $\partial\Omega$ and in $\mathbb{C}\setminus\overline{\Omega}$. The procedure for constructing such singular functions has been studied already in detail in earlier papers by Papamichael et al [3, 6, 9, 12] in connection with the mapping of finite N-connected domains. Since this procedure extends immediately to the additional cases considered in this paper (*i.e.* the cases where Ω is infinite), we do not repeat the details concerning the construction of basis sets here. Instead, we refer the reader to the references cited above and, in each numerical example, we list the basis functions that we actually use.

Regarding the other computational details, these have already been described in detail in the above earlier papers and are not repeated here. We merely note that in the present paper we made use of the numerical procedures summarized in [6] in order to:

(a) perform the quadratures needed for computing, by means of (2.3) and (2.4), the inner products $\langle \eta_s, \eta_r \rangle$ and $\langle H, \eta_r \rangle$,

(b) estimate the errors in the computed approximation to the mapping function f and the associated domain functionals,

(c) determine the optimum number n_{opt} of basis functions needed for obtaining the "best" approximation $f_{n_{\text{opt}}}$ to f.

§3 Applications

As was previously remarked, the numerical method described in §2 can be used to approximate the conformal maps of finite and infinite N-connected domains. Naturally, in each case, the canonical domain S has the same connectivity as the corresponding physical domain Ω. Furthermore, in each case, the canonical domain S is bounded by circles and concentric circular arcs and, in each case, its actual form depends on whether Ω is finite or infinite and on whether the origin 0 is in or outside Ω, *i.e.* $0 \in \Omega$ or $0 \notin \Omega$. Thus, in each of the four cases considered below we indicate the following:

(a) the physical domain Ω,

(b) the position of the origin 0,

(c) the form of the resulting canonical domain S.

In addition to the above, the radii $r_1, r_2, \ldots, r_N$ of the N circles or circular arcs that form the boundary of the canonical domains S are of particular interest. In fact, in the finite case $N-1$ of these radii and in the infinite case all N of these are unknowns of the conformal mapping problem and must also be determined by the numerical process. Thus, in each case, we also give the formulae for computing the unknown $N-1$ or N radii of S. For the cases where Ω is finite these formulae have already been given in [3] and [6]. The derivation of the corresponding formulae for the cases where Ω is infinite follows in a similar manner.

3.1 Slit circular ring ($\mathrm{N} \geq 2$)

The physical domain Ω*:* This is a finite domain bounded by N closed piecewise analytic Jordan curves Γ_ν, $\nu = 1, 2, \ldots, N$. Here we assume that:

(a) Γ_1 is the outer component of $\partial\Omega$, *i.e.* Γ_1 encloses all the other Γ_ν,

(b) the origin 0 lies in the interior of Γ_N, *i.e.* $0 \in \mathrm{int}(\Gamma_\mathrm{N})$.

The canonical domain $S_{1,N}$*:* This consists of a circular ring

$$R := \{w : r_N < |w| < r_1\}$$

slit along $N-2$ arcs a_ν, $\nu = 2, 3, \ldots, N-1$, of circles

$$|w| = r_\nu, \quad \text{where } r_N < r_\nu < r_1, \qquad \nu - 2, 3, \ldots, N-1.$$

The mapping function f is normalized by imposing a condition of the form

$$A(\zeta) = 0, \quad \zeta \in \Gamma_1, \qquad i.e.\ f(\zeta) = \zeta, \quad \zeta \in \Gamma_1, \tag{3.1.1}$$

so that f maps conformally Ω onto $S_{1,N}$ in such a way that

$$\begin{aligned} f(\Gamma_1) &= \{w : |w| = r_1\}, \\ f(\Gamma_N) &= \{w : |w| = r_N\}, \\ f(\Gamma_\nu) &= a_\nu, \qquad \nu = 2, 3, \ldots, N-1. \end{aligned}$$

The radii r_ν, $\nu = 1, 2, \ldots, N$*:* The radius r_1 is known from the normalizing condition (3.1.1), *i.e.*

$$r_1 = |\zeta|.$$

The $N-1$ ratios

$$M_\nu := \frac{r_1}{r_\nu}, \qquad \nu = 2, 3, \ldots, N,$$

are important domain functionals known as the radial moduli of Ω. These are given by the formulae

$$\begin{aligned} 2\pi \log M_\nu &= \frac{1}{2i} \int_{\partial\Omega} \frac{1}{z - \beta_\nu} (\overline{A(z)} + 2 \log |z|)\, dz, \\ & \qquad \nu = 2, 3, \ldots, N-1, \qquad (3.1.2) \\ 2\pi \log M_N &= \frac{1}{2i} \int_{\partial\Omega} \frac{1}{z} \log |z|\, dz - ||H||^2, \end{aligned}$$

where the points $\beta_2, \beta_3, \ldots, \beta_{N-1}$ are arbitrarily chosen points in the interiors of the boundary curves $\Gamma_2, \Gamma_3, \ldots, \Gamma_{N-1}$ (see [2] and [3]). Thus, the corresponding approximations $M_\nu^{(n)}$ to the moduli M_ν are given by

$$\begin{aligned} M_\nu^{(n)} &= \exp\left\{ \frac{1}{2\pi} \left(\frac{1}{2i} \int_{\partial\Omega} \frac{1}{z - \beta_\nu} (\overline{A(z)} + 2 \log |z|)\, dz \right) \right\}, \\ & \qquad \nu = 2, 3, \ldots, N-1, \qquad (3.1.3) \\ M_N^{(n)} &= \exp\left\{ \frac{1}{2\pi} \left(\frac{1}{2i} \int_{\partial\Omega} \frac{1}{z} \log |z|\, dz - ||H_n||^2 \right) \right\}, \end{aligned}$$

and the radii r_ν, $\nu = 2, 3, \ldots, N$, of the circular arcs corresponding to the boundary curves Γ_ν, $\nu = 2, 3, \ldots, N$, are approximated respectively by

$$r_\nu^{(n)} = \frac{|\zeta|}{M_\nu^{(n)}}, \qquad \nu = 2, 3, \ldots, N.$$

The special case $N = 2$ corresponds to the well-known problem of mapping a doubly-connected domain

$$\Omega := (\text{int}\Gamma_1) \cap (\text{ext}\Gamma_2)$$

onto a circular ring

$$R = S_{1,2} := \{w : r_1 < |w| < r_2\}.$$

Here, the single modulus

$$M = M_2 := \frac{r_2}{r_1}$$

determines uniquely the conformal equivalence class of the doubly-connected domain Ω, *i.e.* two doubly-connected domains can be mapped onto each other only if they have the same modulus.

3.2 Slit disc (N $\geq$ 1)

The physical domain Ω*:* This is as in Section 3.1 except that in this case the origin lies in Ω, *i.e.* $0 \in \Omega$ rather than in the interior of one of the boundary curves.
The canonical domain $S^{(d)}$*:* This consists of a disc

$$D := \{w : |w| < r_1\}$$

slit along $N-1$ arcs a_ν, $\nu = 2, 3, \dots, N$, of circles

$$w = |r_\nu|, \quad \text{where } r_\nu < r_1, \qquad \nu = 2, 3, \dots, N.$$

The mapping f maps conformally Ω onto $S^{(d)}$ in such a way that

$$f(\Gamma_1) = \{w : |w| = r_1\}$$

and

$$f(\Gamma_\nu) = a_\nu, \qquad \nu = 2, 3, \dots, N.$$

In this case the mapping function f can be either normalized by imposing a condition of the form (3.1.1), so that $r_1 = |\zeta|$, or by requiring that

$$A(0) = 0, \qquad \text{s.t. } f(0) = 0 \quad \text{and } f'(0) = 1,$$

in which case r_1 is given by the formula

$$2\pi \log r_1 = \frac{1}{i} \int_{\partial\Omega} \frac{1}{z} \log |z| \, dz - ||H||^2. \tag{3.2.1}$$

Regarding the moduli

$$M_\nu := \frac{r_1}{r_\nu}, \qquad \nu = 2, 3, \dots, N,$$

and their approximationg formulae these are given again by means of the formulae (3.1.2) and (3.1.3).

The special case $N = 1$ leads to the best known problem of conformal mapping, *i.e.* the problem of mapping a simply-connected domain Ω onto a disc

$$D_R := \{w : |w| < R = r_1\}$$

so that

$$f(0) = 0 \quad \text{and } f'(0) = 1.$$

In this case the radius R of the disc is called the conformal radius of the domain Ω with respect to the origin and can be approximated by means of (3.2.1).

3.3 Exterior of a circle slit along N − 1 arcs

The physical domain Ω*:* This is an infinite domain bounded by N closed piecewise analytic Jordan curves Γ_ν, $\nu = 1, 2, \ldots, N$. Here, we assume that the origin 0 lies in the interior of Γ_1, *i.e.* $0 \in \text{int}(\Gamma_1)$.
The canonical domain $S_e^{(d)}$*:* This consists of the exterior

$$D_e := \{w : |w| > r_1\}$$

of a circle of radius r_1, slit along $N - 1$ arcs a_ν, $\nu = 2, 3, \ldots, N$, of circles

$$|w| = r_\nu, \qquad \text{where } r_1 < r_\nu, \quad \nu = 2, 3, \ldots, N.$$

The mapping is normalized by imposing the condition

$$\lim_{z\to\infty} A(z) = 0, \qquad i.e. \quad \lim_{z\to\infty} f(z) = \infty \quad \text{and} \quad \lim_{z\to\infty} \frac{f(z)}{z} = 1. \quad (3.3.1)$$

Then, f maps conformally Ω onto $S_e^{(d)}$ in such a way that

$$f(\Gamma_1) = \{w : |w| = r_1\}.$$

The radii r_ν, $\nu = 1, 2, \ldots, N$*:* These are given by the formulae

$$2\pi \log r_1 = ||H||^2 - \frac{1}{i} \int_{\partial\Omega} \frac{1}{z} \log|z|\, dz, \quad (3.3.2)$$

$$2\pi \log r_\nu = \frac{i}{2} \int_{\partial\Omega} \frac{1}{z - \beta_\nu} (\overline{A(z)} + 2\log|z|)\, dz, \quad (3.3.3)$$

$$\nu = 2, 3, \ldots, N,$$

where the points $\beta_2, \beta_3, \ldots, \beta_N$ are arbitrarily chosen points in the interiors of the boundary curves $\Gamma_2, \Gamma_3, \ldots, \Gamma_N$ (see [2] and [3]). The corresponding approximations $r_\nu^{(n)}$ to these radii are given respectively by

$$r_1^{(n)} = \exp\left\{\frac{1}{2\pi}\left(||H_n||^2 - \frac{1}{i}\int_{\partial\Omega}\frac{1}{z}\log|z|\,dz\right)\right\}, \tag{3.3.4}$$

$$r_\nu^{(n)} = \exp\left\{\frac{1}{2\pi}\left(\frac{i}{2}\int_{\partial\Omega}\frac{1}{z-\beta_\nu}(\overline{A(z)} + 2\log|z|)\,dz\right)\right\}, \tag{3.3.5}$$
$$\nu = 2, 3, \ldots, N,$$

The special case $N = 1$ corresponds the classical problem of mapping the exterior of a closed analytic Jordan curve onto the exterior

$$D_c := \{w : |w| > c = r_1\}$$

of a circle of radius $c = r_1$, in such a way that

$$\lim_{z\to\infty} f(z) = \infty \quad \text{and} \quad \lim_{z\to\infty}\frac{f(z)}{z} = 1.$$

Here, the radius c of the circle is called the capacity (or transfinite diameter) of $\partial\Omega = \Gamma_1$.

3.4 Circular slit domain (N $\geq$ 1)

The physical domain Ω: This is as in Section 3.3, except that in this case the origin lies in Ω, *i.e.* $0 \in \Omega$, rather than in the interior of one of the boundary curves.
The canonical domain $S^{(c)}$: This consists of the entire plane slit along N arcs of circles

$$|w| = r_\nu, \qquad \nu = 1, 2, \ldots, N.$$

The mapping is normalized, as in Section 3.3, by imposing the condition (3.3.1).
The radii r_ν, $\nu = 1, 2, \ldots, N$: These are given by (3.3.3) with $\nu = 1, 2, \ldots, N$ and are approximated by (3.3.4) and (3.3.5) with $\nu = 1, 2, \ldots, N$.

§4 Numerical examples

Several examples illustrating the application of the ONM to cases where Ω is finite and $N \geq 2$ can be found in [9, 3, 10]. More specifically, [9] and [3] contain examples illustrating the application of the method in cases where the mapping function f may have singularities only on the boundary $\partial\Omega$ (corner singularities) whilst [10] concerns the doubly-connected case $f : \Omega \to R := \{w : r_1 < |w| < r_2\}$ and contains examples for dealing with singularities that f may have not only on $\partial\Omega$ but also in $\mathbb{C}\setminus\overline{\Omega}$. Finally, [3] includes one example illustrating the application of the "classical" ONM (*i.e.* the ONM with monomial basis) to the problem $f : \Omega \to S^{(d)}$.

The purpose of the examples given below is to show that the ONM can be applied successfully to all the mappings (finite or infinite) that were presented in §3, even when the mapping function f has singularities. In each example, we state:

(a) the physical domain Ω,

(b) the conformal mapping problem under consideration,

(c) the normalizing condition,

(d) the special points $\beta_\nu \in \text{int}(\Gamma_\nu)$ that appear in formulae (3.1.2) and (3.3.5), and which are also used for the construction of the basis set $\{\eta_j\}$,

(e) the basis functions $\{\eta_j\}$ used.

Our numerical results consist of the number n_{opt} of basis functions that gives the "best" approximation $f_{n_{\text{opt}}}$ to f, the corresponding estimate $E_{n_{\text{opt}}}$ of the maximum error in modulus

$$\max_{z \in \partial\Omega} \left| |f_{n_{\text{opt}}}(z)| - |f(z)| \right|,$$

and the corresponding approximations $r_\nu^{n_{\text{opt}}}$ to the radii r_ν. Further, in the examples that involve slit canonical domains we also list the ONM estimates $\theta_\nu^{n_{\text{opt}}}$ of the angles θ_ν that the arcs a_ν subtend at the origin. These are computed by estimating the arguments of the end points of the approximate arcs.

4.1 Example 1

Domain Ω: L-shaped region Ω of Figure 1, *i.e.* Ω is finite and $N = 1$.
Mapping problem: $f : \Omega \to D_R := \{w : |w| < R\}$.
Normalizing condition: $f(0) = 0$ and $f'(0) = 1$.
Basis set: The basis set $\{\eta_j\}$ is formed by introducing into the monomial set

$$z^j, \qquad j = 0, 1, 2, \ldots,$$

the following singular functions:

$$(z - z_c)^{2l/3-1}, \qquad l = 1, 2, 4, 5, 7, \tag{4.1.1}$$

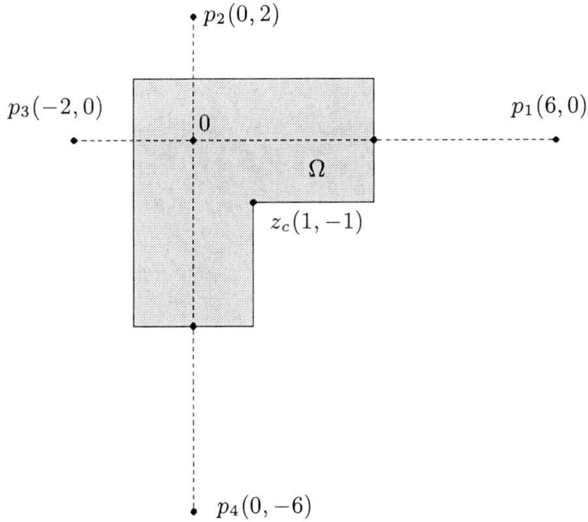

Figure 1.

where $z_c = 1 - i$. These are needed for dealing with the corner singularity of f at the re-entrant corner z_c. We also add

$$\frac{1}{z - p_j}, \qquad j = 1, 2, 3, 4, \tag{4.1.2}$$

where $p_1 = 6$, $p_2 = 2i$, $p_3 = -2$, $p_4 = 6i$. These are needed for dealing with the simple pole singularities that f has at the points p_j, $j = 1, 2, 3, 4$.
Numerical results:

$$n_{\text{opt}} = 25, \quad E_{25} = 6.0 \times 10^{-6}, \quad R_{25} = 1.307158877.$$

These should be compared with the results

$$n_{\text{opt}} = 27, \quad E_{27} = 1.8 \times 10^{-5}, \quad R_{27} = 1.307158877.$$

obtained by Levin et al [4], using the closely related Bergman kernel method. (In fact, the idea of using singular functions of the form (4.1.1) and (4.1.2) for dealing with corner and pole singularities was first used in [4]).

4.2 Example 2

Domain Ω*:* Exterior of the L-shape of Figure 2, *i.e.* Ω is infinite, $N = 1$ and $0 \in \mathrm{int}(\partial\Omega)$.

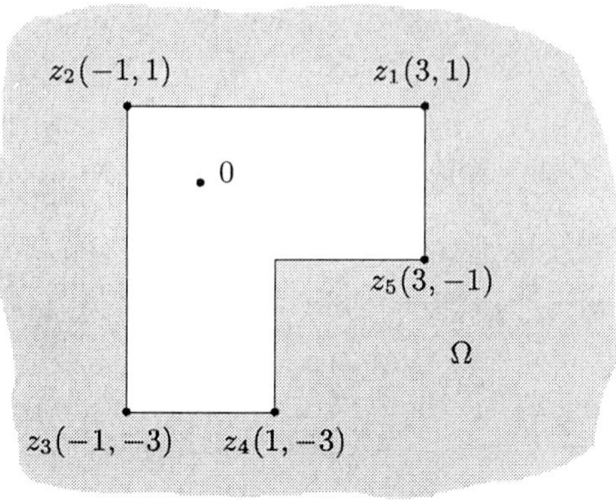

Figure 2.

Mapping problem: $f : \Omega \to D_c := \{w : |w| > c\}$.
Normalizing condition: $\lim_{z\to\infty} f(z) = \infty$ and $\lim_{z\to\infty} \frac{f(z)}{z} = 1$.
Basis set: The basis set $\{\eta_j\}$ is formed by introducing into the "monomial" set

$$z^{-j}, \qquad j = 2, 3, 4, \dots,$$

singular functions for dealing with the singularities at each of the five re-entrant corners z_j, $j = 1, 2, 3, 4, 5$. The actual singular functions used are

$$\left(\frac{1}{z} - \frac{1}{z_j}\right)^q \frac{1}{z^2}, \qquad j = 1, 2, 3, 4, 5,$$

with $q = -\frac{1}{3}, \frac{1}{3}, \frac{2}{3}, \frac{4}{3}$.
Numerical results:

$$n_{\mathrm{opt}} = 26, \quad E_{26} = 8.1 \times 10^{-5}, \quad R_{26} = 2.169780610.$$

These should be compared with the results

$$n_{\mathrm{opt}} = 29, \quad E_{29} = 4.3 \times 10^{-5}, \quad R_{29} = 2.169780749.$$

obtained by Papamichael and Kokkinos [8] using the Bergman kernel method.

4.3 Example 3

Domain Ω*:* 3-connected domain consisting of a square with 2 symmetric square holes; see Figure 3(a). The components of $\partial\Omega$ are:

$$\begin{aligned}\Gamma_1 &= \{(x,y): |x|=2, |y|\le 2\} \cup \{(x,y): |x|\le 2, |y|=2\},\\ \Gamma_\nu &= \{(x,y): |x+(-1)^\nu| = 0.5, |y|\le 0.5\} \cup \\ &\quad \{(x,y): |x+(-1)^\nu| \le 0.5, |y| = 0.5\}, \qquad \nu = 2,3.\end{aligned}$$

That is Ω is finite, $N = 3$ and $0 \in \partial\Omega$.

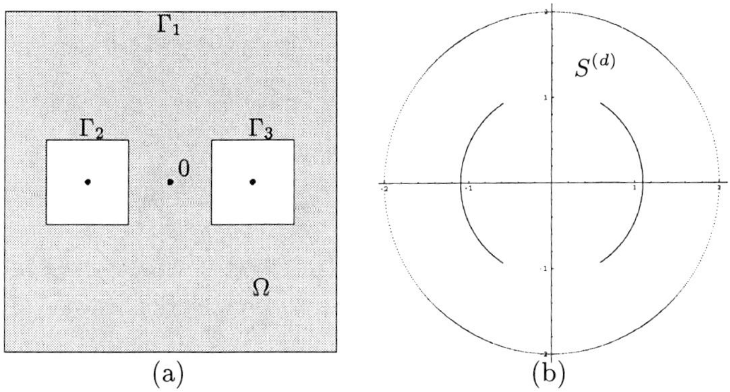

Figure 3.

Mapping problem: $f : \Omega \to S^{(d)}$, see Section 3.2.
Normalizing condition: $f(2) = 2$.
Special points: $\beta_\nu = (-1)^\nu$, $\nu = 2,3$.
Basis set: We take advantage of the two-fold rotational symmetry of Ω, see [11], and form the basis set $\{\eta_j\}$ by introducing into the "monomial" set

$$z^{2j-1}, \quad (z-\beta_2)^{-(j+1)} + e^{i\theta}(z-\beta_3)^{-(j+1)}, \quad \theta = j\pi \qquad j = 1,2,3,\ldots,$$

singular functions for dealing with the singularities at each of the corners z_c of Γ_2 and Γ_3 (*i.e.* at each of the corners of the square holes). The singular functions are of the form

$$\left(\frac{1}{z-\beta_\nu} - \frac{1}{z_c-\beta_\nu}\right)^q \frac{1}{(z-\beta_\nu)^2} \tag{4.3.1}$$

with $q = -\frac{1}{3}, \frac{1}{3}, \frac{2}{3}, \frac{4}{3}$. (In fact, because of the symmetry the 32 singular functions can be combined into 16 "symmetric" functions, see [3], Ex. 5.)

Numerical results:

$$n_{\text{opt}} = 44, \quad E_{44} = 3.3 \times 10^{-5},$$

$$r_1 = 2, \quad r_2^{(44)} = r_3^{(44)} = 1.16837584, \quad \theta_2^{(44)} = \theta_3^{(44)} = 2.334382.$$

The plot of the approximate image domain is shown in Figure 3(b). This image domain was obtained by plotting the images of 500 equally spaced points on each of the components Γ_1, Γ_2 and Γ_3 of $\partial\Omega$. The plot shows that there is considerable crowding of the image points on the lower and upper parts of the outer circle bounding the canonical domain $S^{(d)}$.

4.4 Example 4

Domain Ω*:* 5-connected domain consisting of the exterior of a circle and 4 square holes, see Figure 4(a). The components of $\partial\Omega$ are

$$\begin{aligned}
\Gamma_1 &= \{z : z = 0.8e^{it}, 0 \le t \le 2\pi\}, \\
\Gamma_{2\nu} &= \{(x,y) : |x - (-1)^\nu 1.5| = 0.5, |y| < 0.5\} \cup \\
&\quad \{(x,y) : |x - (-1)^\nu 1.5| \le 0.5, |y| = 0.5\}, \\
\Gamma_{2\nu+1} &= \{(x,y) : |x| \le 0.5, |y - (-1)^\nu 1.5i| = 0.5, |y| < 0.5\} \cup \\
&\quad \{(x,y) : |x| = 0.5, |y - (-1)^\nu 1.5i| \le 0.5\},
\end{aligned}$$

with $\nu = 1, 2$. That is Ω is infinite, $N = 5$ and $0 \in \text{int}(\Gamma_1)$.

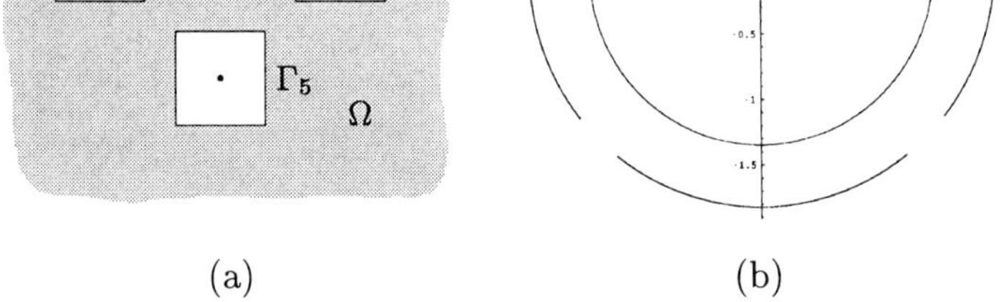

Figure 4.

Mapping problem: $f : \Omega \to S_e^{(d)}$, see Section 3.3.

Normalizing condition: $\lim_{z\to\infty} f(z) = \infty$ and $\lim_{z\to\infty} \frac{f(z)}{z} = 1$.
Special points: $\beta_{2\nu} = -(-1)^{\nu}1.5$, $\beta_{2\nu+1} = -(-1)^{\nu}1.5i$, $\nu = 1, 2$.
Basis set: We take advantage of the four-fold rotational symmetry of Ω and form the basis set $\{\eta_j\}$ by introducing into the "monomial" set

$$z^{-(4j+1)}, \quad \sum_{\nu=2}^{5} e^{(\nu-2)i\theta}(z-\beta_\nu)^{-(j+1)}, \quad \theta = \frac{j\pi}{2} \qquad j = 1, 2, 3, \ldots, \quad (4.4.1)$$

singular functions of the form (4.3.1) for dealing with the singularities at each of the corners z_c of Γ_ν, $\nu = 2, 3, 4, 5$ (*i.e.* at each of the corners of the square holes). Because of the symmetry, the 64 singular functions can be combined into 16 "symmetric" functions.
Numerical results:

$$n_{\text{opt}} = 26, \quad E_{26} = 5.1 \times 10^{-5},$$

$$r_1^{(26)} = 1.34724949, \quad r_j^{(26)} = 1.82031060, \quad j = 2, 3, 4, 5,$$

$$\theta_j^{(26)} = 1.340691, \quad j = 2, 3, 4, 5.$$

The plot of the approximate image domain is shown in Figure 4(b).

4.5 Example 5

Domain Ω: 2-connected domain consisting of the exterior of 2 circles, see Figure 5(a). The components of $\partial\Omega$ are

$$\begin{aligned} \Gamma_1 &= \{z : z = 1 + 0.5e^{it}, 0 \le t \le 2\pi\}, \\ \Gamma_2 &= \{z : z = -1 + 0.8e^{it}, 0 \le t \le 2\pi\}. \end{aligned}$$

That is Ω is infinite, $N = 2$ and $0 \in \Omega$.
Mapping problem: $f : \Omega \to S^{(c)}$, see Section 3.4.
Normalizing condition: $\lim_{z\to\infty} f(z) = \infty$ and $\lim_{z\to\infty} \frac{f(z)}{z} = 1$.
Special points: $\beta_\nu = (-1)^{\nu+1}$, $\nu = 1, 2$.
Basis set: The basis set $\{\eta_j\}$ used is

$$(z - \beta_1)^{-(j+1)}, \quad (z - \beta_2)^{-(j+1)} \qquad j = 1, 2, 3, \ldots.$$

(Here the domain does not contain corners and, for this reason, we do not use singular functions.)
Numerical results:

$$n_{\text{opt}} = 70, \quad E_{70} = 4.5 \times 10^{-5}$$

$$r_1^{(70)} = 1.43964885, \qquad r_2^{(70)} = 1.10892369,$$

$$\theta_1^{(70)} = 1.605811, \quad \theta_2^{(70)} = 3.497673.$$

The plot of the approximate image domain is shown in Figure 5(b).

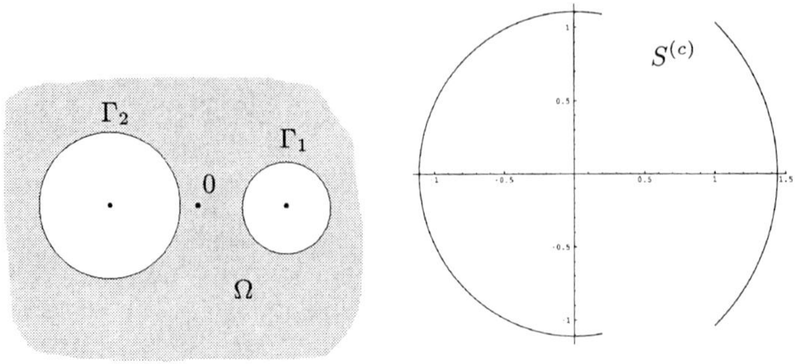

(a) (b)

Figure 5.

4.6 Example 6

Domain Ω*:* 4-connected domain consisting of the exterior of 4 square holes, see Figure 6(a). The components of $\partial\Omega$ are

$$\begin{aligned}\Gamma_{2\nu-1} &= \{(x,y) : |x-(-1)^\nu 1.5| = 0.2, |y| < 0.2\} \cup \\ &\quad \{(x,y) : |x-(-1)^\nu 1.5| \leq 0.2, |y| = 0.2\}, \\ \Gamma_{2\nu} &= \{(x,y) : |x| = 0.2, |y-(-1)^\nu 1.5i| < 0.2\} \cup \\ &\quad \{(x,y) : |x| \leq 0.2, |y-(-1)^\nu 1.5i| = 0.2\}\end{aligned}$$

with $\nu = 1, 2$. That is Ω is infinite, $N = 4$ and $0 \in \Omega$.
Mapping problem: $f : \Omega \to S^{(c)}$, see Section 3.4.
Normalizing condition: $\lim_{z\to\infty} f(z) = \infty$ and $\lim_{z\to\infty} \frac{f(z)}{z} = 1$.
Special points: $\beta_{2\nu-1} = -(-1)^\nu 1.5$, $\beta_{2\nu} = -(-1)^\nu 1.5i$, $\nu = 1, 2$.
Basis set: The set $\{\eta_j\}$ consists of the "monomial" set (4.4.1) and the 16 symmetric singular functions used in Example 4.
Numerical results:

$$n_{\text{opt}} = 19, \quad E_{19} = 4.9 \times 10^{-5},$$

$$r_j^{(19)} = 1.555085, \quad \theta_j^{(19)} = 0.612909, \quad j = 1, 2, 3, 4.$$

The plot of the approximate image domain is shown in Figure 6(b).

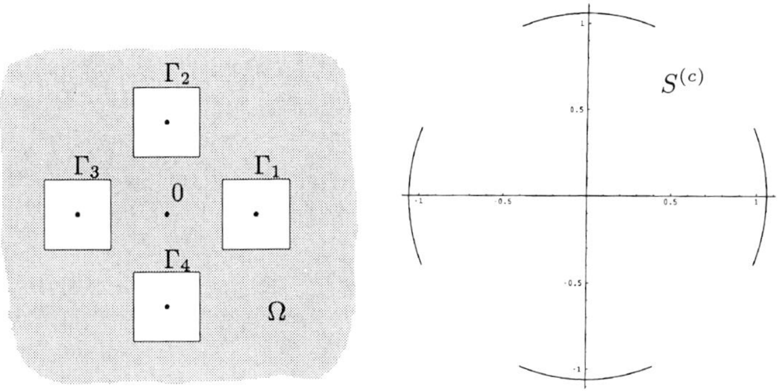

(a) (b)

Figure 6.

Acknowledgement. I am grateful to Professor N. Papamichael for many helpful discussions.

Most of this work was carried out while the author was on a sabbatical leave at the Department of Mathematics and Statistics of the University of Cyprus.

References

[1] Bergman, S., *The Kernel Function and Conformal Mapping*, Math. Surveys, No. 5, Providence, R. I.: Am. Math. Soc., 1970.

[2] Gaier, D., *Konstruktive Methoden der konformen Abbildung*, Springer, Berlin, Göttingen, Heidelberg, 1964.

[3] Kokkinos, C. A., N. Papamichael and A. B. Sideridis, An orthonormalization method for the approximate conformal mapping of multiply-connected domains, *IMA J. Numer. Anal.* **9** (1990), 343–359.

[4] Levin, D., N. Papamichael and A. B. Sideridis, The Bergman kernel methods for the numerical conformal mapping of simply-connected domains, *J. Inst. Math. Applics* **22** (1978), 171–187.

[5] Nehari, Z., *Conformal Mapping*, McGraw-Hill, New York, 1952.

[6] Papamichael N., Two numerical methods for the conformal mapping of simply and multiply-connected domains, *Bulletin Greek Math. Soc.* **32** (1991), 21–43.

[7] Papamichael, N. and Kokkinos, C. A., Two numerical methods for the conformal mapping of simply-connected domains, *Comput. Meth. Appl. Mech. Engrg* **28** (1981), 285–307.

[8] Papamichael, N. and Kokkinos, C. A., Numerical conformal mapping of exterior domains, *Comput. Meth. Appl. Mech. Engrg* **31** (1982), 189–203.

[9] Papamichael, N. and Kokkinos, C. A., The use of singular functions for the approximate conformal mapping of doubly-connected domains, *SIAM J. Sci. Stat. Comp.* **5** (1984), 684–700.

[10] Papamichel, N. and M. K. Warby, Pole-type singularities and the numerical conformal mapping of doubly-connected domains, *J. Comput. Appl. Math.* **10** (1984), 93–106.

[11] Papamichel, N. and M. K. Warby, Stability and convergence properties of Bergman kernel method in numerical conformal mapping, *Numer. Math.* **48** (1986), 639–669.

[12] Papamichel, N., M. K. Warby and D. M. Hough, The treatment of corner and pole-type singularities in numerical conformal mapping techniques, *J. Comp. Appl. Math.* **14** (1986), 163–191.

[13] Warby, M. K., Bergman kernel methods for the numerical conformal mapping of simply and doubly-connected domains, dissertation, Department of Mathematics and Statistics, Brunel University, 1984.

Christodoulos A. Kokkinos
Department of Mathematics
National Technical University of Athens
Zographou 157–173
15780 Athens
GREECE
chkokki@math.ntua.gr

Positive trigonometric sums in the theory of univalent functions

Stamatis Koumandos

Abstract. We give several positive sine sums associated with certain problems on the logarithmic coefficients of univalent functions. In particular, we give a class of positive convex sine sums.

§1 Introduction

Let $\mathcal{S}$ be the class of functions $f(z) = z + a_2 z^2 + \dots$ analytic and univalent in the unit disk $\mathbb{D}$. The Koebe function

$$k(z) = z(1-z)^{-2} = z + 2z^2 + 3z^3 + \dots$$

often plays the role of extremal function in $\mathcal{S}$. The logarithmic coefficients γ_n of f in $\mathcal{S}$, are defined by

$$\log \frac{f(z)}{z} = 2\sum_{n=1}^{\infty} \gamma_n z^n.$$

Thus the Koebe function $k(z)$ has logarithmic coefficients $\gamma_n = \frac{1}{n}$.

In 1971, Milin conjectured that

$$\sum_{k=1}^{n} (n-k+1)k\,|\gamma_k|^2 \leq \sum_{k=1}^{n} (n-k+1)\frac{1}{k}, \tag{1.1}$$

which he showed would imply the famous Bieberbach conjecture $|a_n| \leq n$. Subsequently, de Branges proved (1.1) and thus established the Bieberbach conjecture.

In [7], Duren and Leung showed that

$$\sum_{k=1}^{\infty} |\gamma_k|^2 \leq \sum_{k=1}^{\infty} \frac{1}{k^2} = \frac{\pi^2}{6}, \tag{1.2}$$

and conjectured that

$$\sum_{k=1}^{n} |\gamma_k|^2 \leq \sum_{k=1}^{n} \frac{1}{k^2} \qquad \text{for all } n, \tag{1.3}$$

Computational Methods and Function Theory (CMFT'97)
N. Papamichael, St. Ruscheweyh and E. B. Saff (Eds.), pp. 345–357

ISBN 981-02-3626-3

a conjecture which remains open.

The common feature of (1.1), (1.2) and (1.3) is that the Koebe function maximizes the left hand side among all functions in $\mathcal{S}$.

In [12], Milin and Grinshpan considered, among other things, inequalities of the form

$$\sum_{k=1}^{n} \sigma_k k \left|\gamma_k\right|^2 \leq \sum_{k=1}^{n} \sigma_k \frac{1}{k}, \tag{1.4}$$

where $\sigma_k \geq 0$ are given weights. Of course, the inequality (1.4) holds if and only if the Koebe function yields the maximum of the quadratic form on the left hand side. By using a spire variation Milin and Grinshpan derived the following necessary condition. If (1.4) is valid then

$$\sum_{k=1}^{n} \sigma_k \sin k\theta > 0, \qquad \text{for all } n, 0 < \theta < \pi. \tag{1.5}$$

This result has been also proved by Andreev and Duren in [1], using Schiffer's boundary variation.

In the present article, we give several examples of coefficients $\sigma_k \geq 0$ for which the inequality (1.5) is valid for all n and thus we provide some information for the order of magnitude of the coefficients σ_k for which (1.4) may be valid. In particular, we give a new positivity result for a class of convex sine sums (*cf.* §3 and §4). We shall see that the Duren-Leung conjecture (1.3) is a special case of a more general problem.

§2 Positive sine sums

For Milin's inequality (1.1), $\sigma_k = (n-k+1)$, the corresponding sum (1.5), reduces to

$$\frac{(n+1)\sin\theta - \sin(n+1)\theta}{4(n+1)\sin^2\frac{\theta}{2}} > 0, \qquad 0 < \theta < \pi,$$

which represents the positivity of the classical conjugate Fejér kernel (*cf.* [15]).

For the Duren-Leung sum (1.3) the corresponding sine sum is

$$\sum_{k=1}^{n} \frac{\sin k\theta}{k} > 0, \qquad 0 < \theta < \pi, \tag{2.1}$$

which shows that all the partial sums of the Fourier series of the function $(\pi - \theta)/2$ are positive in $(0, \pi)$. Inequality (2.1) had been conjectured by Fejér in 1910 and was first proved by Jackson [11], then by Gronwall [10], by Fejér himself [9] and many other authors who have offered different proofs (*cf.* [15], p. 62).

A dramatic improvement on (2.1) was published by Vietoris [14], in 1958. Vietoris's theorem is the following. Let (a_k), $k = 0, 1, \ldots,$ be any nonincreasing sequence of nonnegative real numbers such that $a_0 > 0$ and $2ka_{2k} \leq (2k-1)a_{2k-1}$ for $k \geq 1$. Then one has

$$\sum_{k=1}^{n} a_k \sin k\theta > 0, \qquad \text{for all } n, 0 < \theta < \pi. \tag{2.2}$$

Vietoris observed that (2.2) follows from the corresponding assertions for the specific case in which $a_k = c_k$, c_k being a certain sequence having order of magnitude $k^{-1/2}$, as opposed to the order of magnitude k^{-1} for the coefficients in (2.1).

In [4], Brown and Hewitt found an interesting refinement of (2.2). They considered a nonincreasing, nonnegative sequence (a_k) satisfying the relations $a_{2k} = a_{2k+1}$, $(2k+1)a_{2k} \leq 2ka_{2k-1}$. Then

$$\sum_{k=1}^{n} a_k \sin k\theta > 0, \tag{2.3}$$

for $0 < \theta < \pi$, when n is odd and $0 < \theta < \pi - \frac{\pi}{n}$, when n is even. We see that the sine sums in this generalization are not positive everywhere, nevertheless negative values for the sine sum (2.3) occur only in the interval $[\pi - \frac{\pi}{n}, \pi]$ for n even. In any case, (2.3) is stronger than (2.2) for its range of validity.

Recently, Brown and Wang [6] have obtained the following result. Suppose that λ_0 is the solution of the equation $(1 + \lambda)\pi = \tan(\lambda\pi)$, for $0 < \lambda < 1/2$ and α_0 is the solution of the equation

$$\sum_{k=1}^{\infty} \frac{2k}{(2k-1+\alpha)(2k+\alpha)(2k+1+\alpha)} = \frac{\sin \lambda_0\pi}{2(1+\lambda_0)\pi}, \tag{2.4}$$

if $-1 < \alpha \leq \alpha_0$, then

$$\sum_{k=1}^{n} \frac{\sin k\theta}{k+\alpha} > 0, \tag{2.5}$$

for $0 < \theta < \pi$ when n is odd and $0 < \theta \leq \pi - \frac{\pi}{n+\frac{1}{2}}$ when n is even. There is no direct connection between (2.3) and (2.5), however both results are best possible in a certain sense (see details in [4] and [6]).

We refer to Askey's monograph [2] for an historical account of positivity results for trigonometric sums and their applications. See also [3] for some interesting connection of these questions to function theory.

In the next section, we shall show that a trifling modification of the coefficients in both (2.3) and (2.5) yields an everywhere positive sine sum in $(0, \pi)$.

§3 Convex sine sums

Let (a_k) be a positive, monotonically decreasing and convex sequence of real numbers. We consider the sine sums

$$S_n(\theta) = \sum_{k=1}^{n} a_k \sin k\theta, \quad 0 < \theta < \pi.$$

A convex sine sum is a sum of the form

$$\begin{aligned} S_n^b(\theta) &= bS_n(\theta) + (1-b)S_{n-1}(\theta) \\ &= \sum_{k=1}^{n-1} a_k \sin k\theta + ba_n \sin n\theta, \qquad 0 < b < 1. \end{aligned} \tag{3.1}$$

It is well-known that

$$S_n^{\frac{1}{2}}(\theta) > 0, \qquad \text{for all } n, 0 < \theta < \pi,$$

and this need not be true for the sums $S_n(\theta)$ (see [8], Theorem II and also [15], Ch. V, p. 228.)

In this section, we let $a_k = \frac{1}{k+\alpha}$, $\alpha > -1$ and seek to determine the maximum range of b in [0, 1] for which

$$S_n^b(\theta) > 0, \qquad \text{for all } n, 0 < \theta < \pi. \tag{3.2}$$

The special case where $a_k = \frac{1}{k+1}$, has been considered by Brown and Wilson in [5]. In particular, it is shown in [5] that

$$\sum_{k=1}^{n-1} \frac{\sin k\theta}{k+1} + \beta \frac{\sin n\theta}{n+1} > 0, \qquad \text{for all } n, 0 < \theta < \pi, \tag{3.3}$$

for $\beta = \log 2$ and this inequality fails to hold for $\beta > \log 2$. It is, of course, obvious that (3.3) cannot hold for all β in $[0, \log 2]$ because it is false when $\beta = 0$, n is odd and θ near π (see Theorem A in [5]).

In the case where $a_k = \frac{1}{k}$, we have

$$\sum_{k=1}^{n-1} \frac{\sin k\theta}{k} + b\frac{\sin n\theta}{n} = b\sum_{k=1}^{n} \frac{\sin k\theta}{k} + (1-b)\sum_{k=1}^{n-1} \frac{\sin k\theta}{k} > 0,$$

for all $0 \le b \le 1$, $n = 1, 2, \ldots$, $0 < \theta < \pi$, and this is an immediate consequence of the Fejér-Jackson inequality (2.1).

It is easy to see that if $0 \le a < b < c \le 1$, then

$$S_n^b(\theta) = \frac{c-b}{c-a} S_n^a(\theta) + \frac{b-a}{c-a} S_n^c(\theta). \tag{3.4}$$

So if (3.2) holds for $b = b_1$ and $b = b_2$, where $0 \le b_1 < b_2 \le 1$, then it is true for all b in $[b_1, b_2]$.

We observe that if

$$S_n^b(\theta) = \sum_{k=1}^{n-1} \frac{\sin k\theta}{k+\alpha} + b\frac{\sin n\theta}{n+\alpha} > 0,$$

for all n and $0 < \theta < \pi$, then $1 - p(\alpha) \le b \le p(\alpha)$, where

$$p(\alpha) = \sum_{k=0}^{\infty} (-1)^k \frac{\alpha}{k+\alpha}.$$

Indeed, the derivative of $S_n^b(\theta)$ is positive at π for $b > p(\alpha)$ (resp. $b < 1 - p(\alpha)$) when n is even (resp. odd) and sufficiently large. In this case the sum must assume negative values near π.

We can now state our main result.

Theorem 1. *The inequality*

$$\sum_{k=1}^{n-1} \frac{\sin k\theta}{k+\alpha} + b\frac{\sin n\theta}{n+\alpha} > 0, \tag{3.5}$$

holds for all n and $0 < \theta < \pi$, when

$$1 - p(\alpha) \le b \le p(\alpha) \qquad \text{and } 0 \le \alpha \le \alpha_0.$$

The constant α_0 is defined as follows:

Let λ_0 be the solution of the equation $(1 + \lambda)\pi = \tan(\lambda\pi)$ in the interval $0 < \lambda < \frac{1}{2}$. ($\lambda_0 = 0.4302967\ldots$). Then α_0 is the solution of the equation

$$p(\alpha) = \cos^2 \frac{\lambda_0 \pi}{2}. \tag{3.6}$$

The numerical value of α_0 is $\alpha_0 = 2.11023396\ldots$.

§4 Proof of the main result

For the proof of this Theorem we need the following elementary lemma.

Lemma 1. *Let*

$$p(\alpha) = \sum_{k=0}^{\infty} (-1)^k \frac{\alpha}{k+\alpha}.$$

Then we have

(i) $1/2 < p(\alpha) < 1$ *for all* $\alpha > 0$.

(ii) $\lim_{\alpha\to\infty} p(\alpha) = \frac{1}{2}$.

(iii) $p(\alpha)$ *is a strictly decreasing function in* $[0, \infty)$.

(iv) $p(\alpha) = \frac{\alpha}{2}\left[\psi\left(\frac{\alpha+1}{2}\right) - \psi(\frac{\alpha}{2})\right]$, *where* $\psi(z) = \frac{\Gamma'(z)}{\Gamma(z)}$.

Proof: It follows directly from the definition of $p(\alpha)$ that $p(\alpha) < 1$. Next, we observe that $p(\alpha) = \alpha\beta(\alpha)$, where

$$\beta(\alpha) = \sum_{k=0}^{\infty} (-1)^k \frac{1}{k+\alpha}.$$

It is known that (see [13] Vol. 1, p.775)

$$\beta(\alpha) = \int_0^1 \frac{t^{\alpha-1}}{t+1}\, dt.$$

Then, an integration by parts gives

$$p(\alpha) = \frac{1}{2} + \int_0^1 \frac{t^\alpha}{(t+1)^2}\, dt. \tag{4.1}$$

From this it follows that $p(\alpha) > \frac{1}{2}$. Assertions (ii) and (iii) follow immediately from (4.1). We also have from [13], p. 775, that

$$\beta(\alpha) = \frac{1}{2}\left[\psi\left(\frac{\alpha+1}{2}\right) - \psi\left(\frac{\alpha}{2}\right)\right],$$

which establishes (iv). The proof of the Lemma is complete. ■

Proof of Theorem 1: We first show that α_0 is the solution of the equation (3.6) which is simpler than (2.4). Since $p(\alpha)$ is strictly decreasing for $\alpha \geq 0$, the equation (3.6) has a unique solution. We write

$$\begin{aligned} F(\alpha) &:= \sum_{k=1}^{\infty} \frac{2k}{(2k-1+\alpha)(2k+\alpha)(2k+1+\alpha)} \\ &= \frac{\alpha+1}{2} \sum_{k=0}^{\infty} \frac{1}{(2k+\alpha)(2k+\alpha+1)} \\ &\quad - \frac{\alpha-1}{2} \sum_{k=0}^{\infty} \frac{1}{(2k+\alpha)(2k+\alpha-1)}. \end{aligned}$$

Using the formula

$$\sum_{k=0}^{\infty} \frac{1}{(k+a)(k+b)} = \frac{\psi(b)-\psi(a)}{b-a},$$

(*cf.* [13], Vol 1, p.655), we get

$$\begin{aligned} F(\alpha) &= \frac{\alpha+1}{4} \left\{ \psi\left(\frac{\alpha+1}{2}\right) - \psi\left(\frac{\alpha}{2}\right) \right\} \\ &\quad - \frac{\alpha-1}{4} \left\{ \psi\left(\frac{\alpha}{2}\right) - \psi\left(\frac{\alpha-1}{2}\right) \right\}. \end{aligned}$$

Since

$$\psi(z+1) = \psi(z) + \frac{1}{z},$$

it follows from the above and (iv) of the Lemma that

$$\begin{aligned} F(\alpha) &= -\frac{1}{2} + \frac{\alpha}{2} \left\{ \psi\left(\frac{\alpha+1}{2}\right) - \psi\left(\frac{\alpha}{2}\right) \right\} \\ &= -\frac{1}{2} + \sum_{k=0}^{\infty} (-1)^k \frac{\alpha}{k+\alpha} = -\frac{1}{2} + p(\alpha). \end{aligned} \tag{4.2}$$

Since

$$\frac{\cos \lambda_0 \pi}{2} = \frac{\sin \lambda_0 \pi}{2(1+\lambda_0)},$$

it follows from this and (4.2) that equation (2.4) is equivalent to the equation (3.6).

In view of (3.4) it suffices to prove inequality (3.5) only for $b = p(\alpha)$ and $b = 1 - p(\alpha)$. In fact, it follows from (3.4) that

$$\frac{1}{2} S_n^{p(\alpha)}(\theta) = \left(p(\alpha) - \frac{1}{2} \right) T_n^{\alpha}(\theta) + (1 - p(\alpha)) S_n^{\frac{1}{2}}(\theta), \tag{4.3}$$

where

$$T_n^\alpha(\theta) = \sum_{k=1}^{n} \frac{\sin k\theta}{k+\alpha}.$$

As it has already been mentioned

$$S_n^{\frac{1}{2}}(\theta) = \sum_{k=1}^{n-1} \frac{\sin k\theta}{k+\alpha} + \frac{1}{2}\frac{\sin n\theta}{n+\alpha} > 0, \qquad \text{for } 0 < \theta < \pi.$$

Then using Lemma 1 and the inequality (2.5) it suffices to prove that

$$S_n^{p(\alpha)}(\theta) > 0, \qquad \text{when } n \text{ is even and } \pi - \frac{\pi}{n+\frac{1}{2}} < \theta < \pi. \tag{4.4}$$

We shall prove that the sum $S_n^{p(\alpha)}(\theta)$ is strictly decreasing for this range of θ. Since clearly $S_n^{p(\alpha)}(\pi) = 0$, positivity follows. In fact, differentiating and summing by parts we get

$$\begin{aligned}
\frac{d}{d\theta} S_n^{p(\alpha)}(\theta) &= \sum_{k=0}^{n-1} \frac{k}{k+\alpha} \cos k\theta + p(\alpha)\frac{n\cos n\theta}{n+\alpha} \\
&= \sum_{k=0}^{n-2} \left(\frac{k}{k+\alpha} - \frac{k+1}{k+1+\alpha}\right) \sum_{j=0}^{k} \cos j\theta \\
&\quad + \frac{n-1}{n-1+\alpha} \sum_{j=0}^{n-1} \cos j\theta + p(\alpha)\frac{n\,\cos n\theta}{n+\alpha} \\
&= \sum_{k=0}^{n-2} \left(\frac{k}{k+\alpha} - \frac{k+1}{k+1+\alpha}\right) \left(\frac{1}{2} + \frac{\sin(k+\frac{1}{2})\theta}{2\sin\frac{\theta}{2}}\right) \\
&\quad + \frac{n-1}{n-1+\alpha} \left(\frac{1}{2} + \frac{\sin(n-\frac{1}{2})\theta}{2\sin\frac{\theta}{2}}\right) + p(\alpha)\frac{n\cos n\theta}{n+\alpha} \\
&= -\sum_{k=0}^{n-2} \frac{\alpha}{(k+\alpha)(k+1+\alpha)} \frac{\sin(k+\frac{1}{2})\theta}{2\sin\frac{\theta}{2}} \\
&\quad + \frac{n-1}{n-1+\alpha} \frac{\sin(n-\frac{1}{2})\theta}{2\sin\frac{\theta}{2}} + p(\alpha)\frac{n}{n+\alpha}\cos n\theta \\
&= -\frac{1}{2\sin\frac{\theta}{2}} \left\{ \sum_{k=0}^{n-2} \frac{\alpha}{(k+\alpha)(k+1+\alpha)} \sin\left(k+\frac{1}{2}\right)\theta \right.
\end{aligned}$$

$$-\frac{n-1}{n-1+\alpha}\sin\left(n-\frac{1}{2}\right)\theta$$
$$-p(\alpha)\frac{n}{n+\alpha}\left(\sin\left(n+\frac{1}{2}\right)\theta-\sin\left(n-\frac{1}{2}\right)\theta\right)\Bigg\}.$$

Next, we show that

$$\sum_{k=0}^{n-2}\frac{\alpha}{(k+\alpha)(k+1+\alpha)}\sin(k+\frac{1}{2})\theta$$
$$+\left(\frac{p(\alpha)n}{n+\alpha}-\frac{n-1}{n-1+\alpha}\right)\sin\left(n-\frac{1}{2}\right)\theta \tag{4.5}$$
$$-\frac{p(\alpha)n}{n+\alpha}\sin\left(n+\frac{1}{2}\right)\theta>0.$$

We have

$$\sum_{k=0}^{n-2}\frac{\alpha}{(k+\alpha)(k+1+\alpha)}\sin\left(k+\frac{1}{2}\right)\theta>0,$$

and this follows by a summation by parts and using the fact that

$$\sum_{k=0}^{N}\sin(k+\frac{1}{2})\theta=\frac{1-\cos(N+1)\theta}{2\sin\frac{\theta}{2}}\geq 0,\qquad \text{for all } N, 0<\theta<\pi.$$

We observe that

$$\frac{p(\alpha)n}{n+\alpha}-\frac{n-1}{n-1+\alpha}<0,\qquad \text{for } n\geq 2$$

and consider two cases:

The case $\pi-\pi/(n+\frac{1}{2})<\theta<\pi-\pi/(2n+1)$. Since n is even, for θ lying in this interval we have $\sin(n+1/2)\theta<0$, $\cos(n+1/2)\theta>0$, whence $\sin(n-1/2)\theta<-\sin(n+1/2)\theta$, so that

$$\left(\frac{p(\alpha)n}{n+\alpha}-\frac{n-1}{n-1+\alpha}\right)\sin\left(n-\frac{1}{2}\right)\theta-\frac{p(\alpha)n}{n+\alpha}\sin\left(n+\frac{1}{2}\right)\theta$$
$$>-\sin\left(n+\frac{1}{2}\right)\theta\left(\frac{2p(\alpha)n}{n+\alpha}-\frac{n-1}{n-1+\alpha}\right). \tag{4.6}$$

By (i) of Lemma 1, we easily get

$$\frac{2p(\alpha)n}{n+\alpha}-\frac{n-1}{n-1+\alpha}>0,$$

hence the second part of (4.6) is positive and this completes the proof of (4.5) for the interval under consideration.

The case $\pi - \pi/(2n+1) < \theta < \pi$. In this interval we set $\phi = \pi - \theta$ so that $0 < \phi < \pi/(2n+1)$. We let $a_k = \frac{\alpha}{(k+\alpha)(k+1+\alpha)}$ and observe that for this range of ϕ we have

$$\begin{aligned}\sum_{k=0}^{n-2} a_k \sin(k+\frac{1}{2})\theta &= \sum_{k=0}^{n-2}(-1)^k a_k \cos\left(k+\frac{1}{2}\right)\phi \\ &> \cos\left((n-2)+\frac{1}{2}\right)\phi \sum_{k=0}^{N-1}(a_{2k}-a_{2k+1}),\end{aligned} \tag{4.7}$$

where $N = n/2$. It follows from (4.7) that the left side of (4.5) exceeds

$$\begin{aligned}&\cos\left((n-2)+\frac{1}{2}\right)\phi \sum_{k=0}^{N-1}(a_{2k}-a_{2k+1}) \\ &-\left(\frac{p(\alpha)n}{n+\alpha}-\frac{n-1}{n-1+\alpha}\right)\cos\left(n-\frac{1}{2}\right)\phi - \frac{p(\alpha)n}{n+\alpha}\cos\left(n+\frac{1}{2}\right)\phi \\ &> \cos\left(n+\frac{1}{2}\right)\phi\left\{\sum_{k=0}^{N-1}(a_{2k}-a_{2k+1}) + \frac{n-1}{n-1+\alpha} - \frac{2p(\alpha)n}{n+\alpha}\right\}.\end{aligned} \tag{4.8}$$

Since n is even, we have in this case $\cos(n+1/2)\phi > 0$. We need also to prove that the expression inside the brackets in the last part of (4.8) is also positive. For this purpose we let

$$\begin{aligned}U_N(\alpha) &= \sum_{k=0}^{N-1}\frac{2\alpha}{(2k+\alpha)(2k+\alpha+1)(2k+\alpha+2)} \\ &\quad + \frac{2N-1}{2N-1+\alpha} - \frac{4Np(\alpha)}{2N+\alpha}.\end{aligned}$$

It follows by a direct computation that $U_N(\alpha) > 0$, for $N = 1, 2, \dots 5$, $0 < \alpha \le \alpha_0$. In order to establish the positivity of $U_N(\alpha)$ for $N \ge 6$ and the same range of α, we observe that

$$\begin{aligned}&\sum_{k=0}^{\infty}\frac{1}{(2k+\alpha)(2k+\alpha+1)(2k+\alpha+2)} \\ &= \frac{1}{2\alpha}\left(1+\alpha\psi\left(\frac{\alpha+1}{2}\right)-\alpha\psi\left(\frac{\alpha}{2}+1\right)\right) \\ &= \frac{1}{2\alpha}\left(\alpha\psi\left(\frac{\alpha+1}{2}\right)-\alpha\psi\left(\frac{\alpha}{2}\right)-1\right).\end{aligned}$$

Hence

$$\sum_{k=0}^{\infty}\frac{2\alpha}{(2k+\alpha)(2k+\alpha+1)(2k+\alpha+2)} = \alpha\psi\left(\frac{\alpha+1}{2}\right)-\alpha\psi\left(\frac{\alpha}{2}\right)-1.$$

It follows from this and (iv) of Lemma 1 that

$$\lim_{N\to\infty} U_N(\alpha) = 0. \tag{4.9}$$

Moreover,

$$\begin{aligned}
&U_N(\alpha) - U_{N+1}(\alpha)\\
&\quad = \frac{2N-1}{2N-1+\alpha} - \frac{2N+1}{2N+1+\alpha} + p(\alpha)\left(\frac{4N+4}{2N+2+\alpha} - \frac{4N}{2N+\alpha}\right)\\
&\qquad - \frac{2\alpha}{(2N+\alpha)(2N+\alpha+1)(2N+\alpha+2)}\\
&\quad = \frac{(8p(\alpha)-4)N^2 - (6+4\alpha-8\alpha p(\alpha))N + (\alpha^2-1)(2p(\alpha)-1) - 3\alpha}{(2N+\alpha-1)(2N+\alpha)(2N+\alpha+1)(2N+\alpha+2)}\\
&\quad > 0,
\end{aligned}$$

for $N \geq 6$, $0 < \alpha \leq \alpha_0$ and this in combination with (4.9) gives $U_N(\alpha) > 0$, for the same range of N and α, as desired. At this stage the proof of (3.5) for $b = p(\alpha)$ is complete. In the same way we can establish (3.5) for $b = 1 - p(\alpha)$ and complete the proof of the Theorem. ■

Remark 1: Using (ii) of Lemma 1 we get

$$\lim_{\alpha\to\infty} \alpha S_n^{p(\alpha)}(\theta) = \sum_{k=1}^{n-1} \sin k\theta + \frac{1}{2}\sin n\theta > 0.$$

To prove the last inequality we write

$$\sum_{k=1}^{n-1} \sin k\theta + \frac{1}{2}\sin n\theta = \cos\frac{\theta}{2}\sum_{k=0}^{n-1}\sin\left(k+\frac{1}{2}\right)\theta = \cos\frac{\theta}{2}\,\frac{1-\cos n\theta}{2\sin^2\frac{\theta}{2}} > 0,$$

for all n and $0 < \theta < \pi$, which is to say that $S_n^{p(\alpha)}(\theta) > 0$ for sufficiently large values of α. The same result, of course, holds true for the sums $S_n^{q(\alpha)}(\theta)$, where $q(\alpha) = 1 - p(\alpha)$. Thus we are led to the conjecture that Theorem 1 holds for all $\alpha > 0$. This conjecture appears to be supported by numerical experimentation.

Remark 2: It follows from Theorem 1 that (3.3) holds for $1 - \log 2 \leq \beta \leq \log 2$. Let the sequence (a_k) be as in (2.3). Then using part (iii) of Theorem B in [5], a small adaptation of the proof of our Theorem gives

$$\sum_{k=1}^{n-1} a_k \sin k\theta + \alpha\, a_n \sin n\theta > 0, \qquad \text{for all } n, 0 < \theta < \pi,$$

when $1/4 \leq \alpha \leq 3/4$.

Remark 3: It would be interesting to see for which of the positive sine sums discussed in this paper, the corresponding inequality (1.4) is true. This is, of course, a much more difficult problem than the Duren-Leung conjecture (1.3).

References

[1] Andreev, V. V. and P. L. Duren, Inequalities for logarithmic coefficients of univalent functions and their derivatives, *Indiana Univ. Math. J.* **37** no. 4 (1988), 721–733.

[2] Askey, R., *Orthogonal Polynomials and Special Functions*, Regional Conference Series in Applied Mathemetics 21, SIAM, Philadelphia, 1975.

[3] Askey, R. and G. Gasper, Inequalities for polynomials, in *The Bieberbach Conjecture: Proceedings of the Symposium on the Occasion of the Proof*, A. Baernstein, D. Drasin, P. Duren and A. Marden (eds.), Amer. Math. Soc., Providence, 1986, 7–32.

[4] Brown, G. and E. Hewitt, A class of positive trigonometric sums, *Math. Ann.* **268** (1984), 91–122.

[5] Brown, G. and D. C. Wilson, A class of positive trigonometric sums II, *Math. Ann.* **285**, (1989), 57–74.

[6] Brown, G. and K. Wang, An extension of the Fejér-Jackson inequality. *J. Austral. Math. Soc. Ser. A* **62** no. 1 (1997), 1–12.

[7] Duren, P. L. and Y. J. Leung, Logarithmic coefficients of univalent functions, *J. Analyse Math.* **36** (1979), 36–43.

[8] Fejér, L., Über die Positivität von Summen die nach trigonometrischen oder Legendreschen Funktionen fortschreiten (erste Mitteilung), *Acta Litt. ac Scientarum Szeged* **2** (1925), 75–86. Also in *Leopold Fejér Gesammelte Arbeiten*, Vol. II, Birkhäuser, Basel, Suttgart, 1970, 128–138.

[9] Fejér, L. , Einige Sätze, die sich auf das Vorzeichen einer ganzen rationalen Funktion beziehen, *Monatsh. Math. Physik* **35** (1928), 305–344. Also in *Gesammelte Arbeiten II*, Birkhäuser-Verlag, Basel, Stuttgart, 1970.

[10] Gronwall, T. H. , Über die Gibbssche Erscheinung und die trigonometrischen Summen $\sin x + \frac{1}{2}\sin 2x + \ldots + \frac{1}{n}\sin nx$, *Math. Ann.* **72** (1912), 228–243.

[11] Jackson, D., Über eine trigonometrische Summe, *Rend. Circ. Math. Palermo* **32** (1911), 257–262.

[12] Milin, I. M. and A. Z. Grinshpan, Logarithmic coefficients means of univalent functions, *Complex Variables Theory Appl.* **7** (1986), 143–151.

[13] Prudnikov, A. P. , Yu. A. Brychkov, and O. I. Marichev, *Integrals and Series*, Gordon & Breech Science Publishers, 1990.

[14] Vietoris, L., Über das Vorzeichen gewisser trigonometrischer Summen, *S. B. Öster Akad. Wiss.* **167** (1958), 125–135; Teil II: *Anzeiger Öster. Akad. Wiss.* (1959), 192–193.

[15] Zygmund, A., *Trigonometric Series*, 2nd ed. Cambridge University Press, 1959.

Stamatis Koumandos
Department of Mathematics and Statistics
The University of Cyprus
P.O. Box 537
1678 Nicosia
CYPRUS
skoumand@pythagoras.mas.ucy.ac.cy

On computing zeros and poles of meromorphic functions

Peter Kravanja

Marc Van Barel

and Ann Haegemans

Abstract. Given a meromorphic function f, we present an accurate numerical method that computes all the zeros and poles of f that lie inside a Jordan curve γ, together with their respective multiplicities and orders. An upper bound for the total number of poles of f that lie inside γ is assumed to be known. Our algorithm is based on numerical integration along γ and formal orthogonal polynomials. It uses the logarithmic derivative f'/f. Initial approximations for the zeros and poles are not needed. Numerical examples illustrate the effectiveness of our approach.

§1 Introduction

Let $\mathcal{P}$ be the linear space of polynomials. Let $\langle\cdot,\cdot\rangle : \mathcal{P}\times\mathcal{P}\to\mathbb{C}$ be any symmetric and bilinear form such that

$$\langle\phi,\psi\rangle = \sum_{k=1}^{m}\lambda_k\phi(x_k)\psi(x_k) \tag{1.1}$$

for any $\phi,\psi\in\mathcal{P}$, where $m\in\mathbb{N}_0$, $\lambda_1,\ldots,\lambda_m\in\mathbb{C}_0$ and $x_1,\ldots,x_m\in\mathbb{C}$ are unknown. Suppose that the points $x_1,\ldots,x_m$ are known to be mutually distinct. Assume that we have an 'oracle' at our disposal that provides us with the value of $\langle\phi,\psi\rangle$ for any $\phi,\psi\in\mathcal{P}$ upon simple request. Let an upper bound M for m be given. We will present an accurate numerical algorithm for computing the unknowns in (1.1). It is based on formal orthogonal polynomials and proceeds by evaluating $\langle\cdot,\cdot\rangle$ and by solving certain generalized eigenvalue problems.

Our interest in this problem is motivated by the following application. Let W be a simply connected region in $\mathbb{C}$, $f: W\to\mathbb{C}$ meromorphic in W, and γ a positively oriented Jordan curve in W. Suppose that f has neither zeros nor poles on γ. We consider the problem of computing *all* the zeros and poles of f that lie in the interior of γ, together with their

Computational Methods and Function Theory (CMFT'97)
N. Papamichael, St. Ruscheweyh and E. B. Saff (Eds.), pp. 359–369

ISBN 981-02-3626-3

respective multiplicities and orders. Let N denote the total number of zeros of f that lie inside γ (*i.e.*, the number of zeros where each zero is counted according to its multiplicity), and let P denote the total number of poles of f that lie inside γ (*i.e.*, the number of poles where each pole is counted according to its order). Suppose from now on that $N+P>0$. Let n denote the number of mutually distinct zeros of f that lie inside γ. Let $z_1,\dots,z_n$ be these zeros and $\nu_1,\dots,\nu_n$ their respective multiplicities. Let p denote the number of mutually distinct poles of f that lie inside γ. Let $y_1,\dots,y_p$ be these poles and $\mu_1,\dots,\mu_p$ their respective orders. We define a symmetric and bilinear form $\langle\cdot,\cdot\rangle : \mathcal{P}\times\mathcal{P}\to\mathbb{C}$ by setting

$$\langle\phi,\psi\rangle := \frac{1}{2\pi i}\int_\gamma \phi(z)\psi(z)\frac{f'(z)}{f(z)}\,dz \tag{1.2}$$

for any two polynomials $\phi,\psi\in\mathcal{P}$. An easy calculation shows that $f'(z)/f(z)$ has a simple pole at z_k with residue ν_k for $k=1,\dots,n$ and a simple pole at y_l with residue $-\mu_l$ for $l=1,\dots,p$. It follows that

$$\langle\phi,\psi\rangle = \sum_{k=1}^{n}\nu_k\phi(z_k)\psi(z_k) - \sum_{l=1}^{p}\mu_l\phi(y_l)\psi(y_l).$$

The form $\langle\cdot,\cdot\rangle$ can be evaluated via numerical integration along γ. It is clearly of type (1.1). Our algorithm requires an upper bound M for $m=n+p$. Assume that an upper bound $\hat{P}$ for P is known, and define $s_r := \langle 1, z^r\rangle$ for $r\in\mathbb{N}_0$. Then $s_r = \nu_1 z_1^r+\cdots+\nu_n z_n^r-\mu_1 y_1^r-\cdots-\mu_p y_p^r$. In particular, $s_0 = N-P$ and we may assume that the value of s_0 has been computed. As $n+p\le N+P = s_0+2P\le s_0+2\hat{P}$, it follows that we may take $M=s_0+2\hat{P}$. In case γ is the unit circle, an upper bound for P can be obtained by using the heuristic approach of Gleyse and Kaliaguine [7].

For previous attempts to tackle this problem, we refer to Abd-Elall, Delves and Reid [1] and also Ioakimidis [9, 10].

In case $P=0$, our problem reduces to that of computing *all* the zeros, together with their respective multiplicities, of an analytic function that lie inside a Jordan curve. Methods for the determination of zeros of analytic functions that are based on the numerical evaluation of integrals are called *quadrature methods.* A review of such methods is given in [11]. The integrals s_r are equal to the Newton sums of the unknown zeros, $s_r = \sum_{k=1}^{n}\nu_k z_k^r$, and thus the problem is transformed into that of solving a system of polynomial equations. This approach was taken by Delves and Lyness [5] and was reconsidered by Li [12]. In our opinion, what is wrong with these approaches is that they are based on integrals involving polynomials from the standard monomial basis (ordinary moments). The mapping between the data and the unknowns is usually ill-conditioned, especially in the case of clustered zeros. This leads to algorithms that may produce very inaccurate results. We will present an algorithm that gives very accurate results. Instead of the standard monomial basis, it uses the formal orthogonal polynomials associated with $\langle \cdot, \cdot \rangle$. The key to our success lies in the fact that we consider the zeros (and poles) and the multiplicities (and orders) as *separate* unknowns, an idea that was overlooked in the past. Note that our algorithm does not require initial approximations for the zeros and poles.

§2 Theoretical solution

Our unknowns m, $x_1, \ldots, x_m$ and $\lambda_1, \ldots, \lambda_m$ in (1.1) can be calculated by applying the form $\langle \cdot, \cdot \rangle$ to polynomials from an arbitrary basis for $\mathcal{P}$. Indeed, let ψ_k be a monic polynomial of degree k for $k = 0, 1, 2, \ldots$. Define the $k \times k$ matrices F_k and $F_k^{(1)}$ as

$$F_k := \left[\ \langle \psi_r, \psi_s \rangle \ \right]_{r,s=0}^{k-1} \qquad \text{and} \qquad F_k^{(1)} := \left[\ \langle \psi_r, \psi_1 \psi_s \rangle \ \right]_{r,s=0}^{k-1}$$

for $k = 1, 2, \ldots$.

Theorem 1. $m = \operatorname{rank} F_{m+r}$ *for every nonnegative integer* r. *In particular,* $m = \operatorname{rank} F_M$.

Proof: Let r be a nonnegative integer. The matrix F_{m+r} can be written as

$$\begin{aligned} F_{m+r} &= \sum_{k=1}^{m} \lambda_k \begin{bmatrix} \psi_0(x_k)\psi_0(x_k) & \cdots & \psi_0(x_k)\psi_{m+r-1}(x_k) \\ \vdots & & \vdots \\ \psi_{m+r-1}(x_k)\psi_0(x_k) & \cdots & \psi_{m+r-1}(x_k)\psi_{m+r-1}(x_k) \end{bmatrix} \\ &= \sum_{k=1}^{m} \lambda_k \begin{bmatrix} \psi_0(x_k) \\ \vdots \\ \psi_{m+r-1}(x_k) \end{bmatrix} \left[\ \psi_0(x_k) \ \cdots \ \psi_{m+r-1}(x_k) \ \right]. \end{aligned}$$

This implies that $\operatorname{rank} F_{m+r} \le m$. However, F_m is regular. Indeed, one can easily verify that F_m can be factorized as $F_m = V_m D_m V_m^T$ where V_m is the Vandermonde-like matrix $V_m := \left[\psi_r(x_s) \right]_{r=0,s=1}^{m-1,m}$ and D_m is the diagonal matrix $D_m := \operatorname{diag}(\lambda_1, \ldots, \lambda_m)$. Therefore $\operatorname{rank} F_{m+r} \ge m$. It follows that $\operatorname{rank} F_{m+r} = m$. ■

Theorem 2. *The eigenvalues of the pencil* $F_m^{(1)} - \lambda F_m$ *are the numbers* $\psi_1(x_1), \ldots, \psi_1(x_m)$.

Proof: Define V_m as the Vandermonde-like matrix $V_m := [\,\psi_r(x_s)\,]_{r=0,s=1}^{m-1,m}$ and let D_m and $D_m^{(1)}$ be the diagonal matrices $D_m := \mathrm{diag}(\lambda_1, \ldots, \lambda_m)$ and $D_m^{(1)} := \mathrm{diag}(\lambda_1\psi_1(x_1), \ldots, \lambda_m\psi_1(x_m))$. Then F_m and $F_m^{(1)}$ can be factorized as $F_m = V_m D_m V_m^T$ and $F_m^{(1)} = V_m D_m^{(1)} V_m^T$. Let $\lambda^\star$ be an eigenvalue of the pencil $F_m^{(1)} - \lambda F_m$ and x a corresponding eigenvector. Then

$$\begin{aligned} & F_m^{(1)} x = \lambda^\star F_m x \\ \Leftrightarrow\quad & (V_m D_m^{(1)} V_m^T) x = \lambda^\star (V_m D_m V_m^T) x \\ \Leftrightarrow\quad & D_m^{(1)} y = \lambda^\star D_m y \quad \text{if } y := V_m^T x \\ \Leftrightarrow\quad & \mathrm{diag}(\psi_1(x_1), \ldots, \psi_1(x_m)) y = \lambda^\star y. \end{aligned}$$

This proves the theorem. ■

As $\psi_1(z)$ is a polynomial of degree one, $x_1, \ldots, x_m$ can be calculated by applying the previous theorem. Once m and $x_1, \ldots, x_m$ have been found, the problem becomes linear and $\lambda_1, \ldots, \lambda_m$ can be computed by solving a Vandermonde-like system of linear equations.

The question remains, of course, which polynomials $\psi_k(z)$ to choose. We will obtain very accurate numerical results by using the formal orthogonal polynomials associated with $\langle \cdot, \cdot \rangle$.

§3 Formal orthogonal polynomials

A monic polynomial φ_t of degree $t \geq 0$ that satisfies

$$\langle z^k, \varphi_t(z) \rangle = 0, \qquad k = 0, 1, \ldots, t-1, \tag{3.1}$$

is called a *formal orthogonal polynomial* (FOP) of degree t. (Observe that condition (3.1) is void for $t = 0$.) The adjective *formal* emphasizes the fact that, in general, the form $\langle \cdot, \cdot \rangle$ does not define a true inner product. An important consequence of this fact is that, in contrast to polynomials that are orthogonal with respect to a true inner product, FOPs φ_t need not exist or need not be unique for every degree t. (For details, see for example [8] and the references cited therein.) If (3.1) is satisfied and φ_t is unique, then φ_t is called a *regular* FOP and t a *regular index*. If we set

$$\varphi_t(z) =: u_{0,t} + u_{1,t} z + \cdots + u_{t-1,t} z^{t-1} + z^t$$

then condition (3.1) translates into the Yule-Walker system

$$\begin{bmatrix} s_0 & s_1 & \cdots & s_{t-1} \\ s_1 & & \cdot^{\cdot^{\cdot}} & \vdots \\ \vdots & \cdot^{\cdot^{\cdot}} & & \vdots \\ s_{t-1} & \cdots & \cdots & s_{2t-2} \end{bmatrix} \begin{bmatrix} u_{0,t} \\ u_{1,t} \\ \vdots \\ u_{t-1,t} \end{bmatrix} = - \begin{bmatrix} s_t \\ s_{t+1} \\ \vdots \\ s_{2t-1} \end{bmatrix} \tag{3.2}$$

where $s_r := \langle 1, z^r \rangle$ for $r = 0, 1, 2, \ldots$. Hence, the regular FOP of degree $t \geq 1$ exists if and only if the matrix $H_t := [s_{k+l}]_{k,l=0}^{t-1}$ is regular. Thus, the rank profile of $H := [s_{k+l}]_{k,l\geq 0}$ determines which regular FOPs exist.

If $s_0 = \langle 1, 1 \rangle = \sum_{k=1}^{m} \lambda_k \neq 0$, then the regular FOP of degree 1 exists and is given by $\varphi_1(z) = z - \mu$ where $\mu := s_1/s_0$. Theorem 1 implies that the regular FOP φ_m of degree m exists and tells us also that regular FOPs of degree larger than m do not exist. The polynomial φ_m is easily seen to be

$$\varphi_m(z) = (z - x_1) \cdot \cdots \cdot (z - x_m).$$

It is the monic polynomial of degree m that has $x_1, \ldots, x_m$ as simple zeros.

If H_m is strongly regular, *i.e.*, if all its leading principal submatrices are regular, then we have a full set $\{\varphi_0, \varphi_1, \ldots, \varphi_m\}$ of regular FOPs.

What happens if H_m is not strongly regular? By filling up the gaps in the sequence of existing regular FOPs it is possible to define a sequence $\{\varphi_t\}_{t=0}^{\infty}$, with φ_t a monic polynomial of degree t, such that if these polynomials are grouped into blocks according to the sequence of regular indices, then polynomials belonging to different blocks are orthogonal with respect to $\langle \cdot, \cdot \rangle$. More precisely, define $\{\varphi_t\}_{t=0}^{\infty}$ as follows. If t is a regular index, then let φ_t be the regular FOP of degree t. Else define φ_t as $\varphi_r \psi_{t,r}$ where r is the largest regular index less than t and $\psi_{t,r}$ is an arbitrary monic polynomial of degree $t - r$. In the latter case φ_t is called an *inner polynomial.* If $\psi_{t,r}(z) = z^{t-r}$ then we say that φ_t is defined *by using the standard monomial basis.* These polynomials $\{\varphi_t\}_{t=0}^{\infty}$ can be grouped into blocks. Each block starts with a regular FOP and the remaining polynomials are inner polynomials. Note that the last block has infinite length.

The block orthogonality property is expressed by the fact that the Gram matrix $G_m := [\langle \varphi_r, \varphi_s \rangle]_{r,s=0}^{m-1}$ is block diagonal. The diagonal blocks are regular, symmetric and zero above the main antidiagonal. If all the inner polynomials in a certain block are defined by using the standard monomial basis, then the corresponding diagonal block has Hankel structure. The matrix $G_m^{(1)} := [\langle \varphi_r, \varphi_1 \varphi_s \rangle]_{r,s=0}^{m-1}$ is block tridiagonal. The diagonal blocks are symmetric and lower anti-Hessenberg (*i.e.*, its entries are equal to zero along all the antidiagonals that lie above the main antidiagonal, except for the antidiagonal that precedes the main antidiagonal). Again, if all the inner polynomials in a certain block are defined by using the standard monomial basis, then the corresponding diagonal block is a Hankel matrix. The entries of the off-diagonal blocks are all equal to zero, except for the entry in the south-east corner. For proofs and further details, we refer to [3].

Corollary 1. *The eigenvalues of the pencil $G_m^{(1)} - \lambda G_m$ are the numbers $\varphi_1(x_1), \ldots, \varphi_1(x_m)$. In particular, if $s_0 = \sum_{k=1}^m \lambda_k \neq 0$, then they are given by $x_1 - \mu, \ldots, x_m - \mu$ where $\mu := s_1/s_0$.*

§4 The algorithm

If m and $\varphi_0, \varphi_1, \ldots, \varphi_{m-1}$ are known, then we can apply the previous theorem to obtain $x_1, \ldots, x_m$. By Theorem 1 the value of m can be computed as the rank of G_M, but this is not a very practical approach. Instead, we will start to compute the polynomials $\{\varphi_t\}_{t=0}^{\infty}$ one by one and determine the value of m as the degree of the last existing regular FOP. The upper bound M will play a crucial role in the computations.

Corollary 1 can be interpreted as follows: the zeros of the mth degree regular FOP can be calculated by solving a generalized eigenvalue problem. This property holds for *all* regular FOPs. This will enable us to compute regular FOPs in their product form, which is numerically very stable. Define $G_k := [\langle \varphi_r, \varphi_s \rangle]_{r,s=0}^{k-1}$ and $G_k^{(1)} := [\langle \varphi_r, \varphi_1 \varphi_s \rangle]_{r,s=0}^{k-1}$ for $k = 1, 2, \ldots$.

Theorem 3. *Let $t \geq 1$ be a regular index and let $z_{t,1}, \ldots, z_{t,t}$ be the zeros of the regular FOP φ_t. Then the eigenvalues of the pencil $G_t^{(1)} - \lambda G_t$ are given by $\varphi_1(z_{t,1}), \ldots, \varphi_1(z_{t,t})$. In particular, if $s_0 \neq 0$, then they are given by $z_{t,1} - \mu, \ldots, z_{t,t} - \mu$ where $\mu := s_1/s_0$.*

Proof: Define the Hankel matrix $H_t^{<}$ as $H_t^{<} := [s_{1+k+l}]_{k,l=0}^{t-1}$. We will first show that the zeros of φ_t are given by the eigenvalues of the pencil $H_t^{<} - \lambda H_t$. The zeros of φ_t are given by the eigenvalues of its companion matrix C_t. Let $\lambda^\star$ be an eigenvalue of C_t and x a corresponding eigenvector. As H_t is regular, we may conclude that $C_t x = \lambda^\star x \Leftrightarrow H_t C_t x = \lambda^\star H_t x$. Using (3.2) one can easily verify that $H_t C_t = H_t^{<}$.

Let A_t be the unit upper triangular matrix that contains the coefficients of $\varphi_0, \varphi_1, \ldots, \varphi_{t-1}$. Then G_t can be factorized as $G_t = A_t^T H_t A_t$. Suppose that $\varphi_1(z) =: z - \beta$. (If 1 is a regular index, then $\beta = s_1/s_0$.) Then $G_t^{(1)}$ is given by $[\langle \varphi_r, z\varphi_s \rangle]_{r,s=0}^{t-1} - \beta G_t$. The matrix $[\langle \varphi_r, z\varphi_s \rangle]_{r,s=0}^{t-1}$ can be written as $A_t^T H_t^{<} A_t$ and thus $G_t^{(1)} = A_t^T (H_t^{<} - \beta H_t) A_t$. Now let $\lambda^\star$ be an eigenvalue of the pencil $H_t^< - \lambda H_t$ and x a corresponding eigenvector. Then

$$\begin{aligned} & H_t^< x = \lambda^\star H_t x \\ \Leftrightarrow \quad & (H_t^< - \beta H_t)x = (\lambda^\star - \beta) H_t x \\ \Leftrightarrow \quad & A_t^T (H_t^< - \beta H_t) A_t y = \varphi_1(\lambda^\star) A_t^T H_t A_t y \quad \text{if } y := A_t^{-1} x \\ \Leftrightarrow \quad & G_t^{(1)} y = \varphi_1(\lambda^\star) G_t y. \end{aligned}$$

This proves the theorem. ■

Regular FOPs are characterized by the fact that the determinant of a Hankel matrix is different from zero, while inner polynomials correspond to singular Hankel matrices. By using an explicit determinant expression for regular FOPs, one can show that $\langle \varphi_t, \varphi_t \rangle = \det H_{t+1} / \det H_t$. Therefore, if t is a regular index, then $t+1$ is a regular index if and only if $\langle \varphi_t, \varphi_t \rangle \neq 0$. However, from a numerical point of view a test "is equal to zero" does not make sense. Because of rounding errors in the evaluation of $\langle \cdot, \cdot \rangle$ we would encounter only regular FOPs. Strictly speaking one could say that inner polynomials are not needed in numerical calculations. However, the opposite is true! Let us call a regular FOP *well-conditioned* if its corresponding Yule-Walker system (3.2) is well-conditioned, and *ill-conditioned* otherwise. To obtain a numerically stable algorithm, it is crucial to generate only well-conditioned regular FOPs and to replace ill-conditioned regular FOPs by inner polynomials. Stable look-ahead solvers for linear systems of equations that have Hankel structure are based on this principle [6, 4, 2]. In this approach the diagonal blocks in G_m are taken (slightly) larger than strictly necessary to avoid ill-conditioned blocks. A disadvantage is that part of the structure of G_m and $G_m^{(1)}$ gets lost and that there is some additional fill-in.

We will ask the user for two thresholds, ϵ_{stop} and ϵ_{cond} with $\epsilon_{\text{stop}} < \epsilon_{\text{cond}}$, to decide whether the algorithm may stop or not, and to determine the size of a block. Suppose that the algorithm has just generated a well-conditioned regular FOP φ_r. If $r = M$, then we may stop. Else we proceed to calculate $\langle \varphi_r, \varphi_r \rangle$. If $|\langle \varphi_r, \varphi_r \rangle| \geq \epsilon_{\text{cond}}$ then we generate φ_{r+1} as a regular FOP. Else we scan the sequence $\left(|\langle z^t \varphi_r, \varphi_r \rangle| \right)_{t=0}^{M-1-r}$. If $|\langle z^t \varphi_r, \varphi_r \rangle| < \epsilon_{\text{stop}}$ for $t = 0, 1, \ldots, M-1-r$ then we conclude that $m = r$ and stop. Else we search for the first element that is larger than ϵ_{cond}. The corresponding value of t then determines the size of the block of polynomials. If all the elements are less than ϵ_{cond} then we use the value of t that corresponds to the maximum to determine the block size and warn the user that we could not obtain the level of well-conditioning that he or she requested.

§5 Numerical examples

We implemented our algorithm in Matlab. The m-files are available from the first author. In the following examples the computations were done with floating point relative accuracy $\approx 2.2204\,10^{-16}$.

Example 1: Suppose that $m = 10$ and $x_k = k$, $\lambda_k = 1$ for $k = 1, \ldots, m$. This corresponds to the problem of computing the zeros of the Wilkinson-like polynomial $\prod_{k=1}^{m}(z-k)$. We evaluate the form (1.1) explicitely. By calculating the eigenvalues of the pencil $H_m^< - \lambda H_m$ (which corresponds to the classical approaches of [5] and [12]) we obtain the approximations

$$
\begin{array}{ll}
0.99997610636668, & 1.99812692712904,\\
2.96964756470803, & 3.85164221027260,\\
4.76929486601830, & 5.83891586633208,\\
6.93895160469148, & 7.98903115901926,\\
8.99931401487053, & 9.99999170411583.
\end{array}
$$

(The imaginary parts are not shown — they are $\mathcal{O}(10^{-9})$.) Our algorithm (with $M = 15$, $\epsilon_{\text{stop}} = 10^{-12}$ and $\epsilon_{\text{cond}} = 1$) generates only regular FOPs, concludes that $m = 10$ and obtains the (very accurate!) approximations

$$
\begin{array}{ll}
1.00000000000001, & 2.00000000000000,\\
3.00000000000000, & 4.00000000000000,\\
4.99999999999996, & 5.99999999999998,\\
7.00000000000001, & 8.00000000000000,\\
9.00000000000000, & 9.99999999999999.
\end{array}
$$

(The imaginary parts are $\mathcal{O}(10^{-15})$.) The mapping between the ordinary moments and the zeros is particularly ill-conditioned in the case of clustered zeros. However, also in this case our algorithm obtains comparatively very good results. Of course, the problem of computing zeros of polynomials is currently receiving a lot of attention and there exist specific algorithms that are better suited than our more general approach. See, *e.g.*, the paper by Pan [13] for a recent review.

In the following examples the form (1.2) was evaluated via numerical integration. If γ is a circle, then (1.2) is the integral of a periodic function over a complete period, and therefore the trapezoidal rule is an appropriate quadrature rule.

To increase the accuracy of our algorithm, the approximations for the zeros or poles were refined iteratively via Newton's method,

$$
z_k^{(\alpha+1)} = z_k^{(\alpha)} - \nu_k \frac{f(z_k^{(\alpha)})}{f'(z_k^{(\alpha)})}, \qquad y_l^{(\alpha+1)} = y_l^{(\alpha)} + \mu_l \frac{f(y_l^{(\alpha)})}{f'(y_l^{(\alpha)})}, \qquad \alpha \geq 0.
$$

Example 2: Define the function f by $f(z) := e^{3z} + 2z\cos z - 1$ and let γ be the circle $\{z \in \mathbb{C} : |z| = 2\}$. Our algorithm (with $\hat{P} = 0$ and thus $M = s_0 = \langle 1, 1\rangle = 4$, $\epsilon_{\text{stop}} = 10^{-12}$ and $\epsilon_{\text{cond}} = 1$) finds that $N = 4$. It generates φ_1 as a regular FOP, defines φ_2 as an inner polynomial, generates φ_3 and φ_4 as regular FOPs and concludes that $m = n = 4$. The approximations for the zeros and the corresponding multiplicities are shown in the following table.

$-1.84423395326221 + i\ 0.00000000000000$	1
$0.53089493029293 + i\ 1.33179187675112$	1
$0.53089493029293 - i\ 1.33179187675112$	1
$0.00000000000000 + i\ 0.00000000000000$	1

The correct digits are underlined. The calculated multiplicities were at a distance of $\mathcal{O}(10^{-15})$ to integers. If we put $\epsilon_{\text{cond}} = 0.1$ then our algorithm generates only regular FOPs. Now the approximations for the zeros are less accurate.

$-1.84423395326194 - i\ 0.00000000000499$	1
$0.53089493028783 + i\ 1.33179187675442$	1
$0.53089493029794 - i\ 1.33179187674729$	1
$-0.00000000000071 - i\ 0.00000000000211$	1

This shows the importance of generating only well-conditioned regular FOPs. The calculated multiplicities were at a distance of $\mathcal{O}(10^{-11})$ to integers.

Example 3: Let $f(z) := 1/(z^2(z-1)(z^2+9)) + z\sin z + e^{-3z} + 4$ and let γ be the circle $\{z \in \mathbb{C} : |z| = 2\}$. We set the upper bound $\hat{P} := 5$. It turns out that $\langle 1, 1\rangle = 0$ (in other words, $N = P$) and thus $M = s_0 + 2\hat{P} = 10$. The approximations for the zeros and poles obtained by our algorithm (again with $\epsilon_{\text{stop}} = 10^{-12}$ and $\epsilon_{\text{cond}} = 1$) had about 12 correct significant digits. After one step of iterative refinement, we obtain the following results:

x_k	λ_k
0.97843635600921	1
0.16974891913248	1
−0.13327146070751	1
1.00000000000000	−1
0.00000000000000	−2

The computed values of the λ_k's were at a distance of $\mathcal{O}(10^{-12})$ to integers. Note that zeros and poles can be distinguished easily by checking the signs of the λ_k's.

Acknowledgement. This research was partially supported by the Fund for Scientific Research-Flanders (FWO-V), project "Orthogonal Systems and their Applications", grant #G.0278.97.

References

[1] Abd-Elall, L. F., L. M. Delves, and J. D. Reid, A numerical method for locating the zeros and poles of a meromorphic function, in *Numerical Methods for Nonlinear Algebraic Equations*, P. Rabinowitz (ed.), Gordon and Breach, 1970, 47–59.

[2] Bojanczyk, A. W. and G. Heinig, A multi-step algorithm for Hankel matrices, *J. Complexity* **10** no. 1 (1994), 142–164.

[3] Bultheel, A. and M. Van Barel, *Linear Algebra, Rational Approximation and Orthogonal Polynomials*, Studies in Computational Mathematics, vol. 6, North-Holland, 1997.

[4] Cabay, S. and R. Meleshko, A weakly stable algorithm for Padé approximants and the inversion of Hankel matrices, *SIAM J. Matrix Anal. Appl.* **14** no. 3 (1993), 735–765.

[5] Delves, L. M. and J. N. Lyness, A numerical method for locating the zeros of an analytic function, *Math. Comput.* **21** (1967), 543–560.

[6] Freund, R. W. and H. Zha, A look-ahead algorithm for the solution of general Hankel systems, *Numer. Math.* **64** (1993), 295–321.

[7] Gleyse, B. and V. Kaliaguine, On algebraic computation of number of poles of meromorphic functions in the unit disk, in *Nonlinear Numerical Methods and Rational Approximation II*, A. Cuyt (ed.), Kluwer Academic Publishers, 1994, 241–246.

[8] Gragg, W. B. and M. H. Gutknecht, Stable look-ahead versions of the Euclidean and Chebyshev algorithms, in *Approximation and Computation: A Festschrift in Honor of Walter Gautschi*, R. V. M. Zahar (ed.), Birkhäuser, 1994, 231–260.

[9] Ioakimidis, N. I., Application of the generalized Siewert-Burniston method to locating zeros and poles of meromorphic functions, *J. Appl. Math. Phys.* **36** (1985), 733–742.

[10] Ioakimidis, N. I., Determination of poles of sectionally meromorphic functions, *J. Comput. Appl. Math.* **15** (1986), 323–327.

[11] Ioakimidis, N. I., Quadrature methods for the determination of zeros of transcendental functions — a review, in *Numerical Integration: Recent Developments, Software and Applications*, P. Keast and G. Fairweather (eds.), Reidel, Dordrecht, The Netherlands, 1987, 61–82.

[12] Li, T.-Y., On locating all zeros of an analytic function within a bounded domain by a revised Delves/Lyness method, *SIAM J. Numer. Anal.* **20** no. 4 (1983), 865–871.

[13] Pan, V. Y., Solving a polynomial equation: some history and recent progress, *SIAM Review* **39** no. 2 (1997), 187–220.

Peter Kravanja
Katholieke Universiteit Leuven
Department of Computer Science
Celestijnenlaan 200 A
B-3001 Heverlee
BELGIUM
Peter.Kravanja@na-net.ornl.gov

Marc Van Barel
Katholieke Universiteit Leuven
Department of Computer Science
Celestijnenlaan 200 A
B-3001 Heverlee
BELGIUM
Marc.VanBarel@cs.kuleuven.ac.be

Ann Haegemans
Katholieke Universiteit Leuven
Department of Computer Science
Celestijnenlaan 200 A
B-3001 Heverlee
BELGIUM
Ann.Haegemans@cs.kuleuven.ac.be

Fast decreasing rational functions and their applications

Eli (A. L.) Levin

Abstract. Let r_n be a rational function of the form $p_n(x)/p_n(-x)$, where p_n is a polynomial of degree n. Let φ be continuous and increasing on [0,1], with $\varphi(0) = 0$. Under what conditions on φ, does a sequence r_n, $n \geq 1$, exist such that

$$|r_n(x)| \leq C \exp\big(-n\varphi(x)\big), \quad x \in [0,1] \ ?$$

We present a survey of recent results concerning this problem. Some applications to rational approximation with locally geometric rates are also given.

§1 Introduction

Rational functions of the form

$$r_n(x) = p_n(x)/p_n(-x) \tag{1.1}$$

(p_n stands for a polynomial of degree $\leq n$) play an important role in rational approximation of $|x|$ and related functions. Indeed, set

$$R_n(x) := x\, \frac{1 - r_n(x)}{1 + r_n(x)} \ . \tag{1.2}$$

Since $r_n(x)r_n(-x) = 1$, we see that R_n is an *even* rational function (of degree $\leq n + 1$) and the following identity holds:

$$|x| - R_n(x) = 2|x|\frac{r_n(|x|)}{1 + r_n(|x|)}. \tag{1.3}$$

Therefore, if we construct r_n of form (1.1) that is fast decreasing on $[0, 1]$ (note that $r_n(0) = 1$), we automatically get a good rational approximant to $|x|$ on $[-1, 1]$.

Computational Methods and Function Theory (CMFT'97)
N. Papamichael, St. Ruscheweyh and E. B. Saff (Eds.), pp. 371–386

ISBN 981-02-3626-3

Example 1: Let $p_n(x) = (1-x)^n$, and let R_n be defined by (1.1), (1.2). Then (1.3) yields:

$$\big||x| - R_n(x)\big| \le 2|x|\Big(\frac{1-|x|}{1+|x|}\Big)^n, \quad x \in [-1,1] \, .$$

Hence, as $n \to \infty$, $R_n(x) \to |x|$ *geometrically fast* on any subinterval $[a,b] \subset [-1,1]\backslash\{0\}$. This is a nice feature, and one would expect this from good rational approximants, since $|x|$ is analytic on such $[a,b]$. On the other hand, it is easy to see that

$$cn^{-1} \le \max_{[-1,1]} \big||x| - R_n(x)\big| \le Cn^{-1}, \quad n \ge 1 \, ,$$

so that the *global rate* of approximation is poor.

Example 2: Let

$$p_n(x) = \prod_{j=0}^{n-1}(\zeta^j - x), \quad \zeta = \exp(-1/\sqrt{n}) \, .$$

Then (D. Newman [11])

$$|r_n(x)| \le e^{-\sqrt{n}}, \quad x \in [e^{-\sqrt{n}}, 1] \quad , \quad n \ge 5 \, .$$

Combined with an obvious inequality

$$0 < r_n(x) \le 1, \quad x \in [0, e^{-\sqrt{n}}] \, ,$$

this yields, for $x \in [0,1]$,

$$x|r_n(x)| \le e^{-\sqrt{n}} \quad \text{and} \quad 1 + r_n(x) \ge 1 - e^{-\sqrt{n}} \, . \tag{1.4}$$

Therefore, the corresponding R_n satisfy, by (1.3):

$$\max_{[-1,1]} \big||x| - R_n(x)\big| \le 3e^{-\sqrt{n}} \, ,$$

so that the global rate of approximation is incomparably better than in Example 1. However, we have now lost the geometric convergence!

Indeed, for any fixed $a \in (0,1)$, we have by (1.3)

$$\big||x| - R_n(x)\big| \ge a|r_n(x)|, \quad x \in [a,1] \, . \tag{1.5}$$

Next, split r_n as

$$r_n = \prod_{j\in J} \frac{\zeta^j - x}{\zeta^j + x} \prod_{j \notin J} \frac{\zeta^j - x}{\zeta^j + x} =: r_{n1} r_{n2},$$

where $J := \{j : \zeta^j < an^{-1}\}$. Since

$$\left|\frac{\zeta^j - x}{\zeta^j + x}\right| \geq 1 - \frac{2}{n}, \quad x \in [a,1], \quad j \in J,$$

we have

$$|r_{n1}(x)| \geq (1 - 2n^{-1})^n, \quad x \in [a,1] .$$

Furthermore, $\zeta^j \geq an^{-1}$ implies $j \leq \sqrt{n}\log(n/a)$, and thus $\deg r_{n2} = O(\sqrt{n}\log n)$. Now, for $0 < a < b \leq 1$,

$$\begin{aligned} \max_{[a,b]} \log |r_{n2}| &\geq \frac{1}{\log b/a} \int_a^b \log |r_{n2}(x)| \frac{dx}{x} \\ &\geq -\frac{\pi^2}{2\log b/a} \cdot \deg r_{n2}, \end{aligned}$$

by *Newman's inequality* (*cf.* [11])

$$\int_a^b \log \left|\frac{\zeta - x}{\zeta + x}\right| \frac{dx}{x} \geq -\frac{\pi^2}{2}, \quad 0 < a < b < \infty , \quad \zeta \in \mathbb{C} . \tag{1.6}$$

Collecting the above estimates we obtain that

$$\max_{[a,b]} \big||x| - R_n(x)\big| \geq C\exp\{-c\sqrt{n}\log n\}, \quad n \geq 5 ,$$

hence R_n does not converge geometrically to $|x|$ on any $[a,b] \subset [0,1]$.

The same conclusion is true for rational approximants constructed by N. Vyacheslavov [23], and for the best ones studied by H. Stahl [17]. In both cases we have a relation (1.5), with a suitable r_n of form (1.1), and the zeros of these r_n behave like Newman's nodes, ζ^j – they are extremely crowded near 0. An inevitable consequence of this is (as the above argument reveals) the lack of geometric convergence.

Example 3: For n even, let

$$r_n := r^{(1)}_{\frac{n}{2}} r^{(2)}_{\frac{n}{2}},$$

where $r^{(1)}_{n/2}, r^{(2)}_{n/2}$ are rational functions of Examples 1 and 2, respectively, and define R_n by (1.2). Since $|r^{(1)}_{n/2}| \leq 1$ on [0,1], we see that (1.4) holds (with $\sqrt{n}$ replaced by $\sqrt{n/2}$) for the above r_n as well. Also, $|r^{(2)}_{n/2}| \leq 1$ on $[0,1]$. Hence (1.3) yields:

$$\big||x| - R_n(x)\big| \leq 3\Big(\frac{1-|x|}{1+|x|}\Big)^{\frac{n}{2}} e^{-\sqrt{n/2}}, \quad x \in [-1,1] .$$

Thus, we now have both effects – R_n converge geometrically to $|x|$ on any $[a,b] \subset [-1,1]\backslash\{0\}$, and an $\exp(-c\sqrt{n})$ global rate is retained.

An obvious question is to what extent this last estimate can be improved? More precisely:

- Let φ be a continuous nondecreasing function on [0,1], with $\varphi(0) = 0$, and let $c > 0$ be a constant. Under what conditions on φ, c does a sequence $\{R_n\}$ exist such that, for some $C > 0$

$$\big||x| - R_n(x)\big| \leq Ce^{-n\varphi(x)}e^{-c\sqrt{n}}, \quad x \in [-1,1], \quad n \geq 1? \tag{1.7}$$

(Recall that the error in best approximation of $|x|$ by rational functions of order n behaves like $\exp(-\pi\sqrt{n})$ (*cf.* [17], [23]).) Hence, in any event, the exponent c in (1.7) must satisfy $c \leq \pi$).

It is clear from the above examples that we must first settle a companion problem:

- Let φ be as above. Under what conditions on φ does a sequence $\{r_n\}$ of rational functions of form (1.1) exist, such that

$$|r_n(x)| \leq Ce^{-n\varphi(x)}, \quad x \in [0,1], \quad n \geq 1? \tag{1.8}$$

§2 Fast decreasing rationals – crude estimates

In this section we present an elementary condition on φ, that is necessary and sufficient for (1.8) to hold in a *weaker form*, namely:

$$|r_n(x)| \leq Ce^{-cn\varphi(x)}, \quad x \in [0,1], \quad n \geq 1, \tag{2.1}$$

with some $C, c > 0$.

The following result is due to V. Maimeskul, E. Saff and the author.

Theorem 1. ([6]) *Let φ be a right continuous, nondecreasing function on [0,1] with $\varphi(0) = 0$. There exists a sequence $\{r_n\}$ of rational functions of form (1.1) with property (2.1), if and only if*

$$\int_0^1 \frac{\varphi(x)}{x}dx < \infty\ . \tag{2.2}$$

Sketch of the proof. Assume (2.1) holds. By Newman's inequality (1.6), we have for any $0 < \varepsilon < 1$:

$$-\frac{\pi^2}{2} \leq \frac{1}{n}\int_\varepsilon^1 \log|r_n(x)|\frac{dx}{x} \leq \frac{1}{n}\log\frac{1}{\varepsilon}\log C - c\int_\varepsilon^1 \frac{\varphi(x)}{x}dx\ .$$

On letting first $n \to \infty$ and then $\varepsilon \to 0$, we obtain

$$c\int_0^1 \frac{\varphi(x)}{x}dx \le \frac{\pi^2}{2} . \tag{2.3}$$

Conversely, assume (2.2) holds and set

$$s := \sum_0^\infty \varphi(2^{-k}) .$$

The monotonicity of φ implies:

$$\frac{1}{\log 2}\int_0^1 \frac{\varphi(x)}{x}dx \le s \le \varphi(1) + \frac{1}{\log 2}\int_0^1 \frac{\varphi(x)}{x}dx < \infty . \tag{2.4}$$

Fix $n \ge 1$, and let

$$m_k := \left[\frac{n}{s}\varphi(2^{-k})\right], \quad k = 0, 1, \dots ,$$

where $[\cdot]$ denotes the greatest integer function. Then

$$p_n(x) := \prod_{k\ge 0}(2^{-k} - x)^{m_k}$$

is a polynomial of degree $\le n$, and it is not hard to see that the corresponding $r_n(x) = p_n(x)/p_n(-x)$ satisfy (2.1) with $c = s^{-1}$. Note that, by (2.4),

$$c\int_0^1 \frac{\varphi(x)}{x}dx \le \log 2 . \quad \blacksquare \tag{2.5}$$

Remark:

(a) Since the conditions (2.3), (2.5) are not matching, the above result does not settle the original problem (1.7). The reason is (as we shall see) that the integral in (2.2) is the "wrong one" to consider.

(b) One should compare Theorem 1 with a result by K. Ivanov and V. Totik:

Theorem 2. ([4]) *Let φ be as in Theorem 1. There exists a sequence $\{p_n\}$ of even polynomial of degree n, $n = 1, 2, \dots$, with the property*

$$|p_n(x)| \le Ce^{-cn\varphi(x)}, \quad x \in [0,1], \quad n \ge 1, \tag{2.6}$$

if and only if

$$\int_0^1 \frac{\varphi(x)}{x^2}dx < \infty. \tag{2.7}$$

§3 Applications

We have already seen that fast decreasing rationals are intimately related to rational approximation of $|x|$. They also prove useful in approximation of many other functions, such as piecewise analytic ones. Other applications include digital filter design (we briefly touch on this below), numerical integration in the Hardy spaces H^p (*cf.* [8]), etc. In all these applications one usually needs to approximate the signum function as a first step.

(a) Approximation of the signum function

The case of polynomial approximation was considered by E. Saff and V. Totik:

Theorem 3. ([15]) *Let φ be as in Theorem 1. There exists a sequence $\{P_n\}$ of polynomials of degree n, $n = 1, 2, \dots,$ with the property*

$$|\mathrm{sgn} x - P_n(x)| \leq C \exp\big(-cn\varphi(|x|)\big), \quad x \in [-1, 1], \quad n \geq 1, \tag{3.1}$$

if and only if φ satisfies (2.7).

(Note that $\varphi(x) = x^\alpha$ was considered in [15], but the same proof works in the general case. See also [4].)

A similar result holds for the case of rational approximation:

Theorem 4. ([6], [7]) *Let φ be as in Theorem 1. There exists a sequence $\{R_n\}$ of rational functions of degree n, $n = 1, 2, \dots$ with the property*

$$|\mathrm{sgn} x - R_n(x)| \leq C \exp\big(-cn\varphi(|x|)\big), \quad x \in [-1, 1], \quad n \geq 1, \tag{3.2}$$

if and only if φ satisfies (2.2).

Sketch of the proof. Let such $\{R_n\}$ exist. It is not hard to see that, at the expense of C, we may assume that R_n is odd. Then, by the classical Newman's argument (*cf.* [12]) one deduces that a sequence $\{r_n\}$ of functions of form (1.1) exists, with property (2.1). Hence (Theorem 1) φ must satisfy (2.2).

Conversely, if φ satisfies (2.2), one can find a sequence $\{r_n\}$ with property (2.1). We may assume that all zeros of r_n lie in (0,1]. Indeed, if it is not the case, replace each factor $(x - \zeta)/(x + \zeta)$, with $\zeta \notin (0, 1]$, by $(x - \zeta')/(x + \zeta')$, with $\zeta' := \min\{|\mathrm{Re}\,\zeta|, 1\}$, thereby making $|r_n|$ smaller on [0,1].

Next, for n large enough, define $\varepsilon_n \in (0, 1)$ by

$$C \exp\big(-cn\varphi(\varepsilon_n)\big) = 1/2$$

(with C, c as in (2.1)), and replace each zero ζ of r_n lying in $(0, \varepsilon_n)$ by ε_n. Then the resulting r_n still satisfies (2.1), with C replaced by $\max\{2, C\}$, and we also have

$$1 + r_n(x) \geq 1/2, \quad x \in [0, 1]. \tag{3.3}$$

Finally, set

$$R_n = \frac{1 - r_n}{1 + r_n} .$$

R_n is odd, and (1.2), (3.3) ensure that R_n satisfies (3.2). ∎

(b) Digital filter design

Various problems of signal processing can be reduced to the following one. Consider an *ideal lowpass filter*, that is characterized by the function

$$u(x) = \begin{cases} 1, & 0 \leq x < \omega_c \\ 0, & \omega_c < x \leq \pi \end{cases}, \tag{3.4}$$

where ω_c is the so-called cutoff frequency. Let us choose some passband cutoff frequency ω_p and some stopband cutoff frequency ω_s, such that $0 < \omega_p < \omega_c < \omega_s < \pi$. We seek a complex rational function $R_n(z)$, such that $|R_n(e^{ix})|$ approximates $u(x)$ within given errors $\pm\delta_1$ and $\pm\delta_2$ in $[0, \omega_p]$ and $[\omega_s, \pi]$, respectively, and such that $|R_n|$ drops smoothly in the transition band $[\omega_p, \omega_s]$. For this purpose one may use the classical Zolotarev's construction, which leads to the so-called elliptic filters. A drawback of such filters is that the difference $|R_n| - u$ equiripples in the passband and the stopband. To make this difference more "flat" away from the transition band, a damping geometric factor can be introduced (compare with Example 3, Section 1). The following result (which is a variant of Theorem 4) then comes in handy.

Theorem 5. ([6]) *Let φ be as in Theorem 1, and let u be defined by (3.4). Then, given $\varepsilon \in (0, 1)$ and $n \geq 1$, there exists R_n such that for $x \in [0, \pi]$ there holds*

$$|u(x) - R_n(x)| \leq C \exp\left\{-cn\left(\frac{1}{\log\frac{1}{\varepsilon}} + \varphi(|x - \omega_p|)\right)\right\}, \quad \text{if} \quad |x - \omega_p| > \varepsilon ,$$

and

$$|R_n(x)| \leq C, \quad \text{if} \quad |x - \omega_p| < \varepsilon .$$

The constants C, c are independent of n, ω_p and ε.

(c) Rational approximation with locally geometric rates

Let $\mathbf{P}_n$ denote the class of all real polynomials of degree $\le n$, and let $\mathbf{R}_n$ be the class of all rational functions P/Q, $P, Q \in \mathbf{P}_n$. For any $f \in C[-1,1]$ we denote by $e_n(f)$ and $\rho_n(f)$ the errors in best uniform approximation of f on $[-1,1]$ by elements of $\mathbf{P}_n$ and $\mathbf{R}_n$, respectively. Let P_n^*, R_n^* be the best approximants, that is

$$e_n(f) = \|f - P_n^*\|_{L_\infty[-1,1]}, \quad \rho_n(f) = \|f - R_n^*\|_{L_\infty[-1,1]} \,.$$

Assume now that f is piecewise analytic on $[-1,1]$, by which we mean that there exists a partition

$$-1 = x_0 < x_1 < \cdots < x_s < x_{s+1} = 1 \tag{3.5}$$

such that the restriction of f to each $[x_j, x_{j+1}]$, $0 \le j \le s$, has an analytic continuation to a neighborhood of this *closed* interval, but f itself is not analytic at $x_1, \ldots, x_s$. We call these points the *singularities* of f on $(-1,1)$ and assume that f has at least one such singularity (that is, $s \ge 1$). Then $f \in C^{k-1}[-1,1] \backslash C^k[-1,1]$, for some $k \ge 1$, and hence (*cf.* [18, Chapter 7])

$$c_1 n^{-k} \le e_n(f) \le c_2 n^{-k}, \quad n = 1, 2, \ldots \tag{3.6}$$

for some $c_1, c_2 > 0$. Also (*cf.* [3], [22])

$$e^{-c_3\sqrt{n}} \le \rho_n(f) \le e^{-c_4\sqrt{n}}, \quad n = 1, 2, \ldots \tag{3.7}$$

for some $c_3, c_4 > 0$.

It seems reasonable to expect, however, that $f - P_n^*$ and $f - R_n^*$ tend to zero much faster (say, geometrically) on intervals of analyticity of f. Unfortunately, P_n^* never enjoys this property (see [13], [15] for a more detailed discussion of this phenomenon) and R_n^* may not enjoy it (see [17]). Thus it is natural to seek "near best" approximants, that converge to f with a global rate close to the best one, and at the same time converge geometrically outside singularities of f.

This problem was investigated, for the polynomial case, in [2], [5], [15], [20]. In particular, the following result was obtained by E. Saff and V. Totik.

Theorem 6. ([15]) *Let φ be as in Theorem 1 and satisfy (2.7). Let f be piecewise analytic on [-1,1]. Then there exist constants $c, C > 0$ and polynomials $P_n \in \mathbf{P}_n$, $n = 1, 2, \ldots$, such that for every $x \in [-1,1]$*

$$|f(x) - P_n(x)| \le C e_n(f) \exp\big(-cn\varphi(d_x)\big), \tag{3.8}$$

where d_x denotes the distance from x to the nearest singularity of f in (-1,1).

Remark: For $\varphi(x) = x^\beta$, $\beta > 1$, this theorem follows from Theorem 3 in [15], if we take into account estimates (3.6). For a general φ, the proof is the same, but one applies the general result (Theorem 2) instead of Theorem 2 of [15].

A similar result holds for the case of rational approximation, and it is due to V. Maimeskul, E. Saff and the author.

Theorem 7. *Let φ be as in Theorem 1 and satisfy (2.2). Let f be piecewise analytic on [-1,1]. Then there exist constants $c > 0$, $0 < \alpha < 1$, and rational functions $R_n \in \mathbf{R}_n$, $n = 1, 2, \ldots$, such that for all $x \in [-1, 1]$*

$$|f(x) - R_n(x)| \leq \rho_{\alpha n}(f) \exp\big(- cn\varphi(d_x)\big), \tag{3.9}$$

where d_x denotes the distance from x to the nearest singularity of f in (-1,1).

Remarks:

(a) This result follows from Theorem 4.1 in [6], combined with estimates (3.7).

(b) The restriction $\alpha < 1$ is important and, in general, one cannot replace $\rho_{\alpha n}$ by $C\rho_n$, with some constant $C > 0$ (see Theorem 11 below).

(c) Note that the present φ can grow much faster near 0 than φ of Theorem 6 (compare the conditions (2.2) and (2.7)). Hence the geometric rate of convergence is better (near singularities of f) than in the polynomial case.

(d) Theorem 7 can be extended to more general functions. For example, one may assume that the restriction of f to each *open* interval (x_j, x_{j+1}) has an analytic and bounded continuation to some open rhombus with opposite vertices x_j, x_{j+1}, $0 \leq j \leq s$. See Theorem 5.1 in [6].

§4 Sharp estimates

In this section we answer two questions that were posed in Section 1. We look for conditions on φ that ensure the existence of r_n, $n \geq 1$, of the form

$$r_n(x) = \prod_{j=1}^{n} \frac{x - \zeta_{jn}}{x + \zeta_{jn}}, \tag{4.1}$$

that satisfy, with some $C > 0$,

$$|r_n(x)| \le Ce^{-n\varphi(x)}, \quad x \in [0,1], \quad n = 1, 2, \ldots. \tag{4.2}$$

We also consider the property

$$|r_n(x)| \le e^{o(n)-n\varphi(x)}, \quad x \in [0,1], \quad n = 1, 2, \ldots, \tag{4.3}$$

where $o(n)$ is *uniform in* x.

Note that (4.3) is weaker than (4.2) but stronger (in a sense) than (2.1). We have already obtained (see in the proof of Theorem 2.1) that the condition

$$\frac{2}{\pi^2}\int_0^1 \frac{\varphi(x)}{x}dx \le 1 \tag{4.4}$$

is necessary for r_n satisfying (4.2) to exist. The same proof actually shows that (4.4) is also necessary for (4.3) to hold. The proof of (4.4) was based on Newman's inequality (1.6), that can be written as

$$\frac{2}{\pi^2}\int_a^b \log\left|\frac{\zeta+t}{\zeta-t}\right|\frac{dt}{t} \le 1, \quad 0 \le a < b \le \infty, \quad \zeta \in \mathbb{C}\,.$$

This is sharp, since equality holds provided $a = 0$, $b = \infty$ and $\zeta \in (0,\infty)$ (*cf.* [11]). It turns out that, on replacing dt/t by a suitable measure, a stronger inequality can be obtained for $0 \le a < b \le 1$, and this will lead to a stronger necessary condition on φ.

Lemma 1. ([7]). *For* $0 \le a < b \le 1$,

$$\frac{2}{\pi^2}\int_a^b \log\left|\frac{\zeta+t}{\zeta-t}\right|\frac{dt}{t\sqrt{1-t^2}} \le 1, \quad \zeta \in \mathbb{C}, \tag{4.5}$$

and

$$\frac{2}{\pi^2}\int_0^1 \log\left|\frac{\zeta+t}{\zeta-t}\right|\frac{dt}{t\sqrt{1-t^2}} = 1, \quad \zeta \in (0,1]\,. \tag{4.6}$$

Proof: For $\zeta > 0$, the integrand in (4.6) is positive. Hence (4.6) implies (4.5) for such ζ. Then the maximum principle (note that the integral in (4.5) is a subharmonic function of ζ, for $\zeta \in \overline{\mathbb{C}}\backslash[a,b]$) yields (4.5) for all $\zeta \in \mathbb{C}$. Thus, it remains to establish (4.6).

Let $0 < \varepsilon < 1$. It is well-known (*cf.* [1]) that, for $\zeta \in [\varepsilon, 1]$,

$$\frac{1}{\pi}\int_\varepsilon^1 \log\left|\frac{\zeta+t}{\zeta-t}\right|\frac{dt}{\sqrt{(1-t^2)(t^2-\varepsilon^2)}} = K(\varepsilon),$$

where $K(\varepsilon)$ is the complete elliptic integral with modulus ε. Because $K(\varepsilon) \to \pi/2$, $\varepsilon \to 0$, we obtain (4.6). ■

Now, assuming $\{r_n\}$ satisfies (4.3) and proceeding as in the proof of Theorem 1, we obtain that the condition

$$\frac{2}{\pi^2} \int_0^1 \varphi(x) \frac{dx}{x\sqrt{1-x^2}} \leq 1 \tag{4.7}$$

is necessary for (4.3) (hence, for (4.2)) to hold.

This is *sharp* in the sense that, under some restrictions on φ, (4.7) is sufficient for (4.3) to hold.

Theorem 8. ([7]) *Assume φ is continuous, increasing and concave on [0,1]. Then there exist r_n, $n \geq 1$, with property (4.3) if and only if φ satisfies (4.7).*

The necessity of (4.7) has just been established. The proof of sufficiency is more involved and it employs potential theoretic methods, in the spirit of Totik's paper [19] on fast decreasing polynomials. We only indicate the main steps.

Step 1. Given a measure $\mu \geq 0$ in $D := \{z \in \mathbb{C}, \text{Re } z > 0\}$, let

$$U_g^\mu(z) := \int \log \left| \frac{z+\bar{t}}{z-t} \right| d\mu(t), \quad z \in D, \tag{4.8}$$

denote the Green potential of μ. Let $\varphi \in C[0,1]$, $\varphi(0) = 0$. We show that there exists μ^* of total mass $\|\mu^*\| = 1$ with support $S(\mu^*) \subseteq [0,1]$, that satisfies, with some constant c^*

$$U_g^{\mu^*} - c^* = \varphi, \quad \text{on } S(\mu^*), \tag{4.9}$$

$$U_g^{\mu^*} - c^* \geq \varphi, \quad \text{on } [0,1] \,. \tag{4.10}$$

Note that this is a non-standard "singular" equilibrium problem, since [0,1] touches the boundary of D (*cf.* [16, Chapter II.5]). We also show (with the aid of Lemma 1) that $U_g^{\mu^*} - c^*$ is the *least superharmonic majorant for φ on* [0,1], in the class of all superharmonic functions of the form $U_g^\nu +$ const, $\|\nu\| \leq 1$, $S(\mu) \subset D$, thereby proving the *uniqueness* of the above μ^*. (Incidentally, the equilibrium measure μ^* may have a mass at 0; *e.g.* if $\varphi = 0$ then $S(\mu^*) = \{0\}$ and $U_g^{\mu^*} \equiv 0$.)

The relevance of the above equilibrium problem is obvious: given r_n as in (4.1), we may write

$$|r_n(z)| = \exp\big(-nU_g^{\nu_n}(z)\big),$$

where

$$\nu_n := \frac{1}{n}\sum_{j=1}^{n} \delta_{\zeta_{jn}}, \quad \|\nu_n\| = 1, \tag{4.11}$$

and $\delta_{\zeta_{jn}}$ denotes the unit point mass at ζ_{jn}. Then the property (4.3) can be formulated as

$$U_g^{\nu_n}(x) - o(1) \geq \varphi(x), \quad x \in [0,1], \tag{4.12}$$

(with $o(1)$ uniform in x). Thus, our goal is to describe those φ for which c^* in (4.9), (4.10) is zero. Then, on discretizing μ^*, we may hope to get the desired r_n.

Step 2. Define, for any compact $K \subseteq [0,1]$,

$$F(K) := -\frac{1}{\operatorname{cap}_g K} + \int \varphi d\omega_K,$$

where $\operatorname{cap}_g K$ is the Green capacity of K relative to D (*cf.* [24]) and ω_K denotes the equilibrium distribution of K. We show that

$$\max_K F(K) = F\big(S(\mu^*)\big) = -c^* \tag{4.13}$$

(compare with the F-functional introduced by Mhaskar and Saff in [10]).

Step 3. Assume now that φ is increasing and concave on [0,1] and satisfies

$$0 < \rho := \frac{2}{\pi^2}\int_0^1 \varphi(x)\frac{dx}{x\sqrt{1-x^2}} \leq 1\ . \tag{4.14}$$

We show, by virtue of (4.13), that $c^* = 0$ and $S(\mu^*) = [0,1]$. This yields (see (4.9))

$$U_g^{\mu^*} = \varphi, \quad \text{on} \quad [0,1]\ .$$

Furthermore,

$$\mu^*(\{0\}) = 1 - \rho\ .$$

Step 4. Assume $\rho = 1$ and define for a given natural number n, the partition $0 = \zeta_{0n} < \zeta_{1n} < \cdots < \zeta_{nn} = 1$ by

$$\int_{\zeta_{j-1,n}}^{\zeta_{jn}} d\mu^* = \frac{1}{n}, \quad j = 1, \ldots, n,$$

and let ν_n be as in (4.11). Proceeding as in [9], we deduce (4.12) which is equivalent to (4.3). If $\rho < 1$, we have (4.3) even with $\rho^{-1}\varphi$ instead of φ. ■

To get r_n with a stronger property (4.2), one needs (4.12) with $o(1)$ replaced by $O(n^{-1})$. Here we have to analyze the density of μ^*. In doing so we impose extra conditions on φ, that ensure regular behaviour of this density near 0. Assume that for x small enough,

(a) $x\varphi'(x)$ is increasing

(b) for some $\delta > 0$, $x\varphi'(x) \leq (1-\delta)\varphi(x)$.

Then we have:

Theorem 9. ([7]) *Let φ be as in Theorem 8, and assume that φ satisfies (a), (b) for x small enough. Then there exist r_n, $n \geq 1$ with property (4.2) provided strict inequality holds in (4.7), i.e.*

$$\frac{2}{\pi^2}\int_0^1 \varphi(x)\frac{dx}{x\sqrt{1-x^2}} < 1 \; . \tag{4.15}$$

Remarks:

(i) Many important functions, like cx^α, $\alpha < 1$ or $[\log(c/x)]^{-\alpha}$, $\alpha > 1$, satisfy (a), (b) above. Moreover, applying a finer discretization technique (*cf.* [21], [16, Chapter VI.4]) one can replace (4.15) by (4.7), at least for the case $\varphi(x) = cx^\alpha$, $\alpha < 1$.

(ii) One should compare Theorems 8, 9 with a result of V. Totik:

Theorem 10. ([19]) *Let φ be continuous and increasing on [0,1], and assume that $\varphi(\sqrt{x})$ is concave. Then there exist polynomials $p_n \in \mathbf{P}_n$, $n \geq 1$, such that*

$$|p_n(x)| \leq e^{o(n)-n\varphi(|x|)}, \quad x \in [-1,1] \; , \tag{4.16}$$

if and only if

$$\frac{2}{\pi}\int_0^1 \varphi(x)\frac{dx}{x^2\sqrt{1-x^2}} \leq 1 \; . \tag{4.17}$$

Moreover, if strict inequality holds in (4.17), then (4.16) holds with $O(1)$ instead of $o(n)$.

Finally, we apply Theorem 9 to the rational approximation of $|x|$. We look for conditions on c, φ to ensure the existence of $R_n \subset \mathbf{R}_n$, $n \geq 1$, such that, with some $C > 0$,

$$\big||x| - R_n(x)\big| \leq C\exp\big(-c\sqrt{n} - n\varphi(|x|)\big), \quad x \in [-1,1] \; . \tag{4.18}$$

Theorem 11. ([7]) *Let φ be a continuous increasing function on [0,1], with $\varphi(0) = 0$. If a sequence $\{R_n\}$ with property (4.18) exists, then*

$$\frac{2}{\pi^2}\int_0^1 \varphi(x)\frac{dx}{x\sqrt{1-x^2}} + \frac{c^2}{\pi^2} \le 1 \, . \tag{4.19}$$

In particular, if $c = \pi$, geometric convergence is not possible.

Conversely, if strict inequality holds in (4.19) and φ is as in Theorem 9, then there exist $\{R_n\}$ satisfying (4.18).

References

[1] Akhiezer, N. I., *Elements of the Theory of Elliptic Functions*, Translations of Amer. Math. Soc. v. 79, Providence 1990.

[2] Gaier, D., Polynomial approximation of piecewise analytic functions, *Journal of Analysis* **4** (1996), 67–79.

[3] Gonchar, A. A., Estimates of the growth of rational functions and some of their applications, *Mat. Sb.* **72(114)** (1967), 489–503; Engl. transl. in *Math. USSR Sb.* **1** (1967), 445–456.

[4] Ivanov, K. G. and V. Totik, Fast decreasing polynomials, *Constr. Approx.* **6** (1990), 1–20.

[5] Ivanov, K. G., E. B. Saff and V. Totik, Approximation by polynomials with locally geometric rates, *Proc. Amer. Math. Soc.* **106** (1989), 153–161.

[6] Levin, A. L., V. V. Maimeskul and E. B. Saff, Rational approximation with locally geometric rates, *J. Approx. Th.* **92** (1998), 308–330.

[7] Levin, A. L. and E. B. Saff, Fast decreasing rational functions, *Israel J. Math.*, 1998 to appear.

[8] Levin, A. L. and E. B. Saff, Exact convergence rates for best L_p-rational approximation and for the optimal quadrature in H^p, in: *Methods of Approximation Theory in Complex Analysis and Mathematical Physics*, Moscow, Nauka (1992), 98–109.

[9] Lubinsky, D. S. and E. B. Saff, *Strong asymptotic for extremal polynomials associated with weights on* $\mathbb{R}$, in Lecture Notes in Math., v. 1305, Springer Verlag, 1988.

[10] Mhaskar, H. N. and E. B. Saff, Where does the sup norm of a weighted polynomial live? (A generalization of incomplete polynomials), *Constr. Approx.* **1** (1985), 71-91.

[11] Newman, D. J., Rational approximation to $|x|$, *Michigan Math. J.* **11** (1964), 11–14.

[12] Petrushev, P. P. and V. A. Popov, *Rational Approximation of Real Functions*, Cambridge Univ. Press, Cambridge, 1987.

[13] Saff, E. B., A principle of contamination in best polynomial approximation, Proc. of Cuba seminar in approximation and optimization (Habana, 1987), *Lecture Notes in Math.*, v. 1354, Springer (1988), 79–97.

[14] Saff, E. B. and H. Stahl, Asymptotic distribution of poles and zeros of best rational approximants to x^α on [0,1], in: *Topics in Complex Analysis, Banach Center Publications*, v. 31, Institute of Mathematics, Polish Academy of Warsaw, 1995.

[15] Saff, E. B. and V. Totik, Polynomial approximation of piecewise analytic functions, *J. London Math. Soc.* **2(39)** (1989), 487–498.

[16] Saff, E. B. and V. Totik, *Logarithmic Potentials with External Fields*, Springer, 1997.

[17] Stahl, H., Best uniform rational approximation of $|x|$ on [-1,1], *Mat. Sb.* **183** (1992), 85–112; Engl. transl.: *Russian Acad. Sci. Sb. Math.* **76** (1993), 461–487.

[18] Timan, A. F., *Theory of Approximation of Functions of a Real Variable*, Pergamon Press, 1963.

[19] Totik, V., Fast decreasing polynomials via potentials, *J. d'Analyse Math.* **62** (1994), 131–154.

[20] Totik, V., Polynomial approximation with locally geometric rate, In: *Colloquia Mathematica Societatis János Bolyai*, v. 58, Approximation Theory, Kecskemet (Hungary), 1990.

[21] Totik, T., *Weighted Approximation with Varying Weights*, Lecture Notes in Math., v. 1569, Springer, 1994.

[22] Turán, P., On the approximation of piecewise analytic functions by rational functions, *Contemporary Problems of Theory of Analytic Functions*, Nauka, Moscow, 1966, 296–299.

[23] Vyacheslavov, N. S., On the approximation of $|x|$ by rational functions, *Dokl. Akad. Nauk SSSR* **220** (1975), 512–515; Engl. transl.: *Soviet Math. Dokl.* **16** (1975), 100–104.

[24] Widom, H., Rational approximation and n-dimensional diameter, *J. Approx. Th.* **5** (1972), 343–361.

Eli (A. L.) Levin
Department of Mathematics
The Open University of Israel
Tel Aviv 61392
ISRAEL
elile@oumail.openu.ac.il

Harmonic univalent mappings of the unit disc onto a vertical strip

Wiesław Majchrzak

Abstract. In this paper the class $\mathcal{P}_H$ of harmonic, univalent, sense-preserving and normalized mappings of the unit disc onto a vertical strip is considered, as well its closure $\mathcal{F}_H$. An integral representation for functions in $\mathcal{F}_H$ is determined and several important properties are studied. In particular, the coefficient problem is partially solved.

§1 Statement of results

Let $\mathbb{D}$ denote the unit disc $\{z : |z| < 1\}$ and $\Omega = \Omega(m, M)$ the strip $\{w : m < \operatorname{Re} w < M\}$ where m, M are arbitrary real constants, $m < 1 < M$.

We consider the class $\mathcal{P}_H = \mathcal{P}_H(\mathbb{D}, \Omega)$ of all harmonic, univalent, orientation-preserving mappings $f = h + \overline{g}$ from $\mathbb{D}$ onto Ω, where h and g are holomorphic in $\mathbb{D}$ and

$$h(z) = 1 + a_1 z + \sum_{n=2}^{\infty} a_n z^n, \qquad g(z) = \sum_{n=2}^{\infty} \overline{a}_{-n} z^n, \tag{1.1}$$

with $a_1 > 0$.

If $f \in \mathcal{P}_H$, then $f = u + iv$ where u, v are real harmonic functions defined in $\mathbb{D}$. Taking account of the normalizing conditions in (1.1), we have

$$u(0) = 1, \qquad u'_x(0) = v'_y(0) > 0, \qquad u'_y(0) = v'_x(0) = 0. \tag{1.2}$$

Let u_∂ denote the step function

$$u_\partial(z) = \begin{cases} M & \text{for } z \in \Gamma, \\ m & \text{for } z \in \operatorname{Int}(\mathbb{T} \setminus \Gamma), \end{cases} \tag{1.3}$$

defined on the unit circle $\mathbb{T}$, where $\Gamma \subset \mathbb{T}$ is an open arc. The function u_∂ will be called a circle mapping.

For the constants m, M denote $\alpha := (1 - m)/(M - m)$.

Computational Methods and Function Theory (CMFT'97)
N. Papamichael, St. Ruscheweyh and E. B. Saff (Eds.), pp. 387–396

ISBN 981-02-3626-3

Lemma 1. *For a given circle mapping u_∂, there exists a unique solution $u = u(z)$ of the Dirichlet problem in $\mathbb{D}$ under conditions (1.2). It has the form*

$$u(z) = 2m - 1 + \frac{M-m}{\pi} \arg \frac{e^{i\pi\alpha} - z}{e^{-i\pi\alpha} - z}, \qquad z \in \mathbb{D},$$

where $\Gamma = \Gamma(\alpha) = \{e^{it} : t \in (-\pi\alpha, \pi\alpha)\}$ *and* $\arg\{\frac{e^{i\pi\alpha}-z}{e^{-i\pi\alpha}-z}\}\big|_{z=0} = 2\pi\alpha$.

Proof: Write $\Gamma = \{e^{it} : t \in (t_1, t_2)\}$. It is well known that a solution of the Dirichlet problem in $\mathbb{D}$ with given u_∂ is

$$u(z) = \frac{1}{2\pi} \int_0^{2\pi} u_\partial(e^{it}) \operatorname{Re} \frac{e^{it} + z}{e^{it} - z}\, dt.$$

Thus, using (1.3), we have

$$u(z) = m + \frac{M-m}{2\pi} \int_{t_1}^{t_2} \operatorname{Re} \frac{e^{it} + z}{e^{it} - z}\, dt.$$

Hence, after integration, we obtain

$$u(z) = m + \frac{M-m}{2\pi} \left[t_2 - t_1 + 2 \arg \frac{e^{it_1}(e^{it_2} - z)}{e^{it_2}(e^{it_1} - z)} \right], \qquad z \in \mathbb{D},$$

where $\arg 1 = 0$. Finally, taking account of the normalizing conditions (1.2), we obtain $t_1 = -\pi\alpha$ and $t_2 = \pi\alpha$, $\alpha = (1-m)/(M-m)$. Thus, u as well as Γ are as asserted. ■

For $z \in \mathbb{D}$, $|\eta| = 1$ and $\alpha \in (0,1)$ define the function

$$K(z,\eta) := \int_0^z \frac{1+\eta z}{1-\eta z} \frac{dz}{(e^{i\pi\alpha} - z)(e^{-i\pi\alpha} - z)} \tag{1.4}$$

$$= \begin{cases} \dfrac{e^{i\pi\alpha} + e^{-i\pi\alpha}}{(e^{-i\pi\alpha} - e^{i\pi\alpha})^2} \log \dfrac{1 - ze^{i\pi\alpha}}{1 - ze^{-i\pi\alpha}} \\ \quad - \dfrac{i}{\sin \pi\alpha} \dfrac{z}{e^{-i\pi\alpha} - z}, & \text{for } \eta = e^{i\pi\alpha} \\ \dfrac{e^{i\pi\alpha} + e^{-i\pi\alpha}}{(e^{-i\pi\alpha} - e^{i\pi\alpha})^2} \log \dfrac{1 - ze^{-i\pi\alpha}}{1 - ze^{i\pi\alpha}} \\ \quad + \dfrac{i}{\sin \pi\alpha} \dfrac{z}{e^{i\pi\alpha} - z}, & \text{for } \eta = e^{-i\pi\alpha} \\ \dfrac{-2\overline{\eta}}{(\overline{\eta} - e^{i\pi\alpha})(\overline{\eta} - e^{-i\pi\alpha})} \log(1 - \eta z) \\ \quad + \dfrac{e^{i\pi\alpha} + \overline{\eta}}{(\overline{\eta} - e^{i\pi\alpha})(e^{i\pi\alpha} - e^{-i\pi\alpha})} \log(1 - ze^{-i\pi\alpha}) \\ \quad + \dfrac{e^{-i\pi\alpha} + \overline{\eta}}{(\overline{\eta} - e^{-i\pi\alpha})(e^{-i\pi\alpha} - e^{i\pi\alpha})} \log(1 - ze^{i\pi\alpha}), & \text{for } \eta \neq e^{\pm i\pi\alpha}, \end{cases}$$

where $\log 1 = 0$. Let $\mathcal{M}$ denote the set of all probability measures μ defined on Borel subsets of the circle $|\eta| = 1$. Consider the class $\mathcal{F}_H$ of all harmonic mappings $f : \mathbb{D} \to \mathbb{C}$ of the form

$$\begin{aligned} f(z) &= 2m - 1 + \frac{M-m}{\pi} \arg \frac{e^{i\pi\alpha} - z}{e^{-i\pi\alpha} - z} \\ &+ i\frac{2(M-m)\sin\pi\alpha}{\pi} \operatorname{Im}\left\{\int_{|\eta|=1} K(z,\eta)\, d\mu\right\}, \end{aligned} \tag{1.5}$$

where $\mu \in \mathcal{M}$.

Theorem 1. *If $f \in \mathcal{P}_H$, then there is a measure $\mu \in \mathcal{M}$ such that f may be written as in (1.5).*

Proof: If $f = h + \overline{g} \in \mathcal{P}_H$, then $a = g'/h'$ satisfies the assumptions of Schwarz's Lemma. Let $u = \operatorname{Re} f = \operatorname{Re}(h+g)$ and $v = \operatorname{Im} f = \operatorname{Im}(h-g)$. Since we can apply Lemma 1 to $\operatorname{Re} f$, we have

$$h'(z) + g'(z) = 2u'(z) = \frac{2(M-m)}{\pi}\frac{d}{dz} \arg \frac{e^{i\pi\alpha} - z}{e^{-i\pi\alpha} - z}$$

and

$$h'(z) + g'(z) = \frac{2(M-m)\sin\pi\alpha}{\pi(e^{i\pi\alpha} - z)(e^{-i\pi\alpha} - z)}. \tag{1.6}$$

Note that

$$h'(z) - g'(z) = [h'(z) + g'(z)]\frac{1 - a(z)}{1 + a(z)}.$$

Thus

$$h'(z) - g'(z) = \frac{2(M-m)\sin\pi\alpha p(z)}{\pi(e^{i\pi\alpha} - z)(e^{-i\pi\alpha} - z)}, \tag{1.7}$$

where the function p is holomorphic in $\mathbb{D}$ with $\operatorname{Re} p(z) > 0$, $z \in \mathbb{D}$, $p(0) = 1$. Since, by (1.7),

$$v(z) = \operatorname{Im} f(z) = \operatorname{Im}\left\{\int_0^z [h'(z) - g'(z)]\, dz\right\},$$

we get

$$v(z) = \frac{2(M-m)\sin\pi\alpha}{\pi} \operatorname{Im}\left\{\int_{|\eta|=1} K(z,\eta)\, d\mu\right\}$$

from the well-known Herglotz integral formula for the function p, where $K(z,\eta)$ is defined by (1.4), which was asserted. ■

Therefore each function $f \in \mathcal{P}_H$ is representable by formula (1.5). This implies $f \in \mathcal{F}_H$ and therefore $\mathcal{P}_H \subset \mathcal{F}_H$.

Theorem 2. *$\mathcal{F}_H$ is a compact and convex class of harmonic, univalent and orientation-preserving mappings from $\mathbb{D}$ into Ω.*

Proof: It is easily noticed that $\mathcal{M}$ is a compact convex set of normalized measures. Hence, the class $\mathcal{F}_H$, generated by $\mathcal{M}$, is compact and convex.

We shall justify the univalence of mappings in $\mathcal{F}_H$. If $f \in \mathcal{F}_H$ then $\operatorname{Re} f = \operatorname{Re} \varphi^{-1}$ where

$$\varphi^{-1}(z) = 2m - 1 - i\frac{M-m}{\pi}\log\frac{e^{i\pi\alpha} - z}{e^{-i\pi\alpha} - z}, \qquad \varphi^{-1}(0) = 1, \tag{1.8}$$

maps the disc $\mathbb{D}$ conformally onto the strip Ω. Hence, $\operatorname{Re}(f \circ \varphi)$ is the identity mapping and $f \circ \varphi$ maps each vertical line in Ω into itself. Therefore, it is sufficient to verify the univalence of function in $\mathcal{F}_H$ on such lines.

Let $w = \delta + i\tau \in \Omega$, $\tau \in \mathbb{R}$, δ fixed. Then, for $f = u + iv$, we get

$$\frac{\partial}{\partial\tau}(f\circ\varphi)(w) = i\frac{\partial}{\partial\tau}(v\circ\varphi)(w).$$

If we decompose $f = h + \overline{g}$, then $v = \operatorname{Im}(h-g)$ and

$$\frac{\partial}{\partial\tau}(f\circ\varphi)(w) = i\frac{\partial}{\partial\tau}[\operatorname{Im}(h-g)\circ\varphi](w).$$

Next, from (1.7) we get

$$\begin{aligned}&\frac{\partial}{\partial\tau}(f\circ\varphi)(w) = \\ &\quad -\frac{2(M-m)}{\pi}\sin\pi\alpha\,\operatorname{Im}\left\{\frac{1}{(e^{i\pi\alpha}-\varphi)(e^{-i\pi\alpha}-\varphi)}(p\circ\varphi)\right\}\varphi'_w,\end{aligned} \tag{1.9}$$

where

$$\varphi'_w = \frac{\pi}{2(M-m)\sin\pi\alpha}(e^{i\pi\alpha} - \varphi(w))(e^{-i\pi\alpha} - \varphi(w)).$$

Since $\operatorname{Re}\frac{\partial}{\partial\tau}(f\circ\varphi) = 0$, equations (1.8) and (1.9) imply that $\operatorname{Re}\{(e^{i\pi\alpha} - \varphi(w))(e^{-i\pi\alpha} - \varphi(w))\} = 0$ and, in consequence,

$$\frac{\partial}{\partial\tau}(f\circ\varphi)(w) = i\operatorname{Re}(p\circ\varphi) \tag{1.10}$$

where $\operatorname{Re} p(z) > 0$, $z \in \mathbb{D}$. From (1.10) we immediately deduce

$$\operatorname{Im}\frac{\partial}{\partial\tau}(f\circ\varphi)(w) = \operatorname{Re}(p\circ\varphi) > 0,$$

for $\tau \in \mathbb{R}$ and $\delta \in (m, M)$ arbitrarily fixed. By the conformality of φ, this shows that f is univalent and orientation-preserving in $\mathbb{D}$. ■

A function f will be called convex in $\mathbb{D}$ if f is univalent in $\mathbb{D}$ and $f(\mathbb{D})$ is a convex set.

Theorem 3. *All functions in the class $\mathcal{F}_H$ are convex in $\mathbb{D}$.*

Proof: Note that a function $f = h + \overline{g} \in \mathcal{F}_H$ is convex in $\mathbb{D}$ if and only if, for any $\theta \in [0, \pi)$, the holomorphic function $F = -e^{-i\theta}(h + e^{2i\theta}g)$ is convex in the vertical direction. To check this we apply the Robertson condition ([2], p. 203). Namely, by (1.6) and (1.7), we have

$$\mathrm{Re}\{-i(e^{-i\pi\alpha} - z)(e^{i\pi\alpha} - z)F'(z)\} = a_1 \sin\theta \,\mathrm{Re}\, p(z), \tag{1.11}$$

where $a_1 > 0$ and $\mathrm{Re}\, p(z) > 0$ for $z \in \mathbb{D}$. Thus, equation (1.11) implies $\mathrm{Re}\{-i(e^{-i\pi\alpha} - z)(e^{i\pi\alpha} - z)F'(z)\} > 0$ for any $\theta \in (0, \pi)$ and $\alpha \in (0, 1)$. Moreover, if $\theta = 0$, then $F = -(h + g) = -\varphi^{-1}$, where φ^{-1} is the function in (1.8). Hence F is convex in the vertical direction for any $\theta \in [0, \pi)$. ■

Proposition 1. *If $f = h + \overline{g} \in \mathcal{F}_H$ and $\sup_{z\in\mathbb{D}} |g'/h'| < 1$ then $f(\mathbb{D}) = \Omega$ and $f \in \mathcal{P}_H$.*

Proof: Is is well known that any function $f \in \mathcal{F}_H$ is a normalized univalent orientation-preserving harmonic mapping of the disc $\mathbb{D}$ into the strip Ω. If $\sup_{z\in\mathbb{D}} |g'(z)/h'(z)| = \sup_{z\in\mathbb{D}} |f'_{\overline{z}}/f'_z| < 1$, then there is a $R \in (0, 1)$ such that $\sup_{z\in\mathbb{D}} |f'_{\overline{z}}/f'_z| \leq R$, so f is a K_R-quasiconformal mapping in $\mathbb{D}$, where $K_R = (1 + R)/(1 - R)$ (*cf.* [4]). Since the set $f(\mathbb{D})$ is convex, $f(\mathbb{D}) \neq \Omega$ would imply that $\partial f(\mathbb{D})$ would be a Jordan curve in $\mathbb{C} \cup \{\infty\}$. Consequently, the quasiconformal mapping f would extend (*cf.* [4]) to a homeomorphism between $\overline{\mathbb{D}}$ and $\overline{f(\mathbb{D})}$. However, this is not possible because, at the points $e^{\pm i\alpha} \in \mathbb{T}$, each function $f \in \mathcal{F}_H$ is discontinuous, which implies $f(\mathbb{D}) = \Omega$, so $f \in \mathcal{P}_H$. ■

To solve extremal problems involving continuous functionals we need to determine the closure of $\mathcal{P}_H$.

Proposition 2. *The closure of $\mathcal{P}_H$ in the topology of locally uniform convergence in $\mathbb{D}$ is $\mathcal{F}_H$.*

Proof: We know that $\mathcal{P}_H \subset \mathcal{F}_H$. Moreover, if $f = h + \overline{g} \in \mathcal{F}_H$, then g'/h' is a Schwarz function. Hence, there is a function p holomorphic in $\mathbb{D}$, $p(0) = 1$, $\mathrm{Re}\, p(z) > 0$, $z \in \mathbb{D}$, such that

$$\frac{g'(z)}{h'(z)} = \frac{p(z) - 1}{p(z) + 1}, \qquad z \in \mathbb{D}. \tag{1.12}$$

Any function $f \in \mathcal{F}_H$ corresponds to a measure $\mu \in \mathcal{M}$ which can be approximated by measures $\mu_n \in \mathcal{M}$ generating functions $f_n = h_n + \overline{g}_n \in \mathcal{F}_H$ such that (*cf.* (1.12))

$$\sup_{z\in\mathbb{D}}\left|\frac{g_n'(z)}{h_n'(z)}\right| = \sup_{z\in\mathbb{D}}\left|\frac{p_n(z)-1}{p_n(z)+1}\right| < 1 \tag{1.13}$$

since $\operatorname{Re} p_n(z) > 0$ for $z \in \mathbb{D}$.

Consequently, from Proposition 1 and (1.13) we obtain $f_n \in \mathcal{P}_H$. In turn if μ is a pointwise limit of $\{\mu_n\}$ as $n \to \infty$, then $f_n \to f$ in the sense of locally uniform convergence in $\mathbb{D}$. This gives the assertion. ■

Theorem 4. *The functions*

$$\begin{aligned} f(z) &= 2m - 1 + \frac{M-m}{\pi}\arg\frac{e^{i\pi\alpha}-z}{e^{-i\pi\alpha}-z} \\ &\quad + i\frac{2(M-m)\sin\pi\alpha}{\pi}v_\eta(z), \qquad z \in \mathbb{D}, \end{aligned} \tag{1.14}$$

where $v_\eta = \operatorname{Im} K(\cdot,\eta)$, $|\eta| = 1$, $\alpha = (1-m)/(M-m)$, *and* $K(\cdot,\eta)$ *is defined by (1.4), form the set of extreme points of the class* $\mathcal{F}_H$.

Proof: It is well known that extreme points in the class $\mathcal{M}$ are unit point masses. The mapping

$$\mathcal{M} \ni \mu \mapsto \operatorname{Im}\left\{\int_{|\eta|=1} K(z,\eta)\,d\mu\right\}$$

is a linear homeomorphism. Hence, from the integral representation (1.5) of functions belonging to the class $\mathcal{F}_H = \overline{\mathcal{P}_H}$ we obtain formula (1.14) for extreme points of this class. ■

We shall consider the coefficient problems for the class $\mathcal{P}_H$ and its closure. It is clear that the extremal functions of $\mathcal{P}_H$ are of the form (1.14).

For any function $f = h + \overline{g} \in \mathcal{P}_H$ formula (1.6) (or (1.7)) implies

$$a_1 = \frac{2(M-m)\sin\pi\alpha}{\pi}, \qquad \alpha = \frac{1-m}{M-m}, \tag{1.15}$$

and the coefficient depends on Ω only. Thus (1.15) is valid in the class $\mathcal{F}_H$, too. Next, since

$$\frac{1}{(e^{i\pi\alpha}-z)(e^{-i\pi\alpha}-z)} = 1 + \sum_{n=1}^{\infty}\frac{\sin(n+1)\pi\alpha}{\sin\pi\alpha}z^n, \qquad z \in \mathbb{D},$$

for any function $f \in \mathcal{P}_H$, we have by (1.6)

$$a_n + \overline{a}_{-n} = \frac{2(M-m)}{n\pi} \sin n\pi\alpha, \qquad n = 2, 3 \dots. \tag{1.16}$$

Formula (1.16) is true in the class $\mathcal{F}_H$ as well.

Theorem 5. *If $f \in \mathcal{F}_H$ then, for $n = 2, 3, \dots$,*

$$|\overline{a}_{-n}| \leq \frac{2(M-m)}{\pi n} \sum_{k=1}^{n-1} |\sin k\pi\alpha|, \tag{1.17}$$

$$|a_n| \leq \frac{2(M-m)}{\pi n} \sum_{k=1}^{n} |\sin k\pi\alpha|, \tag{1.18}$$

where $\alpha \in (0,1)$, $\alpha = (1-m)/(M-m)$.

Proof: Let $f = h + \overline{g} \in \mathcal{P}_H$. From (1.6) and (1.7) we obtain

$$g'(z) = -\frac{2(M-m)}{\pi} \frac{\sin \pi\alpha}{(e^{i\pi\alpha} - z)(e^{-i\pi\alpha} - z)} \int_{|\eta|=1} \frac{\eta z}{1 - \eta z}\, d\mu, \qquad z \in \mathbb{D},$$

and, consequently,

$$g'(z) = -\frac{2(M-m)}{\pi} \left[\sin \pi\alpha + \sum_{n=1}^{\infty} \sin(n+1)\pi\alpha z^n \right] \sum_{n=1}^{\infty} c_n z^n, \tag{1.19}$$

where

$$c_n = \int_{|\eta|=1} \eta^n \, d\mu, \qquad n = 1, 2, \dots.$$

From (1.19) we get

$$\overline{a}_{-n} = -\frac{2(M-m)}{n\pi} \sum_{k=1}^{n-1} c_{n-k} \sin k\pi\alpha, \tag{1.20}$$

where

$$|c_k| \leq 1, \qquad k = 1, 2, \dots, n-1.$$

Estimate (1.17) results from (1.20) if we take account of the fact that $\mathcal{P}_H$ is a dense subclass of $\mathcal{F}_H$. Estimate (1.18) follows from (1.16) and (1.17). ■

Corollary 1. *For $n = 2$ with $\alpha \in (0,1)$ and for $n \geq 3$ with $\alpha \in (0, 1/(n-1)]$, the following exact estimates hold:*

$$|\overline{a}_{-n}| \leq \frac{2(M-m)}{\pi n} \sum_{k=1}^{n-1} \sin k\pi\alpha, \tag{1.21}$$

$$|a_n| \leq \frac{2(M-m)}{\pi n} \sum_{k=1}^{n} \sin k\pi\alpha. \tag{1.22}$$

The only extremal function for (1.21) and (1.22) is the function in (1.14) where $\eta = 1$.

In the cases $n = 3$ and $n = 4$, we can obtain exact estimates of $|\overline{a}_{-n}|$ and $|a_n|$ in the class $\mathcal{F}_H$ for any $\alpha \in (0,1)$. We shall precede these results by some general remarks.

As is asserted in Theorem 4, extremal functions for coefficient problems in the class $\mathcal{F}_H$ are generated by unit point masses belonging to the class $\mathcal{M}$. Thus we need to find a point $\eta = e^{it}$ on the unit circle, such that the function (*cf.* (1.20))

$$W_n^{(\alpha)}(t) = \left| \sum_{k=1}^{n-1} e^{i(n-k)t} \sin k\pi\alpha \right|, \qquad t \in [-\pi, \pi],\ \alpha \in (0,1),\ n = 3, 4, \ldots,$$

attains its maximum. For this purpose, consider

$$\left[W_n^{(\alpha)}(t)\right]^2 = A_0^{(n)}(\alpha) + 2\sum_{l=1}^{n-2} A_l^{(n)}(\alpha) \cos lt, \tag{1.23}$$

where

$$A_0^{(n)}(\alpha) = \sum_{k=1}^{n-1} \sin^2 k\pi\alpha,$$

$$A_l^{(n)}(\alpha) = \sum_{k=1}^{n-l-1} \sin k\pi\alpha \sin(k+l)\pi\alpha, \qquad l = 1, 2, \ldots, n-2.$$

Note that (1.23) is an even function. Also, a calculation shows that

$$W_n^{(1-\alpha)}(\pi - t) = W_n^{(\alpha)}(t) \tag{1.24}$$

holds for $t \in [0,\pi]$ and $n = 3, 4, \ldots$. From (1.24), for $n \geq 3$ and $\alpha \in [1/2, 1)$, we get

$$\max_{t\in[0,\pi]} W_n^{(\alpha)}(t) = \max_{t\in[0,\pi]} W_n^{(1-\alpha)}(\pi - t). \tag{1.25}$$

By (1.25), it is sufficient to find $\max_{t\in[0,\pi]} W_n^{(\alpha)}(t)$ only for $\alpha \in [0, 1/2)$ in order to determine this maximum for any $\alpha \in (0,1)$. For this reason, from (1.25) and (1.21), (1.22) we immediately obtain

Corollary 2. *For $n = 3$, estimates (1.17) and (1.18) are still exact for $\alpha \in (1/2, 1)$. In this case, function (1.14) is extremal with $\eta = -1$ only.*

Now, we can prove:

Theorem 6. *If $f \in \mathcal{F}_H$, then*

$$|\bar{a}_{-4}| \leq \frac{M-m}{2\pi} \max_{t\in[-\pi,\pi]} W_4^{(\alpha)}(t), \tag{1.26}$$

where

$$\max_{t\in[-\pi,\pi]} W_4^{(\alpha)}(t) = \begin{cases} W_4^{(\alpha)}(0) & \text{for } \alpha \in (0, \alpha_0], \\ W_4^{(\alpha)}(\pm t_0) & \text{for } \alpha \in (\alpha_0, 1/2], \\ W_4^{(\alpha)}(\pm\pi \mp t_0) & \text{for } \alpha \in (1/2, 1-\alpha_0], \\ W_4^{(\alpha)}(\pi) & \text{for } \alpha \in (1-\alpha_0, 1), \end{cases}$$

and $\alpha_0 = 0,3508\ldots$ is the unique solution of the equation

$$2\cos^3 \pi\alpha + 4\cos^2 \pi\alpha - 1 = 0, \qquad \alpha \in (1/3, 1/2], \tag{1.27}$$

while

$$t_0 = \arccos\left[\frac{(\cos 4\pi\alpha - 1)\cos\pi\alpha}{4\sin\pi\alpha\sin 3\pi\alpha}\right], \qquad t_0 \in (0, \pi/2]. \tag{1.28}$$

Estimate (1.26) is sharp.

Proof: By (1.23), for $n = 4$, we have

$$[W_4^{(\alpha)}(t)]^2 = A_2^{(4)}(\alpha)\cos 2t + A_1^{(4)}(\alpha)\cos t + A_0^{(4)}(\alpha), \tag{1.29}$$

$t \in [-\pi, \pi]$, $\alpha \in (1/3, 1/2]$, where

$$\begin{aligned} A_2^{(4)}(\alpha) &= \cos 2\pi\alpha - \cos 4\pi\alpha, \\ A_1^{(4)}(\alpha) &= 2\cos\pi\alpha - \cos 3\pi\alpha - \cos 5\pi\alpha, \\ A_0^{(4)}(\alpha) &= \sin^2\pi\alpha + \sin^2 2\pi\alpha + \sin^2 3\pi\alpha. \end{aligned}$$

Consider

$$\frac{d}{dt}[W_4^{(\alpha)}(t)]^2 = 4(\cos 4\pi\alpha - \cos 2\pi\alpha)(\cos t - \nu(\alpha))\sin t,$$

where

$$\nu(\alpha) = \frac{(\cos 4\pi\alpha - 1)\cos\pi\alpha}{4\sin 3\pi\alpha\sin\pi\alpha}, \qquad \alpha \in (1/3, 1/2].$$

Since $\nu(\alpha)$ is decreasing in $(1/3, 1/2]$ and $\lim_{\alpha\to 1/3}\nu(\alpha) = +\infty$, $\nu(1/2) = 0$, there is exactly one α_0 such that $\nu(\alpha_0) = 1$. It is easy to check that α_0 is a solution of equation (1.27), belonging to $(1/3, 1/2]$. If $\alpha \in (1/3, \alpha_0)$ then $\nu(\alpha) > 1$ and, in consequence, function (1.29) attains its upper bound for $t = 0$. In this way, taking account of Corollary 1, we get estimate (1.26) for $\alpha \in (0, \alpha_0]$. For $\alpha \in (\alpha_0, 1/2]$, the function $\nu(\alpha)$ takes its values in $(0, 1]$. Thus $\max_t [W_4^{(\alpha)}(t)]^2 = [W_4^{(\alpha)}(\pm t_0)]^2$ where t_0 is given (1.28). Hence estimate (1.26) holds in this case. For $\alpha \in (1/2, 1)$ estimate (1.26) is a consequence of (1.25) and the results concerning $\alpha \in (1/2, 1)$. ■

Remark: Estimate (1.26) and relation (1.16) easily imply an exact estimate of the coefficient $|a_4|$ in the class $\mathcal{F}_H$.

As an open problem, it remains to determine the upper bound for $|\overline{a}_{-n}|$ and $|a_n|$ in the class $\mathcal{F}_H$ for $n \geq 5$ and $\alpha \in (1/(n-1), (n-2)/(n-1))$.

The paper generalizes some results due to Hengartner and Schober ([3]) and it refers to [5] where functions $p : \mathbb{D} \to \Omega(O, M)$, $p(0) = 1$, holomorphic in $\mathbb{D}$, were studied.

References

[1] Clunie, J. and T. Sheil-Small, Harmonic univalent functions, *Ann. Acad. Sci. Fenn.* Ser. A I **9** (1984), 3–25.

[2] Goodman, A. W., *Univalent Functions*, vol. I, Mariner Publ. Comp., Tampa, 1983.

[3] Hengartner, W., and G. Schober, Univalent harmonic mappings, *Trans. Amer. Math. Soc.* **299** (1987), 1–31.

[4] Lehto, O., and K. I. Virtanen, *Quasiconformal Mappings in the Plane*, Springer-Verlag, Berlin Heidelberg New York, 1973.

[5] Olejniczak, E., and S. Walczak, Fonctions holomorphes aux parties réelles limitées, *Acta. Univ. Lodziensis Ser. II* **10** (1977), 201–208.

Wiesław Majchrzak
Faculty of Mathematics
University of Lodz
Ul. S. Banacha 2
90-238 Lodz
POLAND
wmajch@imul.uni.lodz.pl

Euclidean circles of curvature for geodesics of conformal metrics

David Minda

Abstract. The purpose of this note is to establish a simple property of euclidean circles of curvature for geodesic arcs relative to a conformal metric. For the special case of the hyperbolic metric on a Nehari region which is not Möbius equivalent to a strip, this property gives an elegant geometrical interpretation for the Ahlfors-Weill quasiconformal reflection in the boundary of the region. The interpretation reveals that Ahlfors-Weill reflection is a natural extension of the classical notion of Schwarz reflection in a circle or line.

§1 Introduction

Suppose $\rho(w)|dw|$ is a $\mathcal{C}^2$ conformal metric on a plane region Ω. For a point $a \in \Omega$ and a geodesic arc Γ (relative to $\rho(w)|dw|$) through a, let C be the euclidean circle of curvature for Γ at a. We show that as Γ varies over all geodesic arcs through a, the associated collection of all euclidean circles of curvature at a have precisely two points in common, a and

$$r(a) = a + \frac{1}{\frac{\partial \log \rho}{\partial w}(a)}. \tag{1.1}$$

Note that $r(a) \in \mathbb{P}$, the Riemann sphere. In fact, $r(a) = \infty$ if and only if a is a critical point for the density function ρ. In the particular case that Ω is a disk or half-plane and the conformal metric is the hyperbolic metric, the point $r(a)$ is the ordinary reflection of a in the circle or line $\partial\Omega$. To see this, recall that points a and a^* are symmetric about a circle or line if and only if every circle through both a and a^* is orthogonal to the circle or line. Also, in a disk or half-plane the hyperbolic geodesics through a point a are arcs of circles orthogonal to the boundary; these circles all go through a^*.

In the general situation the function r does not define a reflection in $\partial\Omega$; that is, an orientation reversing homeomorphism of $\mathbb{P}$ which is an involution and fixes $\partial\Omega$ pointwise. In the special case that the conformal metric is the hyperbolic metric $\lambda_\Omega(w)|dw|$ on a hyperbolic regoin Ω, we write r_Ω in place of r. A simply connected region Ω is called a Nehari region provided

$$(1 - |z|^2)^2 |S_f(z)| \le 2, \qquad z \in \mathbb{D}, \tag{1.2}$$

Computational Methods and Function Theory (CMFT'97)
N. Papamichael, St. Ruscheweyh and E. B. Saff (Eds.), pp. 397–403

ISBN 981-02-3626-3

for any conformal mapping $f : \mathbb{D} \to \Omega$, where

$$S_f = \frac{f'''}{f'} - \frac{3}{2}\left(\frac{f''}{f'}\right)^2$$

is the Schwarzian derivative of f. It is elementary that $S_f \equiv 0$ if and only if f is a Möbius transformation. Bounds of the type (1.2) give a measure of how much f differs from being a Möbius transformation. Gehring and Pommerenke [5] proved that unless Ω is a Möbius equivalent to a strip, then a function f satisfying inequality (1.2) extends to an injective function from $\overline{\mathbb{D}}$ into $\mathbb{P}$. In case Ω is a Nehari region that is not Möbius equivalent to a strip, Chuaqui and Osgood [2] proved that the analytically defined mapping r_Ω is a reflection in $\partial\Omega$. If the upper bound "2" in (1.2) is replaced by "$2t$" for some $t \in [0, 1)$, then this reflection is $\frac{1+t}{1-t}$-quasiconformal. When r_Ω is a reflection in $\partial\Omega$, they showed that

$$F(z) = \begin{cases} f(z) & z \in \overline{\mathbb{D}}, \\ r_\Omega(f(1/\bar{z})) & z \in \mathbb{P}\backslash\mathbb{D}, \end{cases} \tag{1.3}$$

gives a homeomorphic extension of f to $\mathbb{P}$ which is $\frac{1+t}{1-t}$-quasiconformal when f satisfies the stronger form of (1.2). In the latter case F is precisely the quasiconformal extension of f to $\mathbb{P}$ defined by Ahlfors and Weill [1]. Of course, in the special case $t = 0$ the extension (1.3) is the classical Schwarz reflection principle. Our geometric interpretation of r_Ω reveals that (1.3) is a natural geometric extension of Schwarz reflection. Also, it gives a simple explanation for the known Möbius invariance of the extension (1.3).

Later, Chuaqui and Osgood [3] presented a unified differential geometric treatment of univalence criteria and homeomorphic extensions relating to Schwarzians relative to various conformal metrics. In these cases the formula (1.1) provides a geometric interpretation for the analytically defined reflections they use.

§2 Preliminaries

Suppose $\rho(w)|dw|$ is a conformal metric on Ω. For a path $\gamma : w = w(t)$, $t \in I$, the geodesic curvature of γ at $w(t)$ relative to the metric $\rho(w)|dw|$ is

$$\kappa_\rho(w(t), \gamma) = \frac{\kappa_e(w(t), \gamma) + 2\,\mathrm{Im}\left\{\frac{\partial \log \rho}{\partial w}(w(t))\frac{w'(t)}{|w'(t)|}\right\}}{\rho(w(t))},$$

where

$$\kappa_e(w(t), \gamma) = \frac{1}{|w'(t)|}\,\mathrm{Im}\left\{\frac{w''(t)}{w'(t)}\right\}$$

is the euclidean curvature of γ at $w(t)$. Observe that if a is a critical point for the density ρ, then the geodesic curvature at a is just the fixed scalar $1/\rho(a)$ times the euclidean curvature at a. If $f : \Delta \to \Omega$ is a locally univalent holomorphic function, then

$$\sigma(z)|dz| = f^*(\rho(w)|dw|) = \rho(f(z))|f'(z)||dz|$$

is the conformal metric on Δ obtained by pulling-back $\rho(w)|dw|$ to Δ via f. Geodesic curvature is invariant under the pull-back operation:

$$\kappa_\sigma(z,\delta) = \kappa_\rho(f(z), f\circ\delta).$$

For each point $a \in \Omega$ and any direction $e^{i\theta}$, there is a unique geodesic arc Γ (relative to $\rho(w)|dw|$) through a with unit tangent $e^{i\theta}$ at a. These local geodesic arcs do not extend to global geodesics unless the conformal metric is complete. Geodesic arcs have vanishing geodesic curvature. Thus, if Γ is a geodesic arc through a with unit tangent $e^{i\theta}$ at a, then $\kappa_\rho(a,\Gamma) = 0$ and so

$$\kappa_e(a,\Gamma) = -2\,\mathrm{Im}\left\{\frac{\partial \log \rho}{\partial w}(a)e^{i\theta}\right\}. \tag{2.1}$$

We briefly recall some facts about the (signed) euclidean curvature. For a path $\delta : z = z(t)$, $t \in I$, and a locally univalent function f, the change of curvature formula is [4]

$$\kappa_e(f(z(t)), f\circ\delta)|f'(z)| = \kappa_e(z(t),\gamma) + \mathrm{Im}\left\{\frac{f''(z(t))}{f'(z(t))}\,\frac{z'(t)}{|z'(t)|}\right\}. \tag{2.2}$$

The euclidean circle of curvature for δ at the point $z(t)$ is the circle through $z(t)$ which has the same unit tangent $z'(t)/|z'(t)|$ as δ at $z(t)$ and the same euclidean curvature as δ at $z(t)$. This is a line if $\kappa_e(z(t),\delta) = 0$ and a circle with radius $1/|\kappa_e(z(t),\delta|$ otherwise. A more detailed description is given in [6].

§3 Main results

We now establish our main result and present some examples.

Theorem 1. *Suppose $\rho(w)|dw|$ is a C^2 conformal metric on a region Ω. For each $a \in \Omega$ the point*

$$r(a) = a + \frac{1}{\frac{\partial \log \rho}{\partial w}(a)}$$

lies on the euclidean circle of curvature at a for any geodesic arc through a. The points a and $r(a)$ are the unique points on the Riemann sphere common to all the euclidean circles of curvature for geodesic arcs through a.

Proof: Fix $a \in \Omega$. For each direction $e^{i\theta}$ let Γ_θ be the geodesic arc through a in the direction $e^{i\theta}$. Suppose C_θ is the euclidean circle of curvature for Γ_θ at a. We want to show $\cap C_\theta = \{a, r(a)\}$.

We first consider the case in which $\frac{\partial \log \rho}{\partial w}(a) = 0$. Then from (2.1), $\kappa_e(a, \Gamma_\theta) = 0$ for all θ. Thus, C_θ is the straight line through a in the direction $e^{i\theta}$ and so $\cap C_\theta = \{a, \infty\}$. Since $r(a) = \infty$, we are done in this special situation.

The idea of the proof is to convert the general case to the special situation by using a suitable Möbius transformation. Let T be the unique Möbius transformation such that $T(a) = 0$, $T'(a) = \rho(a)$ and $T''(a) = 2\frac{\partial \rho}{\partial w}(a)$. Set $S = T^{-1}$ and $\Delta = T(\Omega)$. Take $\sigma(z)|dz|$ to be the pull-back to Δ of $\rho(w)|dw|$ via S. Then $\sigma(0) = \rho(a)|S'(0)| = \rho(a)/T'(a) = 1$. From $\sigma(z) = \rho(S(z))|S'(z)|$ we obtain

$$\frac{\partial \log \sigma}{\partial z}(z) = \frac{\partial \log \rho}{\partial w}(S(z))S'(z) + \frac{S''(z)}{2S'(z)}.$$

Therefore,

$$\begin{aligned}\frac{\partial \log \sigma}{\partial z}(0) &= \frac{\partial \log \rho}{\partial w}(a)S'(0) + \frac{S''(0)}{2S'(0)} \\ &= \frac{T''(a)S'(0)}{2T'(a)} + \frac{S''(0)}{2S'(0)} \\ &= 0\end{aligned}$$

because $S(T(w)) \equiv w$. Now, $T(\Gamma_\theta)$ is a geodesic arc relative to $\sigma(z)|dz|$ through the origin and in the direction $e^{i\theta}$ since $T'(a) > 0$. Because the origin is a critical point for σ, we conclude that $\kappa_e(0, T(\Gamma_\theta)) = 0$. Then L_θ, the straight line through 0 in the direction $e^{i\theta}$, is the euclidean circle of curvature for $T(\Gamma_\theta)$ at the origin. We show that $S(L_\theta)$ is the euclidean circle of curvature for Γ_θ at a. Clearly, $S(L_\theta)$ passes through $a = S(0)$. Also, $S(L_\theta)$ has unit tangent $e^{i\theta}$ at a since $S'(0) > 0$. All that remains is to demonstrate $\kappa_e(a, S(L_\theta)) = \kappa_e(a, \Gamma_\theta)$. From (2.1)

$$\kappa_e(a, \Gamma_\theta) = -2\,\mathrm{Im}\left\{\frac{\partial \log \rho}{\partial w}(a)e^{i\theta}\right\},$$

while (2.2) yields

$$\kappa_e(a, S(L_\theta))S'(0) = \mathrm{Im}\left\{\frac{S''(0)}{S'(0)}e^{i\theta}\right\},$$

or

$$\kappa_e(a, S(L_\theta)) = \mathrm{Im}\left\{\frac{S''(0)}{S'(0)^2}e^{i\theta}\right\},$$

since L_θ has vanishing euclidean curvature. From $T(S(z)) \equiv z$, we find

$$\frac{S''(0)}{S'(0)^2} = -\frac{T''(a)}{T'(a)} = -\frac{2\frac{\partial \rho}{\partial w}(a)}{\rho(a)} = -2\frac{\partial \log \rho}{\partial w}(a).$$

Consequently, Γ_θ and $S(L_\theta)$ have the same euclidean curvature at a, so $S(L_\theta)$ is C_θ, the euclidean circle of curvature for Γ_θ at a. Then

$$\cap C_\theta = \cap S(L_\theta) = S(\cap L_\theta) = \{S(0), S(\infty)\}.$$

Note that

$$S(z) = a + \frac{S'(0)z}{1 - \frac{S''(0)}{2S'(0)}z}$$

since both sides are Möbius transformations which have the same value $(S(0))$, first derivative $(S'(0))$ and second derivative $(S''(0))$ at the origin. Because

$$S(\infty) = a - \frac{2S'(0)^2}{S''(0)} = a + \frac{1}{\frac{\partial \log \rho}{\partial w}(a)},$$

the proof is complete. ∎

Corollary 1. *Let $\rho(w)|dw|$ be a conformal metric on a region Ω. If there exists a Möbius transformation T such that $\sigma(z)|dz| = T^*(\rho(w)|dw|)$ then*

$$T \circ r_\sigma = r_\rho \circ T. \tag{3.1}$$

Proof: The Möbius invariance expressed by (3.1) can be verified directly by using the definitions of r_σ and r_ρ, but the computations involved are not enlightening. The invariance is clear from the geometric interpretation given by the theorem. Because $w = T(z)$ gives an isometry from $T^{-1}(\Omega)$ with the metric $\sigma(z)|dz|$ to Ω with the metric $\rho(w)|dw|$. T maps geodesic arcs relative to $\sigma(z)|dz|$ through a point $a \in T^{-1}(\Omega)$ to geodesic arcs relative to $\rho(w)|dw|$ through $T(a)$. Because Möbius transformations map circles to circles, it follows that if C is a euclidean circle of curvature at a for a geodesic arc Γ through a, then $T(C)$ will be the euclidean circle of curvature at $T(a)$ for the geodesic arc $T(\Gamma)$ through $T(a)$. Hence, $T(C)$ will contain the points $T(a)$ and $T(r_\sigma(a))$, so $r_\rho(T(a)) = T(r_\sigma(a))$. ∎

The hyperbolic metric is conformally invariant. In particular,

$$T^*(\lambda_\Omega(w)|dw|) = \lambda_{T^{-1}(\Omega)}(z)|dz|$$

for any Möbius transformation T. Thus, the preceding corollary yields $T \circ r_{T^{-1}(\Omega)} = r_\Omega \circ T$. In words, r_Ω is Möbius invariant, a result established analytically in [2].

§4 Remarks

In case $r(a) \neq \infty$, the locus of the centers of the euclidean circles of curvature for all geodesic arcs through a is the line which is the perpendicular bisector of the line segment $[a, r(a)]$. For a convex region Ω it is known that the center of any euclidean circle of curvature for a hyperbolic geodesic lies in $\mathbb{P}\backslash\Omega$, *cf.* [6]. Thus, for a convex region Ω and a point $a \in \Omega$ with $r_\Omega(a) \neq \infty$, the centers of the euclidean circles of curvature at a for all hyperbolic geodesics through a lie on a line which determines a half-plane containing the region.

Suppose Ω is a hyperbolic region and $f : \mathbb{D} \to \Omega$ is a holomorphic universal covering projection. We write r in place of r_Ω. Straightforward computations reveal that

$$\left|\frac{r_w(f(z))}{r_{\bar{w}}(f(z))}\right| = \frac{1}{2}(1-|z|^2)^2|S_f(z)|.$$

If f satisfies (1.2) with "$2t$" in place of "2" for some $t \in [0,1)$, then this shows that r is an orientation reversing quasiconformal mapping with dilatation at most $\frac{1+t}{1-t}$. If Ω is simply connected, then f satisfies (1.2) with "2" replaced by "6". In this situation r need not be injective or quasiconformal, but does satisfy an "alternating" Beltrami equation in the sense of Srebro and Yakubov [7]. It would be interesting to know whether r has properties that follow from geometric properties of Ω.

If $\rho(w)|dw|$ has zero Gauss curvature, then r is meromorphic unless it is identically infinite. The Gauss curvature of a conformal metric $\rho(w)|dw|$ is

$$-\frac{\Delta \log \rho(w)}{\rho^2(w)} = -\frac{4\frac{\partial^2 \log \rho}{\partial \bar{w}\partial w}(w)}{\partial^2(w)}.$$

Thus, if $\rho(w)|dw|$ has zero curvature, then $\log \rho$ is a harmonic function on Ω. The critical points of a harmonic function are isolated unless it is constant. Since

$$\frac{\partial r}{\partial \bar{w}}(w) = -\frac{\frac{\partial^2 \log \rho}{\partial \bar{w}\partial w}(w)}{\left[\frac{\partial \log \rho}{\partial w}(w)\right]^2},$$

$r_{\bar{w}} = 0$ when $\rho(w)|dw|$ has Gauss curvature zero. Hence, for ρ nonconstant, r is meromorphic on Ω. The logarithmic metric on $\mathbb{C}\backslash\{0\}$ is $\frac{|dz|}{|z|}$. For this metric $r(z) = -z$ which is reflection in the origin.

It is tempting to conjecture that the property of euclidean circles of curvature in Theorem 1 characterizes conformal metrics. This is not the case. For instance, consider the Klein model of hyperbolic geometry on the unit disk $\mathbb{D}$. The Riemannian metric on $\mathbb{D}$ which gives the Klein model is not a conformal metric. The geodesics are arcs of euclidean lines in $\mathbb{D}$. Thus, for each point $a \in \mathbb{D}$ the euclidean circles of curvature for geodesics through a are all straight lines. These "circles" of curvature have the two common points a and ∞.

Acknowledgement. The author acknowledges support from the C.P. Taft Memorial Fund for a travel grant to partially cover local expenses at CMFT '97.

References

[1] Ahlfors, L. V. and G. Weill, A uniqueness theorem for Beltrami equations, *Proc. Amer. Math. Soc.* **13** (1962), 975–978.

[2] Chuaqui, M. and B. Osgood, Ahlfors-Weill extensions of conformal mappings and critical points of the Poincaré metric, *Comment. Math. Helv.* **69** (1994), 659–668.

[3] Chuaqui, M. and B. Osgood, General univalence criteria in the disk: extensions and extremal functions, to appear.

[4] Flinn, B. and B. Osgood, Hyperbolic curvature and conformal mapping, *Bull. London Math. Soc.* , **18** (1986), 272–276.

[5] Gehring, F. W. and Ch. Pommerenke, On the Nehari univalence criterion and quasicircles, *Comment. Math. Helv.* **59** (1984), 226–242.

[6] Minda, D., Applications of hyperbolic convexity to euclidean and spherical convexity, *J. Analyse Math.* **49** (1987), 90–105.

[7] Srebro, U. and E. Yakubov, Branched folded maps and alternating Beltrami equations, *J. Analyse Math.* **70** (1996), 65–90.

David Minda
Department of Mathematical Sciences
University of Cincinnati
Cincinnati, OH 45221-0025
U.S.A.
David.Minda@math.uc.edu

Some integral equations related to the Riemann map

A. H. M. Murid

M. Z. Nashed

and M. R. M. Razali

Abstract. Each of the Szegö and the Bergman kernel functions is known to satisfy a certain integral equation of the second kind. In this paper, based on a certain boundary relationship, an integral equation is formulated such that both the integral equations for the Szegö and the Bergman kernels can be derived from that equation. New integral equations related to the Riemann map are also derived.

§1 Introduction

Let Ω be a simply connected domain in the complex plane whose boundary Γ is a Jordan curve. The Szegö and the Bergman kernel functions of Ω are well-known reproducing kernels and are connected to the Riemann mapping function which maps Ω onto the unit disk $\mathbb{D}$. These kernels are also related to orthonormal polynomials, where the orthogonality is defined with respect to a boundary integral (Szegö kernel) or an area integral (Bergman kernel), and several studies about the use of Szegö and Bergman kernel orthonormalization methods for numerical conformal mapping are available in the literature; see *e.g.* [4, 5, 6, 11, 14, 16, 17, 18]. The discovery by Kerzman and Trummer [13] that the Szegö kernel satisfies a boundary integral equation of the second kind gives an alternative method of computing the Szegö kernel (provided the boundary Γ is of class C^2). The integral equation is derived from the concrete realization of an operator equation given by Kerzman and Stein [12]. A markedly different way of its derivation has been given by Henrici [9, pp. 560–563] for analytic boundary Γ. Henrici's approach is function-theoretic and is based on a boundary relationship of the Riemann map. This boundary relationship can also be rewritten in terms of the Szegö kernel. The authors have shown recently [19] that Henrici's approach can also be adapted to develop an integral equation of the second kind for the Bergman kernel. Surveys and various results on integral equation methods for computing the Riemann map can be found in [1, 6, 9, 11, 20, 22].

Computational Methods and Function Theory (CMFT'97)
N. Papamichael, St. Ruscheweyh and E. B. Saff (Eds.), pp. 405–419

ISBN 981-02-3626-3

In this paper, an integral equation is formulated such that both the integral equations for the Szegö and the Bergman kernels can be derived as special cases. New integral equations of the second kind related to the Riemann map are also derived.

In Section 2 we give a brief review of the Szegö and the Bergman kernels as well as their relationships with the Riemann mapping function. In Section 3 we derive an integral equation based on a certain boundary relationship. In Section 4 we give special cases of this new formulation that lead to several integral equations related to the Szegö kernel, the Bergman kernel and the Riemann map.

§2 The Riemann map, the Szegö kernel, and the Bergman kernel

Let Ω be any simply connected bounded region in the complex plane whose boundary Γ is assumed to be an analytic Jordan curve. Suppose Γ admits a counterclockwise parametrization $z(t)$, $0 \leq t \leq \beta$, with $z'(t) = dz/dt \neq 0$ for all t. The unit tangent to Γ at the point $z(t)$ will be denoted by $T(z) = z'(t)/|z'(t)|$.

Let $a \in \Omega$ be a fixed point and let R be the Riemann map $R : \Omega \to \mathbb{D}$ normalized at a, that is

$$R(a) = 0, \qquad R'(a) > 0. \tag{2.1}$$

The existence and uniqueness of R is well-known (see *e.g.* [9], pp. 324–328). Since Γ is analytic, the function R can be extended to a function that is analytic on the closure $\overline{\Omega} = \Omega \cup \Gamma$.

Closely connected to the Riemann mapping function R are the Szegö kernel (briefly, S) and the Bergman kernel (briefly, B) of Ω. For definitions and basic properties of S and B, see [2, 3, 9, 10, 15, 21]. The Szegö kernel $S(z,a)$ and the Bergman kernel $B(z,a)$ are continuous with respect to z on $\overline{\Omega}$, and

$$R'(z) = \frac{2\pi}{S(a,a)} S(z,a)^2, \qquad z \in \overline{\Omega}, \tag{2.2}$$

$$R'(z) = \sqrt{\frac{\pi}{B(a,a)}} B(z,a), \qquad z \in \overline{\Omega}. \tag{2.3}$$

It can also be shown [13] that the Riemann mapping function can be computed on the boundary Γ without any integration by means of

$$R(z) = \frac{1}{i} T(z) \frac{R'(z)}{|R'(z)|}, \qquad z \in \Gamma. \tag{2.4}$$

Hence to solve the conformal mapping problem it is sufficient to compute the boundary values of either the Szegö or the Bergman kernel. These kernels can be computed as the solutions of second kind integral equations as stated in the following theorems:

Theorem 1 ([9, 13]). *The Szegö kernel $S(z,a)$ is the unique continuous solution of the Kerzman-Stein-Trummer integral equation (KST integral equation)*

$$S(z,a) + \int_{\Gamma} A(z,w)S(w,a)\,|dw| = \overline{H(a,z)}, \qquad z \in \Gamma, a \in \Omega, \tag{2.5}$$

where

$$A(z,w) = \begin{cases} \overline{H(w,z)} - H(z,w), & \text{if } w,z \in \Gamma, w \neq z \\ 0 & \text{if } w = z \in \Gamma \end{cases}, \tag{2.6}$$

and

$$H(w,z) = \frac{1}{2\pi i}\frac{T(z)}{z-w}, \qquad w \in \overline{\Omega}, z \in \Gamma, w \neq z. \tag{2.7}$$

The kernel A is known as the Kerzman-Stein kernel and is smooth, skew-Hermitian, and is identically zero if and only if Ω is a disc. The kernel H is usually referred to as the Cauchy kernel.

Theorem 2 ([19]). *The Bergman kernel $B(z,a)$ is the unique continuous solution of the integral equation*

$$B(z,a) + \int_{\Gamma} M(z,w)B(w,a)\,|dw| = -\frac{1}{\pi}\frac{\overline{T(z)^2}}{(\overline{z}-\overline{a})^2}, \qquad z \in \Gamma, a \in \Omega, \tag{2.8}$$

where

$$M(z,w) = \begin{cases} \dfrac{T(w)}{2\pi i}\left[\dfrac{\overline{T(z)^2}}{\overline{w}-\overline{z}} - \dfrac{1}{w-z}\right], & \text{if } w,z \in \Gamma, w \neq z, \\[2ex] \dfrac{1}{2\pi}\dfrac{\operatorname{Im}\left[z''(t)\overline{z'(t)}\right]}{|z'(t)|^3}, & \text{if } w = z \in \Gamma. \end{cases} \tag{2.9}$$

This integral equation for the Bergman kernel B takes a more pleasant form if we multiply both sides by $T(z)$ and let $\tilde{B}(z,a) = T(z)B(z,a)$, in which case we obtain:

Theorem 3 ([19]). *The function $\tilde{B}(z,a) = T(z)B(z,a)$ is the unique continuous solution of the integral equation*

$$\tilde{B}(z,a) + \int_\Gamma N(z,w)\tilde{B}(w,a)\,|dw| = -\frac{1}{\pi}\frac{\overline{T(z)}}{(\overline{z}-\overline{a})^2}, \qquad z \in \Gamma, a \in \Omega, \quad (2.10)$$

where

$$N(z,w) = \begin{cases} \dfrac{1}{\pi}\mathrm{Im}\left[\dfrac{T(z)}{z-w}\right], & \text{if } w, z \in \Gamma, w \neq z, \\[2ex] \dfrac{1}{2\pi}\dfrac{\mathrm{Im}\left[z''(t)\overline{z'(t)}\right]}{|z'(t)|^3}, & \text{if } w = z \in \Gamma. \end{cases} \quad (2.11)$$

The real kernel N is the Neumann kernel which arises frequently in the integral equations of potential theory and conformal mapping (see *e.g.* [9]). The Neumann kernel N is continuous at all points $(z,w) \in \Gamma \times \Gamma$ and the value of $2\pi N(z,z)$ also has a geometrical interpretation as the curvature of Γ at the point $z = z(t)$ (see [8, p. 186]).

§3 An integral equation related to a boundary relationship

Reflecting on Henrici's approach [9, pp. 560–563] in deriving the KST integral equation for the Szegö kernel S and our adaptation of the approach [19] in deriving the integral equation for the Bergman kernel B, we found that both cases depend on a key observation, namely that these kernels admit the following boundary relationships:

$$S(z,a) = i\left[\frac{T(z)S(z,a)}{R(z)}\right]^-, \qquad z \in \Gamma, \quad (3.1)$$

$$B(z,a) = -\left[\frac{T(z)^2B(z,a)}{R(z)^2}\right]^-, \qquad z \in \Gamma, \quad (3.2)$$

where $a \in \Omega$ and R is the Riemann map of Ω with the usual normalization $R(a) = 0$ and $R'(a) > 0$. Here the minus sign in the superscript denotes complex conjugation.

Motivated by these results we seek a general boundary relationship which embodies both of these boundary relationships as special cases. For this purpose we let $D(z)$ be a function which is analytic with respect to $z \in \Omega$, is continuous on $\overline{\Omega}$, and Hölder continuous on Γ. We suppose further that D satisfies the following boundary relationship:

$$D(z) = c\left[\frac{T(z)Q(z)D(z)}{P(z)}\right]^-, \qquad z \in \Gamma, \quad (3.3)$$

where c is a complex constant, Q and P are complex-valued functions with the following properties:

(1) $P(z)$ is analytic with respect to $z \in \Omega$, continuous on $\overline{\Omega}$, and Hölder continuous on Γ.

(2) $P(z) \neq 0, Q(z) \neq 0, z \in \Gamma$.

Note that the boundary relationship (3.3) also has the following two equivalent forms:

$$\frac{D(z)}{\overline{D(z)}} = \frac{P(z)}{\overline{c}\,T(z)Q(z)}, \qquad z \in \Gamma, \tag{3.4}$$

and

$$P(z) = \overline{c}\,\frac{T(z)Q(z)D(z)^2}{|D(z)|^2}, \qquad z \in \Gamma. \tag{3.5}$$

With these assumptions and notations, we derive an integral equation satisfied by $D(z)$ which plays a pivotal rôle in this paper.

Theorem 4. *Let u and v be complex-valued functions that are defined on a smooth Jordan curve Γ. If P has a finite number of zeros at $a_1, a_2, \ldots, a_N$ in Ω, then, for $z \in \Gamma$, we have*

$$\begin{aligned} &\frac{1}{2}\left[v(z) + \frac{u(z)}{\overline{T(z)Q(z)}}\right] D(z) + \mathrm{PV}\int_\Gamma K(z,w)D(w)\,|dw| \\ &= -c\,u(z)\left[\sum_{j=1}^{N} \operatorname*{Res}_{w=a_j} \frac{D(w)}{(w-z)P(w)}\right]^{-}, \end{aligned} \tag{3.6}$$

where

$$K(z,w) = \frac{1}{2\pi i}\left[\frac{u(z)}{(\overline{w}-\overline{z})\overline{Q(w)}} - \frac{v(z)T(w)}{w-z}\right]. \tag{3.7}$$

Proof: We write

$$\mathrm{PV}\frac{1}{2\pi i}\int_\Gamma \left[\frac{u(z)}{(\overline{w}-\overline{z})\overline{Q(w)}} - \frac{v(z)T(w)}{w-z}\right] D(w)\,|dw|$$

$$= \mathrm{PV}\frac{1}{2\pi i}\int_\Gamma \frac{u(z)D(w)}{(\overline{w}-\overline{z})\overline{Q(w)}}\,|dw| - \mathrm{PV}\frac{1}{2\pi i}\int_\Gamma \frac{v(z)T(w)D(w)}{w-z}\,|dw|,$$

and evaluate the two integrals on the right separately. In the second integral, we use $T(w)|dw| = dw$ and the formula [7, p. 27]

$$\mathrm{PV}\frac{1}{2\pi i}\int_{\Gamma}\frac{h(w)}{w-z}\,dw = \frac{1}{2}h(z), \qquad z\in\Gamma, \tag{3.8}$$

for a function h that is analytic in Ω, continuous in $\overline{\Omega}$, and Hölder continuous on Γ. Hence we obtain

$$\mathrm{PV}\frac{1}{2\pi i}\int_{\Gamma}\frac{v(z)T(w)D(w)}{w-z}\,|dw| = \frac{1}{2}v(z)D(z).$$

As for the first integral, we use (3.3) and $\overline{T(w)}|dw| = \overline{dw}$ to get

$$\mathrm{PV}\frac{1}{2\pi i}\int_{\Gamma}\frac{u(z)D(w)}{(\overline{w}-\overline{z})\overline{\overline{Q(w)}}}\,|dw| = -c\,u(z)\left[\mathrm{PV}\frac{1}{2\pi i}\int_{\Gamma}\frac{D(w)}{(w-z)P(w)}\,dw\right]^{-}.$$

By the calculus of residues and formula (3.8), we obtain

$$\mathrm{PV}\frac{1}{2\pi i}\int_{\Gamma}\frac{u(z)D(w)}{(\overline{w}-\overline{z})\overline{\overline{Q(w)}}}\,|dw| = -c\,u(z)\left[\sum_{j=1}^{N}\operatorname*{Res}_{w=a_j}\frac{D(w)}{(w-z)P(w)} + \frac{1}{2}\frac{D(z)}{P(z)}\right]^{-},$$

which on using (3.3) gives

$$\begin{aligned}\mathrm{PV}\frac{1}{2\pi i}\int_{\Gamma}&\frac{u(z)D(w)}{(\overline{w}-\overline{z})\overline{\overline{Q(w)}}}\,|dw|\\ &= -c\,u(z)\left[\sum_{j=1}^{N}\operatorname*{Res}_{w=a_j}\frac{D(w)}{(w-z)P(w)}\right]^{-} - \frac{1}{2}\frac{u(z)D(z)}{\overline{T(z)}\overline{\overline{Q(z)}}}.\end{aligned}$$

Summarizing,

$$\begin{aligned}&\mathrm{PV}\frac{1}{2\pi i}\int_{\Gamma}\left[\frac{u(z)}{(\overline{w}-\overline{z})\overline{\overline{Q(z)}}} - \frac{v(z)T(w)}{w-z}\right]D(w)\,|dw|\\ &= -c\,u(z)\left[\sum_{j=1}^{N}\operatorname*{Res}_{w=a_j}\frac{D(w)}{(w-z)P(w)}\right]^{-} - \frac{1}{2}\left[\frac{u(z)}{\overline{T(z)}\overline{\overline{Q(z)}}} + v(z)\right]D(z),\end{aligned}$$

which is equivalent to (3.6). ■

For (3.6) to be an integral equation of the second kind it is necessary and sufficient that

$$\frac{1}{2}\left[v(z)+\frac{u(z)}{\overline{T(z)Q(z)}}\right]=l(z), \qquad z\in\Gamma,$$

where $l(z)\neq 0$ for all $z\in\Gamma$. Without loss of generality we take $l(z)=1$, and seek conditions under which the kernel $K(z,w)$ can be made continuous for all $z,w\in\Gamma$. We next give necessary and sufficient conditions under which the integral equation (3.6) reduces to an integral equation of the second kind with a continuous kernel.

Theorem 5. *Let* $u(z)=(2-v(z))\overline{T(z)Q(z)}$. *Let the smooth Jordan curve* Γ *have the parametric representation* $z(t)$, $0\le t\le\beta$, *and* $W(t)=Q(z(t))$ *be such that* $W'(t)$ *exists and is continuous on* $[0,\beta]$. *Then for every fixed* $z\in\Gamma$,

$$\lim_{w\to z}K(z,w)=-\frac{1}{2\pi i}\frac{\overline{W'(t)}}{|z'(t)|\overline{W(t)}}, \tag{3.9}$$

if and only if $v(z)=1$.

Proof: In (3.7) let $z=z(t)$ and $w=z(s)$. Recall $T(z(t))=z'(t)/|z'(t)|$. Then, from (3.7), we have

$$\begin{aligned}2\pi i|z'(s)|K(z,w)= &\quad (2-v(z))\left|\frac{z'(s)}{z'(t)}\right|\left[\frac{z'(t)}{z(s)-z(t)}\frac{W(t)}{W(s)}\right]^{-}\\ &-v(z)\frac{z'(s)}{z'(t)}\frac{z'(t)}{z(s)-z(t)}.\end{aligned}\tag{3.10}$$

Using

$$z(s)=z(t)+z'(t)(s-t)+\frac{1}{2}z''(t)(s-t)^2+O((s-t)^3),$$

we have for $s\neq t$,

$$\frac{z'(t)}{z(s)-z(t)}=\frac{1}{s-t}\left\{1+\frac{1}{2}\frac{z''(t)}{z'(t)}(s-t)+O((s-t)^3)\right\}^{-1}. \tag{3.11}$$

Using the fact that

$$\frac{1}{1+\xi}=1-\xi+O(\xi^2)$$

for ξ close to zero, (3.11) becomes

$$\frac{z'(t)}{z(s)-z(t)} = \frac{1}{s-t}\left\{1 - \frac{1}{2}\frac{z''(t)}{z'(t)}(s-t) + O((s-t)^3)\right\}. \tag{3.12}$$

Next we observe that

$$\left|\frac{z'(s)}{z'(t)}\right|^2 = \left(\frac{z'(s)}{z'(t)}\right)\overline{\left(\frac{z'(s)}{z'(t)}\right)} = 1 + 2\mathrm{Re}\left(\frac{z''(t)}{z'(t)}\right)(s-t) + O((s-t)^2).$$

Hence

$$\left|\frac{z'(s)}{z'(t)}\right| = 1 + \mathrm{Re}\left(\frac{z''(t)}{z'(t)}\right)(s-t) + O((s-t)^2). \tag{3.13}$$

By hypothesis we also have

$$W(s) = Q(z(s)) = W(t) + W'(t)(s-t) + O((s-t)^2). \tag{3.14}$$

Substituting (3.12), (3.13), and (3.14) into (3.10), and simplifying, we obtain

$$2\pi i|z'(s)|K(z,w) = \frac{2-2v(z)}{t-s} + (2-v(z))\left[\frac{1}{2}\frac{z''(t)}{z'(t)} - \frac{\overline{W'(t)}}{\overline{W(t)}}\right]$$
$$-\frac{1}{2}v(z)\frac{z''(t)}{z'(t)} + O(s-t).$$

To get rid of the singular term we have no other choice but to take $v(z) = 1$. With this choice, we get

$$2\pi i|z'(s)|K(z,w) = -\frac{\overline{W'(t)}}{\overline{W(t)}} + O(s-t),$$

which implies (3.9). ■

Combining the results of Theorems 4 and 5 leads to:

Theorem 6. *Let the smooth Jordan curve Γ have the parametric representation $z(t)$, $0 \le t \le \beta$, and $W(t) = Q(z(t))$ be such that $W'(t)$ exists and continuous on $[0,\beta]$. If P has a finite number of zeros, say at $a_1, a_2, \ldots, a_N$ in Ω, then, for $z \in \Gamma$, we have*

$$D(z) + \int_\Gamma K(z,w)\,D(w)\,|dw|$$
$$= -c\,\overline{T(z)Q(z)}\left[\sum_{j=1}^{N}\operatorname*{Res}_{w=a_j}\frac{D(w)}{(w-z)P(w)}\right]^{-}, \tag{3.15}$$

where

$$K(z,w)=\begin{cases}\dfrac{1}{2\pi i}\left[\dfrac{\overline{T(z)Q(z)}}{(\overline{w}-\overline{z})\overline{Q(w)}}-\dfrac{T(w)}{w-z}\right], & \text{if } w,z\in\Gamma, w\neq z,\\[2ex] -\dfrac{1}{2\pi i}\dfrac{\overline{W'(t)}}{|z'(t)|\overline{W(t)}}, & \text{if } w=z\in\Gamma.\end{cases}\tag{3.16}$$

Note that, in Theorem 5, if Γ is assumed to have the arc length parametrization $z(\sigma)$, then $|z'(\sigma)|=1$ and therefore (3.9) becomes

$$K(z(\sigma),z(\sigma))=-\frac{1}{2\pi i}\frac{\overline{W'(\sigma)}}{\overline{W(\sigma)}}.\tag{3.17}$$

Theoretically it is simpler to work with arc length parametrization. Thus to determine $K(z(t),z(t))$, where $z(t)$ is any parametrization of Γ, one may use

$$K(z(t),z(t))=\frac{K(z(\sigma),z(\sigma))|_{\sigma=t}}{|z'(t)|}.\tag{3.18}$$

§4 Applications of Theorem 6

In this section, we give several applications of Theorem 6, which give rise to some integral equations related to the Szegö kernel, the Bergman kernel and the Riemann map. We assume that the boundary Γ is an analytic Jordan curve.

Application 1

Comparison of (3.1) and (3.3) leads to a choice of $D(z)=S(z,a)$, $c=i$, $Q(z)=1$, $P(z)=R(z)$. Applying Theorem 6, we obtain the KST integral equation (2.5).

Application 2

Suppose now we compare equation (3.2) with (3.3). This leads to a choice of $D(z)=B(z,a)$, $c=-1$, $Q(z)=T(z)$, $P(z)=R(z)^2$. Applying Theorem 6, we obtain the integral equation (2.8) for the Bergman kernel.

Application 3

Comparison of (2.4) with (3.5) leads to a choice of $D(z)=\sqrt{R'(z)}$, $c=i$, $Q(z)=1$, and $P(z)=R(z)$. Note that since $R'(z)\neq 0$ in $\overline{\Omega}$, a single-valued and analytic branch of $\sqrt{R'(z)}$ can be defined. Applying Theorem 6, we obtain

$$U_1(z,a)+\int_\Gamma A(z,w)U_1(w,a)\,|dw|=2\pi\,\overline{H(a,z)},\qquad z\in\Gamma,\tag{4.1}$$

where $U_1(z,a) = \sqrt{R'(a)R'(z)}$ and $A(z,w)$ is the Kerzman-Stein kernel (2.6). Notice the close resemblance of this integral equation with that of the KST integral equation (2.5). But no knowledge of the Szegö kernel is required here. Using (2.4), the boundary values of the Riemann map R are evaluated in terms of U_1 by

$$R(z) = \frac{1}{i}T(z)\frac{U_1(z,a)^2}{|U_1(z,a)^2|}, \qquad z \in \Gamma. \tag{4.2}$$

Application 4
From (2.4) we obtain

$$R(z)^2 = -T(z)^2\frac{R'(z)^2}{|R'(z)|^2}, \qquad z \in \Gamma. \tag{4.3}$$

Comparison with (3.5) leads to $D(z) = R'(z)$, $c = -1$, $Q(z) = T(z)$, and $P(z) = R(z)^2$. Applying Theorem 6, we obtain

$$U_2(z,a) + \int_\Gamma M(z,w)U_2(w,a)\,|dw| = -\frac{\overline{T(z)^2}}{(\overline{z}-\overline{a})^2}, \qquad z \in \Gamma, \tag{4.4}$$

where $U_2(z,a) = R'(a)R'(z)$ and $M(z,w)$ is the kernel (2.9). If we multiply both sides by $T(z)$ and let $U_3(z,a) = T(z)U_2(z,a)$, we obtain

$$U_3(z,a) + \int_\Gamma N(z,w)U_3(w,a)\,|dw| = -\frac{\overline{T(z)}}{(\overline{z}-\overline{a})^2}, \qquad z \in \Gamma, \tag{4.5}$$

where N is the Neumann kernel (2.11). Notice the close resemblance of the integral equations (4.4) and (4.5) with the integral equations (2.8) and (2.10) respectively. But no knowledge of the Bergman kernel is required here. Using (2.4), the boundary values of the Riemann map R are evaluated in terms of U_3 by

$$R(z) = \frac{1}{i}\frac{U_3(z,a)}{|U_3(z,a)|}, \qquad z \in \Gamma. \tag{4.6}$$

Application 5
Observe that (4.3) may also be written as

$$1 = -\overline{T(z)}^2\frac{R(z)^2/R'(z)^2}{|R(z)/R'(z)|^2}, \qquad z \in \Gamma. \tag{4.7}$$

Comparison with (3.5) leads to a choice of $D(z) = R(z)/R'(z)$, $c = -1$, $Q(z) = \overline{T(z)}^3$, and $P(z) = 1$. Applying Theorem 6, we obtain

$$\frac{R(z)}{R'(z)} + \int_\Gamma M_1(z,w)\frac{R(w)}{R'(w)}\,|dw| = 0, \qquad z \in \Gamma, \tag{4.8}$$

where

$$M_1(z,w) = \begin{cases} \dfrac{1}{2\pi i}\left[\dfrac{T(z)^2}{T(w)^3(\overline{w}-\overline{z})} - \dfrac{T(w)}{w-z}\right], & \text{if } w, z \in \Gamma, w \neq z, \\ -\dfrac{3}{2\pi|z'(t)|}\mathrm{Im}\left[\dfrac{z''(t)}{z'(t)}\right], & \text{if } w = z \in \Gamma. \end{cases} \tag{4.9}$$

Dividing both sides of (4.8) by $T(z)$ and applying (2.4), we get

$$|R'(z)|^{-1} + \int_\Gamma M_2(z,w)|R'(w)|^{-1}\,|dw| = 0, \qquad z \in \Gamma, \tag{4.10}$$

where

$$M_2(z,w) = \begin{cases} \dfrac{1}{\pi}\mathrm{Im}\left[\dfrac{T(w)^2}{T(z)(z-w)}\right], & \text{if } w, z \in \Gamma, w \neq z, \\ -\dfrac{3}{2\pi|z'(t)|}\mathrm{Im}\left[\dfrac{z''(t)}{z'(t)}\right], & \text{if } w = z \in \Gamma. \end{cases} \tag{4.11}$$

Assuming Γ has the parametric representation $z(t)$, $0 \leq t \leq \beta$, the integral equation (4.10) can be written in terms of $\theta'(t)$, where $\theta(t)$ is the boundary correspondence function. Using the relationship

$$R'(z(t))z'(t) = i\theta'(t)e^{i\theta(t)},$$

the integral equation (4.10) becomes

$$\theta'(t)^{-1} + \int_0^\beta M_3(t,s)\theta'(s)^{-1}\,ds = 0, \qquad z \in \Gamma, \tag{4.12}$$

where

$$M_3(t,s) = \begin{cases} \dfrac{1}{\pi}\mathrm{Im}\left[\dfrac{z'(s)^2}{z'(t)(z(t)-w)}\right], & \text{if } t \neq \varepsilon, \\ -\dfrac{3}{2\pi}\mathrm{Im}\left[\dfrac{z''(t)}{z'(t)}\right], & \text{if } t = s. \end{cases} \tag{4.13}$$

Thus the reciprocal of the derivative of the boundary correspondence function $\theta(t)$ satisfies the homogeneous integral equation (4.12).

Application 6
Let R be a Riemann mapping of the simply connected region Ω onto the unit disk. We assume that the boundary Γ of Ω is an analytic Jordan curve. Let $0 \in \Omega$ and let R be normalized by

$$R(0) = 0, \qquad R'(0) > 0. \tag{4.14}$$

Since R has a simple zero at 0, R can be represented by

$$R(z) = z\,g(z), \qquad z \in \Omega, \tag{4.15}$$

where g is analytic in Ω, $g(0) > 0$, and $g(z) \neq 0$. One method of determining R having this form is by solving an associated boundary value problem of Dirichlet type. By seeking a solution of this problem in the form of a logarithmic potential, one arrives at the Symm's integral equation which is an integral equation of the first kind [9, Sections 16.5–16.6].

Since R is analytic in $\overline{\Omega}$, from (4.15) and using the fact that $|R(z)| = 1$ for $z \in \Gamma$, we have

$$R(z) = \frac{z\,g(z)}{|z||g(z)|}, \qquad z \in \Gamma. \tag{4.16}$$

A comparison of (4.16) and (3.5) leads to $a = 0$, $c = 1$, $P(z) = R(z)$, $Q(z) = z\overline{T(z)}/|z|$, and $D(z) = \sqrt{g(z)}$. Applying Theorem 6, gives

$$\sqrt{g(z)} + \int_\Gamma J(z,w)\sqrt{g(w)}\,|dw| = \frac{1}{|z|}\left[\frac{\sqrt{g(0)}}{R'(0)}\right]^{-}, \qquad z \in \Gamma, \tag{4.17}$$

where the kernel J is given by

$$J(z,w) = \begin{cases} \dfrac{1}{2\pi i}\left[\dfrac{\overline{z}\,|w|\overline{T(w)}}{\overline{w}\,|z|(\overline{w}-\overline{z})} - \dfrac{T(w)}{w-z}\right], & \text{if } w, z \in \Gamma, w \neq z, \\[2ex] \dfrac{1}{2\pi i\,|z'(t)|}\left[\dfrac{\rho'(t)}{\rho(t)} - \dfrac{\overline{z'(t)}}{\overline{\overline{z(t)}}} - i\,\mathrm{Im}\left(\dfrac{z''(t)}{z'(t)}\right)\right], & \text{if } w = z \in \Gamma, \end{cases} \tag{4.18}$$

where $\rho(t) = |z(t)|$.

If we write $U(z) = \sqrt{g(0)g(z)}$ and use the fact that $R'(0) = g(0)$, the integral equation (4.17) can be rewritten as

$$U(z) + \int_\Gamma J(z,w)U(w)\,|dw| = \frac{1}{|z|}, \qquad z \in \Gamma. \tag{4.19}$$

If this integral equation admits a unique solution, then using (4.16) we can express the Riemann mapping function R in terms of U as

$$R(z) = \frac{z\,U(z)^2}{|z||U(z)^2|}, \qquad z \in \Gamma. \tag{4.20}$$

We have demonstrated that the integral equation (4.19) has a solution. But is the solution that we have found the only one? Our numerical experiments have indicated that we do have uniqueness. In any case, the question of uniqueness is not important here. Even if it can be proven that the integral equation (4.19) has a unique solution, it might not be the preferable integral equation for numerical conformal mapping because of its complicated looking kernel $J(z, w)$. What we wish to emphasize here is how Henrici's method for proving the KST integral equation can be adapted to derive the integral equation (3.15) from which various integral equations related to the Riemann map follow.

Acknowledgement. A.H.M. Murid is indebted to the organizers of the CMFT'97 meeting for the financial support he received. The authors acknowledge the valuable suggestions of Prof. N. Papamichael and the referee, which helped improve the presentation of the paper.

References

[1] Andersen, C., S. E. Christiansen, O. Møller and H. Tornehave, Conformal mapping, in *Selected Numerical Methods*, C. Gram (ed.), Regnecentralen, Coppenhagen, 1962, 114–261.

[2] Bell, S. R., *The Cauchy Transform, Potential Theory, and Conformal Mapping*, CRC Press, Boca Raton, 1992.

[3] Bergman, S., *The Kernel Function and Conformal Mapping*, AMS Math. Survey No. 5, New York, 1950.

[4] Bergman, S. and J. G. Herriot, Application of the method of the kernel function for solving boundary-value problem, *Numer. Math.* **3** (1961), 209–225.

[5] Burbea, J., A procedure for conformal maps of simply-connected domains by using the Bergman function, *Math. Comp.* **24** (1970), 821–829.

[6] Gaier, D., *Konstruktive Methoden der Konformen Abbildung*, Springer, Heidelberg, 1964.

[7] Gakhov, F. D., *Boundary Value Problems*, English translation of Russian edition 1963 ed. by I. N. Sneddon, Dover, New York, 1990.

[8] Henrici, P., *Applied and Computational Complex Analysis, vol. 1*, Wiley, New York, 1974.

[9] Henrici, P., *Applied and Computational Complex Analysis, vol. 3*, Wiley, New York, 1986.

[10] Hille, E., *Analytic Function Theory, vol. 2*, Ginn, Boston, 1962.

[11] Kantorovich, L. V. and V. I. Krylov, *Approximate Methods of Higher Analysis*, English translation of Russian edition 1941 ed. by C.D. Benster, Noordhoff, Groningen, 1964.

[12] Kerzman, N. and E. M. Stein, The Cauchy kernel, the Szegö kernel, and the Riemann mapping function, *Math. Ann.* **236** (1978), 85–93.

[13] Kerzman, N. and M. R. Trummer, Numerical conformal mapping via the Szegö kernel, *J. Comp. Appl. Math.* **14** (1986), 111–123.

[14] Levin, D., N. Papamichael and A. Sideridis, The Bergman kernel method for numerical conformal mapping of simply connected domains, *J. Inst. Maths. Applics.* **22** (1978), 171–187.

[15] Nehari, Z. *Conformal Mapping*, Dover, New York, 1952.

[16] Papamichael, N. and C. A. Kokkinos, Two numerical methods for the conformal mapping of simply-connected domains, *Computer Methods in Applied Mechanics and Engineering* **28** (1981), 285–307.

[17] Papamichael, N., M. K. Warby and D. M. Hough, The determination of poles of the mapping function and their use in numerical conformal mapping, *J. Comp. Appl. Math.* **9** (1983), 155–166.

[18] Rabinowitz, P., Numerical experiments in conformal mapping by the method of orthonormal polynomials, *J. Assoc. Comput. Mach.* **13** no. 2 (1966), 296–303.

[19] Razali, M. R. M., M. Z. Nashed and A. H. M. Murid, Numerical conformal mapping via the Bergman kernel, *J. Comp. Appl. Math.* **82** no. 1-2 (1997), 333–350.

[20] Schinzinger, R. and P. A. A. Laura, *Conformal Mapping: Methods and Applications*, Elsevier, Amsterdam, 1991.

[21] Szegö, G., *Orthogonal Polynomials*, Amer. Math. Soc. Colloq. Publ. 23, American Mathematical Society, Providence, 1939.

[22] Trefethen, L. N., (ed.), *Numerical Conformal Mapping*, North-Holland, Amsterdam, 1986.

A. H. M. Murid
Department of Mathematics
Faculty of Science
Universiti Teknologi Malaysia
Locked Bag 791
80990 Johor Bahru
MALAYSIA
alihmm@math.fs.utm.my

M. Z. Nashed
Department of Mathematical Sciences
University of Delaware
Delaware 19716
U.S.A.
nashed@math.udel.edu

M. R. M. Razali
Department of Mathematics
Faculty of Science
Universiti Teknologi Malaysia
Locked Bag 791
80990 Johor Bahru
MALAYSIA
s-rshidi@fs.utm.my

An extension of the notion of universal Taylor series

Vassili Nestoridis

Abstract. Let Ω be a simply connected domain $\Omega \subset \mathbb{C}$, $\Omega \neq \mathbb{C}$ and $\zeta \in \Omega$. For $f \in H(\Omega)$ we denote by $S_N(f,\zeta)(z)$ the sequence of partial sums of the Taylor development of f with center ζ. We prove that generically for $f \in H(\Omega)$ the following holds:

For every compact set K, $K \cap \Omega = \emptyset$ with K^c connected and every function: $\phi : K \to \mathbb{C}$ continuous on K and holomorphic in K^o, there exists a strictly increasing sequence $\lambda_n \in \{0,1,2,\ldots\}$, such that, for every compact set $L \subset \Omega$ we have

$$\sup_{\zeta \in L} \sup_{z \in K} |S_{\lambda_n}(f,\zeta)(z) - \phi(z)| \to 0, \quad \text{as } n \to +\infty.$$

§1 Definitions and results

We denote by Ω any simply connected domain $\Omega \subset \mathbb{C}$, $\Omega \neq \mathbb{C}$. If $f : \Omega \to \mathbb{C}$ is holomorphic in Ω ($f \in H(\Omega)$) and $\zeta \in \Omega$, then f has a Taylor development with center ζ:

$$f(z) = \sum_{n=0}^{\infty} a_n(f,\zeta)(z-\zeta)^n, \quad \text{where } a_n(f,\zeta) = \frac{f^{(n)}(\zeta)}{n!}.$$

We denote by $S_N(f,\zeta)(z) = \sum_{n=0}^{N} a_n(f,\zeta)(z-\zeta)^n$ the partial sums.

First we define the class $U(\Omega,\zeta)$ ($\zeta \in \Omega$) of universal functions in Ω with center ζ.

Definition 1. *Let Ω and $\zeta \in \Omega$ be as above. A function $f \in H(\Omega)$ belongs to $U(\Omega,\zeta)$, iff for every compact set $K \subset \mathbb{C}$, $K \cap \Omega = \emptyset$ with K^c connected and for every function $h : K \to \mathbb{C}$ continuous on K and holomorphic in K^0, there exists a sequence $\lambda_n \in \{0,1,2,\ldots\}$, $n = 1,2,\ldots$, such that $\sup_{z \in K} |S_{\lambda_n}(f,\zeta)(z) - h(z)| \to 0$, as $n \to +\infty$.*

As a complement to this definition we notice that we may easily obtain $\lambda_n < \lambda_{n+1}$ and that the radius of convergence of $\sum_{n=0}^{\infty} a_n(f,\zeta)(z-\zeta)^n$ is exactly dist $(\zeta, \Omega^c) < +\infty$.

Computational Methods and Function Theory (CMFT'97)
N. Papamichael, St. Ruscheweyh and E. B. Saff (Eds.), pp. 421–430

ISBN 981-02-3626-3

Remark: The assumption "K^c connected" is natural, because if K^c has a bounded component W, then for $b \in W$ the function $\frac{1}{z-b}$ cannot be approximated by polynomials uniformly on K (maximum principle). We also notice that the compact set K may contain segments of the boundary of Ω.

Definition 2. *Let Ω be as above and let $L \subset \Omega$ be compact. A function $f \in H(\Omega)$ belongs to $U(\Omega, L)$, iff for every compact set $K \subset \mathbb{C}$, $K \cap \Omega = \emptyset$ with K^c connected and for any function $h : K \to \mathbb{C}$ continuous on K and holomorphic in K^0, there exists a sequence $\lambda_n \in \{0, 1, 2, \ldots\}$, $n = 1, 2, \ldots$, such that $\sup_{\zeta \in L} \sup_{z \in K} |S_{\lambda_n}(f, \zeta)(z) - h(z)| \to 0$, as $n \to +\infty$.*

We notice that in this definition we may always have $\lambda_n < \lambda_{n+1}$. Finally we define the class $U(\Omega)$ of universal functions (Taylor series) in Ω.

Definition 3. *Let Ω be as above. A function $f \in H(\Omega)$ belongs to $U(\Omega)$, iff for every compact set $K \subset \mathbb{C}$, $K \cap \Omega = \emptyset$ with K^c connected and for any function $h : K \to \mathbb{C}$ continuous on K and holomorphic in K^0, there exists a sequence $\lambda_n \in \{0, 1, 2, \ldots\}$, $n = 1, 2, \ldots$, such that for every compact set $L \subset \Omega$ we have $\sup_{\zeta \in L} \sup_{z \in K} |S_{\lambda_n}(f, \zeta)(z) - h(z)| \to 0$, as $n \to +\infty$.*

We notice that in this definition we may always arrange $\lambda_n < \lambda_{n+1}$. Our results are the following.

Theorem 1. *The classes $U(\Omega, \zeta), U(\Omega, L)$ and $U(\Omega)$ are G_δ-dense in $H(\Omega)$ endowed with the topology of uniform convergence on compacta. In particular, they are non void.*

Theorem 1 provides new generic properties of holomorphic functions (see also [15], [7], [8], [9]).

Proposition 1. *For every $f \in H(\Omega)$, there exist $u_1, u_2 \in U(\Omega)$ such that $f = u_1 + u_2$.*

Proposition 2. *Let $f \in U(\Omega, \zeta)$, for some $\zeta \in \Omega$. Let μ be any Randon measure on $\partial\Omega$ and let $h : \partial\Omega \to \mathbb{C}$ be any μ-measurable function. Then there exists a strictly increasing sequence $\lambda_n \in \{0, 1, 2, \ldots\}$, $n = 1, 2, \ldots$, such that $S_{\lambda_n}(f, \zeta)(z) \to h(z)$, μ-almost everywhere, as $n \to +\infty$.*

Proposition 2 relates to universal trigonometric series of Menchoff ([34], [1], [37]).

Proposition 3. *Let $f \in U(\Omega, \zeta)$, for some $\zeta \in \Omega$. Then f is not a rational function.*

Theorem 2. *Let $f(z) = \sum_{n=0}^{\infty} a_n (z-\zeta)^n$, $f \in U(\Omega,\zeta)$ for some $\zeta \in \Omega$. Then for every $z_0 \in \Omega^c$ the series $\sum_{n=0}^{\infty} a_n (z_0-\zeta)^n$ is not (C,k) summable for any $k = 1, 2, \ldots$.*

A consequence of Theorem 2 is the following.

Proposition 4. *Let $f \in U(\Omega, \zeta)$, for some $\zeta \in \Omega$. Then f does not extend continuously on $\overline{\Omega}$.*

Our last result is:

Proposition 5. *The classes $U(\Omega)$ and $\bigcap_{\zeta\in\Omega} U(\Omega,\zeta)$ contain only holomorphic functions in Ω, which are non extendable.*

§2 Sketch of the proof of Theorem 1

For the proof of Theorem 1 we need the following.

Lemma 1. *There exists a sequence $K_m \subset \mathbb{C}$, $m = 1, 2, \ldots$, of compact sets, $K_m \cap \Omega = \emptyset$ with K_m^c connected, such that for every compact set $K \subset \mathbb{C}$, $K \cap \Omega = \emptyset$ with K^c connected, there exists $m = 1, 2, \ldots$, with $K \subset K_m$.*

In the case where Ω is the open unit disk, the previous lemma has been proven in [37]. The proof in the general case is analogous.

We fix a sequence K_m, $m = 1, 2, \ldots$, as in the lemma. Let also f_j, $j = 1, 2, \ldots$, be an enumeration of the polynomials with coefficients in $\mathbb{Q} + i\mathbb{Q}$. For $\zeta \in \Omega$, $m, j, s = 1, 2, \ldots$, and $n = 0, 1, 2, \ldots$, we consider the set

$$E(\Omega, \zeta, K_m, f_j, s, n) = \left\{ f \in H(\Omega) : \sup_{z\in K_m} |S_n(f,\zeta)(z) - f_j(z)| < \frac{1}{s} \right\}.$$

Using Mergelyan's Theorem we obtain the following.

Lemma 2.

1) $U(\Omega,\zeta) = \bigcap_m \bigcap_j \bigcap_s \bigcup_{n=0}^{\infty} E(\Omega, \zeta, K_m, f_j, s, n).$

2) $U(\Omega, L) = \bigcap_m \bigcap_j \bigcap_s \bigcup_{n=0}^{\infty} \left[\bigcap_{\zeta \in L} E(\Omega, \zeta, K_m, f_j, s, n) \right]$.

3) If L_ρ, $\rho = 1, 2, \ldots$, is an exhausting sequence of compact sets in Ω, then

$$U(\Omega) = \bigcap_{\rho=1}^{\infty} U(\Omega, L_\rho).$$

Since $H(\Omega)$ is complete metrizable, we have Baire's Theorem at our disposal. Combining this with the previous lemma, we see that Theorem 1 follows if we prove that

$$\bigcap_{\zeta \in L} E(\Omega, \zeta, K_m, f_j, s, n)$$

is open in $H(\Omega)$ and

$$\bigcup_{n=0}^{\infty} \left[\bigcap_{\zeta \in L} E(\Omega, \zeta, K_m, f_j, s, n) \right]$$

is dense in $H(\Omega)$.

The first fact has a proof similar to the proof of Lemma 2.3 of [37]. A difference in the proof is that, if we have the inequalities

$$\sup_{z \in K_m} |S_n(f, \zeta)(z) - f_j(z)| < \frac{1}{s}$$

for every $\zeta \in L$, then by continuity on the compact set $L \times K_m$ we obtain

$$\sup_{\zeta \in L} \sup_{z \in K_m} |S_n(f, \zeta)(z) - f_j(z)| = \beta < \frac{1}{s}.$$

The fact that the set

$$\bigcup_{n=0}^{\infty} \left[\bigcap_{\zeta \in L} E(\Omega, \zeta, K_m, f_j, s, n) \right]$$

is dense in $H(\Omega)$ follows from Mergelyan's Theorem and the simple connectivity of Ω. The proof is similar to that of Lemma 2.4 in [37]. One also uses the trivial fact that, if g is a polynomial and $n \geq \deg g$, then for any center ζ we have $S_n(g, \zeta) \equiv g$.

Remark: Let S be any infinite subset of $\{0, 1, 2, \ldots\}$. If in definitions 1, 2, 3 we require that $\lambda_n \in S$ for all $n = 1, 2, \ldots$, then we have new classes $U(\Omega, \zeta, S), U(\Omega, L, S)$ and $U(\Omega, S)$ respectively. These classes are also G_δ-dense in $H(\Omega)$; the proof is similar to that of Theorem 1.

§3 Previous results

Regarding the origin of universal Taylor series we should mention the universal trigonometric series of Menchoff ([34],[1]), a result of Chui and Parnes ([2]) and the works of W. Luh (see especially [16]). Luh considers also universality in other senses. In the results of Luh and Chui and Parnes the approximation does not take place on the boundary of Ω (the compact set K must be disjoint from the topological closure $\overline{\Omega}$ of Ω). In our definitions ([37] and definitions 1, 2, 3 of the present paper) the compact set K is allowed to contain segments of the boundary $\partial\Omega$; thus, the approximation also holds on the points of the boundary. This fact produces essential differences between the corresponding classes of functions (see [33]). We also notice that the fact that the approximation is uniform with respect to the center ζ, when ζ varies on compact sets $L \subset \Omega$, seems to be a new element (see Definitions 2 and 3). Furthermore, the fact that universality of Taylor series is a generic property of holomorphic functions had, as far as we know, passed unnoticed (see [37], [33], [7], [8], [9] and Theorem 1 of the present paper). The use of Baire's theorem simplifies the proofs and allows to realize simultaneously several generic properties. Finally we mention that some properties of universal Taylor series in our sense have already been investigated (see [37], [33], [14] and the results of the present paper).

Initially we have introduced the class $U(D,0) = U$, where D is the open unit disk ([37]). What led us to give this definition and to include segments of the boundary in the compact set K, is a question of S. K. Pichorides, who had attracted the attention of E. S. Katsoprinakis and myself on a theorem of Marcinkiewicz and Zygmund ([32], [39], [6]). A series of papers have been published jointly or independently ([10], [11], [12], [13], [35], [36], [38]). In some of these papers geometric assumptions on the partial sums led to rational functions. The question of S. K. Pichorides was related to approximation by the partial sums on the boundary and to rational functions (for details see [37]).

During the last year in Greece several colleagues of mine worked on universal Taylor series. We have a Master's thesis by G. Costakis at the University of Athens, which is in preparation. The main results are obtained in Katsoprinakis-Papadimitrakis ([14]) and Melas-Nestoridis-Papadoperakis ([33]). In [14] the authors obtain an extension of Rogosinski's formula, realizing a program suggested by J.-P. Kahane, and show that every element of $U(D,0)$ is nowhere (C,k) summable on the unit circle. As a consequence of the extended Rogosinski's formula, they also obtain an extension of Marcinkiewicz-Zygmund theorem. In [33] we compare the properties of $U(D,0)$ and the class of Chui and Parnes and we investigate the growth of the coefficients a_n of any element $\sum_{n=0}^{\infty} a_n z^n$ of $U(D,0)$. The best known result on the growth of a_n is due to Melas. It states that we cannot have $|a_n| \leq Ce^{nb_n}$ for all $n = 0, 1, 2, \ldots$, where $C < +\infty$ and b_n is any decreasing sequence such that $\sum_1^\infty \frac{b_n}{n}$ converges. As a corollary we prove that $U(D, 0)$ is disjoint from the Nevanlinna class. The first result towards this direction was obtained with different methods by M. Papadimitrakis. It states that a_n can not converge to zero. The second result was established by I. Papadoperakis.

Proposition 6. *Let* $\sum_{n=0}^\infty a_n z^n \in U(D, 0)$, $\tau < 1$, *then* $\limsup \frac{|a_n|}{n^\tau} = +\infty$.

We sketch the simple proof of this result of Papadoperakis.

Proof: Assume $|a_n| \leq Cn^\tau$ for all $n = 1, 2, \ldots$, where $C < +\infty$ (and $\tau < 1$), to arrive at a contradiction.

For any N we find $M \equiv M(N) > N$, such that

$$\left|\sum_N^M a_n e^{in\theta} - 1\right| < \frac{1}{2} \quad \text{for} \quad -\frac{\pi}{2} \leq \theta \leq \frac{\pi}{2}.$$

This is possible because $\sum_{n=N}^\infty a_n z^n \in U(D, 0)$. It follows that

$$\frac{1}{2} < \operatorname{Re} \sum_N^M a_n e^{in\theta}.$$

Let $x_n = \operatorname{Re}\, a_n$ and $y_n = I_m a_n$; then $|x_n| \leq Cn^\tau$ and

$$\frac{1}{2} < \sum_N^M (x_n \cos n\theta - y_n \sin n\theta)$$

for all $\theta \in \left[-\frac{\pi}{2}, \frac{\pi}{2}\right]$. Considering also the analogous inequality for $-\theta$ and adding the two inequalities, we obtain

$$\frac{1}{2} < \sum_N^M x_n \cos n\theta, \qquad 0 \leq \theta \leq \frac{\pi}{2}.$$

Integrating on $[0, \theta]$, $0 \leq \theta \leq \frac{\pi}{2}$, we take

$$\frac{\theta}{2} \leq \sum_N^M \frac{x_n}{n} \sin n\theta, \qquad 0 \leq \theta \leq \frac{\pi}{2}.$$

Integrating again on $[0, \theta]$, $0 \leq \theta \leq \frac{\pi}{2}$, we obtain

$$\frac{\theta^2}{4} \leq \sum_N^M \frac{x_n}{n^2}(1 - \cos n\theta) \leq 2C \sum_N^M \frac{n^\tau}{n^2}.$$

Thus,

$$\frac{1}{2C}\frac{\pi^2}{16} \le \sum_{N}^{M} \frac{1}{n^{2-\tau}}.$$

Since this holds for infinitely many pairs $N < M$, the series $\sum_1^\infty n^{\tau-2}$ should diverge. This is absurd, because $\tau < 1$. The proof is completed. ■

According to Proposition 1 the classes $U(D,0)$ and $U(D)$ are not closed under addition. We show by a counterexample that they are neither closed under multiplication.

Counterexample: Let $u_1(z) = \sum_{n=0}^{\infty} a_n z^n$ be any element of $U(D,0)$ (or of $U(D)$). We set $u_2(z) = \sum_{n=0}^{\infty} \overline{a}_n z^n$; then it is easy to check that u_2 also belongs to $U(D,0)$ (or $U(D)$, respectively). We consider the product $(u_1 \cdot u_2)(z) \equiv \sum_{n=0}^{\infty} b_n z^n$. Then the coefficients b_n are all real. This easily implies that $u_1 \cdot u_2$ is not universal, because on the real axis the partial sums are real and they can not approximate the constant i. This shows that $U(D,0)$ and $U(D)$ are not closed under multiplication.

On extending some of the previous results from the disk to any simply connected domain, it was helpful to me that I read an unpublished note ([3]), in which A. Córdova obtains an extension of the Chui-Parnes result to more general domains simultaneously for a denumerable set of centers. At that time I did not know the result in [16], where the Chui-Parnes result was extended for all centers simultaneously.

Acknowledgements. I wish to express my thanks to J.-P. Kahane, S. Ruscheweyh and M. Zinsmeister.

References

[1] Bary, N. K., *A Treatise on Trigonometric Series, Vol. I, II*, Pergamon Press, 1964.

[2] Chui, C. and M. N. Parnes, Approximation by overconvergence of power series, *J. Math. Anal. Appl.* **36** (1971), 693–696.

[3] Córdova, A., Final examination of approximation theory, Departamento de Matemáticas, Universidad Técnica Federico Santa María, unpublished.

[4] Dienes, P., *The Taylor Series*, Dover Pub. Inc., New York, 1957.

[5] Faulstich, K., W. Luh and L. Tomm, Universelle Approximation durch Riesz-Transformierte der Geometrischen Reihe, *Manuscripta Math.* **36** (1981), 309–321.

[6] Kahane, J.-P., Sur la structure des ensembles de points limites des sommes partielles d'une série de Taylor, *Acta Sci. Math.* **45** (1983), 247–251.

[7] Kahane, J.-P., Baire Theory in Fourier and Taylor series, *Conference in the honor of Donald Newman*, Philadelphia, March 1996.

[8] Kahane, J.-P., General properties of Taylor series, 1896–1996, *J. Fourier Anal. Appl.* **3** Spec. Iss. (1997), 907–911.

[9] Kahane, J.-P., A few generic properties of Fourier and Taylor series, Taiwan, November 1996.

[10] Katsoprinakis, E. S., On a theorem of Marcinkiewicz and Zygmund for Taylor series, *Ark. Mat.* **27** 1 (1989), 105–126.

[11] Katsoprinakis, E. S., Taylor series with limit-points on a finite number of circles, *Trans. Am. Math. Soc.* **337** 1 (1993), 437–450.

[12] Katsoprinakis, E. S. and V. Nestoridis, Partial sums of Taylor series on a circle, *Ann. Inst. Fourier* **39** 3 (1989), 715–736.

[13] Katsoprinakis, E. S. and V. Nestoridis, An application of Kronecker's theorem to rational functions, *Math. Ann.* **298** 1 (1994), 145–166.

[14] Katsoprinakis, E. S. and M. Papadimitrakis, Extensions of a theorem of Marcinkiewicz-Zygmund and of Rogosinski's formula and an application to universal Taylor series, to appear.

[15] Kierst, S. and E. Szpirajn, Sur certaines singularités des fonctions analytiques uniformes, *Fundam. Math.* **21** (1933), 267–294.

[16] Luh, W., Universal approximation properties of overconvergent power series on open sets, *Analysis* **6** (1986), 191–207.

[17] Luh, W., Universalfunktionen in einfach zusammenhängenden Gebieten, *Aequationes Math.* **19** (1979), 183–193.

[18] Luh, W. and R. Trautner, Summierbarkeit der geometrischen Reihe auf vorgeschriebenen Mengen, *Manuscr. Math.* **18** (1976), 317–326.

[19] Luh, W., On universal functions, in *Fourier Analysis and Approximation Theory*, Budapest (Hungary), Colloq. Math. Soc. János Bolyai **19** 1976, 503–511.

[20] Luh, W., Multiply universal holomorphic functions, *J. Approximation Theory* **89** 2 (1997), 135–155.

[21] Luh, W., Über cluster sets analytischer Funktionen, *Acta Math. Acad. Sci. Hung.* **33** (1979), 137–142.

[22] Luh, W., Approximation by antiderivatives, *Constructive Approximation* **2** (1986), 179–187.

[23] Luh, W., On the "universality" of any holomorphic function, *Result. Math.* **10** (1986), 130–136.

[24] Luh, W., Holomorphic monsters, *J. Approximation Theory* **53** 2 (1988), 128–144.

[25] Luh, W., Universal functions and conformal mappings, *Serdica* **19** 2–3 (1993), 161–166.

[26] Luh, W., Entire functions with various universal properties, *Complex Variables, Theory Appl.* **31** (1996), 87–96.

[27] Luh, W., Approximation analytischer Funktionen durch überkonvergente Potenzreihen und deren Matrix-Transformierten, *Mitt. Math. Sem. Giessen* **88** (1970), 56.

[28] Luh, W., Über die Summierbarkeit der geometrischen Reihe, *Mitt. Math. Sem. Giessen* **113** (1977), 70.

[29] Luh, W., V. A. Martirosian and J. Müller, Universal entire functions with gap power series, *Indagationes math.*, to appear.

[30] Luh, W., V. A. Martirosian and J. Müller, T-universal functions with lacunary power series, *Acta Sci. math. (Szeged)* **64** (1998), 51–63.

[31] Luh, W. and R. Trautner, Punktweise Überkonvergenz von Potenzreihen, in *Constructive Function Theory '77*, Sofia, 1980, 373–380.

[32] Marcinkiewicz, J. and A. Zygmund, On the behaviour of trigonometric series and power series, *Trans. Am. Math. Soc.* **50** (1941), 407–453.

[33] Melas, A., V. Nestoridis and I. Papadoperakis, Growth of coefficients of universal Taylor Series and comparison of two classes of functions, *J. Anal. Math*, **73** (1997), 187–207.

[34] Menchoff, D., Sur les series trigonometriques universelles, *C. R. (Dokl.) Acad. Sci. URSS, n. Ser.* **49** (1945), 79–82.

[35] Nestoridis, V., Limit points of partial sums of Taylor series, *Mathematika* **38** 2 (1991), 239–249.

[36] Nestoridis, V., Distribution of partial sums of the Taylor development of rational functions, *Trans. Am. Math. Soc.* **346** 1 (1994), 283–295.

[37] Nestoridis, V., Universal Taylor series, *Ann. Inst. Fourier* **46** 5 (1996), 1293–1306.

[38] Nestoridis, V. and S. K. Pichorides, The circular structure of the set of limit points of partial sums of Taylor series, *Semin. Anal. Harmonique* Annee 1989/90, 71–77.

[39] Zygmund, A., *Trigonometric Series*, second edition reprinted, vol. I, II, Cambridge University Press, 1979.

Vassili Nestoridis
Department of Mathematics
University of Athens
Panepistimiopolis
15784 Athens
GREECE

Starlikeness of certain integral transforms

Milutin Obradović

Abstract. By using the method of differential-subordination we consider starlikeness of certain integral transforms in the unit disc.

§1 Introduction and preliminaries

As usual, let A denote the class of functions f analytic in the unit disc $\mathbb{D} = \{z : |z| < 1\}$ and normalized by $f(0) = f'(0) - 1 = 0$. Furthermore, let $S^* \subset A$ be the class of *starlike* functions in $\mathbb{D}$, *i.e.* those $f \in A$ which satisfy

$$\operatorname{Re} \frac{zf'(z)}{f(z)} > 0, \qquad z \in \mathbb{D}$$

(see, *cf.* [2]).

Let $B(\alpha, \beta, h, g)$ denote the class of functions $f \in A$ which admit the representation

$$f(z) = \left[(\alpha + i\beta) \int_0^z h(t) g^\alpha(t) t^{i\beta - 1}\, dt \right]^{\frac{1}{\alpha + i\beta}}, \tag{1.1}$$

where α and β are real, $\alpha > 0$, $zh(z) \in A$, $\operatorname{Re} h(z) > 0$, $z \in \mathbb{D}$, and $g \in S^*$. This is the class of Bazilevič functions [1], which is well-known to form a subclass of the class of normalized univalent functions in $\mathbb{D}$. The case $g(z) = z$, $\beta = 0$, which is equivalent to either of the following conditions:

$$f(z) = \left[\alpha \int_0^z h(t) t^{\alpha - 1}\, dt \right]^{\frac{1}{\alpha}}, \tag{1.2}$$

$$\operatorname{Re} f'(z) \left(\frac{f(z)}{z} \right)^{\alpha - 1} > 0, \qquad \alpha > 0, \quad z \in \mathbb{D}, \tag{1.3}$$

was considered by many authors. Recently, Ponnusamy [7], and later the author [6] considered starlikeness of functions satisfying (1.3).

In this paper we consider starlikeness of functions of the form (1.1), but with other assumptions on g and h than those in the Bazilevič case. We shall also consider starlikeness of certain other integral transforms which have been studied earlier in related papers. We need the following lemma [3].

Computational Methods and Function Theory (CMFT'97)
N. Papamichael, St. Ruscheweyh and E. B. Saff (Eds.), pp. 431–436

ISBN 981-02-3626-3

Lemma 1. *Let Ω be a set in the complex plane $\mathbb{C}$ and suppose that the function $\psi : \mathbb{C}^2 \times \mathbb{D} \to \mathbb{C}$ satisfies $\psi(ix, y, z) \notin \Omega$, for all real x, y with $y \leq -(1+x^2)/2$ and all $z \in \mathbb{D}$. If the function $zp \in A$ satisfies*

$$\psi(p(z), zp'(z); z) \in \Omega, \qquad z \in \mathbb{D},$$

then $\operatorname{Re} p(z) > 0$ *holds for all* $z \in \mathbb{D}$.

§2 Integral transforms and starlikeness

We start with the integral transform (2.2) which appeared in the paper of Mocanu [4] in connection with the study of α-convex functions.

Theorem 1. *Let $f \in A$ and $\alpha \in \mathbb{C}$, $\operatorname{Re} \alpha > 0$. If*

$$\left| \alpha \frac{zf'(z)}{f(z)} - 1 \right| < 1, \qquad z \in \mathbb{D} \tag{2.1}$$

then

$$F(z) = \left(\alpha \int_0^z \frac{f^\alpha(t)}{t}\, dt \right)^{\frac{1}{\alpha}} \tag{2.2}$$

is starlike in $\mathbb{D}$.

Proof: Differentiating F^α we obtain $\alpha F^{\alpha-1}(z)F'(z) = \alpha f^\alpha(z)/z$, and from there, upon logarithmic differentiation,

$$(\alpha - 1)\frac{zF'(z)}{F(z)} + \frac{zF''(z)}{F'(z)} = \alpha \frac{zf'(z)}{f(z)} - 1. \tag{2.3}$$

(2.3) and (2.1) imply

$$\left| (\alpha - 1)\frac{zF'(z)}{F(z)} + \frac{zF''(z)}{F'(z)} \right| < 1, \qquad z \in \mathbb{D}, \tag{2.4}$$

so that we must prove that (2.4) implies $F \in S^*$. Let $zF'(z)/F(z) = p(z)$. Then $zF''(z)/F'(z) = zp'(z)/p(z) + p(z) - 1$ and (2.4) becomes

$$\left| \frac{zp'(z)}{p(z)} + \alpha p(z) - 1 \right| < 1, \qquad z \in \mathbb{D}.$$

For an application of Lemma 1 we consider the function

$$\psi(ix, y) = \left| \frac{y}{ix} + \alpha ix - 1 \right|^2 - 1, \tag{2.5}$$

where x, y are real ($x \neq 0$) and $y \leq -(1+x^2)/2$. If we prove that under these conditions $\psi(ix, y) \geq 0$, then Lemma 1 will give $\operatorname{Re} p(z) > 0$, $z \in \mathbb{D}$, which implies $F \in S^*$.

Writing $\alpha = \alpha_1 + i\alpha_2$, α_1, α_2 real, (2.5) becomes

$$\begin{aligned} \psi(ix, y) &= \left|(-\alpha_2 x - 1) + i\left(\alpha_1 x - \frac{y}{x}\right)\right|^2 - 1 \\ &= (\alpha_1^2 + \alpha_2^2)x^2 + 2\alpha_2 x + \frac{y^2}{x^2} - 2\alpha_1 y. \end{aligned} \tag{2.6}$$

Since $y \leq -(1+x^2)/2$ and $\alpha_1 = \operatorname{Re}\alpha > 0$ we get $-2\alpha_1 y \geq \alpha_1(1+x^2)$. Also, $y^2/x^2 \geq (1+x^2)^2/(4x^2) \geq 1$. Now, from (2.6) we obtain

$$\psi(ix, y) \geq (\alpha_1^2 + \alpha_2^2 + \alpha_1)x^2 + 2\alpha_2 x + 1 + \alpha_1. \tag{2.7}$$

The trinom on the right hand side of (2.7) has the discriminant

$$D = 4\left[\alpha_2^2 - (1+\alpha_1)(\alpha_1^2 + \alpha_2^2 + \alpha_1)\right] = -4\alpha_1\left[\alpha_2^2 + (1+\alpha_1)^2\right] < 0$$

which shows that it is positive for all real x. This implies our assertion. ■

Remark 1: In [8] the author proved that, under the stronger condition $|\alpha - 1| < 1$, the function F given by (2.2) is univalent. Our present result is better.

Theorem 2. *Let $f \in A$ and $\alpha \in \mathbb{C}$, $\operatorname{Re}\alpha > 0$. If*

$$\left|\alpha - 1 + \frac{zf''(z)}{f'(z)}\right| < 1, \qquad z \in \mathbb{D}, \tag{2.8}$$

then the function

$$F(z) = \left[\alpha \int_0^z t^{\alpha-1} f'(t)\, dt\right]^{\frac{1}{\alpha}}, \tag{2.9}$$

is starlike in $\mathbb{D}$.

Proof: As in the proof of Theorem 1, we obtain from (2.9) and (2.8)

$$\left|(\alpha - 1)\frac{zF'(z)}{F(z)} + \frac{zF''(z)}{F'(z)}\right| < 1, \qquad z \in \mathbb{D},$$

which is the same as relation (2.4). Hence, we get the same conclusion. ■

The univalence problem, but not the starlikeness, for the integral transformation (2.9) has earlier been studied in [5].

Next we wish to consider starlikeness in the general case given by (1.1). However, first we quote two results which deal with the special cases given by (1.2) and (1.3).

Theorem 3 ([7]). *If $\alpha > 0$, $0 < r \le 1$, $f \in A$ and*

$$\left| f'(z)\left(\frac{f(z)}{z}\right)^{\alpha-1} - 1\right| < \frac{r}{1+(1+r)(\alpha/(1+\alpha))}, \qquad z \in \mathbb{D},$$

then $f \in S^*$ and $\left|\frac{zf'(z)}{f(z)} - 1\right| < r$.

Theorem 4 ([6]). *If $f \in A$ is of the form (1.2), where $zh \in A$, $h(z) \neq 0$ for all $z \in \mathbb{D}$, and satisfies the condition*

$$\left|\frac{zh'(z)}{h(z)}\right| < \sqrt{\alpha^2+\alpha+1}, \qquad z \in \mathbb{D}.$$

Then $f \in S^*$.

Theorem 5. *Let the function f have the form (1.1), where α, β, are real, $\alpha > 0$, $zh(z) \in A$, $h(z) \neq 0$ for all $z \in \mathbb{D}$, and satisfies*

$$\left|\beta + \operatorname{Im}\frac{zh'(z)}{h(z)}\right| \le \sqrt{\alpha+1} - 1, \tag{2.10}$$

while $g \in A$ satisfies

$$\left|\frac{zg'(z)}{g(z)} - 1\right| < \frac{1}{\alpha}, \qquad z \in \mathbb{D}, \tag{2.11}$$

then $f \in S^*$.

Proof: Similar as in the proof of Theorem 1 we obtain from (1.1)

$$(\alpha+i\beta-1)\frac{zf'(z)}{f(z)} + \frac{zf''(z)}{f'(z)} = \frac{zh'(z)}{h(z)} + \alpha\frac{zg'(z)}{g(z)} + i\beta - 1,$$

which implies

$$\begin{aligned}(\alpha+i\beta-1)\frac{zf'(z)}{f(z)} + \frac{zf''(z)}{f'(z)} - \frac{zh'(z)}{h(z)} - (\alpha+i\beta-1) &= \\ = \alpha\left(\frac{zg'(z)}{g(z)} - 1\right).&\end{aligned} \tag{2.12}$$

From (2.12) and (2.11) we deduce that

$$\left|(\alpha+i\beta-1)\frac{zf'(z)}{f(z)} + \frac{zf''(z)}{f'(z)} - \frac{zh'(z)}{h(z)} - (\alpha+i\beta-1)\right| < 1 \tag{2.13}$$

for $z \in \mathbb{D}$.

Let $p(z) := zf'(z)/f(z)$, so that $zf''(z)/f'(z) = zp'(z)/p(z) + p(z) - 1$, and relation (2.13) becomes

$$\begin{aligned} &\psi_1(p(z), zp'(z); z) := \\ &\left|\frac{zp'(z)}{p(z)} + (\alpha + i\beta)p(z) - \frac{zh'(z)}{h(z)} - (\alpha + i\beta)\right| < 1. \end{aligned} \tag{2.14}$$

To prove $\operatorname{Re} p(z) > 0$, $z \in \mathbb{D}$ using Lemma 1, it is enough to prove that

$$\psi_1(ix, y; z) \geq 1 \tag{2.15}$$

for real x, y, $y \leq -(1+x^2)/2$ and for all $z \in \mathbb{D}$. Writing $zh'(z)/h(z) = u+iv$, we get

$$\begin{aligned} \psi_1(ix, y; z) &= \left|[-\beta x - (u + \alpha)] + i\left(\alpha x - \frac{y}{x}\right) - i(v + \beta)\right| \\ &\geq \left|[-\beta x - (u + \alpha)] + i\left(\alpha x - \frac{y}{x}\right)\right| - |v + \beta|. \end{aligned}$$

Hence, to prove (2.15) it is sufficient to show that

$$\left|[-\beta x - (u + \alpha)] + i\left(\alpha x - \frac{y}{x}\right)\right| \geq |v + \beta| + 1,$$

or equivalently

$$\left|[-\beta x - (u + \alpha)] + i\left(\alpha x - \frac{y}{x}\right)\right|^2 - (|v + \beta| + 1)^2 \geq 0. \tag{2.16}$$

If we denote the left hand side of (2.16) by $\psi_2(x, y; z)$, then we get

$$\psi_2(x, y; z) = (\alpha^2 + \beta^2)x^2 + 2\beta(u+\alpha)x + \frac{y^2}{x^2} - 2\alpha y + (u+\alpha)^2 - (|v+\beta|+1)^2.$$

As in the proof of Theorem 1, we have $-2\alpha y \geq \alpha(1 + x^2)$ and $y^2/x^2 \geq 1$, which implies

$$\psi_2(x, y; z) \geq (\alpha^2 + \beta^2 + \alpha)x^2 + 2\beta(u+\alpha)x + (u+\alpha)^2 + \alpha + 1 - (|v+\beta|+1)^2.$$

The discriminant of the last trinom is equal to

$$D = 4[-(\alpha^2 + \alpha)(u + \alpha)^2 - (\alpha^2 + \beta^2 + \alpha)(\alpha + 1 - (|v + \beta| + 1)^2)],$$

and since $\alpha > 0$ and (2.10) are satisfied, we see that $D \leq 0$. Hence $\psi_2(x, y; z) \geq 0$ for all real x, as required.

By Lemma 1 we conclude $\operatorname{Re} p(z) > 0$, $z \in \mathbb{D}$, *i.e.* $f \in S^*$. ■

Remark 2: If we put $g(z) = z$ and $\beta = 0$ in Theorem 5, then we get

$$|\operatorname{Im} zh'(z)/h(z)| \leq \sqrt{\alpha + 1} - 1, \quad z \in \mathbb{D},$$

instead of the appropriate condition of Theorem 4. But since $\sqrt{\alpha^2 + \alpha + 1} > \sqrt{\alpha + 1} - 1$ for $\alpha > 0$, we cannot obtain Theorem 4 from Theorem 5.

Acknowledgement. This work was supported by Grant No. 04M03 of MNTRS through Math. Institute SANU. I am grateful to the referee for the careful reading and the useful comments.

References

[1] Bazilevič, I. E., On a case of integrability in quadratures of the Loewner-Kufarev equation, *Mat.Sb.* **37 (79)** (1955) 471–476 (Russian).

[2] Duren, P. L., *Univalent Functions*, Springer-Verlag, 1983.

[3] Miller, S. S. and P. T. Mocanu, Second order differential inequalities in the complex plane, *J. Math. Anal. Appl.* **65** (1978), 289–305.

[4] Mocanu, P. T., Uné propriété de convexité généralisée dans la théorie de la représentation conforme, *Mathematica (Cluj)*, **11 (34)**, 1 (1969), 127–133.

[5] Moldoveanu, S. and N. N. Pascu, Sufficient conditions for univalence of regular functions in the unit disc, "Babes-Bolyai" University, Seminar on Geometric Function Theory, preprint No.5 (1986), 111–114.

[6] Obradović, M., Starlikeness of certain class of analytic functions, Second joint Conference on Modern Applied Mathematics, Ilieni (Romania), June 3-7, 1997 (communicated).

[7] Ponnusamy, S., Pólya-Schoenberg conjecture by Carathéodory functions, *J. London Math. Soc.* **(2) 51** (1995), 93–104.

[8] Radomir, I., Sufficient conditions for univalence of a large class of functions, Univ. N. Sad, *Zb. Rad. Prirod.-Mat. Fak. Ser. Mat.* **25,2** (1995), 23–30.

Milutin Obradović
Department of Mathematics
Faculty of Technology and Metallurgy
4 Karnegijeva Street,
11000 Belgrade
YUGOSLAVIA
obrad@elab.tmf.bg.ac.yu

Rational approximation with varying weights in the complex plane

Igor E. Pritsker

and Richard S. Varga

Abstract. Given an open bounded set G in the complex plane and a weight function $W(z)$ which is analytic and different from zero in G, we consider the problem of locally uniform rational approximation of any function $f(z)$, which is analytic in G, by particular ray sequences of weighted rational functions of the form $W^{m+n}(z)R_{m,n}(z)$, where $R_{m,n}(z) = P_m(z)/Q_n(z)$, with $\deg P_m \leq m$ and $\deg Q_n \leq n$. The main result of this paper is a necessary and sufficient condition for such an approximation to be valid. We also consider a number of applications of this result to various classical weights, and find explicit criteria for the possibility of weighted approximation in these cases.

§1 Introduction and General Results.

In this paper, we shall develop the ideas of [11] and apply them to the study of the approximation of analytic functions in an open set G by weighted rationals $W^{m+n}(z)R_{m,n}(z)$. Specifically, we examine triples of the form

$$(G, W, \gamma) \tag{1.1}$$

where

$$\left.\begin{array}{ll}
i) & G \text{ is an open bounded set in the complex plane } \mathbb{C}, \\
 & \text{which can be represented as a finite or countable} \\
 & \text{union of disjoint simply connected domains, } i.e., \\
 & \qquad G = \bigcup_{\ell=1}^{\sigma} G_\ell \qquad \text{where } 1 \leq \sigma \leq \infty, \\
ii) & W(z), \text{ the weight function, is analytic in } G \text{ with} \\
 & W(z) \neq 0 \text{ for any } z \in G, \text{ and} \\
iii) & \gamma \text{ satisfies } 0 \leq \gamma \leq 1.
\end{array}\right\} \tag{1.2}$$

Computational Methods and Function Theory (CMFT'97)
N. Papamichael, St. Ruscheweyh and E. B. Saff (Eds.), pp. 437–448

ISBN 981-02-3626-3

We say that the triple (G, W, γ) has the *rational approximation property* if,

$$\left.\begin{array}{l}\text{for any } f(z) \text{ which is analytic in } G \text{ and for any compact subset } E \text{ of } G, \text{ there exists a sequence of rational functions } \{R_{m_i,n_i}(z)\}_{i=0}^{\infty}, \text{ where } R_{m_i,n_i}(z) = P_{m_i}(z)/Q_{n_i}(z), \text{ with } \deg P_{m_i} \leq m_i \text{ and } \deg Q_{n_i} \leq n_i \text{ for all } i \geq 0, \text{ and where } (m_i+n_i) \to \infty \text{ as } i \to \infty, \text{ such that} \\ i) \quad \lim_{i\to\infty} \dfrac{m_i}{m_i+n_i} = \gamma, \\ \quad\ \text{and} \\ ii) \quad \lim_{i\to\infty} \|f - W^{m_i+n_i} R_{m_i,n_i}\|_E = 0,\end{array}\right\} \tag{1.3}$$

where all norms throughout this paper are the uniform (Chebyshev) norms on the indicated sets.

Given a triple (G, W, γ), as in (1.1) which satisfies the conditions of (1.2), we state below our main result, Theorem 1, which gives a characterization, in terms of potential theory, for the triple (G, W, γ) to have the rational approximation property. Let $\mathcal{M}(E)$ be the set of all positive unit Borel measures on $\mathbb{C}$ which are supported on a compact set E, *i.e.*, for any $\mu \in \mathcal{M}(E)$, we have $\mu(\mathbb{C}) = 1$ and $\operatorname{supp} \mu \subset E$. Also, ∂G denotes the boundary of the set G, and the logarithmic potential of an arbitrary compactly supported signed measure μ is defined (see Tsuji [18], p. 53) by

$$U^{\mu}(z) := \int \log \frac{1}{|z-t|}\, d\mu(t).$$

Theorem 1. *A triple (G, W, γ), satisfying (1.2), has the rational approximation property (1.3) if and only if there exist a signed measure*

$$\mu(G, W, \gamma) = \gamma\mu^{+}(G, W, \gamma) - (1-\gamma)\mu^{-}(G, W, \gamma),$$

with $\mu^{+}(G, W, \gamma), \mu^{-}(G, W, \gamma) \in \mathcal{M}(\partial G)$, and a constant $F(G, W, \gamma)$ such that

$$U^{\mu(G,W,\gamma)}(z) - \log |W(z)| = F(G, W, \gamma), \qquad \text{for any } z \in G. \tag{1.4}$$

Below, we state some consequences of Theorem 1, while in §2, we state applications of Theorem 1 in a number of specific cases.

Remark 1: Results on weighted rational approximation of analytic functions in open sets with *multiply* connected components (as opposed, in (1.2i), to unions of simply connected domains) will be considered elsewhere.

Remark 2: The condition in (1.2ii) that $W(z) \neq 0$ for all $z \in G$ cannot be dropped, for if $W(z_0) = 0$ for some $z_0 \in G_k$, where $G = \bigcup_{\ell=1}^{\sigma} G_\ell$, then the necessarily null sequence

$$\left\{W^{m_i+n_i}(z_0)R_{m_i,n_i}(z_0)\right\}_{i=0}^{\infty}$$

trivially fails to converge to any $f(z)$, analytic in G, with $f(z_0) \neq 0$; whence, the rational approximation property fails.

Corollary 1. *A triple (G, W, γ), satisfying (1.2), has the rational approximation property (1.3) if and only if (1.3) holds for $f(z) \equiv 1$, i.e., if and only if this single function is locally uniformly approximable on compact subsets of G by a corresponding sequence of the weighted rational functions.*

Remark 3: The function $f(z) \equiv 1$ in Corollary 1 can be replaced by any function which is analytic in G and not equal identically to 0 in G.

Corollary 2. *Given a triple (G, W, γ), which satisfies (1.2) with σ finite, assume that there exist a constant F and a signed measure μ with*

$$\operatorname{supp} \mu \subset \partial G \quad \text{and} \quad \mu(\mathbb{C}) = 2\gamma - 1, \tag{1.5}$$

such that

$$U^{\mu}(z) - \log|W(z)| = F, \qquad \text{for any } z \in G. \tag{1.6}$$

Then, the triple (G, W, γ) has the rational approximation property (1.3) if and only if the signed measure μ can be decomposed as

$$\mu = \gamma\mu^{+} - (1-\gamma)\mu^{-}, \tag{1.7}$$

with $\mu^+, \mu^- \in \mathcal{M}(\partial G)$.

Furthermore, let a Jordan decomposition of the signed measure μ, satisfying (1.5) and (1.6), be given by

$$\mu = \tau^{+} - \tau^{-},$$

where τ^+ and τ^- are positive measures with

$$\operatorname{supp} \tau^{+},\ \operatorname{supp} \tau^{-} \subset \partial G \quad \text{and} \quad \mu(\operatorname{supp} \tau^{+} \cap \operatorname{supp} \tau^{-}) = 0.$$

Then, the triple (G, W, γ) has the rational approximation property (1.3) if and only if

$$\tau^{+}(\mathbb{C}) \leq \gamma. \tag{1.8}$$

If (1.7) or (1.8) holds true for a signed measure μ satisfying (1.5) and (1.6), then

$$\mu(G, W, \gamma) = \mu \quad \text{and} \quad F(G, W, \gamma) = F.$$

The study of weighted *rational* approximation has recently been introduced in papers by Borwein and Chen [1], Borwein, Rakhmanov and Saff [2], and Rakhmanov, Saff, and Simeonov [12]. The last two papers deal with weighted rational approximation only on the real line. Certain special cases of the triples (G, W, γ), in the notation of (1.1), were considered in the complex plane in [1], but that research did not attack the general question of necessary and sufficient conditions for (G, W, γ) to have the rational approximation property of (1.3), as in Theorem 1.

§2 Applications.

Finding the signed measure $\mu(G, W, \gamma)$ of Theorem 1, or verifying its existence, is a nontrivial problem in general. Since $U^{\mu(G,W,\gamma)}(z)$ is harmonic in $\mathbb{C}\backslash\text{supp}\ \mu(G, W, \gamma)$ and, since it can be shown from (1.4), if $\log|W(z)|$ is continuous on $\overline{G}$ and if G is a *finite* union of $G_\ell, \ell = 1, 2, \ldots, \ell_0$, that $U^{\mu(G,W,\gamma)}(z)$ is equal to $\log|W(z)| + F(G, W, \gamma)$ on $\text{supp}\,\mu(G, W, \gamma) \subset \partial G$, then $U^{\mu(G,W,\gamma)}(z)$ can be found as the solution of the corresponding Dirichlet problems. The signed measure $\mu(G, W, \gamma)$ can be recovered from its potential, using the Fourier method described in Section IV.2 of Saff and Totik [13].

However, we next consider a different method, dealing with specific weight functions, which allows us to deduce "explicit" expressions for the signed measure $\mu(G, W, \gamma)$ of Theorem 1. For simplicity, we assume throughout this section that G is given as in (1.2i), but with σ finite. We denote the unbounded component of $\overline{\mathbb{C}}\backslash\overline{G}$ by Ω. Let ν^+ and ν^- be two *positive* Borel measures on $\mathbb{C}$, with compact supports satisfying

$$\text{supp}\,\nu^+ \subset \overline{\mathbb{C}}\backslash G \quad \text{and} \quad \text{supp}\,\nu^- \subset \overline{\mathbb{C}}\backslash G,$$

such that

$$\nu^+(\mathbb{C}) = \nu^-(\mathbb{C}) = 1.$$

For real numbers $\alpha \geq 0$ and $\beta \geq 0$, assume that the weight function $W(z)$, satisfying

$$\log|W(z)| = -\left(\alpha U^{\nu^+}(z) - \beta U^{\nu^-}(z)\right) = -U^\nu(z), \qquad z \in G, \tag{2.1}$$

with $\nu := \alpha\nu^+ - \beta\nu^-$ being a signed measure, is analytic in G. Then, we state, as an application of Theorem 1, our next result as

Theorem 2. *Given any pair of real numbers $\alpha \geq 0$ and $\beta \geq 0$, given an open bounded set $G = \bigcup_{\ell=1}^{\sigma} G_\ell$, as in (1.2i) with σ finite, and given the*

weight function $W(z)$ of (2.1), then the triple (G, W, γ) has the rational approximation property (1.3) if and only if the signed measure

$$\mu := (2\gamma - 1 + \alpha - \beta)\omega(\infty, \cdot, \Omega) - \alpha\hat{\nu}^+ + \beta\hat{\nu}^- \tag{2.2}$$

can be decomposed as

$$\mu = \gamma\mu^+ - (1-\gamma)\mu^-, \tag{2.3}$$

where $\mu^+, \mu^- \in \mathcal{M}(\partial G)$. Here, $\omega(\infty, \cdot, \Omega)$ is the harmonic measure at ∞ with respect to Ω, and $\hat{\nu}^+$ and $\hat{\nu}^-$ are, respectively, the balayages of ν^+ and ν^- from $\overline{\mathbb{C}} \backslash \overline{G}$ to $\overline{G}$.

Furthermore, if μ of (2.2) satisfies (2.3), then (see Theorem 1)

$$\mu(G, W, \gamma) = \mu.$$

We point out that the harmonic measure $\omega(\infty, \cdot, \Omega)$ (defined in Nevanlinna [8] or Tsuji [18]) is the same as the equilibrium distribution measure for $\overline{G}$, in the sense of classical logarithmic potential theory [18]. For the notion of balayage of a measure, we refer the reader to Chapter IV of Landkof [6] or Section II.4 of Saff and Totik [13].

In the following series of subsections, we consider various classical weight functions and we find their corresponding signed measures, associated with the weighted rational approximation problem in G, as given in Theorem 1.

2.1 Incomplete rationals

With $\mathbb{N}_0$ and $\mathbb{N}$ denoting respectively the sets of nonnegative and positive integers, the *incomplete polynomials* of Lorentz [7] are a sequence of polynomials of the form

$$\left\{z^{m(i)} P_{n(i)}(z)\right\}_{i=0}^{\infty}, \qquad \deg P_{n(i)} \leq n(i), \qquad m(i), n(i) \in \mathbb{N}_0, \tag{2.1.1}$$

where it is assumed that $\lim_{i\to\infty} m(i)/n(i) =: \alpha$, where $\alpha > 0$ is a real number. The question of the possibility of the approximation of functions by incomplete polynomials is closely connected to that of the approximation of functions by the weighted polynomials

$$\{z^{\alpha n} P_n(z)\}_{n=0}^{\infty}, \qquad \deg P_n \leq n. \tag{2.1.2}$$

The approximation question for the incomplete polynomials of (2.1.1) was completely settled, by Saff and Varga [14] and by v. Golitschek [4], for the real interval $[0, 1]$ (see Totik [17] and Saff and Totik [13] for the associated history and later developments), and by the authors [11] in the complex plane. We consider now the analogous problem for *incomplete rational functions* in the complex plane. A special case of incomplete rational approximation in the complex plane was studied by Borwein and Chen in [1]. The latest such developments, on the real line, are in Borwein, Rakhmanov and Saff [2] and Rakhmanov, Saff and Simeonov [12].

Since the weight $W(z) := z^\alpha$ in (2.1.2) is multiple-valued in $\mathbb{C}$ if $\alpha \notin \mathbb{N}$, we then restrict ourselves to the slit domain $S_1 := \mathbb{C}\backslash(-\infty, 0]$ and the single-valued branch of $W(z)$ in S_1 satisfying $W(1) = 1$. Thus,

$$W(z) := z^\alpha, \qquad z \in S_1 := \mathbb{C}\backslash(-\infty, 0], \tag{2.1.3}$$

where $\alpha > 0$ is a real number.

Theorem 3. *Given an open set G, as in (1.2 i)) with σ finite, such that $\overline{G} \subset S_1$, and given the weight function $W(z)$ of (2.1.3), then the triple (G, z^α, γ) has the rational approximation property (1.3) if and only if the signed measure*

$$\mu = (2\gamma - 1 + \alpha)\omega(\infty, \cdot, \Omega) - \alpha\omega(0, \cdot, \Omega) \tag{2.1.4}$$

can be decomposed as

$$\mu = \gamma\mu^+ - (1-\gamma)\mu^-,$$

where $\mu^+, \mu^- \in \mathcal{M}(\partial G)$. Here, $\omega(\infty, \cdot, \Omega)$ and $\omega(0, \cdot, \Omega)$ are the harmonic measures with respect to the unbounded component Ω of $\overline{\mathbb{C}}\backslash\overline{G}$, respectively, at $z = \infty$ and at $z = 0$.

In special cases where the geometric shape of G is given explicitly, it is possible to determine the explicit form of the signed measure in (2.1.4). As a simple example, we consider below the special case of a disk and $\gamma = 1/2$.

Corollary 3. *Given the disk $D_r(a) := \{z \in \mathbb{C} : |z - a| < r\}$ with center $a \in (0, +\infty)$ and where $\overline{D}_r(a) \subset S_1 = \mathbb{C}\backslash(-\infty, 0]$, i.e., $r < a$, and given the weight function of (2.1.3), then the triple $(D_r(a), z^\alpha, 1/2)$ has the rational approximation property (1.3) if and only if*

$$r \le r_{\max}(a,\ \alpha) := \begin{cases} a, & \alpha \in (0, 1/2], \\ a \sin\frac{\pi}{4\alpha}, & \alpha \in (1/2, +\infty). \end{cases} \tag{2.1.5}$$

Furthermore, if (2.1.5) is satisfied, then the associated signed measure $\mu(D_r(a), z^\alpha, 1/2)$ (see Theorem 1) is given by

$$d\mu(D_r(a), z^\alpha, 1/2) = \frac{\alpha}{2\pi r}\left(1 - \frac{a^2 - r^2}{|z|^2}\right) ds,$$

where ds is the arclength measure on the circle $|z - a| = r$.

Remark 4: Generally, it is possible to show that the triple $(D_r(a),\ z^\alpha,\ \gamma)$, as in Corollary 3 but with *any* $\gamma \in [0,1]$, has the approximation property (1.3) if and only if

$$r \le \begin{cases} a, & \alpha+\gamma \le 1, \\ au_0, & \alpha+\gamma > 1, \end{cases}$$

where $u_0 \in (0,1]$ is the largest solution of the equation

$$\begin{aligned}(2\gamma-1)\arccos&\left[\frac{1-2\gamma-(2\gamma-1+2\alpha)u^2}{2u(2\gamma-1+\alpha)}\right]\\ +2\alpha\arccos&\left[\frac{(2\gamma-1+2\alpha)\sqrt{1-u^2}}{2\sqrt{\alpha(2\gamma-1+\alpha)}}\right] = \gamma\pi,\end{aligned}$$

in the interval $(0,1]$.

2.2 Exponential weight

Consider the weight function

$$W(z) := e^{-z}, \qquad z \in \mathbb{C}. \tag{2.2.1}$$

This section is devoted to the study of weighted rational approximation, with respect to the exponential weight of (2.2.1), in disks centered at the origin and in certain domains, arising in connection with Padé approximations of the exponential function. Our next result treats the case of disks.

Theorem 4. *Given $D_r(0) := \{z \in \mathbb{C} : |z| < r\}$ and given the weight $W(z)$ of (2.2.1), then the triple $(D_r(0), e^{-z}, \gamma)$ has the rational approximation property (1.3) if and only if*

$$r \le r_{\max}(\gamma), \qquad 0 \le \gamma \le 1, \tag{2.2.2}$$

where the number $r_{\max}(\gamma)$ is the unique positive solution, for r in the interval $[|\gamma - 1/2|, +\infty)$, of the following equation:

$$\sqrt{r^2-\left(\gamma-\frac{1}{2}\right)^2}-\left(\gamma-\frac{1}{2}\right)\arccos\left(\frac{\gamma-\frac{1}{2}}{r}\right)=\frac{\pi}{2}(1-\gamma). \tag{2.2.3}$$

Moreover, if inequality (2.2.2) holds, then the associated signed measure $\mu(D_r(0),\, e^{-z}, \gamma)$ is given by

$$d\mu(D_r(0), e^{-z}, \gamma) = (2\gamma - 1 - 2r\cos\theta)\frac{d\theta}{2\pi},$$

where $d\theta$ is the angular measure on $|z| = r$ and where $z = re^{i\theta}$.

In particular, $r_{\max}(1) = 1/2$ (see also [10], Thm. 3.8), $r_{\max}(1/2) = \pi/4$ and $r_{\max}(0) = 1/2$.

We remark that the solution, $r_{\max}(\gamma)$ of (2.2.3), can be verified to be symmetric about $\gamma = 1/2$, as a function of γ in the interval $[0, 1]$.

Next, we again consider the weight function $W(z) := e^{-z}$ of (2.2.1), but we now consider the triple $(G_\gamma, e^{-z}, \gamma)$, where G_γ, a *generalized Szegő domain*, is defined below. To begin, first assume that $0 < \gamma < 1$. Then following [15], the two conjugate complex numbers, defined by

$$z_\gamma^{\pm} := \exp\{\pm\, i \arccos(2\gamma - 1)\}, \tag{2.2.4}$$

have modulus unity, and we consider the complex plane $\mathbb{C}$ slit along the two rays

$$\mathbb{R}_\gamma := \left\{z \in \mathbb{C} : z = z_\gamma^+ + i\tau \text{ or } z = z_\gamma^- - i\tau, \text{ for } \tau \geq 0\right\}.$$

This is shown below in Figure 1. (For readers who are familiar with [15], the quantity $\sigma := \lim_{i\to\infty} n_i/m_i$ in that paper and γ of (1.3i) are related through $\gamma = 1/(1+\sigma)$.)

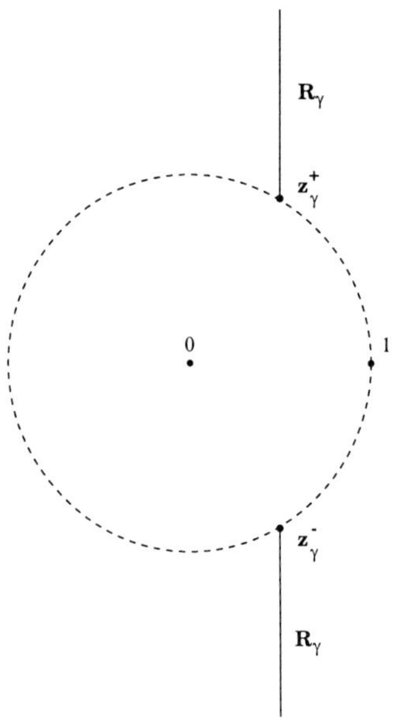

Figure 1. The set $\mathbb{C}\backslash\mathbb{R}_\gamma$.

Next, the function

$$\hat{g}_\gamma(z) := \sqrt{1 + z^2 - 2z(2\gamma - 1)}$$

has z_γ^+ and z_γ^- as branch points, which are the finite extremities of $\mathbb{R}_\gamma$. On taking the principal branch for the square root, *i.e.*, on setting $\hat{g}_\gamma(0) = 1$ and extending g_γ analytically on the doubly slit domain $\mathbb{C}\backslash\mathbb{R}_\gamma$, then g_γ is analytic and single-valued on $\mathbb{C}\backslash\mathbb{R}_\gamma$. It can also be verified that $1 \pm z + \hat{g}_\gamma(z)$ does not vanish on $\mathbb{C}\backslash\mathbb{R}_\gamma$.

Next, we define the functions $(1+z+\hat{g}_\gamma(z))^{2\gamma}$ and $(1-z+\hat{g}_\gamma(z))^{2(1-\gamma)}$ so that their values at $z = 0$ are respectively, $2^{2\gamma}$ and $2^{2(1-\gamma)}$, with remaining values determined by analytic continuation. These functions are also analytic and single-valued on $\mathbb{C}\backslash\mathbb{R}_\gamma$. With these definitions, we then set

$$w_\gamma(z) := \frac{4\gamma\left(\frac{1-\gamma}{\gamma}\right)^{1-\gamma} z e^{\hat{g}_\gamma(z)}}{(1+z+\hat{g}_\gamma(z))^{2\gamma}\,(1-z+\hat{g}_\gamma(z))^{2(1-\gamma)}}, \qquad 0 < \gamma < 1, \quad (2.2.5)$$

and it follows that $w_\gamma(z)$ is also analytic and single-valued on $\mathbb{C}\backslash\mathbb{R}_\gamma$. For the omitted cases $\gamma = 0$ and $\gamma = 1$, it can be verified that $w_1(z) = \lim_{\gamma\to 1} w_\gamma(z)$ and $w_0(z) = \lim_{\gamma\to 0} w_\gamma(z)$ satisfy

$$\begin{aligned} w_1(z) &= ze^{1-z} \quad \text{for Re } z < 1, \\ w_0(z) &= ze^{1+z} \quad \text{for Re } z > -1. \end{aligned}$$

(Again, for those familiar with [15], the function $w_\gamma(z)$ of (2.2.5) is exactly the function $w_\sigma(z)$ in [15], eq. (2.5).)

It is known (see [15], Lemma 4.1) that $w_\gamma(z)$ is univalent in $|z| < 1$, for any γ with $0 \le \gamma \le 1$, and this allows us to define the domain

$$G_\gamma := \{z \in \mathbb{C} : |w_\gamma(z)| < 1 \text{ and } |z| < 1\}. \qquad (2.2.6)$$

Its boundary, ∂G_γ, is a well-defined Jordan curve which lies interior to the unit disk, except for its points $z_\gamma^\pm$ of (2.2.4). This is shown in Figure 2. We call G_γ an *extended Szegő domain*, as the special case $\gamma = 1$ corresponds to a domain originally treated by Szegő in [16] in 1924.

We now have all the necessary definitions for the statement of our next result.

Theorem 5. *For any γ with $0 < \gamma \le 1$, let G_γ be the domain of (2.2.6), and let $W(z) = e^{-z}$ be the weight function of (2.2.1). Then, the triple $(G_\gamma, e^{-z}, \gamma)$ has the rational approximation property (1.3).*

To conclude this section, we note that, except for the final result of Theorem 5, all preceding results stated in §1 and §2 are of the "if and only if" type, *i.e.*, these results are by definition *sharp*. The result of Theorem 5, however, leaves open the possibility that for a given γ with $0 \le \gamma \le 1$, there

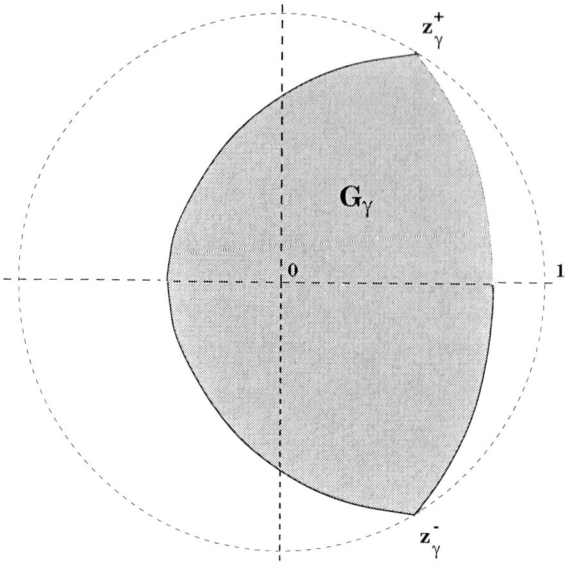

Figure 2. The set G_γ.

could be a *larger* domain H, with $G_\gamma \subset H$, such that the triple (H, e^{-z}, γ) has the rational approximation property (1.3), but we strongly doubt this.

Also of general interest is the extension of the results of this paper to triples (G, W, γ) of (1.1), where one has the *sharpened rational approximation property*, that is, for any $f(z)$, analytic in G and continuous in $\overline{G}$, there is a sequence of rational functions $\{R_{m_i,n_i}\}_{i=0}^{\infty}$ satisfying (1.3i), such that

$$\lim_{i\to\infty} \left\| f - W^{m_i+n_i} R_{m_i,n_i} \right\|_{\overline{G}} = 0.$$

For the essentially polynomial case of $\gamma = 0$ and $W(z) := e^{-z}$, this is treated in part in [10], Theorem 3.2. Some general results in weighted polynomial approximation on compact sets are obtained in [9]. To our knowledge, general results on the sharpened rational approximation property have not as yet been treated in the literature.

References

[1] Borwein, P. B. and W. Chen, Incomplete rational approximation in the complex plane, *Constr. Approx.* **11** (1995), 85–106.

[2] Borwein, P. B., E. A. Rakhmanov, and E. B. Saff, Rational approximation with varying weights I, *Constr. Approx.* **12** (1996), 223–240.

[3] Bruin, de, M. G., E. B. Saff, and R. S. Varga, On the zeros of generalized Bessel polynomials, I and II, *Nederl. Akad. Wetensch. Indag. Math.* **23** (1981), 1–25.

[4] Golitschek, v., M., Approximation by incomplete polynomials, *J. Approx. Theory* **28** (1980), 155–160.

[5] Kuijlaars, A. B. J., The role of the endpoint in weighted polynomial approximation with varying weights, *Constr. Approx.* **12** (1996), 287–301.

[6] Landkof, N. S., *Foundations of Modern Potential Theory*, Springer-Verlag, Berlin, 1972.

[7] Lorentz, G. G., Approximation by incomplete polynomials (problems and results), in *Padé and Rational Approximations: Theory and Applications*, E. B. Saff and R. S. Varga (eds.), Academic Press, New York, 1977, pp. 289–302.

[8] Nevanlinna, R. *Analytic Functions*, Springer-Verlag, New York, 1970.

[9] Pritsker, I. E., Polynomial approximation with varying weights on compact sets of the complex plane, *Proc. Amer. Math. Soc.* **126** (1998), 3283–3292.

[10] Pritsker, I. E. and R. S. Varga, The Szegő curve, zero distribution and weighted approximation, *Trans. Amer. Math. Soc.* **349** (1997), 4085–4105.

[11] Pritsker, I. E. and R. S. Varga, Weighted polynomial approximation in the complex plane, *Constr. Approx.* **14** (1998), 475–492.

[12] Rakhmanov, E. A., E. B. Saff and P. C. Simeonov, Rational approximation with varying weights II, *J. Approx. Theory* **92** (1998), 331–338.

[13] Saff, E. B., and V. Totik, *Logarithmic Potentials with External Fields*, Springer-Verlag, Heidelberg, 1997.

[14] Saff, E. B., and R. S. Varga, On incomplete polynomials, *Numerische Methoden der Approximationstheorie*, L. Collatz, G. Meinardus, H. Werner (eds.), ISNM 42, Birkhäuser-Verlag, Basel, 1978, 281–298.

[15] Saff, E. B. and R. S. Varga, On the zeros and poles of Padé approximants to e^z. III, *Numer. Math.* **30** (1978), 241–266.

[16] Szegő, G., Über eine Eigenschaft der Exponentialreihe, *Sitzungsber. Berl. Math. Ges.* **23** (1924), 50–64.

[17] Totik, V., Weighted Approximation with Varying Weights, *Lecture Notes in Math.*, vol. 1569, Springer-Verlag, Heidelberg, 1994.

[18] Tsuji, M., *Potential Theory in Modern Function Theory*, Maruzen, Tokyo, 1959.

[19] Varga, R. S., and A. J. Carpenter, Asymptotics for zeros and poles of normalized Padé approximants to e^z, *Numer. Math.* **68** (1994), 169–185.

[20] Walsh, J. L., Interpolation and Approximation by Rational Functions in the Complex Domain, *Colloquium Publications*, vol. 20, Amer. Math. Soc., Providence, 1969.

[21] Yosida, K., *Functional Analysis*, Springer-Verlag, Berlin, 1978.

Igor E. Pritsker
Dept. of Math.
Case Western Reserve Univ.
Cleveland, Ohio 44106
U.S.A.
iep@po.cwru.edu

Richard S. Varga
Inst. for Computational Mathematics
Dept. of Math. & Computer Science
Kent State University
Kent, Ohio 44242
U.S.A.
varga@mcs.kent.edu

Radii of neighborhoods for coefficient estimates of functions close to the identity

Dmitri V. Prokhorov

Abstract. We estimate from below the radius $M(\alpha,\beta)$ of a neighborhood S^M of the identity function, where the Pick functions maximize the linear functional $I(\alpha,\beta;f)=\mathrm{Re}\{a_4+\alpha a_3+\beta a_2\}$ for functions $f(z)=z+a_2z^2+\dots$ holomorphic and univalent in the unit disk and such that $|f(z)|<M$.

§1 Introduction

Let S^M be the class of functions $f(z)=z+a_2z^2+\dots$ holomorphic and univalent in the unit disk and such that $|f(z)|<M$ there. We denote by P^M the Pick functions which map the unit disk onto the disk of radius M centered at the origin and slit along a segment on the negative real semiaxis. It is shown in [2] that the Pick function P^M maximizes the linear functional $I(\alpha,\beta;f)=\mathrm{Re}\{a_4+\alpha a_3+\beta a_2\}$ for $(\alpha,\beta)\in D$ where

$$D=\{\alpha\leqslant -2,\beta>-1\}\cup\{-2<\alpha<6,\beta>\alpha^2/4+\alpha\}\cup\{\alpha\geqslant 6,\beta>4\alpha-9\}$$

in the class S^M if $1<M<M(\alpha,\beta)$.

In this paper we estimate the bounds $M(\alpha,\beta)$ from below. The corresponding theorem will appear in the last section.

Recall that a dense subclass of S^M is represented by solutions of the Loewner differential equation

$$\frac{dw}{dt}=-w\frac{e^{iu}+w}{e^{iu}-w},\qquad w|_{t=0}=z,\qquad 0\leqslant t\leqslant \log M. \tag{1.1}$$

In fact, let $w(z,t)$ be an integral of equation (1.1) with a piecewise continuous function $u=u(t)$. Then the set of all functions $Mw(z,\log M)$ is dense in S^M.

Substitute $e^tw(z,t)=z+a_2(t)z^2+\dots$ in (1.1) to obtain a system of differential equations for $a_2(t),\dots,a_n(t)$:

$$\begin{aligned}\frac{da_k(t)}{dt} &= G_{k-1}(t,a_2(t),\dots,a_{k-1}(t),u(t)), \\ a_k(0) &= 0,\qquad a_k(\log M)=a_k,\qquad 2\leqslant k\leqslant n.\end{aligned} \tag{1.2}$$

Computational Methods and Function Theory (CMFT'97)
N. Papamichael, St. Ruscheweyh and E. B. Saff (Eds.), pp. 449–459

ISBN 981-02-3626-3

The functions G_{k-1} may be written explicitly (see, *e.g.* [1]). We restrict ourselves to the case $n = 4$. The solutions of system (1.2) at $t = \log M$ form a dense subset of the coefficient region $V_4(M) = \{(a_2, a_3, a_4) : f \in S^M\}$. Its boundary $\partial V_4(M)$ is the reachable set of the control system (1.2) at $t = \log M$.

Every point of $\partial V_4(M)$ corresponds to an optimal control u in (1.2) which maximizes the Hamilton function $H = \operatorname{Re} \sum_{k=1}^{3} G_k \overline{\Psi}_k$. The conjugate coordinates $\Psi_k(t)$ solve the differential equations

$$\frac{d\overline{\Psi}_k}{dt} = -\frac{\partial H}{\partial a_{k+1}}, \qquad \Psi_k(0) = \xi_{2k-1} + i\xi_{2k}, \qquad 1 \leqslant k \leqslant 3. \tag{1.3}$$

Hence $\Psi_3(t)$ is constant. The vector $(\Psi_1(\log M), \Psi_2(\log M), \Psi_3(\log M))$ is orthogonal to a tangent or supporting hyperplane for $\partial V_4(M)$ if it exists (see, *e.g.* [1]).

To maximize the functional $I(\alpha, \beta; f)$ on S^M it is equivalent to maximize the functional $\operatorname{Re}\{a_4 + \alpha a_3 + \beta a_2\}$ as a linear function of a_2, a_3, a_4 on $V_4(M)$. The maximum is attained at boundary points of $V_4(M)$ where $(\Psi_1(\log M), \Psi_2(\log M), \Psi_3(\log M)) = (\beta, \alpha, 1)$. Thus it is sufficient to put $\Psi_3 = 1$, and it is enough to determine the coefficient region $V_4^1(M) = \{(a_2, a_3, \operatorname{Re} a_4) : f \in S^M\}$ and its boundary hypersurface $\partial V_4^1(M)$ instead of $V_4(M)$ and $\partial V_4(M)$.

$\partial V_4^1(M)$ is the set of points $(a_2(\log M), a_3(\log M), \operatorname{Re} a_4(\log M))$ where a_2, a_3, $\operatorname{Re} a_4$ are solutions of (1.2). The systems (1.2) and (1.3) are connected since the right-hand sides of (1.2)–(1.3) contain a control u which maximizes the Hamilton function H for all $t \in [0, \log M]$. The initial data in (1.3) are arbitrary. So $\partial V_4^1(M)$ is parametrized by $\xi = (\xi_1, \xi_2, \xi_3, \xi_4)$ (see [1] for detailed proofs). The conjugate coordinates $\Psi_1(\log M)$, $\Psi_2(\log M)$ also depend on ξ.

It is easy to verify that the control $u(t) = \pi$, corresponding to the Pick functions, satisfies the necessary extremum conditions for maximizing $I(\alpha, \beta; f)$, $(\alpha, \beta) \in D$, in the classe S^M if M is close to 1. The above mentioned solution of the extremal problem in [2] was based on the fact that the Jacobi matrix of the mapping $\xi \longrightarrow (\Psi_1(\log M), \Psi_2(\log M))$ is close to the unit matrix. This guarantees local existence of an inverse mapping and the uniqueness of optimal controls satisfying the necessary extremum conditions. Thus the necessary conditions prove to be in fact sufficient, *i.e.* the Pick functions maximize the functional $I(\alpha, \beta; f)$.

In this paper we analyse the known proof of the inverse function theorem to estimate radii of local existence.

§2 Preliminary statements

For $f \in S^M$ we put $T = \log M$, $a_k(t) = X_{k-1}(t) = x_{2k-3}(t) + ix_{2k-2}(t)$, $a_k = X_{k-1}(T), \Psi_{k-1}(t) = \psi_{2k-3}(t) + i\psi_{2k-2}(t)$, $k = 2,3,4$. Inserting the explicit formulae for G_k in (1.2) leads to the phase control system for $0 \leqslant t \leqslant T$

$$\frac{dX_1}{dt} = -2e^{-t-iu} = G_1 = g_1 + ig_2, \; X_1(0) = 0, \tag{2.1}$$

$$\frac{dX_2}{dt} = -4e^{-t-iu}X_1 - 2e^{-2t-i2u} = G_2 = g_3 + ig_4, \; X_2(0) = 0, \tag{2.2}$$

$$\frac{dx_5}{dt} = \mathrm{Re}[-2e^{-t-iu}(2X_2+X_1^2) - 6e^{-2t-i2u}X_1 - 2e^{-3t-i3u}] = g_5, \; x_5(0) = 0.$$

The second equation of the conjugate system (1.3)

$$\frac{d\overline{\Psi}_2}{dt} = 4e^{-t-iu}, \qquad \Psi_2(0) = \xi_3 + i\xi_4 \tag{2.3}$$

implies that

$$\overline{\Psi}_2 = -2X_1 + \xi_3 - i\xi_4. \tag{2.4}$$

Hence the first equation in (1.3) may be written as

$$\frac{d\overline{\Psi}_1}{dt} = 2e^{-t-iu}(\overline{\Psi}_2 + \xi_3 - i\xi_4) + 6e^{-2t-i2u}, \qquad \Psi_1(0) = \xi_1 + i\xi_2. \tag{2.5}$$

It follows also from (2.1)–(2.5) that

$$X_2 = (\xi_1 - i\xi_2 - \overline{\Psi}_1 + \overline{\Psi}_2(\overline{\Psi}_2 - \xi_3 + i\xi_4))/3. \tag{2.6}$$

If $f \in S^M$ is represented by an optimal control u in the Loewner differential equation (1.1) and maximizes $I(\alpha, \beta; f)$ in S^M, then u maximizes the Hamilton function

$$H(t, x, \psi, u) = \sum_{k=1}^{5} g_k \psi_k, \qquad \psi_5 = 1,$$

and hence $u = u(t, x, \psi)$ is defined as a root of the equation

$$H_u(t, x, \psi, u) = 0. \tag{2.7}$$

Here x and ψ are 5-dimensional vectors. The transversality conditions $\psi(T) = (\beta, 0, \alpha, 0, 1)$ are also necessary for the extremal function.

Let $F : \xi \longrightarrow y = (\psi_1(T), \psi_2(T), \psi_3(T), \psi_4(T))$ be the mapping of the initial data of the equation onto the solution values at $t = T$ induced by the differential system (2.3), (2.5) with optimal control u. Then

$F(\xi^0) = y^0$, $y^0 = ((7-4\alpha-1/M)(1-1/M)+\beta, 0, -4(1-1/M)+\alpha, 0)$, and $\xi^0 = (\beta, 0, \alpha, 0)$, provided (α, β) and (y_3^0, y_1^0) belong to D. The operator F is differentiable. Its derivative $F'(\xi)$ is a (4×4)-matrix with elements $\psi_{jk}(T)$, where $\psi_{jk}(t)$ is defined as the derivative of $\psi_j(t)$ with respect to ξ_k. For known ψ_j and $u(t, x, \psi)$ the functions ψ_{jk} have the initial data $\psi_{jk}(0) = \delta_{jk}$ (where δ_{jk} is the Kronecker symbol), and solve a system of 16 differential equations obtained after differentiating the equations (2.3), (2.5) for ψ_j with respect to ξ_k.

Thus F' is close to the unit matrix if M is close to 1. Hence it is invertible and F' has a local inversion $(F')^{-1}$. So F maps a neighborhood $U(\xi^0)$ of ξ^0 one-to-one onto a neighborhood $U(y^0)$ of y^0. We wish to estimate the radii of these neighborhoods by applying the method decribed in the proof of Theorem 9.17 in [3]. This method requires the following conditions to hold:

(i) $\|(F')^{-1}(\xi^0)\| \leqslant 2$;

(ii) $\|F'(\xi^0) - F'(\xi)\| \leqslant 1/4$, $\xi \in U(\xi^0)$;

(iii) if the radius of $U(\xi^0)$ is equal to l, then $\|y^0 - y\| \leqslant l/8$.

In addition, we shall assume that

(iv) $\|\xi^0 - y^0\| \leqslant l/8$ and l is so large that $\|\xi - \xi^0\| > l$ implies $F(\xi) \neq \xi^0$.

The matrix norm in (i)–(iv) is supposed to be induced by the euclidean vector norm. Let us remark that condition (iii) restricts neither $U(\xi^0)$ nor M but together with the first part of (iv) it means that $\xi^0 \in U(y^0)$. The second part of (iv) guarantees the uniqueness of a preimage for $\xi^0 \in U(y^0)$ not only in $U(\xi^0)$ but in the whole ξ-space.

§3 Matrix norm estimates

Let A and B be (4×4)-matrices with elements $a_{jk} + \delta_{jk}$ and $b_{jk} + \delta_{jk}$ respectively, $1 \leqslant j, k \leqslant 4$. Assume $|a_{jk}| \leqslant 1/32$, $|b_{jk}| \leqslant 1/32$ for all j, k. Then $\det A > 0.8$. For a minor A_{jk} of order 3 we have: $|A_{jj}| < 1.2$ and $|A_{jk}| < 0.1$, $j \neq k$. Hence $\|A^{-1}\| < 1.5/0.8 < 2$. It is also evident that $\|A - B\| \leqslant 1/4$.

We proved:

Lemma 1. *To provide conditions (i), (ii) it is sufficient to satisfy*

$$|\psi_{jk}(T) - \delta_{jk}| \leqslant 1/32, \qquad 1 \leqslant j, k \leqslant 4, \qquad \xi \in U(\xi^0). \tag{3.1}$$

§4 Phase and conjugate estimates

Now we will estimate x_k and ψ_k, $1 \leqslant k \leqslant 4$, given by (2.1)–(2.3), (2.5). Directly from (2.1), (2.3) we have

$$|x_k(t)| \leqslant 2(1-e^{-t}), \qquad k=1,2, \tag{4.1}$$

$$|\psi_k(t)-\xi_k| \leqslant 4(1-e^{-t}), \qquad k=3,4. \tag{4.2}$$

Using (4.1) in (2.2) we obtain after integration

$$|x_k(t)| \leqslant 9-16e^{-t}+7e^{-2t}, \qquad k=3,4. \tag{4.3}$$

And similarily, using (2.4) and (4.2) in (2.5):

$$|\psi_k(t)-\xi_k| \leqslant 4(|\xi_3|+|\xi_4|+4)(1-e^{-t})-5(1-e^{-2t}), \quad k=1,2. \tag{4.4}$$

§5 Preliminary radii estimates

We need to estimate $|H_{uu}(t,x,\psi,u)|$ with optimal $u=u(t,x,\psi)$ from below. Let $t=0$. Denote

$$h(\xi)=H_{uu}(0,0,\xi,u(0,0,\xi)). \tag{5.1}$$

Since $u(0,0,\xi^0)=\pi$ (see [2]),

$$h_0=h(\xi^0)=-2\beta+8\alpha-18<0 \tag{5.2}$$

if $(\alpha,\beta)\in D$. We wish to estimate a radius l of the neighborhood $U(\xi^0)$, subject to $h(\xi)\leqslant h_0/2$ for $\xi\in U(\xi^0)$.

We set

$$\begin{aligned} l_1 &= (\sqrt{(4|\beta|+32|\alpha|+108-2h_0)^2+18h_0^2} \\ &\qquad -4|\beta|-32|\alpha|-108+2h_0)/144. \end{aligned} \tag{5.3}$$

Lemma 2. *If $\|\xi-\xi^0\|\leqslant l_1$, then $h(\xi)\leq h_0/2$.*

Proof: After differentiating (5.1) we have

$$h'(\xi)=H_{uu\psi}(0,0,\xi,u)+H_{uuu}(0,0,\xi,u)u_\xi. \tag{5.4}$$

Differentiate equation (2.7) at $t=0$ to obtain

$$H_{u\psi}(0,0,\xi,u)+H_{uu}(0,0,\xi,u)u_\xi=0,$$

which leads to

$$u_\xi=-H_{u\psi}(0,0,\xi,u)/h(\xi). \tag{5.5}$$

(5.4) and (5.5) imply the differential equation

$$h'(\xi) = H_{uu\psi}(0,0,\xi,u) - H_{uuu}(0,0,\xi,u)H_{u\psi}(0,0,\xi,u)/h(\xi). \tag{5.6}$$

It is easy to verify that

$$\begin{aligned}
|H_{uuu}(0,0,\xi,u)| &\leqslant 2|\xi_1|+2|\xi_2|+16|\xi_3|+16|\xi_4|+54,\\
|H_{u\psi_j}(0,0,\xi,u)| &\leqslant 2, \qquad j=1,2,\\
|H_{u\psi_j}(0,0,\xi,u)| &\leqslant 4, \qquad j=3,4,\\
|H_{uu\psi_j}(0,0,\xi,u)| &\leqslant 2, \qquad j=1,2,\\
|H_{uu\psi_j}(0,0,\xi,u)| &\leqslant 8, \qquad j=3,4.
\end{aligned} \tag{5.7}$$

Equation (5.6) and inequalities (5.7) lead to the differential inequalities

$$h_{\xi_j} \leqslant 8|H_{uuu}(0,0,\xi,u)/h_0| + 8, \qquad 1 \leqslant j \leqslant 4,$$

and the derivative $h_e(\xi)$, for any direction e, satisfies

$$h_e(\xi) \leqslant 16|H_{uuu}(0,0,\xi,u)/h_0| + 16.$$

There exists l such that $h(\xi) \leqslant h(\xi^0)/2$ if $\|\xi - \xi^0\| \leqslant l$. For $\|\xi - \xi^0\| \le l$ this yields

$$|h(\xi) - h(\xi^0)| \leqslant 16(|H_{uuu}(0,0,\xi,u)/h_0| + 1)l.$$

Therefore to verify $h(\xi) \leqslant h(\xi^0)/2$ it is sufficient to satisfy the condition $l \leqslant |h_0|/(32(|H_{uuu}(0,0,\xi,u)/h_0| + 1))$ which together with the inequality $|H_{uuu}(0,0,\xi,u)| \leqslant 2|\beta| + 16|\alpha| + 54 + 36l$ (implied by (5.7)) is equivalent to the assertion of Lemma 2. ■

Now we are able to restrict M in the light of (iv) and the condition $(y_3^0, y_1^0) \in D$.

Lemma 3. *Condition (iv) is valid with $l = l_1$ if $M \leqslant \min\{M_1, M_2\}$, where*

$$M_1 = 1 + \frac{l_1}{\left|8\sqrt{(7+4|\alpha|)^2+16} - l_1\right|}, \tag{5.8}$$

$$M_2 = 1 + \frac{7l_1}{\left|8\left(\sqrt{(7+4|\alpha|)^2+16} + 8\,(12+|\alpha|)\right) - 7l_1\right|}. \tag{5.9}$$

Proof: The first part of (iv) is equivalent to

$$\|((7-4\alpha-1/M)(1-1/M), 0, -4(1-1/M), 0)\| \leqslant l_1/8$$

which is valid if $M \leqslant M_1$.

To establish the second part of (iv) we note first that, according to (4.2), $F(\xi) \neq \xi^0$ if $|\xi_3 - \alpha| > 4(1-1/M)$ or $|\xi_4| > 4(1-1/M)$.

If $|\xi_3 - \alpha| \leqslant 4(1-1/M)$ and $|\xi_4| \leqslant 4(1-1/M)$ then $F(\xi) \neq \xi^0$ provided that

$$\|F(\xi)-F(\xi^0)\| > \|F(\xi^0)-\xi^0\| = (1-1/M)((7-4\alpha-1/M)^2+16)^{1/2}. \quad (5.10)$$

But it follows from (4.2), (4.4) and the first part of (iv) that

$$\begin{aligned} \|F(\xi) - F(\xi^0)\| &\geqslant \|\xi - \xi^0\| - \|F(\xi^0) - \xi^0\| - \|F(\xi) - \xi\| \\ &\geqslant l_1 - l_1/8 - 8(|\xi_3| + |\xi_4| + 4)(1 - 1/M) \quad (5.11) \\ &> 7l_1/8 - 8(|\alpha| + 12)(1 - 1/M). \end{aligned}$$

(5.10) and (5.11) imply that the second part of (iv) is true if $M \leqslant M_2$. ■

Denote by ρ the distance of ξ^0 from the boundary of D, and let

$$M_3 = 1 + \frac{\rho}{\left|\sqrt{(7+4|\alpha|)^2+16} - \rho\right|}. \quad (5.12)$$

Lemma 4. *If $M \leqslant M_3$, then $(y_3^0, y_1^0) \in D$.*

Proof: The assertion of the lemma is equivalent to $\|y^0 - \xi^0\| < \rho$ or $(1-1/M)((7-4\alpha-1/M)^2+16)^{1/2} < \rho$ which follows from $M \leqslant M_3$. ■

§6 Regularity estimates

We continue to estimate $|H_{uu}(t, x, \psi, u)|$ with optimal u from below. Now we wish to estimate M subject to $H_{uu}(t, x, \psi, u) \leqslant h_0/4$ for $\|\xi - \xi^0\| \leqslant l_1$ and $0 \leqslant t \leqslant T$.

Put $G(t, \xi) = H_{uu}(t, x, \psi, u)$ where x and ψ depend on the initial data ξ of (2.1)–(2.3), (2.5). The optimal control u is a differentiable function of t (see [1]). After differentiating G we have

$$\begin{aligned} G_t(t, \xi) = H_{uut}(t, x, \psi, u) + H_{uux}(t, x, \psi, u)\frac{dx}{dt} \\ + H_{uu\psi}(t, x, \psi, u)\frac{d\psi}{dt} + H_{uuu}(t, x, \psi, u)\frac{du}{dt}. \end{aligned} \quad (6.1)$$

Differentiate (2.7) with respect to t to obtain

$$H_{ut}(t,x,\psi,u) + H_{ux}(t,x,\psi,u)\frac{dx}{dt} + H_{u\psi}(t,x,\psi,u)\frac{d\psi}{dt} + H_{uu}(t,x,\psi,u)\frac{du}{dt} = 0$$

which leads to the following formula for $\frac{du}{dt}$

$$\frac{du}{dt} = -\frac{H_{ut}(t,x,\psi,u) + H_{ux}(t,x,\psi,u)\frac{dx}{dt} + H_{u\psi}(t,x,\psi,u)\frac{d\psi}{dt}}{H_{uu}(t,x,\psi,u)}. \tag{6.2}$$

(6.1) and (6.2) give the differential equation

$$\begin{aligned} G_t(t,\xi) &= H_{uut}(t,x,\psi,u) + H_{uux}(t,x,\psi,u)\frac{dx}{dt} \\ &\quad + H_{uu\psi}(t,x,\psi,u)\frac{d\psi}{dt} - \frac{H_{uuu}(t,x,\psi,u)}{G(t,\xi)} \\ &\quad \times \left(H_{ut}(t,x,\psi,u) + H_{ux}(t,x,\psi,u)\frac{dx}{dt} + H_{u\psi}(t,x,\psi,u)\frac{d\psi}{dt}\right). \end{aligned} \tag{6.3}$$

It is easy to verify that

$$\begin{aligned} |H_{uuu}(t,x,\psi,u)| &< 2e^{-t}(|\xi_1| + |\xi_2| + 16|\xi_3| + 16|\xi_4| + 134) \\ &\leqslant 2e^{-t}(|\beta| + 16|\alpha| + 34l_1 + 134). \end{aligned}$$

$|H_{uut}(t,x,\psi,u)|$ is estimated similarly. Next we verify that

$$\begin{aligned} \left|H_{uux}(t,x,\psi,u)\frac{dx}{dt}\right| &= \left|\sum_{k=1}^{4} H_{uux_k}(t,x,\psi,u)g_k\right| \\ &< 16e^{-2t}(|\xi_3| + |\xi_4| + 20) \\ &\leqslant 16e^{-2t}(|\alpha| + 2l_1 + 20). \end{aligned}$$

The derivatives

$$\begin{aligned} \left|H_{uu\psi}(t,x,\psi,u)\frac{d\psi}{dt}\right| &= \left|\sum_{k=1}^{4} H_{uu\psi_k}(t,x,\psi,u)H_{x_k}(t,x,\psi,u)\right|, \\ \left|H_{ux}(t,x,\psi,u)\frac{dx}{dt}\right| &= \left|\sum_{k=1}^{4} H_{ux_k}(t,x,\psi,u)g_k\right| \end{aligned}$$

and

$$\left|H_{u\psi}(t,x,\psi,u)\frac{d\psi}{dt}\right| = \left|\sum_{k=1}^{4} H_{u\psi_k}(t,x,\psi,u)H_{x_k}(t,x,\psi,u)\right|$$

are estimated similarly.

The last derivative is estimated as

$$|H_{ut}(t,x,\psi,u)| < 4e^{-2t}(|\xi_3| + |\xi_4| + 20) \leqslant 4e^{-2t}(|\alpha| + 2l_1 + 20).$$

Hence putting

$$p = 72(|\beta| + 16|\alpha| + 34l_1 + 134)(|\alpha| + 2l_1 + 20), \quad (6.4)$$

$$q = 2(|\beta| + 32|\alpha| + 66l_1 + 134), \quad (6.5)$$

we have

$$\left|H_{uut}(t,x,\psi,u) + H_{uux}(t,x,\psi,u)\frac{dx}{dt} + H_{uu\psi}(t,x,\psi,u)\frac{d\psi}{dt}\right|$$
$$< 2e^{-t}(|\beta| + 16|\alpha| + 34l_1 + 134) + 32e^{-2t}(|\alpha| + 2l_1 + 20) < e^{-t}q$$

and

$$\left|H_{uuu}(t,x,\psi,u)(H_{ut}(t,x,\psi,u) + H_{ux}(t,x,\psi,u)\frac{dx}{dt}\right.$$
$$\left. + H_{u\psi}(t,x,\psi,u)\frac{d\psi}{dt})\right| < e^{-t}p.$$

From this and (6.3) we obtain the differential inequality

$$G_t(t,\xi) \leqslant e^{-t}(4p/|h_0| + q), \qquad G(0,\xi) < h_0/2,$$

which is easy to integrate:

$$\begin{aligned} G(T,\xi) &\leqslant G(0,\xi) + 4p(1-1/M)/|h_0| + q(1-1/M) \\ &\leqslant h_0/2 + (1-1/M)(4p/|h_0| + q). \end{aligned}$$

We need that the right-hand side of the last inequality is $\leq h_0/4$.

Lemma 5. *The inequality $G(t,\xi) = H_{uu}(t,x,\psi,u) \leqslant h_0/4$ is valid for all $t \in [0,T]$ and $\|\xi - \xi^0\| \leqslant l_1$ if $M \leqslant M_4$, where*

$$M_4 = 4\frac{4p - qh_0}{|16p - 4qh_0 - h_0^2|}. \quad (6.6)$$

The proof follows immediately from the above assumptions.

§7 Estimates of Jacobi matrix elements

Lemma 6. *Condition (3.1) is true if $M \leqslant M_5$, where*

$$M_5 = 1 + \left(4\left(33 + 16\frac{(5|\alpha| + 10l_1 + 737)(|\alpha| + 2l_1 + 4)}{3|h_0|}\right) - 1\right)^{-1}. \quad (7.1)$$

Proof: First, we estimate $\|u_\xi\|$. Differentiate equality (2.7) with respect to ξ and obtain

$$H_{ux}(t,x,\psi,u)x_\xi + H_{u\psi}(t,x,\psi,u)\psi_\xi + H_{uu}(t,x,\psi,u)u_\xi = 0$$

which leads to the formula

$$u_\xi = -(H_{ux}(t,x,\psi,u)x_\xi + H_{u\psi}(t,x,\psi,u)\psi_\xi)/H_{uu}(t,x,\psi,u). \quad (7.2)$$

After a straightforward estimation of (7.2), using (2.4), (2.6), (4.1)–(4.4), the inequalities for H_{ux} and $H_{u\psi}$ of the previous paragraph, the inequalities $\|\xi - \xi^0\| \leqslant l_1$, $H_{uu}(t,x,\psi,u) \leqslant h_0/4$ and $|\psi_{jk}(t) - \delta_{jk}| \leqslant 1/32$, $1 \leqslant j,k \leqslant 4$, $0 \leqslant t \leqslant T$, we obtain

$$|u_{\xi_k}| \leqslant e^{-t}\frac{5|\alpha| + 10l_1 + 737}{6|h_0|}, \qquad 1 \leqslant k \leqslant 4. \quad (7.3)$$

Now we have to prove (3.1). Differentiate equations (2.3), (2.5) for ψ_j with respect to ξ_k and obtain a system of 16 differential equations for ψ_{jk}, $1 \leqslant j,k \leqslant 4$. After a straightforward estimation of the right-hand sides of these equations, using (2.4), (2.6), (4.1)–(4.4), (7.3), $\|\xi - \xi^0\| \leqslant l_1$ and $|\psi_{jk}(t) - \delta_{jk}| \leqslant 1/32$, we obtain

$$\left|\frac{d\psi_{jk}}{dt}\right| \leqslant e^{-t}\left(\frac{33}{8} + 2\frac{(5|\alpha| + 10l_1 + 737)(|\alpha| + 2l_1 + 4)}{3|h_0|}\right), \quad (7.4)$$
$$j = 1,2, \quad 1 \leqslant k \leqslant 4.$$

The inequalities for $j = 3,4$ are weaker and hence the corresponding estimates for M are included in those implied by (7.4).

Estimate (7.4) leads to the conclusion of the lemma. ■

§8 Final estimates

Now we are in a position to summarize the previous estimates. Observe that the class S^M, $M > 1$, is a closed neighborhood of the identity function. So $T = \log M$ may serve as a radius of this neighborhood. Let $M(\alpha, \beta)$

be the radius which determines a neighborhood where the Pick functions maximize the linear functional $I(\alpha, \beta; f)$ in S^M. Put

$$M_0 = \min_{1 \leqslant k \leqslant 5} M_k. \tag{8.1}$$

Theorem 1. *Let* $(\alpha, \beta) \in D$, h_0, $l = l_1$, p, q, M_k, $0 \leqslant k \leqslant 5$, *be given by (5.2), (5.3), (6.4), (6.5), (8.1), (5.8), (5.9), (6.2), (6.6), (7.1) respectively. Then* $M(\alpha, \beta) \geqslant M_0$.

The proof follows directly from the inverse function theorem and Lemmas 1–6.

Acknowledgement. This work was partially supported by a RFBR grant no. 98-01-00842.

References

[1] Prokhorov, D. V., *Reachable Set Methods in Extremal Problems for Univalent Functions*, Saratov Univ., Saratov, 1993.

[2] Prokhorov, D. V. and Z. Vasileva, Linear extremal problems for univalent functions close to identity, *Bull. Soc. Sci. Lettr. Lodz* **45** (1995), 11–17.

[3] Rudin, W., *Principles of Mathematical Analysis*, McGraw–Hill Book Comp., New York, 1964.

Dmitri V. Prokhorov
Department of Mathematics and Mechanics
Saratov State University
Astrakhanskaya Str., 83
410026 Saratov
RUSSIA
prokhor.ma.sgu@oda.ceu.runnet.ru

A remark on the Loewner differential equation

Oliver Roth

Abstract. We give a control-theoretic proof of Pommerenke's result on the parametric representation of normalized univalent functions in the unit disk as solutions of the Loewner differential equation. The method consists in combining a classical result on finite-dimensional control-linear systems with Montel's theorem on normal families.

§1 Introduction

We first recall some important subsets of the vector space $\mathcal{H}(\mathbb{D})$ of all functions analytic in the unit disk $\mathbb{D} = \{z \in \mathbb{C} : |z| < 1\}$. Equipped with the topology of locally uniform convergence, $\mathcal{H}(\mathbb{D})$ becomes a metrizable locally convex Hausdorff space.

For fixed $M \in [1, \infty]$ let $\mathcal{S}^M$ be the class of all functions f analytic and univalent in $\mathbb{D}$ satisfying $|f(z)| < M$ $(|z| < 1)$, normalized by $f(0) = 0$, $f'(0) = 1$. We put $\mathcal{S} := \mathcal{S}^\infty$. Let $\mathcal{P}$ denote the set of functions p that are analytic in $\mathbb{D}$ and satisfy $\operatorname{Re} p(z) > 0$ $(|z| < 1)$ and $p(0) = 1$. It is well-known that

$$\text{extr } \mathcal{P} = \left\{ z \mapsto \frac{1 + xz}{1 - xz} \,\middle|\, |x| = 1 \right\} \tag{1.1}$$

is the set of extreme points of the class $\mathcal{P}$.

Definition 1. *Let $I \subseteq [0, \infty]$ be an interval and $\Omega \subseteq \mathcal{H}(\mathbb{D})$ a normal family. A function $h : \mathbb{D} \times I \to \mathbb{C}$ is called* a Carathéodory function on I with values in Ω *if the following conditions are satisfied:*

(i) The mapping $z \mapsto h(z,t)$ belongs to Ω for almost all $t \in I$.

(ii) The mapping $t \mapsto h(z,t)$ is measurable on I for each fixed $z \in \mathbb{D}$.

The set of all Carathéodory functions on I with values in Ω will be denoted by $L^1(I; \Omega)$.

Using the normality of Ω it is easily seen that $z \mapsto \int_I h(z,s)\, ds$ is a well-defined analytic function in $\mathbb{D}$ for every $h \in L^1(I; \Omega)$. We remark that it is useful to regard the functions in $L^1(I; \Omega)$ as L^1-functions on I taking their values in Ω.

Computational Methods and Function Theory (CMFT'97)
N. Papamichael, St. Ruscheweyh and E. B. Saff (Eds.), pp. 461–469

ISBN 981-02-3626-3

We fix $M \in [1,\infty]$ and consider the Loewner differential equation

$$\begin{aligned} \frac{d}{dt}w(z,t) &= -w(z,t)p(w(z,t),t), \qquad t\in[0,\log M], \\ w(z,0) &= \mathrm{id}(z) := z, \end{aligned} \tag{1.2}$$

where $p \in L^1([0,\log M];\mathcal{P})$. The corresponding solution is denoted by $w(z,t;p)$. For fixed $\Omega \subseteq \mathcal{P}$, the set

$$\mathcal{R}_{\log M}(\mathrm{id},\Omega) := \left\{ z \mapsto w(z,\log M;p) \,\middle|\, p \in L^1([0,\log M];\Omega) \right\} \tag{1.3}$$

of all such solutions for $t = \log M$ generated by all Carathéodory functions on I taking their values in Ω, is called the *time*-$\log M$ *reachable set* of (1.2).

It is easily seen that $M\mathcal{R}_{\log M}(\mathrm{id},\mathcal{P}) \subseteq \mathcal{S}^M$, $M>1$. Hence it is convenient to introduce the sets $\tilde{\mathcal{R}}_{\log M}(\mathrm{id},\Omega) := M\mathcal{R}_{\log M}(\mathrm{id},\Omega)$ if $M \in [1,\infty)$ and $\tilde{\mathcal{R}}_{\infty}(\mathrm{id},\Omega) := \{\lim_{t\to\infty} e^t w(z,t;p) \,|\, p \in L^1([0,\infty);\Omega)\}$ for $\Omega \subseteq \mathcal{P}$.

Loewner [5] showed that $\tilde{\mathcal{R}}_{\infty}(\mathrm{id},\mathrm{extr}\,\mathcal{P})$ is dense in $\mathcal{S}$, *cf.* (1.1), and Pommerenke [6], [8] later established that $\tilde{\mathcal{R}}_{\infty}(\mathrm{id},\mathcal{P}) = \mathcal{S}$. The analogous statements that $\tilde{\mathcal{R}}_{\log M}(\mathrm{id},\mathcal{P}) = \mathcal{S}^M$ and that $\tilde{\mathcal{R}}_{\log M}(\mathrm{id},\mathrm{extr}\,\mathcal{S})$ is dense in $\mathcal{S}^M$ for every $M \in [1,\infty)$ can be proved in a similar fashion as in the case $M=\infty$.

The purpose of this paper is to give a short proof of the following theorem, by combining a similar result for finite-dimensional control-linear systems with Montel's theorem that the locally bounded families in $\mathcal{H}(\mathbb{D})$ are exactly the normal ones.

Theorem 1. *Let $\Omega \subseteq \mathcal{P}$ and let $M \in [1,\infty)$ be fixed. If Ω is compact and convex, then*

(a) $\mathcal{R}_{\log M}(id,\Omega)$ is compact and

(b) $\mathcal{R}_{\log M}(id,\mathrm{extr}\,\Omega)$ is dense in $\mathcal{R}_{\log M}(id,\Omega)$.

Remark 1: If $\Omega = \mathcal{P}$ then Loewner's result, *i.e.* that $M\mathcal{R}_{\log M}(\mathrm{id},\mathrm{extr}\,\mathcal{P})$ is dense in $\mathcal{S}^M$, and (1) imply $M\mathcal{R}_{\log M}(\mathrm{id},\mathcal{P}) = \mathcal{S}^M$, *i.e.* Pommerenke's result. On the other hand, $M\mathcal{R}_{\log M}(\mathrm{id},\mathcal{P}) = \mathcal{S}^M$ and (2) give Loewner's result. The case $M=\infty$ can be dealt with by using an additional limiting argument.

The assertions of Theorem 1 remain valid for more general control systems than the Loewner equation (1.2).

In fact, the essential assumption in Theorem 1 is the convexity and compactness of Ω. The fact that we suppose $\Omega \subseteq \mathcal{P}$ is only necessary to ensure the *existence* of the solutions of the Loewner equation for all $z \in \mathbb{D}$ and all $t \in [0, \log M]$. To make this precise, let us consider the following more general situation.

Let $\Omega \subseteq \mathcal{H}(\mathbb{D})$ be a compact and convex family of functions h analytic in the unit disk $\mathbb{D}$ satisfying $h(0) = 0$. For a fixed initial function $f_0 \in \mathcal{H}(\mathbb{D})$ such that $f_0(0) = 0$, we consider the initial value problem

$$\begin{aligned} \frac{d}{dt} w(z,t) &= h(w(z,t),t), \qquad t \in [0,T], \\ w(z,0) &= f_0(z), \end{aligned} \tag{1.4}$$

with a given Carathéodory function $h \in L^1([0,T];\Omega)$.

Substituting $u := f_0(z)$, this equation constitutes an ordinary initial value problem in $\mathbb{C}$ depending on two complex parameters $z \in \mathbb{D}$ and $u \in \mathbb{C}$. By a classical result of Carathéodory it is known that there exists an open set $\mathcal{D} \subseteq [0,T] \times \mathbb{D} \times \mathbb{C}$ and a function $\mathcal{D} \ni (t,z,u) \mapsto \varphi(t,t_0,z,u)$ such that for fixed $z_0 \in \mathbb{D}$ the function $t \mapsto w(z_0,t;h) := \varphi(t,t_0,z_0,f_0(z_0))$ is the unique absolutely continuous solution of (1.4) defined on some (maximal) subinterval $J(z_0)$ of $[0,T]$, *i.e.* $t \mapsto w(z_0,t;h)$ is differentiable a.e. in $J(z_0)$ and satisfies (1.4), *cf.* [1]. Moreover, w depends analytically on z_0. Usually, the intervall J depends on z_0 and becomes smaller if z_0 approaches the unit circle as simple examples show. Therefore, we assume the following:

Assumption I: *For fixed $T > 0$, the solution $w(z,t;h)$ of (1.4) does exist on $\mathbb{D} \times [0,T]$ for each Carathéodory function $h \in L^1([0,T];\Omega)$.*

Assumption I implies in particular that $w : \mathbb{D} \times [0,T] \to \mathbb{D}$. Consequently, the time-T reachable set

$$R_T(f_0,\Omega) := \{ z \mapsto w(z,T;h) \,|\, h \in L^1([0,T];\Omega) \}$$

of (1.4) is a bounded, *i.e.* normal, subset of $\mathcal{H}(\mathbb{D})$.

Theorem 2. *Let $\Omega \subseteq \mathcal{P}$, let T be a fixed non-negative real number and let Assumption I be satisfied.*

(a) If Ω is compact and convex, then the time-T reachable set $\mathcal{R}_T(f_0,\Omega)$ of (1.4) is compact.

(b) If, moreover, Ω is the set of functions

$$h(z) = \int_{|x|=1} F(xz)\, d\mu(x), \tag{1.5}$$

where μ is any probability measure on $\partial\mathbb{D}$ and $F(z) = \sum_{n=1}^{\infty} A_n z^n \in \mathcal{H}(\mathbb{D})$ satisfies $A_n \neq 0$ $(n = 1, 2, \ldots)$, then $\mathcal{R}_T(f_0, \text{extr}\,\Omega)$ is dense in $\mathcal{R}_T(f_0,\Omega)$.

Remark 2: By the Herglotz representation of functions in $\mathcal{P}$, Theorem 1 is contained in Theorem 2. It is an interesting problem whether Theorem 2 can be sharpened to conclude that $\mathcal{R}_T(f_0, \text{extr } \Omega)$ is already compact. In the case of the Loewner equation (1.2), and $T = \infty$, this question has been studied by Pommerenke [7], but remains unsolved; compare Remark 3 below.

§2 Proof of Theorem 2

The strategy of the proof of Theorem 2 is as follows. Each Carathéodory function $h \in L^1(I; \Omega)$ can be expanded into a power series

$$h(z,t) = \sum_{j=1}^{\infty} c_j(t) z^j, \qquad |z| < 1.$$

It is easy to check that the coefficient functions $c_j : [0,T] \to \mathbb{C}$ are measurable. Because Ω was assumed to be a normal family, the Cauchy integral formula for the coefficients $c_j(t)$ shows that the functions c_j are contained in $L^1([0,T]; \mathbb{C})$, the space of all complex-valued L^1-functions on the interval $I = [0,T]$. We now expand the solution $w(z,t;h)$ of (1.4), generated by $h \in L^1([0,T]; \Omega)$ into a power series

$$w(z,t;h) = \sum_{j=1}^{\infty} a_j(t) z^j, \qquad |z| < 1.$$

The coefficient functions $t \mapsto a_j(t)$, $j \in \mathbb{N}$, are absolutely continuous on $[0,T]$ and thus differentiable a.e.. Comparing coefficients of z^j in (1.4) we obtain the infinite system of initial value problems

$$\begin{aligned} \frac{d}{dt} a_j(t) &= \sum_{s=1}^{j} c_s(t) P_{s,j}(a_1(t), \ldots, a_j(t)), \qquad t \in [0,T], \\ a_j(t) &= a_j^0, \end{aligned} \tag{2.1}$$

where the $P_{s,j}$ are certain polynomials in $a_1, \ldots, a_j$, and the a_j^0's are the Taylor coefficients of f_0. The recursive structure of (2.1) is forced by our normalization $h(0,t) \equiv 0$ and $f_0(0) = 0$ and will be a crucial ingredient in what follows. It enables us to fix $k \in \mathbb{N}$ and consider the vector $a(t) := (a_1(t), \ldots, a_k(t))^T$ which is the solution of a system of the form

$$\begin{aligned} \frac{d}{dt} a(t) &= \sum_{j=1}^{k} c_j(t) B_j(a(t)), \qquad t \in [0,T], \\ a(0) &= a^0 := (a_1^0, \ldots, a_k^0)^T. \end{aligned} \tag{2.2}$$

Here, the $B_j : \mathbb{C}^k \to \mathbb{C}^k$ are polynomials. The finite system (2.2) constitutes an ordinary control system in $\mathbb{C}^k$ where the control functions c_j enter *linearly*. It can be viewed as a k-dimensional projection of the original system (1.4). The vector control function $C(t) := (c_1(t), \ldots, c_k(t))$ is a L^1-function on I taking its values in

$$U_k(\Omega) := \left\{ (c_1, \ldots, c_k)^T \in \mathbb{C}^k \,\middle|\, \exists h \in \Omega : h(z) \equiv \sum_{j=1}^{\infty} c_j z^j \right\},$$

the kth coefficient variability region of Ω, *i.e.* $C \in L^1(I; U_k(\Omega))$. Since Ω is compact and convex, the same is true for $U_k(\Omega) \subseteq \mathbb{C}^k$. In particular, the time-T reachable set $\mathcal{R}_T(a^0; U_k(\Omega))$ of (2.2) coincides with the kth coefficient region $U_k(\mathcal{R}_T(f_0, \Omega))$ of the time-T reachable set $\mathcal{R}_T(f_0, \Omega)$ of (1.4).

For the k-dimensional version (2.2) of equation (1.4) the analogous result to Theorem 2 for (1.4) is a well-known theorem in control theory:

Theorem 3. *If $U_k = U_k(\Omega) \subseteq \mathbb{C}^k$ is compact and convex. then*

(a) $\mathcal{R}_T(a^0, U_k)$ is compact and

(b) $\mathcal{R}_T(a^0, \text{extr } U_k)$ is dense in $\mathcal{R}_T(a^0, U_k)$.

For a proof we refer to [4], p. 177 ff. We shall now use Theorem 3 and the classical theorem of Montel on normal families to prove Theorem 2. To do this, we let $I := [0, T]$ and emphasize the two basic ingredients of the proof of Theorem 3:

If $t \mapsto a_{j,C}(t)$, $j = 1, \ldots, k$, denotes the solution of (2.1) generated by $C \in L^1(I; U_k)$ then

(I) $L^1(I; U_k)$ is weakly sequential compact for each $k \in \mathbb{N}$ and

(II) if $\{C_l\}_{l \in \mathbb{N}} \subseteq L^1(I; U_k)$ converges weakly to $C \in L^1(I; U_k)$, then the corresponding solutions $a_{j,C_l}(T)$ converge to $a_{j,C}(T)$ for each $j = 1, \ldots, k$ as $l \to \infty$.

By (I) we mean that for each sequence $\{C_n\}_{n \in \mathbb{N}} \subseteq L^1(I; U_k)$ there is a subsequence $\{C_{n_l}\}_{l \in \mathbb{N}}$ and a function $C \in L^1(I; U_k)$ such that C_{n_l} converges weakly in $L^1(I; U_k)$ to C, *i.e.*

$$\int_I v(s)^T C_{n_l}(s)\, ds \to \int_I v(s)^T C(s)\, ds, \qquad l \to \infty,$$

for each $v \in L^\infty(I;U_k)$, the dual space of $L^1(I;U_k)$. Let us note that $L^1(I;\mathbb{C})$ is weakly complete: If a sequence $\{c_k\}_{k\in\mathbb{N}}$ in $L^1(I;\mathbb{C})$ converges weakly, *i.e.* $\int_T v(s)^T c_k(s)\,ds$ converges (in $\mathbb{C}$) for every $v \in L^\infty(I;\mathbb{C})$ then $\{c_k\}_{k\in\mathbb{N}}$ is weakly convergent to some $c \in L^1(I;\mathbb{C})$, *cf.* [2], Theorem 4.21.4.

Proof of Theorem 2 (a): Let $\{w_n(z,T)\}_{n\in\mathbb{N}}$ be a sequence in $\mathcal{R}_T(f_0,\Omega)$ such that $w_n(z,T) = w(z,T;h_n)$, $h_n \in L^1(I;\Omega)$ for all $n \in \mathbb{N}$.

(a) Let $h_n(z,t) = \sum_{j=1}^\infty c_{j,n}(t)z^j$, *i.e.* $C_{k,n}(t) := (c_{1,n}(t),\dots,c_{k,n}(t))^T \in L^1(I;U_k)$. Using a diagonal argument, the weak compactness property (I) of $L^1(I;U_k)$ shows that there exists a subsequence $\{n_l\}_{l\in\mathbb{N}} \subseteq \mathbb{N}$ such that $\{C_{k,n_l}\}_{l\in\mathbb{N}}$ converges weakly in $L^1(I;U_k)$ for each $k\in\mathbb{N}$. Let c_j denote the weak limit of $\{c_{j,n_l}\}_{l\in\mathbb{N}}$ in $L^1(I;\mathbb{C})$ for each $j\in\mathbb{N}$.

(b) We fix $v \in L^\infty(I;\mathbb{C})$ and consider for $z\in\mathbb{D}$

$$g_{n_l}(z) := \int_I v(s)h_{n_l}(z,s)\,ds = \sum_{j=1}^\infty \int_I v(s)c_{j,n_l}(s)\,ds\, z^j =: \sum_{j=1}^\infty d_{j,n_l} z^j .$$

Because $z \mapsto h_{n_l}(z,t)$ is contained in the normal family Ω for a.a. $t \in [0,T]$, the sequence $\{g_{n_l}\}_{l\in\mathbb{N}} \subseteq \mathcal{H}(\mathbb{D})$ is a normal family and by (a) $d_{j,n_l} \to d_j = \int_I v(s)c_j(s)\,ds$ for each $j\in\mathbb{N}$ as $l\to\infty$. If g is a locally uniform subsequential limit of $\{g_{n_l}\}_{l\in\mathbb{N}}$ in $\mathcal{H}(\mathbb{D})$, then $g^{(j)}(0) = j!\,d_j$, $j\in\mathbb{N}$, and the function

$$z \mapsto g(z) = \sum_{j=1}^\infty d_j z^j$$

is analytic in $\mathbb{D}$. By normality, $g_{n_l} \to g$ in $\mathcal{H}(\mathbb{D})$ as $l\to\infty$. In particular,

$$g_{n_l}(z,t) := \int_0^t h_{n_l}(z,s)\,ds \longrightarrow G(z,t) := \sum_{j=1}^\infty \int_0^t c_j(s)\,ds\, z^j \in \mathcal{H}(\mathbb{D})$$

locally uniformly in $\mathbb{D}$ as $l\to\infty$ for each $t\in I$. Since $L^1(I;\mathbb{C})$ is weakly complete and the functions $t\mapsto h_{n_l}(z,t) \in L^1(I;\mathbb{C})$, $z\in\mathbb{D}$ fixed, converge weakly in $L^1(I;\mathbb{C})$, there exists a function $h:\mathbb{D}\times I\to\mathbb{C}$ such that $G(z,t) = \int_0^t h(z,s)\,ds$ and $t\mapsto h(z,t)\in L^1(I;\mathbb{C})$ for each fixed $z\in\mathbb{D}$. But then,

$$\frac{1}{\delta}\int_t^{t+\delta} h(z,s)\,ds = \frac{G(z,t+\delta)-G(z,t)}{\delta} = \lim_{l\to\infty}\frac{1}{\delta}\int_t^{t+\delta} h_{n_l}(z,s)\,ds,$$

for all $t, t+\delta \in I$, $\delta>0$, locally uniformly in $\mathbb{D}$. Since Ω is compact and convex and $h_{n_l}(\,.\,,t) \in \Omega$ for all $l \in \mathbb{N}$ and a.a. $t \in [0,T]$, this implies

$$\frac{1}{\delta}\int_t^{t+\delta} h(z,s)\,ds \in \Omega.$$

Letting $\delta \to 0$, we conclude that $h \in L^1(I;\Omega)$ by the following argument. Let $t^* \in I$ be a Lebesgue point of all coefficient functions $t \mapsto c_j(t)$, *i.e.*

$$\lim_{\delta\to 0}\frac{1}{\delta}\int_{t^*}^{t^*+\delta} c_j(t)\,dt = c_j(t^*) \qquad \text{for all } j \in \mathbb{N}.$$

We conclude

$$\frac{1}{\delta}\int_{t^*}^{t^*+\delta} h(z,s)\,ds = \sum_{j=1}^{\infty}\left[\frac{1}{\delta}\int_{t^*}^{t^*+\delta} c_j(t)\,dt\right] z^j \to \sum_{j=1}^{\infty} c_j(t^*)z^j \in \Omega,$$

as $\delta \to 0$, by the compactness of Ω for all Lebesgue points $t^* \in I$, *i.e.* for a.a. $t^* \in I$. Obviously, $\sum_{j=1}^{\infty} c_j(t^*)z^j = h(z,t^*)$ and hence $h \in L^1(I;\Omega)$.

(c) Finally, let w be the solution of (1.4) corresponding to $h \in L^1(I;\Omega)$. If we use the power series expansions $w(z,t) = \sum_{j=1}^{\infty} a_j(t)z^j$ and $w_{n_l}(z,t) = \sum_{j=1}^{\infty} a_{j,n_l}(t)z^j$, we get $a_{j,n_l}(T) \to a_k(T)$ for each $j \in \mathbb{N}$ as $l \to \infty$ by (II). Since the solutions of (1.4) are bounded, a simple normal family argument shows that $z \mapsto w_{n_l}(z,T)$ converges to $z \mapsto w(z,T)$ in $\mathcal{H}(\mathbb{D})$. ■

In order to prove Theorem 2 (b), we first establish the following simple lemma.

Lemma 1. *Let $F(z) = \sum_{k=0}^{\infty} A_k z^k \in \mathcal{H}(\mathbb{D})$ such that $A_n \neq 0$ $(n = 1,2,\ldots)$ and let Ω be the set of functions $h \in \mathcal{H}(\mathbb{D})$ of the form (1.5) where μ is any probability measure on $\partial\mathbb{D}$. Then for each fixed $k \in \mathbb{N}$, the extreme points of the kth coefficient variability region of Ω are contained in the coefficient variability region of the set of extreme points of Ω, i.e.*

$$\mathit{extr}\ U_k(\Omega) \subseteq U_k(\mathit{extr}\ \Omega).$$

Proof: Let $h(z) = \sum_{k=0}^{\infty} c_k z^k \in \Omega$ be such that $C := (c_0,\ldots,c_k) \in \operatorname{extr} U_k(\Omega)$, *i.e.* $C \in \partial U_k(\Omega)$. Applying Theorem 9.10 in [3], we conclude that $h(z) = \sum_{j=1}^{k} \lambda_j F(x_j z)$ where the x_j's are distinct, $|x_j| = 1$, $\lambda_j \geq 0$ and $\sum_{j=1}^{k} \lambda_j = 1$, *i.e.* $c_s = A_s \sum_{j=1}^{k} \lambda_j x_j^s$, $s = 1,\ldots,k$. Thus, C is a convex combination of the vectors $A_s(1,x_s,\ldots,x_s^k)^T \in U_k(\Omega)$, $s = 1,\ldots,k$.

Because the x_j's are distinct, $A_s \neq 0$, and $C \in \text{extr } U_k(\Omega)$, we conclude that $h(z) = F(xz)$ for some $|x| = 1$. Because $\text{extr } \Omega = \{F(xz) \,|\, |x| = 1\}$, *cf.* [3], this implies $h \in \text{extr } \Omega$, *i.e.* $C \in U_k(\text{extr } \Omega)$. ■

Proof of Theorem 2 (b): Let $w(z,T;h) = \sum_{j=1}^{\infty} a_j(T) z^j \in \mathcal{R}_T(f_0, \Omega)$ for $h \in L^1(I;\Omega)$. By Theorem 3 (b) and Lemma 1, $\mathcal{R}_T(a^0, U_k(\text{extr } \Omega))$ is dense in $\mathcal{R}_T(a^0, U_k(\Omega))$. From this we obtain a sequence $\{h_n\}_{n\in\mathbb{N}} \subseteq L^1(I; \text{extr } \Omega)$, $h_n(z,t) = \sum_{j=1}^{\infty} c_{j,n}(t) z^j$ such that the solution $a_{j,n}(t)$ of (2.1) for $t = T$ corresponding to $c_{1,n}(t), \ldots, c_{j,n}(t)$ tends to $a_j(T)$ as $n \to \infty$ for each fixed $j \in \mathbb{N}$. As in part (c) of the proof of Theorem 2 (a), we conclude that $w(z,T;h_n) \to w(z,T;h)$ locally uniformly in $\mathbb{D}$. ■

Remark 3: In view of Remark 2 it would be interesting to know whether $\mathcal{R}_T(a_0, \text{extr } U_k(\Omega))$ is already compact. In the special case of the Loewner equation (here the initial value a^0 in (2.2) is $a^0 = e_1 := (1, 0, \ldots, 0)^T \in \mathbb{C}^k$) it is not difficult to show the weaker result that for each $M \in [1, \infty]$ the so-called $\leq \log M$*-reachable set, i.e.*

$$\mathcal{R}_{\leq \log M}(e_1, \text{extr } U_k(\mathcal{P})) = \bigcup_{1 \leq \tilde{M} \leq M} \mathcal{R}_{\log \tilde{M}}(e_1, \text{extr } U_k(\mathcal{P}))$$

is compact and coincides with $\{1\} \times V_k^M$ where $V_k^M = U_k(\mathcal{S}^M)$ is the kth coefficient region of the class $\mathcal{S}^M$. Therefore, in order to describe the kth coefficient region V_k^M of (bounded) univalent functions, it is enough to use the k-dimensional projection of the classical Loewner differential equation

$$\begin{aligned} \frac{d}{dt} w(z,t) &= -w(z,t) \frac{\kappa(t) + w(z,t)}{\kappa(t) - w(z,t)}, \qquad t \in [0, \log M], \\ w(z,0) &= z, \end{aligned} \tag{2.3}$$

and to consider the solutions $a(t)$ corresponding to all measurable functions $\kappa : [0, \log M] \to \partial\mathbb{D}$ for *all* $t \in [0, \log M]$. The proof relies on the facts that $V_k^M = \bigcup_{1 \leq \tilde{M} \leq M} \partial V_k^{\tilde{M}}$ and that the boundary functions of V_k^M are slit mappings.

References

[1] Coddington, E. A. and N. Levinson, *Theory of Ordinary Differential Equations*, McGraw-Hill, New York, 1955.

[2] Edwards, R. E., *Functional Analysis - Theory and Applications*, Holt, Rinehart and Winston, New York, 1965.

[3] Hallenbeck, D. J. and T. H. MacGregor, *Linear Problems and Convexity Techniques in Geometric Function Theory*, Pitman, Boston, 1984.

[4] Jurdjevic, V., *Geometric Control Theory*, Cambridge University Press, Cambridge, 1997.

[5] Löwner, K., Untersuchungen über schlichte konforme Abbildungen des Einheitskreises, I., *Math. Ann.* **89** (1923), 103–121.

[6] Pommerenke, Chr., Über die Subordination analytischer Funktionen, *J. Reine Angew. Math.* **218** (1965), 159–173.

[7] Pommerenke, Chr., On the Loewner differential equation, *Michigan Math. J.* **13** (1966), 435–443.

[8] Pommerenke, Chr., *Univalent Functions*, Vandenhoeck & Ruprecht, Göttingen, 1975.

Oliver Roth
Mathematisches Institut
Universität Würzburg
D-97074 Würzburg
GERMANY
roth@mathematik.uni-wuerzburg.de

Repeated bivariate rational approximants with prescribed poles

Catarina Rudälv

Abstract. In [19], Walsh considers approximation to analytic functions on compact sets by interpolation with rational functions having prescribed poles. Ambroladze and Wallin studied in [3], [4] and [5] this interpolation problem from a somewhat different point of view. Inspired by their ideas and by the approximants defined by Chaffy-Camus in [7] and Guillaume in [13] we construct rational approximants with prescribed poles to bivariate functions using a repeated interpolation procedure. In this paper we give some convergence results and numerical examples.

§1 Introduction

Generalisation of the one-dimensional Padé approximants (PA) to the multivariate case (MPA) is not straightforward. Several papers have been written about the subject. One of the major problems turns out to be the defining equations, *i.e.* the choices that should be made in order to get the right number of linear equations determining the coefficients of the rational function. Another problem is that it is difficult to show convergence. Chisholm, Karlsson and Wallin, Cuyt, Hughes-Jones and also Levin, among others, have all contributed to the area. Each of them showed that their PA satisfied some algebraic properties and some of them also gave some properties of convergence. In [7], Chaffy-Camus introduced a new class of bivariate approximants using a repeated procedure. The main idea was to create an approximant which was easy to compute, used less equations, was easy to extend to higher dimensions and last but not least, to create an approximant for which we could achieve some convergence results.

The main idea in [7] was to use the univariate PA in the following way. In the first step one constructs the PA with respect to the first variable, z. In the next step one computes the PA, of the resulting function, with respect to the second variable, w. The resulting function is referred to as the $\text{Padé}_w \circ \text{Padé}_z$. In [7] some algebraic properties are proved and also a version of the classical Montessus de Ballore theorem is generalised to these approximants. Another attempt to get convergence results for multivariate approximants, was made by Guillaume in [13]. He defined a multivariate

Computational Methods and Function Theory (CMFT'97)
N. Papamichael, St. Ruscheweyh and E. B. Saff (Eds.), pp. 471–484

ISBN 981-02-3626-3

Padé approximant using a nested procedure. The first step is the same as in [7], but in the second step he constructs the Padé approximants for the coefficients of the resulting rational function, not for the whole resulting function.

In this paper we use the ideas of Chaffy-Camus and Guillaume to construct bivariate Padé-type approximants, *i.e.* approximants with prescribed poles. The extension to the general, multivariate case is straightforward. Since multivariate Padé-type approximants (PTA) do not give rise to the same problems as MPA, the motivation which led Chaffy-Camus to their definition of $\text{Padé}_w \circ \text{Padé}_z$ has in [7] will not be be ours here. We do not have any problems with large and unstable system of equations, nor with the need of symmetric knowledge of coefficients. We are motivated purely by the need to search for new convergence results. The main result of this paper is Theorem 3. Its proof is based on theorems and techniques used in [3] and [4]. An important ingredient is classical potential theory.

§2 Notation

We will use the following notation:

$\bar{\mathbb{C}}$:	the extended complex plane, $\bar{\mathbb{C}} = \mathbb{C} \cup \infty$.
$\bar{\mathbb{C}}^2$:	Cartesian product of two copies of $\bar{\mathbb{C}}$.
Ω:	domain (open, connected set) in $\bar{\mathbb{C}}^2$.
D, D_1, D_2:	domains in $\bar{\mathbb{C}}$.
$D_1 \times D_2$:	product domain in $\bar{\mathbb{C}}^2$, $\{(z,w) : \ z \in D_1, w \in D_2\}$.
K, K_1, K_2:	closed sets in $\bar{\mathbb{C}}$.
$\partial D, \partial D_j, \partial K_j$:	boundary of D, D_j and K_j, respectively, $j = 1, 2$.
Π_n:	the space of univariate polynomials (in the variable z or w) of degree at most n with complex coefficients.
$\Pi_n(w)$:	the space of univariate polynomials (in the variable z) of degree at most n with coefficients being functions of w.
$\mathbb{N}^2$:	Cartesian product of two copies of $\mathbb{N}$, the set of natural numbers.
M_1, M, N:	subsets of $\mathbb{N}^2$.
Π_N:	the space of bivariate polynomials, $\sum_{(i,j)} a_{i,j} z^i w^j$, with complex coefficients, a_{ij}, and $(i,j) \in N$.

§3 Interpolation at zero

Suppose f is analytic in a product domain $D_1 \times D_2 \subset \mathbb{C}^2$, $0 \in D_j \subset \mathbb{C}$, $j = 1, 2$. We shall carry out an interpolation in two steps.

Step 1. Given a set $M_1 \subset \mathbb{N}^2$, a polynomial $Q_1(z,w) \in \Pi_{M_1}$, $Q_1(0,0) \neq 0$, and a number $n_1 \in \mathbb{N}$, we wish to find a polynomial $P_1(z,w) \in \Pi_{n_1}(w)$, *i.e.*

$$P_1(z,w) = \sum_{k=0}^{n_1} a_k(w) z^k, \text{ for } w \in D_2, \tag{3.1}$$

such that, for some function $w \mapsto g(w)$,

$$f(z,w) - \frac{P_1(z,w)}{Q_1(z,w)} = g(w)\mathcal{O}(z^{n_1+1}), \ z \to 0. \tag{3.2}$$

Since $Q_1(0,0) \neq 0$, this is equivalent to the linearised condition

$$f(z,w)Q_1(z,w) - P_1(z,w) = g(w)\mathcal{O}(z^{n_1+1}), \ z \to 0. \tag{3.3}$$

Solving the above system of equations, (3.3), for the coefficients $a_k(w)$, we have the following proposition.

Proposition 1. *$P_1(z,w)$ exists uniquely and its coefficients are determined by*

$$a_k(w) = \sum_{j=0}^{k} b_j(w) c_{k-j}(w), \quad 0 \leq k \leq n_1, \ b_j(w) = 0 \text{ if } j > m_1, \tag{3.4}$$

where, for fixed $w \in D_2$, $c_j(w)$ and $b_j(w)$ are respectively the Taylor coefficients of $f(z,w)$ and $Q_1(z,w)$ with respect to z, and $m_1 = \max\{i : (i,j) \in M_1$ for some $j \in \mathbb{N}\}$. Moreover, $a_k(w)$ are analytic in D_2.

Remark 1: If we choose to interpolate at $a = \infty$, instead of at $a = 0$, then the coefficients $a_k(w)$ will no longer be analytic, since they will have a polynomial part given by $b_j(w)$ which has a pole at infinity. Thus, it will be impossible to proceed to the next step. However, the case $a = \infty$ will be possible if we choose Q_1 to be independent of w, since in this case every $b_j(w)$ will be constant. If, instead of choosing $a = 0$, we choose the interpolation point to be $a \neq 0, \neq \infty$, then the result will be the same as in the case of $a = 0$.

Step 2. Now, given two numbers $m_2, n_2 \in \mathbb{N}$ and a polynomial $Q_2(w) \in \Pi_{m_2}$, $Q_2(0) \neq 0$, we can construct Padé-type approximants to each $a_k(w)$, since every $a_k(w)$ is analytic in D_2. That is, we can find polynomials $P_{2,k}(w) \in \Pi_{n_2}$ such that

$$a_k(w) - \frac{P_{2,k}(w)}{Q_2(w)} = \mathcal{O}(w^{n_2+1}), \ w \to 0. \tag{3.5}$$

Since $Q_2(0) \neq 0$, this is equivalent to

$$a_k(w)Q_2(w) - P_{2,k}(w) = \mathcal{O}(w^{n_2+1}), \ w \to 0. \tag{3.6}$$

Let $M = M_1 \cup (\{0\} \times \{0, \dots, m_2\})$ and $N = \{0, \dots n_1\} \times \{0, \dots, n_2\}$. Then we define:

Definition 1. *The rational function $\tilde{P}(z,w)/\tilde{Q}(z,w)$ constructed above, where $\tilde{P}(z,w) = \sum_{k=0}^{n_1} P_{2,k}(w)z^k$ and $\tilde{Q}(z,w) = Q_1(z,w)Q_2(w)$, will be referred to as a nested Padé-type approximant around zero. To indicate the dependence on the index sets, we will denote it by $\{N/M\}_{f,0}$ or just by $\{N/M\}_f$.*

§4 Interpolation at infinity

If f is analytic in a product domain $D_1 \times D_2 \subset \bar{\mathbb{C}}^2$, $\infty \in D_j$, $j = 1, 2$, and if we interpolate at $a = \infty$ then (as was explained in Remark 1) the polynomial Q_1, which was used in the first step, needs to be univariate. We shall consider the diagonal case, *i.e.* we shall use polynomials of the same degree in the numerator and the denominator. Two items in the construction need to be changed. First, given a number $n_1 \in \mathbb{N}$ and a polynomial $Q_1(z) \in \Pi_{n_1}$, we shall determine a polynomial $P_1(z,w) \in \Pi_{n_1}(w)$ such that (instead of (3.3)) for some function $w \mapsto g(w)$,

$$f(z,w)Q_1(z) - P_1(z,w) = g(w)\mathcal{O}\left(\frac{1}{z}\right), \ z \to \infty. \tag{4.1}$$

Thus, P_1 is the polynomial part of the Laurent expansion of fQ_1 with respect to z. $P_1(z,w)$ still exists uniquely but, instead of (3.4), we will have the following formula for $a_k(w)$,

$$a_k(w) = \sum_{r=0}^{n_1-k} c_r(w)b_{k+r}, \tag{4.2}$$

where, for fixed $w \in D_2$, each $c_j(w)$ is a coefficient of the Laurent expansion of f with respect to z, and every b_j is a coefficient of $Q_1(z)$.

Now, in the second step, given a number $n_2 \in \mathbb{N}$ and a polynomial $Q_2(w) \in \Pi_{n_2}$, we can find polynomials $P_{2,k}(w) \in \Pi_{n_2}$ such that

$$a_k(w)Q_2(w) - P_{2,k}(w) = \mathcal{O}\left(\frac{1}{w}\right), \ w \to \infty. \tag{4.3}$$

Thus, $P_{2,k}$ is the polynomial part of the Laurent expansion of a_kQ_2. Finally, if $N = \{0, \dots, n_1\} \times \{0, \dots, n_2\}$, then define:

Definition 2. *The rational function $\tilde{P}(z,w)/\tilde{Q}(z,w)$ constructed above, where $\tilde{P}(z,w) = \sum_{k=0}^{n_1} P_{2,k}(w)z^k$ and $\tilde{Q}(z,w) = Q_1(z)Q_2(w)$, will be referred to as a nested Padé-type approximant around infinity. To indicate the dependence on the index sets, we will denote it by $\{N/N\}_{f,\infty}$.*

Remark 2: In the two definitions above one can choose Q_2 to depend on k. Different choices will lead to different error formulas and different convergence results. In this paper we will assume that Q_2 is independent of k.

We have the following relation between Definition 1 and Definition 2. The proof can be found in [18].

Proposition 2 (Invariance of value transformation). *Suppose f is analytic in a product domain $D_1 \times D_2 \subset \mathbb{C}^2$, $0 \in D_j$, $j = 1,2$. If $\hat{f}(z,w) = f(\frac{1}{z}, \frac{1}{w})$ and if the relation between the denominators of the nested PTA of f and $\hat{f}$ satisfies*

$$\hat{Q}(z,w) = z^{n_1} w^{n_2} \tilde{Q}\left(\frac{1}{z}, \frac{1}{w}\right), \tag{4.4}$$

then

$$\{N/N\}_{\hat{f},\infty}(z,w) = \{N/N\}_{f,0}\left(\frac{1}{z}, \frac{1}{w}\right). \tag{4.5}$$

§5 Ordinary Padé-type approximants

Let us recall the definition of ordinary Padé-type approximants (PTA) introduced by Abouir and Cuyt in [1]. In this paper the reader can also find a review of multivariate Padé-type approximation methods in existence at that time. The definition is for when $\Omega \subset \mathbb{C}^2$ and $(0,0) \in \Omega$. The expansions used are Taylor expansions around zero.

Definition 3. *Given two index sets $M, N \subset \mathbb{N}^2$ and a polynomial $Q(z,w) \in \Pi_M$, find a polynomial $P(z,w) \in \Pi_N$ such that*

$$\left(f - \frac{P}{Q}\right)(z,w) = \sum_{(i,j)\in\mathbb{N}^2\setminus N} e_{ij} z^i w^j \tag{5.1}$$

or equivalently

$$(fQ - P)(z,w) = \sum_{(i,j)\in\mathbb{N}^2\setminus N} \tilde{e}_{ij} z^i w^j. \tag{5.2}$$

It is easy to see that P exists uniquely. We refer to the approximant P/Q as an ordinary Padé-type approximant *and we denote it by $(N/M)_f$.*

Now, if we choose $Q_1(z)$ and $Q_2(w)$ in the two steps of the construction of the nested PTA and $Q(z,w) = Q_1(z)Q_2(w)$ as the prescribed denominator of the ordinary Padé-type approximant, P/Q, then the following is true (see [18]).

Proposition 3.

$$\frac{\tilde{P}(z,w)}{\tilde{Q}(z,w)} = \frac{P(z,w)}{Q(z,w)} \tag{5.3}$$

i.e. nested Padé-type approximants and ordinary Padé-type approximants are the same if the denominators are chosen as above.

Remark 3: Proposition 3 is true for nested and ordinary Padé-type approximants around infinity too. This follows from Proposition 2. However, Proposition 3 is not true if we choose Q_1 to be a general bivariate polynomial in the variables z and w.

Remark 4: The nested Padé-type approximants defined in §3 satisfy some useful algebraic properties. For instance, it can be proved that they satisfy properties like *consistency, factorisation* and *restriction.* We refer to [18] for further information about algebraic properties and the proofs.

§6 Old convergence results

There are some convergence results that need to be mentioned. The first one concerns a summability method discussed by Eiermann in [10]. He shows that a special case of this summability method can be regarded as ordinary Padé-type approximantion and proves that under certain conditions we get convergence to f on a subset of D, the domain of analyticity of f. A new proof of Eiermann's theorem, which was easier to follow, was given in [6]. Daras [9] generalised this to the multivariate case when $\Omega \subset \mathbb{C}^2$ is a polydisc and Rudälv [16] proved the same result for the case where $\Omega \subset \mathbb{C}^2$ is a more general product domain. Also, a new proof of Daras' theorem was given in [17]. However, neither [10] and [6] nor [9], [16] and [17] contain any uniform convergence result on compact subsets of the whole domain Ω.

Another result was proved by Gonzalez-Vera and Orive in [12]. Given a function f, analytic in Ω, they proved that if the denominator is a product of univariate polynomials, then one can expect geometric convergence on the largest polydisc, centred at the origin and entirely contained in Ω. However, we would like the convergence to be on the whole domain Ω.

Yet another result is given in [8]. This paper deals with homogeneous series expansions of f where uniform convergence is proved for balls.

The first two cases above deal with so-called rectangular PTA, *i.e.* the choice of the index set for the numerator $P(z,w)$ is a rectangular index set. The third case concerns triangular index sets. The last case, which can be found in [2], deals with general index sets. If the index sets and the denominators of the ordinary PTA satisfy some properties, then convergence is obtained for ordinary PTA on polydiscs and balls.

Our goal here is to obtain convergence on Ω, the domain of analyticity, when Ω is a product domain $D_1 \times D_2 \subset \bar{\mathbb{C}}^2$, $\infty \in D_j$, D_j regular, $j = 1,2$.

§7 Convergence results

In order to prove convergence for nested Padé-type approximants we need to look at the remainder after approximation. In the rest of this paper, interpolation is done at infinity. The error term will look almost the same when interpolation is done at zero or at some other point.

$$\begin{aligned}
E\ &(z,w)\\
&= f(z,w) - \frac{\tilde{P}(z,w)}{\tilde{Q}(z,w)}\\
&= f(z,w) - \frac{P_1(z,w)}{Q_1(z)} + \frac{P_1(z,w)}{Q_1(z)} - \frac{\tilde{P}(z,w)}{Q_1(z)Q_2(w)}\\
&= \left(f(z,w) - \frac{P_1(z,w)}{Q_1(z)}\right) + \frac{1}{Q_1(z)}\left(P_1(z,w) - \frac{\tilde{P}(z,w)}{Q_2(w)}\right)\\
&= \underbrace{\left(f - \frac{P_1}{Q_1}\right)(z,w)}_{E_1} + \underbrace{\frac{1}{Q_1(z)}\sum_{k=0}^{n_1}\underbrace{\left(a_k - \frac{P_{2,k}}{Q_2}\right)(w)}_{E_{2,k}} z^k}_{E_2}. \qquad (7.1)
\end{aligned}$$

If we decide to work with interpolation at infinity, then we have the following integral representation of the univariate approximation error. This will be used in the proof of Theorem 3.

Theorem 1 (see for instance [15]). *Suppose g is analytic in $D \subset \bar{\mathbb{C}}$, $\infty \in D$, and suppose $\{Q_n\}$ is a sequence of univariate polynomials given in advance. Let P_n/Q_n be the univariate PTA of g with prescribed poles at the zeros of Q_n defined as in [3]. Then*

$$g(z) - \frac{P_n(z)}{Q_n(z)} = -\frac{1}{2\pi i}\int_\gamma \frac{Q_n(t)}{Q_n(z)}\frac{g(t)}{t-z}dt \qquad (7.2)$$

for all points z outside γ, where $\gamma \in D$ is a simple closed positively oriented curve, or a finite union of such curves, winding once around each point of $K = \bar{\mathbb{C}} \setminus D$.

In this section we will prove that it is possible to obtain uniform convergence of sequences of nested PTA (around infinity) to functions analytic in a product domain $D_1 \times D_2 \subset \bar{\mathbb{C}}^2$, $\infty \in D_j$, $j = 1, 2$. The proof is based on the error formula obtained before and on the following univariate theorem.

Theorem 2 (Ambroladze and Wallin, [3]). *If D is a regular domain in $\bar{\mathbb{C}}$, $\infty \in D$, $K = \bar{\mathbb{C}} \setminus D$ and $\{Q_n\}$ is a sequence of monic polynomials having asymptotically no zeros in D, then the following conditions are pairwise equivalent:*

(1) *For all f, analytic in D, the sequence of PTA with prescribed poles at the zeros of Q_n converges, locally uniformly, to f on D.*

(2) $\lim_{n\to\infty} \max_{z\in K} |Q_n(z)|^{1/n} = \operatorname{cap}(K)$

(3) $\lim_{n\to\infty} |Q_n(z)|^{1/n} = \operatorname{cap}(K) e^{g_D(z)}$ *locally uniformly on D.*

(4) *The zeros of $\{Q_n\}$ have regular asymptotic distribution.*

(5) *For an arbitrary bounded closed set $F \subset D$ there exists an open neighbourhood O of K such that*

$$\lim_{n\to\infty} \left(\sup_{t\in O, z\in F} \left| \frac{Q_n(t)}{Q_n(z)} \right| \right) = 0,$$

where $\operatorname{cap}(K)$ and g_D are respectively the logarithmic capacity and the Green function of D with pole at infinity, and the meaning of regular asymptotic distribution is that every weak limit of the set of zero counting measures $\{\mu_n\}$ is supported by K and its sweeping out onto ∂K coincides with the equilibrium measure. For further definitions see [3].*

Now we are ready to state the main theorem of this paper. Since we work with nested PTA around infinity, the two sequences of polynomials, $\{Q_1\}$ and $\{Q_2\}$, which we will choose in advance, will be univariate in the variables z and w, respectively. We will indicate the dependence on the numbers n_1 and n_2 by using these numbers as second subscripts, Q_{1,n_1} and Q_{2,n_2}.

Theorem 3. *Suppose f is analytic in a product domain $D_1 \times D_2 \subset \bar{\mathbb{C}}^2$, $\infty \in D_j$, D_j regular, $j = 1,2$. Let n_1 and n_2 be positive integers. Let $\{Q_1\} = \{Q_{1,n_1}\}$ and $\{Q_2\} = \{Q_{2,n_2}\}$, $Q_1 \in \Pi_{n_1}$, $Q_2 \in \Pi_{n_2}$, be two sequences of monic polynomials, with asymptotically no zeros in D_1 and D_2, respectively, satisfying one of the conditions (2)–(5) in Theorem 2. Let $\{\tilde{P}/\tilde{Q}\}$ be the corresponding sequence of nested PTA around infinity. Let $\tilde{K}$ be a compact subset of $D_1 \times D_2$. Then there exists $n_2 = n_2(n_1)$, depending on n_1, such that*

$$\lim_{\substack{n_1 \to \infty \\ n_2 = n_2(n_1) \to \infty}} \left(f - \frac{\tilde{P}}{\tilde{Q}} \right)(z,w) = 0 \tag{7.3}$$

uniformly on $\tilde{K}$.

We give a short version of the proof and refer to [18] for details. In the proof we will need the following three lemmas, where we use the notation and assumptions in Theorem 3.

Lemma 1. *Let $\{Q_{j,n_j}\}$, $j = 1,2$, be a sequence of polynomials satisfying the assumptions in Theorem 3. Let K_j be a compact subset of D_j. Then there exists a curve Γ_j satisfying the assumptions in Theorem 1 where $K_j = K$, $D_j = D$, $\Gamma_j = \Gamma$, such that, for some constants $\rho_j < 1$ and $N_j > 0$,*

$$\max_{\substack{u \in K_j \\ t \in \Gamma_j}} \left| \frac{Q_{j,n_j}(t)}{Q_{j,n_j}(u)} \right| < \rho_j^{n_j}, \text{ when } n_j > N_j. \tag{7.4}$$

The next lemma will be used to prove the third lemma.

Lemma 2. *Let $\{Q_{1,n_1}\}$ be a sequence of polynomials satisfying one of the conditions (2)–(5) in Theorem 2 and let γ be a curve in D_1. Then there exist constants $C > 0$ and $N_1 > 0$ such that*

$$\max_{t \in \gamma} |Q_{1,n_1}(t)| \le C^{n_1}, \; n_1 > N_1.$$

Lemma 3. *Let $\{a_k(w)\}$ be the coefficients of the numerator constructed in the first step of the construction of the nested Padé-type approximant around infinity. Let Γ_2 be a curve in D_2. Then there exist constants $A > 0$, $B > 0$ and $N_1 > 0$ such that*

$$\sum_{k=0}^{n_1} \sup_{s \in \Gamma_2} |a_k(s)| \le B \cdot n_1 \cdot A^{n_1}, \; n_1 > N_1.$$

Proof of Theorem 3: We have divided the error formula (7.1) into two parts E_1 and E_2 where E_2 is a sum of smaller errors $E_{2,k}$. We shall take care of them separately.

Let $\tilde{K}$ be an arbitrary compact subset of $D_1 \times D_2$. Let K_1 and K_2 be the projections of $\tilde{K}$ onto the z-plane and w-plane, respectively. Let Γ_1 be the curve which appears in Lemma 1. Then using Theorem 1, we shall get the following estimate for E_1:

$$\max_{(z,w)\in\tilde{K}} |E_1| \le A \cdot \max_{\substack{w\in K_2,\\ t\in\Gamma_1}} |f(t,w)| \cdot \max_{\substack{z\in K_1,\\ t\in\Gamma_1}} \left|\frac{Q_{1,n_1}(t)}{Q_{1,n_1}(z)}\right|. \tag{7.5}$$

Since $\{Q_{1,n_1}\}$ satisfies one of the conditions (2)–(5) in Theorem 2, we know from Lemma 1 that this implies that

$$\max_{(z,w)\in\tilde{K}} |E_1| \le C_1 \cdot M_1 \cdot \rho_1^{n_1}, \text{ when } n_1 > N_1,$$

where C_1 and M_1 are constants which do not depend on n_1. Thus, E_1 will be small if n_1 is large enough. If we use the same technique on $E_{2,k}$ with Γ_2 instead of Γ_1 we get

$$\max_{w\in K_2} |E_{2,k}| \le C_2 \cdot M_k \cdot \rho_2^{n_2}, \text{ when } n_2 > N_2, \tag{7.6}$$

where C_2 is a constant which does not depend on n_2 and M_k is a constant defined as $M_k = \max_{s\in\Gamma_2} |a_k(s)|$.

Next, we will estimate E_2. This involves an estimation of the quotient $z^k/Q_{1,n_1}(z)$. However, it can be proved that there exists a constant c such that

$$\left|\frac{z^k}{Q_{1,n_1}(z)}\right| \le c^{n_1}, \ k = 1, ..., n_1, \ z \in K_1. \tag{7.7}$$

Next, Lemma 3 tells us that the growth of the constants M_k can be estimated by an exponential function. Thus, by combining Lemma 3 with (7.6) and (7.7), we get

$$\max_{(z,w)\in\tilde{K}} |E_2| \le C \cdot \rho_2^{n_2} \cdot A^{n_1} \cdot n_1 \cdot c^{n_1}, \text{ when } n_1 > N_1 \text{ and } n_2 > N_2,$$

where C is a constant which does not depend on n_1 and n_2. This can be made arbitrarily small if we choose n_2 large enough. Since $E(z,w)$ consists of two parts, E_1 and E_2, we can conclude that on each compact set $\tilde{K} \subset D_1 \times D_2$, it is possible to make $E(z,w)$ arbitrarily small by choosing first n_1 and then n_2 large enough. ■

§8 Examples

We shall give some examples to illustrate how Theorem 3 works. The idea is to start with a product domain $D_1 \times D_2 \subset \bar{\mathbb{C}}^2$ and then to choose two sequences of univariate polynomials $\{Q_1\}$, $\{Q_2\}$ in such a way that the assumptions in Theorem 3 are satisfied. Our first choice is to use the so-called Fekete points as zeros of $Q_j, j = 1, 2$. It can be proved (see for instance [14]) that Fekete points, associated to a given compact set K, exist and are located on the outer boundary of K, *i.e.* the boundary of the unbounded component of the complement of K, $\bar{\mathbb{C}} \setminus K$.

Proposition 4 (see for instance [14]). *If $K \subset \mathbb{C}$ is non-polar and compact, $n \geq 2$, and μ_n is the probability measure that puts mass $1/n$ at each of the nth Fekete points of K, then $\mu_n \to \nu$ in the weak*-sense where ν is the equilibrium measure of K.*

Proposition 4 shows that if we interpolate at infinity and choose the sequences $\{Q_1\}$ and $\{Q_2\}$ to be Fekete polynomials, *i.e.* the polynomials which have Fekete points as zeros, then Q_1 and Q_2 satisfy condition (4) in Theorem 2. Now, Fekete points are easy to compute when K is a disc but, unfortunately, are not easy to compute in general. In the following numerical example we have taken K_1 and K_2 to be discs.

Example 1: Let $D_j = \bar{\mathbb{C}} \setminus K_j$ where K_1 and K_2 are discs with radius 2 and 3, respectively. Suppose also that Q_1 and Q_2 are chosen as Fekete polynomials with respect to K_1 and K_2, respectively. If the degrees of Q_1 and Q_2 are at most n_1 and n_2, respectively, then $Q_1(z) = z^{n_1} - 2^{n_1}$ and $Q_2(w) = w^{n_2} - 3^{n_2}$. In the following numerical example we have used $f(z,w) = e^{1/z+1/w}/(z-2)(w-3)$. The calculations have been done with Mathematica 3.0. $|E(z,w)|$ denotes the absolute value of the error term evaluated at the point (z,w).

(n_1, n_2)	$\lvert E(3,4)\rvert$	$\lvert E(100,100)\rvert$
$(10,10)$	$1.50 \cdot 10^{-2}$	$2.49 \cdot 10^{-20}$
$(20,20)$	$5.93 \cdot 10^{-4}$	$1.43 \cdot 10^{-35}$
$(50,50)$	$8.87 \cdot 10^{-8}$	$2.94 \cdot 10^{-81}$

(n_1, n_2)	$\lvert E(3,4)\rvert$	$\lvert E(100,100)\rvert$
$(10,20)$	$6.22 \cdot 10^{-3}$	$6.94 \cdot 10^{-22}$
$(20,30)$	$1.26 \cdot 10^{-4}$	$7.12 \cdot 10^{-39}$
$(50,60)$	$5.48 \cdot 10^{-9}$	$7.64 \cdot 10^{-90}$

Our second choice will be to use a discrete Brownian motion to generate the zeros of Q_1 and Q_2. Inspired by Theorem 3 and the papers by Ambroladze and Wallin, [4] and [5], we will give an example where interpolation is done at zero. However, if we choose to interpolate at a point $a \neq \infty$, then condition (4) in Theorem 2 will be replaced by the following condition. The sweeping out of every weak* limit of the set of zero counting measures $\{\mu_n\}$, where μ_n is the measure that puts mass $1/n$ at each zero of Q_n, counting multiplicity, coincides with the harmonic measure evaluated at a. The following well-known theorem by Kakutani tells us about the connection between harmonic measures and Brownian motions.

Theorem 4 (see for instance [11]).

$$\begin{aligned}\omega(z, E, D) &= Prob(\text{Brownian motion starting at } z \in D \\ &\qquad \text{first hits } \partial D \text{ in the set } E),\end{aligned} \tag{8.1}$$

where $\omega(z, E, D)$ denotes the harmonic measure of $E \subset \partial D$ evaluated at the point z (i.e. the solution to the Dirichlet problem in D with boundary values equal to the characteristic function χ_E) and Prob stands for the probability.

Example 2: In the following numerical example we have taken $f(z, w) = 1/(z-1)(w-1)$ and D_j, $j = 1, 2$, to be the rectangles

$$D_1 = \{z : |\text{Re}z| < 1, |\text{Im}z| < 1\} \qquad D_2 = \{w : |\text{Re}w| < 1, |\text{Im}w| < 1\}.$$

Mathematica 3.0 has been used to simulate the Brownian motion. The steplength is equal to ϵ and in each step we have chosen one of four directions (north, east, south and west) randomly. $|E(z, w)|$ denotes the absolute value of the error term evaluated at the point (z, w). For simplicity, we have chosen $m_1 = n_1$ and $m_2 = n_2$.

(n_1, n_2)	$\lvert E(0.5, 0.5)\rvert, \epsilon = 0.1$	$\lvert E(0.5, 0.5)\rvert, \epsilon = 0.01$
$(10, 10)$	$1.03 \cdot 10^{-4}$	$1.18 \cdot 10^{-3}$
$(20, 20)$	$2.12 \cdot 10^{-6}$	$8.05 \cdot 10^{-9}$
$(30, 30)$	$9.56 \cdot 10^{-10}$	$4.58 \cdot 10^{-8}$
$(40, 40)$	$1.41 \cdot 10^{-11}$	$4.43 \cdot 10^{-11}$

(n_1, n_2)	$\lvert E(0.5, 0.5)\rvert, \epsilon = 0.1$	$\lvert E(0.5, 0.5)\rvert, \epsilon = 0.01$
$(10, 20)$	$5.68 \cdot 10^{-3}$	$7.61 \cdot 10^{-3}$
$(20, 30)$	$4.47 \cdot 10^{-6}$	$9.76 \cdot 10^{-7}$
$(30, 40)$	$4.76 \cdot 10^{-12}$	$1.15 \cdot 10^{-8}$
$(40, 50)$	$4.34 \cdot 10^{-11}$	$8.94 \cdot 10^{-10}$

References

[1] Abouir, J. and A. Cuyt, Multivariate partial Newton-Padé and Newton Padé-type approximants, *Journal of Approxiamtion Theory* **72** (1993), 301–306.

[2] Abouir, J., A. Cuyt, P. Gonzalez-Vera and R. Orive, On the convergence of general order multivariate Padé-type approximants, *Journal of Approximation Theory* **86** (1996), 216–228.

[3] Ambroladze, A. and H. Wallin, Convergence of rational interpolants with prescribed poles, *Journal of Approximation Theory* **89** (1997), 238–256.

[4] Ambroladze, A. and H. Wallin, Rational interpolants with prescribed poles, theory and practice, *Complex Variables* **34** (1997), 399–413.

[5] Ambroladze, A. and H. Wallin, Rational interpolants with preassigned poles, theoretical aspects, *Studia Mathematica*, 1998 to appear.

[6] Cala-Rodriguez, F. and H. Wallin, Padé-type approximants and a summability theorem by Eiermann, *Journal of Computational and Applied Mathematics* **39** (1992), 15–21.

[7] Chaffy-Camus, C., (PADE)y of (PADE)x approximants of F(x,y), in *Nonlinear numerical methods and rational approximation*, A. Cuyt (ed.), D. Reidel Publishing Company, 1988, 155–166.

[8] Cuyt, A., P. Gonzalez-Vera and R. Orive, Convergence of "homogeneous" Padé-type approximants on the unit ball of $\mathbb{C}^n$, preprint.

[9] Daras, N., Domaine de convergence d'une transformation de la suite des sommes partielles d'une fonction holomorphe et applications aux approximants de type Padé, dissertation, Université de Lille Flandres-Artois, 1988.

[10] Eiermann, M.,On the convergence of Padé-type approximants to analytic functions, *Journal of Computational and Applied Mathematics* **10** (1984), 219–227.

[11] Garnett, J., *Applications of Harmonic Measure*, The University of Arkansas lecture notes in the mathematical sciences, vol. 8, A Wiley-Interscience publication, 1986.

[12] Gonzalez-Vera, P. and R. Orive, On convergence of certain multivariate rational approximants, *Journal of Institute of Mathematics and Computer Sciences* **1** (1991), 31–40.

[13] Guillaume, P., A new class of multivariate Padé approximants. Rapport interne No 96.38, UMR 5640, CNRS-Université Paul Sabatier-INSA, Complexe Scientifique de Rangueil, Toulouse, 1996.

[14] Hille, E., *Analytic Function Theory*, Ginn and Company, Boston, vol. 2, 1962.

[15] Karlberg, L. and H. Wallin, Padé-type approximants and orthogonal polynomials for Markov-Stieltjes functions, *Journal of Computational and Applied Mathematics* **32** (1990), 153–157.

[16] Rudälv, C., Multivariate Padé-type approximants-algebraic properties and convergence results, Department of Mathematics, University of Umeå, no.1, 1994.

[17] Rudälv, C., On the extension of Eiermann's theorem to several complex variables, in *Nonlinear Numerical Methods and Rational Approximation II*, A. Cuyt (ed.), Kluwer, Dordrecht, 1994, 219–228.

[18] Rudälv, C., Nested multivariate rational approximants with prescribed poles, Department of Mathematics, University of Umeå, no.1, 1998.

[19] Walsh, J. L., *Interpolation and Approximation by Rational Functions in the Complex Domain*, Coll. Publ., AMS, vol. 20, Providence, R.I., 4th ed., 1995.

Catarina Rudälv
Department of Mathematics
University of Umeå
S-901 87 Umeå
SWEDEN
Catarina.Rudalv@math.umu.se

Real zeros of Bernstein polynomials

Gerhard Schmeisser

Abstract. We describe subsets $\mathcal{U}$ of the real line with the property that, if a polynomial f has all its zeros in $\mathcal{U}$, then so have all the Bernstein polynomials $B_n[f]$ of positive degree. Certain related questions including extensions to entire functions are also studied. Finally, we outline a connection with a recent result on polynomial transformations.

§1 Introduction

The Bernstein polynomials

$$B_n[f](x) := \sum_{\nu=0}^{n} \binom{n}{\nu} f\left(\frac{\nu}{n}\right) x^{\nu}(1-x)^{n-\nu}$$

have an extensive literature. The bibliographies of Stark [11] and Gonska & Meier [1], [2] comprise the contributions from 1912 to 1986. It seems that mainly the approximation and shape-preserving properties of the Bernstein operator and its diverse variants or extensions have been studied, while a classical question on polynomials, namely the location of zeros, has been touched only indirectly.

For instance, it is known [8], p. 87, that, for an entire function f, the Bernstein polynomials $B_n[f]$ converge to f as $n \to \infty$, uniformly on compact subsets of the complex plane. Hence, when f is a polynomial, then, by a theorem of Hurwitz, the zeros of $B_n[f]$ converge to those of f. Another result on zeros can be deduced from a theorem of Schoenberg ([10], Theorem 1), who proved that the Bernstein operator is variation diminishing on the interval $(0, 1)$. Therefore, if a polynomial f with real coefficients has no zeros in $(0, 1)$, then its Bernstein polynomials of positive degree cannot have zeros of odd multiplicities in $(0, 1)$. But the proof of Schoenberg shows that they cannot have zeros of even multiplicities either. Hence the following holds.

Theorem A. *Let f be a polynomial with real coefficients having all its zeros in $\mathbb{C} \setminus (0, 1)$. Then so have all the Bernstein polynomials $B_n[f]$ of positive degree.*

Computational Methods and Function Theory (CMFT'97)
N. Papamichael, St. Ruscheweyh and E. B. Saff (Eds.), pp. 485–495

ISBN 981-02-3626-3

The conclusion of Theorem A is not true when the coefficients are allowed to be complex. For example, the polynomial

$$f(z) = z^2 - z(1+2\mathrm{i}) + \mathrm{i}$$

has all its zeros in $\mathbb{C} \setminus (0,1)$, but $B_1[f](z) = -2\mathrm{i}(z - 1/2)$. Nevertheless, we may ask if there are non-trivial subsets of the complex plane which the Bernstein operator prefers for the location of zeros and others which it tries to avoid. In order to make that question more precise, we first introduce a notion.

Definition 1. *A subset $\mathcal{U}$ of the complex plane is* Bernstein invariant *if it has the following property. Whenever a polynomial f with complex coefficients has all its zeros in $\mathcal{U}$, then so have all the Bernstein polynomials $B_n[f]$ of positive degree.*

In our talk at the CMFT conference in Cyprus, we have given sufficient conditions and also necessary conditions for a subset $\mathcal{U}$ of the complex plane to be Bernstein invariant. In addition, we have established a list of 21 different Bernstein invariant sets, each having at most two connected components. These results will appear elsewhere. Here we investigate Bernstein invariant subsets of the real line by using a different, quite elementary approach.

§2 Statement of results

Theorem 1. *The following subsets of $\mathbb{R}$ are Bernstein invariant:*

$$(-\infty, 0), \quad (-\infty, 0], \quad (1, \infty), \quad [1, \infty), \tag{2.1}$$

$$\mathbb{R} \setminus (0,1), \quad \mathbb{R} \setminus [0,1), \quad \mathbb{R} \setminus (0,1], \quad \mathbb{R} \setminus [0,1]. \tag{2.2}$$

Remark 1: The proof of Theorem 1 is based on Lemma 1 below and allows refinements. For example, if f has all its zeros in one of the sets (2.2), denoted by $\mathcal{U}$, and $n \geq \deg f$, then, in each of the connected components of $\mathcal{U}$, the polynomials f and $B_n[f]$ have the same number of zeros.

Theorem 1 allows an extension from polynomials to certain entire functions. Let us introduce the following terminology [6], chap. 7, § 2.

Definition 2. *An entire function f belongs to the Laguerre–Pólya class if it has a representation of the form*

$$f(z) = cz^m e^{-az^2+bz} \prod_{\nu=1}^{N} \left(1 + \frac{z}{x_\nu}\right) e^{-\lambda z/x_\nu},$$

where $c \in \mathbb{R} \setminus \{0\}$, m is a non-negative integer, $a, b \in \mathbb{R}$, $a > 0$, $N \in \mathbb{N}_0 \cup \{\infty\}$, $\lambda \in \{0,1\}$, and $x_\nu \in \mathbb{R} \setminus \{0\}$ with $\sum_{\nu=1}^{N} |x_\nu|^{-\lambda-1} < \infty$.

Within the Laguerre–Pólya class those functions f with $\lambda = 0$, $a = 0$, $b > 0$, and $x_\nu > 0$ ($\nu = 1, \ldots, N$) are said to be of type I.

The functions of the Laguerre–Pólya class are exactly those which are limits of sequences of polynomials having only real zeros; the subclass of functions of type I consists exactly of those which are limits of sequences of polynomials having only negative zeros. For the approximation of f by polynomials with real or negative zeros it is important to recall that

$$\left(1 + \frac{\xi z^k}{n}\right)^n \longrightarrow e^{\xi z^k} \qquad \text{as} \quad n \to \infty,$$

uniformly on compact subsets of $\mathbb{C}$. When $k = 1$ and $\xi > 0$, the polynomial on the left-hand side has only one (multiple) zero, which is negative and tends to $-\infty$; when $k = 2$ and $\xi < 0$, it has two distinct real zeros which tend to $\pm\infty$. Therefore Theorem 1 implies the following statement.

Corollary 1. *Let $\mathcal{U}$ be one of the sets (2.2). Suppose that f belongs to the Laguerre–Pólya class and has all its zeros in $\mathcal{U}$. Then the Bernstein polynomials $B_n[f]$ of positive degree have all their zeros in $\mathcal{U}$.*

Suppose that f is of type I in the Laguerre–Pólya class. Then the Bernstein polynomials $B_n[f]$ of positive degree have all their zeros in $(-\infty, 0]$.

Remark 2: It should be mentioned that Theorem 1 does not give a complete list of all Bernstein invariant subsets of $\mathbb{R}$. This can be seen as follows. Let $\mathcal{U}$ be any of the sets (2.1)–(2.2), and denote by $\mathbb{A}$ the field of algebraic numbers. If f has all its zeros in $\mathcal{U} \cap \mathbb{A}$, then, by Theorem 1, $B_n[f]$ has all its zeros in $\mathcal{U}$. Besides, since the zeros of f are algebraic numbers, the coefficients of $B_n[f]$ are seen to be also algebraic numbers. But the field of algebraic numbers is algebraically complete. Hence the zeros of $B_n[f]$ are again algebraic numbers, and so $\mathcal{U} \cap \mathbb{A}$ is another Bernstein invariant set.

The set $\mathcal{U} \cap \mathbb{A}$ is totally disconnected, but is dense in $\mathcal{U}$. Apart from sets of such a type, there are no other Bernstein invariant subsets of $\mathbb{R}$ than those of Theorem 1.

Theorem 2. *Let $\mathcal{U}$ be a Bernstein invariant subset of $\mathbb{R}$. Then $\mathcal{U}$ is dense in one of the sets (2.1)–(2.2). Furthermore, if $\mathcal{U}$ is not totally disconnected, then $\mathcal{U}$ is one of the sets (2.1)–(2.2).*

An interesting problem would be to study the location of the zeros of $B_n[f]$ under the assumption that only some of the zeros of f lie in a Bernstein invariant set. The following theorem is a result in that direction.

Theorem 3. *Let g and h be polynomials with real coefficients. Suppose that h has all its zeros in one of the sets (2.1)–(2.2). Then $B_n[gh]$ cannot have more non-real zeros than $B_n[g]$.*

§3 Proofs

The Möbius transform

$$\varphi : z \longmapsto \frac{z}{z-1} \tag{3.1}$$

is an involution, that is, $\varphi(\varphi(z)) \equiv z$, and maps $\mathbb{C} \setminus \{1\}$ bijectively onto itself. Since

$$B_n[f](z) = (z-1)^n \sum_{\nu=0}^{n} \binom{n}{\nu} (-1)^{n-\nu} f\left(\frac{\nu}{n}\right) \left(\varphi(z)\right)^{\nu},$$

we may associate with $B_n[f]$ the polynomial

$$P_n[f](w) := \sum_{\nu=0}^{n} \binom{n}{\nu} (-1)^{n-\nu} f\left(\frac{\nu}{n}\right) w^{\nu}.$$

Then $z \mapsto \varphi(z)$ achieves a one-to-one correspondence between the zeros of $B_n[f]$ in $\mathbb{C} \setminus \{1\}$ and those of $P_n[f]$ in $\mathbb{C} \setminus \{1\}$. Besides, the point 1 is a k-fold zero of one of the two polynomials if and only if the other polynomial is of degree $n-k$. Hence it is enough to study the zeros of $P_n[f]$.

Next we observe that $P_n[f]$ can be generated from the zeros of

$$f(z) := a_m \prod_{\mu=1}^{m} (z - z_\mu)$$

by a recurrence formula. In fact, setting

$$f_0(z) :\equiv 1, \quad f_j(z) = n^j \prod_{\mu=1}^{j} (z - z_\mu) \quad (j = 1, \ldots, m),$$

we easily find that

$$P_n[f_j](w) = w\left(P_n[f_{j-1}]\right)'(w) - nz_j P_n[f_{j-1}](w).$$

Hence, starting with

$$F_0(w) := P_n[f_0](w) = \sum_{\nu=0}^{n} \binom{n}{\nu}(-1)^{n-\nu}w^{\nu} = (w-1)^n$$

and defining recurrently

$$F_j(w) := wF'_{j-1}(w) - nz_j F_{j-1}(w) \qquad (j = 1, \ldots, m), \tag{3.2}$$

we have $F_j(w) = P_n[f_j](w)$ $(j = 1, \ldots, m)$, and so

$$P_n[f](w) = \frac{a_m}{n^m} F_m(w).$$

Now the question about the location of the zeros of $B_n[f]$ relative to those of f has been reduced to a study of the location of the zeros of F_j in (3.2) relative to z_j and the zeros of F_{j-1}. For this aim, we shall use the following lemma.

Lemma 1. *Let F be a polynomial of degree n with real coefficients having a k-fold zero at 1, ℓ zeros in $(0,1)$, and m zeros in $(1,\infty)$, counted with multiplicities. Denote by k_1, ℓ_1, and m_1 the corresponding numbers associated with the polynomial*

$$F_1(x) := xF'(x) - n\xi F(x).$$

Then the following holds.

If $k \geq 1$ and $\xi < 0$, then $k_1 = k-1$, $\ell_1 \geq \ell + 1$, and $m_1 \geq m$.

If $k \geq 1$ and $\xi > 1$, then $k_1 = k-1$, $\ell_1 \geq \ell$, and $m_1 \geq m+1$.

If $k = 0$ and $\xi < 0$, then $k_1 + \ell_1 + m_1 \geq \ell + m$, $\ell_1 \geq \ell$, and $m_1 \geq m-1$.

If $k = 0$ and $\xi > 1$, then $k_1 + \ell_1 + m_1 \geq \ell + m$, $\ell_1 \geq \ell - 1$, and $m_1 \geq m$.

Proof: Consider $\Phi(x) := x^{-n\xi}F(x)$. Obviously, Φ and F have the same zeros in $(0,\infty)$ even with the same multiplicities. Since

$$\Phi'(x) = x^{-n\xi-1}F_1(x)\,,$$

it is enough to count the zeros of Φ' in the intervals $(0,1)$, $(1,\infty)$, and at 1.

Clearly, every positive j-fold zero of Φ is a $(j-1)$-fold zero of Φ'. Besides, by Rolle's theorem, there is a zero of Φ' between any two consecutive zeros of Φ. Finally, we note that $\Phi(0) = 0$ when $\xi < 0$ and $\lim_{x\to\infty}\Phi(x) = 0$ when $\xi > 1$. Hence, denoting by a the smallest positive zero of Φ and by b the largest, we conclude that Φ' has a zero in $(0,a)$ when $\xi < 0$ and a zero in (b,∞) when $\xi > 1$. With these observations the statement of the lemma is easily verified. ■

Proof of Theorem 1: First we observe that $B_n[f](0) = 0$ if and only if $f(0) = 0$ and $B_n[f](1) = 0$ if and only if $f(1) = 0$. More precisely, if $f(z) = zg(z)$, then

$$B_n[f](z) = zB_{n-1}[g_n](z), \quad \text{where} \quad g_n(z) := g\big((n-1)z/n + 1/n\big)\,,$$

and if $f(z) = (z-1)h(z)$, then

$$B_n[f](z) = (z-1)B_{n-1}[h_n](z), \quad \text{where} \quad h_n(z) = g\big((n-1)z/n\big)\,.$$

Furthermore, if g or h has its zeros in one of the sets (2.1)–(2.2), then g_n or h_n has its zeros in the same set and $g_n(0) \neq 0$ or $h_n(1) \neq 0$, respectively. Therefore it is enough to prove that $(-\infty,0)$, $(1,\infty)$, and $\mathbb{R}\setminus[0,1]$ are Bernstein invariant sets.

Using the recurrence formula (3.2) for the construction of $P_n[f]$ and employing Lemma 1 successively, we conclude that,

(i) if f has all its zeros in $(-\infty,0)$, then $P_n[f]$ has all its zeros in $(0,1]$;

(ii) if f has all its zeros in $(1,\infty)$, then $P_n[f]$ has all its zeros in $[1,\infty)$;

(iii) if f has all its zeros in $\mathbb{R}\setminus[0,1]$, then $P_n[f]$ has all its zeros in $(0,\infty)$.

Now, in order to determine the location of the zeros of $B_n[f]$, it is enough to note that the mapping (3.1) maps $(0,1)$ onto $(-\infty,0)$, $(1,\infty)$ onto itself, and $(0,\infty)$ onto $\mathbb{R}\setminus[0,1]$. ■

Remark 3: The Bernstein operator has a symmetry property, which we shall use to our advantage in the subsequent proof. If f has real coefficients and $g(x) := f(1-x)$, then $B_n[g](x) = B_n[f](1-x)$. Hence to each statement concerning the location of zeros of $B_n[f]$ in a subset of $(-\infty,0]$ there is a corresponding statement with respect to a subset of $[1,\infty)$.

Proof of Theorem 2: Let $\mathcal{U}$ be a Bernstein invariant subset of $\mathbb{R}$, and let $f_\xi(x) := (x-\xi)^2$, where ξ will be chosen appropriately. Then, for $n \geq 2$,

$$B_n[f_\xi](x) = x^2\left(1-\frac{1}{n}\right) - x\left(2\xi - \frac{1}{n}\right) + \xi^2. \tag{3.3}$$

The discriminant of $B_n[f_\xi]$ is

$$D(\xi) := \frac{1}{n}\left(\frac{1}{n} - 4\xi(1-\xi)\right).$$

If $\xi \in (0,1)$, then $D(\xi) < 0$ for large n, which means that $B_n[f_\xi]$ has a pair of conjugate non-real zeros. This shows that $\mathcal{U} \cap (0,1) = \emptyset$.

Now suppose that $\mathcal{V} := (-\infty, 0] \cap \mathcal{U} \neq \emptyset$. Assume that $\mathcal{V}$ is not dense in $(-\infty, 0]$. Then there exists a $\lambda \in (-\infty, 0]$ such that

$$d := \inf\left\{|\lambda - \xi| \,:\, \xi \in \mathcal{V}\right\} > 0. \tag{3.4}$$

Clearly, there is a sequence (ξ_j) in $\mathcal{V}$ converging to a point ξ^* in $(-\infty, 0]$, but not necessarily in $\mathcal{V}$, such that $|\lambda - \xi_j| \to |\lambda - \xi^*| = d$. If $\xi^* = 0$, then we conclude that $\xi^* \in \mathcal{V}$ and $\lambda \neq 0$. Consequently, the Bernstein polynomial $B_n[f](x)$ of $f(x) := x^2$, which is

$$\frac{n-1}{n}\, x\left(x + \frac{1}{n-1}\right) \qquad (n \geq 2),$$

must have its zeros in $\mathcal{V}$. But

$$\left|\lambda + \frac{1}{n-1}\right| < |\lambda| = d$$

for large n. A contradiction!

It remains to consider the case that $\xi^* \in (-\infty, 0)$. Then the sequence (ξ_j) can be chosen in $(-\infty, 0)$. This implies that the discriminant $D(\xi_j)$ is positive, and so the polynomial (3.3) for $\xi = \xi_j$ has two distinct real zeros. Denoting the smaller one by $x_{n,1}(\xi_j)$ and the other by $x_{n,2}(\xi_j)$, we find by a short reflection that

$$x_{n,1}(\xi_j) < \xi_j < x_{n,2}(\xi_j) < 0.$$

Now, if j is fixed and $n \to \infty$, then $x_{n,1}(\xi_j)$ and $x_{n,2}(\xi_j)$ both tend to ξ_j; but if n is fixed and $j \to \infty$, then $x_{n,1}(\xi_j) \to x_{n,1}(\xi^*)$, $x_{n,2}(\xi_j) \to x_{n,2}(\xi^*)$, and

$$x_{n,1}(\xi^*) < \xi^* < x_{n,2}(\xi^*).$$

With this we conclude that, for appropriately chosen j and n, we have either

$$|\lambda - x_{n,1}(\xi_j)| < d \quad \text{or} \quad |\lambda - x_{n,2}(\xi_j)| < d.$$

This contradicts (3.4). Hence $\mathcal{V}$ is dense in $(-\infty, 0]$. Now Remark 3 allows us to conclude that, in any case, $\mathcal{U}$ is dense in one of the sets (2.1)–(2.2).

If $\mathcal{U}$ is not totally disconnected, then it must contain an interval. Suppose that it contains a subinterval (c, d) of $(-\infty, 0)$ and that (a, b) is the largest open interval contained in $\mathcal{U} \cap (-\infty, 0)$, such that (a, b) contains (c, d). Assume that $a > -\infty$. Then, in the notation of the previous paragraph, $(x_{n,1}(a), x_{n,1}(b))$ is another interval contained in $\mathcal{U}$. Since $x_{n,1}(a) < a$ and $x_{n,1}(b) \to b$ as $n \to \infty$, we find that, when n is sufficiently large, then $(x_{n,1}(a), b)$ is an interval larger than (a, b), but also contained in $\mathcal{U}$. This is a contradiction. Hence a must be $-\infty$. Similarly, we obtain that b must be 0. In view of Remark 3, we infer that, if a connected component of $\mathcal{U}$ contains two points, then that component is one of the sets (2.1).

In order to complete the proof, it remains to show that $\mathcal{U}$ cannot be the union of an interval $\mathcal{U}_1$ of the form (2.1) and a totally disconnected set $\mathcal{U}_2$ such that $\mathcal{U}_1 \cup \mathcal{U}_2$ is dense in one of the sets (2.2). In view of Remark 3, it is enough to consider the case that $\mathcal{U}_1$ contains the interval $(-\infty, 0)$ and $\mathcal{U}_2$ is dense in $[1, \infty)$. Let

$$f(x) := (x - \xi)(x - \eta) \qquad \text{with} \quad \xi \in \mathcal{U}_1 \text{ and } \eta \in \mathcal{U}_2.$$

For $n \geq 2$, the Bernstein polynomial $B_n[f]$ has a zero in $(-\infty, 0)$ and one in $[1, \infty)$. If we keep η fixed but vary ξ continuously, then both zeros of $B_n[f]$ vary continuously. Hence $\mathcal{U}_2$ must contain an interval. A contradiction! ■

Lemma 2. *Let F be a polynomial of degree n with real coefficients. If $\xi \in \mathbb{R} \setminus (0, 1)$, then*

$$F_1(x) := xF'(x) - n\xi F(x)$$

cannot have more non-real zeros than F.

Proof: It is well known [9], V. Abschnitt, Aufgabe 53, that the derivative of a polynomial G with real coefficients cannot have more non-real zeros than G itself. Hence the conclusion of the lemma holds for $\xi = 0$.

Next, for $\xi = 1$, the polynomial F_1 can be generated as follows

$$G(x) := x^n F(1/x), \quad H(x) := G'(x), \quad F_1(x) = -x^{n-1} H(1/x).$$

Again, under these operations, the number of non-real zeros cannot increase. Hence the conclusion holds for $\xi = 1$ as well.

Finally, for $\xi \in \mathbb{R} \setminus [0,1]$, the lemma is an exercise in [9], V. Abschnitt, Aufgabe 66. ■

Proof of Theorem 3: Let $h(z) = c_m \prod_{\mu=1}^{m}(z - z_\mu)$. In order to generate $P_n[gh]$, we may start the recurrence formula (3.2) with $F_0(w) := P_n[g](w)$. Then Lemma 2 assures that F_m cannot have more non-real zeros than F_0. But

$$B_n[g](z) = (z-1)^n F_0\big(\varphi(z)\big)$$

and

$$B_n[gh](z) = \frac{c_m}{n^m}(z-1)^n F_m\big(\varphi(z)\big),$$

with φ defined in (3.1). Since φ maps $\mathbb{C} \setminus \mathbb{R}$ bijectively onto itself, we conclude that $B_n[gh]$ cannot have more non-real zeros than $B_n[g]$. ■

§4 An approach by a transformation method

In a series of papers, Iserles et al. [3], [4], [5] have investigated a transformation problem whose most general form may be formulated as follows.

Let $\{P_\nu\}$ and $\{Q_\nu\}$ be two bases of the linear space of polynomials. Suppose that those polynomials P_ν and Q_ν which are not a constant have only real zeros. Find a pair of intervals I and J with the following property. If all the zeros of $\sum_{\nu=0}^{n} a_\nu P_\nu(z)$ reside in I, then all the zeros of $\sum_{\nu=0}^{n} a_\nu Q_\nu(z)$ reside in J.

The study of the zeros of $B_n[f]$ in terms of those of f can be seen as such a transformation problem. In fact, introducing the finite differences

$$\Delta^0 f\left(\frac{\nu}{n}\right) := f\left(\frac{\nu}{n}\right),$$

$$\Delta^{k+1} f\left(\frac{\nu}{n}\right) := \Delta^k f\left(\frac{\nu+1}{n}\right) - \Delta^k f\left(\frac{\nu}{n}\right) \quad (k = 0, 1, \ldots),$$

we have [8], p. 13,

$$B_n[f](x) = \sum_{\nu=0}^{n} \binom{n}{\nu} \Delta^\nu f(0)\, x^\nu. \tag{4.1}$$

On the other hand, Newton's interpolation formula allows us to represent f as

$$f(x) = \sum_{\nu=0}^{n} \binom{n}{\nu} \Delta^\nu f(0)\, P_{n,\nu}(x) \qquad (n \geq \deg f), \tag{4.2}$$

where $P_{n,0}(x) \equiv 1$ and

$$P_{n,\nu}(x) = \frac{nx(nx-1)\cdots(nx-\nu+1)}{n(n-1)\cdots(n-\nu+1)} \qquad (\nu = 1, 2, \ldots).$$

Hence $B_n[f]$ can be obtained from f by a change of the basis as in the above problem. The following result of Iserles and Nørsett [3], Theorem 9, will be useful.

Theorem B. *Let $(x)_k$ be the Pochhammer symbols, defined by*

$$(x)_0 := 1, \quad (x)_k := (x+k-1)\cdot(x)_{k-1} \quad (k = 1, 2, \ldots).$$

Then

$$\sum_{\nu=0}^{n} a_\nu (x)_\nu \longmapsto \sum_{\nu=0}^{n} a_\nu x^\nu$$

maps polynomials with positive zeros into polynomials with positive zeros.

This result can be applied as follows. Suppose that f, as given by (4.2), has negative zeros. Then

$$g(x) := f\Big(-\frac{x}{n}\Big) = \sum_{\nu=0}^{n} \frac{(-1)^\nu}{\nu!} \Delta^\nu f(0)\,(x)_\nu$$

has positive zeros. By Theorem B, the polynomial

$$\sum_{\nu=0}^{n} \frac{(-1)^\nu}{\nu!} \Delta^\nu f(0)\, x^\nu$$

has positive zeros. A result of Laguerre [7], sec. 18, pp. 30–31, allows to conclude that

$$\sum_{\nu=0}^{n} \frac{(-1)^\nu}{\nu!\,(n-\nu)!} \Delta^\nu f(0)\, x^\nu,$$

also has positive zeros. Hence $B_n[f]$, as given by (4.1), has negative zeros. This is a partial proof for $(-\infty, 0)$ to be Bernstein invariant. Unfortunately, the transformation method does not apply when $n < \deg f$.

References

[1] Gonska H. H. and J. Meier, A bibliography on approximation of functions by Bernstein-type operators, in *Approximation Theory IV*, C. K. Chui, L. L. Schumaker and J. D. Ward (eds.), Academic Press, New York, 1983, 739–785.

[2] Gonska H. H. and J. Meier, A bibliography on approximation of functions by Bernstein-type operators (Supplement 1986), in *Approximation Theory V*, C. K. Chui, L. L. Schumaker and J. D. Ward (eds.), Academic Press, New York, 1986, 621–654.

[3] Iserles, A. and S. P. Nørsett, Zeros of transformed polynomials, *SIAM J. Math. Analysis* **21** (1990), 483–509.

[4] Iserles, A., S. P. Nørsett and E. B. Saff, On transformations and zeros of polynomials, *Rocky Mountain Journal of Mathematics* **21** (1991), 331–357.

[5] Iserles, A. and E. B. Saff, Zeros of expansions in orthogonal polynomials, *Proc. Cambridge Phil. Soc.* **105** (1989), 559–573.

[6] Karlin, S., *Total Positivity,* I, Stanford University Press, Stanford, 1968.

[7] Laguerre, E., *Œuvres de Laguerre* 1, 2nd edn., Chelsea, Bronx, 1972.

[8] Lorentz, G. G., *Bernstein Polynomials,* 2nd edn., Chelsea Publishing Company, New York, 1986.

[9] Pólya, G. and G. Szegö, *Aufgaben und Lehrsätze aus der Analysis,* 2, 4th edn., Springer-Verlag, Berlin, 1971.

[10] Schoenberg, I. J., On variation diminishing approximation methods, in *On Numerical Approximation*, R. E. Langer (ed.), The University of Wisconsin Press, Madison, 1959, 249–274.

[11] Stark, E. L., Bernstein polynomials 1912–1955, in *Functional Analysis and Approximation*, Proc. Conf. Math. Res. Oberwolfach 1980, P. L. Butzer, B. Sz.-Nagy and E. Görlich (eds.), Birkhäuser, Basel, 1981, 443–461.

Gerhard Schmeisser
Mathematisches Institut
Universität Erlangen Nürnberg
D-91054 Erlangen
GERMANY
schmeisser@mi.uni-erlangen.de

Location of critical points for polynomials

Gerald Schmieder

Abstract. Let some complex polynomial be given in terms of its zeros. We estimate the location of the zeros of the derivative.

§1 Introduction

The zeros of the derivative of a complex polynomial p are functions of the zeros of p itself. In general we do not know explicit expressions for these functions. So we try to locate the derivative-zeros geometrically in terms of the given zeros of p.

As a question of this type we mention the famous Sendov Conjecture [2] which is about 30 years old and still open. This conjecture states:

Conjecture 1 (Sendov Conjecture). *Let $p \in \mathbb{C}[z]$ be some polynomial of degree $n > 1$ having all its zeros $z_1, \ldots, z_n$ in the closed unit disk. Does there exist to every z_j some w_j with $|z_j - w_j| \leq 1$ and $p'(w_j) = 0$?*

This conjecture has been confirmed in several special cases (for references see [1]).

Here we shall discuss the problem of locating the zeros of the derivative for general complex polynomials, without restrictions to the zeros. At the end we consider the results in view of the Sendov Conjecture. We observe two different heuristic principles for the distribution of the derivative-zeros:

Zeros of p' prefer the neighborhood of:

- a cluster of zeros of p, as well as
- a lone zero of p.

The second principle can be obtained by moving some zero of p to ∞. Then also some derivative-zero has to disappear because the degree of the resulting polynomial is one less than that of p. The first principle follows from the fact that zeros of order ≥ 2 are zeros of the derivative, combined with an obvious continuity argument.

It is clear that these considerations are only of qualitative nature and contain no quantitative statement about the distribution of the zeros of the derivative. The aim of this paper is to give more precise descriptions and consequences of the mentioned principles. A crucial tool for our considerations is the following generalized version of Rouché's Theorem:

Computational Methods and Function Theory (CMFT'97)
N. Papamichael, St. Ruscheweyh and E. B. Saff (Eds.), pp. 497–504

ISBN 981-02-3626-3

Theorem 1 (Rouché). *Let f and g be meromorphic functions defined on the domain $G \subset \mathbb{C}$. We assume the cycle Γ to be homologous to zero in G. Let $n(z,\Gamma)$ denote the winding number of Γ with respect to z and $0 \le n(z,\Gamma) \le 1$ be fulfilled for all $z \in G \setminus \Gamma$. The strict inequality*

$$|f(z)+g(z)| < |f(z)| + |g(z)|$$

should be satisfied everywhere on Γ (which implies that Γ meets neither a zero nor a pole of f or g). Then $N(f) - P(f) = N(g) - P(g)$, where $N(h)$ resp. $P(h)$ stand for the number of zeros resp. poles (counting multiplicity) of the meromorphic function $h : G \to \overline{\mathbb{C}}$ in the set $\{z \in G : n(z,\Gamma) = 1\}$.

This theorem is a consequence of the Argument Principle applied to the meromorphic function f/g. The conclusion follows immediately if the cycle $\Theta := (f/g) \circ \Gamma$ has winding number 0 with respect to the origin. Therefore it suffices to obtain some half-line from 0 to ∞ which does not meet Θ. Exactly this is guaranteed by the provided inequality, where the half-line is taken as the positive real axis.

§2 Strip conditions

We start with:

Lemma 1. *Let $p(z) = \prod_{j=1}^{n}(z - z_j)$ be a complex polynomial of degree $n > 1$ and let some point $a \in \mathbb{C}$ and some real number $r > 0$ be given. Assume that the circle $|z - a| = r$ contains neither a zero of p nor of p'. Furthermore let $\operatorname{Re}(a - z_j) > r$ be fulfilled for (at least) $j = \mu+1, \ldots, n$ and assume that the inequality*

$$\sum_{j=\mu+1}^{n} \frac{\operatorname{Re}(a - z_j) - r}{|a - z_j|^2 - r^2} \ge \max_{|z-a|=r} \left| \sum_{k=1}^{\mu} \frac{1}{z - z_k} \right| \tag{2.1}$$

holds. If M denotes the total number of $k \in \{1, \ldots, n\}$ with $|a - z_k| < r$, then there are exactly M zeros of p' in the disk $|z - a| < r$ (counting multiplicity).

Proof: Let some complex number a be fixed. Each function

$$f_j(t) := \frac{1}{re^{it} + a - z_j}, \qquad t \in [0, 2\pi[, \quad j \in \{\mu+1, \ldots, n\}$$

describes a circular line in the right halfplane. Therefore we have a lower estimate for $\left|\sum_{j=\mu+1}^{n} f_j(t)\right|$ by $\sum_{j=\mu+1}^{n} \min_{0 \le t < 2\pi} \operatorname{Re} f_j(t)$. The logarithmic derivative $g(z)$ of the polynomial

$$q_\mu(z) = \prod_{j=\mu+1}^{n} (z - z_j)$$

has neither a pole nor (as follows from the Gauß-Lucas-Theorem [3]) a zero in $|z - a| \le r$. We may estimate as follows:

$$\begin{aligned} |g(a + re^{it})| &\ge \sum_{j=\mu+1}^{n} \min_{0 \le t < 2\pi} \operatorname{Re} \frac{1}{re^{it} + a - z_j} \\ &= \sum_{j=\mu+1}^{n} \frac{\operatorname{Re}(a - z_j) - r}{|a - z_j|^2 - r^2}, \qquad 0 \le t \le 2\pi. \end{aligned}$$

If the inequality (2.1) is fulfilled, we can apply Rouché's Theorem with g as above and $f := -p'/p$ (note that $|z - a| = r$ contains no zero of p' by assumption). If M stands for the total number of $k \in \{1, \dots, n\}$ with $|a - z_k| < r$, Rouché's Theorem shows that there are exactly M zeros of p' in $|z - a| < r$. This completes the proof. ∎

Now we establish a sufficient condition for inequality (2.1), which is much easier to handle. First we define

$$\theta := \frac{1}{n - \mu} \sum_{j=\mu+1}^{n} \operatorname{Re}(a - z_j)$$

and

$$d := \max_{j=\mu+1,\dots,n} |a - z_j|.$$

It can be easily seen that the inequality

$$\frac{n - \mu}{\mu} \frac{\theta - r}{d^2 - r^2} \ge \frac{1}{||a - z_k| - r|}, \qquad k = 1, \dots, \mu,$$

has the desired properties. Thus we have:

Theorem 2. *Let $p(z) = \prod_{j=1}^{n}(z - z_j)$ be a complex polynomial of degree $n > 1$ and let some point $a \in \mathbb{C}$ and some real number $r > 0$ be given. If*

$$|a - z_k| < r - \frac{\mu}{n - \mu} \frac{d^2 - r^2}{\theta - r}, \qquad k = 1, \dots, \mu, \tag{2.2}$$

and

$$\operatorname{Re}(a - z_j) > r, \qquad j = \mu + 1, \dots, n,$$

is fulfilled, then exactly μ zeros of p' (counting multiplicity) are in the disk $|z - a| < r$.

§3 Disk conditions

The main result of this section (Theorem 3) can also be derived from Walsh's two-circle Theorem [3], which is a consequence of the Grace Apolarity Theorem. Therefore it may be of interest to give another proof by completely different methods.

Lemma 2. *Let $p(z) = \prod_{j=1}^{n}(z - z_j)$ be a complex polynomial of degree $n > 1$. We choose some point $a \in \mathbb{C}$ and some real number $r > 0$ such that the circle $|z - a| = r$ contains no zero of p. For all $j = 1, \ldots, m$ we assume $|a - z_j| < r$ and moreover*

$$\left| \sum_{k=m+1}^{n} \frac{1}{\zeta - z_k} \right| < \left| \sum_{j=1}^{m} \frac{1}{\zeta - z_j} \right| + \left| \sum_{\nu=1}^{n} \frac{1}{\zeta - z_\nu} \right| \tag{3.1}$$

for all ζ which fulfill $|\zeta - a| = r$. Let $M\,(\geq m)$ be the total number of $j \in \{1, \ldots, n\}$ with $|a - z_j| < r$. Then the disk $K_r(a) = \{z : |z - a| \leq r\}$ contains exactly $M - 1$ zeros of p' (counting multiplicity). The boundary $\partial K_r(a)$ contains no zero of p'.

Proof: We have

$$\frac{p'}{p}(z) = \sum_{k=1}^{m} \frac{1}{z - z_k} + \sum_{k=m+1}^{n} \frac{1}{z - z_k}.$$

From the well-known Gauß-Lucas-Theorem [3] we obtain that the rational function

$$L(z) := \sum_{j=1}^{m} \frac{1}{z - z_j}$$

as the logarithmic derivative of the polynomial

$$p_m(z) := (z - z_1)(z - z_2) \ldots (z - z_m)$$

has exactly $m - 1$ zeros (counting multiplicity) in the disk $K_r(a)$. Assumption (3.1) allows to apply Rouché's Theorem with

$$f := \frac{p'}{p}, \quad g := -\frac{p'_m}{p_m}, \qquad G = \mathbb{C}, \quad \Gamma = \partial K_r(a),$$

(from (3.1) we obtain that there is neither a zero of p' nor of p'_m on the circle $|z - a| = r$) and therefore we have

$$N(p') - N(p) = N(p'_m) - N(p_m),$$

where N stands for the number of zeros in $K_r(a)$ respectively. This gives the desired conclusion. ■

We maintain the notations of the preceding proof and consider

$$|g(a+re^{it})| = \left|\sum_{j=1}^{m} \frac{1}{r+e^{i(-t+\omega_j)}|a-z_j|}\right|$$

where $\omega_j := \arg(a-z_j)$ and $t \in [0, 2\pi[$. The image of each function

$$g_j(t) = \frac{1}{r+e^{i(-t+\omega_j)}|a-z_j|}, \qquad j = 1, \ldots, m,$$

describes some segment of a circle, in the right half-plane, centered at a real-positive number. This leads to a lower estimate of $|g(a+re^{it})|$ by

$$|g(a+re^{it})| \geq \sum_{j=1}^{m} \frac{1}{r+|a-z_j|} \geq \frac{m}{r+r}$$

Note that

$$\left|\sum_{\nu=1}^{n} \frac{1}{\zeta - z_\nu}\right| = \left|\frac{p'}{p}(\zeta)\right| > 0$$

for $\zeta = a+re^{it}$ in the case that $|z-a| = r$ contains neither a zero of p nor of p'. Thus (under this assumption) we obtain the following inequality which is sufficient for (3.1):

$$\frac{m}{(n-m)2r} \geq \frac{1}{||a-z_j|-r|}, \qquad j = m+1, \ldots, n.$$

Our result allows a simpler formulation if we omit the assumption that $|z-a| = r$ contains neither a zero of p nor a zero of p'. This can be arranged as follows: Our considerations are valid for all $\rho < r$ with the property that the zeros of p and of p' inside the disks $|z-a| < \rho$ and $|z-a| < r$ are the same. Then the circle $|z-a| = \rho$ contains neither a zero of p nor of p'. Now $\rho \to r$ gives the conclusion. Thus we have proved:

Theorem 3. *Let $p(z) = \prod_{j=1}^{n}(z-z_j)$ be a complex polynomial of degree $n > 1$ and some point $a \in \mathbb{C}$ as well as some $r > 0$ be given. We assume that $|a-z_j| < r$ holds for $j = 1, \ldots, m$. For all $j = m+1, \ldots, n$ let the estimate*

$$|a-z_j| \geq r\left(1 + 2 \cdot \frac{n-m}{m}\right)$$

be fulfilled. Then there are exactly $m-1$ (counting multiplicity) zeros of p' in the disk $|z-a| < r$.

Now we apply this result in the case that the zeros z_1 and z_2 of p are close. For this purpose we take $a = \frac{z_1+z_2}{2}$ and $m = 2$. Our Theorem 3 gives the following. If

$$\frac{2}{n-1}\left|\frac{z_1+z_2}{2} - z_j\right| > |z_1 - z_2|$$

holds for all $j = 3, \ldots, n$ then the disk $\left|z - \frac{z_1+z_2}{2}\right| < r$ contains exactly one zero of p' whenever

$$\left|\frac{z_1-z_2}{2}\right| < r \leq \frac{1}{n-1} \min_{3\leq j\leq n} \left|\frac{z_1+z_2}{2} - z_j\right|$$

We conclude that there is exactly one zero of p' in the disk[1]

$$\left|z - \frac{z_1+z_2}{2}\right| \leq \left|\frac{z_1-z_2}{2}\right|.$$

Similar calculations give that there is *no* zero of p' in the disk

$$|z - z_1| < \frac{1}{2n-1} \min_{2\leq j\leq n} |z_1 - z_j|$$

(which is empty in the case $z_1 = z_2$). But we can improve this result by use of (3.1) instead of Theorem 3. We obtain that there is actually *no* zero of p' in the disk

$$|z - z_1| < \frac{1}{n} \min_{2\leq j\leq n} |z_1 - z_j|.$$

§4 An application of Theorem 2

Now we give an application of Theorem 2 in view of the Sendov Conjecture.

Let p be a complex polynomial of degree $n \geq 4$ with zeros $z_1, \ldots, z_n$ in $K_1(0) = \{z \in \mathbb{C} : |z| \leq 1\}$. Without loss of generality we assume $0 \leq z_1 \leq 1$. We will apply Theorem 2 with the data $a = z_1$ and $\mu = 1$. Let some $R > 0$ be given such that $\operatorname{Re} z_j < a - R$ for all $j \geq 2$. Then no zero of p is an inner point of the strip

$$S = \{z \in \mathbb{C} : a - R \leq \operatorname{Re} z \leq a\}.$$

We define $r := \lambda \cdot R$ where $0 < \lambda < 1$ is treated at first as a free parameter. By an obvious continuity argument we may assume that no zero of p' lies on the circle $|z - a| = r$ (otherwise we take for the following consideration the

[1]The Grace-Heawood Theorem [4] gives *some* zero of p' in the closed disk centered at $\frac{z_1+z_2}{2}$ with radius $\frac{|z_1-z_2|}{2} \cot \frac{\pi}{n}$.

number r suitably smaller or larger). Then the first part of the assumption of Theorem 2 is fulfilled. For d and θ from Theorem 2 we have the estimates (using $|z_j| \leq 1$)

$$d \leq 1 + a \leq 2 \quad , \qquad \theta \geq R.$$

Then (2.2) becomes

$$0 = |a - z_1| \leq \lambda R - \frac{1}{n-1} \frac{4 - \lambda^2 R^2}{R(1-\lambda)}.$$

The right side is non-negative for $\lambda_0 \leq \lambda \leq \lambda_1$ where

$$\lambda_{0/1} := \frac{n-1}{2(n-2)} \left(1 \pm \sqrt{1 - \frac{16(n-2)}{R^2(n-1)^2}} \right),$$

provided that (in order to get real roots):

$$R \geq \frac{4}{n-1} \sqrt{n-2} =: R_0. \tag{4.1}$$

It is easy to verify that $\lambda_0 > 0$ and $\lambda_1 < 1$ (using $n \geq 4$ and $R < 2$). From Theorem 2 we obtain: *Under the assumption of (4.1) every disk $K_r(a)$ with $r = \lambda \cdot \rho$, $\lambda_0 < \lambda < \lambda_1$ and $R_0 \leq \rho \leq R$ contains exactly one zero of p'.* We summarize:

Theorem 4. *Let $z_1, \ldots, z_n \in K_1(0)$ be the zeros of the complex polynomial $p(z)$ of degree $n \geq 4$ and let $z_1 > 0$. For all $j > 1$ we assume* $\operatorname{Re} z_j \leq z_1 - R$, *where $R \geq \frac{4\sqrt{n-2}}{n-1}$. Then the closed disk centered at z_1 with radius*

$$r(n) := \frac{2}{\sqrt{n-2}}$$

contains exactly one zero of p'. Moreover, the annulus

$$\frac{2}{\sqrt{n-2}} < |z - z_1| < \frac{R(n-1)}{2(n-2)} \left(1 + \sqrt{1 - \frac{16(n-2)}{R^2(n-1)^2}} \right)$$

is free of zeros of p'.

References

[1] Pflug, P. and G. Schmieder, Remarks on the Ilieff-Sendov problem, *Ann. UMCS*, **48** (1994), 98–105.

[2] Hayman, W. K., *Research Problems in Function Theory*, Athlone Press, University of London, 1967.

[3] Marden, M., *The Geometry of the Zeros of a Polynomial in a Complex Variable*, Amer. Math. Soc. (Math. Surveys), New York, 1949.

[4] Szegő, G., Bemerkungen zu einem Satz von J.H. Grace über die Wurzeln algebraischer Gleichungen, *Math. Z.* **13** (1922), 28–55.

Gerald Schmieder
Fachbereich Mathematik
Universität Oldenburg
Postfach 2503
D-26111 Oldenburg
GERMANY
schmieder@mathematik.uni-oldenburg.de

Conformal maps via Sinc methods

Frank Stenger

and Ross Schmidtlein

Abstract. In this paper we derive a *Sinc procedure* for the construction of a conformal map, f, of a simply connected domain, or Riemann surface B in the complex plane to the unit disc U. The construction is based on the solution of a boundary integral equation which always has a unique solution. We assume that ∂B, the boundary of B, consists of a finite number of analytic arcs, with well defined angles at the junctions. We also give an explicit procedure for evaluating f in the interior of B. We furthermore give a brief description of an explicit *Sinc construction* which enables the computation of the inverse map $F = f^{-1}$ from U to B, based on the computed f on ∂B. Given any $\varepsilon > 0$, the time complexity of sequential computation of f_ε on ∂B such that $\sup_{\zeta \in \partial B} |f(\zeta) - f_\varepsilon(\zeta)| < \varepsilon$ is $\mathcal{O}\left((\log(\varepsilon))^6\right)$.

§1 Introduction and summary

This paper deals with the construction of a conformal map, f, from a simply connected domain or Riemann surface $B \subset \mathbb{C}$ to the unit disc U in the complex plane $\mathbb{C}$, as well as the construction of the inverse map, F, from U to B, under the assumption that the boundary of B consists of a finite number of analytic arcs that have well defined angles of intersection at their junctions. Assuming that $f(0) = 0$, we define $\mathcal{G}$ in B by $f(z) = z \exp\{\mathcal{G}(z)\}$, and we then construct an approximation to $\mathcal{G}$ via the solution of a linear boundary integral equation on ∂B, the boundary of B. The boundary integral equation that we use for the determination of $\mathcal{G}$ on ∂B is not new; it is, in fact, a double layer potential boundary integral equation for solving Dirichlet's problem in B. We also describe a highly efficient procedure for computing f and F in the interior of their domains of analyticity once they are known on the boundary. The complexity of our method, *i.e.* the amount of work required to construct an approximation f_ε to f, such that $\sup_{\zeta \in \partial B} |f(\zeta) - f_\varepsilon(\zeta)| \leq \varepsilon$ is $\mathcal{O}(\log(\varepsilon)^6)$.

Computational Methods and Function Theory (CMFT'97)
N. Papamichael, St. Ruscheweyh and E. B. Saff (Eds.), pp. 505–549

ISBN 981-02-3626-3

1.1 Perspective

In general, there are two classes of integral equation methods for constructing conformal maps: linear boundary integral equation methods based on constructing f, such as those of the present paper, as well as those in the excellent publications [30], [31], [21], [17], [2], and (nonlinear) methods based on the construction of F (see [11], [10], [32], [7], [1], [34], [36], [37], [20], [8], [9]). Numerical computations based on the Schwarz–Christoffel approach are also nonlinear (see [35], [33]), in that they need the solution of a system of nonlinear algebraic equations for the required parameters. For many applications, both f and F are required, and for this reason, the present paper describes a procedure for constructing both.

Let us cite some well known methods of constructing conformal maps, in order to get a better perspective of how the method of this paper relates to these.

To date, there are two classes of boundary integral equation methods for the conformal map f of B onto U:

1. *Linear integral equations for Θ', the derivative of Θ, where $f = |f|e^{i\Theta}$.* Included among these from the standpoint of numerical computation and under our assumptions on ∂B are Warschawski's equation ([16]: §16.7), and Berrut's equation [2], [3]. Warschawski's and Berrut's equations are linear equations of the second kind, that have bounded kernels. Both take the form $(I + K)\,\Theta' = g$, with different operators K and with different right hand sides g, both have unique inverses. The $K = K_W$ of Warschawski's kernel (discussed in §2.2 below) having an eigenvalue 1, which is the eigenvalue of largest modulus, while other eigenvalues of K_W are less than 1 in absolute value. Moreover, numerical evidence has shown that the eigenvalues of Berrut's kernel $K = K_B$ appear to be less than 1 in absolute value, so that Berrut's equation appears to have an advantage over Warschawski's for solution by successive approximation. Both equations yield excellent results when the boundary of B is smooth, but the accuracy of approximation is poor when the region has corners. Furthermore, Θ' is unbounded at re-entrant corners.

   Symm's equation [30], [11] is a first kind integral equation for Θ', with unbounded logarithmic kernel; it does not always have a solution [11]. However, Hough [18] has derived an excellent numerical algorithm based on Symm's equation under assumptions on ∂B that are similar to ours. We shall make some comparisons of Hough's and our method in §4.4 of this paper.

2. *Linear, second kind integral equations for* Θ, such as Lichtenstein's equation ([10]: Ch. 1, §2) and Gersgorin's ([16]: §16.7); each of these has the form $(I - K)\,\Theta = g_2$, with K as for Warschawski's equation, so that $(I - K)$ does not have an inverse, which also gives rise to numerical problems.

The equation which we shall base our construction on has the form $(I+K)u = g$, with K as for Warschawski's equation. We show, moreover, that this equation always has a unique and bounded solution.

Theodorsen [11] and Wegmann [36], [37], [38] have derived nonlinear boundary integral equation procedures for approximating the map F of U onto B. Of these, the FFT procedure of Wegmann is the more efficient, since it requires $\mathcal{O}(n \log(n))$ operations to construct a Fourier polynomial approximation of degree n. If the boundary of B is an analytic contour, *i.e.* if F is analytic in $\{w \in \mathbb{C} : |w| < R\}$, with $R > 1$, then the error of approximation is $\mathcal{O}(R^{-n})$, and thus the complexity of Wegmann's method is $\mathcal{O}(\log(1/\varepsilon) \log\log(1/\varepsilon))$, whereas the complexity reduces to $\mathcal{O}(\varepsilon^{-\alpha} \log(1/\varepsilon))$ in the case when ∂B has a corner with interior angle π/α.

The above complexity results show that it is difficult to compare the methods cited above either among themselves or with ours, since they all excel in different situations. We also remark that whereas Berrut's and Wegmann's methods do not produce accurate approximations in the neighborhoods of corners, they nevertheless are quite accurate away from the corners, so that Berrut's, or Wegmann's approximations may be satisfactory for many users, even when ∂B has corners.

Numerical computations based on the Schwarz–Christoffel (S–C) approach to construct a map of the unit disc U to the interior of a polygonal region are also nonlinear (see [19], [35], [33]), in that they need the solution of a system of nonlinear algebraic equations for the required parameters. Programs for these methods have been shown to be very efficient.

Although the present paper gives an explicit description for the construction of f when ∂B consists of an arbitrary number of analytic arcs, the complexity of construction of the method of this paper as a function of the number of arcs n is $\mathcal{O}(n^3)$, which grows rapidly as a function of n. We thus expect that future research combining the work of this paper with the excellent domain splitting approach of [12], [13], [24], [25] (with n relatively small on each sub-domain) would make it possible to make the method of this paper more efficient for large n.

1.2 Layout of the paper

In this paper we present the following:

- At the outset, in §1.3 below, we state explicit assumptions about the make up of the boundary of B, which consists of a finite number of analytic arcs.

- In §2.1 we present some definitions and properties of Cauchy type integrals, which we shall require in this paper.

- In §2.2 we derive two integral equations for conformal maps of B onto U: (i) Lichtenstein's equation, and (ii) the integral equation on which the construction of this paper is based.

- In §3.1 we deduce some requisite boundary behaviors of the map f, to enable us to identify the spaces of analytic functions that house f.

- In §3.2 we deduce the boundary behavior and the spaces of analytic functions for the inverse map F.

- In §4.1 we give an explicit algorithm based on *Sinc approximation* for the construction of an approximation f_ε to f on the boundary of B, with $\sup_{\zeta \in B} |f(\zeta) - f_\varepsilon(\zeta)| < \varepsilon$.

- In §4.2 we give an explicit algorithm for approximating f in the interior of B, using the computed values of f at the *Sinc points* of the arcs Γ_j on the boundary of B.

- In §4.3 we sketch a procedure for the construction of the inverse map F by means of the forward map, based on determining the required values of F on ∂U using the computed values of f on ∂B.

- In §4.4 we give two explicit examples of our procedure, by constructing maps of B onto U, with B a wedge of a disc and a "pac–man".

- In §5 we give a proof of convergence of the algorithm described in §4.1.

- The content of §6 is an appendix, in which we present a summary of methods of *Sinc approximation* that we require in earlier sections of the paper.

1.3 The boundary of B

Let us make some explicit assumptions about the boundary of B. We use the notation "arg(a)" to denote the principal value of the argument of a complex number a, *i.e.* $-\pi < \arg(a) \leq \pi$, whereas "Arg(a)" is a continuous on $\mathbb{R}$, which may take on values outside of this range.

Assumption I: Assumptions on the Boundary of B: *An* arc Γ *is here defined by a mapping* $\zeta(\tau) = \xi(\tau) + i\,\eta(\tau)$, *with* ξ *and* η *real valued functions that are continuous on* $[0,1]$, *such that the derivative* $\dot{\zeta}(\tau) = \dot{\xi}(\tau) + i\,\dot{\eta}(\tau)$ *exists and is continuous on* $[0,1]$ *and such that* $\dot{\xi}(\tau)^2 + \dot{\eta}(\tau)^2 > 0$ *for all* $\tau \in [0,1]$. *For some positive integer* n, a contour *is here defined to be the union of* n *arcs* $\Gamma_j = \{\zeta_j(\tau) : 0 \leq \tau \leq 1\}$ *such that* $\zeta_j(1) = \zeta_{j+1}(0)$, *for* $j = 1, 2, \ldots, n$, *and where we make the identification* $\zeta_{n+1} \equiv \zeta_1$.

We make the following assumptions about ∂B, *the boundary of* B:

1. *We assume that* ∂B *does not intersect itself, although we do allow* B *to be a Riemann surface, and that* ∂B *can be represented as a union of* analytic arcs, *in the form*

$$\partial B = \bigcup_{j=1}^{n} \Gamma_j, \tag{1.3.1}$$

where we encounter the arcs $\Gamma_1, \Gamma_2, \ldots, \Gamma_n$ *consecutively as we traverse* ∂B *in a counterclockwise manner.*

2. *We assume that we are given an explicit function* ζ_j *for each* $j = 1, 2, \ldots, n$, *such that*

$$\Gamma_j = \{\zeta_j(\tau) : 0 \leq \tau \leq 1\}, \tag{1.3.2}$$

with each $\zeta_j \in \mathbf{H}^\infty(\mathcal{E}_d)$, *where, for some* $d > 0$ *and independent*[1] *of* j, $\mathcal{E}_d$ *is the lens shaped region [28], pp. 67–68,*

$$\mathcal{E}_d \equiv \left\{\tau \in \mathbb{C} : \left|\arg\left(\frac{\tau}{1-\tau}\right)\right| < d\right\}, \tag{1.3.3}$$

and where $\mathbf{H}^\infty(\mathcal{E}_d)$ *denotes the* Hardy space *of all functions that* are *analytic and bounded in* $\mathcal{E}_d$.

[1]For simplicity, we have taken the same value "d" for each arc. A more efficient algorithm may result by using a more optimal value of d for each arc.

We remark here, that whereas each arc Γ_j was originally defined as a function of τ on the interval $[0, 1]$, we will now consider the functions ζ_j as analytic functions of the complex variable τ.

In addition, it will be assumed that there are points $a_0, a_1, \ldots, a_n$, with $a_n = a_0$, which one encounters consecutively as one traverses ∂B in a counterclockwise manner, and that, as τ increases from 0 to 1, $\zeta_j(\tau)$ traverses Γ_j from $a_{j-1} = \zeta_j(0)$ to $a_j = \zeta_j(1)$, for $j = 1, \ldots, n$, with (as stated above) $\zeta_{n+1} \equiv \zeta_1$.

3. *We assume that for $j = 1, 2, \ldots, n$, ζ_j provides a conformal map of the domain $\mathcal{E}_d$ in (1.3) onto a simply connected domain D_j, and that the inverse map, $\zeta_j^{[-1]}$, can be explicitly expressed*[2]*. It will thus be assumed that $\dot{\zeta}$ is uniformly bounded and non–vanishing in $\overline{\mathcal{E}_d}$.*

4. *We assume that the size of the interior angle of intersection π/β_j' of the arcs Γ_j and Γ_{j+1} at the point a_j, namely*

$$\frac{\pi}{\beta_j'} = \pi + \operatorname{Arg}\left(\frac{\dot{\zeta}_{j+1}(0)}{\dot{\zeta}_j(1)}\right), \tag{1.3.4}$$

satisfies the relation $0 < \pi/\beta_j' < \infty$.

Remark:

- Under #3 of the above assumptions, it is possible to give an explicit expression for a function φ_j, that maps D_j conformally onto the domain

$$\mathcal{D}_d = \{w \in \mathbb{C} : |\operatorname{Im} w| < d\} \tag{1.3.5}$$

with d the fixed positive number referred to above. Indeed, we have the following identities:

[2] In practice it will suffice to be able to compute $\zeta_j^{[-1]}(z)$ to desired accuracy for any given $z \in \Gamma_j$.

$$
\begin{aligned}
z = \zeta_j(\tau) &\iff \tau = \zeta_j^{[-1]}(z) \\
w = \log\left(\frac{\tau}{1-\tau}\right) &\iff \tau = \frac{e^w}{1+e^w} \\
\varphi_j(z) = \log\left(\frac{\zeta_j^{[-1]}(z)}{1-\zeta_j^{[-1]}(z)}\right) &\iff \varphi_j^{[-1]}(w) = \zeta_j\left(\frac{e^w}{1+e^w}\right) \\
\frac{1}{\varphi_j'(z)} &= \zeta_j'\left(\frac{e^w}{1+e^w}\right)\frac{e^w}{(1+e^w)^2}.
\end{aligned}
\tag{1.3.6}
$$

- We note that for the case of a slit at one or more of the points a_j, the interior angle of the map at these points is 2π, so that $\beta_j' = 1/2$. However, we also allow $0 < \beta_j' < 1/2$, in which case we have a Riemann surface emanating from such branch points a_j.

§2 Integral equations

In this section we review some properties of Cauchy and Hilbert transforms, and we derive some integral equations for constructing the requisite conformal map f.

2.1 Some properties of Hilbert and related transforms

Let us establish some standard notations and definitions, which will be convenient throughout this paper.

We use the notation

$$
\mathcal{S}g(\zeta) = \frac{P.V.}{\pi i}\int_\Gamma \frac{g(t)}{t-\zeta}\,dt, \quad \zeta \in \Gamma, \tag{2.1.1}
$$

for the *Hilbert transform* of g over the arc or contour Γ.

Let Φ be defined for $z \in \mathbb{C}$ and for a given function g that is Hölder continuous on Γ by the equation

$$
\Phi(z) = \frac{1}{2\pi i}\int_\Gamma \frac{g(t)}{t-z}\,dt, \quad z \notin \Gamma. \tag{2.1.2}
$$

Let $\Phi^+(\zeta)$ (resp., $\Phi^-(\zeta)$) denote the (non–tangential) limit as $z \to \zeta$ from the right hand side (resp., left hand side) of Γ. Then, the *Sokhotski formulas* [14]

$$\begin{aligned} \Phi^+(\zeta) &= \frac{1}{2}g(\zeta) + \frac{1}{2}\mathcal{S}g(\zeta) \\ \Phi^-(\zeta) &= -\frac{1}{2}g(\zeta) + \frac{1}{2}\mathcal{S}g(\zeta) \end{aligned} \tag{2.1.3}$$

are satisfied whenever ζ is a point of smoothness of the arc, *i.e.* not a corner point. We shall ignore the "exceptional" corner points in our derivation below of the integral equations, since they are "sets of measure zero", although we shall have to consider them again in the choice of our approximating bases.

It thus follows for the case that $\Gamma = \partial B$, *i.e.* when Γ is the boundary of B, that if $g \in \mathbf{Hol}(B)$, and if g is Hölder continuous on $\overline{B}$, the closure of B, then, by Cauchy's formula, we have for $z \in B$,

$$g(z) = \frac{1}{2\pi i}\int_{\partial B} \frac{g(t)}{t-z}\,dt, \quad z \in B. \tag{2.1.4}$$

Upon letting $z \to \zeta \in \partial B$, we get the equation

$$g(\zeta) = \frac{1}{2}g(\zeta) + \frac{1}{2}\,\mathcal{S}g(\zeta), \tag{2.1.5}$$

where this formula is certainly valid at all points of smoothness of ∂B. It thus follows that *if $g \in \mathbf{Hol}(B)$, and if g is Hölder continuous on $\overline{B}$, then at every point of smoothness of ∂B we have the equation*

$$g(\zeta) = \mathcal{S}g(\zeta). \tag{2.1.6}$$

Next [14], §5.1, *if Γ is a contour, and if for some constant $\alpha \in (0,1)$ we have $u \in \mathbf{Lip}_\alpha(\Gamma)$, then the function v defined by*

$$v = \mathcal{S}u \tag{2.1.7}$$

also belongs to $\mathbf{Lip}_\alpha(\Gamma)$. Moreover, we then have the dual inverse Hilbert transform expression *[14], §7.3,*

$$u = \mathcal{S}v. \tag{2.1.8}$$

The expressions (2.1.7) and (2.1.8) are valid at all points of smoothness of ∂B.

It is straight forward to show [14], that if $u \in \mathbf{Lip}_\alpha(\Gamma)$, with Γ a closed contour, then at all points of smoothness of ∂B we have $\mathcal{S}^2 u = u$.

2.2 Integral equations for conformal maps

Let f denote a conformal map of a bounded region B onto U, where we assume that $0 \in B$ and $f(0) = 0$. We start with the convenient representation

$$f(z) = z \exp\{\mathcal{G}(z)\}, \tag{2.2.1}$$

in which $\mathcal{G}$ is well defined, *e.g.* as $\mathcal{G}(z) = \text{Log}\{f(z)/z\}$, where $\text{Log}(\cdot) = \log|\cdot| + i\text{Arg}(\cdot)$ is defined in Assumption I.

Now for ζ on the boundary of B, we set

$$\mathcal{G} = g + i\,h \tag{2.2.2}$$

with g and h real valued functions, and $g(\zeta) = -\log|\zeta|$. If h were also known, we could, of course, compute $\mathcal{G}$ at any point of B via (2.1.4). A number of integral equations are known for the determination of h – see, for example, [10], Ch. I, §2.

2.2.1 *Lichtenstein's Equation.* Applying the operator $\mathcal{S}$ to both sides of (2.2.2), using the equation (2.1.6), *i.e.* $\mathcal{G} = \mathcal{S}\mathcal{G}$, and then taking the imaginary part of each side, we get

$$h - \text{Re}\,\mathcal{S}\,h = \text{Im}\,\mathcal{S}\,g. \tag{2.2.3}$$

This is, in fact, *Lichtenstein's equation*, [10], Eqs. (2.4)–(2.5).

2.2.2 *Another Integral Equation.* Let us now derive the integral equation on which the method of this paper is based for the determination of the conformal map f. By [16], Thm. 14.5d, we may start with the representation

$$\mathcal{G}(z) = \frac{1}{\pi i}\int_{\partial B}\frac{u(t)}{t-z}\,dt + i\,\delta, \quad z \in \mathbb{C}, \tag{2.2.4}$$

where u is an unknown real valued function and δ is a real–valued constant. Letting $z \to \zeta$, with ζ a point of smoothness of ∂B, we deduce, via the *Sokhotski formulas* that

$$\mathcal{G}(\zeta) = u(\zeta) + (\mathcal{S}u)(\zeta) + i\,\delta. \tag{2.2.5}$$

By taking the real part of each side of this equation, we get the following integral equation, which may also be found in [16], pp. 280 & 370,

$$u(\zeta) + \text{Re}(\mathcal{S}u)(\zeta) = g(\zeta). \tag{2.2.6}$$

The integral equation operator K defined by $K\,u = \text{Re}\;\mathcal{S}\,u$ also arose for the case of Lichtenstein's equation above, and moreover, it arises for nearly every other integral equation that is used for constructing the conformal map f (see *e.g.* [10], Ch. I., §2.1). It is shown in [10], Ch. I, §3, that this operator K has a simple eigenvalue 1, for which the corresponding eigenfunction is also 1. Furthermore, the other eigenvalues λ such that the equation $K\,v = \lambda\,v$ has non–trivial solutions v are all less than 1 in absolute value.

It is thus evident that the moduli of the eigenvalues of the operator $I + K$ all lie in the interval $1 - |\lambda_2| \le |\mu| \le 2$, with λ_2 the eigenvalue of largest modulus less than 1, *i.e.* $|\lambda_2| < 1$, and therefore, the operator $I + K$ has a condition number that is bounded by $2/(1 - |\lambda_2|)$. We may therefore expect that the condition number of the matrix obtained via application of a sufficiently accurate Galerkin or collocation method to Eq. (2.2.6) is also uniformly bounded (although it may be large, if λ_2 is close to 1). The integral equation (2.2.6) therefore has a unique solution. After solving this equation for the unique solution u, we can (if we so desire) get h from the identity

$$h = \text{Im}\;\mathcal{S}\,u. \tag{2.2.7}$$

We can then compute $\mathcal{G}$ in the interior of B by use of (2.2.4). Note that, whereas other integral equations yield either h or h' (the derivative of h) directly, we propose to first determine u by solving the stable integral equation (2.2.6), and then to determine h, if necessary, via use of (2.2.7). We do not actually require h for the computation of f, in view of (2.2.4).

We will of course replace (2.2.6) and (2.2.7) by systems of m linear equations which approximate the solution to these operator equations. Thus, the operator K as defined above will be replaced by an $m \times m$ matrix $\mathbf{K}$, the right hand side g of (2.2.6) will be replaced by a vector $\mathbf{g}$ of order m, and we will end up with the matrix equation

$$(\mathbf{I} + \mathbf{K})\,\mathbf{u} = \mathbf{g} \tag{2.2.8}$$

for the vector $\mathbf{u}$. The matrix of this system of equations will have a uniformly bounded condition number for all m sufficiently large, enabling the direct soluiton of (2.2.8), via LU factorization. We will then be able to construct a vector $\mathbf{h}$ (with the entries of $\mathbf{h}$ being evaluations of the function h at points of ∂B) via replacement of the operation $\text{Im}\;\mathcal{S}u$ by its discretized matrix–vector approximation $\mathbf{Lu}$ (see (4.1.24)).

We note in passing that the Fredholm alternative must be satisfied for solutions of (2.2.3) to exist, and this is, in fact, the case. Thus Lichtenstein's equation, as well as several of the other equations considered in [10],

Ch. I, §2, do not have unique solutions. Nevertheless, it may be shown, for example ([10], Ch1, §3), that although the operator $I - \text{Re}\ \mathcal{S}$ is singular, due to the special forms of the right hand sides of (2.2.3), the series produced by applying successive approximation to the equation (2.2.3) converges.

§3 Boundary behavior of the maps

In this section we deduce some smoothness properties of the function $\mathcal{G}$ on ∂B, where $f(z) = z \exp\{\mathcal{G}(z)\}$, as well as of the inverse map F on ∂U.

3.1 Boundary behavior of $\mathcal{G}$

Consider the junction point a_k between the two arcs, Γ_k and Γ_{k+1}, which meet at an *interior angle* π/β_k' as defined in (1.3.4). It is well known, [4], pp. 91–92, that as $z \to a_k$, we have

$$\begin{aligned} f(z) - f(a_k) &\sim c\,(z - a_k)^{\beta_k'} && \text{if } \ \beta_k' \neq 1 \\ f(z) - f(a_k) &\sim c\,(z - a_k)\,w(z) && \text{if } \ \beta_k' = 1, \end{aligned} \qquad (3.1.1)$$

with $w(z)$ either 1 or $\log(z)$, and with c a constant.
That is, since $f(z) = z \exp(\mathcal{G}(z))$, we get

$$\begin{aligned} e^{\mathcal{G}(z)} - e^{\mathcal{G}(a_k)} &= \frac{f(z)}{z} - \frac{f(a_k)}{a_k} \\ &= \frac{f(z) - f(a_k)}{z} + f(a_k)\left(\frac{1}{z} - \frac{1}{a_k}\right) \\ &\sim \frac{c}{a_k}(z - a_k)^{\beta_k'} - \frac{f(a_k)}{a_k^2}(z - a_k). \end{aligned} \qquad (3.1.2)$$

We thus deduce that

$$\mathcal{G}(z) - \mathcal{G}(a_k) \sim \begin{cases} \dfrac{c}{f(a_k)}(z - a_k)^{\beta_k'}, & \text{if } \ \beta_k' < 1, \\ -\dfrac{1}{a_k}(z - a_k), & \text{if } \ \beta_k' > 1. \end{cases} \qquad (3.1.3)$$

If $\beta_{k+1}' = 1$, it may happen [4] that $\mathcal{G}(z) - \mathcal{G}(a_k) = \mathcal{O}(|z - a_k|\,|\log|z - a_k|\,|) = \mathcal{O}(|z - a_k|^{1-\varepsilon'})$, as $z \to a_k$, with ε' any number in the range $0 < \varepsilon' < 1/2$.

Corresponding to an arbitrary positive number $\varepsilon' < 1/2$ it is thus convenient, at this point, to set[3]

$$\beta_k = \begin{cases} \beta_k', & \text{if} \quad 0 < \beta_k' < 1; \\ 1 - \varepsilon', & \text{if} \quad 1 \leq \beta_k', \end{cases} \tag{3.1.4}$$

for $k = 0, 1, \ldots, n$. We then have

$$\mathcal{G}(z) - \mathcal{G}(a_k) = \mathcal{O}((z - a_k)^{\beta_k}) \quad \text{as} \quad z \to a_k. \tag{3.1.5}$$

We now assume that the reader is familiar with the definition of the space $\mathbf{M}_{\alpha,\beta}(\mathcal{D})$, as described in the paragraph of Eqs. (6.2)–(6.3) of the Appendix of this paper. In *Sinc notation*, the functions $\rho_k(z) = \exp(\varphi_k(z))$, with φ_k defined as in (1.3.6), are used to define the *Sinc spaces* $\mathbf{L}_{\alpha,\beta}(D_k)$ and $\mathbf{M}_{\alpha,\beta}(D_k)$. We thus need a knowledge of the behavior of $\rho_k(z)$ as $z \to a_{k-1}$ and also as $z \to a_k$.

From the definitions of ζ_k in §1.3, we find that, as $\tau \to 0$,

$$z = \zeta_k(\tau) \sim \zeta_k(0) + \zeta_k'(0)\,\tau = a_{k-1} + \zeta_k'(0)\,\tau, \tag{3.1.6}$$

so that

$$\tau \sim \frac{z - a_{k-1}}{\zeta_k'(0)} \quad \text{as} \quad z \to a_{k-1}. \tag{3.1.7}$$

Similarly, as $\tau \to 1$, we have

$$1 - \tau \sim \frac{a_k - z}{\zeta_k'(1)} \quad \text{as} \quad z \to a_k. \tag{3.1.8}$$

That is, in view of the definition (6.2) of $\rho(z)$ and (1.3.6), we have $\rho_k(z) = \tau/(1 - \tau)$, and therefore

$$\rho_k(z) \sim \begin{cases} \dfrac{z - a_{k-1}}{\zeta'(0)}, & z \to a_{k-1} \\ \dfrac{\zeta_k'(1)}{a_k - z}, & z \to a_k. \end{cases} \tag{3.1.9}$$

We have thus shown, recalling (3.1.5), that

$$\begin{aligned} \mathcal{G}(z) - \mathcal{G}(a_{k-1}) &= \mathcal{O}((\rho_k(z))^{\beta_{k-1}}), \qquad z \to a_{k-1}, \\ \mathcal{G}(z) - \mathcal{G}(a_k) &= \mathcal{O}((\rho_k(z))^{-\beta_k}), \qquad z \to a_k. \end{aligned} \tag{3.1.10}$$

[3] It is convenient to take $\varepsilon' = 0$ in the algorithm of this paper.

We next discuss the analytic continuation of f from B to $B \cup_{j=1}^{n} D_j$, for the case of $n > 1$, the case of $n = 1$ being trivial. To this end, consider the mapping $\Phi = f \circ \zeta_j$ that maps the interval $(0,1)$ to an arc S_j of the boundary of U. Since $f(z)$ is analytic in D_j^+, the part of D_j to the left of Γ_j, it follows that the function Φ is clearly an analytic function in $\mathcal{E}_d^+$, the part of $\mathcal{E}_d$ to the left of $(0,1)$. Hence, if ω is a unimodular constant such that ω does not belong to the closure of S_j, the map $\Psi \equiv i\,(\omega + \Phi)/(\omega - \Phi)$ transforms S_j to a finite interval (s_1, s_2) of $\mathbb{R}$. In addition, Ψ is analytic in $\mathcal{E}_d^+$, Ψ has a continuous extension from $\mathcal{E}_d^+$ to the interval $(0,1)$ and moreover Ψ is real valued on $(0,1)$. Hence, by the Schwarz reflection principle, Ψ may be continued analytically from $\mathcal{E}_d^+$ to all of $\mathcal{E}_d$. That is, Φ may be continued analytically to all of $\mathcal{E}_d$ and therefore f may be continued analytically to all of D_j. The analytic continuation of f from B to $B \cup_{j=1}^{n} D_j$ therefore follows.

The following theorem thus follows, in view of (3.1.10) and the definition of $\mathbf{M}_{\alpha,\beta}(\mathcal{D})$ given in the appendix of this paper.

Theorem 1. *Let the function $\mathcal{G}$ be defined as in (2.2.1) and let D_j be defined as in Assumption I. Then $\mathcal{G}$ has an analytic continuation from B to $B \cup_{j=1}^{n} D_j$. Moreover, $\mathcal{G}$ restricted to D_j belongs to $\mathbf{M}_{\beta_{j-1},\beta_j}(D_j)$, $j = 1, \ldots, n$, with $\beta_0 \equiv \beta_n$.*

3.2 Boundary behavior of F

At the outset we deduce analyticity properties of the inverse map F in a neighborhood of ∂U, for ∂B as described in Assumption I. These will enable us to determine F in U to arbitrary accuracy, under the assumption that we know f on ∂B.

The determination of the points $b_k = e^{i\theta_k} = f(a_k)$ on the boundary of U is straight forward and simple, once f is known on ∂B.

By setting $w = f(z) \Leftrightarrow z = F(w)$, and then setting $b_k = \exp\{i\,\theta_k\}$ $(= f(a_k))$, we see from (3.1.1) that

$$z - a_k \sim c\,(w - b_k)^{1/\beta'_k}, \quad w \to b_k. \tag{3.2.1}$$

Hence, we set $\alpha'_k = 1/\beta'_k$, for $1 \le k \le n$, and then (by proceeding as in §3.1) we define α_k by

$$\alpha_k = \begin{cases} \alpha'_k & \text{if } 0 < \alpha'_k < 1; \\ 1 - \varepsilon', & \text{if } 1 \le \alpha'_k, \end{cases} \tag{3.2.2}$$

for $k = 1, \ldots, n$, and with $\alpha_0 = \alpha_n$, and arbitrary $\varepsilon' \in (0, 1/2)$.

Just as we used arcs $\Gamma_k \in \partial B$, we now use circular arcs $\Gamma_k^T \in \partial U$, which we define by

$$\Gamma_k^T = \left\{ w = e^{i\theta} \in \mathbb{C} : \theta_{k-1} < \theta < \theta_k \right\}, \tag{3.2.3}$$

and we define the conformal maps

$$\psi_k(w) = i \, \frac{\theta_k - \theta_{k-1}}{2} + \log \left(\frac{w - e^{i\theta_{k-1}}}{e^{i\theta_k} - w} \right), \tag{3.2.4}$$

which map the "lens shaped" regions

$$\mathcal{Q}_k = \{ w \in \mathbb{C} : |\mathrm{Im} \; \psi_k(w)| < d \}. \tag{3.2.5}$$

onto $\mathcal{D}_d$, where $\mathcal{D}_d$ is defined as in (1.3.5) above.

We then have, by proceeding as for f in §3.1 above:

Theorem 2. *Let ∂B be determined as in Assumption I. Then the mapping F has an analytic continuation from U to $U \cup_{j=1}^{n} \mathcal{Q}_j$. The function F restricted to $\mathcal{Q}_k$ belongs to $\mathbf{M}_{\alpha_{k-1}, \alpha_k}(\mathcal{Q}_k)$.*

§4 Computational details

In §4.1, we study the behavior of the principal value of a certain integral taken over the boundary of B, and we show that this does, in fact, stay bounded at all points of the closure of B. From that, we deduce a *Sinc approximation scheme* for accurately replacing the integral equation (2.2.6) by the system of algebraic equations (2.2.8), and we thus give an explicit description of the entries in the system of equations (2.2.8). We also describe in this section an explicit algorithm for approximating f at the *Sinc points* of each arc Γ_j. In §4.2, we describe an algorithm for computing f at any point in the interior of B, using the already computed approximation of f on ∂B. In §4.3, we sketch an algorithm for approximating $F : U \to B$ on the boundary of U as well as in the interior of U to arbitrary accuracy.

4.1 Approximation of f on the boundary of B

Let w be a function that is Hölder continuous on the boundary of B, and let us consider the approximation of v defined by $v(\zeta) = \mathcal{S}w(\zeta)$, for $\zeta \in \partial B$. That is, by Assumption I, we have

$$v(\zeta) = \sum_{q=1}^{n} \frac{P.V.}{\pi i} \int_{\Gamma_q} \frac{w(t)}{t - \zeta}\, dt. \tag{4.1.1}$$

Evidently, the integral over Γ_q in (4.1.1) will become infinite as ζ approaches an endpoint of Γ_q. However, the total sum in (4.1.1) remains bounded, at all points $\zeta \in \Gamma_q$, as we shall now show by means of a sequence of n linear expressions that interpolate w at the endpoints of Γ_q.

Recalling that Γ_q has endpoints a_{q-1} and a_q, we define a linear expression $u_q(t)$ for $t \in \Gamma_q$ by

$$u_q(t) = \frac{(a_q - t)\, w(a_{q-1}) + (t - a_{q-1})\, w(a_q)}{a_q - a_{q-1}}. \tag{4.1.2}$$

Clearly, u_q interpolates $w(t)$ at the endpoints of Γ_q. Then, using the notation

$$Q(\Gamma_q, \zeta) = \frac{P.V.}{\pi i} \int_{\Gamma_q} \frac{dt}{t - \zeta}, \tag{4.1.3}$$

we find that

$$\frac{P.V.}{\pi i} \int_{\Gamma_q} \frac{u_q(t)}{t - \zeta}\, dt = \frac{1}{\pi i} \{w(a_q) - w(a_{q-1})\} + u_q(\zeta)\, Q(\Gamma_q, \zeta). \tag{4.1.4}$$

The evaluation of the real part of $Q(\Gamma_q, \zeta)$ may be difficult for complicated contours Γ_q, since the evaluation of the argument of the logarithm might then be non–trivial. We shall discuss the evaluation of $Q(\Gamma_q, \zeta)$ at the end of §4.2.

Lemma 1. *Let χ_q denote the characteristic function of Γ_q, and let u_q , and $Q(\Gamma_q, \zeta)$ be defined as above. Then, for $\zeta \in \Gamma_p$, we have*

$$\sum_{q=1}^{n} \frac{1}{\pi i} \int_{\Gamma_q} \frac{u_q(t)}{t - \zeta}\, dt = U_p(\zeta) + V_p(\zeta), \tag{4.1.5}$$

where

$$U_p(\zeta) =$$

$$\frac{1}{\pi i}\left\{(\zeta - a_{p-1})\mathrm{Log}(\zeta - a_{p-1})\left(\frac{w(a_p) - w(a_{p-1})}{a_p - a_{p-1}} - \frac{w(a_{p-1}) - w(a_{p-2})}{a_{p-1} - a_{p-2}}\right)\right.$$

$$\left. - (a_p - \zeta)\mathrm{Log}(a_p - \zeta)\left(\frac{w(a_{p+1}) - w(a_p)}{a_{p+1} - a_p} - \frac{w(a_p) - w(a_{p-1})}{a_p - a_{p-1}}\right)\right\},$$

$$V_p(\zeta) = \frac{1}{\pi\, i}\left\{\mathrm{Log}(\zeta - a_{p-2})\, u_{p-2}(\zeta) + \mathrm{Log}(a_{p+1} - \zeta)\, u_{p+1}(\zeta)\right\}$$

$$- \sum_{q, q \neq p-1, p, p+1} Q(\Gamma_q, \zeta)\, u_q(\zeta). \tag{4.1.6}$$

Proof: The proof is straight forward, using (4.1.4). Taking $\zeta \in \Gamma_p$, and summing both sides of (4.1.4) over q from 1 to n, or equivalently, substituting (4.1.4) into (4.1.5) we find that the sums of the "$w(a_q) - w(a_{q-1})$" terms on the right hand side of (4.1.4) telescopes to zero. The term U_p consists of all the terms in this sum that contain either the coefficient $\log(\zeta - a_{p-1})$ or the coefficient $\log(a_p - \zeta)$, whereas the term V_p is analytic at all points of the closure of Γ_p. We thus arrive at (4.1.6). ■

Remark: Inspection of (4.1.6) shows that the expression (4.1.5) does, in fact, stay bounded for all $\zeta \in \partial B$.

At this point we set

$$W_q(\zeta) = \frac{P.V.}{\pi i}\int_{\Gamma_q} \frac{w(t) - u_q(t)}{t - \zeta}\, dt, \tag{4.1.7}$$

so that, for $\zeta \in \Gamma_p$,

$$g(\zeta) = U_p(\zeta) + V_p(\zeta) + \sum_{q=1}^{n} W_q(\zeta). \tag{4.1.8}$$

The evaluation of U_p and V_p may be carried out via the explicit expressions (4.1.6) while approximation of the terms W_q as defined in (4.1.7) are carried out via *Sinc methods*. We now describe the details.

Let us assume that $\zeta \in \Gamma_p$, and let us then perform the evaluation of the integral (4.1.7) over Γ_q. By assumption, $w \in \mathbf{Lip}_\mu(\partial B)$, whereas

$u_q \in \mathbf{Lip}_1(\Gamma_q)$. Hence it follows, upon setting $v_q(t) = w(t) - u_q(t)$ that $v_q(t) = \mathcal{O}(|t - a_q|^\mu)$ as $t \to a_q$, with μ denoting the exponent of continuity of the Hölder continuous function w, and similarly, $v_q(t) = \mathcal{O}(|t - a_{q-1}|^\mu)$ as $t \to a_{q-1}$. Hence we use the formula (6.18) to compute accurate values of $W_p(\zeta)$ at the Sinc points $\zeta = z_\ell^{(p)} = \varphi_p^{[-1]}(\ell h_p)$, and we use ordinary *Sinc quadrature* over Γ_q to approximate $W_q(\zeta)$, for $\zeta \in \Gamma_p$, $p \neq q$.

Let us now consider the a priori determination of the parameters for explicit Sinc approximation on each arc Γ_p *in order to achieve a final approximation* $\mathcal{G}_\varepsilon$ *of* $\mathcal{G}$ *in* B *such that* $\|\mathcal{G} - \mathcal{G}_\varepsilon\|_B = \sup_{z\in B} |\mathcal{G}(z) - \mathcal{G}_\varepsilon(z)| = \mathcal{O}(\varepsilon)$. To this end we know, by the maximum principle for analytic functions, that, if we have an analytic (or harmonic) function approximation of $\mathcal{G}$ to within ε on ∂B then the same approximation is within ε of $\mathcal{G}$ in all of B, *i.e.* $\|\mathcal{G} - \mathcal{G}_\varepsilon\|_B = \|\mathcal{G} - \mathcal{G}_\varepsilon\|_{\partial B} = \sup_{z\in\partial B} |\mathcal{G}(z) - \mathcal{G}_\varepsilon(z)|$. Hence it suffices to determine $\mathcal{G}_\varepsilon$ accurately merely on ∂B. Let us therefore consider the equations

$$\begin{aligned} u + K\,u &= g \\ u_m + K_m\,u_m &= g_m. \end{aligned} \tag{4.1.9}$$

The first equation in (4.1.9) is just (2.2.6). The terms of the second equation of (4.1.9) are described in detail in (5.1.3) below.

We see below that the result $K_m\,u_m$ is an $\mathcal{O}(\varepsilon)$ approximations of $K\,u_m$ that is obtained via application of approximations described in §6. By subtracting the second of (4.1.9) from the first, we get

$$\|u - u_m\|_{\partial B} \le \|(I + K_m)^{-1}\| \left(\|g - g_m\|_{\partial B} + \|(K - K_m)\,u\|_{\partial B}\right). \tag{4.1.10}$$

Now it is proved in §5 that since $\|(I + K)^{-1}\|$ is uniformly bounded, so is $\|(I + K_m)^{-1}\|$, for all m sufficiently large. In addition, if $\|g - g_m\|$ is of the order of ε, then $\|(K - K_m)\,u\|$ is also of the order of ε. We can estimate the error of *Sinc approximation on* Γ_p using (6.8). For, suppose that in view of (6.8) and (6.10), we have

$$\sup_{z\in\Gamma_p} \left| g(z) - \sum_{k=-N_{p-1}}^{N_p} g(z_k^{(p)})\,\omega_k^{(q)}(z) \right| = \mathcal{O}(\varepsilon), \tag{4.1.11}$$

with $(*)$ $\varepsilon = N_p^{1/2} \exp\left\{-(\pi d \beta_p N_p)^{1/2}\right\}$. By solving this equation $(*)$ for N_p and recalling the definition of h given in (6.5), we get the estimates

$$\begin{aligned} N_p &= \left[\frac{(\log(1/\varepsilon)+\log|\log(1/\varepsilon)|)^2}{\pi d\,\beta_p}\right], \\ h_p &= \left(\frac{\pi d}{\beta_p N_p}\right)^{1/2}, \end{aligned} \tag{4.1.12}$$

where $[\cdot]$ denotes the greatest integer function. It then follows from Theorem 8 that all errors at the *Sinc points* are of the order of $\varepsilon/\log(1/\varepsilon)$, and the errors in the final approximations will be of the order of ε.

Having determined h_p and N_p as above, we now compute *Sinc points*, $z_k^{(p)} = \varphi_p^{[-1]}(k\,h_p)$, as well as the derivatives, $\varphi_p'(z_k^{(p)})$. We select the following basis for approximating on Γ_q a function $u \in \mathbf{M}_{\beta_{q-1},\beta_q}(D_q)$, for $q = 1, \ldots, n$.

$$\begin{aligned} \sigma_j^{(q)} &= \operatorname{sinc}\{[\varphi_q - j\,h_q]/h_q\}, \quad = -N_{q-1}, \ldots, N_q, \\ w_j^{(q)} &= \sigma_j^{(q)}, \quad j = -N_{q-1}+1, \ldots, N_q - 1, \\ w_{-N_{q-1}}^{(q)}(z) &= \frac{a_q - z}{a_q - a_{q-1}} - \sum_{j=-N_{q-1}+1}^{N_q} \frac{a_q - z_j^{(q)}}{a_q - a_{q-1}}\,\sigma_j^{(q)}(z), \\ w_{N_q}^{(q)}(z) &= \frac{z - a_{q-1}}{a_q - a_{q-1}} - \sum_{j=-N_{q-1}}^{N_q-1} \frac{z_j^{(q)} - a_{q-1}}{a_q - a_{q-1}}\,\sigma_j^{(q)}(z). \end{aligned} \tag{4.1.13}$$

We remark here, that the functions $\sigma_j^{(q)}$ vanish at the endpoints of γ_q, whereas the class of functions $\mathbf{M}_{\beta_{q-1},\beta_q}(D_q)$ contains functions that do not vanish at the endpoints of Γ_q. The functions $w_{-N_{q-1}}^{(q)}$ and $w_{N_q}^{(q)}$ are basis functions which we use to interpolate non–zero functions at the endpoints of Γ_q. For example, $w_{N_q}^{(q)}$ accurately approximates the values 0 and 1 at the endpoints a_{q-1} and a_q respectively.

Our approximation of u on ∂B thus takes the form

$$u(\zeta) \approx u_\varepsilon(\zeta) = \sum_{q=1}^{n} \sum_{j=-N_{q-1}}^{N_q} c_j^{(q)}\, w_j^{(q)}(\zeta)\,\chi_q(\zeta), \tag{4.1.14}$$

where the $c_j^{(q)}$ are constants, and where χ_q denotes the characteristic function of Γ_q.

Corresponding to the basis functions $\omega_\ell^{(q)}$, we determine the *Hilbert transform bases* (compare the end of §6 of this paper), in which

$$\zeta = z_k^{(p)}, \tag{4.1.15}$$

i.e. ζ is a *Sinc point.* Letting $\mathcal{S}$ denote the Hilbert transform operator, we thus set

$$\begin{aligned} v_\ell^{(q)}(\zeta) &= \mathcal{S}\sigma_\ell^{(q)}(\zeta) \\ d_\ell^{(q)}(\zeta) &= \mathcal{S}\omega_\ell^{(q)}(\zeta), \end{aligned} \tag{4.1.16}$$

where, based on Theorems 9, 11, and Eq. (6.16), we have the following accurate approximations for these functions:

$$v_\ell^{(q)}(\zeta) = \begin{cases} \dfrac{h_q}{\pi\, i\, \varphi_q'\left(z_\ell^{(q)}\right)\left(z_\ell^{(q)} - \zeta\right)} & \text{if} \quad q \neq p \\ \left.\begin{array}{ll} 0, & \text{if} \quad \ell = k + 2s, \\ \dfrac{2\, h_p}{\pi\, i\, \varphi_p'\left(z_\ell^{(p)}\right)\left(z_\ell^{(p)} - \zeta\right)} & \text{if} \quad \ell = k + 2s + 1, \end{array}\right\} & \text{if} \quad q = p. \end{cases} \tag{4.1.17}$$

Here s is an arbitrary integer. Although the right hand sides of (4.1.16) are only approximations based on Theorem 11, we take them to be identities, and we set

$$\begin{aligned} d_\ell^{(q)}(\zeta) &= v_\ell^{(q)}(\zeta) \ \text{if} \ -N_{q-1} < \ell < N_q; \\ d_{-N_{q-1}}^{(q)}(\zeta) &= -\frac{1}{\pi\, i} + \frac{a_q - \zeta}{a_q - a_{q-1}}\, Q(\Gamma_q, \zeta) \\ &\quad - \sum_{j=-N_{q-1}+1}^{N_q} \frac{a_q - z_j^{(q)}}{a_q - a_{q-1}}\, v_j^{(q)}(\zeta) \\ d_{N_q}^{(q)}(\zeta) &= \frac{1}{\pi\, i} + \frac{\zeta - a_{q-1}}{a_q - a_{q-1}}\, Q(\Gamma_q, \zeta) \\ &\quad - \sum_{j=-N_{q-1}}^{N_q - 1} \frac{z_j^{(q)} - a_{q-1}}{a_q - a_{q-1}}\, v_j^{(q)}(\zeta), \end{aligned} \tag{4.1.18}$$

where $Q(\Gamma_q, \zeta)$ is defined in (4.1.3).

It therefore follows that if $\zeta = z_m^{(\ell)}$, then an accurate approximation of the Hilbert transform of the approximation in (4.1.14) taken over the boundary of B is given by

$$\mathcal{S}\, u_\varepsilon(\zeta) = \sum_{q=1}^{n} \sum_{\ell=-N_{q-1}}^{N_q} c_\ell^{(q)}\, d_\ell^{(q)}(\zeta). \tag{4.1.19}$$

We are now in position to describe the explicit matrix set up, in the form of an *algorithm*:

1. Form the numbers $v_\ell^{(q)}(z_k^{(p)})$ as defined in (4.1.17), for

$$k = -N_{p-1}, \ldots, N_p, \quad p = 1, \ldots, n$$

$$\ell = -N_{q-1}, \ldots, N_q, \quad q = 1, \ldots, n.$$

2. Form the numbers $d_\ell^{(q)}(z_k^{(p)})$ as defined in (4.1.18), for

$$k = -N_{p-1}, \ldots, N_p, \quad p = 1, \ldots, n$$

$$\ell = -N_{q-1}, \ldots, N_q, \quad q = 1, \ldots, n,$$

   where as already mentioned above, the evaluation of $Q(\Gamma_q, \zeta)$ is based on the use of (4.1.3).

3. Form blocks of matrices $\mathbf{A}_{p,q}$ of size $(N_{p-1}+N_p+1) \times (N_{q-1}+N_q+1)$, with $d_\ell^{(q)}(z_k^{(p)})$ denoting the $(k,\ell)^{th}$ element of $\mathbf{A}_{p,q}$. Then form a comprehensive matrix $\mathbf{A} = [\mathbf{A}_{p,q}]$ with $\mathbf{A}_{p,q}$ denoting the $(p,q)^{th}$ block of $\mathbf{A}$, $p, q = 1, \ldots, n$. Recall that $N_0 = N_n$.

4. Form the vectors $\mathbf{g}_p$, $p = 1, \ldots, n$, with

$$\mathbf{g}_p = -\left(\log\left|z_{-N_{p-1}}^{(p)}\right|, \ldots, \log\left|z_{N_p}^{(p)}\right|\right), \tag{4.1.20}$$

   and then form the vector

$$\mathbf{g} = (\mathbf{g}_1, \ldots, \mathbf{g}_n)^T, \tag{4.1.21}$$

   where "T" denotes the transpose.

5. Form the matrices

$$\mathbf{K} = \mathrm{Re}\ \mathbf{A}, \qquad \mathbf{L} = \mathrm{Im}\ \mathbf{A}. \tag{4.1.22}$$

6. Solve the system of equations

$$(\mathbf{I} + \mathbf{K})\,\mathbf{u} = \mathbf{g} \tag{4.1.23}$$

for the vector $\mathbf{u}$, and (if h is desired) form the vectors

$$\mathbf{h} = \mathbf{L}\,\mathbf{u}, \qquad \mathbf{c} = \mathbf{g} + i\,\mathbf{h}. \tag{4.1.24}$$

7. At this point we have

$$\begin{aligned} \mathbf{c} &\equiv (\mathbf{c}_1, \ldots, \mathbf{c}_n)^T \\ \mathbf{c}_p &\equiv \left(c^{(p)}_{-N_{p-1}}, \ldots, c^{(p)}_{N_p}\right), \quad p = 1, \ldots, n. \end{aligned} \tag{4.1.25}$$

It is convenient for sake of continuity of our final approximation to assign the same value to the coefficients of the end bases at the junctions of Γ_{p-1} and Γ_p. We accomplish this by averaging the two computed end coefficients, *i.e.* for $p = 1, 2, \ldots, n$, we perform the following operations in order:

$$\begin{aligned} a &= \frac{c^{(p)}_{-N_{p-1}} + c^{(p-1)}_{N_{p-1}}}{2} \\ c^{(p)}_{-N_{p-1}} &= c^{(p-1)}_{N_{p-1}} = a, \end{aligned} \tag{4.1.26}$$

and where $c^{(0)}_{N_0} = c^{(n)}_{N_n}$.

8. Our final approximation of the function $\mathcal{G}$ as defined in (2.2.1) then takes the form

$$\mathcal{G}(\zeta) \approx \sum_{q=1}^{n} \chi_q(\zeta) \sum_{\ell=-N_{q-1}}^{N_q} c^{(q)}_\ell\, w^{(q)}_\ell(\zeta). \tag{4.1.27}$$

4.2 Approximation of f in B

Our basic equation for approximating $f(z) = z\,\exp\{\mathcal{G}(z)\}$ for $z \in B$ is Cauchy's integral formula,

$$\mathcal{G}(z) = \sum_{j=1}^{n} \frac{1}{2\pi i} \int_{\Gamma_j} \frac{\mathcal{G}(t)}{t - z}\, dt. \tag{4.2.1}$$

We proceed similarly for the evaluation of this integral as we did for (4.1.1) at the outset of §4.1.

Introducing only a negligible error, we have identified the values of $\mathcal{G}$ at the junctions a_p of the arcs with the computed values at the nearest *Sinc points.*

Next, we set

$$u_p(t) = \frac{(a_p - t)\,\mathcal{G}(a_{p-1}) + (t - a_{p-1})\,\mathcal{G}(a_p)}{a_p \;-\; a_{p-1}}. \tag{4.2.2}$$

Then, as in (4.1.3), for $z \in B$ we set

$$R(\Gamma_p, z) = \frac{1}{2\pi i} \int_{\Gamma_p} \frac{dt}{t - z}, \tag{4.2.3}$$

so that, setting

$$U_p(z) = \frac{1}{2\pi i} \int_{\Gamma_p} \frac{u_p(t)}{t - z}\, dt, \tag{4.2.4}$$

we find that

$$U_p(z) = \frac{1}{2\,\pi\, i}\,\{u(a_p) - u(a_{p-1})\} + u_p(z)\, R(\Gamma_p, z). \tag{4.2.5}$$

We must excercise care in the evaluation of the real part of $R(\Gamma_p, z)$, just as for the case of $Q(\Gamma_p, \zeta)$ in (4.1.3) above. Assume for the moment that we know how to accurately compute this quantity. We shall derive a suitable method of evaluating the real part of $Q(\Gamma_p, \zeta)$ and $R(\Gamma_p, z)$ at the end of this section.

Set

$$w_p(t) = u(t) - u_p(t), \tag{4.2.6}$$

and then consider for fixed $z \in B$ the approximation of the integrals

$$W_p(z) = \frac{1}{2\pi i} \int_{\Gamma_p} \frac{w_p(t)}{t - z}\, dt. \tag{4.2.7}$$

Let $J(z)$ denote the set of all integers p such that $z \in D_p$, and set $K(z) = \{1, 2, \ldots, n\} \setminus J(z)$. We remark here that from §1.3 if $|\operatorname{Im} \varphi_p(z)| < d$ then $p \in J(z)$. Then we set

$$\begin{aligned} G_J(z) &= \sum_{p \in J(z)} \sum_{j=-N_{p-1}}^{N_p} w_p(z_j^{(p)})\, c_j^{(p)}(z) \\ G_K(z) &= \sum_{p \in K(z)} \frac{h_p}{2\pi i} \sum_{j=-N_{p-1}}^{N_p} \frac{w_p(z_j^{(p)})}{\varphi'(z_j^{(p)})(z_j^{(p)} - z)}, \end{aligned} \tag{4.2.8}$$

with w_p defined as in (4.2.6), and $c_j^{(p)}$ as in (6.15). Thus, we use the formula (6.19) to approximate W_p whenever $z \in D_p$, and the quadrature formula (6.11) whenever z does not belong to D_p.
Our approximation of $\mathcal{G}$ to within a uniform error of the order of ε then takes the form

$$\mathcal{G}(z) \approx G_J(z) + G_K(z) + \sum_{p=-N_{p-1}}^{N_p} u_p(z)\, R(\Gamma_p, z). \tag{4.2.9}$$

Since the system of equations (4.1.23) has a bounded condition number that is independent of ε for all ε sufficiently small, by (4.1.10), the resulting error in the approximation of f via $f(z) = z \exp\{\mathcal{G}(z)\}$ is then also of the order of ε.

The numerical evaluation of $Q(\Gamma_p, \zeta)$ defined in (4.1.3) and $R(\Gamma_p, z)$ defined in (4.2.3) may be non–trivial for complicated contours Γ_p. For this reason we shall consider in detail the evaluation of these quantities. To this end, recall here the differences between the terms "Arg" and arg as given at the outset of §1.3.

For the case when ζ is a *Sinc point* of Γ_p, *i.e.* when $\zeta = z_k^{(p)}$, π times the real part of $Q(\Gamma_p, \zeta)$ is given by the identity

$$\operatorname{Im}\left(\int_{\Gamma_p} \frac{dt}{t - z_k^{(p)}}\right) = \sum_{k=-N_{p-1}-1, k \neq k-1, k}^{N_p} \arg\left(\frac{y_{k+1} - y_k}{y_k - y_k}\right) + \arg\left(\frac{y_{k+1} - y_k}{y_k - y_{k-1}}\right), \tag{4.2.10}$$

in which we have set $y_k = z_k^{(j)}$ for $-N_{j-1} \le k \le N_j$, $y_{-N_{j-1}-1} = a_{j-1}$, and $y_{N_j+1} = a_j$. The identity (4.2.10) is readily verified by inspection; it is valid provided that the *Sinc point* spacing on Γ_j is sufficiently refined (a necessary requirement for the method of this paper to be accurate) so that each of the argument terms in the imaginary part of the integral on the left taken over the two consecutive values, y_{k-1} and y_k is in fact less than π in magnitude. The formula (4.2.12) can then be used for evaluating the term $Q(\Gamma_p, \zeta)$ defined in (4.1.3) for $\zeta \in \Gamma_p$.

Next, we describe simple techniques for approximating the real parts of $Q(\Gamma_p, \zeta)$, $\zeta \notin \Gamma_p$, and $R(\Gamma_p, z)$, $z \notin \Gamma_p$. The main difficulty is the approximation of

$$\Theta \equiv \operatorname{Im}\left(\int_{\Gamma_p} \frac{dt}{t - z}\right). \tag{4.2.11}$$

To this end, let us first set

$$S \equiv \sum_{k=-N_{p-1}}^{N_p+1} \arg\left(\frac{y_k - z}{y_{k-1} - z}\right), \tag{4.2.12}$$

where, as above, $y_k = z_k^{(j)}$ for $-N_{j-1} \le k \le N_j$, and where $y_{-N_{p-1}-1} = a_{p-1}$, and $y_{N_p+1} = a_p$.
We define an integer ℓ by

$$\ell = \left[\frac{1}{h_q} \operatorname{Re} \varphi_q(z)\right], \tag{4.2.13}$$

where $[\cdot]$ denotes the greatest integer function, and where φ_q is defined as in (1.3.6).
We then state the following rules for the evaluation of Θ:

- If $\ell \notin [-N_{q-1} - 1, N_q]$, or else if $\varphi_q(z)$ is undefined, then Θ is given exactly by (4.2.12), *i.e.*

$$\Theta = S. \tag{4.2.14}$$

- Otherwise, with y_j defined as in (4.2.12), we set

$$\begin{array}{rclrcl} u_1 &=& \mathrm{Re}(z-y_\ell), & v_1 &=& \mathrm{Im}(z-y_\ell), \\ u_2 &=& \mathrm{Re}(y_{\ell+1}-y_\ell), & v_2 &=& \mathrm{Im}(y_{\ell+1}-y_\ell), \end{array} \tag{4.2.15}$$

and we define yet another integer, "s", by

$$s = \begin{cases} 1, & \text{if} \ \ u_1 v_2 - u_2 v_1 > 0, \\ \\ 0, & \text{if} \ \ u_1 v_2 - u_2 v_1 \le 0. \end{cases} \tag{4.2.16}$$

Then it follows, recalling the definition of "arg" and the expression for the area of the triangle subtended by the three points $0, u_1 + i\, v_1$, and $u_2 + i\, v_2$, namely $\frac{1}{2}(u_1 v_2 - u_2 v_1)$, that this area is positive if and only if $\arg(u_1 + i\, v_1) < \arg(u_2 + i\, v_2)$. Therefore

$$\mathrm{Im}\left(\int_{y_\ell}^{y_{\ell+1}} \frac{dt}{t-z}\right) = \arg\left(\frac{y_{\ell+1}-z}{y_\ell - z}\right) + 2\pi s, \tag{4.2.17}$$

where the integral is taken along Γ_p.

Hence it then follows, in view of the definition of S above, that

$$\Theta = S + 2\pi s. \tag{4.2.18}$$

4.3 Approximation of F

A straightforward and simple scheme is the construction of F by first constructing f.

Having determined the mappings ψ_k as described in (3.2.4) above, we then determine parameters N_k and h_k as in (4.1.12), but with β_k replaced by α_k, and we then determine the *Sinc points* on each arc Γ_k^T. The approximation of $F(w)$ for $w \in U$ then proceeds in the same manner as the evaluation of $f(z)$ for $z \in B$, provided that we know the values of F at the *Sinc points* of each arc Γ_k^T. By assumption we already have an accurate approximation of f on ∂B. We therefore use the expression (4.2.9) to compute an accurate approximation of $\mathcal{G}$ for any $\zeta \in \Gamma_k$. The determination of the values of F at the *Sinc points* of Γ_k^T is thus carried out in a straight forward manner, via the well known method of *false position* applied to the argument of f. We omit the details.

4.4 Examples

In this section we illustrate the computation of conformal maps via the methods of this paper, and we make some comparisons of computation time required by Hough's [18] and our method. We thus present two examples, the case when B is a half disc, and that of a "pac–man". Comparisons are not so straight forward, since our method sets up the problem size according to a specified input error, while Hough's method is less flexible. We found that Hough's method is at times more efficient than ours for certain regions with corners that are multiples of right angles. Also, our timed outputs were not always the same, even when we performed the same computations, since we did our computations on a multi–tasking computer. As we already mentioned, our method works for interior angles greater than 2π (*i.e.* Riemann surfaces), whereas Hough's does not, although we believe that Hough's method could be easily modified to be able to handle interior angles greater than 2π. We did not present an example of a map with an interior angle greater than 2π, nor one for which Symm's equation cannot be solved (see [11]), and for which Hough's method therefore fails. For the examples that we present, compilation time of our program took about 10 times as long as computation after compilation. Finally, Hough's method requires about 5 times as long to compile as our method. A copy of our *Fortran Program* may be obtained by sending an e-mail message to *stenger@sinc.utah.edu.*

Example 1: A Half Disc. Let us take as an example $\partial B = \Gamma_1 \cup \Gamma_2$, with

$$\begin{aligned} \Gamma_1 &= \left\{ z = \zeta_1(\tau) = \frac{1}{2} + i(2\tau - 1),\ 0 \le \tau \le 1 \right\} \\ \Gamma_2 &= \left\{ z = \zeta_2(\tau) = \frac{1}{2} + \exp\left\{ i\left(\frac{\pi}{2} + \pi\tau\right)\right\},\ 0 \le \tau \le 1 \right\}. \end{aligned} \tag{4.4.1}$$

That is, B shown in Fig. 1 is the left half of a disc, shifted 1/2 unit to the right. Upon solving these equations for τ, we get

$$\begin{aligned} \zeta_1^{[-1]}(z) &= \frac{1}{2i}\left(z + i - \frac{1}{2}\right) \\ \zeta_2^{[-1]}(z) &= \frac{1}{\pi i}\log\left(\frac{1}{2} - z\right) + \frac{1}{2}. \end{aligned} \tag{4.4.2}$$

In view of (1.3.6), the mapping functions φ_k, the *Sinc points* $z_j^{(k)}$ and the derivatives of φ_k are thus given by

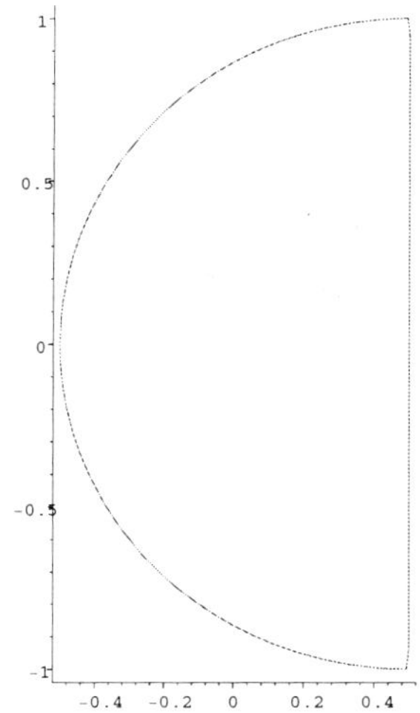

Figure 1. The region B of this example

$$\begin{aligned}
\varphi_1(z) &= \log\left(\frac{i+(z-\frac{1}{2})}{i-(z-\frac{1}{2})}\right) \\
\varphi_2(z) &= \log\left(\frac{\frac{1}{2}+\frac{1}{\pi i}\log(\frac{1}{2}-z)}{\frac{1}{2}-\frac{1}{\pi i}\log(\frac{1}{2}-z)}\right) \\
z_j^{(1)} &= \frac{1}{2}-i+\frac{2\,i\,e^{jh_1}}{1+e^{jh_1}} \\
z_j^{(2)} &= \frac{1}{2}+i\exp\left(\frac{i\,\pi\,e^{jh_2}}{1+e^{jh_2}}\right) \\
\{\varphi_1\}'(z) &= -\frac{2\,i}{1+(z-\frac{1}{2})^2} \\
\{\varphi_2\}'(z) &= \frac{\pi\,i}{(\frac{1}{2}-z)\left(\log^2(\frac{1}{2}-z)+\frac{\pi^2}{4}\right)}.
\end{aligned} \tag{4.4.3}$$

Since the interior angles at the junctions are both less than π, we can take $\beta_1 = \beta_2 = 1$, so that, according to (4.1.12), $N_1 = N_2$, and furthermore

$h_1 = h_2$.

Finally, an explicit conformal map of B onto U is given by

$$f(z) = z \, \frac{2\,z - 5}{(z-1)(2\,z+3)}. \tag{4.4.4}$$

The next figure, Fig. 2, illustrates a comparison of the amount of computer time (not counting compilation) it takes to compute the mapping f on ∂B, to within a certain accuracy, for Hough's method $(+)$ and the method of our paper (o).

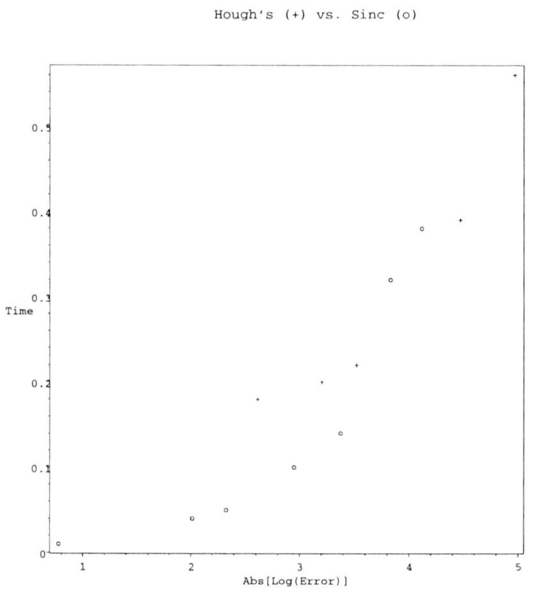

Figure 2. Computation time as a function of maximal error

Example 2: A Pac–Man. For this example we applied the method of this paper to a "pac–man", with boundary curves $\Gamma_1 = \{z \in \mathbb{C} : z = 1/2 + \tau\, e^{i\,\theta_0}, 0 \le \tau \le 1\}$, $\Gamma_2 = \{z \in \mathbb{C} : z = 1/2 + e^{i(\theta_0 + \tau(\pi - \theta_0))}, 0 \le \tau \le 1\}$, and $\Gamma_3 = \{z \in \mathbb{C} : z = 1/2 + e^{-i\theta_0} - \tau\, e^{-i\,\theta_0}, 0 \le \tau \le 1\}$. Our run time comparisons were similar to those for the half–disc above.

Figs. 3 & 4 illustrate evaluations of the maps in the interior of B.

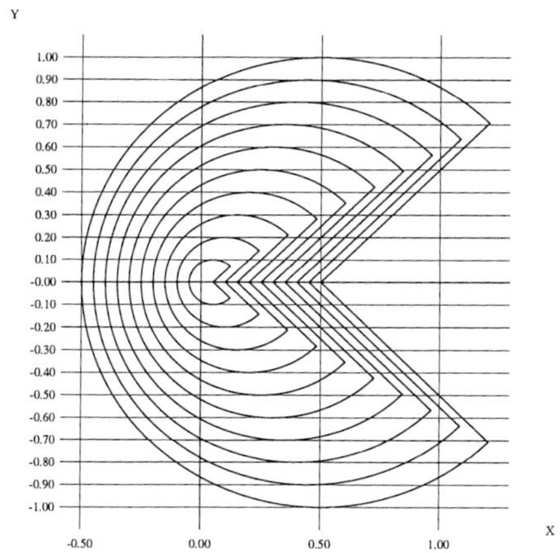

Figure 3. Pac-Man pre-map curves

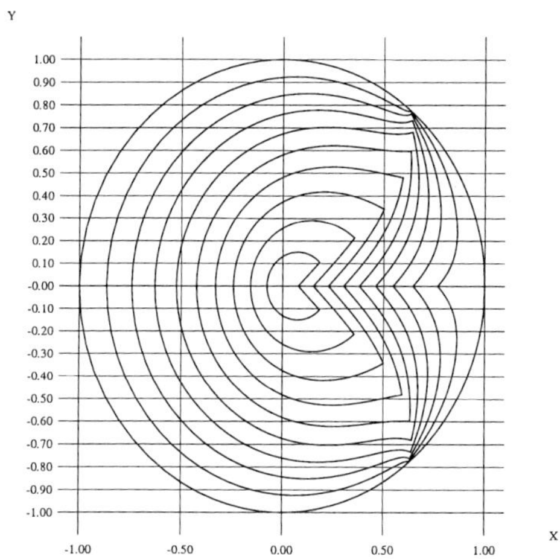

Figure 4. Pac-Man mapped curves

§5 Convergence

In this section we discuss the convergence of the solution to (2.14) of the approximation constructed in §4.1–4.2.

5.1 Nyström's Approach

The exact equation to be solved is

$$u + K\,u = g, \qquad K\,u \equiv \mathrm{Re}\;\mathcal{S}\,u, \tag{5.1.1}$$

whereas the approximate equation is

$$u_m + K_\varepsilon\,u_m = g_m, \tag{5.1.2}$$

with $m = n + 2\sum_{p=1}^{n} N_p$, and with N_p defined in terms of ε as in (4.1.12). Letting $\zeta \in \partial B$, the entries in this approximating equation are defined as follows:

$$\begin{aligned} u_m(\zeta) &= \sum_{q=1}^{n}\sum_{\ell=-N_{q-1}}^{N_q} c_\ell^{(q)}\,\omega_\ell^{(q)}(\zeta)\,\chi_q(\zeta), \\ g_m(\zeta) &= \sum_{q=1}^{n}\sum_{\ell=-N_{q-1}}^{N_q} g\left(z_\ell^{(q)}\right)\,\omega_\ell^{(q)}(\zeta)\,\chi_q(\zeta), \\ K_m\,u_m(\zeta) &= \sum_{q=1}^{n}\sum_{\ell=-N_{q-1}}^{N_q} c_\ell^{(q)}\;\mathrm{Re}\left(d_\ell^{(q)}(\zeta)\right), \end{aligned} \tag{5.1.3}$$

where the $c_\ell^{(q)}$ are real, and where the first two equations in (5.1.3) stem from (4.1.11) and the third from (4.1.19). We assume here that the numbers $c_\ell^{(q)}$ have been determined by solving the linear system of equations (2.2.8). That is, for $\zeta \in \partial B$ we have the *Nyström approximation*

$$u_N(\zeta) \approx g(\zeta) - \sum_{q=1}^{n}\sum_{\ell=-N_{q-1}}^{N_q} c_\ell^{(q)}\;\mathrm{Re}\left(d_\ell^{(q)}(\zeta)\right). \tag{5.1.4}$$

5.2 Function Spaces

Definition: *Let D_j and φ_j be defined as at the end of §1, for $j = 1, 2, \ldots, n$. Let $\mathbf{W}$ denote the family of all functions w that are defined on each region $\mathcal{D}_j$, such that w is continuous on ∂B and such that, in the notation of Theorem 1, $w \in \mathbf{M}_{\beta_{j-1},\beta_j}(D_j)$ for $j = 1, 2, \ldots, n$. Let us set*

$$\mathbf{W}_R = \{\mathrm{Re}\,w : w \in \mathbf{W}\}, \tag{5.2.1}$$

and $D = \cup_{j=1}^{n} D_j$. It is readily seen that the following three spaces of functions are Banach spaces, each normed by $\|w\| = \sup_{z\in\partial B}|w(z)|$:

1. *The space* $\mathbf{C}(\partial B)$ *of functions that are continuous on* ∂B*;*

2. *the space of functions* $\mathbf{W}$*; and*

3. *the space of functions* $\mathbf{W}_R$.

Lemma 2. *Let* $\mathcal{G} = g + i\,h$ *be defined as in (2.2.1)–(2.2.2). Then* $\mathcal{G} \in \mathbf{W}$*, so that both* g *and* h *belong to* $\mathbf{W}_R$.

Proof: By Theorem 1, $\mathcal{G} \in \mathbf{W}$. ■

Theorem 3. *Eq. (5.1.1) has a unique solution* $u \in \mathbf{W}_R$*. Moreover,* $Ku \in \mathbf{W}_R$.

Proof: By our discussion in §2, the equation (5.1.1) (which is the same as Eq. (2.2.6)) has a unique real valued solution u that is Hölder continuous at each point of smoothness of ∂B (see [10], Ch. 1, §3). By the Sokhotski formulas we may therefore set

$$u(\zeta) = w_+(\zeta) - w_-(\zeta), \qquad (Su)(\zeta) = w_+(\zeta) + w_-(\zeta), \quad \zeta \in \partial B, \quad (5.2.2)$$

with $w_\pm$ Hölder continuous on ∂B. By substituting this expression into the equation $(I + K)\,u = g$, or equivalently, by using (2.2.4), we find that $2\,w_+ = g + i\,h - i\,\delta$. Hence by Theorem 1 the function w_+ has an analytic extension into $B \cup D$, with D defined as above. It therefore follows that $w_+ \in \mathbf{W}$.
Let us next show that $w_- \in \mathbf{W}$. To this end, let us set $B^* = B \setminus \overline{D}$. Starting with the assumed known restriction of the function $w_+ \in \mathbf{W}$ to ∂B^*, where it is Hölder continuous, let us *define* w_-^* by the solution to the *Riemann problem*

$$\operatorname{Re}\left\{i\,w_-^*(\zeta)\right\} = \operatorname{Re}\left\{i\,w_+(\zeta)\right\}, \quad \zeta \in \partial B^*. \qquad (5.2.3)$$

It then follows that this function w_-^* has an analytic extension into $\mathbb{C} \setminus \overline{B^*}$, and moreover that the boundary condition $w_-^*(\infty) = 0$ uniquely determines w_-^* (see [16], p. 132). Furthermore, by *Point #3* of Assumption I, ∂B^* is Hölder continuous, and therefore w_-^* is Hölder continuous on $\mathbb{C} \setminus B^*$. That is, $w_-^* \in \mathbf{W}$.
Clearly, if we now solve a second Riemann problem similar to (5.2.3), this time with ∂B^* replaced by ∂B, with w_-^* replaced by w_-, with the function

w_+ defined on ∂B as in (5.2.2), and subject to $w_-(\infty) = 0$, we must deduce as a consequence of the existence and uniqueness of the solution of both of these Riemann problems that $w_- = w_-^*$ on ∂B.
It then follows by (5.2.2) that the real valued solution u of (5.1.1) belongs to $\mathbf{W}_R$.
Finally, since $u \in \mathbf{W}_R$, and $g \in \mathbf{W}_R$, it follows that $K\,u = g - u \in \mathbf{W}_R$. ■

In the next Lemma, the space $\mathbf{L}_{\beta_{q-1},\beta_q}(D_q)$ is defined in the paragraph of Eqs. (6.2) and (6.3) of the Appendix.

Lemma 3. *Let $w \in \mathbf{W}$, so that w restricted to D_q belongs to the space of functions $\mathbf{M}_{\beta_{q-1},\beta_q}(D_q)$, and let u_q be defined as in (4.1.2). Then:*

(i) $w - u_q \in \mathbf{L}_{\beta_{q-1},\beta_q}(D_q)$;

(ii) For $\zeta \in \Gamma_p$, $p \neq q$, $w_q^* \in \mathbf{L}_{\beta_{q-1},\beta_q}(D_q)$, where $w_q^*(t) = (w(t) - u_q(t))/\left(\varphi_q'(t)(t-\zeta)\right)$.

Proof: *Part (i):* This part of the proof is immediate, since both w and u_q belong to $\mathbf{M}_{\beta_{q-1},\beta_q}(D_q)$, and since $w - u_q$ vanishes at the endpoints of Γ_q.

Part (ii): Since $\zeta \notin \Gamma_q$, we immediately have for some constant C_0 that $|t-\zeta| \geq (1/C_0)|t-a_{q-1}|\,|t-a_q|$, and so

$$\left|\frac{w(t)-u_q(t)}{(t-\zeta)\varphi_q'(t)}\right| \leq \frac{C_0|w(t)-u_q(t)|}{|\varphi_q'(t)|\,|t-a_{q-1}|\,|a_q-t|}. \tag{5.2.4}$$

Since by *Part (i)* of the lemma we have $w - u_q \in \mathbf{L}_{\beta_{q-1},\beta_q}(D_q)$, we only need to show that there exists a constant C_1 such that

$$\frac{1}{|\varphi_q'(t)|\,|t-a_{q-1}|\,|a_q-t|} \leq C_1. \tag{5.2.5}$$

But by (3.1.7)–(3.1.8) it follows, with $t = \zeta_q(\tau)$, that

$$\frac{1}{|t-a_{q-1}|\,|a_q-t|} \leq \frac{C_2}{|\tau|\,|1-\tau|}, \quad \tau \in \mathcal{E}_d. \tag{5.2.6}$$

Next, by (1.3.6) we have

$$\begin{aligned}\frac{1}{\varphi_q'(t)} &= \frac{\zeta_q^{[-1]}(t)[1-\zeta_q^{[-1]}(t)]}{\dot\zeta(\tau)}\\ &= \frac{\tau(1-\tau)}{\dot\zeta(\tau)}.\end{aligned} \tag{5.2.7}$$

But by Assumption #3, the term $\dot{\zeta}$ is non–vanishing in $\overline{\mathcal{E}_d}$, and hence the inequality (5.2.5) follows. ■

Theorem 4. *Let the numbers $d_\ell^{(q)}(\zeta)$ be defined as in (4.1.18). Let ε, N_p, and h_p be related as in (4.1.12). If $u \in \mathbf{W}_R$, then for every $p = 1, \ldots, n$ and $k = -N_{p-1}, \ldots, N_p$, as $\varepsilon \to 0$,*

$$\left| (K\,u)\left(z_k^{(p)}\right) - \sum_{q=1}^{n} \sum_{\ell=-N_{q-1}+1}^{N_q} \operatorname{Re}\left\{ d_\ell^{(q)}\left(z_k^{(p)}\right)\right\} u\left(z_k^{(p)}\right) \right| = \mathcal{O}\left(\varepsilon\right). \tag{5.2.8}$$

Proof: Let $w \in \mathbf{W}$, and for $\zeta = z_k^{(p)}$ a *Sinc point* of Γ_p, consider the expression

$$W_q^*(\zeta) = \frac{P.V.}{\pi i} \int_{\Gamma_q} \frac{w(t)}{t-\zeta}\, dt. \tag{5.2.9}$$

Upon defining u_q as in (4.1.2), we have, by (4.1.4), that

$$\begin{aligned} W_q^*(\zeta) &= \frac{1}{\pi i}\left\{w(a_q) - w(a_{q-1}\right\} + u_q(\zeta)\, Q(\Gamma_q, \zeta) \\ &\quad + \frac{1}{\pi i} \int_{\Gamma_q} \frac{w(t) - u_q(t)}{t-\zeta}\, dt. \end{aligned} \tag{5.2.10}$$

If $q = p$, then *Part (i)* of Lemma 3 applies, and we can therefore use (6.18) to obtain an accurate approximation to the integral in (5.2.10). By taking $x = \zeta = z_m^{(p)}$ in (5.2.10) and using (6.16) and the notation of (4.1.16)–(4.1.17), we thus obtain

$$\begin{aligned} W_p^*(\zeta) &\approx \frac{1}{\pi i}\left\{w(a_p) - w(a_{p-1}\right\} + u_p(\zeta)\, Q(\Gamma_p, \zeta) \\ &\quad + \sum_{\ell=-N_{p-1}}^{N_p} \left(w\left(z_\ell^{(p)}\right) - u_p\left(z_\ell^{(p)}\right)\right) v_\ell^{(p)}(\zeta), \end{aligned} \tag{5.2.11}$$

with an error of the order of $\varepsilon/\log(1/\varepsilon)$ as $\varepsilon \to 0$.
On the other hand, if $q \neq p$, then *Part (ii)* of Lemma 3 applies, and in this case we may use the *Sinc quadrature formula* (6.11), to get

$$\begin{aligned} W_q^*(\zeta) &\approx \frac{1}{\pi i}\{w(a_q) - w(a_{q-1}\} + u_q(\zeta)\, Q(\Gamma_q, \zeta) \\ &\quad + \sum_{\ell=-N_{p-1}}^{N_p} \left(w\left(z_\ell^{(q)}\right) - u_q\left(z_\ell^{(p)}\right)\right) v_\ell^{(q)}(\zeta), \end{aligned} \tag{5.2.12}$$

with once again, $v_\ell^{(q)}(\zeta)$ defined as in (4.1.17), and with an error of the order of $\varepsilon/\log(1/\varepsilon)$, $\varepsilon \to 0$.

It thus follows, in the notation of (4.1.8), that

$$\begin{aligned} \mathcal{S}\, w(\zeta) &= \frac{1}{\pi i}\int_{\partial B} \frac{w(t)}{t-\zeta}\, dt = \sum_{q=1}^{n} W_q^*(\zeta) \\ &\approx U_p(\zeta) + V_p(\zeta) \\ &\quad + \sum_{q=1}^{n} \sum_{\ell=-N_{q-1}}^{N_q} \left(w\left(z_\ell^{(q)}\right) - u_q\left(z_\ell^{(q)}\right)\right) v_\ell^{(q)}(\zeta) \end{aligned} \tag{5.2.13}$$

with an error of the order of $\varepsilon/\log(1/\varepsilon)$, $\varepsilon \to 0$.

We next wish to show that we also have

$$\mathcal{S}\, w(\zeta) \approx \sum_{q=1}^{n} \sum_{\ell=-N_{q-1}}^{N_q} w\left(z_\ell^{(q)}\right) d_\ell^{(q)}(\zeta) \tag{5.2.14}$$

with an error of the order of $\varepsilon/\log(1/\varepsilon)$, $\varepsilon \to 0$, where the $d_\ell^{(q)}(\zeta)$ are defined in (4.1.18).

In view of (4.1.6) and (4.1.18), we see that the main discrepancy between the two expressions (5.2.13) and (5.2.14) is that (5.2.13) contains values of w at the corner points of ∂B, whereas in (5.2.14) these have been replaced by values at the *Sinc points* $z_{N_{q-1}}^{(q)}$ and $z_{N_q}^{(q)}$. Since, however, $w \in \mathbf{M}_{\beta_{q-1},\beta_q}(D_q)$, it follows using (6.3) and the definition (6.2) of ρ that there are constants C, C_1 and C_2 such that

$$\begin{aligned}\left|w\left(a_q\right)-w\left(z_{N_q}^{(q)}\right)\right| &\leq C\left|\rho\left(z_{N_q}^{(q)}\right)\right|^{-\beta_q} \\ &= C\, e^{-N_q\, h_q\, \beta_q}=\frac{C}{h_q}\, h_q\, e^{-N_q\, h_q\, \beta_q} \\ &\leq \frac{C_1}{h_q} \frac{\varepsilon}{\ln\left(\frac{1}{\varepsilon}\right)} \\ &\leq C_2\, \varepsilon .\end{aligned} \tag{5.2.15}$$

Similarly, we have

$$w\left(a_{q-1}\right)-w\left(z_{-N_{q-1}}^{(q)}\right)=\mathcal{O}(\varepsilon). \tag{5.2.16}$$

We then form the difference between the two expressions (5.2.13) and (5.2.14) and, based on these estimates, we can show that this difference is of the order of the right hand sides of (5.2.15) and (5.2.16). We omit the details. ■

Let us recall that the spaces $\mathbf{C}(\partial B)$, $\mathbf{W}$, and $\mathbf{W}_R$ are Banach spaces under $\|\cdot\|$, the uniform norm on ∂B. Observing that the basis functions $\left\{\omega_m^{(p)}\right\}_{-N_{p-1}}^{N_p}$ are continuous on the closure of Γ_p, let us determine a strictly decreasing sequence of positive numbers ε_s, $s=1,2,\ldots$. Corresponding to a given $\varepsilon=\varepsilon_s$, keeping in mind the definition (4.1.12) of N_p in terms of ε, we set $m=n+2\sum_{p=1}^{n} N_p$ and we define an operator $K_m : \mathbf{W}_R \to \mathbf{C}(\partial B)$ by

$$\begin{aligned}&(K_m\, u)(\zeta) \\ &=\sum_{p=1}^{n} \sum_{k=-N_{p-1}}^{N_p}\left\{\sum_{q=1}^{n} \sum_{\ell=-N_{q-1}+1}^{N_q} \operatorname{Re}\left\{d_\ell^{(q)}\left(z_k^{(p)}\right)\right\}\, u\left(z_\ell^{(q)}\right)\right\} \chi_p(\zeta)\, \omega_k^{(p)}(\zeta).\end{aligned} \tag{5.2.17}$$

By Theorem 7, we have

Corollary 1. *Let u and u_m denote the solutions to (5.1.1) and (5.1.2) respectively. Then $\|K\, u-K_m\, u\| \to 0$ as $s \to \infty$.*

Lemma 4. *Given any compact family of functions $U \in \mathbf{W}_R$, the set $\mathcal{A}(U) \equiv \{K_m u : u \in U\}$ is relatively compact. That is, the set $\mathcal{A}(U)$ is collectively compact.*

Proof: In view of the uniform convergence of $K_m u \to K u \in \mathbf{W}_R$, the closure of the set $\mathcal{A}(U)$ is compact, *i.e.* $\mathcal{A}(U)$ is relatively compact. ■

Theorem 5. *If K_m and K are defined as above, then*

$$\left\|(K_m - K)\,(I + K)^{-1}\right\| \to 0 \tag{5.2.18}$$

as $s \to \infty$.

Proof: It follows from Theorem 3 above that, since $I + K$ is a continuous operator defined on $\mathbf{W}_R$ with uniformly bounded inverse, a compact set $U \in \mathbf{W}_R$ is mapped by $(I+K)^{-1}$ onto another compact set, call it $V \in \mathbf{W}_R$. Hence it follows, by proceeding as in the proof of Theorem 10.1 of [23], that given any $\varepsilon > 0$, there exists $m(\varepsilon)$ such that when $m > m(\varepsilon)$, then $\|(K_m - K)(I+K)^{-1}\, u\| < \varepsilon$ for all $u \in U$. Hence $\|(K_m - K)(I+K)^{-1}\| \leq \varepsilon$ for all $m \geq m(\varepsilon)$. ■

We therefore have

Corollary 2. *The inverse, $(I+K_m)^{-1}$ exists for all m sufficiently large such that $\|(K_m - K)(I + K)^{-1}\| < 1$.*

Proof: Set

$$T \equiv I + (K_m - K)(I + K)^{-1}. \tag{5.2.19}$$

Then T^{-1} exists, by Banach's theorem, and since

$$I + K_m = T\,(I + K), \tag{5.2.20}$$

the existence of $(I + K_m)^{-1}$ follows from

$$(I + K)^{-1}\, T^{-1} = (I + K_m)^{-1}. \quad \blacksquare \tag{5.2.21}$$

The following result now follows from the inequality (10.2) of [23].

Theorem 6. *Let u denote the exact solution of the boundary integral equation (5.1.1), and let u_m denote the solution determined by solving the system (2.2.8). Then*

$$\|u - u_m\| = \mathcal{O}(\varepsilon), \quad s \to \infty. \tag{5.2.22}$$

Having determined an accurate approximation u_m to the solution of the integral equation (5.1.1), we can then accurately determine $\mathcal{G}$ on ∂B. The procedure described in §4.2 then enables us to determine $\mathcal{G}$ and hence f in the interior of B, the approximations being readily justified by the results of Theorem 11 of the appendix.

§6 Appendix: Sinc spaces of approximation

We shall describe here some notions about approximation by *Sinc methods* that we shall need for the approximate determination of the functions f and F that are defined at the outset of this paper. Proofs of the results in this section can be found in [22].

Let $\mathcal{D}$ denote a simply connected domain in the complex plane $\mathbb{C}$, let $1 \le p \le \infty$, and let $\mathbf{H}^p(\mathcal{D})$ denote the family of all functions f that are analytic in $\mathcal{D}$ and such that

$$N_p(f, \mathcal{D}) \equiv \begin{cases} \left(\int_{\partial \mathcal{D}} |f(z)|^p |dz| \right)^{1/p} < \infty & \text{if } 1 \le p < \infty, \\ \sup_{z \in \mathcal{D}} |f(z)| < \infty & \text{if } p = \infty. \end{cases} \tag{6.1}$$

Let φ denote a function which provides a conformal mapping of a simply connected region $\mathcal{D}$ onto $\mathcal{D}_d$, where $\mathcal{D}_d$ is defined as in (1.3.5). Moreover, let us assume that φ is also a one-to-one map of Γ onto $\mathbb{R}$, with Γ an arc in $\mathcal{D}$, having endpoints a and b, so that $\varphi(a) = -\varphi(b) = \infty$.

Let us now consider some simple *Sinc spaces* which go hand-in-hand with *Sinc methods*. Let $\alpha \in (0, 1]$, $\beta \in (0, 1]$, and $0 < d < \pi$. Let $\mathbf{M}_{\alpha,\beta}(\mathcal{D})$ denote the family of all functions $f \in \mathbf{H}^\infty(\mathcal{D})$, which have finite limits $f(a)$ and $f(b)$ at the endpoints a and b of Γ, such that if we set

$$\rho(z) = \exp(\varphi(z)), \tag{6.2}$$

then

$$\begin{aligned} f(z) - f(a) &= \mathcal{O}(|\rho(z)|^{\alpha}) \text{ as } z \to a, \\ f(z) - f(b) &= \mathcal{O}(|\rho(z)|^{-\beta}) \text{ as } z \to b, \end{aligned} \tag{6.3}$$

where the limits are taken with respect to z from within $\mathcal{D}$. The notation $\mathbf{L}_{\alpha,\beta}(\mathcal{D})$ will be used for that subset of functions $f \in \mathbf{M}_{\alpha,\beta}(\mathcal{D})$ for which $f(a) = f(b) = 0$.

Let $\mathcal{D}_d$ be defined as in (1.3.5) and, for $0 < d' < d$, let φ map $\mathcal{D}'$ to $\mathcal{D}_{d'}$. The most important properties of these spaces are, in essence, that if $f \in \mathbf{M}_{\alpha,\beta}(\mathcal{D})$, then $f'/\varphi' \in \mathbf{L}_{\alpha,\beta}(\mathcal{D}')$; if $f \in \mathbf{M}_{\alpha,\beta}(\mathcal{D})$, and if $(1/\varphi')'$ is uniformly bounded in $\mathcal{D}$ then $f^{(n)}/(\varphi')^n \in \mathbf{L}_{\alpha,\beta}(\mathcal{D}')$, $n = 1, 2, 3, \ldots$. Conversely, if $f'/\varphi' \in \mathbf{L}_{\alpha,\beta}(\mathcal{D})$, then $f \in \mathbf{M}_{\alpha,\beta}(\mathcal{D})$. For example, if (a, b) is a finite interval, if f is of class $\mathbf{Lip}_{\alpha}$ on [a,b], and if f is analytic in some region containing (a, b), then $f \in \mathbf{M}_{\alpha,\alpha}(\mathcal{D})$.

Next, we define some bases for Sinc approximation.

Corresponding to a positive integer N, let us define integers M and N by the equations

$$\begin{aligned} N &= \text{positive integer}; \\ M &= [\beta N/\alpha]; \\ m &= M + N + 1, \end{aligned} \tag{6.4}$$

where $[\cdot]$ denotes the greatest integer function. Letting $\mathbb{Z}$ denote the set of all integers, let us now set

$$\begin{aligned} \operatorname{sinc}(z) &= \frac{\sin(\pi z)}{\pi z}, \\ h &= \left(\frac{\pi d}{\beta N}\right)^{1/2}, \\ z_j &= \varphi^{-1}(jh), \quad j \in \mathbb{Z} \\ \sigma_j &= \operatorname{sinc}\{[\varphi - jh]/h\}, \quad j = -M, \ldots, N, \\ \omega_j &= \sigma_j, \quad j = -M+1, \ldots, N-1, \\ \omega_{-M} &= \frac{1}{1+\rho} - \sum_{j=-M+1}^{N} \frac{1}{1+e^{jh}} \sigma_j, \\ \omega_N &= \frac{\rho}{1+\rho} - \sum_{j=-M}^{N-1} \frac{e^{jh}}{1+e^{jh}} \sigma_j, \\ \varepsilon_N &= N^{1/2} e^{-(\pi d \beta N)^{1/2}}. \end{aligned} \tag{6.5}$$

The points z_j defined here are called the *Sinc points.* It is more convenient in the special case when we know *a priori* that f vanishes at the endpoint $a = \varphi^{-1}(-\infty)$ of Γ, to take $\omega_{-M} = \sigma_{-M}$, and it is similarly more convenient

to take $\omega_N = \sigma_N$ in the case when f vanishes at the right–hand end point b of Γ.
Let us also define a *row vector,* $\mathbf{w}$ *of basis functions*, a diagonal matrix $D(u)$, and an operator V by the equations

$$\begin{aligned} \mathbf{w}(z) &= (\omega_{-M}(z), \ldots, \omega_N(z)) \\ D(u) &= \operatorname{diag}[u(z_{-M}), \ldots, u(z_N)]; \\ V(u) &= (u(z_{-M}), \ldots, u(z_N))^T, \end{aligned} \tag{6.6}$$

with ω_j defined as in (6.5), with u an arbitrary function defined on Γ, and with "$\mathbf{c}^T$" denoting the transpose of the vector $\mathbf{c}$.

We shall also define a norm by

$$\|f\| = \sup_{x \in \Gamma} |f(x)|. \tag{6.7}$$

Next, we describe *interpolation, definite and indefinite integration, and the approximation of Hilbert transforms* via *Sinc methods.*

- *Sinc Interpolation*

  This basic procedure is described in the following theorem.

Theorem 7. *If $f \in \mathbf{M}_{\alpha,\beta}(\mathcal{D})$, then as $N \to \infty$,*

$$\|f - \mathbf{w}\, V\, f\| = \mathcal{O}(\varepsilon_N). \tag{6.8}$$

- *Sinc Collocation*

  While the above defined *Sinc basis* completely determines the formulas of *Sinc approximation*, the role of the basis is primarily in the evaluation or plotting of the final solution. This is the point of the theorem of this section, which enables us to reduce all *Sinc computations* to the solution of (linear or nonlinear) algebraic equations.

Theorem 8. *If the conditions of Theorem 7 are satisfied, and if $\mathbf{c} = (c_{-M}, \ldots, c_N)^T$ is a vector of complex numbers, such that*

$$\max_{j=-M,\cdots,N} |f(z_j) - c_j| = \delta, \tag{6.9}$$

then, as $N \to \infty$,

$$\|f - \mathbf{w} \cdot \mathbf{c}\| = \mathcal{O}\left(\varepsilon_N + \delta\left(\log(\varepsilon)\right)^2\right). \tag{6.10}$$

Proof: Using the triangle inequality, we get $\|f - \mathbf{w}\,\mathbf{c}\| \le \|f - \mathbf{w}\,V\,f\| + \|\mathbf{w}\|\max_j |c_j - f(z_j)| \le \varepsilon_N + \|\mathbf{w}\|\,\delta$. But since $\|\mathbf{w}\| = \mathcal{O}(\log(N))$ ([28], p. 142), and since $\varepsilon_N = \mathcal{O}\left(\exp(-c' N^{1/2})\right)$, we have $N = \mathcal{O}\left((\log(\varepsilon_N))^2\right)$, from which (6.10) follows. ∎

- *Sinc Quadrature*

This simple *Sinc quadrature* approximation is described as follows.

Theorem 9. *If $f/\varphi' \in \mathbf{L}_{\alpha,\beta}(\mathcal{D})$, then, as $N \to \infty$,*

$$\left| \int_\Gamma f(x)\,dx - (V(h/\varphi'))^T\,V f \right| = \mathcal{O}(\varepsilon_N). \tag{6.11}$$

Indeed, it may be shown, that if $h = (2\pi d/(\beta N))^{1/2}$, then the right hand side of this equation may be replaced by $\mathcal{O}\left(\varepsilon_N^{\sqrt{2}}\right)$.

- *Sinc Indefinite Integration*

At the outset, we define numbers σ_k and e_k, by

$$\begin{aligned} \sigma_k &= \int_0^k \operatorname{sinc}(x)\,dx, \quad k \in \mathbb{Z}, \\ e_k &= \frac{1}{2} + \sigma_k. \end{aligned} \tag{6.12}$$

It is then convenient to define indefinite integration operators $\mathcal{J}$ and $\mathcal{J}'$ by $(\mathcal{J}f)(x) = \int_a^x f(t)\,dt$, $(\mathcal{J}'f)(x) = \int_x^b f(t)\,dt$. We define a Toeplitz matrix $I^{(-1)}$ of order m by $I^{(-1)} = [e_{i-j}]$, with e_{i-j} denoting the $(i,j)^{th}$ element of $I^{(-1)}$, and we then define the operators $\mathcal{J}_m$ and $\mathcal{J}'_m$ by $\mathcal{J}_m = \mathbf{w}\,A\,V$, and $\mathcal{J}'_m = \mathbf{w}\,B\,V$, with $A = h\,I^{(-1)}\,D(1/\varphi')$ and $B = h\,\left(I^{(-1)}\right)^T\,D(1/\varphi')$. We then have:

Theorem 10. *If $f/\varphi' \in \mathbf{L}_{\alpha,\beta}(\mathcal{D})$, then, as $N \to \infty$,*

$$\begin{aligned} \|\mathcal{J}\,f - \mathcal{J}_m\,f\| &= \mathcal{O}(\varepsilon_N) \\ \|\mathcal{J}'\,f - \mathcal{J}'_m\,f\| &= \mathcal{O}(\varepsilon_N). \end{aligned} \tag{6.13}$$

- *Cauchy and Hilbert Transforms*

We take $x \in \Gamma$, and $z \in \mathcal{D}^+$, the part of $\mathcal{D}$ to the right of Γ, and define the functions

$$s_k(x) = \frac{h}{\pi} \frac{\sin\{\pi[\varphi(x) - kh]/h\}}{\varphi'(z_k)(x - z_k)} \tag{6.14}$$

$$c_k(z) = \frac{h}{2\,\pi\, i} \frac{\exp\left\{\frac{\pi i}{h}(\varphi(z) - kh)\right\} - 1}{\varphi'(z_k)(z - z_k)}. \tag{6.15}$$

Since $c_k(x) = \frac{1}{2}(s_k(x) + t_k(x))$, we have, for k and ℓ integers, that

$$s_k(z_\ell) = \begin{cases} 0 & \text{if} \quad k \neq \ell \\ 1 & \text{if} \quad k = \ell, \end{cases}$$

$$t_k(z_\ell) = \begin{cases} \dfrac{h}{\pi\, i} \dfrac{1 - (-1)^{k-\ell}}{\varphi'(z_k)\,(z_k - z_\ell)} & \text{if} \quad k \neq \ell \\ 0 & \text{if} \quad k = \ell, \end{cases} \tag{6.16}$$

$$c_k(z_\ell) = \begin{cases} \dfrac{h}{2\pi\, i} \dfrac{1 - (-1)^{k-\ell}}{\varphi'(z_k)\,(z_k - z_\ell)} & \text{if} \quad k \neq \ell \\ \frac{1}{2} & \text{if} \quad k = \ell. \end{cases}$$

The proof of the following theorem may be found in [6], [29].

Theorem 11. *Let α and β denote positive numbers such that $0 < \alpha \leq 1$, $0 < \beta \leq 1$, let $g \in \mathbf{L}_{\alpha,\beta}(\mathcal{D})$, let N be a positive integer and let us select M as in (6.4), h, z_j and ε_N as in (6.5). Let ∂D^+ be Hölder continuous, with $\mathcal{D}^+$ denoting the part of $\mathcal{D}$ to the right of Γ as we traverse Γ from a to b. Then there exist positive constants C_i, $i = 1, 2, 3$, that are independent of N, such that*

$$\sup_{x \in \Gamma} \left| g(x) - \sum_{j=-M}^{N} g(z_j)\, s_j(x) \right| < C_1\, \varepsilon_N, \quad x \in \Gamma, \tag{6.17}$$

$$\sup_{x \in \Gamma} \left| \frac{P.V.}{\pi i} \int_\Gamma \frac{g(t)}{t - x}\, dt - \sum_{j=-M}^{N} g(z_j)\, t_j(x) \right| < C_2\, \varepsilon_N, \quad x \in \Gamma, \tag{6.18}$$

$$\sup_{z \in \mathcal{D}^+} \left| \frac{1}{2\pi i} \int_\Gamma \frac{g(t)}{t-z}\, dt - \sum_{j=-M}^{N} g(z_j)\, c_j(z) \right| \quad < \quad C_3\, \varepsilon_N\,, \quad z \in \mathcal{D}^+. \tag{6.19}$$

Acknowledgement. Part of this research was carried out while Stenger was visiting the University of New South Wales during February and March of 1997. The hospitality of the UNS and the support of the Australian Research Council are gratefully acknowledged. In addition, Stenger also gratefully acknowledges partial support of this work by NSF grant # CCR-9307602, as well as the support the University of Aizu, Japan, during the summer of 1997. The authors are grateful to Jean–Paul Berrut, to Dieter Gaier, Nick Papamichael, and to the referee for valuable criticisms.

References

[1] Bergström, H., An approximation of the analytic function mapping a given domain inside and outside of the unit circle, *Mem. Publ. Soc. Arts Lettr. Hainant* Vol. Hors Série (1958), 193–198.

[2] Berrut, J.–P., A Fredholm integral equation of the second kind for conformal mapping, *J. Comp. Appl. Math.* **14** (1986) 99–110.

[3] Berrut, J.–P., Integralgleichungen und Fourier–Methoden zur numerischen konformen Abbildung, dissertation, Eidgenössische Technische Hochschule Zürich, 1985.

[4] Carathéodory, C., *Conformal Representation*, Cambridge Tracts in Mathematics and Mathematical Physics, 1958.

[5] Davis, P. J., Orthogonalization codes in numerical analysis, in *Survey of Numerical Analysis*, J. Todd (ed.), McGraw–Hill, 1962, Ch. 10.

[6] Elliott E. and F. Stenger, Sinc method of solution of singular integral equations, in *IMACS Conference on CSIE*, PA, Philadelphia, 1984, 155–166.

[7] Friberg, M. S., A new method for the effective determination of conformal maps, dissertation, Univ. of Minn., 1951.

[8] Fornberg, B., A numerical method for conformal mapping of doubly connected regions, *SIAM J. Sci. Stat. Comput.* **1** (1980), 386–400.

[9] Fornberg, B., A numerical method for conformal mappings, *SIAM J. Sci. Stat. Comput.* **5** (1984), 771–783.

[10] Gaier, D., *Konstruktive Methoden der konformen Abbildung*, Springer–Verlag, 1964.

[11] Gaier, D., Integralgleichungen erster Art und konforme Abbildung, *Math. Z.* **147** (1976), 113–129.

[12] Gaier, D. and W. K. Hayman, Moduli of long quadrilaterals and thick ring domains, *Rend. Mat. Appl.* **10** (1990), 809–834.

[13] Gaier, D. and W. K. Hayman, On the computation of modules of long quadrilaterals, *Constr. Approx.* **7** (1991), 453–467.

[14] Gakhov, F. D., *Boundary Value Problems*, Pergamon Press, Oxford, 1966, (in Russian).

[15] Gutknecht, M. H., Numerical conformal mapping methods based on function conjugation, *J. Comp. Appl. Math.* **14** (1986), 31–77.

[16] Henrici, P., *Applied and Computational Complex Analysis*, Vol. III, Wiley, 1986.

[17] Hough, D. M. and N. Papamichael, The use of splines and singular functions in an integral equation method for conformal mapping, *Numer. Math.* **37** (1981), 133–147.

[18] Hough, D. M., User's Guide to CONFPAC, IPS Research Report No. 90–11, ETH Zürich, 1990.

[19] Howe, D., The application of conformal methods to the conformal transformation of polygonal boundaries, *J. Inst. Maths Applics* **12** (1973), 125–136.

[20] Hübner, O., The Newton method for solving Theodorsen's equation, *J. Comp. Appl. Math.* **14** (1967), 19–30.

[21] Jawson, M. A. and G. T. Symm, *Integral Equation Methods in Potential Theory*, Academic Press, London, 1976.

[22] Kowalski, M., K. Sikorski and F. Stenger, *Selected Topics in Approximation and Computation*, Oxford, 1995.

[23] Kress, R., *Linear Integral Equations*, Springer–Verlag, 1989.

[24] Papamichael, N. and N. S. Stylianopoulos, A domain decomposition method of conformal mapping onto a rectangle, *Constr. Approx.* **7** (1991), 63–83.

[25] Papamichael, N. and N. S. Stylianopoulos, The asymptotic behaviour of conformal modules of quadrilaterals with applications to the estimation of resistance values, *Constr. Approx.* **15** (1999), 109–134.

[26] Reifenberg, M., Numerical solution of a boundary integral equation for conformal mapping by means of attenuation factors, dissertation, University of Fribourg (Switzerland), 1997.

[27] Reifenberg, M. and J.–P. Berrut, Numerical solution of boundary integral equations by means of attenuation factors, Universität Freiburg (Schweiz), Mathematisches Institut Report No. 96–2 (1996).

[28] Stenger, F., *Numerical Methods Based on Sinc and Analytic Functions*, Springer-Verlag, New York, 1993.

[29] Stenger, F., Sinc approximation of Cauchy–type integrals over arcs, 1998 to appear in *Proceedings of a meeting in Honor of David Elliott's* 65th *Birthday.*

[30] Symm, G. T., An integral equation method in conformal mapping, *Numer. Math.* **9** (1966), 250–258.

[31] Symm, G. T., Numerical mapping of exterior domains, *Numer. Math.* **10** (1967), 437–445.

[32] Timman, R., The direct and the inverse problem of airfoil theory, a method to obtain numerical solutions, Nat. Luchtiv. Labor., Report #F16, Amsterdam, 1951.

[33] Trefethen, L. N., Numerical computation of the Schwarz–Christoffel transformation, *SIAM J. Sci. Stat. Comput.* **1** (1980), 82–102.

[34] Vertgeim, B. A., Approximation construction of some conformal mappings, *Doklad. Akad Nauk SSSR* **119** (1958), 12–14, (in Russian).

[35] Vecheslavov, V. V. and V. I. Kokoulin, Determination of the parameters of the conformal mapping of simply connected polygonal regions, *USSR Comp. Math. and Math. Phys.* **13** (1974), 57–65.

[36] Wegmann, R., Ein Iterationsverfahren zur konformen Abbildung, *Numer. Math.* **30** (1978), 453–466.

[37] Wegmann, R., Ein Iterationsverfahren zur konformen Abbildung, *ZAMM* **59** (1979), T85–T86.

[38] Wegmann, R., Conformal mapping by the method of alternating projections, *Numer. Math.* **56** (1989), 291–307.

Frank Stenger
Department of Computer Science
University of Utah
Salt Lake City, UT 84112
U.S.A.
stenger@cs.utah.edu

Ross Schmidtlein
Nuclear Engineering Program
University of Utah
Salt Lake City, UT 84112
U.S.A.
schmidtl@eng.utah.edu

The approximation of conformal structures via circle packing

Kenneth Stephenson

Abstract. This is a pictorial tour and survey of circle packing techniques in the approximation of classical conformal objects. It begins with numerical conformal mapping and the conjecture of Thurston which launched this topic, moves to approximation of more general analytic functions, and ends with recent work on the approximation of conformal tilings and conformal structures.

§1 Introduction

A *circle packing* is a configuration of circles with a specified pattern of tangencies. The regular hexagonal or "penny" packing in the plane — every circle tangent to six others — is certainly familiar to everyone, and the literature contains a smattering of other examples stretching back to the ancient Greeks. But I offer this as a first illustration of the type of packings we will discuss.

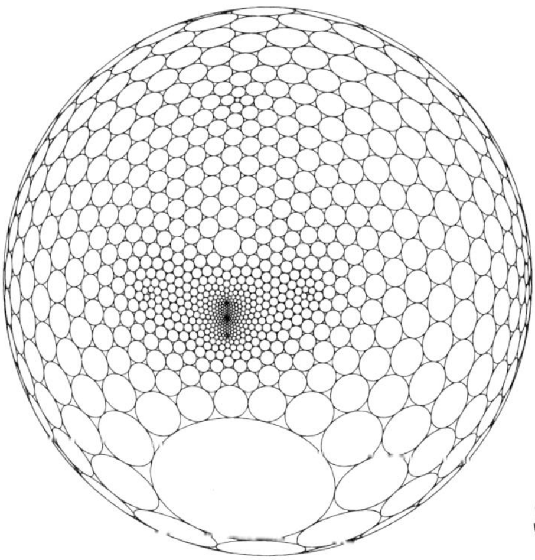

S^2

Computational Methods and Function Theory (CMFT'97)
N. Papamichael, St. Ruscheweyh and E. B. Saff (Eds.), pp. 551–582

ISBN 981-02-3626-3

Circle packing has its beginning as a distinct topic with applications to 3-manifolds in Thurston's Notes [20]. Its connections to analytic functions, the subject of this survey, can be traced precisely to a conjecture of Thurston at the 1985 Bieberbach conference concerning the use of circle packings to approximate conformal mappings and to its subsequent proof by Rodin and Sullivan [16].

I must strongly emphasis right at the beginning that our topic has almost **no contact** with the more familiar "sphere packing" studies, which address density and coverage issues — how many ping pong balls fit in a boxcar. Look to our first figure: it is the **pattern of tangencies** and the triangular interstices which are central. These circles have had to assume widely varied sizes in order to fit together in a prescribed pattern. This packing and the source of its combinatorics will come up later.

When I outline the basic notions of circle packing in a moment, you will find that these objects present an intriguing blend of local geometric rigidity and global geometric flexibility. Mappings between one configuration of circles realizing a certain tangency pattern and another configuration realizing the same pattern turn out to behave in fundamental ways like analytic mappings and have become central to the theory. Two lines of development have emerged regarding these maps. The first, following the lines of Thurston's 1985 conjecture, concerns their use in the **approximation** of analytic functions — that is the subject of this survey. However, the second, which addresses the many **analogies** between these maps and analytic functions, also plays a role here since discoveries in that theory present us with several new targets for approximation.

This will be a decidedly pictoral survey, and I will rely heavily on the reader's geometric feeling for analytic functions. We start with traditional "conformal mapping", where circle packing has little to recommend it, in a numerical sense, over classical methods. However, we move on to more general analytic functions, where it begins to show some advantages, and finally reach quite new, unbroken ground in the approximation of conformal structures. In recent applications to Riemann surfaces and conformal tiling which I survey here, circle packing may stand as the only technique available.

My thanks to the organizers of Computational Methods in Function Theory '97 and to the University of Cyprus for their hospitality. At the end of the paper I point the reader to sources for a circle packing bibliography and for the software package `CirclePack` used for the numerical experiments illustrated here.

1.1 Classical conformal geometry

Our view of analytic functions will be almost exclusively geometric — you will see no power series or integral formulae here, forget complex arithmetic. Think in terms of conformality (angle preserving), curvature and metrics, surfaces and coverings, branching and projections, 'mappings' not 'functions' — the rubber-sheet imagery suggested by Figure 1.

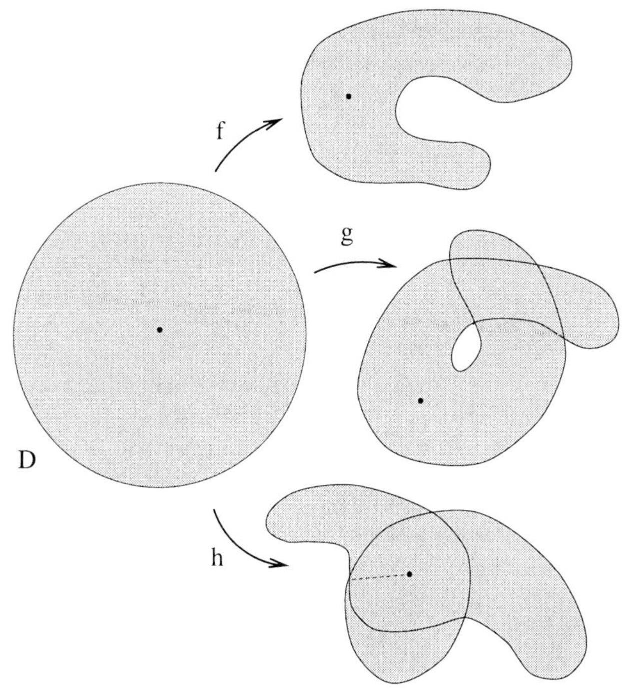

Figure 1. Geometry of classical analytic functions on the disc

In asking you to rely on geometric themes and intuition, it might help to recall a saying which goes to the heart of conformal geometry: namely, *analytic functions f map infinitesimal circles to infinitesimal circles.* Moreover, $|f'|$ is the stretching of these infinitesimal circles. Taking this image seriously and "zooming" in on points z and $f(z)$, one can imagine f mapping a pattern of infinitesimal circles in the domain to the pattern of infinitesimal image circles in the range. If you can get your mind around this image, you will not be far from the truth. In the discrete theory, we simply choose to work with actual circles!

1.2 Circle packing basics

The theory began on the sphere with this theorem, proven successively by Koebe, Andreev, and Thurston: *Let K be an abstract triangulation of a topological sphere. There exists a circle packing P on the Riemann sphere $\mathbb{S}^2$ having the combinatorics of K and P is unique up to Möbius transformations and inversions of $\mathbb{S}^2$.* The conditions on P are that each vertex of $v \in K$ is represented by a circle $c_v \in P$ and that if vertices v and w are connected by an edge in K, then c_v and c_w are tangent.

The richness of circle packings may be traced to their dual natures: *combinatoric* in the pattern of prescribed tangencies and *geometric* in the realization by actual circles. Until §2.4, our circle packings will lie in the sphere $\mathbb{S}^2$, the plane $\mathbb{C}$, or the hyperbolic plane, represented here as $\mathbb{D}$ with the Poincaré metric. Working with them requires a rather modest system of bookkeeping:

• *Complex:* K denotes an abstract simplicial 2-complex triangulating a topological surface $\mathcal{D}$.

• *Label:* R is a collection $\{R(v)\}$ of positive numbers associated with vertices v of K. If we are working in a metric space $\mathcal{D}$, these represent putative radii. We restrict attention to settings with Riemannian metrics of constant curvature $0, 1$, or -1, described as *euclidean*, *spherical*, or *hyperbolic*, respectively.

• *Angle sums:* A label R determines an angle sum $\theta_R(v)$ at each vertex of K. In particular, for each face $\langle v, u, w \rangle \in K$, let $\alpha_R(v; u, w)$ denote the angle at c_v in a triangle formed by the centers of a mutually tangent triple $\langle c_v, c_u, c_w \rangle$ of circles of radii $\langle R(v), R(u), R(w) \rangle$; the law of cosines appropriate to the geometry of R lets us compute α. Then $\theta_R(v) = \sum_{\langle v,u,w \rangle} \alpha_R(v; u, w)$, where the sum is over all faces $\langle v, u, w \rangle \in K$ containing v.

• *Packing:* P is a configuration of circles in the metric space $\mathcal{D}$ satisfying the tangency pattern of K. If the radii are listed in the label R, we would write $P \leftrightarrow K(R)$.

• *Carrier:* $\text{carr}(P)$ denotes the concrete geometric complex in $\mathcal{D}$ formed by connecting the centers of tangent circles of P with geodesic segments; this is simplicially equivalent to the abstract complex K.

Figure 2 illustrates these basics with a simple finite complex labeled K; (b) is a rather generic hyperbolic packing for K in $\mathbb{D}$, (c) a more deliberate euclidean packing with rectangular carrier, (d) is the maximal packing (in $\mathbb{D}$) to be discussed shortly. The carriers are shown for reference. Note that the *radii* and even the geometry change — it is the **combinatorics** which these packings share.

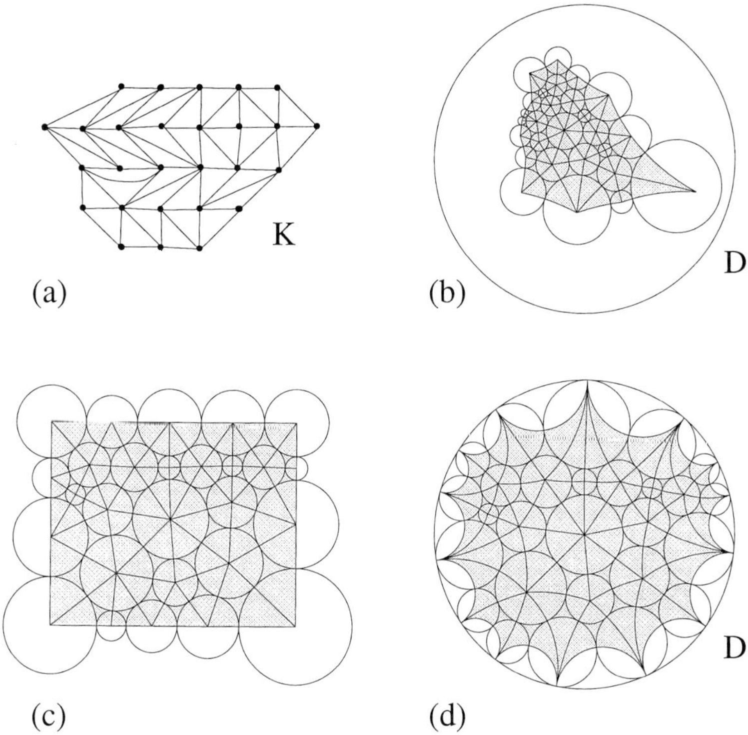

Figure 2. Circle packings for a simple complex K

The surprise, of course, is that for a given complex K there exist *any* packings whatsoever, much less a potentially huge variety of packings. Computationally, the labels are the key.

Definition 1. *A label R is a packing label for K if the angle sum $\theta_R(v)$ is an integral multiple of 2π for every interior vertex v.*

This is clearly a necessary condition for R to represent the radii of a packing P, since $\theta_R(v) = 2\pi n, n \geq 1$, simply reflects the fact that the circles c_{v_i} for the neighbors of v must reach precisely n times around c_v. It is also a sufficient condition if there are no topological obstructions:

Lemma 1. *If K triangulates a simply connected surface, then a label R for K represents the radii for a circle packing of K if and only if R is a packing label.*

We say little about circle centers because they play a secondary role: once a packing label R is found, the circles are laid out in $\mathcal{D}$ ($\mathbb{S}^2, \mathbb{C}$, or $\mathbb{D}$, as appropriate) using a process akin to analytic continuation. The resulting circle packing $P \leftrightarrow K(R)$ is *essentially unique* — that is, it is unique up to a conformal automorphism of $\mathcal{D}$.

A bit of terminology: A packing is *univalent* if its circles have mutually disjoint interiors. Univalence fails locally at a *branch circle* (or *vertex*), that is, at an interior circle whose neighbors wrap two or more times around it (angle sum $4\pi, 6\pi, \ldots$). Figure 3 illustrates a branch circle; the same nine circles which wrap once about the central circle in Figure 3(a) wrap exactly twice around the smaller central circle of Figure 3(b). The *branch set* for a

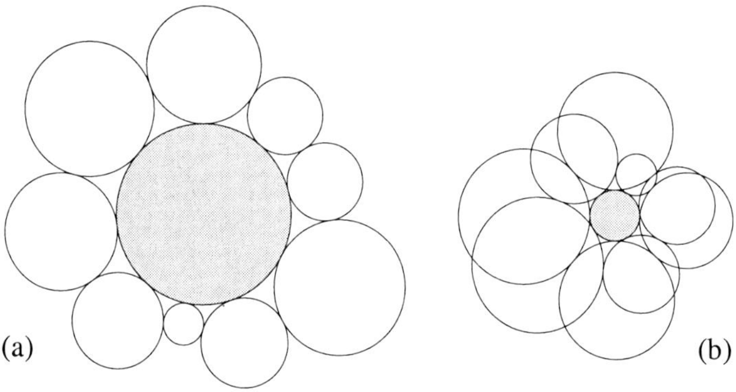

(a) (b)

Figure 3. Local univalence *versus* branching

packing P, denoted $br(P)$, is the set of branch circles, repeated according to multiplicities. If all interior angle sums are 2π, then $br(P) = \emptyset$ and the packing is called *locally univalent.* Univalence can fail, even without branch circles, when circles from one part of a packing overlap those from another.

Packing variety: Given a complex K, there are numerous results on the existence and variety of packings for K; we need two main types. Let's begin with the *extreme rigidity* displayed by maximal packings, proven by Koebe in the finite case [14], Beardon and Stephenson in the bounded degree case [2], and He and Schramm in general [11]:

Theorem 1. *Let K triangulate a simply connected surface. Then for $\mathcal{D}$ equal to precisely one of the spaces $\mathbb{S}^2, \mathbb{C}$, or $\mathbb{D}$, there exists a circle packing $\mathcal{P}_\mathcal{K}$ for K in $\mathcal{D}$ with the property that $\mathcal{P}_\mathcal{K}$ is univalent and has a carrier filling $\mathcal{D}$. Furthermore, $\mathcal{P}_\mathcal{K}$ is essentially unique.*

This circle packing $\mathcal{P}_\mathcal{K}$ is called a *maximal packing* and K is said to be *spherical*, *parabolic*, or *hyperbolic* as $\mathcal{P}_\mathcal{K}$ lies in $\mathbb{S}^2, \mathbb{C}$, or $\mathbb{D}$, respectively.

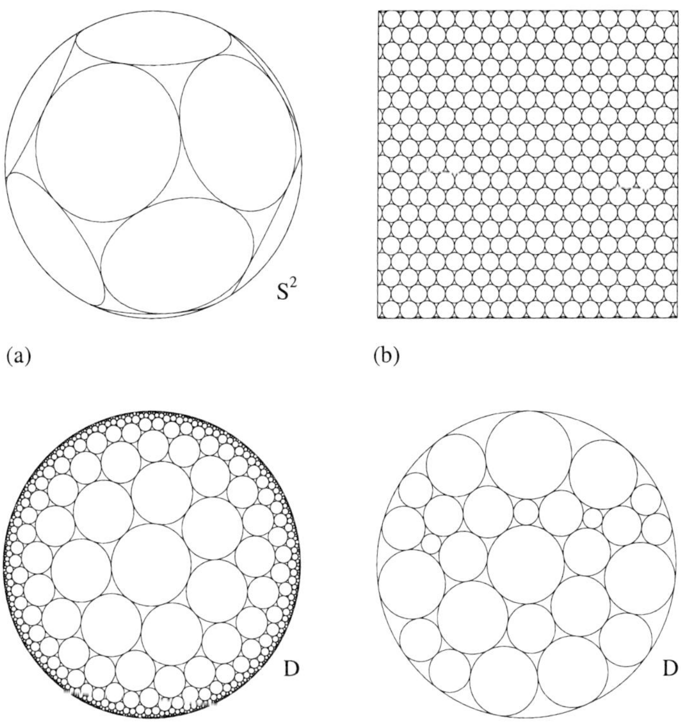

(a) (b) (c) (d)

Figure 4. Maximal Packings

Figure 4 illustrates the four basic situations. A complex K is spherical iff it triangulates a topological sphere; in (a) we see the regular pentagonal packing (every circle with five neighbors). The familiar regular hexagonal or "penny" packing (b) is parabolic, the regular "septagonal" packing (c) is hyperbolic. Computationally, packings such as that of Figure 4(d) are the most important. Here K triangulates a closed topological disc (in fact, this is the maximal packing for the complex of Figure 2). Note that $\text{carr}(\mathcal{P}_K)$ "fills" $\mathbb{D}$ in the sense that its boundary circles are *horocycles* (euclidean circles internally tangent to $\partial\mathbb{D}$) and the boundary vertices have been pushed to the ideal boundary. Spherical packings are computed by puncturing at a vertex, packing the rest of the complex in $\mathbb{D}$, then projecting back to $\mathbb{S}^2$. Infinite packings are handled as appropriately rescaled limits of packings in $\mathbb{D}$ for finite subcomplexes.

In contrast to the rigidity of maximal packings, let's move to the *variety* available when we construct packings meeting prescribed boundary value and branching conditions.

Let K triangulate a closed topological disc. Let $\beta = \{b_1, \cdots, b_k\}$ be a legal branch set for K; this is a (perhaps empty) list of interior vertices of K, with possible repetitions. (Necessary/sufficient combinatorial conditions on β have been proven by Dubejko [9] and Bowers [4].) ∂K denotes the boundary vertices of K.

Theorem 2. *Let K, β, and the boundary label $g : \partial K \longrightarrow (0, \infty)$ be given. Then there exists a unique euclidean packing label R for K with branch set β and satisfying $R(w) = g(w), w \in \partial K$. The same result holds in hyperbolic geometry, where g is also allowed to assume the value $+\infty$.*

In other words, there exists an essentially unique circle packing having a prescribed branch set and prescribed boundary radii. In hyperbolic geometry, infinite radii correspond to horocycles. Thus the maximal packing for K is the packing in $\mathbb{D}$ having prescribing branch set $\beta = \emptyset$ and prescribed boundary radii $g(w) = \infty, w \in \partial K$.

Don't let this global malleability of circle packings obscure the *local rigidity* implied by the packing condition at each vertex. The local-to-global linkage is moderated by the combinatorics of K.

Discrete analytic functions: We can attempt to impose some order on the numerous packings for a given complex K by using a *function theory* paradigm. Define *discrete analytic functions* as maps between circle packings which preserve tangency and orientation. Various packings of K may now be associated with discrete analytic functions $f : \mathcal{P}_K \longrightarrow \mathcal{P}$. Compare the univalent, locally univalent, and branched discrete analytic functions of Figure 5 to their classical models in Figure 1.

As classical analytic functions map infinitesimal circles to infinitesimal circles, so our discrete versions operate with *real* circles. Fundamental

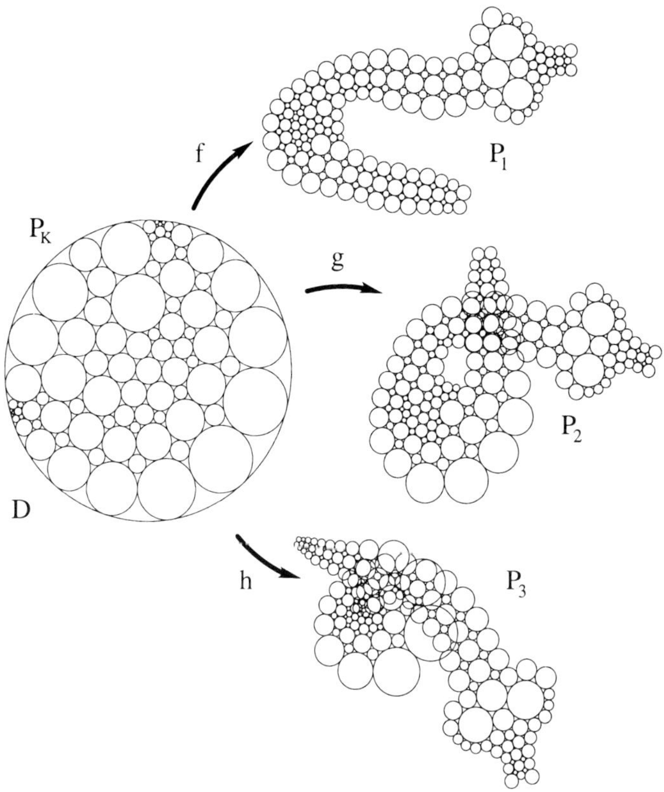

Figure 5. Discrete analytic functions on the disc

behavior is surprisingly similar. There is a *ratio* function $f^{\#}$ defined on vertices of K by $f^{\#}(v) = \text{radius}(f(c_v))/\text{radius}(c_v)$; this reflects the amount that f stretches or shrinks c_v and is the analog of $|f'|$. Hidden in P_3 is a branch circle like that of Figure 3(b); the topological behavior of the mapping is precisely that at a branch point of an analytic function.

Computation: Let me point out that the computational problem centers on finding packing labels. Thurston suggested an iterative algorithm which works beautifully in hyperbolic and euclidean geometry to solve the boundary value problems described above. Here, we will take the algorithm for granted; see [8] for a description and an efficient implementation.

§2 Approximation

Let's proceed now to the principal approximation issues. I will break my survey into four topics: the original conformal mapping setting of Thurston's Conjecture, generalizations to other analytic functions, then on to new uses in conformal tiling and conformal structures on surfaces.

2.1 Conformal mapping

Figure 6 illustrates Thurston's 1985 conjecture. Given the Jordan domain Ω, cut out a finite piece of a regular hexagonal packing of circles having radii $1/n$. Write P_n for this finite circle packing and let K_n encode its combinatorics. If P_{K_n} is the maximal packing for K_n, then $f_n : P_{K_n} \longrightarrow P_n$ represents for us a discrete conformal mapping from $\mathbb{D}$ to Ω. The illustration captures domains and ranges for three such mappings (appropriate normalizations are assumed).

Thurston conjectured and Rodin and Sullivan proved that the discrete conformal maps f_n converge uniformly on compact subsets of $\mathbb{D}$ to the classical conformal mapping $F : \mathbb{D} \longrightarrow \Omega$ as $n \to \infty$. From that convergence one can also show that the ratio functions $f_n^\#$ converge uniformly on compact sets to $|F'|$.

I will not go into the proof of Thurston's conjecture, but I would like to highlight the two key ingredients, which reappear in one guise or another in every setting we consider: namely, *refinement* and *distortion control.* "Refinement" refers to the process of conveniently generating sequences of successively finer circle packings (that is, more numerous and smaller circles) appropriate to a setting — hexagonal packings with successively smaller radii serve that purpose in Figure 6. "Distortion" concerns the comparison of a face in (the carrier of) one packing with the corresponding face in another packing of the same complex, normally quantified in terms of quasiconformal distortion. Less distortion of carrier faces converts to "more conformal" mapping behavior. In Figure 7, observe the distortion of various triangles on the right compared to the corresponding equilateral triangles on the left. Distortion near the center is smaller.

The important Hexagonal Packing Lemma of Rodin and Sullivan and its descendants imply, roughly speaking, that in comparing two univalent packings the amount of distortion that a given face can manifest decreases as its combinatorial distance from the boundary increases. The Rodin/Sullivan Theorem has been considerably extended to accommodate more general combinatorics and to quantify the convergence. The deepest and most general results are due to Z-X. He and O. Schramm [12, 13]; see also [19, 18].

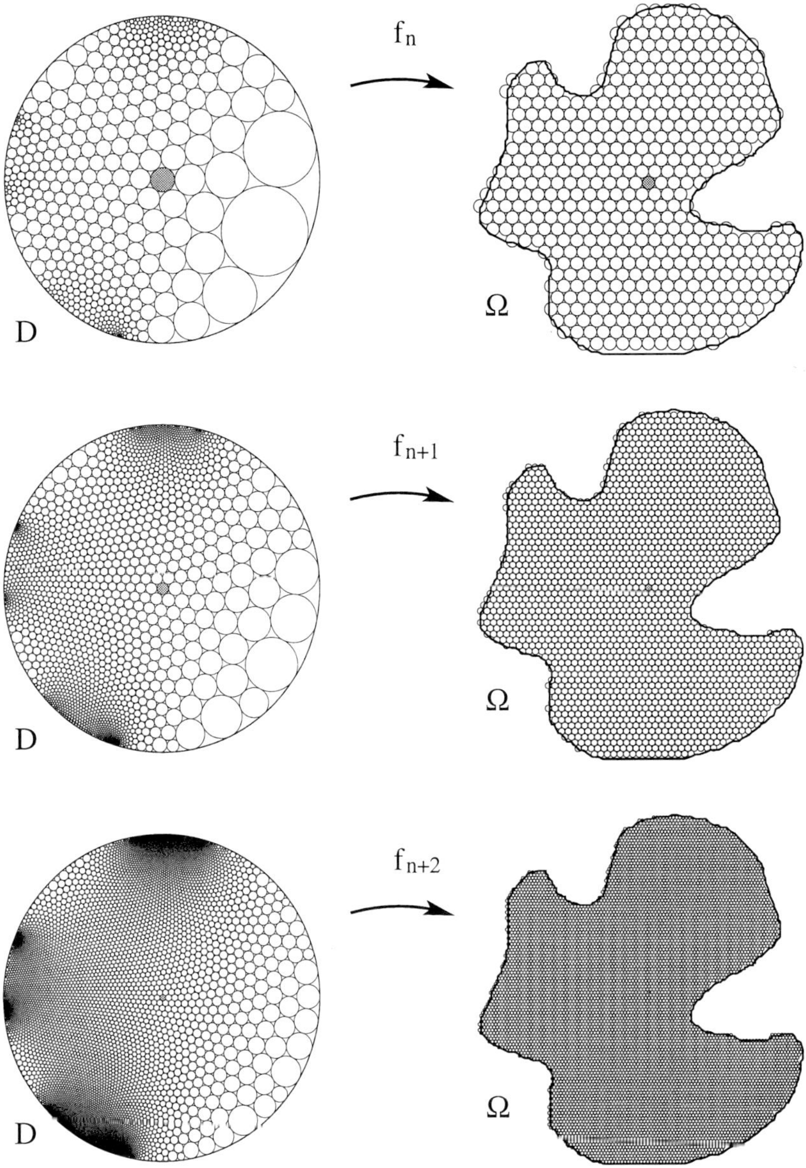

Figure 6. Illustrating Thurston's Conjecture

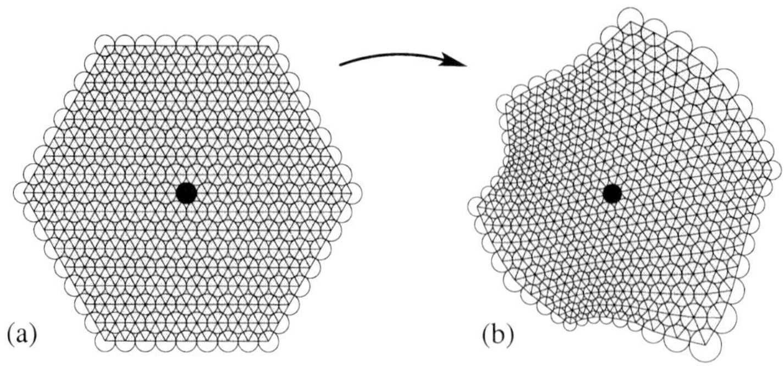

Figure 7. The Hexagonal Packing Lemma

Numerical methods: On the more practical side, for those involved in numerical conformal mapping circle packing may appear to be a disappointment: the "packing" process itself — *e.g.* , computing the radii for P_{K_n} — turns out to be rather slow, data sets are bulky, and the process of laying out the packing itself introduces additional error. Circle packing certainly cannot compete with the classical numerical methods for speed and accuracy.

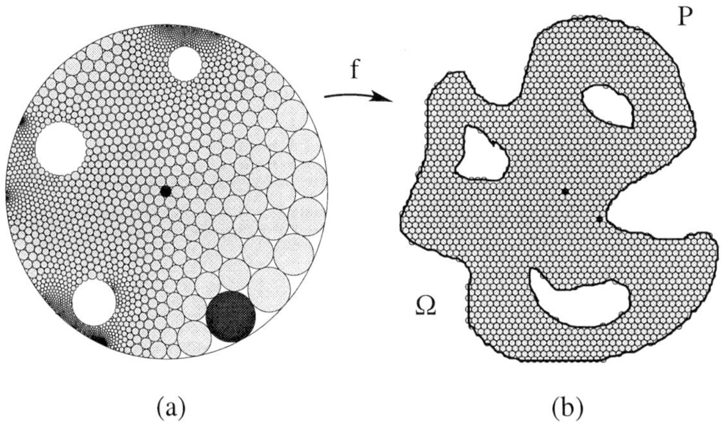

Figure 8. A multiply connected example

Circle packing does have some strengths. For instance, it quite easily handles multiply connected regions. Figure 8(b) illustrates a generic 4-

connected region Ω and the packing P it cuts from a regular hex packing. Augment the complex underlying P with an "ideal" vertex for each hole; that is, a common neighbor for the circles bounding that hole. The augmented complex is simply connected, and displaying its maximal packing with the circles corresponding to the ideal vertices removed leaves the "circle domain" on the left in Figure 8. This then is (discretely) conformally equivalent to the original domain Ω. (Z-X. He and O. Schramm in [11] applied circle packing techniques in a major advance on Koebe's Conjecture regarding infinitely connected *circle domains*; their work is far deeper than our simple numerical example here.)

Prototyping: In the settings of traditional conformal mapping, circle packing may be helpful at what I will call *prototyping*. Suppose one plans to apply Schwarz-Christoffel to approximate a conformal map of $\mathbb{D}$ to the polygon R of Figure 9(b). One might start with a discrete conformal map f

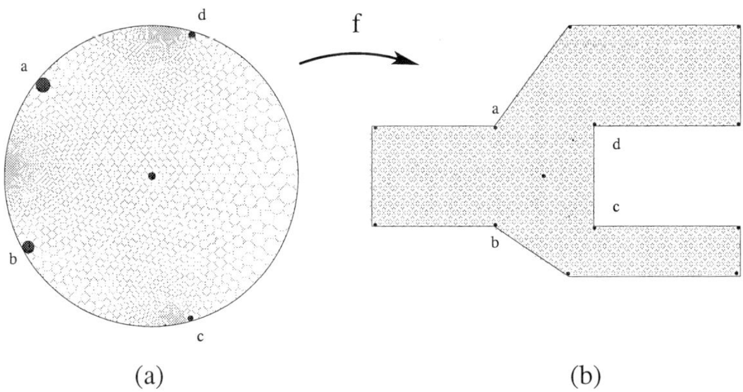

Figure 9. Prototyping a Schwarz-Christoffel Map

involving a modest number of circles (just over 5000 in this instance). I've marked only the four most visible circles associated with the indicated corners of the polygon; these and the other corner preimages can serve as the initial guesses for the Schwarz-Christoffel parameters. Prototyping could be valuable in other situations as well, such as the multiply connected setting discussed earlier.

2.2 Analytic functions

Classical analytic function theory involves more than just conformal maps. So it is, too, with the discrete theory. The numerous parallels are not the aim in this survey, but in passing note that the discrete theory involves fairly general functions on $\mathbb{D}$, polynomials and a smattering of entire and meromorphic functions on the plane, rational functions on the sphere, and discrete maps on Riemann surfaces. The theory has its own Schwarz-Pick, Uniformization, Liouville and Picard theorems, and its own notions of covering theory, extremal length, harmonic measure, and Brownian motion (random walks). It is approaching something like a "full service" function theory, though there remain many gaps.

I have chosen two examples to illustrate the computational and approximation side of this more general discrete theory: one involves a function in the disc algebra, the other a finite Blaschke product.

A disc algebra example: Consider the closed curve γ of Figure 10, with a finite number of transverse self-intersections. Using winding numbers alone, one can verify that γ is the image of the unit circle under a disc algebra function (continuous on $\overline{\mathbb{D}}$ and analytic on $\mathbb{D}$); one can also determine the distribution of branch values (images of branch points) among the components of $\mathbb{C} - \gamma$. (See [15].)

The actual construction of a disc algebra solution is another matter; Schwarz-Christoffel methods apply, but I'm not aware of implementations. We can construct the analogous discrete solutions by introducing branch points to our earlier univalent methods.

Our aim is construction of a multi-sheeted image packing; we use a "ball bearing" rather than a hexagonal pattern as our starting material and overlay the region defined by γ as in the univalent case. However, several "sheets" (up to three deep inside the smallest loop) may be inferred from γ and the branch structure, so appropriate pieces are cut out and attached as in the upper right of Figure 10. The resulting packing P has a complex K which is simply connected. Our discrete disc algebra function F results from mapping $\mathcal{P}_K$ to P with f and projecting with π to a packing in $\mathbb{C}$. The pasting resulted in two branch points, one simple and one of order 2. These are the shaded circles in $\mathcal{P}_K$; the former has 8 neighbors which map 2-to-1 onto the 4 neighbors of its projected image circle, the latter has 12 neighbors mapped 3-to-1 onto the 4 neighbors of its projected image circle.

As in the conformal setting, a sequence of refined packings constructed in this manner will (with suitable normalization) converge uniformly on compact subsets of the disc to the classical disc algebra solution. We can routinely generate the necessary refinements from this initial coarse con-

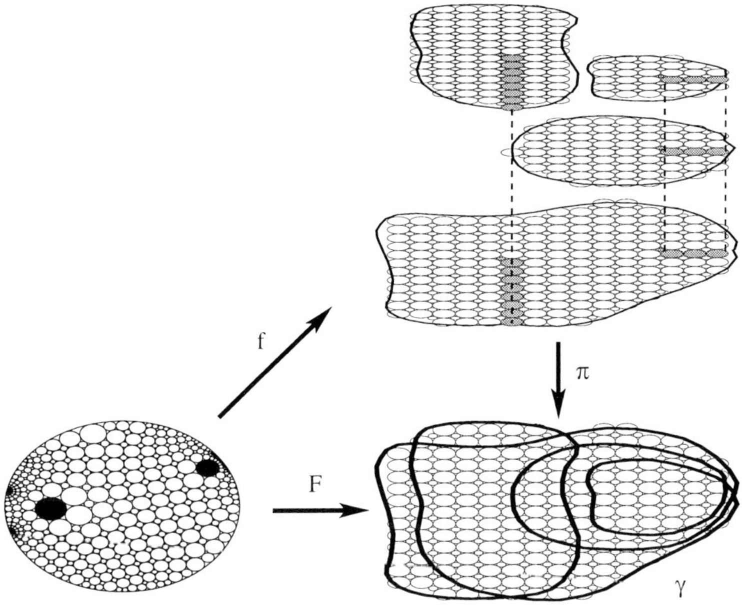

Figure 10. Approximating a disc algebra function

struction, as we will see later, so this is in fact a fairly practical procedure.

Finite Blaschke products: Classical finite Blaschke products arise not only in function theory on the disc and its applications, but also in studies of rational functions, iteration theory, and topological covering theory. An n-fold Blaschke product is typically represented as a product of n Möbius factors, one for each zero, multiplied by a unimodular constant. It assumes each value in the disc n times, counting multiplicities, and maps the unit circle n times around itself.

Lacking arithmetic, we need a more geometric description for the discrete construction. The finite Blaschke products are precisely the "proper" analytic maps of the disc to itself, and the argument principle in turn implies that an n-fold Blaschke product has $n-1$ branch points, counting multiplicities. These geometric features are enough for our construction, as I illustrate with a 4-fold example.

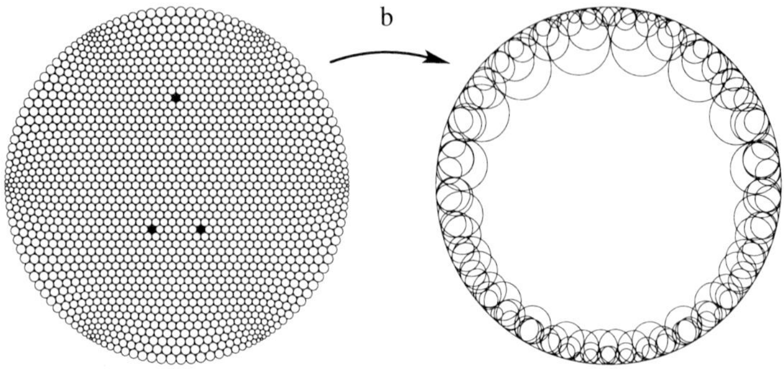

Figure 11. Boundary circles for 4-fold discrete Blaschke product

Take as the domain a maximal packing $\mathcal{P}_{\mathcal{K}}$ in the unit disc, shown on the left in Figure 11. I have identified three shaded interior circles for branching. We repack the complex in hyperbolic geometry, specifying that: (1) boundary circles get infinite hyperbolic radii; (2) designated branch circles get angle sum 4π rather than the usual 2π. The resulting packing label allows us to lay out the range packing. Displaying all its circles on the right would lead to visual meltdown since points are covered four times, so I have drawn only the chain of boundary horocycles and the 3 branch circles (they appear as extremely small dots towards the middle) in the image packing. A careful survey would confirm that those boundary circles wrap 4 times around the unit circle. The construction is complete, aside from applying a Möbius transformation of $\mathbb{D}$ for the sake of normalization.

To approximate a particular classical 4-fold Blaschke product B, we would first need to know its branch points. Then we would run this construction with successively finer complexes K_n, with branch circles approximating the desired branch points, and with appropriate normalizations. The sequence $\{b_n\}$ of discrete finite Blaschke products would converge uniformly on compact subsets of $\mathbb{D}$ to B.

The simplicity of the discrete construction is quite impressive; the repacking process converts the local geometric conditions on boundary radii and branch points to the inevitable global behavior. Classical methods would have difficulty coping with prescribed branch points. On the other hand, constructing a discrete finite Blaschke product with specified *zeros* is tough.

2.3 Conformal tilings

Conformal tilings, introduced by Phil Bowers and me in [7], involve tiles with specified "conformal" rather than euclidean shapes. Though classical objects, they were inspired by experiments with circle packings and circle packings remain the only avenue for approximating their true conformal shapes.

I'll illustrate with a finite example based on the simple schematic pattern of Figure 12(a). Pasting together seven equilateral triangles in the

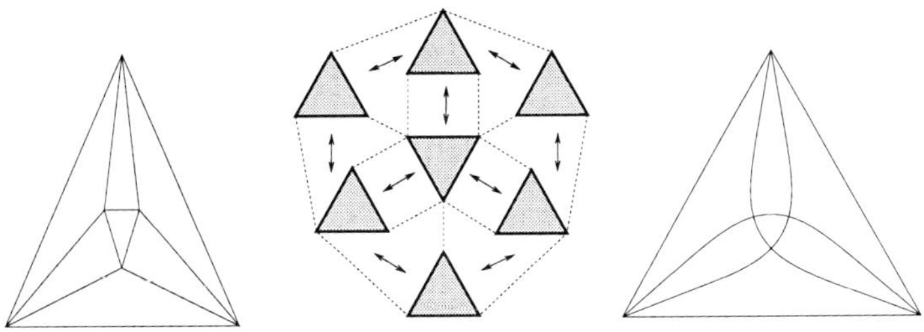

(a) (b) (c)

Figure 12. Combinatorial, affine, and conformal stuctures

prescribed pattern, as suggested in Figure 12(b), yields a topological disc with a piecewise affine structure; there are three "cone" points, each of angle $4\pi/3$. There is a standard (and very classical) procedure for defining a conformal structure compatible with such an affine structure; its atlas involves appropriate power maps $z \mapsto z^{3/2}$ at the cone points (see, *e.g.* Beardon [1]). We arrive, then, at a Riemann surface, and if we map it conformally onto an equilateral triangle, carrying along the markings of the original seven triangles, we arrive at the triangle T of Figure 12(c).

The curves marked on T are uniquely determined by the original combinatorics and by what Bowers and I call a *reflective* structure — namely, any two of the seven triangular pieces having a common edge are anticonformal reflections of one another across that edge. How can one approximate these curves in T? I am not aware of any classical numerical methods which apply. However, circle packing provides *discrete* curves which do the trick.

The discrete construction mimics the original. Conceptually, we take seven copies of the packing of Figure 13(a) and attach them to one another by identifying circles along appropriate edges, just as in Figure 12(b). (In fact, only the resulting combinatorics are important.) When the results are repacked to form an equilateral carrier, the repacking process forces the

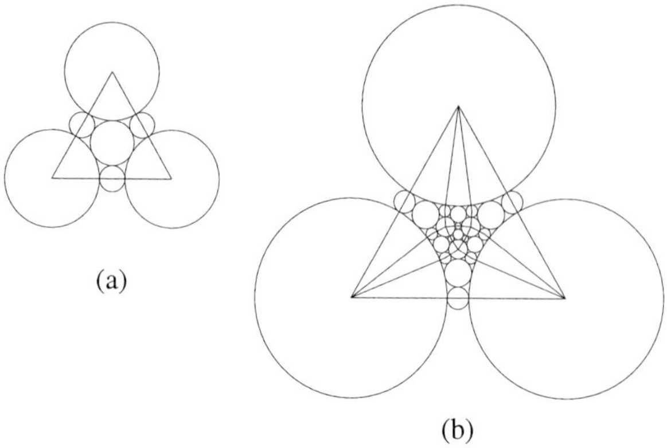

Figure 13. The discrete construction

circle sizes to adjust to flatten out the cone points, as in Figure 13(b).

Figure 14 shows the same process carried out with two successively more refined packings; one can see the marked curves approaching limit shapes.

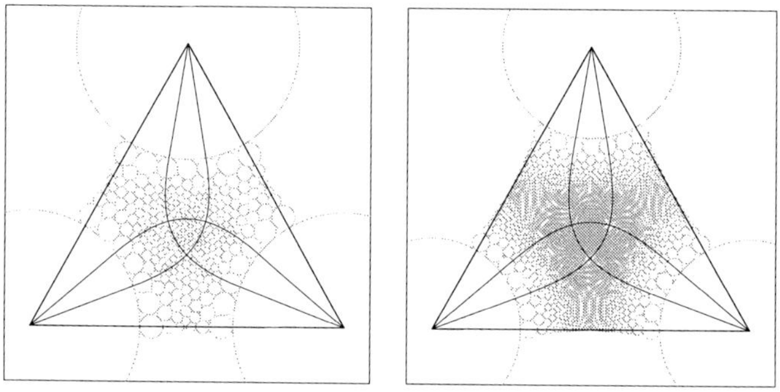

stage-2 refinement stage-4 refinement

Figure 14. Circle packings at two additional refinement levels

I want to take a moment to comment on the refinement process used here. It is called "hex refinement" and operates as shown in Figure 15.

Hex refinement can be applied to any triangulation and has several valuable features: • New combinatoric data is very easy to generate. • The original vertices of a triangulation remain, with their degrees unchanged.

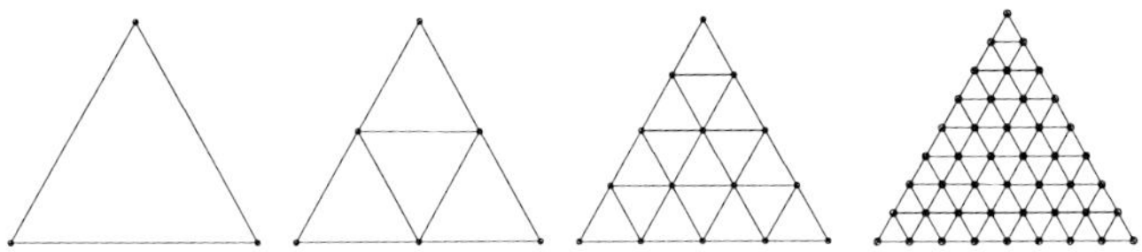

Figure 15. Hex refinement process

• New (interior) vertices are degree six. This means that under successive hex refinement, one is in good position to deploy the Hex Packing Lemma.
• "Refinement" is actually a combinatorial process — one must always compute new radii for the refined complex. The self-scaling nature of hex refinements may provide *a priori* estimates which will save effort in the repacking computation.

Infinite tilings: The processes involved in our simple example, both the classical construction and its discrete analogue, work with only minor adjustment in much more general circumstances. We will look at some infinite patterns here and move to patterns on compact surfaces in the next section.

Figure 16 illustrates four infinite conformally regular tilings of $\mathbb{C}$. Each tile can be mapped conformally onto a regular euclidean n-gon, corners going to corners (and there are additional conditions which we won't go into). These particular patterns are quite special; they arise from "subdivision rules" being applied by Jim Cannon, William Floyd, and Walter Parry to the study of Thurston's Geometrization Conjecture. (See [7] and references therein.) My thanks to Bill Floyd for the data needed to generate these tilings.

What is our interest in tilings? All of these pictures were generated *via* circle packing — the circles themselves aren't shown, but only certain edges of their carriers. Packings bring an experimental component to this research and they provide approximations (like these pictures) of the true conformal shapes of the tiles — capabilities not otherwise available.

I have space to investigate only one example, so let's look at the "twisted pentagonal" tiling in the upper right of Figure 16. In Figure 17(a), denote the three dark curves as $\gamma_0, \gamma_1, \gamma_2$, from inner to outer. These mark central

Figure 16. Four examples of infinite conformal tilings of the plane

"aggregate tiles" associated with the first three stages of the subdivision rule which generated this pattern. Stare at the picture hard enough and you might guess at some scaling relationship among the γ_i's. In fact, using $\lambda \approx 0.55e^{-0.23\pi}$, I have overlaid $\lambda^2 \cdot \gamma_2$, $\lambda \cdot \gamma_1$, and γ_0 in Figure 17(b). Our hunch was right, the vertices $\lambda \cdot \gamma_i$ coincide almost perfectly with vertices of γ_{i-1}. You can probably anticipate what the curves $\gamma_3, \gamma_4, \cdots$ look like; it appears that $\lim_{n\to\infty}(\lambda^n \cdot \gamma_n)$ converges to some fractal curve Γ, that is, the aggregate tiles converge to a fractal tile. The same holds for other tiles in the pattern (not marked), leading to a fractal pentagonal tiling of the plane. Motivated by these very experiments, Walter Parry has used classical methods to prove this scaling property, though the precise value of λ is not known.

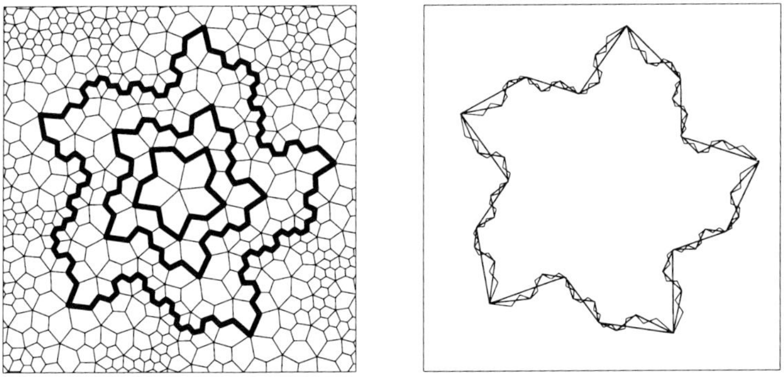

(a) (b)

Figure 17. Aggregates in the twisted pentagonal tiling

Central issues in conformal tiling include the "type" problem – does a pattern tile the plane or the hyperbolic plane – questions about tile shapes, sizes, distortions, and questions about symmetries and self-similarities within the overall pattern. But this is a new topic, and new phenomena appear in nearly every experiment. Numerical computation and display are currently the pivotal tools.

2.4 Grothendieck dessins

Let me begin this section by introducing a certain type of Riemann surface construction. Suppose S denotes a compact, orientable topological surface and K is a triangulation of S. One can paste euclidean equilateral triangles together in the pattern of K to impose a piecewise affine structure on S. Precisely as described in the seven-triangle example of Figure 13, that affine structure determines a conformal structure on S, making it into a Riemann surface. Surfaces constructed this way will be called *equilateral* surfaces.

We quote this surprising result of G. Belyĭ [3], for which we need one other bit of terminology. A nonconstant meromorphic function on a compact Riemann surface which branches only over points $\{0, 1, \infty\}$ is termed a *Belyĭ map*.

Theorem 3. *For a Riemann surface R of positive genus, the following statements are equivalent:*

(a) *R is equilateral.*

(b) *R supports a Belyĭ map.*

(c) *There exists a defining equation for R whose coefficients lie in an algebraic number field F over the rationals.*

In other words, in the Teichmüller space Teich(S) for surfaces of genus $g > 0$ there is a set of points which simultaneously enjoys combinatorial, function theoretic, and algebraic characterizations. By (a) this set is countable, and by (c) it is dense in Teich(S).

Let's turn now to the discrete parallels. We find that S and K determine a point in Teichmüller space in a second way by this result (see [20, 2]).

Theorem 4. *Let K be a triangulation of a compact oriented topological surface S. Then there exists a unique Riemann surface R homeomorphic to S that supports a univalent circle packing $\mathcal{P}_K$ with the combinatorics of K.*

In other words, the combinatorics prescibed for a circle packing will determine the geometry of the space in which the packing can be realized. (The metric for this packing $\mathcal{P}_K$ is the *intrinsic metric* on R; that is, the metric of constant curvature 0 or ± 1 which R inherits from its universal covering surface.) When S has positive genus, the surface R which supports the packing for K is called a *packable* surface of Teich(S). The packable surfaces are clearly countable, and by a result of Bowers and me [6], also dense. With a certain technical condition that we may assume for K, one can also define a discrete meromorphic map from $\mathcal{P}_K$ to a packing in $\mathbb{S}^2$ which branches only over $\{0, 1, \infty\}$; a *discrete* Belyĭ map. In other words, we have discrete parallels for the classical objects described earlier.

Dessins: Equilateral surfaces are at the core of a program initiated by Grothendieck to study the absolute Galois group Gal($\overline{\mathbb{Q}}/\mathbb{Q}$) and to address what is known as the Inverse Galois Problem. This amazing theory begins with a "child's drawing" D on S and ends with an algebraic number field. Briefly, the drawing leads to a triangulation of S, the triangulation leads to an equilateral Riemann surface S_D and a Belyĭ map B_D on S_D, giving the so-called "Belyĭ pair" (S_D, B_D). Schematically,

$$D \longrightarrow K \longrightarrow (S_D, B_D).$$

Of course this is all quite abstract and in only limited special circumstances are any concrete details about the surface and the Belyĭ map known. (See [17].)

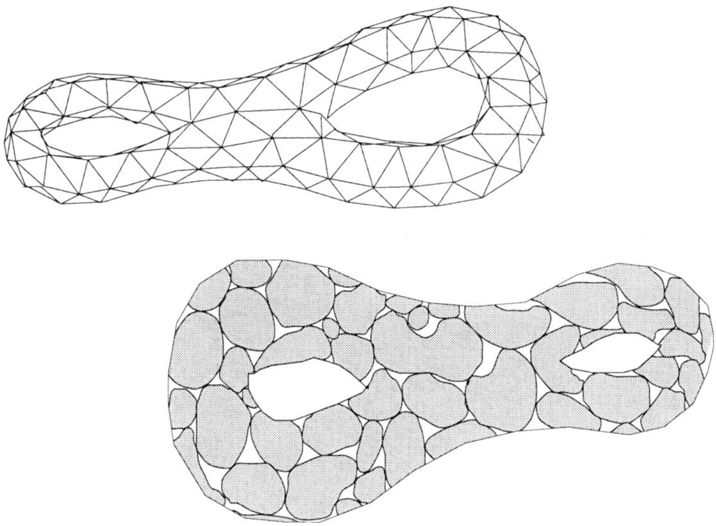

Figure 18. "Equilateral" *versus* "packable" 2-tori

On the other hand, the parallel discrete development can be carried out at various refinement stages and has the advantage of being computable. The idea is to apply n-stages of hex refinement to K, giving a complex $K^{(n)}$, then to construct the associated *discrete Belyĭ pair* $(s_D^{(n)}, b_D^{(n)})$ consisting of $s_D^{(n)} \in \text{Teich}(S)$ and discrete Belyĭ map $b_D^{(n)} : P_{K^{(n)}} \longrightarrow \mathbb{S}^2$. Schematically,

$$D \longrightarrow K \longrightarrow K^{(n)} \longrightarrow (s_D^{(n)}, b_D^{(n)}).$$

Phil Bowers and I have proven the following result [5]:

Theorem 5. *If D is a drawing on a compact orientable surface S then the discrete Belyĭ pairs $(s_D^{(n)}, b_D^{(n)})$ converge to the classical Belyĭ pair (S_D, B_D) as $n \to \infty$. In other words, $s_D^{(n)}$ converges to S_D in the Teichmüller metric of Teich(S) and the functions $b_D^{(n)}$ converge uniformly to B_D on S.*

As examples, let me discuss the two dessins of Figure 19. The first, a very simple genus 0 drawing D on the sphere, is shown in Figure 19(a) along with its canonical triangulation, four shaded and four unshaded faces. Here $s_D^{(n)} = S_D$ for all n since there is only one conformal structure on $\mathbb{S}^2$.

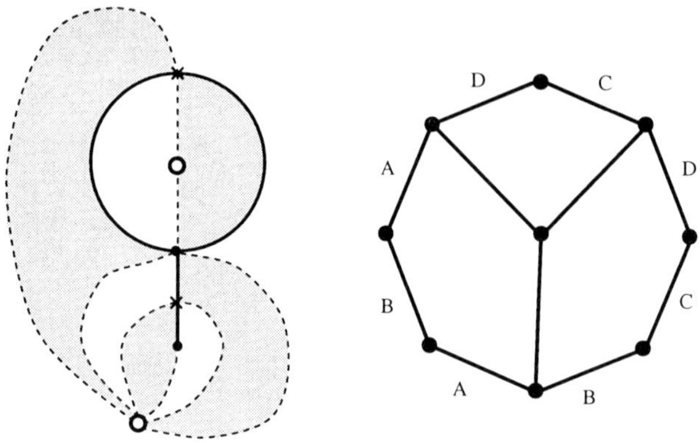

Figure 19. Sample dessins of genus 0 and genus 2

Nevertheless, the convergence of $b_D^{(n)}$ to B_D allows us to approximate the conformal shapes of the faces of K. The stage $n = 3$ construction is shown in Figure 20; the packing is the one shown at the beginning of the paper, but the faces of K are marked in its carrier as the shaded and unshaded regions, and the drawing D itself is shown as the heavy curve. The discrete Belyĭ map $b_D^{(3)}$ maps this packing to the spherical packing on the right with multiplicity 4-to-1, the shaded faces mapping to the shaded hemisphere.

Our second example is the dessin D of Figure 19(b); the indicated side-pairings show this to be a genus 2 case. I have packed this at the original "coarse" stage and at a stage-3 hex refinement and display the results in Figure 21. The packings are laid out in $\mathbb{D}$ as fundamental domains for the universal covering maps from $\mathbb{D}$ onto $s_D^{(0)}$ and $s_D^{(3)}$, respectively. The heavy curve in each case is D and the faces of K are alternately shaded and unshaded.

The packing computations are carried out directly in hyperbolic geometry with the combinatorics of the 2-torus in tact. The results are then laid out as a fundamental domain where numerical information, such as side-pairing maps, can be read off. (Note the side-pairings designated on the upper packing.) The fundamental domains converge as $n \to \infty$ to the fundamental domain of the universal covering of S_D. One thing to note is the surprising accuracy of even the coarse packing — its eight boundary cusps differ by less than 1/500 from those of the stage-3 refinement.

By the way, have our discrete objects lost their connection to number fields? In the original sense, yes; but in fact, number fields reenter in a new

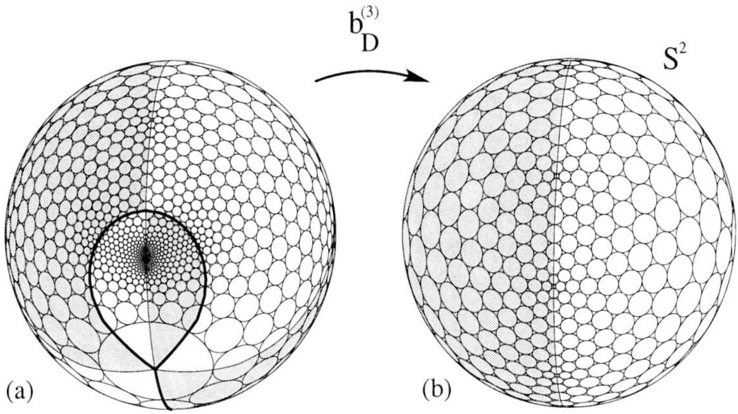

Figure 20. A genus 0 discrete Belyĭ map

and intriguing way as entries in the associated covering maps.

Classical surfaces: Let's wrap up our look at surfaces with two classical examples which (in hindsight) are associated with dessins. The quite famous image in Figure 22 is Klein's surface. He was able to publish the original image more than a century ago because of the surface's ubiquitous symmetries (an order 168 symmetry group) — the triangles are all geodesic. The image here, on the other, generated *via* circle packing, took approximately two seconds for ten digit accuracy starting from purely combinatorial data.

Another famous classical example is Picard's curve $y^3 = x^4 - 1$, a genus three surface. It is actually associated with two distinct dessins. One is shown in Figure 23(a); this is a fundamental domain for a first stage coarse approximation and the circles have been left in for reference. This classical surface also has many symmetries, but the edges of its faces are not all geodesic. As far as I know, aside from circle packing methods, there would be no way to approximate the surface or to create these images.

§3 Concluding remarks

The survey examples suggest a general paradigm: given a classical object of interest — a function, a Riemann surface, or a conformal shape — proceed as follows:

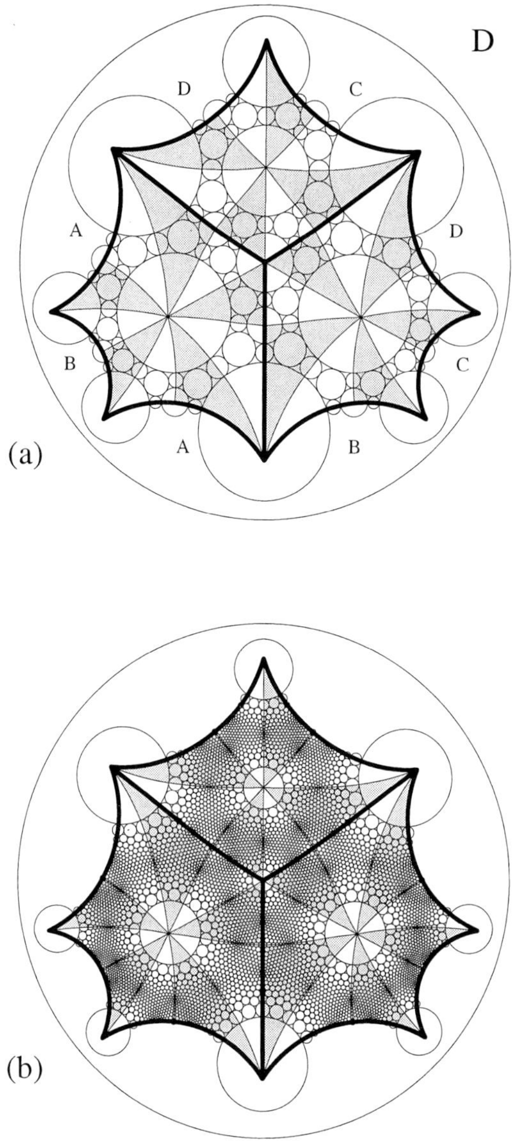

Figure 21. Fundamental domains for genus 2 dessins

Figure 22. Klein's surface

• Construct a combinatorial abstraction — that is, a triangulation and (possibly) a branch structure.

• Endow the triangulation with an appropriate affine structure. Currently we need an "equilateral" model, but certain recent developments are expected to allow more general decompositions.

• Circle pack the result. This is the local-to-global step; nature endows the abstraction with a global discrete conformal geometry compatible with the local packing conditions.

• Apply a process of repeated refinement and repacking to construct a sequence of discrete conformal objects.

If the local geometry is faithfully encoded, then faithful global consequences

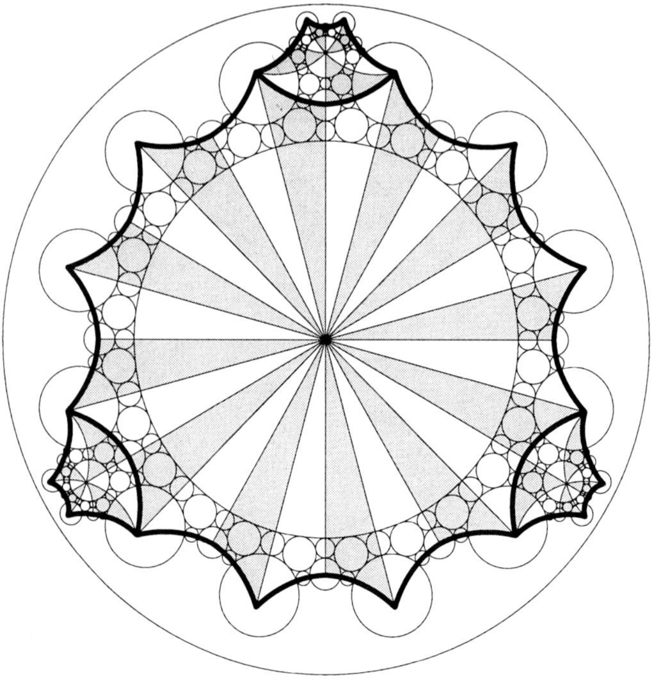

Figure 23. Picard's surface

appear to be inevitable and the discrete conformal objects converge to their classical counterpart. The methods clearly have broad applicability, and the equilateral models, hex refinement, and the hexagonal packing lemma fit so well together that one can contemplate a rather comprehensive and fairly automated "analytic engine" (to steal a phrase) for approximating classical surfaces and functions in a variety of situations.

The methods are also geometrically honest: the discrete objects are generally very faithful to their classical models, so one's intuition remains in tact; the computations take place within the intrinsic geometry; and the packings provide a natural means of visualization.

3.1 Numerical issues

Of course, the devil is in the details. Standard issues of accuracy and rates of convergence remain almost totally open for circle packing methods. These may be quite difficult; the original discrete model, the computed packing labels, and the laying-out process, all introduce errors. A beginning has been made with estimates of the quasiconformal distortion associated with discrete analytic functions, primarily in the hexagonal setting; for the deepest results, see the work of Z.X. He and O. Schramm [13]. The packing algorithm has some beautiful geometry and can be modeled probabilistically, so there is hope of eventual convergence estimates [18].

Nonetheless, we seem to be stuck with the heuristic approach for now — experiments, checks for internal consistency, and, when possible, comparisons to classical examples. In situations where circle packing is the only available tool, we may be forgiven for using it in experiments without first settling the various details of convergence.

3.2 Technical issues

Visit my homepage at *www.math.utk.edu/~kens* for a circle packing bibliography, further packing examples, and directions for acquiring the software package `CirclePack`, which runs under X-Windows.

I have based my circle packing algorithms on an iterative scheme suggested by Thurston; recent improvements are detailed in [8]. Moderate packings, say 5,000–10,000 circles, typically require less than a minute; packings of 100,000+ circles are routine, though data sets become bulky. Hex refinement and "multi-grid" methods open the way for parallelization and potentially significant efficiency improvements. See [10] for discriptions of several experiments and [5] for sample timings.

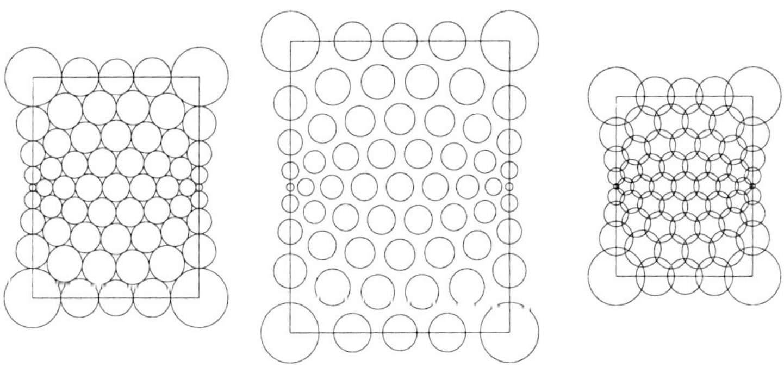

Figure 24. Packing options: tangency, inversive distances, overlaps

3.3 Conclusion

Circle packing is a new and, I think, very exciting topic. It comes with substantial and interesting problems, new capabilities, some available for the first time, an experimental and visual environment, and many directions for further development. As merely one example, Figure 24 illustrates *inversive distance* and *overlap* packings; these generalizations of "tangency" bode even greater flexibility for discrete conformal methods in the future. Moreover, I have left out emerging uses of this geometry inspired by circle packing in areas such as graph embedding and modeling of various physical phenomena.

The traditional applications of conformal mapping — areas which have motivated some truly beautiful mathematics over the years — are increasingly falling to numerical PDE and other methods. At the same time, *conformal geometry* in the broader sense seems to be growing in importance, both in theory and applications. We are confronted with substantial opportunities for numerical approximation, but with flexibility, convenience, visualization, and the intuition gained from real-time experiments taking precedence over traditional speed and precision. Incongruous as it may seem, "circle packing" might provide just the thing in this new approximation environment.

Finally, I admit to loving these pictures, so I want to leave with this last one: a "dodecahedral" tiling from a division rule of Jim Cannon, data from Bill Floyd. None of the circles actually used in the construction are shown, but an intricate pattern of "circles" seems to be emerging from the edge pattern itself!

Acknowledgement. The author gratefully acknowledges support of the National Science Foundation and the Tennessee Science Alliance.

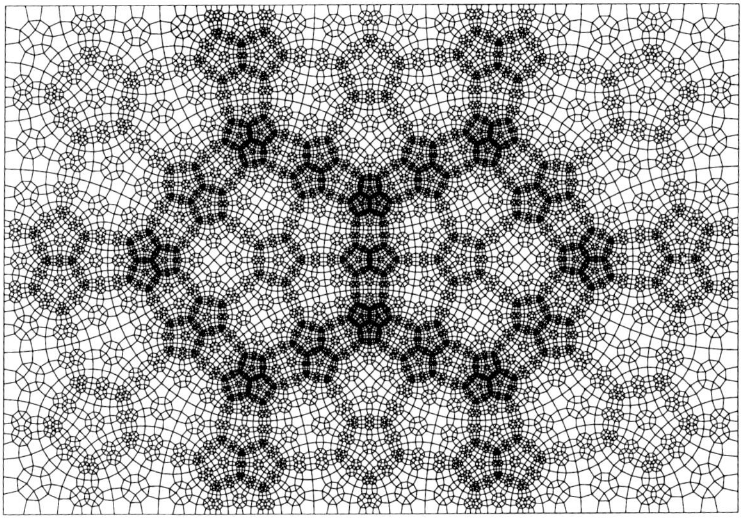

Figure 25. Cannon's dodecahedral pattern

References

[1] Beardon, A. F., *A primer on Riemann Surfaces*, London Math. Soc. Lecture Note Series, vol. 78, Cambridge University Press, Cambridge, 1984.

[2] Beardon, A. F. and K. Stephenson, The uniformization theorem for circle packings, *Indiana Univ. Math. J.* **39** (1990), 1383–1425.

[3] Belyĭ, G., On Galois extensions of a maximal cyclotomic field, *Math. USSR Izv.* **14** (1979), 247–256 (English).

[4] Bowers, P. L., The upper Perron method for labelled complexes with applications to circle packings, *Proc. Camb. Phil. Soc.* **114** (1993), 321–345.

[5] Bowers, P. L. and K. Stephenson, Uniformizing dessins and Belyi maps via circle packing, 1997 to appear.

[6] Bowers, P. L. and K. Stephenson, Circle packings in surfaces of finite type: An in situ approach with applications to moduli, *Topology* **32** (1993), 157–183.

[7] Bowers, P. L. and K. Stephenson, A "regular" pentagonal tiling of the plane, *Conformal Geometry and Dynamics* **1** (1997), 58–86.

[8] Collins, C. and K. Stephenson, A circle packing algorithm, preprint.

[9] Dubejko, T., Branched circle packings and discrete Blaschke products, *Trans. Amer. Math. Soc.* **347** no. 10 (1995), 4073–4103.

[10] Dubejko, T. and K. Stephenson, Circle packing: Experiments in discrete analytic function theory, *Exp. Math.* **4** no. 4 (1995), 307–348.

[11] He, Z.-X. and O. Schramm, Fixed points, Koebe uniformization and circle packings, *Ann. of Math.* **137** (1993), 369–406.

[12] He, Z.-X. and O. Schramm, On the convergence of circle packings to the Riemann map, *Invent. Math.* **125** (1996), 285–305.

[13] He, Z.-X. and O. Schramm, The C^∞-convergence of hexagonal disk packings to the Riemann map, *Acta Mathematica*, 1997 to appear.

[14] Koebe, P., Kontaktprobleme der Konformen Abbildung, *Ber. Sächs. Akad. Wiss. Leipzig, Math.-Phys. Kl.* **88** (1936), 141–164.

[15] Marx, M. L., Extensions of normal immersions of S^1 into $\mathbb{R}^2$, *Trans. Amer. Math. Soc.* **187** (1974), 309–326.

[16] Rodin, B. and D. Sullivan, The convergence of circle packings to the Riemann mapping, *J. Differential Geometry* **26** (1987), 349–360.

[17] Schneps, L. (ed.), *The Grothendieck Theory of Dessins d'Enfants*, London Math. Soc. Lecture Note Series, vol. 200, Cambridge Univ. Press, Cambridge, 1994.

[18] Stephenson, K., A probabilistic proof of Thurston's conjecture on circle packings, *Rendiconti del Seminario Mate. e Fisico di Milano*, to appear.

[19] Stephenson, K., Circle packings in the approximation of conformal mappings, *Bulletin, Amer. Math. Soc.* **23** no. 2 (1990), 407–415.

[20] Thurston, W., The geometry and topology of 3-manifolds, Princeton University Notes, preprint.

Kenneth Stephenson
Dept. of Math.
University of Tennessee
Knoxville, TN 37996
U.S.A.
kens@math.utk.edu

Nonlinear Riemann-Hilbert problems
– history and perspectives –

Elias Wegert

Abstract. This contribution is devoted to a nonlinear boundary value problem for holomorphic functions in the complex unit disc which can be traced back to Bernhard Riemann. We do not intend to give an exhaustive survey of the problem; rather, the paper is meant as a shop window that presents several selected aspects. In order to give the reader a flavour of the subject we comment on the history, sketch some basic results, discuss a number of applications, indicate interrelations to other questions in complex analysis, and address a few open problems.

§1 History

1.1 The origin

The beginning of the story dates back to the year 1851, when 25-year-old Bernhard Riemann published his famous *Inauguraldissertation* "Grundlagen für eine allgemeine Theorie der Functionen einer veränderlichen complexen Grösse" [35]. Though this paper is well known as the origin of one of the most celebrated results in complex analysis, the Riemann mapping theorem, it has widely not been recognized for a long time that Riemann considered conformal mapping just as an example of a more fundamental question. At the beginning of Section 21 Riemann wrote: *"Es wird jedoch zur Erläuterung unserer allgemeinen Sätze ein ausgeführtes Beispiel ihrer Anwendung von Nutzen sein."*

If conformal mapping serves as an illustration, what is the general problem? The answer is given in Section 19, where we learn that Riemann is mainly interested in finding conditions which determine an analytic function in a given domain (more or less) uniquely. He pays special attention to conditions involving boundary values of the unknown function. To say it briefly: What are well-posed boundary value problems for analytic functions?

After some heuristic arguments Riemann came up with the following conclusion: *"Die Bedingungen, welche so eben zur Bestimmung der Function hinreichend und nothwendig befunden worden sind, beziehen sich auf*

Computational Methods and Function Theory (CMFT'97)
N. Papamichael, St. Ruscheweyh and E. B. Saff (Eds.), pp. 583–615

ISBN 981-02-3626-3

ihren Werth ... in Begrenzungspunkten, ... und zwar geben sie für jeden Begrenzungspunkt Eine Bedingungsgleichung. Bezieht sich jede derselben nur auf Einen Begrenzungspunkt, so werden sie durch eine Schaar von Curven repräsentirt, von denen für jeden Begrenzungspunkt Eine den geometrischen Ort bildet."

(The conditions which where just found to be sufficient and necessary for determining the function are related to its values ... at boundary points ..., namely, they give one equation for each boundary point. If each one is related to only one boundary point, they are represented by a family of curves, and for each boundary point one of those [curves] forms the locus [of the corresponding boundary value]).

Riemann's investigations give rise to the following boundary value problem which will be the hero of the present paper:

Let $\{M_t\}_{t\in\mathbb{T}}$ be a given family of curves in the complex plane. Find all functions w holomorphic in the unit disk $\mathbb{D}$ which satisfy the boundary condition

$$w(t) \in M_t$$

for all t on the unit circle $\mathbb{T}$.

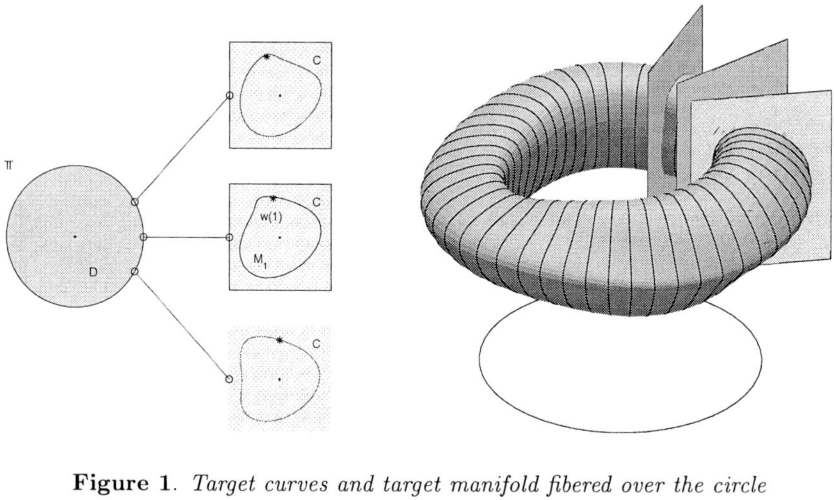

Figure 1. *Target curves and target manifold fibered over the circle*

The problem goes with several names; here it is referred to as the *Riemann-Hilbert problem* (RHP for short). We follow Riemann's geometric approach and set up a framework which is also convenient for visualization.

The curves M_t are said to be the *target curves* or the *restriction curves* of the problem. Since each curve M_t is associated with a point t on the complex unit circle $\mathbb{T}$, the target curves can be viewed as fibers of the *target manifold*

$$M := \{(t, y) \in \mathbb{T} \times \mathbb{C}\colon\ t \in \mathbb{T},\ \ y \in M_t\}.$$

It is clear that the boundary values $w(t)$ have to be understood in an appropriate sense. For the time being we assume that the solutions w belong to the space $H^\infty \cap C$ of functions holomorphic in $\mathbb{D}$ which extend continuously onto the closed disk $\overline{\mathbb{D}}$. Denoting by

$$\operatorname{graph} w := \{(t, w(t)) \in \mathbb{T} \times \mathbb{C}\colon\ t \in \mathbb{T}\},$$

the graph of the boundary function of $w \in H^\infty \cap C$, we write the boundary condition as

$$\operatorname{graph} w \subset M.$$

Any continuous solution of the Riemann-Hilbert problem can be visualized as a line on the target manifold which hits each target curve exactly once.

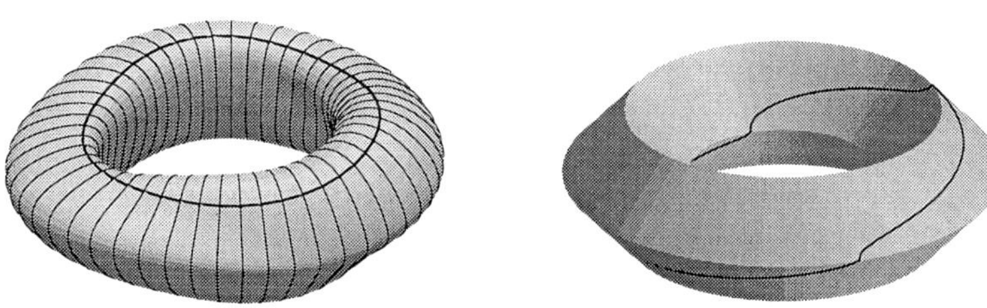

Figure 2. *Solutions on compact target manifolds*

Riemann treated the problem only heuristically, the first general rigorous existence proof was given more than a hundred years later. However, there are two special cases which were investigated in the early years after Riemann: conformal mapping and linear problems.

Conformal mapping of $\mathbb{D}$ onto a Jordan domain G can be formulated as a Riemann-Hilbert problem by setting $M_t := \partial G$ for all $t \in \mathbb{T}$. The background is the Carathéodory-Osgood theorem, which tells us that $\mathbb{T}$ is mapped onto ∂G and hence conformal mapping amounts to solve the Riemann-Hilbert problem $w(t) \in \partial G$, $\forall t \in \mathbb{T}$.

The right part of Figure 2 shows a conformal map of the unit disk onto a square S, interpreted as a solution of the Riemann-Hilbert problem with the target $M := \mathbb{T} \times \partial S$.

The history of conformal mapping is a subject on its own; here we only comment briefly on the development of linear Riemann-Hilbert problems.

1.2 Linear Riemann-Hilbert problems

If all target curves are straight lines, the boundary condition is linear. The simplest problem of this kind consists in finding a holomorphic function w from the boundary values of its real part,

$$\operatorname{Re} w(t) = u(t) \qquad \forall t \in \mathbb{T}. \tag{1.2.1}$$

If the unknown function w is multiplied by a given (continuous nonvanishing) function $\overline{f}\colon\ \mathbb{T} \to \mathbb{C} \setminus \{0\}$ before the real part is taken, we get the general linear Riemann-Hilbert problem

$$\operatorname{Re}\left(\overline{f(t)}\, w(t)\right) = c(t) \qquad \forall t \in \mathbb{T}. \tag{1.2.2}$$

A method for solving the linear problem (1.2.2) was first indicated by David Hilbert. In a lecture at the 3rd International Mathematical Congress in Heidelberg 1904 [22], he introduced the conjugation operator

$$Hf(\mathrm{e}^{\mathrm{i}\tau}) := \frac{1}{2\pi} \int_0^{2\pi} f(\mathrm{e}^{\mathrm{i}\sigma}) \cot \frac{(\sigma - \tau)}{2}\, \mathrm{d}\sigma,$$

and transformed the problem to a (Fredholm) integral equation with a continuous kernel. Unfortunately, Hilbert did not remark that the transformed problem is not always equivalent to the original one. So he missed one of the great discoveries, the 'index' of a singular integral operator.

Though Hilbert published a modification of his earlier approach in 1912 (see [23]), it took several years till Fritz Noether pointed out the gap in Hilbert's paper and understood its consequences ([30], 1921): "... *die Durchführung ist aber nicht korrekt, da sie im Grunde die unzutreffende Annahme enthält, daß die homogenen Gleichungen* $K(\omega) = 0$ *und* $\overline{K}(\omega) = 0$ *gleiche Lösungszahl hätten;...* ". So it was Noether who found that mapping properties of certain operators can be derived from simple geometric characteristics of their 'symbols'.

Noether's work initiated the search for so-called 'index formulas'. Starting with more general classes of singular integral operators (B. Bojarski, S. G. Mikhlin) this development led to the theory of pseudo-differential operators and to general index theory.

In a sense, Noether gave a complete solution of the linear Riemann-Hilbert problem, including explicit solution formulas and solvability conditions for the inhomogenous problem with a negative index (see loc. cit., p. 45). However, he did not (and could not) work in function spaces, which would have been necessary to validate the formulas.

This gap was closed some years later by the investigations of N. I. Muskhelishvili (1935), F. D. Gakhov [13] (1937), and I. N. Vekua [44] (1942), who studied (scalar) problems with Hölder continuous coefficients in Hölder spaces of holomorphic functions.

If the coefficient f is merely a continuous function, the choice of the spaces is more delicate, since the conjugation operator is not continuous in $C(\mathbb{T})$. One possibility is to consider right-hand sides c in the Lebesgue space $L^2_{\mathbb{R}}(\mathbb{T})$ of square-integrable real-valued functions, and to look for solutions in the Hardy space $H^2(\mathbb{D})$ of holomorphic functions in $\mathbb{D}$ for which

$$\|w\|_2 := \Big\{ \sup_{0<r<1} \int_{\mathbb{T}} |w(rt)|^2 \, |dt| \Big\}^{1/2}$$

is finite. Since functions in $H^2(\mathbb{D})$ have angular limits almost everywhere on $\mathbb{T}$, the boundary condition makes sense.

In this setting the essence of Noether's theory is a description of the mapping properties of the Riemann-Hilbert operator

$$R_f \colon \, H^2(\mathbb{T}) \to L^2_{\mathbb{R}}(\mathbb{T}), \quad w \mapsto \operatorname{Re}\big(\overline{f}\, w\big).$$

If the symbol $f\colon \mathbb{T} \to \mathbb{C} \setminus \{0\}$ is continuous, the dimensions of the kernel and the cokernel of R_f are given by

$$\dim \ker R_f = \max\,(0, 2 \cdot \operatorname{wind} f + 1),$$

$$\operatorname{codim} \operatorname{im} R_f = \max\,(0, -2 \cdot \operatorname{wind} f - 1),$$

where $\operatorname{wind} f$ denotes the winding number of f about zero. This implies the index formula

$$\operatorname{ind} R_f := \dim \ker R_f - \operatorname{codim} \operatorname{im} R_f = 2 \cdot \operatorname{wind} f + 1.$$

If the symbol f has winding number zero there is a one parameter family of solutions. If, moreover, f is Hölder continuous, the graphs of all solutions cover the target manifold in a schlicht manner.

With an appropriate definition of the winding number the above result remains in force for problems with piecewise continuous coefficients (I. B. Simonenko [37] 1960).

In the following years a rich theory emerged from the interplay between Riemann-Hilbert problems, singular integral equations, Toeplitz operators, and Birkhoff-Wiener-Hopf factorization. For a comprehensive and up-to-date treatment we recommend the recent monograph of A. Böttcher and Ju. Karlovič, [6]. We shall not (and cannot) follow these developments here, but turn our attention towards nonlinear problems.

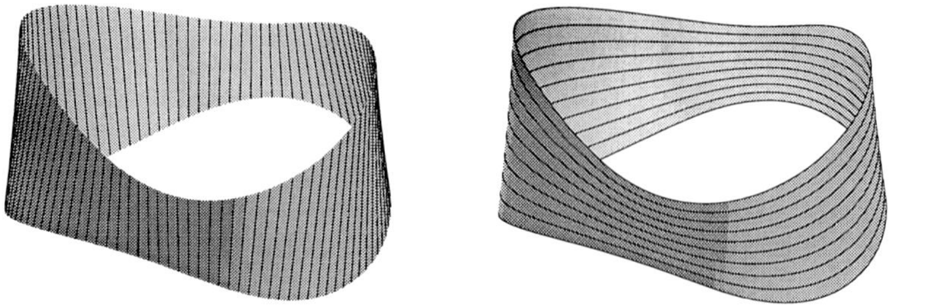

Figure 3. *Target manifold of linear RHP with fibers and solutions*

1.3 Nonlinear Riemann-Hilbert problems

While linear Riemann-Hilbert problems with their different forms had a strong impact on the development of functional analysis and operator theory, nonlinear Riemann-Hilbert problem have not been in the mainstreem of research for a long time.

The investigations started in the fourties with problems of *special structure* which allowed reduction to the linear case ([1], [12], [14], [18], [29], [31], [32], [41]).

At the same time problems involving sufficiently *small nonlinearities* have been studied by means of Banach's and Schauder's fixed point theorem and implicit function theorems ([3], [15], [16], [17], [28], [29], [33]).

Refined fixed point techniques and methods from monotone operator theory allowed to attack problems with *strong nonlinearities* ([9], [40], [52], [53]).

The real breakthrough came in 1972, when Alexander Shnirel'man [36] proved the solvability of Riemann-Hilbert problems under merely *topological* assumptions. This paper marks the beginning of a new period which brought new insights into the geometric aspects of the problem. In the next section we outline some basic results.

§2 Results

2.1 Existence of solutions for compact target manifolds

It is worth to mention that Shnirel'man seems to be the first who returned to Riemann's geometric considerations. He considered Riemann-Hilbert problems with simple closed smooth target curves M_t which depend smoothly on t. In what follows we work with slightly weaker assumptions.

Definition 1. *A (compact) target manifold* $M \subset \mathbb{T} \times \mathbb{C}$ *is said to be* admissible *if it has a parametrization*

$$M = \{(t, \mu(t,s)) \colon \ s, t \in \mathbb{T}\}$$

where $\mu \in \mathbb{T}^2 \to \mathbb{C}$ *is continuously differentiable,* $s \mapsto \mu(t,s)$ *is injective for all* $t \in \mathbb{T}$, *and* $0 < 1/C \leq |\partial_s \mu(t,s)| \leq C$. *The set of all admissible target manifolds is denoted by* $\mathcal{M}$.

The following three examples show that Riemann-Hilbert problems with admissible target manifolds can have distinct solvability properties.

Example 1: Let $M_t := \mathbb{T}$ for all $t \in \mathbb{T}$. Then the solutions $w \in H^\infty \cap C$ are the finite Blaschke products

$$w(z) = c \prod_{k=1}^{n} \frac{z - z_k}{1 - \overline{z}_k z},$$

where $z_k \in \mathbb{D}$ and $c \in \mathbb{T}$ are arbitrarily chosen numbers.

Example 2: If $M_t := \{w \in \mathbb{C} : |w - t^{-1}| = 1\}$, then the only solution is $w \equiv 0$.

Example 3: If $M_t := \{w \in \mathbb{C} : |w - 2\,t^{-1}| = 1\}$, then the Riemann-Hilbert problem has no solution.

The examples stand for three general classes of Riemann-Hilbert problems with a compact target manifold.

Definition 2. *The target manifold* $M \in \mathcal{M}$ *is called* regularly traceable, *(written as* $M \in \mathcal{R}$*) if there exists a holomorphic function* $w_M \in H^\infty \cap C$ *with* graph $w_M \subset \operatorname{int} M$. *The target manifold* $M \in \mathcal{M}$ *is said to be* singularly traceable, $M \in \mathcal{S}$, *if there exists a holomorphic function* $w_M \in H^\infty \cap C$ *with* graph $w_M \subset$ clos int M, *but* $M \notin \mathcal{R}$. *Finally, the set* $\mathcal{N}$ *of* non-traceable *target manifolds is defined by* $\mathcal{N} := \mathcal{M} \setminus (\mathcal{R} \cup \mathcal{S})$.

Here int M denotes the bounded component of $(\mathbb{T} \times \mathbb{C}) \setminus M$ and "clos" denotes the topological closure of a set.

Shnirel'man considered target manifolds with $0 \in M_t$ for all t. The general regularly traceable case can be reduced to this one by subtracting the function w_M.

Before we give the existence result, we remark that any continuous function $w \colon \mathbb{T} \to \mathbb{C}$ with graph $w \subset M$ has a *winding number about* M,

$$\operatorname{wind}_M w := \operatorname{wind}(w - m).$$

Here $m \colon \mathbb{T} \to \mathbb{C}$ is an arbitrary continuous function with graph $m \subset \operatorname{int} M$. Geometrically, $\operatorname{wind}_M w$ counts the number of turns of graph w around M.

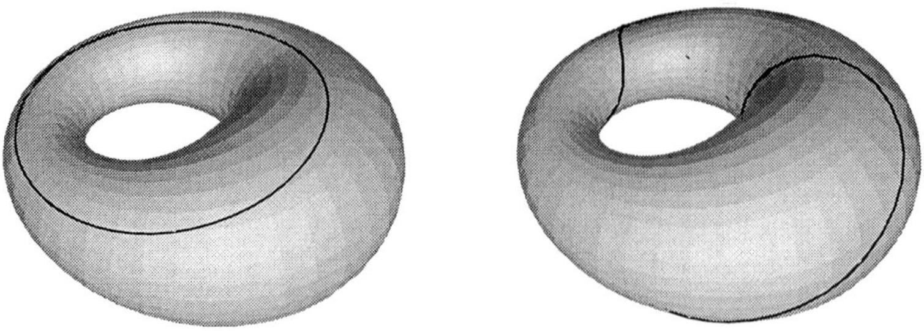

Figure 4. *Solutions with winding numbers zero and two*

Theorem 1. (A. I. Shnirel'man [36]) *If $M \in \mathcal{R}$ the following assertions hold.*

(i) *The target manifold M is foliated by the graphs of solutions of the Riemann-Hilbert problem with winding number zero about M.*

(ii) *If $0 \in \operatorname{int} M_t$ for all $t \in \mathbb{T}$, then for all points $(t_0, w_0) \in M$ and $z_1, \ldots z_n \in \mathbb{D}$, there exists exactly one solution $w \in H^\infty \cap C$ with*

$$\operatorname{wind}_M w = n, \qquad w(t_0) = w_0, \qquad w(z_k) = 0 \; (k = 1, \ldots, n).$$

(iii) *There are no solutions with a negative winding number about M.*

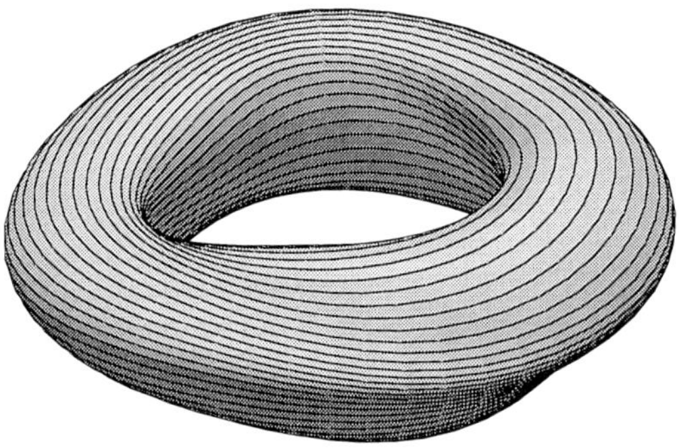

Figure 5. *Regularly traceable target manifold foliated by solutions*

The structure of the solution set of singularly traceable problems is completely different from the regular case.

Theorem 2. *If $M \in \mathcal{S}$, then the following assertions hold.*

(i) *The Riemann-Hilbert problem* graph $w \subset M$ *has exactly one solution* w_0 *in* $H^\infty \cap C$.

(ii) *The winding number of* w_0 *about* M *is negative,* $\operatorname{wind}_M w_0 < 0$.

It is obvious that Riemann-Hilbert problems with $M \in \mathcal{N}$ have no solution.

2.2 Solvability for non-compact target manifolds

Target manifolds of linear Riemann-Hilbert problems do not belong to the class $\mathcal{M}$ since they are not compact.

Considering the target manifold as a subset of $\mathbb{T} \times \widehat{\mathbb{C}}$ would help in the linear case, because this results in smooth compact target curves on the Riemann sphere $\widehat{\mathbb{C}}$. However, for nonlinear problems with a smooth non-compact target in $\mathbb{T} \times \mathbb{C}$, the resulting manifolds M are in general not smooth at $\mathbb{T} \times \{\infty\}$.

Using a special technique for removing singularities and a refinement of Shnirel'man's results it was shown in [45] that the solvability properties of linear Riemann-Hilbert problems extend to a fairly wide class of problems with non-compact target manifold.

Definition 3. *A non-compact target manifold M is said to be* admissible, *if it has a parametrization*

$$M = \{ (t, \mu(t,s)) : t \in \mathbb{T}, s \in \mathbb{R} \} \tag{2.2.1}$$

where $\mu \in C^1(\mathbb{T} \times \mathbb{R})$ is such that

(i) *for each $t \in \mathbb{T}$ the mapping $s \mapsto \mu(t,s)$ is injective,*

(ii) $0 < 1/C \le |D_s \mu| \le C$ *on* $\mathbb{T} \times \mathbb{R}$,

(iii) *the limits*

$$\lim_{s \to \pm\infty} D_s \mu(t,s) := \mu^\pm(t), \qquad \lim_{s \to \pm\infty} s^{-2} D_t \mu(t,s) = 0$$

exist uniformly with respect to $t \in \mathbb{T}$.

The number $\operatorname{ind} M := \operatorname{wind} \mu^+$ *is said to be the* index *of the problem.*

Geometrically, the index appears as a 'twisting number' of the target manifold M.

Condition (iii) guarantees a certain regular behaviour of the target curves M_t at infinity. While the condition implies that the direction of the target curves stabilizes for $s \to \pm\infty$, the existence of a (linear) asymptotics does not follow. Note that there is no interrelation between the behaviour at 'both ends' of M_t.

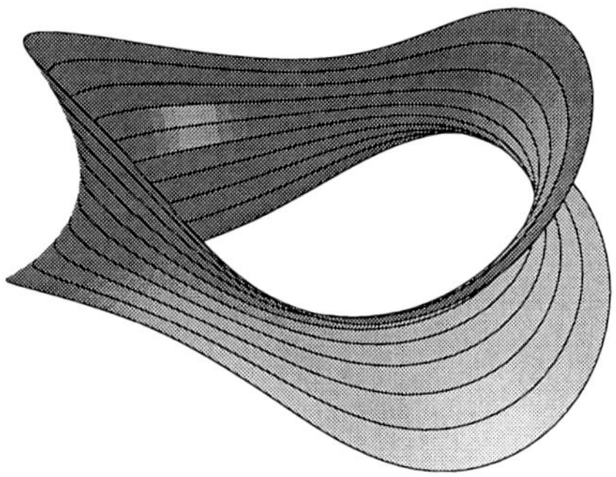

Figure 6. *Non-compact target manifold with ind=0 foliated by solutions*

Theorem 3. (E. Wegert [45], [49]) *Let M be a non-compact admissible target manifold. Then the following assertions hold.*

(i) *If* $\operatorname{ind} M = 0$ *the target manifold is foliated by the graphs of solutions* $w \in H^\infty \cap C$ *to the Riemann-Hilbert problem* $\operatorname{graph} w \subset M$.

(ii) *If* $\operatorname{ind} M < 0$ *there exists at most one solution* $w \in H^\infty \cap C$ *of* $\operatorname{graph} w \subset M$.

(iii) *If* $\operatorname{ind} M = n \geq 0$ *then, for arbitrarily given* $t_0 \in \mathbb{T}$, $W_0 \in M_{t_0}$, $z_1, \ldots, z_n \in \mathbb{D}$, $W_1, \ldots, W_n \in \mathbb{C}$, *there exists exactly one solution of* $\operatorname{graph} w \subset M$ *which satisfies the interpolation conditions*

$$w(t_0) = W_0, \qquad w(z_k) = W_k \quad (k = 1, \ldots, n). \tag{2.2.2}$$

The case of non-orientable non-compact target manifolds can be dealt with by considering the doubled manifold $\widetilde{M}$ having the fibers $\widetilde{M}_t := M_{t^2}$; for details see [49], Section 2.6. It is remarkable that Riemann-Hilbert problems with a unique solution are exactly those with a nonorientable manifold with $\operatorname{ind} M = -1/2$.

2.3 Extremality of solutions

Conformal maps are characterized by a number of extremal properties, so one may wonder if there is also an extremal principle for solutions of Riemann-Hilbert problems.

In order to illustrate the idea, we start with a classical result of Carathéodory, which characterizes conformal mapping of a simply connnected domain $G \ni 0$ onto the disk $\mathbb{D}$.

Lemma 1. *Let $\mathbb{A}^*$ be the set of univalent holomorphic functions $w\colon G \to \mathbb{D}$ with $w(0) = 0$. Fix $z_0 \in G \setminus \{0\}$. Then the modulus $|w(z_0)|$ is maximal in the class $\mathbb{A}^*$ if and only if w maps G conformally onto $\mathbb{D}$.*

This result does not immediately extend to Riemann-Hilbert problems for several reasons, but what can be extended to the general situation is an appropriate reformulation of Carathéodory's principle for the inverse map $w\colon \mathbb{D} \to G$.

The key is to consider a set of bounded holomorphic functions with restricted boundary values

$$\mathbb{A} := \{w \in H^\infty\colon\ w(t) \in \operatorname{clos}\operatorname{int} M_t \quad \text{a.e. on } \mathbb{T}\}$$

and to study the range of values at $z_0 \in \mathbb{D}$

$$R(z_0) := \{w(z_0)\colon\ w \in \mathbb{A}\}.$$

Note that the definition of the set $\mathbb{A}$ does neither involve values at interior points of $\mathbb{D}$ nor univalence of the functions.

Theorem 4. (Extremal Principle [46], [47]) *Let $M \in \mathcal{R}$ and $z_0 \in \mathbb{D}$.*

(i) *The range of values $R(z_0)$ is homeomorphic to the closed disk $\overline{\mathbb{D}}$.*

(ii) *A function w is a solution of the Riemann-Hilbert problem $w(t) \in M_t$ with vanishing winding number about M if and only if $w \in \mathbb{A}$ and $w(z_0) \in \partial R(z_0)$.*

The result yields an effective formula for the range of values.

Corollary 1. *Let $\mathbb{W}_0$ denote the set of solutions of the Riemann-Hilbert problem $w(t) \in M_t$ with winding number zero about M. Then*

$$R(z_0) = \operatorname{clos}\operatorname{int}\{w(z_0)\colon\ w \in \mathbb{W}_0\}.$$

For the special target manifold $M = \mathbb{T}^2$ Theorem 4 is just the maximum modulus principle.

A similar characterization holds for solutions with positive winding numbers satisfying additional interpolation conditions. Schwarz' lemma pertains to the special case where $M = \mathbb{T}^2$.

The next theorem shows that extremal problems over the set $\mathbb{A}$ are not interesting if $M \in \mathcal{S}$ or $M \in \mathcal{N}$.

Theorem 5.

(i) *If $M \in \mathcal{S}$, then $\mathbb{A}$ is a singleton.*

(ii) *If $M \in \mathcal{N}$, then $\mathbb{A}$ is empty.*

The result is not as trivial as it appears. Here are two consequences, which will be applied in Section 3.5.

Corollary 2.

(i) *If $\mathbb{A}$ contains only one element, this function is a continuous solution of the Riemann-Hilbert problem with a negative winding number about M.*

(ii) *If $\mathbb{A}$ contains at least two elements, then $M \in \mathcal{R}$.*

2.4 The geometry of Riemann-Hilbert problems

Recall that we have three classes of compact admissible target manifolds. In order to explore the structure of these classes we equip $\mathcal{M}$ with the Hausdorff distance

$$\operatorname{dist}(M, N) := \max_{p \in M, q \in N} \{\operatorname{dist}(p, N), \operatorname{dist}(q, M)\}.$$

It turns out that the set $\mathcal{S}$ of singularly traceable manifolds is a 'thin set' which forms the common boundary of the two open sets $\mathcal{R}$ and $\mathcal{N}$.

Moreover, if $M \in \mathcal{S}$ the RHP has a unique solution w_0 with negative winding number about M. The definition

$$\mathcal{S}_k := \{M \in \mathcal{M}\colon \operatorname{wind}_M w_0 \leq -k\},$$

endows $\mathcal{S}$ with a hierarchical structure

$$\mathcal{S} = \mathcal{S}_1 \supset \mathcal{S}_2 \supset \mathcal{S}_3 \supset \dots .$$

Theorem 6. (Wegert [48],[49])

(i) *$\mathcal{R}$ and $\mathcal{N}$ are open, $\mathcal{S}$ is closed, $\partial\mathcal{R} = \partial\mathcal{N} = \mathcal{S}$.*

(ii) *The (real) codimension of $\mathcal{S}_{k+1}$ in $\mathcal{S}_k$ equals 2.*

Corollary 3. *Any continuous path in the set of Riemann-Hilbert problems that connects $\mathcal{R}$ and $\mathcal{N}$ necessarily meets $\mathcal{S}$.*

This result has an interesting application to a problem in H^∞-optimization which we discuss in Section 3.5.

§3 Applications

One objective of this paper is to show Riemann-Hilbert problems in action. We start with only a few words about 'direct applications'. Thereafter we will address at some length several topics in complex analysis which are linked with Riemann-Hilbert problems.

3.1 Direct applications

Riemann-Hilbert problems are adequate mathematical models for a number of problems in physics. In most of these applications they appear as boundary value problem for either some physical quantities or their Fourier transforms.

In order to give at least one concrete example, we briefly describe a free boundary problem in electrochemical machining which leads to a nonlinear Riemann-Hilbert problem with a noncompact nonorientable target manifold ([50]).

In the underlying physical process metal is eroded from a workpiece by electrolysis. The workpiece acts as anode whereas the tool consists of the cathode and an insulated part; it is fed towards the workpiece at a constant speed (in Figure 7 to the right). The gap between workpiece and tool is filled with electrolyte.

After an initial period the process becomes quasi-stationary, *i.e.* tool and free surface move at the same horizontal speed and the shape of the latter remains unchanged.

In order to determine the unknown shape of the free surface, we introduce a conformal map of the gap onto a rectangular domain. The real part of this mapping function is the electric potential. The equilibrium conditions on the boundary of the gap result in boundary conditions of Riemann-Hilbert type for the inverse mapping. However, the conditions on the four sides of the rectangle do not fit together continuously. Some special

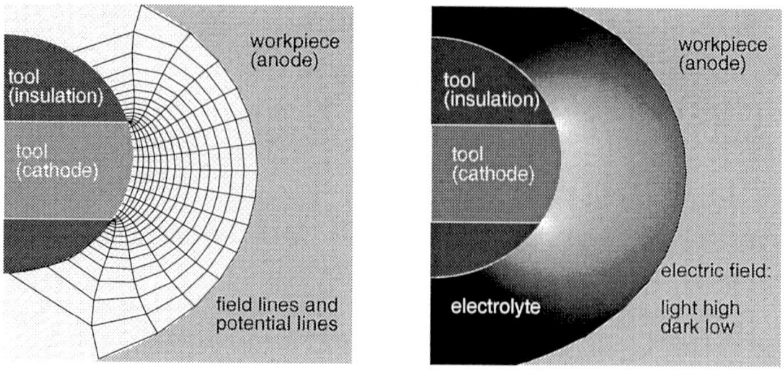

Figure 7. *A free boundary problem in electrochemical machining*

tricks were used to transform the problem to a Riemann-Hilbert problem with a non-orientable non-compact target manifold in the disk for which existence and uniqueness of a solution could be proved. Computational results are shown in Figure 7.

We note that applications which result in 'pure' scalar Riemann-Hilbert problems are rather the exception than the rule.

On the other hand, a number of applications lead to nonlinear Riemann-Hilbert problems for *systems* of holomorphic functions. Those problems play a prominent role, for instance, in models of Keplerian disks around magnetized neutron stars (Wegmann [51]), in inverse problems of electric impedance tomography (Isakov [25], Powell [34]), in connection with identification of memory kernels (v. Wolfersdorf [54]), and in scattering theory for layered media (Sylvester [43]), to mention at least a few. Recently, the interest in Riemann-Hilbert problems for vector-valued functions increased rapidly in connection with investigations of so-called analytic disks. This, however, is another story.

Other applications give rise to nonlinear problems of Riemann-Hilbert-Poincaré type, where the boundary condition involves the derivative of the unknown function

$$f\big(t, w(t), w'(t)\big) = 0, \quad \forall t \in \mathbb{T}.$$

In particular, all boundary value problems for *harmonic* functions u in a simply connected (smoothly bounded) domain $G \to \mathbb{R}^2$ with a nonlinear Robin condition

$$f\big(t, u(t), \partial_n u(t)\big) = 0, \quad \forall t \in \partial G$$

belong to this class. Here ∂_n denotes the normal derivative of u. Since these problems can model more complicated systems which sometimes suffer from bifurcation phenomena, their mathematics have to be more sophisticated. At present the theory is far from beeing complete and probably one should not expect final results, since many problems have their own peculiarities.

Here we shall not enter into a detailed discussion but shall turn our attention to applications where the relation to Riemann-Hilbert problems is somewhat hidden.

3.2 Holomorphic motions

Roughly speaking, a holomorphic motion is a deformation $X \mapsto X_z$ of a set $X \subset \mathbb{C}$ depending on a (complex) parameter $z \in \mathbb{D}$, such that any point of X moves holomorphically with the parameter z. More precisely:

Definition 4. *A holomorphic motion is a mapping $f\colon \mathbb{D}\times X \to \mathbb{C}$ such that*

(i) *the function*

$$f^x\colon \mathbb{D} \to \mathbb{C},\ z \mapsto f(z,x)$$

is holomorphic in $\mathbb{D}$ for any fixed $x \in X$,

(ii) *the function*

$$f_z\colon X \to \mathbb{C},\ x \mapsto f(z,x)$$

is an injection for any fixed $z \in \mathbb{D}$,

(iii) *the mapping f_0 is the identity on X.*

One principle to generate holomorphic motions makes use of Riemann-Hilbert problems. In a sense, holomorphic motions of curves lead just to another approach to Riemann-Hilbert problems.

In order to establish this connection we consider a compact admissible regular target manifold M and recall that $\mathbb{W}_0$ denotes the set of solutions with winding number zero. It is easy to see that the mapping

$$g\colon \mathbb{W}_0 \to \mathbb{C},\ w \mapsto w(0)$$

is a homeomorphism between $\mathbb{W}_0$ and the Jordan curve

$$X = \{w(0)\colon\ w \in \mathbb{W}_0\}.$$

The function $f\colon \mathbb{D} \times X \to \mathbb{C}$ defined by

$$f(z,x) := \big(g^{-1}(x)\big)(z)$$

is a holomorphic motion of X with the Jordan curves $X_z := \{w(z) : w \in \mathbb{W}_0\}$ as fibers.

A basic problem pertains to the extendibility of a given holomorphic motion onto a larger domain. One result of this type is the celebrated λ-Lemma.

Proposition 1. (Mañé, Sad, Sullivan [27]).
Each holomorphic motion $f\colon \mathbb{D}\times X \to \mathbb{C}$ *has an extension to a holomorphic motion* $\widehat{f}\colon \mathbb{D}\times \overline{X} \to \mathbb{C}$ *such that each* $\widehat{f}_z\colon \overline{X} \to \mathbb{C}$ *is quasiconformal, and* $\widehat{f}$ *is continuous on* $\mathbb{D}\times\overline{X}$.

A consequence of the λ-Lemma is the continuity of any holomorphic motion f with respect to the pair (z, x).

Sullivan and Thurston [42] posed the question of whether or not each holomorphic motion of X cannot only be extended to $\overline{X}$ but to a holomorphic motion of $\mathbb{C}$. They gave a partial answer to this question by proving that if $f\colon \mathbb{D}\times X$ is a holomorphic motion of X, then there exists a universal positive constant a such that the restriction of f to $(a\mathbb{D})\times X$ has an extension.

Bers and Royden [5] could show that $a \geq 1/3$. Using ideas from several complex variables, including Forstnerič's results about polynomial hulls and Riemann-Hilbert problems [11], Slodkovski [38] proved that a can be taken to be equal to 1 (*i.e.* f itself admits such an extension). A somewhat simpler proof of this so-called extended λ-Lemma is given by Astala and Martin [2]. Their proof essentially relies on the above construction of holomorphic motions via Riemann-Hilbert problems with a family of target manifolds exhausting $\mathbb{T}\times\mathbb{C}$.

3.3 Generalized Nevanlinna-Pick interpolation

In a sense solutions of Riemann-Hilbert problems generalize the concept of Blaschke products. For illustration we consider interpolation problems for holomorphic functions with additional constraints.

Given $z_j \in \mathbb{D}$ and $W_j \in \mathbb{C}$, the classical Nevanlinna-Pick interpolation problem consists in finding holomorphic functions w in the closed unit ball of H^∞, which satisfy the interpolation conditions

$$w(z_j) = W_j \quad (j = 1, 2, \ldots, n). \tag{3.3.1}$$

It is well known, that a solution exists if and only if the Pick-Matrix

$$P := \left(\frac{1 - W_j\overline{W_k}}{1 - z_j\overline{z}_k}\right)_{j,k=1}^{n}$$

is non-negative. If so, there exist solutions which are Blaschke products of degree not greater than n.

Here we are interested in a more general interpolation problem (3.3.1) with restricted boundary values, where the unit ball of H^∞

$$\mathbb{B} = \{w \in H^\infty\colon\ w(t) \in \overline{\mathbb{D}} \text{ a.e. on } \mathbb{T}\}$$

is replaced by the set

$$\mathbb{A} := \{w \in H^\infty \colon\ w(t) \in \overline{G_t} \text{ a.e. on } \mathbb{T}\ \},$$

and $(G_t)_{t\in\mathbb{T}}$ is a given family of domains bounded by Jordan curves M_t. In order to indicate that $\mathbb{A}$ depends on the target manifold M we sometimes write $\mathbb{A}(M)$.

The extremal principle tells us what to do for solving the one-point interpolation problem

$$w \in \mathbb{A},\ \ w(0) = W.$$

Step 1. Find the solution set $\mathbb{W}_0$ of the RHP.
Step 2. Determine the range of values

$$R = \operatorname{clos}\operatorname{int}\{w(0)\colon\ w \in \mathbb{W}_0\}.$$

Step 3. Consider the location of the point W with respect to R.

3.1. If $W \notin R$ there is no solution.

3.2. If $W \in \partial R$ there exists a unique solution $w_0 \in \mathbb{W}_0$.

3.3. If $W \in \overset{\circ}{R}$ there are several solutions; in particular, for each $k \geq 1$, there exist solutions with

$$\operatorname{graph} w \subset M, \qquad \operatorname{wind}_M w = k.$$

In order to treat multiple interpolation conditions we associate with $(z_0, W_0) \in \mathbb{D} \times \mathbb{C}$ two transformations $S\colon\ \mathcal{M} \to \mathcal{M}$ and $T\colon\ \mathcal{M} \to \mathcal{M}$ of target manifolds, defined by

$$(SM)_t := M_{s(t)}, \qquad (TM)_t := t^{-1}\,(M_t - W_0),$$

with

$$s(t) = \frac{t + z_0}{1 + \overline{z}_0 t}.$$

These transformations are designed such that the solutions of the interpolation problem

$$w(z_0) = W_0, \qquad w \in \mathbb{A}(M)$$

in $\mathbb{A}(M)$ correspond with functions in $\mathbb{A}(TSM)$ without interpolation condition.

Applying the transformations recursively, the number of interpolation conditions is reduced in each step by one. This generates a sequence of sets (closed regions, points, or empty sets)

$$R_1, R_2, \ldots, R_n \subset \mathbb{C}$$

and a sequence of transformed interpolation values

$$W^1, W^2, \ldots, W^n \in \mathbb{C}.$$

Theorem 7. *Let $M \in \mathcal{R}$.*

(i) *There exist K and N with $0 \le K < N \le n+1$, such that*

$$\begin{array}{lll} W^j \in \overset{\circ}{R}_j & \text{if} & j \le K, \\ W^j \in \partial R_j & \text{if} & K < j < N, \\ W^j \notin R_j & \text{if} & N \le j. \end{array}$$

(ii) *The interpolation problem is solvable if and only if $N = n+1$.*

(iii) *The solution is unique if and only if $K < n$.*

(iv) *If $K < n$ the solution belongs to $\mathbb{W}_K$.*

(v) *If $k \ge K = n$ there exist solutions in $\mathbb{W}_k$.*

The case $N = K + 1$ is generic.

Corollary 4. *If the interpolation problem is solvable, there exist solutions with $w(t) \in M_t$ for all $t \in \mathbb{T}$.*

3.4 Polynomial hulls

Let $\mathcal{P}$ denote the set of polynomials in n variables. The *polynomial hull* of a (compact) subset M in $\mathbb{C}^n$ is the set

$$\operatorname{pol} M := \{z \in \mathbb{C}^n : \ |p(z)| \le \max_{t \in M} |p(t)| \ \forall p \in \mathcal{P}\}.$$

If $M \subset \mathbb{T} \times \mathbb{C}$ is an admissible compact target manifold the solutions of the related Riemann-Hilbert problem play an essential role in an explicit description of $\operatorname{pol} M$.

Recall that we denote by $\mathbb{W}_0$ the set of solutions with winding number zero about a regularly traceable target manifold. The family of curves

$$M_z := \{w(z) : \ w \in \mathbb{W}_0\}, \qquad z \in \overline{\mathbb{D}}$$

can be thought of as a holomorphic prolongation of the family of target curves. From the extremal principle it is known that the Jordan curve M_z bounds the range of values,

$$R(z) = \operatorname{clos} \operatorname{int} M_z.$$

Theorem 8. (Forstnerič, [11], Slodkowski [38])
The polynomial hull of $M \in \mathcal{R}$ is given by

$$\operatorname{pol} M = \{(z, w) \in \mathbb{C}^2 \colon\ z \in \overline{\mathbb{D}},\ w \in R(z)\}.$$

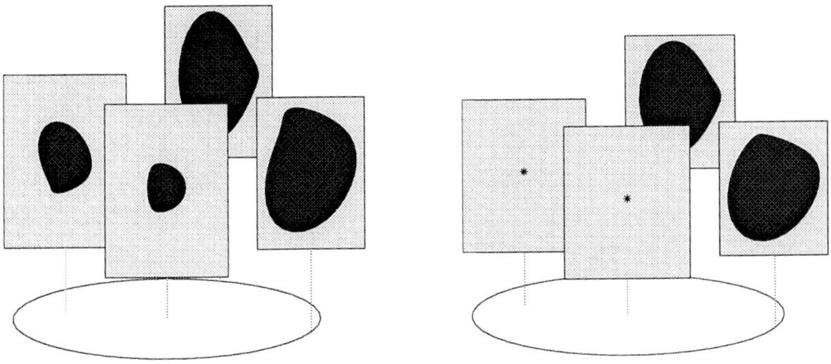

Figure 8. *Polynomial hulls of manifolds in $\mathcal{R}$ and $\mathcal{S}$*

If the target manifold M is singularly traceable, the part of the polynomial hull lying over $\mathbb{D}$ collapses. For $M \in \mathcal{S}$ we denote by $w^* \in H^\infty \cap C$ the unique solution of the Riemann-Hilbert problem $\operatorname{graph} w \subset M$.

Theorem 9. *For each $M \in \mathcal{S}$ the polynomial hull of M is given by*

$$\operatorname{pol} M = \operatorname{clos} \operatorname{int} M \cup \big\{ \big(z, w^*(z)\big) \,:\, z \in \mathbb{D} \big\}.$$

Finally, if $M \in \mathcal{N}$, then the polynomial hull of M lies completely over the circle.

Theorem 10. *For each $M \in \mathcal{N}$ the polynomial hull of M coincides with* $\operatorname{clos} \operatorname{int} M \subset \mathbb{T} \times \mathbb{C}$.

3.5 $\mathbf{H^\infty}$-optimization

Another question which is intimately (but not obviously) related to Riemann-Hilbert problems was raised in the eighties by J. W. Helton. Its engineering background is the design of dynamical systems via optimization of the frequency response function.

Recall that the *transfer function* of a (causal, linear, time invariant, stable) system is a bounded holomorphic function in the complex right-half plane. The boundary function on the imaginary axis is said to be the *frequency response* $\omega \mapsto f(\mathrm{i}\omega)$.

Since the frequency response is the boundary function of a holomorphic function, it cannot by prescribed arbitrarily; design of systems is an optimization problem.

The values $|f(\mathrm{i}\omega)|$ and $\arg f(\mathrm{i}\omega)$ describe gain and phase of the system operating at the frequency ω, and the designer knows what he should estimate as a 'good performance' $f(\mathrm{i}\omega)$ of the system at a certain frequency ω. So he should be able to evaluate this performance in terms of a *penalty function*

$$F\colon \mathbb{R} \times \mathbb{C} \to \mathbb{R}_+, \quad (\omega, w) \mapsto F(\omega, f(\mathrm{i}\omega)),$$

where $F(\omega, f(\mathrm{i}\omega))$ attains small values if $f(\mathrm{i}\omega)$ is 'good' and big values if it is 'bad'. The standard choice of F is the euclidean distance from an 'ideal' response.

Once the penalty function F is fixed, the over-all rating of a system, in the sense of a worst case analysis over all frequencies, is the real number

$$P(f) := \sup \{F(\omega, f(\mathrm{i}\omega))\colon \ \omega \in \mathbb{R}\}.$$

Optimal design of a system then requires to find the *optimal performance*

$$P^* := \inf \{P(f)\colon \ f \in H^\infty(\mathbb{C}_+)\},$$

and all bounded holomorphic functions (systems) for which the infimum (if it exists) is attained. Here $H^\infty(\mathbb{C}_+)$ denotes the Hardy space of bounded holomorphic functions in the complex right-half plane.

In order to relate the optimization problem to Riemann-Hilbert problems we transform it from the right half plane to the unit disk and introduce the sets of bounded analytic functions with restricted boundary values

$$\mathbb{A}(p) := \{w \in H^\infty(\mathbb{D})\colon \ F(t, w(t)) \le p \ \text{ a.e. on } \mathbb{T}\}.$$

The set $\mathbb{A}(p)$ is built from the transfer functions of all systems with performance not worse than p, and the optimization problem is equivalent to finding the smallest possible value of p for which $\mathbb{A}(p)$ is not empty.

Sets of the type $\mathbb{A}(p)$ have already been considered above in conjunction with the extremal principle for Riemann-Hilbert problems. The difference is that we now have a family of target manifolds

$$M(p) := \{(t, w) \in \mathbb{T} \times \mathbb{C}\colon \ F(t, w) = p\}.$$

We call the optimization problem *strictly nested* if it satisfies the following assumptions (i)–(iii):

(i) The $M(p)$ are compact admissible target manifolds which depend continuously on $p \in \mathbb{R}_+$.

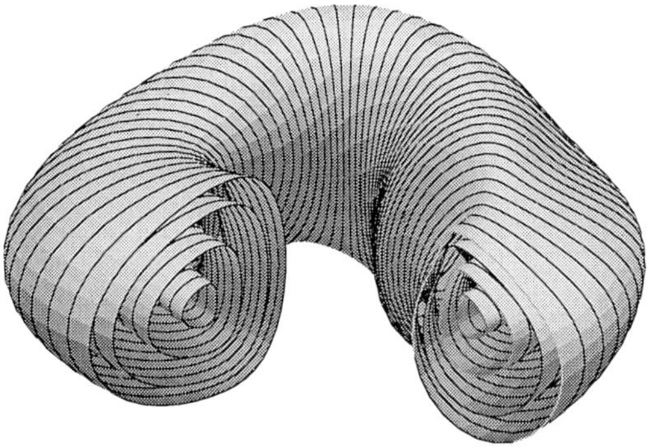

Figure 9. *A nested set of target manifolds with fibers*

(ii) The family $\{M(p)\}$ is strictly nested,

$$0 < p < q \quad \Rightarrow \quad M(p) \subset \operatorname{int} M(q).$$

(iii) $M(p) \in \mathcal{R}$ if p is sufficiently large; $M(p) \in \mathcal{N}$ if p is sufficiently small.

Theorem 11. (Helton and Marshall [21], Wegert [48]) *For any strictly nested optimization problem the following assertions hold:*

(i) *The optimization problem has a unique solution.*

(ii) *A function w is a solution of the optimization problem if and only if*

$$w \in H^\infty \cap C,$$

$$F\big(t, w(t)\big) = \text{const} \quad \text{on } \mathbb{T},$$

$$\operatorname{wind} \partial_w F(.,w) =: k \geq 1.$$

(iii) *Generically the winding number k of a solution equals one.*

Note that $\partial_w F(.,w) = -\operatorname{wind}_{M(p)} w$.

The second assertion of (ii) has a special meaning: Optimal systems have one and the same performance over all frequencies. This 'self flattening property' of continuous solutions was known since 1986 (Helton and Howe [20]). The *existence* of continuous solutions, however, has been an open problem for several years. It was shown independently by Helton and Marshall [21], and under a weaker hypothesis and by different means by Wegert [48].

It is an important feature of this approach that the optimization problem is transformed into a boundary value problem; we get equations instead of inequalities. This allows, for instance, the construction of efficient numerical methods for solving Helton's H^∞-optimization problem.

§4 Perspectives

4.1 Existence of solutions

Even for smooth target manifolds some questions cannot, apparently, be solved if we admit solutions that are not necessarily continuous on $\overline{\mathbb{D}}$. First of all, there are at least two reasonable concepts for understanding the boundary condition.

1. For bounded holomorphic functions $w \in H^\infty(\mathbb{D})$, the angular limits $w(t)$ exist for almost all $t \in \mathbb{T}$, and the classical boundary condition makes sense almost everywhere on the circle,

$$w(t) \in M_t \qquad \text{a.e. on } \mathbb{T}. \tag{4.1.1}$$

2. If we admit the class $\mathcal{O}(\mathbb{D})$ of all holomorphic functions in $\mathbb{D}$, the *cluster set*

$$\operatorname{clus} w(t) := \{y \in \widehat{\mathbb{C}}\colon\ \exists\ (z_k) \subset \mathbb{D}\colon\ z_k \to t,\ w(z_k) \to y\},$$

of w at t may be used as a substitute for the boundary values:

$$\operatorname{clus} w(t) \subset M_t \qquad \forall t \in \mathbb{T}. \tag{4.1.2}$$

In contrast to the first concept, the second one does not permit some types of irregular solutions. In particular, all solutions are continuous on $\overline{\mathbb{D}}$ if the target manifold is smooth. For the first approach, however, there may exist solutions with irregular behaviour near the boundary. For instance, infinite Blaschke products and singular inner functions are solutions of the standard example $|w(t)| = 1$ a.e. on $\mathbb{T}$ (and the latter do not even have zeros in $\mathbb{D}$). It seems not to be known if such irregular solutions do always exist.

Open Problem: Examine which Riemann-Hilbert problems (4.1.1) (with smooth target manifold) possess solutions of the type of infinite Blaschke products and singular inner functions.

This question is closely related to Nevanlinna parametrization for classes of holomorphic functions with restricted boundary values. This problem will be addressed in Section 4.3.

An appropriate understanding of the boundary condition is still more important when *non-smooth target manifolds* are considered. Till recently, all existence proofs utilized to some extent linearization techniques, which cannot be applied to non-smooth target manifolds. Also, it is clear that the results from Section 2 are not valid in full generality if smoothness is omitted.

Example 4: Let φ: $\mathbb{T} \to \mathbb{R}$ be a continuous function with an unbounded image $H\varphi$ under the Hilbert transform. The Riemann-Hilbert problem with the compact regularly traceable target manifold

$$M := \{(t, w) \in \mathbb{T} \times \mathbb{C}\colon\ |w| = \exp \varphi(t)\}$$

has no continuous solution.

In order to get reasonable results for topological target manifolds we are forced to enlarge the class of admitted solutions and to interpret the boundary condition appropriately, for instance in the sense of (4.1.1) or (4.1.2). In both cases Example 4 has again a one-parametric set of zero-free solutions,

$$w(t) = \exp\,\big(\varphi(t) + \mathrm{i}H\varphi(t) + \mathrm{i}c\big), \qquad c \in \mathbb{R}.$$

In what follows we are going to describe an approach proposed by R. Belch who introduced a concept of 'topological target manifolds' that allows to consider a rather large class of geometric objects.

According to Belch [4], a subset M of $\mathbb{T} \times \widehat{\mathbb{C}}$, is said to be a *topological target manifold* if there exist two open subsets A and B of $\mathbb{T} \times \widehat{\mathbb{C}}$ (called the *restriction domains* associated with M) with the following properties:

(i) $\mathbb{T} \times \widehat{\mathbb{C}}$ is the disjoint union of A, B and M,

(ii) the fibers

$$A_t := \{w \in \widehat{\mathbb{C}}\colon\ (t, w) \in A\}, \quad B_t := \{w \in \widehat{\mathbb{C}}\colon\ (t, w) \in B\}$$

are nonempty and (simply) connected for any $t \in \mathbb{T}$,

(iii) $\partial A = \partial B = M$.

The restriction domains A and B are the two connected components of $(\mathbb{T} \times \widehat{\mathbb{C}}) \setminus M$. Since the topology is borrowed from $\mathbb{T} \times \widehat{\mathbb{C}}$, in general, the fibers

$$M_t := \{w \in \widehat{\mathbb{C}}\colon\ (t, w) \in M\},$$

do not even need to be curves.

Example 5: A 'discontinuous' topological target manifold is

$$M := \{(t,w)\colon\ |w(t)| = 1 + h(\operatorname{Im} t)\},$$

where h stands for the Heaviside step function. The two fibers M_1 and M_{-1} are annuli, the other fibers are circles (see Figure 10). The set of zero-free solutions is

$$w(z) = c \cdot \exp\left(\frac{\log 2}{\pi \mathrm{i}} \cdot \log \mathrm{i}\,\frac{1+z}{1-z}\right), \qquad c \in \mathbb{T}.$$

For $t = \pm 1$ the cluster set of each solution coincides with $M_{\pm 1}$.

Belch is not only interested in holomorphic but also in meromorphic solutions. This results in additional difficulties, which he overcomes with the help of so-called p-functions. Here we confine ourselves to holomorphic solutions; the interested reader is referred to [4].

For $w \in \mathcal{O}(\mathbb{D})$ we define

$$\operatorname{graph} w := \{(t,y)\colon\ t \in \mathbb{T},\ \ y \in \operatorname{clus} w(t)\}$$

and interprete the boundary condition as $\operatorname{graph} w \subset M$.

In the following we focus attention to compact topological target manifolds in $\mathbb{T} \times \mathbb{C}$.

A compact topological target manifold is said to be *regularly traceable* if there exist (bounded) holomorphic functions a and b such that

$$\operatorname{graph} a \subset A, \qquad \operatorname{graph} b \subset B.$$

For topological target manifolds the winding number of a (possibly discontinuous) solution w about M makes no sense and we have to look for a substitute.

If two holomorphic functions w and v have disjoint graphs, then their difference has a finite number of zeros in $\mathbb{D}$ and this number (counting multiplicity) is said to be the *intersection number* $\#\,[w = v]$ of v and w.

The set of all solutions satisfying the boundary condition (4.1.2) is denoted by $\mathbb{W}$,

$$\mathbb{W} := \{w \in \mathcal{O}(\mathbb{D})\colon\ \operatorname{graph} w \subset M\}.$$

For compact target manifolds one of the restriction domains is bounded. We assume this to be A. If $w \in \mathbb{W}$ and $a \in \mathcal{O}(\mathbb{D})$ satisfies $\operatorname{graph} a \subset A$, the graphs of w and a do not intersect. The intersection number $\#\,[w = a]$ is well defined and does not depend on the special choice of a. The solutions for which this number equals zero form the most important class

$$\mathbb{W}_0 := \{w \in \mathbb{W}\colon\ \#\,[w = a] = 0\}.$$

In order to keep the symmetry between A and B we fix a holomorphic function b with $\operatorname{graph} b \subset B$ and $\#[b = a] = 0$ and consider the following families of holomorphic functions with restricted cluster sets

$$\begin{aligned}
\mathcal{F}_0 &:= \{w\colon \operatorname{graph} w \subset A, & \#[w = b] = 0\},\\
\mathcal{G}_0 &:= \{w\colon \operatorname{graph} w \subset B, & \#[w = a] = 0\},\\
\mathbb{F}_0 &:= \{w\colon \operatorname{graph} w \subset \overline{A}, & \#[w = b] = 0\},\\
\mathbb{G}_0 &:= \{w\colon \operatorname{graph} w \subset \overline{B}, & \#[w = a] = 0\}.
\end{aligned}$$

Since $M = \overline{A} \cap \overline{B}$, we have $\mathbb{W}_0 = \mathbb{F}_0 \cap \mathbb{G}_0$. The crucial idea is to examine the structure of the *ranges of values*

$$\mathbb{X}(z) := \{w(z)\colon\ w \in \mathbb{X}\},$$

where $\mathbb{X}$ stands for any of the sets $\mathcal{F}_0$, $\mathcal{G}_0$, $\mathbb{F}_0$, $\mathbb{G}_0$, and $\mathbb{W}_0$.

Lemma 2. *If the topological target manifold is regularly traceable, then for each $z \in \mathbb{C}$,*

(i) *$\mathcal{F}_0(z)$ is homeomorphic to $\mathbb{D}$, $\mathcal{G}_0(z)$ is homeomorphic to $\mathbb{C} \setminus \overline{\mathbb{D}}$,*

(ii) *the plane is the disjoint union of $\mathcal{F}_0(z)$ and $\mathbb{G}_0(z)$, as well as of $\mathcal{G}_0(z)$ and $\mathbb{F}_0(z)$,*

$$\mathcal{F}_0(z) \cap \mathbb{G}_0(z) = \emptyset, \qquad \mathcal{F}_0(z) \cup \mathbb{G}_0(z) = \mathbb{C},$$

$$\mathcal{G}_0(z) \cap \mathbb{F}_0(z) = \emptyset, \qquad \mathcal{G}_0(z) \cup \mathbb{F}_0(z) = \mathbb{C}.$$

The existence of a sufficiently large set of solutions of the Riemann-Hilbert problem follows from the next result.

Theorem 12. (Belch [4]). *If the topological target manifold is regularly traceable and $z \in \mathbb{D}$, then for each point $y \in \partial\mathbb{F}_0(z) \cup \partial\mathbb{G}_0(z)$ there exists a unique solution $w \in \mathbb{W}_0$ with $w(z) = y$.*

Belch's proof relies on an approximation technique (smooth exhaustion and inclusion of topological target manifolds) and the results for smooth target manifolds, especially a clever application of the extremal principle in conjunction with a localization technique.

Recall that for smooth target manifolds $\mathbb{W}_0 \cong \mathbb{T}$. According to Theorem 12 we have one solution for each point of $\partial\mathbb{F}_0(z) \cong \mathbb{T}$ and one solution for each point of $\partial\mathbb{G}_0(z) \cong \mathbb{T}$. So there would be more solutions than expected if $\partial\mathbb{F}_0(z)$ and $\partial\mathbb{G}_0(z)$ do not coincide. If we would know that

$$\overline{\mathcal{F}_0(z)} = \mathbb{F}_0(z), \tag{4.1.3}$$

then it would follow from Lemma 2 that $\mathbb{F}_0(z) \cap \mathbb{G}_0(z) = \partial\mathbb{F}_0(z) = \partial\mathbb{G}_0(z)$, and the boundaries of $\mathbb{F}_0(z)$ and $\mathbb{G}_0(z)$ would parametrize the same set of solutions.

Further, even though $\mathcal{F}_0(z)$ is homeomorphic to $\mathbb{D}$ it is in general not clear if $\partial\mathcal{F}_0(z)$ is homeomorphic to $\mathbb{T}$. Belch calls a topological target manifold *normal*, if (for one and then for all $z \in \mathbb{D}$)

$$\partial\mathbb{F}_0(z) = \partial\mathbb{G}_0(z) \cong \mathbb{T}.$$

According to the results of Section 2 all smooth regularly traceable target manifolds are normal. It is conjectured that smoothness is not essential at all.

Main open problem: Prove that *all* regularly traceable topological target manifolds are normal.

It should be mentioned that Belch's class of topological target manifolds still does not cover all cases emerging from conformal mapping $\mathbb{D} \to G$. For example, if G is a slit domain the condition $\partial A = \partial B$ is violated.

Several other problems for general Riemann-Hilbert problems can be found by looking at related results for conformal mapping. For instance, according to the Carathéodory-Osgood theorem, a conformal mapping onto a Jordan domain extends to a homeomorphism of $\overline{\mathbb{D}}$ onto the closure of the domain. One could expect that a similar result holds for Riemann-Hilbert problems. However, Example 4 shows that there are target manifolds M which are homeomorphic to the torus $\mathbb{T}^2$ and have circular fibers but posess no solution that extends continuously onto the closed disk.

Open problem: Find (sufficiently weak) conditions on M that guarantee the existence of *continuous* solutions.

For some classes it is known that $M \in \text{Lip}$ is sufficient for an affirmative answer [8].

Another challenging problem is to describe the 'prime ends' of a target manifold.

Open problem: Give a geometric description (in terms of the target manifold) of the cluster sets of solutions.

For Example 5 of a discontinuous target manifold the cluster sets clus $w(t)$ are points if $t \neq \pm 1$, but clus $w(\pm 1)$ are the annuli $M_{\pm 1}$ (see Figure 10).

In connection with Warschawski's theorem about the asymptotic behaviour of conformal mapping at corner points of the image domain it is interesting to study Riemann-Hilbert problems with target manifolds that have singularities of special types. Since Riemann-Hilbert problems are more general than conformal mapping there is an ample list of specific singularities which could be studied, for example jumps, crossing jumps, different types of edges, point singularities, etc. Figures 10–12 illustrate some typical effects.

The problem in Figure 10 is Example 5. Only one solution of the class $\mathbb{W}_0$ is shown. The right-hand part is a close-up of the jump with logarithmically scaled horizontal axis.

Figure 11 displays solutions from the class $\mathbb{W}_0$ on a target manifold with circular fibers (left from 'below', right from 'above'). At $t = 1$ the limiting circles are shifted against each other. The problem is a transformation of a linear Riemann-Hilbert problem with a discontinuous symbol.

Figure 12 shows two target manifolds with edges. Since the edges are rotating (in different directions in the left and the right part of the figure) the solutions in $\mathbb{W}_0$ are forced to cross them. The background of this example is conformal mapping onto the interior resp. the exterior of a triangle.

These figures (like all other figures in this text) are generated by MATLAB routines. The solutions in Figure 12 are computed using Tobin Driscoll's SCPACK.

There are heuristical arguments for the behaviour of the solutions in more general situations, but in most cases strict proofs are still missing.

4.2 Extremality of solutions

According to Section 2, the solutions of Riemann-Hilbert problems are at the same time solutions of an extremal problem. While it is not hard to show the existence of extremal functions (even under much weaker hypotheses about the target manifold), some effort is necessary to prove there coincidence with solutions of the Riemann-Hilbert problem. In particular, there seems to be no proof which does not require the *a-priori* knowledge of existence of solutions to the Riemann-Hilbert problem.

Since one of the shortest existence proofs for conformal mapping just goes the other way around (a conformal map is constructed as solution of an extremal problem), a similar approach for general Riemann-Hilbert problems would be of great value. What is known at present in this direction seems to be limited to the case of convex target curves, where S. Hui [24] developed an appropriate concept of duality.

4.3 Parametrization

The set $\mathbb{A}$ of bounded holomorphic functions with restricted boundary values

$$\mathbb{A} = \{w \in H^\infty : \ \mathrm{graph}\, w \subset A\}$$

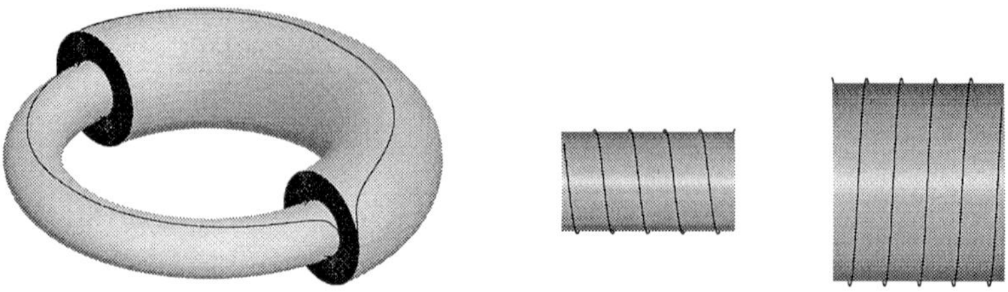

Figure 10. *Jumps produce oscillatory behaviour of logarithmic type*

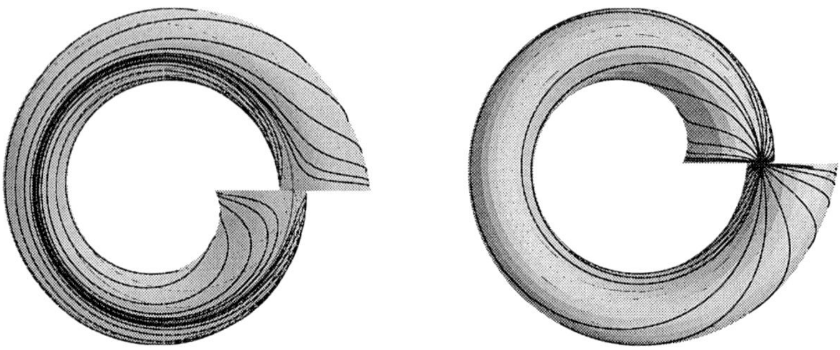

Figure 11. *Crossing jumps have attracting and repelling intersections*

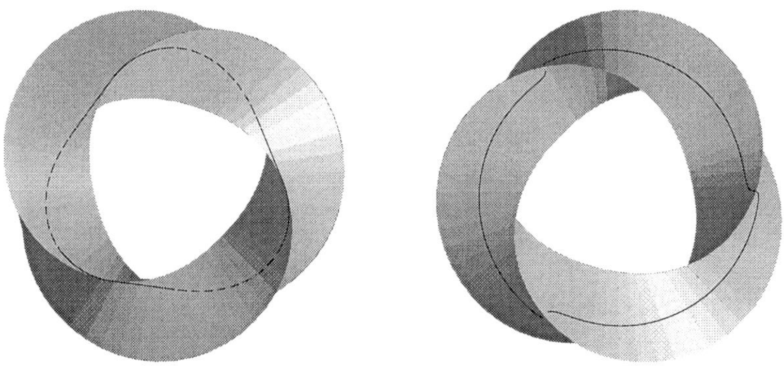

Figure 12. *Edges can be attracting or repelling*

shares several properties with the closed unit ball $\mathbb{B}$ of H^∞. This leads to the question of whether the elements of $\mathbb{B}$ can be used to parametrize the set $\mathbb{A}$ in 'a natural way'.

If all target curves M_t are circles of one and the same size,

$$M_t := \{y \in \mathbb{C}\colon\ |y - f(t)| = 1\}$$

the so-called *Nevanlinna parametrization* gives an affirmative answer,

$$\mathbb{A} = \left\{\frac{P + Qw}{R + Sw} : w \in \mathbb{B}\right\}.$$

Here P, Q, R, S are certain functions in H^∞ depending on f.

Open problem: Is there a 'similar' (whatever this means) parametrization of general sets $\mathbb{A}$ with the closed unit ball $\mathbb{B}$ in H^∞ as the parameter space?

There are strong arguments (using local deformations of the restriction manifold) that 'similar' has to be understood in a fairly wide sense, because otherwise such a parametrization cannot exist.

A solution of the parametrization problem would answer at once the question for solutions with infinite winding number (infinite Blaschke products) and solutions of the type of singular inner functions.

Acknowledgement. The author is supported by grant We 1704/2 of the Deutsche Forschungsgemeinschaft

References

[1] Arzhanov, G. B., On a nonlinear boundary value problem of the Riemann type (Russian), *Sib. Mat. Zh.* **2** (1961) 4, 481–504.

[2] Astala, K. and G. J. Martin, *Holomorphic Motions*, University of Helsinki, Reports of the Department of Mathematics No. 8 (1992).

[3] Bedford, E., Stability of the polynomial hull of $\mathbb{T}^2$, *Ann. Sc. Norm. Super. Pisa, Cl. Sci. IV Ser.* **8** (1981), 311–315

[4] Belch, R., *The Methods of Geometric Function Theory applied to Nonlinear Riemann-Hilbert Problems*, Essay for the J.T. Knight Price, Cambridge University 1996.

[5] Bers, L. and H. Royden, Holomorphic families of injections, *Acta Math.* **157** (1986), 259–286.

[6] Böttcher, A. and Yu.I. Karlovich, *Carleson Curves, Muckenhoupt Weights, and Toeplitz Operators*, Birkhäuser, 1997.

[7] Douady, A., Prolongement de mouvements holomorphes, *Asterisque* **227** (1995), 7–20.

[8] Efendiev, M. A. and E. Wegert, Nonlinear Riemann-Hilbert problems with Lipschitz continuous boundary condition, preprint 94-2 Universität Stuttgart, 1994, 10pp.

[9] Finnilia, R. and J. M. Sloss, Existence and uniqueness theorems for a class of non-linear singular integral equations with application to a hydroelastic problem, *J. Math. Mech.* **16** (1966) 6, 509–534.

[10] Forstnerič, F., Analytic disks with boundaries in a maximal real submanifold of $\mathbb{C}^2$, *Ann. Inst. Fourier* **37** (1987), 1–44.

[11] Forstnerič, F., Polynomial hulls of sets fibered over the circle, *Indiana Univ. Math. J.* **37** 4 (1988), 869–889.

[12] Gabib-Zade, A. Sh., Investigation of a nonlinear Hilbert problem (Russian), *Dokl. Akad. Nauk Azerb. SSR* **14** 4 (1958), 275–278.

[13] Gakhov, F. D., On the Riemann boundary value problem (Russian), *Mat. Sbornik* **2** (**44**) 4 (1937), 673–683.

[14] Gakhov, F. D., On a nonlinear boundary value problem which generalizes the Riemann boundary value problem (Russian), *Dokl. Akad. Nauk SSSR* **181** (1968) 2, 271–274.

[15] Gekht, V. I., Solvability of nonlinear singular integral equations by an iterative method, *Uchen. Zapiski Kaz. Gos. Univ., Matematika* **116** (1956), 111–139.

[16] Gusejnov, A. I., On a nonlinear boundary value problem in the theory of analytic functions (Russian), *Mat. Sbornik* **26 68** (1950), 237–246.

[17] Gusejnov, A. I. and Kh. S. Mukhtarov, *Introduction to the Theory of Nonlinear Singular Integral Equations*, Nauka, Moskow, 1980.

[18] Hatcher, J. R., A nonlinear boundary value problem, *Proc. Amer. Math. Soc.* **95** (1985), 441–448.

[19] Helton, J. W., *Operator Theory, Analytic Functions, Matrices, and Electrical Engineering*, Conference Board of the Mathematical Sciences by the AMS, Providence RI 1987.

[20] Helton, J. W., and R. E. Howe, A bang-bang theorem for optimization over spaces of analytic functions, *J. Approx. Theory* **47** 2 (1986), 101–121.

[21] Helton, J. W. and D. E. Marshall, Frequency domain design and analytic selections, *Indiana Univ. Math. J.* **39** 1 (1990), 157–184.

[22] Hilbert, D., Über eine Anwendung der Integralgleichungen auf ein Problem der Funktionentheorie, *Verh. d. Intern. Mathematiker-Kongr.*, Heidelberg 1904, 233–240.

[23] Hilbert, D., *Grundzüge einer allgemeinen Theorie der linearen Integralgleichungen*, Leipzig 1912.

[24] Hui, S., Qualitative properties of solutions to H^∞-optimization problems, *J. Functional Anal.* **75** (1987), 323–348.

[25] Isakov, V., Prospecting discontinuities by boundary measurements, in: *Inverse Problems: Principles and Applications in Geophysics, Technology, and Medicine*, Akademie Verlag 1993, 215–223.

[26] Kiyasov, S. N., On the solution of a fractional linear Riemann boundary value problem (Russian), *Neklass. Uravn. Mat. Fiz.*, Novosibirsk 1985, 180–183.

[27] Mañé, R., P. Sad, and D. Sullivan, On the dynamics of rational maps. *Ann. Sci. Ec. Norm. Super. IV* **16** (1983), 193–217.

[28] Mukhtarov, Kh. Sh., Investigation of a nonlinear singular integral equation with Hilbert kernel (Russian), *Izv. Vyssh. Ucheb. Zaved., Matematika* **2** (1965), 118–125.

[29] Natalevich, V. K., On a nonlinear singular integral equation and a nonlinear boundary value problem of the theory of analytic functions (Russian), *Dokl. Akad. Nauk SSSR* **83** 1 (1952), 19–22.

[30] Noether, F., Über eine Klasse singulärer Integralgleichungen, *Math. Ann.* **82** (1921) 42–63.

[31] Obnosov, Yu. V., Some nonlinear boundary value problems of the theory of analytic functions, solvable in closed form (Russian), in *Trudy Jubilejnogo Seminara po Kreavym Zadacham*, Minsk, Izd.vo Universitetskoje 1985, 86–95.

[32] Peters, A. S., A solution of certain nonlinear Riemann-Hilbert problems with an application, *Comm. Pure Appl. Math.* **26** (1973), 87–104.

[33] Pogorzelski, W., *Integral Equations and their Applications*, Pergamon Press Oxford and PWN–Pol. Scient. Publ. Warsaw, 1966.

[34] Powell, J., On a small perturbation in the two dimensional inverse conductivity problem, *J. Math. Anal. Appl.* **175** 1 (1993), 292–304.

[35] Riemann, B., Grundlagen für eine allgemeine Theorie der Functionen einer veränderlichen complexen Grösse, in: *Gesammelte mathematische Werke und wissenschaftlicher Nachlaß* (Ed. R. Dedekind and H. Weber) Leipzig 1876, 3–47.

[36] Shnirel'man, A. I., The degree of quasi-linearlike mapping and the nonlinear Hilbert problem (Russian), *Mat. Sb.* **89** 3 (1972), 366–389; English transl.: Math. USSR Sbornik **18** (1974), 373–396.

[37] Simonenko, I. B., The Riemann boundary value problem with measurable coefficients, *Dokl. Akad. Nauk SSSR* **135** (1960) 3, 538–541.

[38] Slodkowski, Z., Polynomial hulls in $\mathbb{C}^2$ and quasicircles, *Ann. Sc. Norm. Super. Pisa, Cl. Sci. IV Ser.* **16** 3 (1989), 367–391.

[39] Slodkowski, Z., Holomorphic motions and polynomial hulls, Proc. Amer. Math. Soc. **111** (1991) 347–355. **16** 3 (1989), 367–391.

[40] Sloss, J.M., Global solution by parameter continuation for a class of nonlinear singular integral equations, *Comm. Pure Appl. Math.* **23** (1970), 841–865.

[41] Solovev, P. V., On a nonlinear boundary value problem in the theory of analytic functions (Russian), *Dokl. Akad. Nauk SSSR* **33** 3 (1941), 190–192.

[42] Sullivan, D. P. and W. P. Thurston, Extending holomorphic motions. *Acta Mathematica* **157** (1986), 243–257.

[43] Sylvester, J., Linear and nonlinear inverse scattering. Manuscript, University of Washington (1997), 47pp.

[44] Vekua, I. N., On a nonlinear Riemann boundary value problem, *Trudy Tbilissk. Mat. Inst.* **11** (1942), 109–139.

[45] Wegert, E., Topological methods for strongly nonlinear Riemann-Hilbert problems for holomorphic functions, *Math. Nachr.* **134** (1987), 201–230.

[46] Wegert, E., Nonlinear Riemann-Hilbert problems and their relationship to extremal problems for holomorphic functions, *Math. Nachr.* **137** (1988), 144–157.

[47] Wegert, E., Boundary value problems and extremal problems for holomorphic functions, *Complex Variables* **11** (1989), 223–256.

[48] Wegert, E., Boundary value problems and best approximation by holomorphic functions, *J. Approx. Theory* **16** 3 (1990), 322–334.

[49] Wegert, E., *Nonlinear Boundary Value Problems for Holomorphic Functions and Singular Integral Equations*, Akademie Verlag Berlin, 1992.

[50] Wegert, E. and D. Oestreich, On a non-symmetric problem in electrochemical machining, *Math. Meth. Appl. Sci.* **20** (1997), 841–854.

[51] Wegmann, R., Keplerian disks around magnetized neutron stars – a free boundary problem, *Methoden Verfahren Math. Phys.* **37** (1991), 233–253.

[52] Wolfersdorf, L. v., On the theory of the nonlinear Riemann-Hilbert problem for holomorphic functions, *Complex Variables* **3** (1984), 323–346.

[53] Wolfersdorf, L. v., A class of nonlinear Riemann-Hilbert problems with monotone nonlinearity, *Math. Nachr.* **130** (1987), 111–119.

[54] Wolfersdorf, L. v., Inverse problems for memory kernels in heat flow and viscoelasticity, J. Inverse Ill-Posed Problems **4** (1996), 341–354.

Elias Wegert
TU Bergakademie Freiberg
D-09596 Freiberg
GERMANY
wegert@math.tu-freiberg.de

Two plane free boundary problems with surface tension

Rudolf Wegmann

Abstract. We study the flow around a bubble and along a wall with a gap. In both problems a free surface is balanced by pressure forces and surface tension. The flow covers a region which is described by an analytic function f in the exterior of the unit disc. The equilibrium condition yields an equation for the derivative f' which may be interpreted as an equation in a Hoelder space. This equation depends on two nondimensional parameters which measure the pressure difference and the surface tension. The implicit function theorem gives the existence of solutions for bubbles with sufficiently large inside pressure. For flow protruding through a gap solutions exist when the surface tension is large enough and the pressure difference is small. In these cases the Newton method converges. In each step of the Newton iteration a function theoretic boundary problem has to be solved. This can be used for numerical calculations. We present some numerical results.

§1 Introduction

We study two problems of irrotational and incompressible flow in the plane, namely the flow around a bubble and the flow along a straight wall with a gap. In both cases the flow occupies a certain region G with a free boundary which is determined by the balance of pressure forces and surface tension. We show that both problems are governed by the same equation. Therefore, both problems can be treated by the same methods.

The bubble with surface tension has been completely treated numerically by Vanden-Broeck and Keller [8] in 1980 and by Shankar [6] in 1992. For a particular parameter combination an exact solution has been found by McLeod [3] in 1955. Analytical properties have been studied by Tanveer [7]. Numerical calculations for the axisymmetric bubble have been performed by Miksis *et al.* [2]. For the gap without surface tension Shankar [5] has found an exact solution. There are no results for the gap with surface tension.

Vanden-Broeck and Keller [8], and Miksis *et al.* [2] use integral equation methods. Shankar [6] describes the problem by a conformal mapping function. The coefficients of this function satisfy a system of equations which was solved by Shankar numerically by a Newton method.

Computational Methods and Function Theory (CMFT'97)
N. Papamichael, St. Ruscheweyh and E. B. Saff (Eds.), pp. 617–633

ISBN 981-02-3626-3

Following Shankar we describe the flow around a bubble by a function f analytic in the exterior D^- of the unit disc D. In the case of the bubble, f is a conformal mapping from D^- to G. In the case of the gap, the function f maps the lower half of D^- conformally to G. The derivative f' has two real zeros which are the preimages of the endpoints of the gap.

From the equilibrium conditions a functional equation $A(f') = 0$ for the derivative f' of f is derived. We interpret this equation as an equation in a Hoelder-space. This equation depends on two nondimensional parameters: γ which measures the difference of the pressure p_b outside the flow and the total pressure of the undisturbed flow, and τ which measures the surface tension. Starting from sufficiently good approximations, which are available for both problems, we show that for certain parameter intervals solutions of the equation exist: The equation for the bubble is solvable when the pressure inside the bubble is sufficiently high. For the gap, solutions exist when the surface tension is large and the outside pressure nearly balances the total pressure of the undisturbed flow. These solutions can be obtained by the Newton method which is shown to converge. In each iterative step a function theoretic boundary value problem must be solved.

§2 Equations

The bubble

The bubble covers a bounded simply connected region B. The flow region G is the exterior of B. The boundary $\partial B = \partial G$ of the bubble must be determined together with the flow field.

We work conveniently in the complex plane $\mathbb{C}$ and combine the two coordinates x, y into the complex variable $z = x + iy$. Since the flow is a potential flow there is an analytic function $W = \Phi + i\Psi$ in G with velocity potential Φ and stream-function Ψ. The velocity vector $\mathbf{U} := (u, v)$ can be obtained either as $(u, v) = \operatorname{grad} \Phi$ or from the components of the derivative $W' = u - iv$.

Far away from the bubble, for $|z| \to \infty$, the flow approaches the undisturbed flow conditions with velocity $\mathbf{U}_0 = (u_0, 0)$, and pressure p_0. Inside the bubble the pressure p_b is constant. The fluid has constant density ρ.

In the flow the pressure p and the velocity $U := \sqrt{u^2 + v^2}$ are connected by Bernoulli's law

$$\frac{\rho U^2}{2} + p = \frac{\rho u_0^2}{2} + p_0. \tag{2.1}$$

On the interface between bubble and flow the pressure difference is balanced by surface tension

$$p_b - p = \sigma \kappa \tag{2.2}$$

with a coefficient $\sigma \geq 0$. The curvature κ of the boundary curve is positive when the curve is concave towards the bubble. Then the force $\sigma\kappa$ acts towards the bubble and compensates an overpressure in the bubble. The sign of κ is negative when the boundary curve is concave towards the flow.

One can use (2.2) to eliminate p from (2.1). This leads to the equation

$$q^2 = \tau\kappa - \gamma \tag{2.3}$$

for the non-dimensional velocity $q = U/u_0$. The parameter

$$\gamma := \frac{p_b - p_0}{\rho u_0^2/2} - 1 \tag{2.4}$$

is a measure of the overpressure in the bubble compared with the total pressure of the flow. The normalized surface tension $\tau := 2\sigma/\rho u_0^2$ has the dimension of a length. Since there is no other length scale in the problem one can achieve $\tau = 1$ by choosing the appropriate unit of length. Therefore, the bubble shape is determined only by the parameter γ. We retain τ in the equations since this allows later a unified treatment with the gap.

There exists a uniquely determined conformal mapping f from the exterior D^- of the unit disc to the flow region G such that

$$f(\infty) = \infty, \qquad f'(\infty) > 0. \tag{2.5}$$

The geometry is shown in the upper part of Figure 1. The flow field W in G is transported to a flow field $w(\zeta) := W(f(\zeta))$ in D^- which describes the flow around the disc. Therefore,

$$w(\zeta) = a_0 \left(\zeta + \frac{1}{\zeta}\right) \tag{2.6}$$

and

$$w'(\zeta) = a_0 \left(1 - \frac{1}{\zeta^2}\right) = W'(f(\zeta))f'(\zeta). \tag{2.7}$$

In the normalization of (2.3), $u_0 = 1$ and $W'(\infty) = 1$. Since $w'(\infty) = a_0 = W'(\infty)f'(\infty)$, it follows that

$$w'(\infty) = f'(\infty) = a_0. \tag{2.8}$$

From equation (2.3) follows the boundary condition for the velocity potential

$$|W'(z)|^2 = \tau\kappa - \gamma \tag{2.9}$$

for $z \in \partial B$.

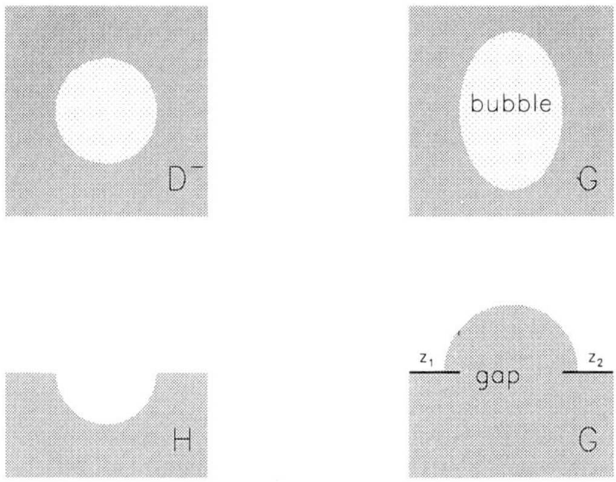

Figure 1. *Geometry of the bubble (upper right) and the gap (lower right) with the canonical regions D^- and H (left)*

Since the problem is invariant with respect to translations it is sufficient to know the derivative f' of the mapping function f. We use the parametrization $\eta(s) = x(s) + iy(s) = f(e^{is})$ to express W' and κ in (2.9) in terms of f'. This yields the condition

$$\frac{2a_0^2(1-\cos 2s)}{|f'(e^{is})|} + \gamma|f'(e^{is})| - \tau - \tau \operatorname{Re}\left(\frac{e^{is}f''(e^{is})}{f'(e^{is})}\right) = 0 \tag{2.10}$$

which must be satisfied for all s. This equation is the equilibrium condition which determines the free surface. It is a necessary condition which the unknown conformal mapping function must satisfy. When f' is calculated the flow field $W'(f(\zeta))$ can be calculated using (2.7) and $a_0 = f'(\infty)$.

Since only the derivative f' enters into (2.10) we write this equation in the form of a functional equation

$$A(f') = 0 \tag{2.11}$$

with a nonlinear operator A which assigns to each analytic function h in D^-, such that $h(e^{is}) \neq 0$ and $h(\infty)$ is real, the real function $\phi = A(h)$ defined by

$$\phi(s) := \frac{2(1-\cos 2s)}{|h(e^{is})|}h(\infty)^2 + \gamma|h(e^{is})| - \tau - \tau \operatorname{Re}\left(\frac{e^{is}h'(e^{is})}{h(e^{is})}\right). \tag{2.12}$$

The gap

Now we consider the flow along a fixed wall $\{z : \operatorname{Im} z = 0, |\operatorname{Re} z| \geq d\}$ with a gap $[-d, d]$. We assume that far away from the gap the flow is in the lower halfplane $\{\operatorname{Im} z < 0\}$ and approaches a flow parallel to the wall with velocity u_0 and pressure p_0. The flow protrudes through the gap, streams a certain distance back on the rear side of the wall and then separates at a point z_1 from the wall. The free streamline hits the other component of the wall at a point z_2 and then again follows the wall.

As before, we describe the velocity by an analytic function W which is defined in the region G occupied by the flow. The condition, that the flow must be parallel to the wall can be expressed by the condition that $W'(z)$ must be real for all real z with $|z| > d$. Therefore, one can choose W in such a way that also W is real for real z outside the gap.

The balance of pressure forces and surface tension along the free boundary leads again to equation (2.3) for the normalized quantities.

Let $H := \{z : |z| > 1, \operatorname{Im} z < 0\}$ be the lower half of D^-. Then there exists a unique conformal mapping from H to G normalized by the conditions

$$f(-1) = z_1, \qquad f(1) = z_2, \qquad f(\infty) = \infty. \tag{2.13}$$

The geometry is shown in the lower part of Figure 1. Since f must be real for real $\zeta \in \partial H$ it can be continued by Schwarz reflection to an analytic function in D^-.

The function $w(\zeta) := W(f(\zeta))$ is defined in H. It is real for real ζ. Therefore, w can also be continued to an analytic function in D^-. This extended function describes the flow around the disc. Therefore, it is given by (2.6).

The free boundary curve is parametrized by $\eta(s) = f(e^{is})$, $-\pi \leq s \leq 0$. One derives from (2.3) that $A(f')(s) = 0$ must hold for $-\pi \leq s \leq 0$. Since the values of $f'(e^{is})$ for $0 \leq s \leq \pi$ are simply obtained by reflection, $A(f')(s) = A(f')(-s)$ holds. This gives for the gap the same equation (2.11) as for the bubble.

§3 The Newton method

For a discussion of the equation $A(f') = 0$ we have to specify more precisely the domain of definition of the operator A. To this end we use for $0 < \mu < 1$ the Hoelder space C^μ of 2π-periodic real functions ϕ which satisfy a Hoelder condition with exponent μ. With the usual norm C^μ is a Banach space.

Let $F^{1+\mu}$ be the space of functions g analytic in D^- such that g' is continuous in the closure of D^-, the real and imaginary parts of $g'(e^{is})$ are in C^μ and $g(\infty)$ is real. Following Vekua ([4], p.193), we assign to each function $\alpha \in C^\mu$ the function g_α which is defined by the integral representation

$$g_\alpha(z) := \int_0^{2\pi} \alpha(t) \log\left(1 - \frac{e^{it}}{z}\right) dt + \int_0^{2\pi} \alpha(t) dt \tag{3.1}$$

for all $z \in D^-$. For each fixed t the branch of $\log(1 - e^{it}/z)$ is understood to be that which vanishes at infinity. Then $\alpha \to g_\alpha$ defines a one-to-one correspondence between C^μ and $F^{1+\mu}$. We choose on $F^{1+\mu}$ the norm which is induced by this mapping from C^μ.

The non-linear operator A is defined by (2.12) for all $h \in F^{1+\mu}$ which satisfy $h(e^{is}) \neq 0$. It assigns to h a function $A(h) \in C^\mu$. The mapping A is composed only by algebraic operations. Therefore, it has derivatives of all orders. The derivative at a function h is a linear operator $DA(h)$ which maps $F^{1+\mu}$ to C^μ. It assigns to a function $g \in F^{1+\mu}$ the function

$$DA(h)g(s) = \mathrm{Re}(A_1(s)g'(e^{is})) + \mathrm{Re}(A_2(s)g(e^{is})) + A_3(s)g(\infty) \tag{3.2}$$

with

$$\begin{aligned} A_1(s) &= -\frac{\tau e^{is}}{h(e^{is})}, \\ A_2(s) &= \gamma\frac{|h(e^{is})|}{h(e^{is})} - \frac{2(1-\cos 2s)h(\infty)^2}{|h(e^{is})|h(e^{is})} + \frac{\tau e^{is}h'(e^{is})}{h(e^{is})^2}, \\ A_3(s) &= \frac{4(1-\cos 2s)h(\infty)}{|h(e^{is})|}. \end{aligned} \tag{3.3}$$

The second derivative is a bilinear operator which maps a pair of functions $g_1, g_2 \in F^{1+\mu}$ to C^μ. For brevity we omit the argument e^{is} in the following representation

$$\begin{aligned} D^2A(h)(g_1,g_2) &= \tau\,\mathrm{Re}\,\frac{e^{is}(g_1g_2' + g_1'g_2)}{h^2} \\ &+ \left(\gamma|h| - \frac{2(1-\cos 2s)}{|h|}h(\infty)^2\right)\left(\mathrm{Re}\,\frac{g_1}{h}\,\mathrm{Re}\,\frac{g_2}{h} - \mathrm{Re}\,\frac{g_1g_2}{h^2}\right) \\ &-4(1-\cos 2s)\frac{h(\infty)}{|h|}\left(g_1(\infty)\,\mathrm{Re}\,\frac{g_2}{h} + g_2(\infty)\,\mathrm{Re}\,\frac{g_1}{h}\right) \\ &+4(1-\cos 2s)\frac{g_1(\infty)g_2(\infty)}{|h|} - 2\tau\,\mathrm{Re}\,\frac{e^{is}h'g_1g_2}{h^3}. \end{aligned} \tag{3.4}$$

We solve the equation (2.11) by the Newton method. In each step f' is changed to $f' + g$ where g is a solution of the equation

$$DA(f')g = -A(f'). \tag{3.5}$$

In view of the representation (3.2) this is a Riemann-Hilbert-Poincaré boundary problem for the analytic function g ([4], p.208). It can be transformed by the representation (3.1) into a singular integral equation for the function $\alpha \in C^\mu$ associated with g. Therefore, one can think of $DA(f')$ as a singular integral operator. It follows that it is a Fredholm operator. This means by definition that its null-space has finite dimension N_1, and its range is closed with finite codimension N_2. The index, defined as the difference $N_1 - N_2$, depends only on the principal part, hence only on the function A_1. From [4], p.208, we get

$$\operatorname{ind}(DA(f')) = -2\operatorname{wind}(f'(e^{is})) \tag{3.6}$$

where $\operatorname{wind}(f'(e^{is}))$ denotes the winding-number of $f'(\zeta)$ as ζ traverses the unit circle in the positive direction. The minus sign in (3.6) occurs since the function g in (3.5) is required to be analytic outside the unit circle, while [4] considers functions which are analytic inside the unit circle.

In view of the different topologies, we get for the bubble the index 0 and for the gap the index 4. Therefore, the two problems must from now on be discussed separately.

§4 The bubble

For large γ the overpressure in the bubble is mainly balanced by the surface tension. The radius of curvature (measured in units τ) must be small. The bubble shape is close to a disc with radius τ/γ.

Theorem 1. *There is a number γ_0 such that for all $\gamma > \gamma_0$ and all $\tau > 0$ there exists a solution $f' \in F^{1+\mu}$ of the equation $A(f') = 0$ with $f'(e^{is}) \neq 0$ and winding number $\operatorname{wind}(f'(e^{is})) = 0$. The Newton method starting from*

$$f_0' = \frac{\tau}{\gamma + 2} \tag{4.1}$$

converges to this solution.

Proof: a) In the first step we show that $DA(f_0')$ has a bounded inverse Γ_0. We insert f_0' into (3.2) and (3.3) and get the following representation

$$DA(f_0')g = \gamma M_1 g + M_2 g \tag{4.2}$$

with two operators

$$M_1 g := -\operatorname{Re}\Big(e^{is}g'(e^{is}) - g(e^{is})\Big), \tag{4.3}$$

$$M_2 g := -2\operatorname{Re}(e^{is}g'(e^{is})) - 2(1 - \cos 2s)\Big(\operatorname{Re} g(e^{is}) - 2g(\infty)\Big) \tag{4.4}$$

which are independent of γ and τ. The equation

$$M_1 g = \psi \tag{4.5}$$

can be explicitly solved for every $\psi \in C^\mu$ by conjugation and integration. This gives a unique solution $g \in F^{1+\mu}$. Therefore, the operator $M_1 : F^{1+\mu} \to C^\mu$ has an inverse which is bounded $||M_1^{-1}|| \leq C_1$. For $\gamma \geq 2||M_2||C_1$ the operator $DA(f_0')$ is also invertible in view of (4.2). The inverse Γ_0 satisfies

$$||\Gamma_0|| \leq \frac{2C_1}{\gamma}. \tag{4.6}$$

b) The operator A maps f_0' to the function

$$A(f_0')(s) = -2a_0 \cos 2s \tag{4.7}$$

with $a_0 := \tau/(\gamma+2)$. This implies $||A(f_0')|| \leq C_2 a_0$ and finally

$$||\Gamma_0 A(f_0')|| \leq \frac{2C_1C_2a_0}{\gamma} \leq C_3\frac{\tau}{\gamma^2} =: \eta_0. \tag{4.8}$$

c) The second derivative has two parts

$$D^2A(h) = \tau B_1(h) + B_2(h) \tag{4.9}$$

with different degrees of homogeneity. For $c > 0$ there holds

$$B_1(ch) = \frac{B_1(h)}{c^2}, \qquad B_2(ch) = \frac{B_2(h)}{c}. \tag{4.10}$$

It follows from the representation (3.4) that the second derivative is a sum of products, where g_1 and g_2 and their derivatives are multiplied by functions which are bounded in C^μ whenever h is uniformly bounded away from zero. Therefore, there are constants such that the estimates $||B_1(h)|| \leq C_4$, $||B_2(h)|| \leq C_5$ hold for $||h-1|| \leq 1/2$. Then for $||h-a_0|| \leq a_0/2$ the second derivative can be estimated by

$$||D^2A(h)|| \leq \frac{\tau C_4}{a_0^2} + \frac{C_5}{a_0} \leq C_6\frac{\gamma^2}{\tau} \tag{4.11}$$

for sufficiently large γ. Finally, this implies that

$$||\Gamma_0 D^2A(h)|| \leq C_7\frac{\gamma}{\tau} =: K_0. \tag{4.12}$$

d) Now choose γ_0 so large that the estimates (4.6) and (4.12) are satisfied for $\gamma \geq \gamma_0$, $\eta_0 K_0 = C_3C_7/\gamma$ is less than $1/2$ and $2\eta_0$ is less than $a_0/2$.

Then a theorem of Kantorovich ([1], p.572), ensures the existence of a solution of $A(f') = 0$ with

$$||f' - a_0|| \leq 2\eta_0 \tag{4.13}$$

and the convergence of the Newton method. It follows from (4.13) that for the solution $f'(e^{is}) \neq 0$, and that the winding number of $f'(e^{is})$ is zero, since $2\eta_0 < a_0/2$. ■

§5 The gap

When the flow protrudes through the gap and returns, then somewhere along the free boundary the curvature must be negative. It follows from $p = p_b - \sigma\kappa > p_b$, the Bernoulli equation (2.1) and the definition (2.4) of γ, that γ must be negative. Shankar [5] found for the case of no surface tension the following exact solutions

$$f'(\zeta) = c\frac{\zeta^2 - 1}{\zeta^2}\frac{\sqrt{-\gamma}\zeta^2 - 1}{\sqrt{-\gamma} - \zeta^2} \tag{5.1}$$

with a constant $c > 0$. These solutions are meaningful only for γ in the interval $-1 < \gamma < 0$. Then f' has zeros for $\zeta_{1,2} = \pm|\gamma|^{-1/4}$. These correspond to the endpoints of the gap where the flow turns by 180°. At these edges the velocity is infinite and $p = -\infty$. Since f' has also zeros at ± 1 the flow separates smoothly from the wall. For γ outside the interval $(-1, 0)$ there are no solutions for $\tau = 0$.

In the case of positive surface tension the solution f' also has two real zeros $\zeta_1 < -1, \zeta_2 > 1$. This implies that $f'(e^{is})$ has winding number -2. Therefore, the derivative $DA(f')$ is an operator of index 4. One has to impose suitable restrictions in order to make the solution unique. This can be done by extending the range space by a space of real dimension 4 and by augmenting the operator A in a suitable way.

We define real functionals ϕ_1, ϕ_2 and a complex functional ϕ_3 on $F^{1+\mu}$ by

$$\phi_1(h) = \frac{1}{2\pi}\int_0^{2\pi} h(e^{is})ds, \quad \phi_2(h) = \frac{1}{2\pi}\operatorname{Im}\left(\int_0^{2\pi} h(e^{is})e^{2is}ds\right), \tag{5.2}$$

$$\phi_3(h) = \frac{1}{2\pi}\int_0^{2\pi} h(e^{is})e^{3is}ds, \tag{5.3}$$

and consider the mapping $B : F^{1+\mu} \to C^\mu \times \mathbb{R}^2 \times \mathbb{C}$ defined by

$$B(h) := (A(h), \gamma\phi_1(h), \gamma\phi_2(h), \gamma\phi_3(h)). \tag{5.4}$$

The functional ϕ_1 is the constant term in the Laurent series of h which is real by definition, ϕ_2 is the imaginary part of the coefficient of ζ^{-2}, and ϕ_3 is the coefficient of ζ^{-3}.

When the surface tension is large the shape of the free surface approaches a semicircle with radius

$$R := -\frac{\tau}{\gamma}. \tag{5.5}$$

The primitive $h_1(\zeta) := R/\zeta$ of the function $h_0(\zeta) := -R/\zeta^2$ maps H to a semidisc of radius R. The next theorem shows that this approximation can be improved to a solution.

Theorem 2. *There is a constant $C_0 > 0$ such that, for all real numbers $\beta_1, \beta_2, \beta_3, \beta_4$ satisfying*

$$\max_l |\beta_l| \leq C_0 \frac{|\gamma|\tau}{|\gamma|+1}, \tag{5.6}$$

there exists a solution $f' \in F^{1+\mu}$ of the equation

$$B(f') = (0, \beta_1, \beta_2, \beta_3 + i\beta_4) \tag{5.7}$$

with $f'(e^{is}) \neq 0$ and winding number $\operatorname{wind}(f'(e^{is})) = -2$. The Newton method starting from

$$f_0' = \frac{\tau}{\gamma\zeta^2} \tag{5.8}$$

converges to this solution.

Proof: a) The function f_0' is a solution of the equation

$$B(f_0') = 0. \tag{5.9}$$

The derivative of A is

$$DA(f_0')g = -\gamma M_3 g. \tag{5.10}$$

The operator

$$M_3 g(s) = \operatorname{Re}\left(e^{3is} g'(e^{is}) + 3e^{2is} g(e^{is})\right) \tag{5.11}$$

is independent of γ and τ. With the coefficients of the Laurent series

$$g(\zeta) = \sum_{k=0}^{\infty} \frac{b_k}{\zeta^k}, \tag{5.12}$$

$M_3 g$ can be represented as a Fourier series

$$\begin{aligned} M_3 g(s) = \quad & \operatorname{Re} b_2 + 2 \operatorname{Re} b_1 \cos s - 2 \operatorname{Im} b_1 \sin s \\ & + (3b_0 - \operatorname{Re} b_4) \cos 2s - \operatorname{Im} b_4 \sin 2s \\ & - \sum_{k=5}^{\infty} (k-3) \Big(\operatorname{Re} b_k \cos(k-2)s + \operatorname{Im} b_k \sin(k-2)s \Big). \end{aligned} \tag{5.13}$$

The equation

$$DB(f_0')g = \gamma(-M_3 g, b_0, \operatorname{Im} b_2, b_3) = (\psi, \alpha_1, \alpha_2, \alpha_3 + i\alpha_4) \tag{5.14}$$

for $\psi \in C^\mu$ and $\alpha_1, .., \alpha_4 \in \mathbb{R}$ can be solved by comparing the coefficients of the Fourier series (5.13) and the Fourier series for ψ. The coefficients $b_0, \operatorname{Im} b_2$ and b_3 cannot be determined from $-\gamma M_3 g = \psi$. They are provided by the additional conditions in (5.14). M_3 is a differential operator. Therefore, the solution of (5.14) is in $F^{1+\mu}$ whenever ψ is in C^μ.

$DB(f_0')$ is γ times an invertible operator which is independent of γ and τ. Therefore, it has an inverse Γ_1 which is bounded by

$$||\Gamma_1|| \leq \frac{C_1}{|\gamma|}. \tag{5.15}$$

b) We use the representation (4.9) for the second derivative. Then there are constants C_4, C_5 such that for $||h + \zeta^{-2}|| \leq 1/2$ the estimates $||B_1(h)|| \leq C_4, ||B_2(h)|| \leq C_5$ hold. Then for $||h - f_0'|| = ||h + R\zeta^{-2}|| \leq R/2$ the second derivative can be estimated by

$$||D^2 A(h)|| \leq \frac{\tau C_4}{R^2} + \frac{C_5}{R} \leq C_6 \frac{|\gamma|(|\gamma|+1)}{\tau}. \tag{5.16}$$

Finally, this implies that

$$||\Gamma_1 D^2 A(h)|| \leq C_7 \frac{|\gamma|+1}{\tau} =: K_0. \tag{5.17}$$

c) Now assume that $\beta_1, \ldots, \beta_4$ satisfy (5.6) with a constant C_0. The operator $\tilde{B}(h) := B(h) - (0, \beta_1, \beta_2, \beta_3 + i\beta_4)$ has the same derivatives as B. The value at f_0' can be estimated by

$$||\tilde{B}(f_0')|| = ||(0, \beta_1, \beta_2, \beta_3 + i\beta_4)|| = \max_l |\beta_l| \leq C_0 \frac{|\gamma|\tau}{|\gamma|+1}. \tag{5.18}$$

In view of (5.15), this implies that

$$||\Gamma_1 \tilde{B}(f_0')|| \leq \frac{C_1 C_0 \tau}{|\gamma|+1} =: \eta_0. \tag{5.19}$$

One can achieve that

$$h := \eta_0 K_0 = C_7 \frac{|\gamma|+1}{\tau} \frac{C_1 C_0 \tau}{|\gamma|+1} \leq \frac{1}{2} \tag{5.20}$$

if the constant C_0 in (5.6) is $\leq 1/(2C_1C_7)$. When $C_0 \leq 1/(4C_1)$, then $\eta_0 \leq R/4$ and the Newton iterates remain in the ball $||h + R\zeta^{-2}|| \leq R/2$ where all the estimates are valid. The Kantorovich theorem is applicable and yields Theorem 2. ■

The solutions of equation (5.7) are physically meaningful not for all parameters. First, one must keep in mind that $f'(\zeta)$ must be real for real ζ. Therefore, all Laurent coefficients of f' must be real. Hence the parameters in (5.7) must be restricted to $\beta_2 = 0, \beta_4 = 0$. Secondly, only functions f' are admitted which are derivatives of univalent functions f. Hence the first coefficient a_1 in the Laurent expansion must vanish. This is certainly guaranteed when f' is an even function and can be achieved by the condition $\beta_3 = 0$ in (5.7).

The remaining family of solutions of (5.7) depends only on the parameter $\beta_1 = a_0$. The solutions have approximately the form

$$f'(\zeta) = a_0 + \frac{\tau}{\gamma\zeta^2} \tag{5.21}$$

with the zeros

$$\zeta_{1,2} = \pm\sqrt{\frac{\tau}{-\gamma a_0}}. \tag{5.22}$$

Only for positive a_0 the zeros are real as required. The values of the mapping $f = a_0\zeta - \tau/\gamma\zeta$ at these zeros, *i.e.* the values

$$f(\zeta_{1,2}) = \pm 2\sqrt{-\tau a_0/\gamma} \tag{5.23}$$

determine the width of the gap.

It follows from Theorem 2 that $f'(e^{is})$ is continuously differentiable on the circle as a function of s. In particular f' is differentiable at ± 1. This means that f preserves the right angles of H at ± 1 since $f'(\pm 1) \neq 0$. The free streamline separates from the wall at right angles.

This is different from the situation of vanishing surface tension. The exact solutions (5.1) for $\tau = 0$ have $f'(\pm 1) = 0$, and the free streamline separates at zero angles from the wall.

When $\gamma > 0$, the outside pressure exceeds the total pressure of the undisturbed flow and the flow cannot protrude through the gap. In this case half of a bubble with the gap inside gives a solution. The bubble is attached to the wall and, supported by high inside pressure, it shields the gap which is no longer felt by the flow. Theorem 1 shows that for fixed γ and τ this solution is locally unique up to a shift along the wall.

A bubble attached to a wall forms with the wall a contact angle β which may be different from 90°. Attached bubbles have been calculated by Vanden-Broeck and Keller [8] for various contact angles. One must expect that also the protrusion through the gap may separate from the wall at an angle β different from 90°. In this case, f' would have either a zero or a singularity at ± 1. We cannot, at present, find such a solution with our method, since we search for f' only in the space $F^{1+\mu}$ which explicitly excludes zeros and singularities.

§6 Numerical results

We solve the equation $A(f') = 0$ by the Newton iteration

$$f'_{l+1} = f'_l + g_l \tag{6.1}$$

for $l \geq 0$ with a Laurent polynomial

$$g_l(\zeta) = \sum_{k=0}^{K} \frac{b_k}{\zeta^k} \tag{6.2}$$

as shift. The conditions

$$DA(f'_l)g_l(s_\nu) = -A(f'_l)(s_\nu) \tag{6.3}$$

for the N grid points $s_\nu = (\nu - 1)2\pi/N$ give a system of N equations for the coefficients b_k in (6.2). This overdetermined system is solved by a least squares method. If f'_l is an even function with real Laurent coefficients a_k, the function $A(f'_l)$ is even and the solution g_l of the Newton equation (3.5) is again an even function with real coefficients b_k. Therefore, one can reduce the number of coefficients b_k in (6.2) accordingly and assume that the b_k are real. This reduces the number of unknowns by a factor of 4.

Shankar [6] used the same equation $A(f') = 0$ as we do. However, he inserted a Laurent polynomial ansatz for f and solved the nonlinear equations for the coefficients numerically with a Newton method. He first discretized and then linearized, we first linearize and then discretize. Our results for the bubble agree with those of Shankar [6] to all decimal places. There are no indications for singularity of the operator $DA(f')$ along the solution branch. Therefore, no bifurcation occurs and the solution branch can be extended by continuation to small values of γ and to $\gamma < 0$.

For the gap we choose $\tau > 0, \gamma < 0$ and $a_0 > 0$ with a_0/τ small. A good choice is $a_0 = -\gamma/4\tau$, since this gives approximately the unit gap $[-1, 1]$ in view of (5.23). We start the iteration from

$$f_0'(\zeta) = a_0 + \frac{\tau}{\gamma\zeta^2}. \tag{6.4}$$

The Newton equation (6.3) is solved with the ansatz (6.2) but with $b_0 = 0$ in order to keep the first coefficient a_0 of f' fixed. With the general ansatz (6.2) one has to impose the additional constraints that b_2 is real and $b_3 = 0$, but this is already implemented if one uses (6.2) only with real coefficients and only with even indices k.

Solutions for other parameter combinations can be obtained by the continuation method, which works as long as $DA(f')$ is nonsingular.

In this way we calculate a solution f' depending on three parameters γ, τ and a_0. The size of the gap $[-d, d]$ is then determined. For each $c > 0$ the function cf' is a solution for the parameters $\gamma, c\tau, ca_0$ with the gap $[-cd, cd]$.

We use this to scale all our results in such a way that the gap is the unit interval $[-1, 1]$. Then $a_0 \approx |\gamma|/4\tau$ in view of (5.23). The condition (5.6) applied to $\beta_1 = a_0$ then yields the constraint

$$|\gamma| + 1 \leq 4C_0\tau^2. \tag{6.5}$$

Therefore, Theorem 2 guarantees solutions only for sufficiently large surface tension τ and only when the outside pressure $p_b > 0$ nearly balances the total pressure of the undisturbed flow. We show results for $\tau = 1$ and $\tau = .05$. The parameters and the first coefficients of the Laurent series

$$f'(\zeta) = \sum_{k=0}^{\infty} a_k \zeta^{-k} \tag{6.6}$$

are listed in Table 1. The corresponding shapes of the free boundary curves are shown in Figure 2.

For small γ the free surface is close to a semi-circle with radius $\tau/|\gamma|$. When the outside pressure decreases, then the pressure forces must be replaced by surface tension. This can only be done by increasing the curvature. The radius must shrink. This effect is clearly visible in the upper panel of Figure 2 showing results for $\tau = 1$. But the radius cannot become less than 1. This puts a natural limit for this family of solutions which cannot be extended to much larger values of $|\gamma|$.

For small surface tension $\tau = .05$ the free surface apparently cannot resist the increasing pressure imbalance $|\gamma|$. Long before the radius shrinks to 1 an indentation forms. The flow becomes too fast and sucks at the free surface. The bump can for a short while be balanced by surface tension, but it must break finally. This branch of solutions ends near $\gamma = -.035$.

Model	1	2	3	4	5	6
γ	−0.5000	−0.7000	−1.0000	−.0200	−.0300	−.0350
τ	1.0000	1.0000	1.0000	.0500	.0500	.0500
ζ_2	4.0129	2.8379	1.7015	5.5720	4.2353	3.8825
x_1	2.0339	1.4838	1.0682	2.8887	2.3305	2.4033
y_i	2.0222	1.4761	1.1654	2.6809	1.8842	1.4777
a_0	.1241	.1737	.2661	.0897	.1184	.1303
a_2	−2.0210	−1.4596	−1.0201	−2.7851	−2.1059	−1.8963
a_4	.3537	.5023	.8543	−0.0423	−.3147	−0.9975
a_6	−.0353	−.1002	−.4374	.0013	−.0072	−.2213
a_8	.0030	.0170	.1902		.0009	−.0318
a_{10}	−.0002	−.0026	−.0757		.0001	−.0038
a_{12}		.0004	.0287			−.0004
a_{14}		−.0001	−.0105			

Table 1. Parameters γ, τ, the positive zero ζ_2 of f', the values $x_1 = f(1)$, $y_i = \operatorname{Im} f(-i)$ and the first Laurent coefficients $a_0, \ldots, a_{14}$ of f' for the six models shown in Figure 2.

These examples show already that the bound for admissible values of γ and τ must be nonlinear as indicated by (6.5)

These two series of models illustrate the theoretical results of Theorem 2. For small $|\gamma|$ and small τ the situation becomes rather complicated. There are several branches of solutions and bifurcation may occur. Since equation (2.10) is quite similar to the equation studied by Tanveer [7], one can use the same method to investigate the analytic continuation of f' to the interior of the unit disc and locate the singularities.

We have not yet found any solution for small surface tension which is close to a solution for $\tau = 0$. Surface tension introduces a singular perturbation of the problem. This can be seen in the function f' of (5.1), which has zeros at ± 1. Therefore, it does not fit into the framework of our theory. The equation (3.5) for the Newton iteration reduces to a Riemann-Hilbert problem but to a singular one, since f' vanishes at two points on the unit circle.

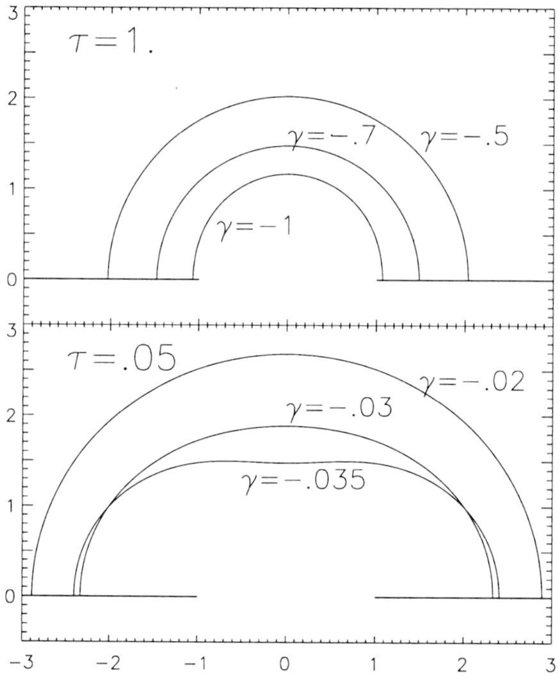

Figure 2. *Shape of the free surface across the gap* $[-1, 1]$ *for* $\tau = 1$ *and* $\tau = .05$ *and different values of* γ

References

[1] Kantorowitsch, L. W. and G. P. Akilow, *Funktionalanalysis in normierten Räumen*, Akademie-Verlag, Berlin, 1964.

[2] Miksis, M., J.-M. Vanden-Broeck and J.B. Keller, Axisymmetric bubble or drop in a uniform flow, *J. Fluid Mech.* **108** (1981), 89–100.

[3] McLeod Jr., E. B., The explicit solution of a free boundary problem involving surface tension, *J. Rat. Mech. Analysis* **4** (1955), 557–567.

[4] Muskhelishvili, N. I., *Singular Integral Equations*, Noordhoff, Groningen, 1953.

[5] Shankar, P. N., Exact solutions for a class of nonlinear free surface flows, *Proc. Roy. Soc. London* **434** (1991), 677–682.

[6] Shankar, P. N., On the shape of a two-dimensional bubble in uniform motion, *J. Fluid Mech.* **244** (1992), 187–200.

[7] Tanveer, S., Some analytical properties of solutions to a two-dimensional steady translating inviscid bubble, *Proc. Roy. Soc. London* **452** (1996), 1397–1410.

[8] Vanden-Broeck, J.-M. and J. B. Keller, Deformation of a bubble or drop in a uniform flow, *J. Fluid Mech.* **101** (1980), 673–686.

Rudolf Wegmann
Max-Planck-Institut für Astrophysik
D–85748 Garching
GERMANY
ruw@mpa-garching.mpg.de

List of Participants

Australia

G. Brown
Vice-Chancellor, The University of Sydney, Sydney, NSW 2006
e-mail: Vice-Chancellor@vcc.usyd.edu.au
Lecture: *A new proof of a theorem of Ruscheweyh*

Belgium

A. Cuyt
Department of Mathematics and Computer Science, University of Antwerp (UIA), Universiteitsplein 1, B-2610 Antwerp
e-mail: cuyt@uia.ua.ac.be
Lecture: *Symbolic versus float–pointing detection of multivariate pole sets*

P. Kravanja
Department of Computer Science, Katholieke Universiteit Leuven, Celestijnenlaan 200 A, B-3001 Heverlee
e-mail: peter.kravanja@cs.kuleuven.ac.be
Lecture: *On computing zeros of analytic functions*

B. Le Bailly
Department of Mathematics, Facultés Universitaires Notre-Dame de la Paix, Rempart de la Vierge, 8, B-5000 Namur
e-mail: bel@math.fundp.ac.be

J. P. Thiran
Department of Mathematics, Facultés Universitaires Notre-Dame de la Paix, Rempart de la Vierge, 8, B-5000 Namur
e-mail: jpt@math.fundp.ac.be
Lecture: *Optimum parameters for the generalized ADI method*

B. Verdonk
Department of Mathematics and Computer Science, University of Antwerp (UIA), Universiteitsplein 1, B-2610 Antwerp
e-mail: verdonk@uia.ua.ac.be
Lecture: *On computational aspects of the qd–algorithm*

Canada

P. Borwein
Department of Mathematics, Simon Fraser University, Burnaby, B.C. V5A 1S6
e-mail: pborwein@cecm.sfu.ca
Lecture: *Littlewood type problems on subarcs of the unit circle*

R. Fournier
Département de Mathématiques, Université de Montréal, Pavillon André-Aisenstadt, 2920 Chemin de la tour, Montréal, H3T1J8
e-mail: fournier@dms.umontreal.ca
Lecture: *Convex univalent functions and omitted values*

P. Gauthier
Département de Mathématiques, Université de Montréal, Pavillon André-Aisenstadt, 2920 Chemin de la tour, Montréal, H3T1J8
e-mail: gauthier@ere.umontreal.ca
Lecture: *Bloch's theorem*

W. Hengartner
Département de Mathématiques, Université Laval, Quebec City, P.Q. G1K 7P4
e-mail: walheng@mat.ulaval.ca

Cyprus

G. Georgiou
Department of Mathematics and Statistics, University of Cyprus, P.O. Box 537, CY-1678 Nicosia
e-mail: georgios@pythagoras.mas.ucy.ac.cy

S. Koumandos
Department of Mathematics and Statistics, University of Cyprus, P.O. Box 537, CY-1678 Nicosia
e-mail: skoumand@pythagoras.mas.ucy.ac.cy
Lecture: *Polynomial inequalities and applications*

G. Kyriazis
Department of Mathematics and Statistics, University of Cyprus, P.O. Box 537, CY-1678 Nicosia
e-mail: kyriazis@pythagoras.mas.ucy.ac.cy

N. Papamichael
Department of Mathematics and Statistics, University of Cyprus, P.O. Box 537, CY-1678 Nicosia
e-mail: nickp@pythagoras.mas.ucy.ac.cy

A. Poullikkas
Electricity Authority of Cyprus, P.O. Box 113, Larnaca
e-mail: mspoul@pythagoras.mas.ucy.ac.cy
Lecture: *Methods of fundamental solutions for elliptic boundary value problems*

N. S. Stylianopoulos
Department of Mathematics and Statistics, University of Cyprus, P.O. Box 537, CY-1678 Nicosia
e-mail: nikos@pythagoras.mas.ucy.ac.cy
Lecture: *The use of domain decomposition for approximating the conformal map of a quadrilateral onto a rectangle*

A. Vidras
Department of Mathematics and Statistics, University of Cyprus, P.O. Box 537, CY-1678 Nicosia
Lecture: *The class of holomorphic functions, representable by Carleman's formula*

Egypt

M. B. Abd-el-Malek
Department of Mathematics and Statistics, Alexandria University, Faculty of Engineering, Alexandria 21544
e-mail: mina@alex.eun.eg
Lecture: *Group methods analysis of electro-static potential between two eccentric conducting cylinders*

N. Moussa
Mathematics Unit, Science Department, The American University in Cairo, P.O. Box 2511, Cairo
e-mail: lsdwnm@acs.auc.eun.eg

Finland

I. Laine
Department of Mathematics, University of Joensuu, FIN-80101 Joensuu
e-mail: laine@joyl.joensuu.fi
Lecture: *A determinant problem related to complex differential equations*

R. Näkki
Department of Mathematics, University of Jyväskylä, FIN-40100 Jyväskylä
e-mail: raimon@math.jyu.fi
Lecture: *How homogeneous is a Swiss cheese?*

M. Vuorinen
Department of Mathematics, University of Helsinki, FIN-00014 Helsinki
e-mail: vuorinen@csc.fi
Lecture: *Quasiconformal images of circles and spheres*

France

L. Baratchart
INRIA, 2004, Route des Lucioles, B.P. 93, F-06902 Sophia-Antipolis Cedex
e-mail: laurent.baratchart@sophia.inria.fr
Lecture: *Rational or meromorphic approximation in Hardy spaces and Morse theory*

B. Beckermann
Laboratoire d'Analyse Numérique et d'Optimisation, Université des Sciences et Technologies de Lille, UFR IEEA - M3, F-59655 Villeneuve d'Ascq Cedex
e-mail: bbecker@ano.univ-lille1.fr
Lecture: *Convergence in capacity of diagonal Padé approximants*

J. D. Fournier
Département G. D. Cassini, Observatoire de Nice, B.P. 4229, F-06364 Nice
e-mail: fournier@obs-nice.fr
Lecture: *Complex roots for random orthogonal polynomials*

J. Leblond
INRIA, 2004, Route des Lucioles, B.P. 93, F-06902 Sophia-Antipolis Cedex
e-mail: leblond@sophia.inria.fr
Lecture: *Problems of Adamjan–Arov–Krein type on subsets of the circle and minimal norm extension*

M. Olivi
INRIA, 2004, Route des Lucioles, B.P. 93, F-06902 Sophia-Antipolis Cedex
e-mail: olivi@sophia.inria.fr
Lecture: *Computing rational H^2 best approximants*

F. Seyfert
INRIA, 2004, Route des Lucioles, B.P. 93, F-06902 Sophia-Antipolis Cedex
e-mail: fseyfert@sophia.inria.fr
Lecture: *An L^p analog to the A.A.K. theory for $2 < p < \infty$*

F. Wielonsky
INRIA, 2004, Route des Lucioles, B.P. 93, F-06902 Sophia-Antipolis Cedex
e-mail: wielonsk@sophia.inria.fr
Lecture: *Some results on Hermite–Padé approximants to e^z*

Germany

V. V. Andrievskii
Mathematisch-Geographische Fakultät, Katholische Universität Eichstätt, Ostenstr. 26-28, D-85072 Eichstätt
e-mail: mgk002@eo-dec-mathsrv.ku-eichstaett.de
Lecture: *Locally convex curves and distribution of zeros of polynomials*

H. P. Blatt
Mathematisch-Geographische Fakultät, Katholische Universität Eichstätt, Ostenstr. 26-28, D-85072 Eichstätt
e-mail: mga009@ku-eichstaett.de
Lecture: *Approximation on unbounded intervals*

A. Böttcher
Fakultät für Mathematik, Technische Universität Chemnitz-Zwickau, D-09107 Chemnitz
e-mail: aboettch@mathematik.tu-chemnitz.de
Lecture: *Lagrange interpolation and approximation by analytic elements in Bergman spaces*

B. Dittmar
Fachbereich Mathematik und Informatik, Martin-Luther-Universität, Halle-Wittenberg, D-06099 Halle (Saale)
e-mail: dittmar@cauchy.mathematik.uni-halle.de
Lecture: *Isoperimetric inequalities for the sums of reciprocal eigenvalues*

H. J. Fischer
Fakultät für Mathematik, Technische Universität Chemnitz-Zwickau, D-09107 Chemnitz
e-mail: hans-juergen.fischer@mathematik.tu-chemnitz.de
Lecture: *On generating orthogonal polynomials with respect to selfsimilar measures*

D. Gaier
Mathematisches Institut, Justus-Liebig-Universität, Arndtstr. 2, D-35392 Giessen
e-mail: dieter.gaier@math.uni-giessen.de
Lecture: *On the relation between $E_n(f)$ and $E_n(f')$*

W. Gehlen
Fachbereich IV - Mathematik, Universität Trier, D-54286 Trier
e-mail: gehlen@merkur.uni-trier.de
Lecture: *Extremal signatures in best polynomial approximation*

R. Greiner
Mathematisches Institut, Universität Würzburg, Am Hubland, D-97074 Würzburg
e-mail: greiner@mathematik.uni-wuerzburg.de

R. Grothmann
Mathematisch-Geographische Fakultät, Katholische Universität Eichstätt, Ostenstr. 26-28, D-85072 Eichstätt
e-mail: grothm@ku-eichstaett.de
Lecture: *Order of strong uniqueness constants, a problem of Poreda*

W. Haussmann
Fachbereich Mathematik, Gerhard Mercator Universität, D-47048 Duisburg
e-mail: haussmann@math.uni-duisburg.de

V. Heuveline
IWR, Universität Heidelberg, Im Neuenheimer Feld 293, D-69120 Heidelberg
e-mail: vincent.heuveline@iwr.uni-heidelberg.de
Lecture: *A block Arnoldi–Faber method for large complex non hermitian eigenvalue problems*

G. Jank
Lehrstuhl II für Mathematik, RWTH Aachen, Templergraben 55, D-52056 Aachen
e-mail: jank@math2.rwth-aachen.de
Lecture: *Nonsymmetric Riccati differential equations*

R. Kühnau
Fachbereich Mathematik und Informatik, Martin-Luther-Universität, Halle-Wittenberg, D-06099 Halle (Saale)
e-mail: kuehnau@mathematik.uni-halle.de

W. Lauf
Mathematisches Institut, Universität Würzburg, Am Hubland, D-97074 Würzburg
e-mail: lauf@mathematik.uni-wuerzburg.de (wolfgang.lauf@lhsystems.com)
Lecture: *The automophism space $\Sigma(G)$ of a domain without punctiform prime ends*

W. Luh
Fachbereich IV - Mathematik, Universität Trier, D-54286 Trier
e-mail: luh@uni-trier.de

G. Opfer
Institut für angewandte Mathematik, Universität Hamburg, Bundesstr. 55, D-20146 Hamburg
e-mail: opfer@math.uni-hamburg.de
Lecture: *The inclusion of the conformal modulus by discrete methods*

H. Renelt
Fachbereich Mathematik und Informatik, Martin-Luther-Universität, Halle-Wittenberg, D-06099 Halle (Saale)
e-mail: renelt@mathematik.uni-halle.de

O. Roth
Mathematisches Institut, Universität Würzburg, Am Hubland, D-97074 Würzburg
e-mail: roth@mathematik.uni-wuerzburg.de

St. Ruscheweyh
Mathematisches Institut, Universität Würzburg, Am Hubland, D-97074 Würzburg
e-mail: ruscheweyh@mathematik.uni-wuerzburg.de

G. Schmeisser
Mathematisches Institut, Universität Erlangen-Nürnberg, D-91054 Erlangen
e-mail: schmeisser@mi.uni-erlangen.de
Lecture: *Zeros of Bernstein polynomials*

G. Schmieder
Fachbereich 6 Mathematik, Universität Oldenburg, Postfach 2503, D-26111 Oldenburg
e-mail: schmieder@mathematik.uni-oldenburg.de
Lecture: *Location of critical points for polynomials*

H. Stahl
Fachbereich 2, Technische Fachhochschule Berlin, Luxemburgerstr. 10, D-13353 Berlin 65
e-mail: chrome@contrib.de
Lecture: *Error estimates for best rational approximants in a H^2–norm*

E. Wegert
Fachbereich Mathematik Informatik, TU Bergakademie Freiberg, Akademiestr. 6, D-09596 Freiberg
e-mail: wegert@math.tu-freiberg.de
Lecture: *Nonlinear Riemann Hilbert problems – history and perspectives*

R. Wegmann
Max-Planck-Institut für Astrophysik, Karl-Schwarzschild-Str. 1, D-85478 Garching
e-mail: ruw@mpa-garching.mpg.de
Lecture: *Two plane free boundary problems with surface tension*

Greece

A. Bacopoulos
Department of Mathematics, National Technical University of Athens, Zografou Campus, GR-15870 Athens
e-mail: abacop@central.ntua.gr
Lecture: *Projections, multicriteria optimization and best approximation*

D. Betsakos
Anatolikis Romylias 7, GR-50100 Kozani
e-mail: betsakos@compulink.gr
Lecture: *Geometric theorems and problems for harmonic measure*

N. Danikas
Department of Mathematics, Aristotle University of Thessaloniki, GR-54006 Thessaloniki
e-mail: danikas@math.auth.gr
Lecture: *Integrals of derivative boundary values and univalence*

V. Dougalis
Department of Mathematics, University of Athens, Panepistemiopolis, GR-15784 Athens
e-mail: doug@eudoxos.dm.uoa.gr
Lecture: *On the numerical solution of the nonlinear Schrödinger equation*

V. Farmaki
Department of Mathematics, University of Athens, Panepistemiopolis, GR-15784 Athens
e-mail: vgeorgil@atlas.uoa.gr

A. Katavolos
Department of Mathematics, University of Athens, Panepistemiopolis, GR-15784 Athens
e-mail: akatavol@atlas.uoa.gr

E. Katsoprinakis
Department of Mathematics, University of Crete, GR-71409 Heraklion
e-mail: katsopr@talos.cc.uch.gr
Lecture: *On the "complex Rolle set" of a polynomial*

C. Kokkinos
Department of Mathematics, National Technical University of Athens, Zografou Campus, GR-15780 Athens
e-mail: chkokki@math.ntua.gr
Lecture: *A unified orthonomalization method for the approximate conformal mapping of simply– and mulitply–connected domains*

G. Kostakis
Department of Mathematics, University of Athens, Panepistemiopolis, GR-15784 Athens

D. Kravvaritis
Department of Mathematics, National Technical University of Athens, Zografou Campus, GR-15780 Athens
e-mail: dkrav@math.ntua.gr

Y. Maroulas
Department of Mathematics, National Technical University of Athens, Zografou Campus, GR-15780 Athens
e-mail: maroulas@math.ntua.gr
Lecture: *On the numerical range of rational matrix function*

S. Negrepontis
Department of Mathematics, University of Athens, Panepistemiopolis, GR-15784 Athens

V. Nestoridis
Department of Mathematics, University of Athens, Panepistemiopolis, GR-15784 Athens
Lecture: *An extension of the notion of universal Taylor series*

S. Notaris
Department of Communications and Mass Media, University of Athens, 5, Stadiou Str., GR-10562 Athens
e-mail: notaris@eudoxos.dm.uoa.gr
Lecture: *An overview of results on the existence or nonexistence and the error of Gauss–Kronrod quadrature formulae*

I. Papadoperakis
Notara 12, GR-10683 Athens
e-mail: doperak@talos.cc.uch.gr

A. Tsarpalias
Department of Mathematics, University of Athens, Panepistemiopolis, GR-15784 Athens

Hong Kong

S. Smale
Mathematics Department, City University of Hong Kong, Tat Chee Ave, Kowloon
e-mail: masmale@sobolev.cityu.edu.hk
Lecture: *Issues of complexity and condition in complex variables*

Hungary

A. Kroó
Mathematical Institute, Hungarian Academy of Sciences, Reáltanoda utca 13-15, H-1053 Budapest
e-mail: kroo@math-inst.hu
Lecture: *Remez–type inequalities for polynomials in* $\mathbb{R}^m$ *and* $\mathbb{C}^m$

S. G. Révész
Mathematical Institute, Hungarian Academy of Sciences, Reáltanoda utca 13-15, H-1053 Budapest
e-mail: revesz@math-inst.hu
Lecture: *On some extremal problems of Landau for nonnegative cosine polynomials*

India

P. Bhattacharyya
Department of Mathematics, IIT, Madras, Madras-600 036
e-mail: pb@luit.iitg.ernet.in
Lecture: *On lacunary interpolation by entire functions*

R. Parvatham
Ramanujan Institute, University of Madras, Chennai 600 005
e-mail: chari@unimad.ernet.in
Lecture: *On a subclass of strongly starlike functions*

S. Ponnusamy
Department of Mathematics, IIT, Madras, Madras-600 036
e-mail: samy@iitg.ernet.in
Lecture: *Some geometric properties and asymptotic approximation of hypergeometric functions*

V. Singh
$\frac{3A}{95}$ Azad Nagar, Kanpur (U.P.) 208002
Lecture: *On some problems of Mocanu type*

Israel

D. Aharonov
Department of Mathematics, Technion, Israel Institute of Technology, Haifa 32000
e-mail: dova@techunix.technion.ac.il

D. Bshouty
Department of Mathematics, Technion, Israel Institute of Technology, Haifa 32000
e-mail: daoud@techunix.technion.ac.il
Lecture: *A constructive method for univalent exterior harmonic mappings with Blaschke dilatation*

E. (A. L.) Levin
Mathematics Department, The Open University of Israel, 16 Klausner Str., Ramat Aviv, Tel-Aviv, 61392
e-mail: elile@oumail.openu.ac.il
Lecture: *Fast decreasing rational functions and their applications*

U. Srebro
Department of Mathematics, Technion, Israel Institute of Technology, Haifa 32000
e-mail: srebro@leeor.math.technion.ac.il

Z. Ziegler
Department of Mathematics, Technion, Israel Institute of Technology, Haifa 32000
e-mail: ziegler@leeor.math.technion.ac.il

Japan

M. Shiba
Department of Applied Mathematics, Hiroshima University, Higashi-Hiroshima 739-8527
e-mail: shiba@amath.hiroshima-u.ac.jp
Lecture: *Area theorems and Rankine ovoids*

Lebanon

A. Lyzzaik
Department of Mathematics, American University of Beirut, Beirut
e-mail: lyzzaik@aub.edu.lb
Lecture: *Recent results for open continuous mappings having two valences between orientable surfaces*

R. N. Maalouf
Department of Mathematics, Notre Dame University, P.O. Box 72, Zouk Mikael
e-mail: rmaalouf@ndu.edu.lb

Malaysia

R. S. Ali
School of Mathematics & Computer Sciences, Universiti Sains Malaysia, 11800 Penang
e-mail: rosihan@cs.usm.my
Lecture: *Coefficient estimates for functions with positive real part and applications*

A. Murid
Department of Mathematics, Faculty of Science, Universiti Teknologi Malaysia, Locked Bag 791, 80990 Johor Bahru, Johor
e-mail: alihmm@math.fs.utm.my
Lecture: *Some integral equations related to the Riemann map*

Norway

K. Hag
Department of Mathematical Sciences, NTNU, N-7034 Trondheim
e-mail: kari@math.ntnu.no
Lecture: *Hyperbolic geometry and disks*

L. Lorentzen
Department of Mathematical Sciences, NTNU, N-7034 Trondheim
e-mail: lisa@math.ntnu.no

O. Njåstad
Department of Mathematical Sciences, NTNU, N-7034 Trondheim
e-mail: njastad@math.ntnu.no
Lecture: *Canonical and extremal solutions of strong moment problems*

V. Petersen
School of Teacher Education, Sor-Trondelag College, Rotvoll Alle, N-7005 Trondheim
e-mail: vigdis.petersen@alu.hist.no

F. Rønning
School of Teacher Education, Sor-Trondelag College, Rotvoll Alle, N-7005 Trondheim
e-mail: frode.ronning@alu.hist.no
Lecture: *Duality for Hadamard products applied to certain integral transforms*

Poland

J. Godula
Institute of Mathematics, UMCS, Plac M. Curie-Sklodowskiej 1, PL-20031 Lublin
e-mail: godula@golem.umcs.lublin.pl

J. Krzyż
Institute of Mathematics, UMCS, Plac M. Curie-Sklodowskiej 1, PL-20031 Lublin
e-mail: krzyz@golem.umcs.lublin.pl
Lecture: *The operator L and Fredholm eigenvalues*

W. Majchrzak
Faculty of Mathematics, University of Lodz, Ul. S. Banacha 2, PL-90238 Lodz
e-mail: kzyskow@imul.uni.lodz.pl
Lecture: *Harmonic univalent mappings of the unit disc onto a vertical strip*

M. Nowak
Institute of Mathematics, UMCS, Plac M. Curie-Sklodowskiej 1, PL-20031 Lublin
e-mail: nowakm@golem.umcs.lublin.pl
Lecture: *Integral means of univalent harmonic maps*

J. Szynal
Institute of Mathematics, UMCS, Plac M. Curie-Sklodowskiej 1, PL-20031 Lublin
e-mail: jsszynal@golem.umcs.lublin.pl

Romania

P. Caraman
Mathematical Institute "Octav Mayer", Romanian Academy - Branch of Iaşi, Str. Palade 12, 6600 Iaşi
e-mail: caraman@tuiasi.ro
Lecture: *Cases of equality between p-modules and p-capacity in $\overline{\mathbb{R}^n}$*

Russia

A. Danielyan
Department of Mathematics, Chair 806, Moscow Aviation Institute, 4, Volokolamskoe Shosse, 125871 Moscow
e-mail: daniel@k806.mainet.msk.su
Lecture: *M. A. Lavrentiev's problems on pointwise polynomial approximation and related questions*

E. Karatsuba
Computer Centre, Russian Academy of Science, Vavilova 40/1, 117967 Moscow
e-mail: ekar@ext1.ccas.ru
Lecture: *The FEE–method for fast evaluation of transcendental functions*

I. Nezhmetdinov
Chair of Mathematical Analysis, Kazan State University, P.O. Box 492, GOS–503, 420503 Kazan
e-mail: iskander.nezhmetdinov@ksu.ru
Lecture: *The dual hull as a counterpart of the closed convex hull*

K. Osipenko
Department of Mathematics, MATI - Russian State University of Technology, Petrovka 27, 103767 Moscow
e-mail: konst@osipenko.mccme.ru
Lecture: *On optimal recovery of periodic analytic functions*

D. Prokhorov
Department of Mathematics and Mechanics, Saratov State University, Astrakhanskaya ul. 83, 410026 Saratov
e-mail: prokhor.ma.sgu@ada.ssu.runnet.ru
Lecture: *Coefficients of functions close to the identity function*

A. Yu. Solynin
Russian Academy of Sciences, V. A. Steklov Mathematical Institute, St. Petersburg Branch, Fontanka 27, 191011 St. Petersburg
e-mail: solynin@pdmi.ras.ru
Lecture: *Conformal invariants: some extremal problems and estimates*

South Africa

K. A. Driver
Department of Mathematics, University of the Witwatersrand, Private Bag 3, Wits 2050
e-mail: 036kad@cosmos.wits.ac.za
Lecture: *Polynomials related with Hermite–Padé approximations to the exponential function*

Spain

A. Martínez Finkelshtein
Dept. Estadística y Matemática Aplicada, Universidad de Almeria, La Canada, E-04120 Almeria
e-mail: andrei@ualm.es
Lecture: *Information entropy for orthogonal polynomials*

Sweden

A. Ambroladze
Department of Mathematics, Umeå University, S-90187 Umeå
e-mail: amiran.ambroladze@math.lth.se
Lecture: *Some open problems in the theory of rational interpolants in the complex plane*

C. Rudälv
Department of Mathematics, Umeå University, S-90187 Umeå
e-mail: catarina.rudalv@mathdept.umu.se
Lecture: *Rational interpolation in* $\mathbb{C}^n$

H. Wallin
Department of Mathematics, Umeå University, S-90187 Umeå
e-mail: hans.wallin@mathdept.umu.se
Lecture: *Rational interpolants with prescribed poles, theory and practice*

Switzerland

J. P. Berrut
Institut de Mathématiques, Université de Fribourg, Perolles, CH-1700 Fribourg
e-mail: jean-paul.berrut@unifr.ch
Lecture: *Numerical solution of a boundary integral equation for conformal mapping by means of attenuation factors*

Ukraine

V. I. Belyĭ †
Institute for Applied Mathematics and Mechanics, Ukrainian Academy of Sciences, 340092 Donetsk
Lecture: *On n-widths of some classes of analytic functions in a bounded quasidisk*

E. Kravchuk
Department of Economics, Donetsk State University, 340012 Donetsk
e-mail: dmitrii@kravchuk.dgtu.donetsk.ua
Lecture: *Generalized Faber sets and applications*

United Kingdom

I. N. Baker
Department of Mathematics, Imperial College, 180 Queen's Gate, London SW7 2AZ
e-mail: i.n.baker@ic.ac.uk
Lecture: *Boundaries of some domains of convergence*

A. F. Beardon
Department of Pure Mathematics and Statistics, University of Cambridge, 16 Mill Lane, Cambridge CB2 1SB
e-mail: afb@dpmms.cam.ac.uk
Lecture: *Some remarks on Schwarz's lemma*

W. K. Hayman
Department of Mathematics, Imperial College, Huxley Building, 180 Queen's Gate, London SW7 2AZ
Lecture: *The growth of solutions of algebraic differential equations*

United States of America

R. W. Barnard
Department of Mathematics, Texas Technical University, Lubbock, TX 79409
e-mail: barnard@math.ttu.edu
Lecture: *Applications of hypergeometric functions*

C. Cao
Department of Mathematics, Boston College, Carney Hall, Chestnut Hill, MA 02167-3806
e-mail: caoch@monet.bc.edu
Lecture: *Packing of horoballs and minmum volume of cusped hyperbolic 3-manifolds*

T. K. DeLillo
Department of Mathematics and Statistics, Wichita State University, Wichita, KS 67260-0033
e-mail: delillo@twsuvm.uc.twsu.edu
Lecture: *Numerical conformal mappings of multiply connected regions by Fornberg-like methods*

M. Dorff
Department of Mathematics and Statistics, University of Missouri - Rolla, Rolla, MO 65409-0020
e-mail: mdorff@umr.edu
Lecture: *Harmonic mappings onto asymmetric vertical strips*

T. Driscoll
Department of Applied Mathematics, University of Colorado, CB 526, Boulder, CO 80309-0526
e-mail: tad@colorado.edu
Lecture: *Schwarz–Christoffel mapping using cross-ratios and Delauny triangulation*

P. Duren
Department of Mathematics, University of Michigan, Ann Arbor, MI 48109-1109
e-mail: duren@umich.edu
Lecture: *Robin capacity*

A. R. Elcrat
Department of Mathematics and Statistics, Wichita State University, 2825 W 17th St., Wichita, KS 67203-1436
e-mail: arelcrat@twsuvm.uc.twsu.edu
Lecture: *Some applications of conformal mapping to inverse problems*

F. W. Gehring
Department of Mathematics, University of Michigan, Ann Arbor, MI 48109-1109
e-mail: fgehring@math.lsa.umich.edu
Lecture: *Möbius groups, iteration and computers*

B. Gragg
Department of Mathematics, Naval Postgraduate School, Monterey, CA 93943
e-mail: gragg@nps.navy.mil
Lecture: *A $O(n^2)$ QR algorithm (shqz) for polynomial zeros*

W. B. Jones
Department of Mathematics, University of Colorado, Boulder, CO 80309-0395
e-mail: wjones@euclid.colorado.edu
Lecture: *Asymptotics for orthogonal polynomials and continued fractions*

L. Miller-Van Wieren
PennState University, Berks Campus, Luerssen 220, Reading, PR 19610
e-mail: leila@math.psu.edu
Lecture: *Univalence criteria for angular sectors*

D. Minda
Department of Mathematics, University of Cincinatti, Cincinatti, OH 45221-0025
e-mail: david.minda@math.uc.edu
Lecture: *Euclidean circles of curvature for geodesics relative to conformal metrics*

I. Pritsker
Department of Mathematics, Case Western Reserve University, 10900 Euclid Avenue, Cleveland, OH 44106-7058
e-mail: pritsker@mcs.kent.edu
Lecture: *Weigthed polynomial and rational approximation in the complex plane*

E. B. Saff
Institute for Constructive Mathematics, University of South Florida, Tampa, FL 33620
e-mail: esaff@math.usf.edu

F. Stenger
Department of Computer Science, University of Utah, Salt Lake City, UT 84112
e-mail: stenger@cs.utah.edu
Lecture: *Sinc construction of conformal maps and Symm's equation*

K. Stephenson
Department of Mathematics, University of Tennessee, Knoxville, TN 37996-1300
e-mail: kens@math.utk.edu
Lecture: *The approximation of conformal structures via circle packing*

R. Thurman
Department of Mathematical Sciences, University of Texas at El Paso, Bell Hall, El Paso, TX 79968-0514
e-mail: thurman@geom.umn.edu
Lecture: *Robin capacity and maximal capacity*

R. S. Varga
Institute for Computational Mathematics, Kent State University, Kent, OH 44242
e-mail: varga@mcs.kent.edu
Lecture: *The Szegő curve and weighted polynomial approximation in the complex plane*

Yugoslavia

M. Obradovič
Department of Mathematics, Faculty of Technology and Metallurgy, 4 Karnegijeva Street, 11000 Beograd
e-mail: obrad@elab.tmf.bg.ac.yu
Lecture: *Starlikeness of certain integral transforms*